PATHOLOGY OF INFECTIOUS DISEASES

Volume I
Helminthiases

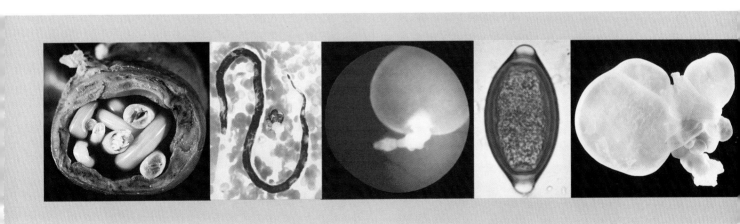

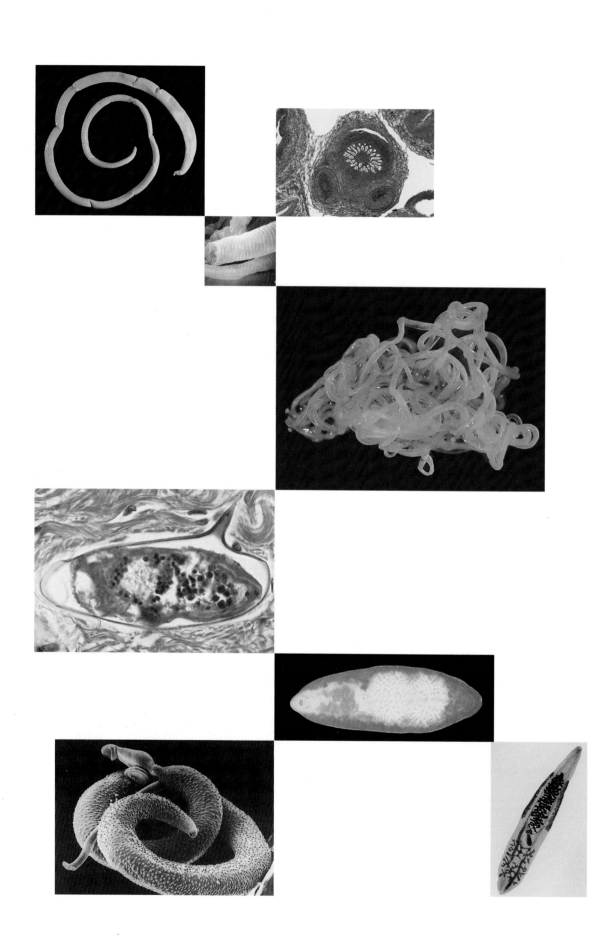

PATHOLOGY
of
INFECTIOUS DISEASES

Volume I
Helminthiases

EDITOR:
Wayne M. Meyers, MD, PhD

COEDITORS:
Ronald C. Neafie, MS
Aileen M. Marty, MD
Douglas J. Wear, MD

Armed Forces Institute of Pathology
American Registry of Pathology

Notes:

1. **Chord vs. cord in the hypodermis of nematodes**: The orthography of these structures differs from source to source. In keeping with 1) *Dorland's Medical Dictionary*, 27th ed, 2) *Clinical Parasitology*, 9th ed, by Beaver, Jung, and Cupp, and 3) the *Commonwealth Institute of Helminthology Keys to the Nematode Parasites of Vertebrates*, we have chosen the spelling "cord" for this volume.

2. **Hematologic values**: White blood cell counts (and platelets, if noted) are in conventional units (eg, $5 \times 10^3/\mu l$), which is equivalent to 5000/cu mm or μl or $5 \times 10^9/l$ International System of Units (SI). Hemoglobin is in conventional units (eg, 14g/dl). (*See Wintrobe's Clinical Hematology*, 9th ed, and *AMA Manual of Style*, 9th ed, p 263.)

3. **Cerebrospinal fluid cell values**: Counts are expressed in conventional units (eg, 5/cu mm), which is equivalent to $5/\mu l$, or $5 \times 10^6/l$ in International System of Units (SI). (*See AMA Manual of Style*, 9th ed, p 263.)

The use of general descriptive names, registered names, trademarks, etc. in this publication does not imply, even in the absence of a specific statement, that such names are exempt from the relevant protective laws and regulations, and therefore free for general use.

Product liability: The publishers cannot guarantee the accuracy of any information contained in this book about therapeutic dosages and their application. In every individual case, the user must check such information by consulting the relevant literature.

The views of the authors and editors do not purport to reflect the positions of the Department of the Army or the Department of Defense.

Armed Forces Institute of Pathology
American Registry of Pathology
Washington, DC

Available from the American Registry of Pathology Bookstore Web site: www.afip.osd

2000
ISBN: 1-881041-65-4

Dedication

This volume is dedicated to the memory of all those
physicians who, while toiling in distant lands, gave their
"last full measure of devotion" to bring healing and comfort
to the sick and impoverished.

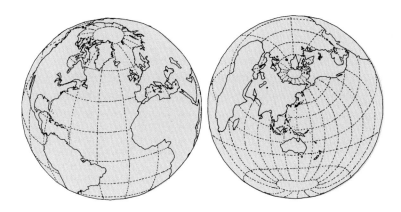

FOREWORD

In a world where mass migrations and world travel are increasingly common, parasitic organisms endemic to a particular region can be widely and rapidly disseminated through a variety of vectors, with enormous social, political, and economic consequences. Parasitologists study this phenomenon, along with the interplay between populations and environments and between parasites and hosts. This new text, *Pathology of Infectious Diseases, Volume 1: Helminthiases*, is a timely and significant contribution to the study of parasitic diseases in humans. The 37 chapters cover a spectrum of helminthic diseases from schistosomiasis and filariasis, to acanthocephaliasis and miscellaneous nematodiases. This new series updates *Pathology of Tropical and Extraordinary Diseases*, edited by C. H. Binford and D. H. Connor, an extremely popular monograph series published in 1976 which was, in turn, the successor to Ash and Spitz's 1945 edition of *Pathology of Tropical Diseases: An Atlas*.

For more than 137 years, the Armed Forces Institute of Pathology has upheld its reputation as the "people's institute" and fulfilled its mission to provide excellence in consultation, education, and research in pathology. The 24 authors who contributed to this text, many of them currently or formerly assigned to the AFIP's Department of Infectious and Parasitic Disease Pathology, have collaborated on a world-class, worldwide effort to support those missions. The book is multidisciplinary and draws upon the Institute's rich collection of nearly 3 million registered cases and the specimens, slides, and blocks that make up our tissue repository and archives.

As the 35th Director of the AFIP, I take great pleasure in presenting our latest effort to provide the most up-to-date discussions and illustrations of helminthic diseases. Our goal, as always, is to provide students, clinicians, researchers, and educators with the knowledge they need to address emerging and re-emerging diseases worldwide.

Glenn N. Wagner, CAPT, MC, USN
The Director

PREFACE AND ACKNOWLEDGMENTS

Before I began writing this preface, I researched the efforts of many others who had faced the same assignment. It soon became apparent that writing an elegant preface is a rarely mastered art that is best approached with considerable trepidation.

I believe David Livingstone, the esteemed explorer, missionary, and physician, best expressed my own sentiments on the making of books. In his preface to *Travels and Researches in South Africa: Including a Sketch of Sixteen Years' Residence in the Interior of Africa*, published in 1858, Livingstone said, "Those who have never carried a book through the press can form no idea of the amount of toil it involves." Commenting on the demands of writing, he further noted, "I think I would rather cross the African continent again than undertake to write another book."

Livingstone did indeed crisscross Africa again, but he did not survive to write an account of those travels. His anatomic heart rests in the village of Chitambo in Zambia, and his other remains are enshrined in London's Westminster Abbey, but the benevolent influence of this selfless physician reverberates throughout Africa to this day. The same can be said of those to whom this book is dedicated, including, along with Livingstone, Paul Carlson, Ralph Kleinschmidt, Albert Schweitzer, and Jeanette Troup. Science and humanity owe a profound debt to their personal sacrifices and generous contributions to our knowledge of tropical diseases.

This volume was conceived as a modest update of *Pathology of Tropical and Extraordinary Diseases*, the history of which is detailed in the Foreword by Captain Wagner. However, not long into our efforts, the Editorial Committee realized that the accumulation of new information and the increasing importance of infectious diseases in the medical sciences demanded that we divide the subject into a series of volumes under the general title *Pathology of Infectious Diseases*. We believe this first volume, while not exhaustive, is a substantial contribution to the study of the pathology of helminthiases, especially the identification of etiologic agents.

The combined efforts of many individuals have brought this book to fruition, and we acknowledge our debt to them with profound appreciation:

— The numerous authors who have waited patiently to see their work in print.

— Those who gave tirelessly of their time and professional expertise in the production of this book, including:
  — Fran Card, publication design and production
  — Bonnie L. Casey, scientific editing
  — Wendy S. Goodman, production coordination and graphic design
  — Ken Stringfellow, photographic scanning and color separation.

— Members of the Visual Information Division, who spent much time and effort providing photographs to illustrate the text.

— Physicians and health care workers around the world who contributed countless numbers of tissue specimens and clinical histories.

We also gratefully acknowledge the administrative support of the following individuals:

— Leslie H. Sobin, MD, SES
  Chief, Division of Gastrointestinal Pathology
  Director, Center for Scientific Publications, Armed Forces Institute of Pathology

—Florabel G. Mullick, MD, SES
   Director, Center for Advanced Pathology
   Principal Deputy Director, Armed Forces Institute of Pathology

—Glenn N. Wagner, CAPT, MC, USN
   Director, Armed Forces Institute of Pathology

Finally, this work would not have gone forward without the interest and support of the American Registry of Pathology, especially that of Dr. Donald West King.

Wayne M. Meyers, MD, PhD
for the Editorial Committee
January 2000

AUTHORS

Ellen M. Andersen, PhD; LCDR, MSC, USN
Director of Field Operations, Naval Medical Research Center Detachment, Lima, Peru.

Lawrence R. Ash, PhD
Professor Emeritus of Infectious and Tropical Diseases, Department of Epidemiology, University of California at Los Angeles School of Public Health.

J. Kevin Baird, PhD; CDR, MSC, USN
Parasitologist, Malaria Program, Naval Medical Research Institute, Bethesda, Maryland.

Maria Dolores Bargues Castello, PhD
Titular Professor of Parasitology, Faculty of Pharmacy, University of Valencia, Valencia, Spain.

Allen W. Cheever, MD
Volunteer Investigator, Biomedical Research Institute, Rockville, Maryland; Laboratory of Parasitic Diseases, National Institute of Allergy and Infectious Diseases, National Institutes of Health, Bethesda Maryland.

John H. Cross, PhD
Professor, Department of Preventive Medicine and Biometrics, Uniformed Services University of the Health Sciences, Bethesda, Maryland.

Brian O.L. Duke, MD, ScD
River Blindness Foundation, Lancaster, England.

Ramon L. Font, MD
Professor of Pathology and Ophthalmology, Director, Ophthalmic Pathology Laboratory, Sara Campbell Blaffer Professor of Ophthalmology, Ophthalmic Pathology Laboratory, Cullen Eye Institute, Baylor College of Medicine, Houston, Texas; Consultant in Pathology, Methodist Hospital, and M.D. Anderson Cancer Center, University of Texas, Houston.

Chris H. Gardiner, PhD; CAPT, MSC, USN
Fleet Hospital Operations and Training Command, Marine Corps Base, Camp Pendleton, California.

Yezid Gutierrez, MD, PhD
Adjunct Professor, Department of Microbiology, Cleveland Clinic Foundation, Cleveland, Ohio.

Parsotam R. Hira, PhD
Professor of Clinical Parasitology, Department of Microbiology, Kuwait University; Consultant Clinical Parasitologist, Mubarak, Farwaniya and Infectious Diseases Hospitals, Division of Laboratories, Ministry of Public Health, Kuwait.

Linda K. Johnson, DVM
Assistant Professor of Pathology, Albert Einstein College of Medicine Institute for Animal Studies, Bronx, New York.

Peter A.S. Johnstone, MD; CDR, MC, USN
Staff Physician, Radiation Oncology Division, Naval Medical Center, San Diego, California; Associate Clinical Professor, University of California, San Diego.

Mary K. Klassen-Fischer, MD; MAJ, MC, USAF
Chief, Fungal Diseases Pathology, Department of Infectious and Parasitic Diseases Pathology, Armed Forces Institute of Pathology, Washington, DC.

Juvady Leopairut, MD
Associate Professor of Pathology, Faculty of Medicine, Ramathibodi Hospital, Bangkok, Thailand; Lecturer, Faculty of Science, Mahidol University, Thailand; Lecturer, Faculty of Medicine, Rangsit University, Thailand.

J. Ralph Lichtenfels, PhD
Research Leader, Biosystematics and National Parasite Collection Unit, Livestock and Poultry Sciences Institute, Agricultural Research Service, US Department of Agriculture, Beltsville, Maryland.

Aileen M. Marty, MD; CDR, MC, USN
  Formerly Chief, Infectious Disease Pathology Branch, Armed Forces Institute of Pathology, Washington, DC; Clinical Associate Professor, Uniformed Services University of the Health Sciences, Bethesda, Maryland.

Santiago Mas-Coma, PhD
  Professor and Director, Department of Parasitology, Faculty of Pharmacy, University of Valencia, Valencia, Spain; Director, International Master Course on Tropical Parasitic Diseases.

Wayne M. Meyers, MD, PhD
  Chief of Mycobacteriology and Registrar for Leprosy, Armed Forces Institute of Pathology, Washington, DC; Research Affiliate, Tulane University, New Orleans, Louisiana.

Pedro Morera, MQC
  Medical Parasitologist, Institute of Health Research and School of Medicine, University of Costa Rica.

Ronald C. Neafie, MS
  Chief, Parasitic Disease Pathology Branch, Department of Infectious and Parasitic Diseases Pathology, Armed Forces Institute of Pathology, Washington, DC.

Eric A. Ottesen, MD; CAPT (Ret), USPHS
  Team Coordinator, Lymphatic Filariasis Elimination (CEE/FIL), Department of Communicable Diseases Eradication and Elimination, World Health Organization, Geneva, Switzerland.

Richard D. Semba, MD
  Assistant Professor of Ophthalmology, Molecular Microbiology and Immunology, and International Health, Wilmer Ophthalmological Institute, Johns Hopkins University School of Medicine, Baltimore, Maryland.

Douglas J. Wear, MD; COL (Ret), USA
  Chairman, Department of Infectious and Parasitic Diseases Pathology, Armed Forces Institute of Pathology, Washington, DC.

CONTENTS

1. **Overview of the Pathogenic Helminths with Discussion of Nonpathogenic Worms, Arthropods, and Other Structures**
   Aileen M. Marty *and* Ronald C. Neafie 1

2. **Schistosomiasis**
   Allen W. Cheever *and* Ronald C. Neafie 23

3. **Paragonimiasis**
   Aileen M. Marty *and* Ronald C. Neafie 49

4. **Hepatic Trematodiases**
   Santiago Mas-Coma, Maria Dolores Bargues Castello,
   Aileen M. Marty, *and* Ronald C. Neafie 69

5. **Fasciolopsiasis and Other Intestinal Trematodiases**
   Aileen M. Marty *and* Ellen M. Andersen 93

6. **Miscellaneous Trematodiases**
   Aileen M. Marty, Santiago Mas-Coma, *and* Maria Dolores Bargues Castello 107

7. **Taeniasis and Cysticercosis**
   Ronald C. Neafie, Aileen M. Marty, *and* Linda K. Johnson 117

8. *Dipylidiasis*
   Aileen M. Marty *and* Ronald C. Neafie 137

9. **Hydatidosis (Echinococcosis)**
   Aileen M. Marty, Linda K. Johnson, *and* Ronald C. Neafie 145

10. **Diphyllobothriasis and Sparganosis**
    Aileen M. Marty *and* Ronald C. Neafie 165

11. **Coenurosis**
    Aileen M. Marty *and* Ronald C. Neafie 185

12. **Hymenolepiasis and Miscellaneous Cyclophyllidiases**
    Aileen M. Marty *and* Ronald C. Neafie 197

13. **Lymphatic Filariasis**
    Eric A. Ottesen, Wayne M. Meyers, Ronald C. Neafie,
    and Aileen M. Marty 215

14. **Mansonelliasis**
    J. Kevin Baird, Ronald C. Neafie, *and* Wayne M. Meyers 245

15. **Loiasis**
    Aileen M. Marty, Brian O.L. Duke, *and* Ronald C. Neafie 261

16. **Dirofilariasis**
    Aileen M. Marty *and* Ronald C. Neafie 275

17. **Onchocerciasis**
    Ronald C. Neafie, Aileen M. Marty, *and* Brian O.L. Duke 287

18 Ocular Onchocerciasis
Ramon L. Font, Yezid Gutierrez, Richard D. Semba,
and Aileen M. Marty 307

19 American Brugian Filariasis
J. Kevin Baird, Mary K. Klassen-Fischer, Ronald C. Neafie,
and Wayne M. Meyers 319

20 Dracunculiasis
Ronald C. Neafie and Aileen M. Marty 329

21 Strongyloidiasis
Wayne M. Meyers, Ronald C. Neafie, and Aileen M. Marty 341

22 Ancylostomiasis
Wayne M. Meyers, Aileen M. Marty, and Ronald C. Neafie 353

23 Creeping Eruption
Douglas J. Wear, Wayne M. Meyers, and Ronald C. Neafie 367

24 Angiostrongyliasis Cantonensis
Aileen M. Marty and Ronald C. Neafie 373

25 Angiostrongyliasis Costaricensis
Pedro Morera, Ronald C. Neafie, and Aileen M. Marty 385

26 Ascariasis
Ronald C. Neafie and Aileen M. Marty 397

27 Toxocariasis
Aileen M. Marty 411

28 Anisakiasis
Ellen M. Andersen and J. Ralph Lichtenfels 423

29 Enterobiasis
Juvady Leopairut, Ronald C. Neafie, Wayne M. Meyers,
and Aileen M. Marty 433

30 Gnathostomiasis
Peter A.S. Johnstone, Parsotam R. Hira, Ronald C. Neafie,
and Mary K. Klassen-Fischer 447

31 Trichuriasis
Aileen M. Marty and Ronald C. Neafie 461

32 Trichinosis
Ronald C. Neafie, Aileen M. Marty, and Ellen M. Andersen 471

33 Capillariasis: Intestinal and Hepatic
John H. Cross and Ronald C. Neafie 481

34 Halicephalobiasis
Chris H. Gardiner, Wayne M. Meyers, and Ronald C. Neafie 493

35 Oesophagostomiasis and Ternidenamiasis
Ronald C. Neafie and Aileen M. Marty 499

36 **Miscellaneous Nematodiases**
Aileen M. Marty, Ronald C. Neafie, Mary K. Klassen-Fischer,
Lawrence R. Ash, *and* Douglas J. Wear 507

37 **Acanthocephaliasis**
Ronald C. Neafie *and* Aileen M. Marty. 519

Index. .... I-XXX

Figure Acknowledgments. XXXI

INDEX OF TABLES

Table A: Classification of parasitic helminths. **xvi**
Table B: Specific morphologic structures useful in identification of etiologic agents. **xx**
Table C: Nematodes seen in human tissue according to maximum diameter. **xxi**
Table D: Features of helminth eggs commonly seen in human tissue. **xxii**
Table E: Morphologic features of trematodes seen in human tissue. **xxiii**
Table F: Morphologic features of cestodes. **xxiv**
Table G: Morphologic features of filarial nematodes seen in human tissue. **xxv**
Table H: Morphologic features of nonfilarial nematodes seen in human tissue. **xxvi**
Table I: Nematodes seen in lumen of human intestine. **xxvii**

Table 1.1: Some comparative morphologic characteristics of the 4 major groups of parasitic helminths. **12-13**
Table 1.2: Terminology of cestode larvae. **14**
Table 1.3: Some morphologic features of gordiids (horsehair worms). **15**
Table 1.4: Structures that mimic helminths. **15**
Table 3.1: Species, range, and distinguishing morphologic features of *Paragonimus* sp. **54-55**
Table 3.2: First intermediate, second intermediate, and definitive hosts of *Paragonimus* sp. **57**
Table 5.1: Epidemiology of intestinal trematodiases. **96-97**
Table 5.2: Host, location, and egg and adult sizes of representative intestinal trematodes. **98**
Table 7.1: Characteristics of the adult tapeworm and the metacestode. **131**
Table 9.1: Morphologic features of adult *Echinococcus* sp. **152**
Table 9.2: Morphologic features of larval stage (metacestode) of *Echinococcus* sp. **153**
Table 13.1: Morphologic features of the most common adult female filariae in humans. **239**
Table 13.2: Morphologic features of the most common adult male filariae in humans. **240**
Table 13.3: Morphologic features of the most common microfilariae in humans. **241**
Table 13.4: Differentiating microfilariae of *Brugia* sp and *W. bancrofti* in blood films. **241**
Table 16.1: Morphologic features of *Dirofilaria* sp infecting humans. **277**
Table 18.1: Comparative prevalence of some clinical signs and ocular lesions among patients in rain forest and savanna regions. **313**
Table 19.1: Dimensions of zoonotic *Brugia* sp. **327**
Table 23.1: Classification, prevalence, and clinical characteristics of nematodes causing creeping eruption. **371**
Table 27.1: Seroprevalence of toxocariasis in humans. **413**
Table 27.2: Comparison of *Toxocara* and other larvae at level of midgut. **418**
Table 28.1: Diagnostic features of *A. simplex*, *P. decipiens*, and *Contracaecum* sp. **428**
Table 36.1: Classification of miscellaneous nematodes, original descriptions, and synonyms. **511**
Table 36.2: Life cycle and epidemiology of miscellaneous nematodes. **515**
Table 36.3: Sizes and morphologic features used to identify adults and eggs of miscellaneous nematodes. **516**
Table 36.4: Clinical findings and pathology in infections of miscellaneous nematodes. **517**

Table A. Classification of parasitic helminths.

PHYLUM: PLATYHELMINTHES
    CLASS: TREMATODA Rudolphi, 1808[a]
        Subclass: Digenea Van Beneden, 1858
            <u>Order: Paramphistomatiformes</u>: Szidat, 1936
                Family: Paramphistomatidae
                    Watsonius
                Family: Gastrodiscidae[b]
                    Gastrodiscus
            <u>Order: Echinostomatiformes (Echinostomida)</u> La Rue, 1957
                Family: Fasciolidae
                    Fasciola
                    Fasciolopsis
                Family: Echinostomatidae
                    Echinostoma
                    Hypoderaeum
                    Echinoparyphium
                    Euparyphium
                    Himasthla
                    Echinochasmus
            <u>Order: Strigeiformes (Strigeatida)</u> La Rue, 1926
                Family: Clinostomatidae
                    Clinostomum
                Family: Schistosomatidae
                    Schistosoma
                Family: Brachylaimidae
                    Brachylaima
                Family: Diplostomatidae
                    Alaria
                    Pharyngostomum
                Family: Gymnophallidae
                    Gymnophalloides
            <u>Order: Opisthorchiformes (Opisthorchiida)</u> La Rue, 1957
                Family: Opisthorchiidae
                    Clonorchis
                    Opisthorchis
                    Amphimerus
                Family: Heterophyidae
                    Heterophyes
                    Metagonimus
                    Centrocestus
                    Pygidiopsis
                    Haplorchis
                    Stellantchasmus
                    Procerovum
                    Diorchitrema
                    Stictodora
            <u>Order: Plagiorchiformes (Plagiorchiida)</u> La Rue, 1957
                Family: Troglotrematidae
                    Paragonimus
                Family: Achillurbainiidae
                    Poikilorchis
                    Achillurbainia

---

[a] This organization is based primarily on the extensive work by Yamaguti (1971), as adapted by Brooks et al (1985) and their revisions (1989, 1990). Other authorities consider that the taxonomic categories listed here as orders are actually superfamilies, and group these superfamilies into either 3 (Gibson, Bray, 1994) or 2 (Singh, 1991) orders.

[b] Some taxonomists consider the Gastrodiscidae and Paramphistomatidae families as synonymous and place the genera *Watsonius* and *Gastrodiscus* under the family Paramphistomatidae.

Table A. Classification of parasitic helminths (continued).

      Family: Nanophyetidae
       Nanophyetus
      Family: Plagiorchiidae
       Plagiorchis
      Family: Lecithodendriidae
       Prosthodendrium
       Phaneropsolus
      Family: Microphallidae
       Spelotrema
      Family: Dicrocoeliidae
       Dicrocoelium
       Eurytrema
 CLASS: CESTODA
  Subclass: Eucestoda
   Order: Pseudophyllidea
    Family: Diphyllobothriidae
     Diphyllobothrium
     Diplogonoporus
     Spirometra
   Order: Cyclophyllidea
    Family: Taeniidae
     Taenia
     Echinococcus
    Family: Dilepididae
     Dipylidium
    Family: Hymenolepididae
     Hymenolepis
    Family: Mesocestoididae
     Mesocestoides
    Family: Davaineidae
     Raillietina
     Inermicapsifer
     Bertiella

PHYLUM: ASCHELMINTHES
 CLASS: NEMATODA
  Subclass: Adenophorea (Aphasmidia)
   Order: Enoplida Chitwood, 1933
    Superfamily: Trichuroidea
     Trichuris
     Capillaria
     Trichinella
    Superfamily: Dioctophymatoidea
     Dioctophyma
     Eustrongylides
    Superfamily: Mermithoidea
     Agamomermis
  Subclass: Secernentea (Phasmidia)
   Order: Rhabditida Chitwood, 1933
    Superfamily: Rhabdiasoidea
     Strongyloides
    Superfamily: Rhabditoidea
     Halicephalobus (*Micronema*)
     Pelodera
     Turbatrix
     Diploscapter
     Rhabditis

Table A. Classification of parasitic helminths (continued).

Order: Tylenchida Thorne, 1949
    Superfamily: Tylenchoidea
        Ditylenchus
        Meloidogyne
        Heterodera
Order: Strongylida Diesing, 1851
    Superfamily: Strongyloidea
        Oesophagostoma
        Ternidens
        Mammomonogamus
    Superfamily: Ancylostomatoidea
        Ancylostoma
        Cyclodontostomum
        Necator
    Superfamily: Trichostrongyloidea
        Trichostrongylus
        Haemonchus
    Superfamily: Metastrongyloidea
        Metastrongylus
        Angiostrongylus
Order: Ascaridida Yamaguti, 1961
    Superfamily: Ascaridoidea
        Ascaris
        Toxocara
        Lagochilascaris
        Baylisascaris
        Anisakis
        Contracaecum
        Pseudoterranova (Phocanema)
Order: Oxyurida Weinland, 1858
    Superfamily: Oxyuroidea
        Enterobius
        Syphacia
Order: Spirurida Chitwood, 1933
Suborder: Spirurina
    Superfamily: Spiruroidea
        Spirocerca
        Gongylonema
    Superfamily: Gnathostomatoidea
        Gnathostoma
    Superfamily: Physalopteroidea
        Physaloptera
    Superfamily: Rictularioidea
        Rictularia
    Superfamily: Thelazoidea
        Thelazia
    Superfamily: Acuarioidea
        Cheilospirura
    Superfamily: Filarioidea
        Brugia
        Mansonella
        Wuchereria
        Dirofilaria
        Loa
        Onchocerca
        Meningonema

Table A. Classification of parasitic helminths (continued).

        Suborder: Camallanina
            Superfamily: Dracunculoidea
              Dracunculus

PHYLUM: ACANTHOCEPHALA

    <u>Order: Archiacanthocephala</u>
        Family: Oligacanthorhynchidae
            Macracanthorhynchus
        Family: Moniliformidae
            Moniliformis
    <u>Order: Palaeacanthocephala</u>
        Family: Polymorphidae
            Bolbosoma

[a]

Table B. Specific morphologic structures useful in identification of etiologic agents.

Morphologic feature	Organisms (page)
Alae	*A. braziliense* (355), *A. costaricensis* (386), *A. lumbricoides* (398), *Baylisascaris* (516), *E. vermicularis* (436), *S. stercoralis* (343), *Toxocara* (412), *Gongylonema* (516)
Areoles	**Paragordius* (3, 4)
Bacillary band	*Capillaria* (482, 487), *T. spiralis* (472), *T. trichiura* (461)
Bosses	*L. loa* (262), *Gongylonema* (516)
Bothria	*D. latum* (167-168), sparganum (171)
Calcareous corpuscles	All adult and larval cestodes
Cephalic gland	Adult *A. duodenale* (355), *N. americanus* (355)
Cuticular combs	*Rictularia* (516)
Excretory columns	*A. lumbricoides* (398), *A. braziliense* (355), *Baylisascaris* (516), *Toxocara* (412)
Excretory gland	Adult *A. duodenale* (355), *N. americanus* (355)
Excretory gland cell	*Anisakis, Contracaecum, Pseudoterranova* (424)
Gynecophoral canal	Adult male *Schistosoma* (24-25)
Head bulb	*Macracanthorhynchus* (521-523), *Moniliformis* (521-524)
Innenkörper	Microfilariae of *Brugia* (221), Microfilariae of *W. bancrofti* (221)
Laminated membrane	*Echinococcus* (148-149)
Lateral body	*A. duodenale* (355), *N. americanus* (355), *O. bifurcum* (500), *T. deminutus* (500)
Microvilli on surface	All adult and larval cestodes
Ovarian balls	*Macracanthorhynchus* (521-523), *Moniliformis* (521-524)
Parauterine organ	*Mesocestoides* (203)
Proboscis	*Gnathostoma* (448)
Pseudocoelomic membranes	*Eustrongylides* (516)
Sclerotized cuticular openings	Pentastomes (10)
Spines	*Echinostoma* (98), *Euparyphium* (98), *Fasciola* (82), *F. buski* (95), *Gnathostoma* (448), *H. heterophyes* (98), *M. yokogawai* (98), *Paragonimus* (52)
Stichosome	*Capillaria* (482, 487), *T. spiralis* (472), *T. trichiura* (461)
Striated muscle	Pentastomes (10), arthropods (11)
Subcuticular glands	Pentastomes (10)
Tracheae	Arthropods (11)
Trophosome	**Mermithids* (2-3)
Ventriculus	*Anisakis, Contracaecum, Pseudoterranova* (424)

*Nonpathogens

Table C. Nematodes seen in human tissue according to maximum diameter.

Nematode (page)	Maximum diameter (μm)	Type of nematode	Usual tissue location
Angiostrongylus costaricensis, larva (386)	11	Nonfilarial	Cecal wall
Toxocara cati, larva (412)	17	Nonfilarial	Eye, viscera
Ancylostoma braziliense, larva (355)	20	Nonfilarial	Skin
Strongyloides stercoralis, filariform larva (343)	20	Nonfilarial	Intestine
Strongyloides stercoralis, rhabditoid larva (343)	20	Nonfilarial	Colonic crypts
Toxocara canis, larva (412)	21	Nonfilarial	Eye, viscera
Capillaria philippinensis, larva (482)	25	Nonfilarial	Intestinal lumen
Capillaria philippinensis, male (482)	30	Nonfilarial	Intestinal lumen
Trichinella spiralis, male (472)	40	Nonfilarial	Skeletal muscle
Capillaria philippinensis, female (482)	50	Nonfilarial	Intestinal lumen
Mansonella streptocerca, male (251)	50	Filarial	Dermis
North American *Brugia* sp, male (320)	55	Filarial	Lymph nodes
Mansonella perstans, male (246)	60	Filarial	Deep soft tissues
South American *Brugia* sp, male (320)	60	Filarial	Lymph nodes
Strongyloides stercoralis, female (343)	60	Nonfilarial	Intestinal lumen
Ascaris lumbricoides, larva (398)	75	Nonfilarial	Lung
Capillaria hepatica, male (487)	78	Nonfilarial	Liver
Mansonella streptocerca, female (251)	85	Filarial	Dermis
Brugia malayi, male (219)	90	Filarial	Lymph nodes, other organs
North American *Brugia* sp, female (320)	90	Filarial	Lymph nodes
South American *Brugia* sp, female (320)	100	Filarial	Lymph nodes
Mansonella perstans, female (246)	150	Filarial	Deep soft tissues
Trichuris trichiura, anterior end, male or female (461)	150	Nonfilarial	Intestinal lumen
Wuchereria bancrofti, male (219)	150	Filarial	Lymph nodes, testis, epididymis
Brugia malayi, female (219)	170	Filarial	Lymph nodes, other organs
Capillaria hepatica, female (487)	184	Nonfilarial	Liver
Enterobius vermicularis, male (436)	200	Nonfilarial	Intestinal lumen
Onchocerca volvulus, male (290)	200	Filarial	Subcutaneous nodules
Wuchereria bancrofti, female (219)	250	Filarial	Lymph nodes, testis, epididymis
Angiostrongylus cantonensis, immature, male or female (375)	260	Nonfilarial	Brain
Dirofilaria tenuis, male (277)	260	Filarial	Subcutis
Dirofilaria immitis, female (277)	300	Filarial	Lung
Dirofilaria immitis, male (277)	300	Filarial	Lung
Necator americanus, male (355)	300	Nonfilarial	Intestinal lumen
Oesophagostomum bifurcum, female (500)	300	Nonfilarial	Mesentery
Angiostrongylus costaricensis, male (386)	310	Nonfilarial	Mesenteric arteries
Angiostrongylus costaricensis, female (386)	350	Nonfilarial	Mesenteric arteries
Oesophagostomum bifurcum, male (500)	350	Nonfilarial	Mesentery
Dirofilaria tenuis, female (277)	360	Filarial	Subcutis
Loa loa, male (262)	400	Filarial	Dermis, conjunctiva
Dirofilaria repens, male (277)	450	Filarial	Subcutis
Necator americanus, female (355)	450	Nonfilarial	Intestinal lumen
Onchocerca volvulus, female (219)	450	Filarial	Subcutaneous nodules
Ancylostoma duodenale, male (355)	500	Nonfilarial	Intestinal lumen
Enterobius vermicularis, female (436)	500	Nonfilarial	Intestinal lumen
Ternidens deminutus, male (500)	560	Nonfilarial	Mesentery
Loa loa, female (262)	600	Filarial	Dermis, conjunctiva
Gnathostoma spinigerum, larva (448)	630	Nonfilarial	Subcutis
Dirofilaria repens, female (277)	650	Filarial	Subcutis
Ancylostoma duodenale, female (355)	700	Nonfilarial	Intestinal lumen
Trichuris trichiura, posterior end, male or female (461)	700	Nonfilarial	Intestinal lumen
Ternidens deminutus, female (500)	730	Nonfilarial	Mesentery
Anisakis, *Contracaecum*, or *Pseudoterranova*, larva (424)	800	Nonfilarial	Mesentery
Dracunculus medinensis, female (330)	1700	Nonfilarial	Skin
Gnathostoma spinigerum, male or female (448)	3000	Nonfilarial	Subcutis
Ascaris lumbricoides, male (398)	4000	Nonfilarial	Intestinal lumen
Ascaris lumbricoides, female (398)	6000	Nonfilarial	Intestinal lumen

Table D. Features of helminth eggs commonly seen in human tissue.

Helminth (page)	Maximum dimension*	Shape	Shell	Egg contents	Usual tissue location
Opisthorchis viverrini (75)	30 μm	Elongate, ovoid	Light yellow-brown, thick, operculate, minute shoulders, abopercular thickening	Miracidium	Intrahepatic bile ducts, intestinal lumen
Clonorchis sinensis (73)	35 μm	Broadly ovoid	Light yellow-brown, thick, operculate, minute shoulders, small protuberance at abopercular end	Miracidium	Intrahepatic bile ducts, intestinal lumen
Taenia sp (119)	40 μm	Spherical	Yellow-brown, thick, radially striated	Oncosphere with 6 hooklets	Intestinal lumen
Capillaria philippinensis (482)	45 μm	Elongate	Thick and striated (unembryonated), or thin (embryonated), bipolar plugs	Unembryonated or embryonated	Intestinal crypts and lumen
Ancylostoma duodenale Necator americanus (358)	55 μm	Ovoid	Thin, usually collapsed and distorted	4-cell stage in stool	Intestinal lumen
Trichuris trichiura (465)	56 μm	Barrel-shaped	Brown-tinged, thick, prominent bipolar plugs	Unembryonated	Intestinal lumen
Enterobius vermicularis (437)	60 μm	Elongate, flattened on one side	Thick	Embryonated	Perianal skin surface, female peritoneum and genital tract
Strongyloides stercoralis (343)	60 μm	Ovoid	Thin, barely perceptible	Embryonated	Intestinal mucosa, crypts, and lumen
Ascaris lumbricoides (fertilized) (401)	70 μm	Round to ovoid	Golden brown, thick, usually mamillated	Unembryonated or in early division	Intestinal lumen, ectopic sites (liver, gallbladder and lung)
Schistosoma japonicum (25)	80 μm	Ovoid	Minute lateral spine (rarely seen in tissue)	Miracidium	Intestinal wall, liver, ectopic sites (lung, CNS, genital tract)
Angiostrongylus costaricensis (388)	85 μm	Ovoid	Thin, barely perceptible	Unembryonated or embryonated	Intestinal wall arteries
Paragonimus westermani (53)	95 μm	Ovoid	Yellow, thick, double-layered, birefringent, operculate	Unembryonated	Lung, ectopic sites (pleura, abdomen)
Schistosoma haematobium (25)	140 μm	Elongate	Small terminal spine	Miracidium	Urinary tract, ectopic sites (genital tract, intestine, liver, lung)
Schistosoma mansoni (25)	140 μm	Elongate	Prominent lateral spine	Miracidium	Intestinal wall, liver, ectopic sites (lung, CNS, genital tract)
Fasciola hepatica (83)	150 μm	Ovoid	Yellow, operculate	Unembryonated or in early division	Intrahepatic bile ducts, intestinal lumen

* Based on observations by Ronald Neafie of eggs in tissue sections.

Table E. Morphologic features of trematodes seen in human tissue.

Blood flukes

Male (page)	Tegumental tuberculations	Ceca with pigment	Size of testes	# of testes	Usual tissue location
Schistosoma haematobium (24)	Moderate	Unite late to form short intestine	Large	4-5	Urinary tract veins
Schistosoma japonicum (24)	None	Unite very late to form very short intestine	Medium	7 in single column	Portal veins
Schistosoma mansoni (24)	Prominent	Unite early to form long intestine	Small	6-9	Portal veins

Female	Tegumental tuberculations	Ceca with pigment	Length of uterus	# of eggs in uterus	Position of ovary	Usual tissue location
Schistosoma haematobium (24)	None	Present	Long	20-30	Posterior half	Urinary tract veins
Schistosoma japonicum (24)	None	Present	Long	50-100	Middle	Portal veins
Schistosoma mansoni (24)	None	Present	Short	1-4	Anterior half	Portal veins

Hermaphroditic flukes (page)	Thickness of tegument (μm)	Features of tegument	Ceca without pigment	Length of uterus	Features of uterus	Position of uterus	Size of testes	Features of testes	Position of testes	Usual tissue location
Clonorchis sinensis (71)	5	Smooth	Unbranched	Long	Coiled	Anterior to middle	Large	Branched	Posterior third	Liver
Opisthorchis viverrini (75)	4	Smooth	Unbranched	Long	Coiled	Middle third	Small	Lobed	Posterior fourth	Liver
Fasciola hepatica (82)	5	Various-sized spines	Highly branched	Short	Coiled	Anterior third	Large	Highly branched	Posterior half	Liver
Paragonimus westermani (52)	20	Scale-like spines	Unbranched	Short	Tightly coiled	Middle	Small	Deeply lobed	Posterior third	Lung

Table F. Morphologic features of cestodes.

Larval cestodes seen in human tissue

Common name (page)	Form	Number of protoscolices	Body wall projections	Body wall microvilli	Special features	Usual tissue location
Cysticercus (120)	Fluid-filled bladder	1 (None in racemose variety)	+	+	4 cup-like suckers per protoscolex (No suckers in racemose variety)	Subcutis, brain, eye
Hydatid cyst (149)	Fluid-filled bladder	Multiple (brood capsules)	−	−	Thick, laminated membrane	Liver, lung, other organs
Sparganum (171)	Solid	None	+/−	+	Bothrium (anterior groove)	Subcutis, muscle
Coenurus (187)	Fluid-filled bladder	Multiple	+	+	4 cup-like suckers per protoscolex	Subcutis, brain, eye

Gravid cestode proglottids found in human stool

Cestode (page)	Maximum size (mm)	Shape	Maximum size of eggs (μm)	Features of eggs	Features of uterus	Genital pore
Taenia saginata (119)	20 x 7	Longer than broad	43	Spherical, thick, radially striated shell	15-20 uterine branches per side	1, lateral
Taenia solium (119)	11 x 5	Longer than broad	43	Spherical, thick, radially striated shell	7-13 uterine branches per side	1, lateral
Dipylidium caninum (138)	23 x 8	Longer than broad with tapered ends	40	Spherical, 4 envelopes	Uterus compartmentalized with eggs in packets	2, one each lateral margin
Diphyllobothrium latum (167)	12 x 4	Broader than long	76	Broadly ovoid, operculate	Uterus coiled in center of proglottid	1, mid-ventral

Table G. Morphologic features of filarial nematodes seen in human tissue.

	Maximum diameter (μm)	Maximum thickness of cuticle (μm)	Features of cuticle	Other features	Usual tissue location
Adult male					
Mansonella streptocerca	50	2	Smooth	Lateral cords pigmented	Dermis
North American Brugia sp	55	2	Fine transverse striations	Somatic muscle mostly contractile (posterior end)	Lymph nodes
South American Brugia sp	60	2	Fine transverse striations	Somatic muscle mostly contractile (posterior end)	Lymph nodes
Mansonella perstans	70	2	Smooth	Lateral cords pigmented	Deep soft tissues
Brugia malayi	90	2	Fine transverse striations	Somatic muscle mostly contractile (posterior end)	Lymph nodes, other organs
Wuchereria bancrofti	150	2	Fine transverse striations	Somatic muscle mostly contractile (posterior end)	Lymph nodes, testis, epididymis
Onchocerca volvulus	200	5	Bilayered, transverse annulations	Somatic muscle equally contractile and sarcoplasmic (posterior end)	Subcutaneous nodules
Dirofilaria tenuis	260	15	Multilayered, fine transverse striations, internal/external longitudinal ridges	Broad lateral cords	Subcutis
Dirofilaria immitis	300	15*	Multilayered, fine transverse striations, internal longitudinal ridges	Immature in humans	Lung
Loa loa	400	10	Irregularly spaced bosses, internal longitudinal ridges	Lateral cords divided into sublaterals	Dermis, conjunctiva
Dirofilaria repens	450	20	Multilayered, coarse transverse striations, internal/external longitudinal ridges	Broad lateral cords	Subcutis
Adult female					
Mansonella streptocerca	85	2	Smooth	Lateral cords pigmented	Dermis
North American Brugia sp	90	2	Fine transverse striations	Usually nongravid	Lymph nodes
South American Brugia sp	100	2	Fine transverse striations	Usually nongravid	Lymph nodes
Mansonella perstans	150	2	Very fine transverse striations	Lateral cords pigmented	Deep soft tissues
Brugia malayi	170	2	Fine transverse striations	2 uteri fill body cavity	Lymph nodes, other organs
Wuchereria bancrofti	250	2	Fine transverse striations	2 uteri fill body cavity	Lymph nodes, testis, epididymis
Dirofilaria immitis	350	15*	Multilayered, fine transverse striations, internal longitudinal ridges	Immature in humans	Lung
Dirofilaria tenuis	360	15	Multilayered, fine transverse striations, internal/external longitudinal ridges	Usually nongravid	Subcutis
Onchocerca volvulus	450	10	Trilayered, regularly spaced transverse ridges	2 uteri do not fill body cavity	Subcutaneous nodules
Loa loa	600	10	Irregularly spaced bosses	3-5 reproductive tubes per cross section	Dermis, conjunctiva
Dirofilaria repens	650	20	Multilayered, coarse transverse striations, internal/external longitudinal ridges	Usually nongravid	Subcutis

*Usually degenerated, distorted, and thicker.

Table H. Morphologic features of nonfilarial nematodes seen in human tissue.

	Maximum diameter (μm)	Thickness of cuticle (μm)	Features of cuticle	Lateral cords	Somatic muscle	Other features	Usual tissue location
Adult male							
Capillaria hepatica	78	2	Fine striations	3 bacillary bands	Poorly developed	Stichosome anterior	Liver
Angiostrongylus cantonensis (immature in humans)	260	6	Multilayered, fine striations	Dome-shaped	Polymyarian, low coelomyarian	Intestine of large diameter	Brain, lung
Angiostrongylus costaricensis	310	2	Smooth	Small, dome-like	Polymyarian, low coelomyarian	Nuclei centered in intestinal cells	Mesenteric arteries
Oesophagostomum bifurcum (immature in humans)	350	10	Striations	Dilated spaces	Meromyarian, platymyarian	Lateral body in cord region	Mesentery
Ternidens deminutus (immature in humans)	560	10	Striations 10 μm apart	Dilated spaces	Meromyarian, platymyarian	Lateral body in cord region	Mesentery
Adult female							
Capillaria hepatica	184	2	Fine striations	3 bacillary bands	Poorly developed	Stichosome anterior	Liver
Angiostrongylus cantonensis (immature in humans)	260	6	Multilayered, fine striations	Dome-like	Polymyarian, low coelomyarian	Intestine of large diameter	Brain, lung
Oesophagostomum bifurcum (immature in humans)	300	10	Striations	Dilated spaces	Meromyarian, platymyarian	Lateral body in cord region	Mesentery
Angiostrongylus costaricensis	350	2	Smooth	Small, dome-like	Polymyarian, low coelomyarian	Nuclei centered in intestinal cells	Mesenteric arteries
Ternidens deminutus (immature in humans)	730	10	Striations 10 μm apart	Dilated spaces	Meromyarian, platymyarian	Lateral body in cord region	Mesentery
Dracunculus medinensis	1700	50	Smooth	Not discernible	Polymyarian, coelomyarian, 2 bands	Atrophic intestine	Skin, subcutis
Larvae							
Angiostrongylus costaricensis (L1)	11	1	Single minute lateral alae	Not evident	Not evident	Notched tail	Cecal wall
Toxocara cati (L2)	17	1	Single minute lateral alae	Small	Coelomyarian	Paired excretory columns	Eye, viscera
Ancylostoma braziliense (L3)	20	1	Double minute lateral alae	Not evident	Coelomyarian	Paired excretory columns	Epidermis
Strongyloides stercoralis (L3, filariform)	20	1	Double minute lateral alae (often not seen)	Not evident	Not evident	Paired excretory columns absent	Colonic wall
Toxocara canis (L2)	21	1	Single minute lateral alae	Small	Coelomyarian	Paired excretory columns	Eye, viscera
Trichinella spiralis (mature L1)	40	1	Fine striations	Readily discernible	Coelomyarian	Stichosome anterior	Skeletal muscle
Ascaris lumbricoides (L3)	75	1	Single prominent lateral alae, striations	Small	Coelomyarian	Paired excretory columns	Lung
Gnathostoma spinigerum (L3)	630	10	Spines	Readily discernible	Polymyarian, coelomyarian	Intestinal cells pigmented	Subcutis
Anisakis, Contracaecum, and *Pseudoterranova* (L3, L4)	800	50	Multilayered	Butterfly or Y-shaped	Polymyarian, coelomyarian	Ventriculus present	Mesentery

Table I. Nematodes seen in lumen of human intestine.

	Maximum diameter (μm)	Maximum thickness of cuticle (μm)	Features of cuticle	Lateral cords	Somatic muscle cells	Other features
Male						
Capillaria philippinensis	30	2	Smooth	Modified into 3 bacillary bands	Poorly developed	Stichosome anterior
Enterobius vermicularis	200	5	Striations, single lateral alae	Vacuolated	Meromyarian, platymyarian	Esophageal bulb
Necator americanus	300	20	Transverse striations	Lateral body at base of each	Meromyarian, platymyarian	Cephalic and excretory glands
Ancylostoma duodenale	500	20	Transverse striations	Lateral body at base of each	Meromyarian, platymyarian	Cephalic and excretory glands
Trichuris trichiura	150 (anterior)	10	Fine transverse striations (anterior)	None	Polymyarian, coelomyarian	Bacillary band and stichosome anterior
Trichuris trichiura	700 (posterior)	10	Annulations, striations (posterior)	None	Polymyarian, coelomyarian	Reproductive organs posterior
Ascaris lumbricoides	4000	40	Multilayered, annulations, striations	Short, narrow	Polymyarian, coelomyarian	Tall columnar intestinal cells
Female						
Capillaria philippinensis	50	2	Smooth	Modified into 3 bacillary bands	Poorly developed	Stichosome anterior
Strongyloides stercoralis	60	2	Fine transverse striations	Poorly developed	Poorly developed	Single row of eggs in uteri
Necator americanus	450	20	Transverse striations	Lateral body at base of each	Meromyarian, platymyarian	Cephalic and excretory glands
Enterobius vermicularis	500	5	Striations, single lateral alae	Vacuolated	Meromyarian, platymyarian	Esophageal bulb
Ancylostoma duodenale	700	20	Transverse striations	Lateral body at base of each	Meromyarian, platymyarian	Cephalic and excretory glands
Trichuris trichiura	150 (anterior)	10	Fine transverse striations (anterior)	None	Polymyarian, coelomyarian	Bacillary band and stichosome anterior
Trichuris trichiura	700 (posterior)	10	Annulations, striations (posterior)	None	Polymyarian, coelomyarian	Reproductive organs posterior
Ascaris lumbricoides	6000	90	Multilayered, annulations, striations	Short, narrow	Polymyarian, coelomyarian	Tall columnar intestinal cells
Larva						
Strongyloides stercoralis (rhabditiform)	20	1	No alae	Poorly developed	Poorly developed	Digestive tube present
Capillaria philippinensis	25	1	No alae	Bacillary bands	Poorly developed	Stichocytes in single or double row

1

Overview of the Pathogenic Helminths
with Discussion of Nonpathogenic Worms, Arthropods, and Other Structures

Aileen M. Marty *and*
Ronald C. Neafie

Introduction

Helminthic diseases affect the health and the social and political development of people worldwide, and have done so since prehistoric times. Helminths are metazoans. Derived from the Greek *meta* (higher) and *zoia* (animal), the term includes all animals more complex than the Protozoa.

Our understanding of the life cycles of helminths and the pathogenesis, host response, and treatment of infections they cause has advanced greatly in recent times. Taxonomic categories are changing because DNA technology[5,7,10,13] and phylogenetic systematics have allowed us to reclassify organisms based on their genetic relatedness.[2,6] Phylogeneticists believe that all living species are mosaics of primitive and derived characteristics; any given organism may carry some very primitive and some relatively recently developed traits. Classifying organisms is a process of synthesizing a complex of characteristics. Many helminths are designated "zoonotic" or "free-living." The distinction between free-living and parasitic helminths is not a natural division. Conditions may permit a free-living helminth to function as a parasite or, as with *Strongyloides*, for a parasitic worm to revert to a free-living stage. Larvae that migrate through human tissue and die within the body, while not properly parasitic, are significant because of possible host reaction in a critical organ. Even dead worms, when ingested, can incite serious allergic reactions.

For the parasitologist and pathologist, whole intact worms received in fixative are most useful for diagnosis. Clearing and staining entire specimens, especially the small worms,[11] permits the examiner to identify most helminths at the least cost. Relationships among various anatomic structures, such as the position of the vulva in relation to the esophagus in a female filarid, can be readily seen. Eggs, larvae, or microfilariae are also more easily observed.

Classifying a portion of a helminth, egg, larva, or microfilaria in tissue is more difficult because of such factors as orientation of the specimen and the possibility that key characteristics are not observable. In addition, tissue processing causes shrinkage which, when profound, leads to distortion. For example, in this book and most parasitol-

Figure 1.1
Coiled adult mermithid *Pheromermis californica* recovered from damp soil in California.*

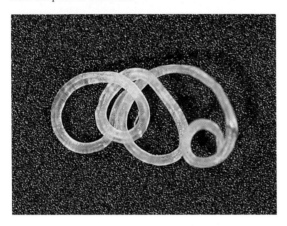

Figure 1.2
Larval mermithid removed from surface of patient's eye. Worm had caused no symptoms and was not considered pathogenic. x16

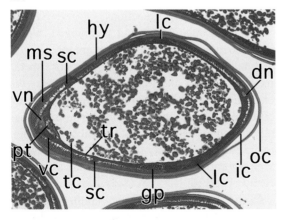

Figure 1.3
Transverse section through postparasitic juvenile mermithid showing outer cuticle (oc), inner cuticle (ic), hypodermis (hy), muscle (ms), lateral cords (lc), subventral cords (sc), ventral cord (vc), ventral nerve (vn), dorsal nerve (dn), pharyngeal tube (pt), gonad primordium (gp), trophosome (tr), and trophocytes (tc). x120

ogy texts, eggs of *Paragonimus westermani* are described as measuring 80 to 120 μm long, but in processed tissue eggs rarely exceed 80 μm. Likewise, microfilariae of *Wuchereria bancrofti* in stained blood films are 7.5 to 10 μm wide, but in tissue sections they are usually 5 μm wide. A gross worm may shrink to half its diameter after processing.

At the very least, effective laboratory analysis requires a knowledge of 1) the kinds of helminths that affect the tissue or body fluid under examination, 2) which stages of the helminth infect that tissue or fluid, 3) the general features of helminths, so as not to confuse the worm with other structures such as arthropods or artifacts, and 4) differences between separate classes of worms, and between nonpathogenic worms submitted as having been passed in stools or from body orifices.

Helminthic infections have several unique clinical and pathologic features. Many helminths migrate through the body, producing disease that varies with their route. Some worms, such as *Toxocara* sp, wander aimlessly and never mature. Others mature as they migrate, causing different clinical symptoms along the way. For example, pneumonitis may develop as schistosomules of *Schistosoma mansoni* travel through the lungs, whereas adult worms in mesenteric vessels release eggs into the liver, obliterating portal venules and leading to pipe stem cirrhosis. Some worms rarely or never migrate out of the organ they first inhabit. *Hymenolepis diminuta* enters the gut as a cysticercoid larva and lodges in the intestine, where it matures into an adult tapeworm and produces eggs that exit in the feces.

Some helminths have very complex life cycles. Some use humans for several stages of their life cycle, provoking different diseases at each stage. For example, consuming eggs of *Taenia solium* leads to cysticercosis, whereas consuming metacestodes of *T. solium* leads only to an adult tapeworm in the intestinal lumen. Well-adapted, healthy adult or larval helminths in tissue usually produce minimal mechanical damage or depletion of essential metabolites. One adult *Ascaris lumbricoides* in the small intestine usually causes no symptoms, but a single worm migrating into a critical site may block a duct, or a bolus of worms may obstruct the intestine. *Diphyllobothrium latum*, which competes with its host for vitamin B_{12}, can

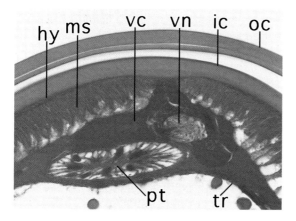

Figure 1.4
Higher magnification of ventral cord area of mermithid illustrated in Figure 1.3 showing outer cuticle (oc), inner cuticle (ic), hypodermis (hy), muscle (ms), ventral cord (vc), ventral nerve (vn), pharyngeal tube (pt), and trophosome (tr). x625

Figure 1.5
Adult female *Paragordius* found in patient's stool. Worm is 25 cm long and is not considered pathogenic. Unstained

cause anemia in patients with limited gastric intrinsic factor. Particularly long worms may produce various reactions in a single biopsy sample, depending on the viability of the parasite in a given area. Such helminths may begin to degenerate in one section but remain viable at other levels. Tissue reaction near the more viable portion of the worm may be minimal or consist of acute inflammation, whereas around the degenerated portion of the worm there may be a chronic inflammatory process, including granuloma formation.

There are 6 basic histopathologic patterns of helminthic infections, and many worms can present several of them at the same time: 1) **Minimal cell destruction**. Well-adapted adults living intravascularly or in the lumen of the intestine; migrating filariae. 2) **Suppurative inflammation**. Formation of abscess, ulceration, or necrotic track, with eosinophilia. 3) **Chronic inflammation**. Can be either diffuse or nodular infiltration of inflammatory cells, usually containing varying mixtures of macrophages and lymphocytes, and sometimes eosinophils. The inflammation may also be perivascular and/or periadnexal, extending a considerable distance from the parasite. Granulomas may develop with or without caseation. 4) **Ischemic necrosis**. Includes obstruction, thrombosis, or destruction of vessels. 5) **Calcification**. Some worms eventually calcify with or without fibrosis. 6) **Cytopathic-cytoproliferative inflammation**. Adult worms irritate tissues and may initiate cell replication and neoplastic change.

Small worms can travel to unexpected sites. *Heterophyes heterophyes*, a tiny intestinal fluke, can penetrate the gut, enter the blood stream, and migrate to the heart. Up to 15% of fatal heart disease in focal areas of the Philippines may result from heterophyid myocarditis.

Helminths have evasive mechanisms that protect them from the human immune system, such as incorporating host antigen. Specific antigens may provide a temporary function for some parasites, then conceal them from immune attack at a crucial moment. *Schistosoma mansoni*, for example, use cytokines of host origin induced by the infectious process for their own development.[4]

Classification of Helminths

Following is a brief discussion of some salient features of the helminths classified in Table A on pages xvi-xix.

I. Acoelomates

Acoelomates are metazoans that lack a body cavity, or coelom. The coelom, when present, is a fluid-filled cavity between the body wall and the internal organs. Acoelomates have solid bodies with internal organs embedded in a loose matrix of connective tissue called parenchyma or mesenchyme. The parenchyma is composed of a network of various fixed cells, smooth muscle fibers, and

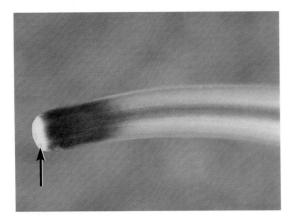

Figure 1.6
Anterior end of *Paragordius* female described in Figure 1.5 demonstrating white calotte (arrow) and pigmented ring. Unstained

Figure 1.7
Posterior end of *Paragordius* female described in Figures 1.5 and 1.6 depicting trilobed tail. Unstained

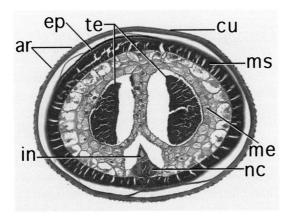

Figure 1.8
Transverse section through adult male *Paragordius* (500 μm in diameter) illustrating cuticular areoles (ar), cuticle (cu), epidermis (ep), muscle (ms), ventral nerve cord (nc), intestine (in), mesenchyme (me), and paired testes (te). x115

fluid-filled spaces. Acoelomate helminths of medical importance to humans belong to the phylum Platyhelminthes.

Phylum: Platyhelminthes

Platyhelminths (flatworms) include 3 classes: Turbellaria, Trematoda, and Cestoda. All are typically soft-bodied, flattened dorsoventrally, and, with rare exceptions, hermaphroditic. Turbellaria are mostly free-living, carnivorous flatworms. Trematodes (flukes) infect many vertebrates. Adult cestodes (tapeworms) are parasites of the small intestine or bile ducts of vertebrates; their larvae parasitize many sites in various vertebrates and invertebrates.

Class: Trematoda (Flukes)

The classification of trematodes, and of the subclass Digenea in particular, is problematic.[9] Trematoda contains 2 subclasses, Aspidocotylea and Digenea. Previous classifications included the Monogenia as an additional subclass, but they are now placed with the Aspidocotylea. Most Aspidocotylea undergo direct development and are parasites of aquatic and amphibious animals. Fish are the principal hosts of both Aspidocotylea and digenetic parasites. Compared to other vertebrates, mammals, including humans, are hosts for relatively few trematodes. All flukes causing disease in humans and other mammals are digenetic.[17]

The 6000 species of digenetic trematodes are common and widespread. They have complex life cycles involving at least 2 different hosts and several generations. The life cycle may require an aquatic phase or be entirely terrestrial. Some flukes use paratenic hosts in which immature parasites undergo no morphologic development or reproduction. Flukes may also use amphiparatenic hosts, which are lactating female mammals that would otherwise serve as definitive hosts, but function instead as paratenic hosts by infecting their offspring with immature parasites via breast milk.

Some morphologic features of trematodes are listed in Table 1.1. Characteristics that help classify specific trematodes include: 1) Body size. Trematode bodies are oval to elongate and range in length from less than 1 mm to 8 cm or longer. Most flukes are a few centimeters long; the largest that parasitizes humans, *Fasciolopsis buski,* measures up to 7.5 cm long. 2) Number, size, and location of tegumental spines or papillae. 3) Morphology of eggs: size, shape, presence and type of operculum, if any, and presence or absence of a miracidium. Eggs are often the only recognizable structures in tissue. 4) Location and relative size of oral and ventral suckers. 5) Arrangement of internal organs. The location and number of internal organs can vary widely. Distinguishing aspects include the length of intestinal ceca and their degree of branching, and the relative size, shape, and location of reproductive organs. Most flukes are hermaphroditic and produce operculate eggs. The schistosomes (blood flukes) are unique among trematodes in that the sexes are separate (Fig 2.6), females produce nonoperculate eggs, and adult worms inhabit the lumens of blood vessels.

Class: Cestoda (Tapeworms)

Unlike trematodes, cestodes have no digestive tract. There are 2 subclasses: Cestodaria and Eucestoda. Cestodaria are intestinal and coelomic parasites of elasmobranchs (eg, sharks) and primitive teleost fish; they use invertebrate intermediate hosts. Cestodaria resemble trematodes in that they lack a scolex and strobila, but are classified with the cestodes because they do not have a digestive tract. Eucestoda, better known as tapeworms, constitute the majority of all cestodes and are responsible for all important cestode diseases in higher vertebrates.

Eucestoda adults have 3 body divisions: an anterior scolex, a neck, and a strobila. The scolex is small, has suckers or bothria for attachment, and may have an armed rostellum. The strobila makes up the greatest part of the worm and consists of linearly arranged, independent, progressively maturing reproductive units called proglottids. Proglottids can be quite motile. The strobila is extremely flat, producing a high surface-area-to-volume ratio—a distinct asset to a parasite that absorbs all nutrients through the tegument.

The 3 orders of Eucestoda are Tetraphyllidea, Cyclophyllidea, and Pseudophyllidea. Worms from 2 orders, Cyclophyllidea and Pseudophyllidea, infect humans. The majority are cyclophyllidean tapeworms. Five of the 13 families of cyclophyllidean tapeworms (Taeniidae, Mesocestoididea, Anoplocephalidae, Dipylidiidae, and Hymenolepididae) contain genera of medical importance.

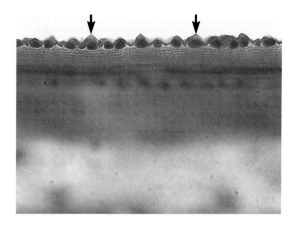

Figure 1.9
Higher magnification of *Paragordius* male depicted in Figure 1.8 demonstrating areoles (arrows) and fibrillar layer of cuticle. x570

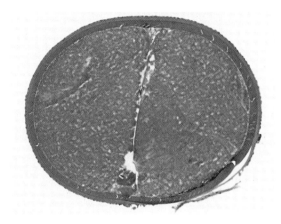

Figure 1.10
Transverse section through larval gordian worm. Body cavity is filled with mesenchymal cells. x145

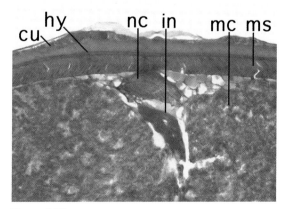

Figure 1.11
Section through ventral nerve cord region of larval gordian worm showing cuticle (cu), hypodermis (hy), muscle (ms), ventral nerve cord (nc), intestine (in), and mesenchymal cells (mc). x245

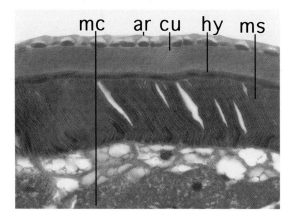

Figure 1.12
Section through body wall of larval gordian worm illustrating areoles (ar), cuticle (cu), hypodermis (hy), muscle (ms), and mesenchymal cells (mc). x590

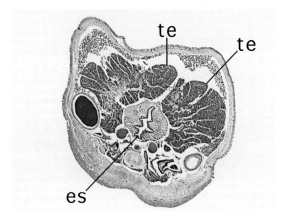

Figure 1.13
Transverse section through earthworm at level of esophagus (es) and testes (te). x12

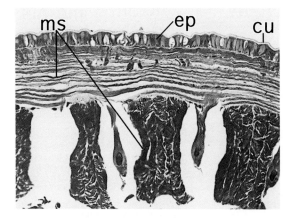

Figure 1.14
Section through body wall of earthworm depicting cuticle (cu), epidermis (ep), and smooth muscle (ms). Movat x120

In the order Pseudophyllidea, only 1 family is medically important: Diphyllobothriidae, of which only 2 genera, *Diphyllobothrium* and *Spirometra,* frequently infect humans.

From a medical or veterinary standpoint, Taeniidae is the most important cyclophyllidean cestode family. Three important pathogens of humans belong to this group: *Echinococcus granulosus*, *Echinococcus multilocularis*, and *T. solium*. All have metacestode stages that produce potentially life-threatening disease.

Some morphologic features of cestodes are listed in Table 1.1. Characteristics that help classify adult cestodes into species include: 1) Size, shape, and decorative features of the scolex, and the presence of grooves (bothria), hooks, or other accessory attachment structures. A scolex with 2 bothria is characteristic of pseudophyllideans. A scolex with 4 suckers is typical of cyclophyllidean tapeworms. The presence or absence of a rostellum and the distribution and size of rostellar hooklets on the scolex distinguish some cyclophyllideans. 2) The length of the neck. 3) The relative size and shape of proglottids and the length of the chain of proglottids. The best morphological indices for differentiating the strobila of cestode species, regardless of host species and worm burden, are a) type of boundary between proglottids, b) position of cirrus sac, c) conjunctive angle of cirrus sac to seminal vesicle, and d) shape of ovary.[1] 4) The size and shape of eggs and the presence or absence of an operculum. Pseudophyllidean tapeworms have ovoid, operculate eggs (Fig 10.19) that are immature when passed in feces. Eggs of cyclophyllidean tapeworms are essentially spherical, nonoperculate, and almost fully embryonated when they exit the proglottid. *Taenia* and *Echinococcus* eggs have striated shells. Larval cestodes have no reproductive system and may be cystic or solid (Table 1.2).[12] Calcareous corpuscles (round to oval laminated bodies) are a normal component of adult and larval cestodes and are diagnostic of this group. They are frequently mistaken for helminth eggs.

II. Pseudocoelomates

Pseudocoelomates have a large, fluid-filled cavity called a pseudocoel between the body wall and the internal organs. Despite the name, there is nothing false about a pseudocoel; it is a structure derived from the blastocoel of the embryo that,

Figure 1.15
Limnatis paluda leech removed from trachea after accidental ingestion by patient living in Saudi Arabia.

Figure 1.16
Section through leech demonstrating prominent bands of smooth muscle oriented in multiple directions. Movat x55

Figure 1.17
Nonencysted mature *Armillifer armillatus* larva attached to abdominal surface of human diaphragm. Note pseudosegmented appearance. x2.2

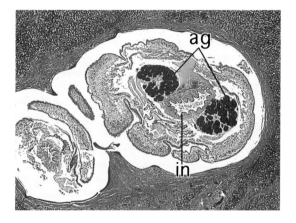

Figure 1.18
Section through encysted *A. armillatus* larva in liver. Intestine (in) and acidophilic glands (ag) are readily identified. x15

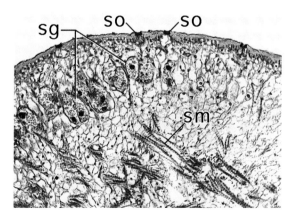

Figure 1.19
Section through encysted *A. armillatus* larva in liver depicting sclerotized openings (so) in cuticle, subcuticular glands (sg), and striated muscle (sm). Movat x165

Figure 1.20
Dermatobia hominus second-stage larva removed from subcutaneous tissue of arm of Guatemalan man with myiasis.

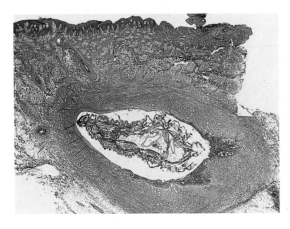

Figure 1.21
Section through skin and subcutaneous tissue of patient with myiasis illustrating larval dipteran in inflammatory lesion. Masson x25

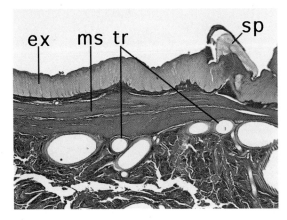

Figure 1.22
Section through *D. hominus* larva showing exoskeleton (ex) with spine (sp), striated muscle (ms), and tracheae (tr). x120

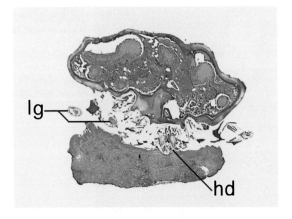

Figure 1.23
Section through nonengorged tick attached to human skin. Note tick's legs (lg), and head (hd) penetrating ulcerated epidermis. x15

unlike a coelom, is not lined by mesoderm. Internal organs are free within the space, since there is no peritoneum bounding the cavity. Two groups of pseudocoelomates cause human disease: nematodes and acanthocephalans.

Phylum: Aschelminthes

Aschelminths comprise a diverse group of organisms that includes nematodes and various free-living groups such as nematomorphs, rotifers, gastrotrichs, and kinorhynchs.

Class: Nematoda (Roundworms)

The Nematoda, or roundworms, are among the largest and most widespread groups of multicellular animals. Free-living nematodes are found in a wide variety of environments. Nematodes belonging to the family Mermithidae are free-living in the adult stage, but are parasitic in juvenile stages in invertebrates, mainly insects. Mermithids have rarely been reported as accidental parasites of man.[14] In most instances, it is difficult to prove they are pathogens since they are usually observed outside the human body, in urine or feces. Clinical histories are usually meager and the identification of the worm is frequently questionable. Mermithids are smooth, filiform worms up to 50 cm long (Fig 1.1). Males are much smaller than females. The body has 6 or 8 longitudinal hypodermal cords, and the intestine is modified into a food-storing organ. Mermithids are illustrated in figures 1.1 to 1.4.

Parasitic nematodes attack virtually all groups of plants and animals in a broad spectrum of parasitism. Some morphologic features of nematodes are listed in Table 1.1. Characteristics that help distinguish species of adult nematodes include: 1) Caudal papillae. 2) Cuticular thickness and markings such as alae, ridges, bosses, and striations (a cuticle regarded as smooth may in fact contain transverse striations). 3) Hypodermis and the shape and size of the lateral cords it forms. 4) Size, type, and amount of somatic musculature. 5) Shape, relative size, number of cells, and specialized features of the digestive tube. 6) Anatomy and arrangement of the male and female reproductive system.

There are 2 subclasses (Secernentea (Phasmida) and Adenophorea (Aphasmida)) and 14 orders of nematodes. The subclass Secernentea contains primarily parasitic nematodes and some free-living forms that inhabit soil. The majority of these parasitic nematodes belong to the orders Rhabditida, Strongylida, Ascaridida, Spirurida, and Camallanina. These nematodes have phasmids (sensory papilla-like structures on the tail). Most members of the subclass Adenophorea are free-living marine and freshwater nematodes. The few parasitic nematodes in this group belong to the superfamilies Trichuroidea and Dioctophymatoidea. Adenophorea do not have phasmids.

Class: Nematomorpha (Horsehair Worms)

Nematomorpha (gordian worms or horsehair worms) (Figs 1.5 to 1.12) are sometimes described in feces, urine, or sputum. The Nematomorpha probably represent a class of the Aschelminthes phylum; however, some taxonomists place them in a phylum of their own. Larvae from the Nematomorpha are parasites of arthropods (insects and crustaceans) and leeches. Adults are free-living. Spurious human encounters result from ingesting adults or larvae in contaminated water, or ingesting an intermediate host containing a fully developed juvenile or immature adult worm.

There are 2 major groups of Nematomorpha: order Nectonematoidea and order Gordioidea. Nectonematoidea includes those that live in the open seas and is represented by a single genus, *Nectonema*. Gordioidea includes all freshwater and terrestrial forms (Table 1.3).

Phylum: Acanthocephala (Thorny-Headed Worms)

Acanthocephala are highly specialized parasites of the digestive tract of vertebrates. Recent molecular evidence indicates that Acanthocephala may be a subtaxon of Rotifera.[8] Rotifera are free-living animals in the phylum Aschelminthes. We choose to retain Acanthocephala as a phylum. Acanthocephala are all endoparasites that require 2 hosts to complete their life cycle.

Some morphologic features of acanthocephalans are listed in Table 1.1. Characteristics that distinguish the Acanthocephala into orders are difficult to recognize in tissue sections. When an intact, unembedded specimen is available, distinguishing features include: 1) Shape, number, and distribution of hooks on the proboscis. 2) Pattern and arrangement of lacunar channels. 3) Presence

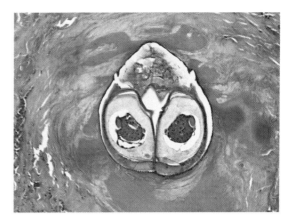

Figure 1.24
Additional sections of tick described in Figure 1.23 revealing mouth parts embedded in necrotic focus of upper dermis. x125

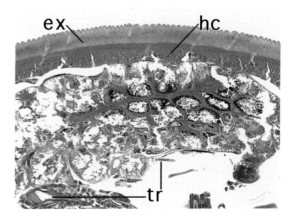

Figure 1.25
Section through tick illustrating thick exoskeleton (ex), hypodermal cells (hc), and tracheae (tr). x125

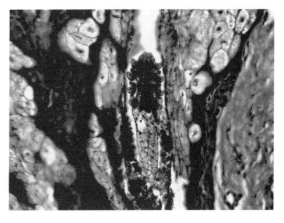

Figure 1.26
Follicle mite, *Demodex folliculorum*, in human skin. Note 4 pairs of short stumpy legs. x260

Figure 1.27
Lipid pseudomembrane in skin with laminations similar to laminated membrane of larval cestode. Patient is 55-year-old man who traveled extensively in Asia and South America. H&E x5.5

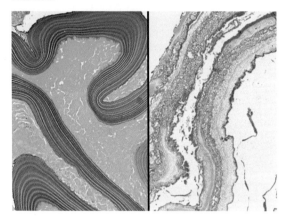

Figure 1.28
Comparison of laminated membrane of hydatid cyst (*left* x22) with lipid pseudomembrane (*right* x55). GMS stain silvers laminations of hydatid cyst; pseudomembrane turns golden with same stain.

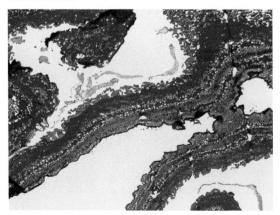

Figure 1.29
Brilliant red staining of lipid pseudomembrane on paraffin-embedded tissue. Oil Red O x22

and distribution of spines on the body. 4) Number of cement glands in males. 5) Morphologic features of eggs in a gravid female.

III. Coelomates

While the coelomates are not helminths, and will be discussed in another volume of this series, we include them here for differential diagnosis. Coelomates are sometimes encountered as pathologic specimens. Coelomates are metazoans with a large, mesoderm-lined, fluid-filled cavity lying between the body wall and the internal organs. The embryogenesis of a coelom is different from that of a pseudocoel; it arises as a cavity within embryonic mesoderm. The mesodermal lining is called the peritoneum. Internal organs are not free within the coelom, but are bounded by peritoneum. Organisms from 3 phyla cause human disease: Annelida, Pentastomida, and Arthropoda.

Phylum: Annelida

Annelida are segmented worms (Figs 1.13 and 1.14) such as earthworms, marine polychaetes, and leeches. The most distinguishing gross characteristic is true segmentation (metamerism), the division of the body into similar segments arranged linearly. Patients sometimes submit earthworms they suspect have passed in feces or other body fluids. Earthworms have a body covered by a thin, moist cuticle over columnar epithelium (Fig 1.14). The class Hirudinea (leeches) does include some parasitic worms. Bloodsucking species attach to the skin or to oropharyngeal or laryngeal mucous membranes. They can cause severe coughing, choking, and ejection of blood. Leeches have suckers (Fig 1.15) and an elaborate, powerful muscular system (Fig 1.16).

Phylum: Pentastomida

There are approximately 95 species of pentastomids, or tongue worms, and all are parasitic. Larval forms develop in humans and are important pathogens (Figs 1.17 and 1.18).

As adults they live within the lungs or nasal passages of vertebrates. Three species commonly infect humans and are discussed in another volume of this series. Pentastomes have a greatly reduced coelom and striated muscle, but are unique among animals in having sclerotized cuticular openings (Fig 1.19).

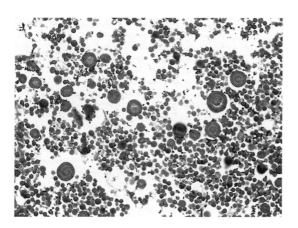

Figure 1.30
Liesegang rings in cystic lesion in kidney. Rings are usually spherical. x105

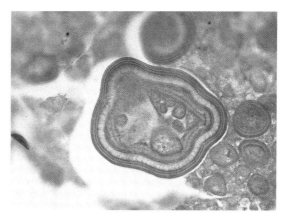

Figure 1.31
Liesegang ring showing conspicuous laminations. x435

Phylum: Arthropoda

The phylum Arthropoda contains over a million described species. It is a very diverse group that includes crustaceans, insects, spiders, ticks, mites, centipedes, millipedes, and other miscellaneous organisms. Arthropods have a hardened exoskeleton containing chitin, striated muscles, and a complete digestive tract. Insects, ticks, and mites contain an elaborate system of air tubes (tracheae). Some arthropods that cause disease in humans may be mistaken for helminths. Several of these are illustrated in figures 1.20 to 1.26. This subject will be discussed in detail in another volume of this series.

Pseudoparasites and Artifacts

Frequently, structures observed in tissue sections are misinterpreted as parasites because they mimic adult worms, larvae, microfilariae, or eggs (Figs 1.27 to 1.65). Often these structures, though not parasites, cannot be specifically identified and remain as "unknowns." Knowledge of parasite anatomy, basic parasitology, and the histopathologic changes parasites cause usually allow the examiner to distinguish between real and false parasites.

Pseudoparasites and artifacts may be endogenous or exogenous. Many of the kinds of structures that may mimic parasites are listed in Table 1.4.

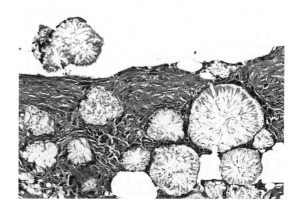

Figure 1.32
Calcium salt crystals in soft tissue. x125

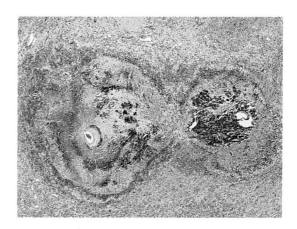

Figure 1.33
Gamma-Gandy body in spleen. Mineralized tissue has haphazard configuration. x25

Table 1.1 Some comparative morphologic characteristics of the 4 major groups of parasitic helminths.

Feature	Trematodes (flukes)	Cestodes (tapeworms)	Nematodes (roundworms)	Acanthocephalans (thorny-headed worms)
Body type	Frequently leaf-shaped (Fig 4.26). Flat, solid, spongy parenchyma with no large cavities. Usually have oral and ventral suckers (Fig 3.1).	Adults have scolex (Figs 7.9 & 10.1), neck, and strobila (proglottids). Flat, solid, spongy parenchyma with calcareous corpuscles (Fig 7.6). Larvae may be cystic (Figs 7.12 & 9.23) or solid (Fig 10.33).	Adults usually elongate and cylindrical (Fig 26.2). Large fluid-filled cavity (pseudocoel) (Fig 26.7). Caudal end curled (Fig 15.3) in usually smaller male.	Short anterior presoma (including a retractile spiny proboscis)(Fig 37.5) and elongated metasoma or trunk (body proper). Trunk surface may appear pseudosegmented (Fig 37.1). Lemnisci thought to function in proboscis extension.
External covering	Tegument: A dynamic living syncytium. Often with spines (Fig 3.3).	Tegument: A dynamic living syncytium. Outer surface with microvilli, +/- folds for increased surface area for absorbing nutrients (Figs 7.6 & 8.15).	Cuticle: A nonliving, acellular, tough covering secreted by the hypodermis (Fig 26.4). Varies greatly in thickness and number of layers; many with 3 main layers (Fig 16.10). Cuticle frequently with striations, annulations, ridges, spines, alae, and other forms of ornamentation.	Tegument is thin (Fig 37.9). Proboscis armed with hooks. Trunk frequently covered with spines.
Hypodermis	Absent	Absent	Hypodermis is a thin cellular layer between the cuticle and muscles (Fig 26.5). Protrudes into body cavity in 4 areas to form longitudinal hypodermal cords. Lateral cords are largest (Fig 26.6), may contain excretory canals, and are useful in diagnosis. Nuclei usually confined to cords.	Hypodermis (epidermis) is a syncytial, fibrous tissue with 3 layers (Figs 37.9 & 37.10): (1) outer thin layer of parallel radial fibers perpendicular to the surface; (2) thick middle layer of fibers running in different directions; (3) thick inner layer of radial fibers containing a lacunar system of channels. A very thin basement layer (dermis) separates the hypodermis from underlying muscles.
Somatic musculature	Below teguement: Smooth muscles (Fig 3.13). Usually consecutive layers of circular, longitudinal and diagonal muscle. (Diagonal layer may be between circular and longitudinal layers or may be absent.)	1. Below tegument: Smooth muscles. Consecutive layers of circular and longitudinal muscle (Fig 7.6). 2. Mesenchymal musculature of longitudinal, transverse, and dorsoventral fibers (Fig 10.24).	Below hypodermis: Smooth muscles. Muscle cells are spindle-shaped and only longitudinally arranged. Each cell is composed of a basement membrane, contractile fibers, and sarcoplasm containing the nucleus. Platymyarian muscle cells have contractile fibers limited to contact with the hypodermis (Fig 22.13). Coelomyarian muscle cells have contractile fibers in contact with the hypodermis but also extending up sides of cell (Fig 26.5). Meromyarian musculature refers to few (2 to 5) muscle cells per quadrant (Fig 29.4). Polymyarian musculature refers to many (6 or more) muscle cells per quadrant (Fig 15.11).	Below hypodermis (epidermis): Smooth muscles. Two layers of somatic muscles (outer circular, inner longitudinal) (Figs 37.9 & 37.10). Other muscles extend from body wall to proboscis and lemnisci.
Digestive tract	Incomplete. Mouth surrounded by oral sucker (Fig 5.14), muscular pharynx; short esophagus that bifurcates into 2 lateral blind ceca (Fig 3.1). Ceca usually straight tubes, but may be highly branched (Fig 4.26).	Absent	Complete. Mouth, stoma, esophagus, intestine, rectum, and anus. Cloaca in males. Esophagus is cuticular-lined and has triradiate lumen (Figs 26.6 & 31.2). Esophagus may be divided into corpus, isthmus, and bulb (Fig 29.3). Esophageal glands may be modified into a stichosome (Figs 31.3 & 33.1). Intestine is a single layer usually of microvilli-lined epithelial cells (Fig 22.14). There is great diversity in structure of esophagus and intestine among species.	Absent

Feature	Trematodes (flukes)	Cestodes (tapeworms)	Nematodes (roundworms)	Acanthocephalans (thorny-headed worms)
Reproductive organs	Hermaphroditic (monoecious) (Fig 4.2) except schistosomes (Fig 2.6). Complex systems. Usually 2 testes and 1 ovary. Sperm are filamentous. Uterus frequently coiled. Yolk glands (vitellaria) lie in lateral fields (Fig 4.38). Sperm and eggs exit into common genital atrium. Male schistosome usually encloses the female in its gynecophoral canal (Fig 2.11).	All hermaphroditic (monoecious) (Fig 12.18). Complex systems. Reproductive organs occupy a major part of each mature proglottid. Testes are numerous but usually 1 ovary. Sperm are filamentous. *Taenia* sp uteri are branched (Fig 7.4). Usually common male and female atrium and gonopore. Male organs may disappear as proglottids mature.	Most dioecious. Gonads tubular and may be long and coiled (Fig 26.7). Males with single reproductive tube (Fig 13.8) usually composed of testis, vas deferens, seminal vesicle, and ejaculatory duct that extends posteriorly. One or 2 copulatory spicules (Fig 17.17). Sperm are round, ameboid, or elongate. Females may have 1 but most have 2 (Fig 17.9) reproductive tubes usually composed of ovary, oviduct, seminal receptacle, uterus, vagina, and vulva. Uteri contain eggs, larvae, or microfilariae.	Dioecious. Complex systems. Reproductive organs enclosed in ligament sacs. Males have 2 testes and produce filamentous sperm that exit through an eversible penis. Males have bursa. Female ovarian tissue breaks up into balls that eventually float freely in the pseudocoel (Fig 37.11). Females have uterine bell. Eggs exit female through genital pore.
Eggs	All except schistosomes have an operculum (Figs 2.10 & 3.6).	All medically important cyclophyllidean tapeworm eggs lack an operculum (Fig 12.12). *Taenia* sp have thick striated shells (Fig 7.11). Pseudophyllidean tapeworms have operculated eggs (Fig 10.19).	Great variability in shell structure (polar plugs (Fig 31.16), mamillations (Fig 26.33), pitting (Fig 36.4), striations (Fig 33.25)). Filarial worms produce microfilariae instead of eggs (Fig 13.5).	Egg development takes place within female pseudocoel (Fig 37.17). Acanthor larva with 3 pairs of larval hooks develops within egg (Figs 37.12 & 37.13) that passes out of vertebrate host in feces.
Excretory system	Bladder and protonephridia (Figs 3.1 & 4.43) (osmoregulatory tubules containing flame cells) usually open posteriorly at nephridiopore (rarely, 2 anterior nephridiopores). Frequently obscure.	Flame cells and tubules in mesenchyme drain into 4 peripheral longitudinal collecting canals. Usually obscure.	May be glandular or tubular. Glandular type consists of 1 or 2 renette glands; usually ventrally near esophagus; a cellular extension runs to the exterior via a midventral excretory pore. Most parasitic nematodes have tubular form consisting of longitudinal canals that run within lateral cords; a transverse canal leads to excretory pore located ventrally at midline. Tubular system of 3 long canals forms an H shape.	Archiacanthocrephala have 2 protonephridia associated with reproductive system.

Table 1.2 Terminology of cestode larvae.

Type of larva	Definition
Hexacanth	Oncosphere: small, spheroid larva that emerges from egg of any Eucestoda (tapeworm) (Fig 12.13). Usually 3 pairs of small hooklets.
Coracidium	Hexacanth with ciliated epithelium, present in the Pseudophyllidea.
Decacanth (lycophore)	Small larva that emerges from egg of cestodarian. Similar to a hexacanth but has 10 hooklets.
Procercoid	Simple, elongated, solid larva, usually with posterior bulb (cercomer) bearing hexacanth hooklets. Cercomer often lost before procercoid is infective to the next host. Body has no apparent segmentation. This stage in first intermediate host usually develops into more advanced stage. Develops in arthropods and annelids. Absent in the Cyclophyllidea.
Plerocercoid	Solid larva that develops from procercoid in second intermediate host. Elongate, solid structure with undifferentiated scolex (with rudimentary ventral groove (bothria) or tentacles) and without true body segmentation (Fig 10.20). Can pass unchanged into paratenic host. Develops into adult worm in definitive host. Termed a sparganum if species unknown.
Plerocercus	Solid larva that develops from procercoid in second intermediate host. Scolex is armed with 4 tentacles. Posterior end of body is bladder (blastocyst) into which rest of body can withdraw. Typical of Trypanorhyncha.
Cysticercus	Fluid-filled, bladder-like cyst containing single armed protoscolex (Figs 7.12 & 7.16). Example: *Cysticercus cellulosae* (larva of *T. solium*). Calcareous corpuscles most numerous in neck region.
Cysticercoid	Small, solid-bodied, cyst-like larva, usually with single protoscolex invaginated into a cavity. Frequently has caudal appendage. Usually develops in invertebrate host, but can develop in mammals. Example: *Hymenolepis nana* (Figs 12.31 & 12.32).
Strobilocercus	Elongated cysticercus that shows strobilization while still in intermediate host (cysticercus with segmentation) (Fig 12.40). *Taenia* sp that regularly form strobiloceri previously grouped under genus *Hydatigera*.
Coenurus	Fluid-filled, bladder-like cyst with numerous invaginated protoscolices (Figs 11.4 & 11.10). *Taenia* sp with these features were previously grouped under genus *Multiceps*.
Hydatid cyst (unilocular)	Spherical, fluid-filled cyst ranging from a few mm to many cm in diameter (Fig 9.23). Tegument has outer laminated wall and inner germinal membrane (Fig 9.6). Mature cyst contains brood capsules, each with many protoscolices (Fig 9.11).
Hydatid cyst (polycystic or multicystic)	Cyst that proliferates by endogenous budding of germinal membrane, forming folds and pockets within primary vesicle (Fig 9.33). In human host, may also have exogenous proliferation.[14] Fluid-filled cysts divide internally to form multichambered growths (Fig 9.29). Protoscolices rare in human infections (Fig 9.30).

Table 1.3 Some morphologic features of gordiids (horsehair worms).

Feature	Gordiids
Body type	Adults and larvae have long, slender, cylindrical bodies generally of uniform diameter (Figs 1.5 and 1.6). They frequently twist and turn upon themselves to form complicated knots. Males usually smaller than females. Posterior end is rounded or has 2 or 3 caudal lobes (Fig 1.7) depending on genus and sex. Pseudocoelom may be spacious or filled with mesenchyme (Fig 1.10).
External covering	Thick, multilayered cuticle. May have conspicuous areoles, inconspicuous areoles, or no areoles. Areoles are contiguous rounded or polygonal areas in the cuticle that often project above the surface as papillae, warts, or bumps (Figs 1.8 and 1.12).
Hypodermis	Thin nucleated layer between cuticle and somatic muscle cells (Fig 1.12). Lateral cords lacking.
Somatic musculature	Smooth. Single layer of longitudinal muscle cells. Contractile fibers completely surround the protoplasm of each cell (circomyarian) (Fig 1.12).
Digestive tract	Reduced and nonfunctional in adults.
Reproductive organs	Dioecious. Paired testes in male (Fig 1.9) and paired ovaries in female. Cloaca in both sexes. Males have no copulatory spicules; sperm are elongate. Ovaries in female develop lateral diverticula. Eggs exit female through cloaca.
Nervous system	Ventral nerve cord present (Fig 1.11).

Table 1.4 Structures that mimic helminths.

Endogenous
    Lipid pseudomembrane (Figs 1.27 to 1.29)
    Liesegang rings (Figs 1.30 and 1.31)
    Mineralization (Figs 1.32 to 1.35)
    Saponification of fat cells (Fig 1.36)
    Hair shafts (Fig 1.37)
    Heterotopic epithelioid nest (Fig 1.38)

Exogenous
    Lycopodium clavatum spore (Figs 1.39 to 1.43)
    Balantidium coli (Fig 1.44)
    Food items
        Plants (seeds) (Figs 1.45 to 1.50)
        Animals (squid) (Figs 1.51 and 1.52)
    Arthropods (whole or fragments) (Figs 1.20 to 1.26)
    Helicosporum spore (Fig 1.53)
    Adiaconidia (fungal conidia) (Fig 1.54)
    Feather (Figs 1.55 and 1.56)
    Splinter (Fig 1.57)
    Staining artifacts (Figs 1.58 to 1.60)
    Unknown (Figs 1.61 to 1.65)

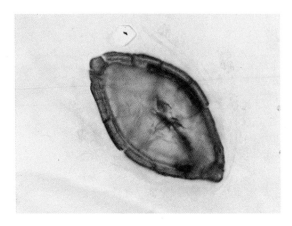

Figure 1.34
Uric acid crystal in urine sediment, easily identified by diamond shape.* Unstained x345

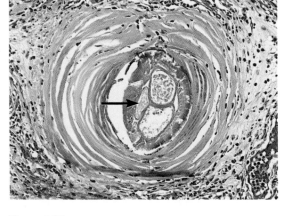

Figure 1.35
Partially mineralized filarial nematode in fibrotic lesion of skin. Spherical, well-circumscribed configuration of parasite is retained, as well as preserved intestine (arrow) and paired uteri. Movat x115

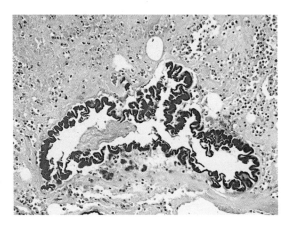

Figure 1.36
Artifact in skin of forearm mimicking larval cestode and believed to be saponified fat cell. Movat x125

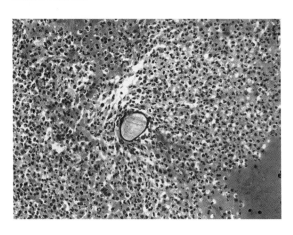

Figure 1.37
Transverse section through hair shaft within exudate of appendix. x180

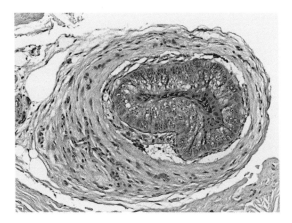

Figure 1.38
Heterotopic epithelial rest in mass from neck. x115

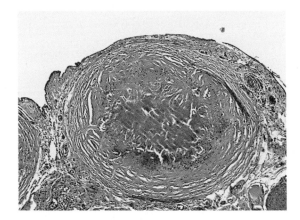

Figure 1.39
Lycopodium granuloma on serosal surface of colon in patient who had had 2 previous abdominal surgeries. x45

The Pathogenic Helminths • 1

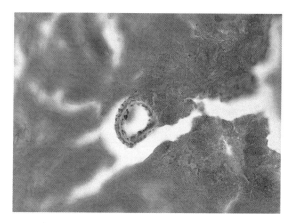

Figure 1.40
Lycopodium clavatum spore within necrotic granuloma described in Figure 1.39. Spores are 20 to 30 μm in diameter and have a spiny outer coat. x450

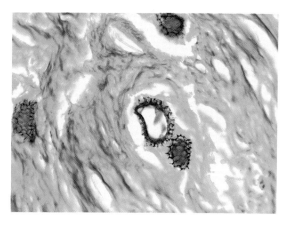

Figure 1.41
GMS stain reveals *L. clavatum* spore's spiny coat. x420

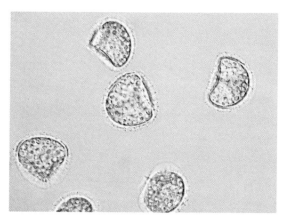

Figure 1.42
Unstained *L. clavatum* spores. x435

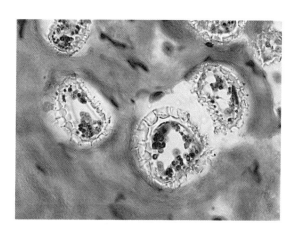

Figure 1.43
Lycopodium clavatum spores in prostate. x600

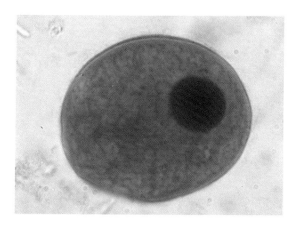

Figure 1.44
Balantidium coli cyst in feces may be mistaken for helminth egg. x575

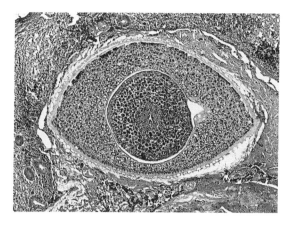

Figure 1.45
Section through seed in lumen of appendix.

17

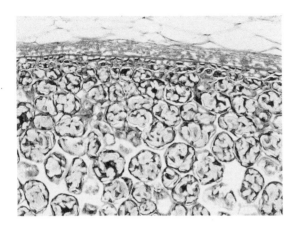

Figure 1.46
Section through lima bean showing outer coat and numerous starch grains. Movat x120

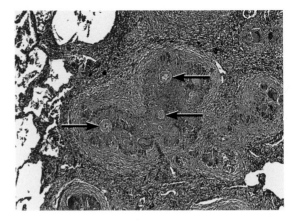

Figure 1.47
Aspiration of leguminous vegetables may result in lentil pneumonia. Lesions begin as abscesses, but are eventually replaced by granulomatous inflammation, as depicted in this patient's lung. Note small islands of vegetable material (arrows) within area of inflammation. x45

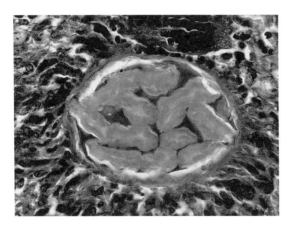

Figure 1.48
Starch grain in granuloma of lung. x450

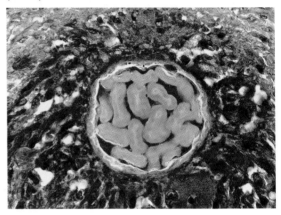

Figure 1.49
Movat-stained section of starch grain in granuloma of lung. Starch grains are usually well-circumscribed, spherical structures with readily identifiable internal, convoluted endosperm. x400

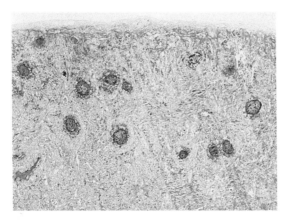

Figure 1.50
Numerous starch grains in wall of gallbladder. PAS x45

Figure 1.51
Section through branching structure removed from Filipino woman during nasal suctioning. Additional clinical history revealed specimen to be regurgitated cooked squid. x11

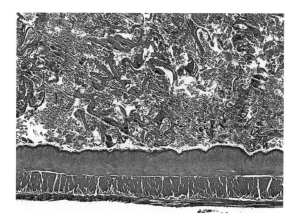

Figure 1.52
Higher magnification of cooked squid described in Figure 1.51 reveals preserved layers of muscle in tentacle. x40

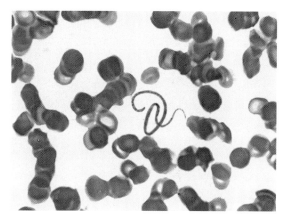

Figure 1.53
Fungal spore of *Helicosporum* in thin peripheral blood film. Spore resembles microfilaria, but lacks definitive diagnostic features. Giemsa x675

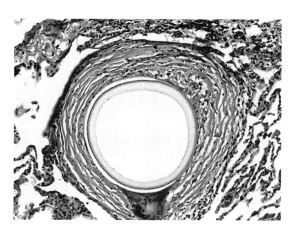

Figure 1.54
Thick-walled adiaconidia of *Emmonsia parva* var *crescens* in fibrotic nodule of lung may be mistaken for helminth or helminth egg. x125

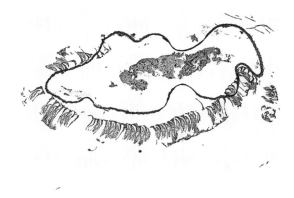

Figure 1.55
Specimen recovered from child's nose, thought to be a worm, was ultimately identified as a feather. x35

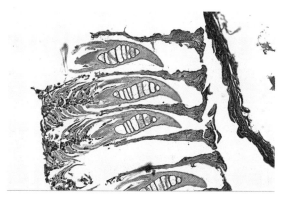

Figure 1.56
Higher magnification of feather described in Figure 1.55. x220

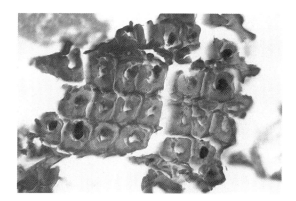

Figure 1.57
Splinter in subcutaneous tissue from patient with phaeomycotic cyst. x215

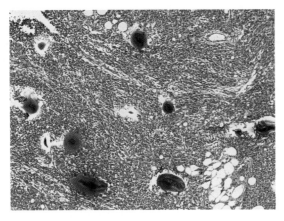

Figure 1.58
Staining artifacts in omental tumor, mistaken for parasites. x45

Figure 1.59
Higher magnification of artifact in Figure 1.58. Although some degree of organization is apparent, characteristic features of parasite are lacking. x165

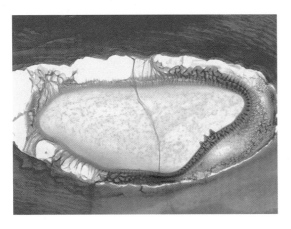

Figure 1.60
Staining artifact mimicking helminth in rectal biopsy. x50

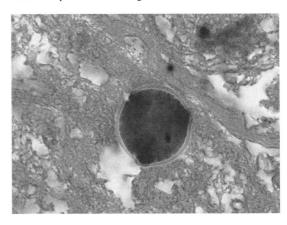

Figure 1.61
Unidentified nonparasitic structure in drainage from cyst in liver. x585

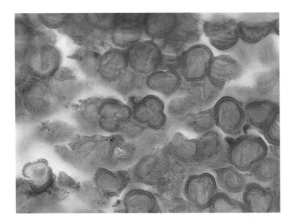

Figure 1.62
Unidentified nonparasitic structures in fallopian tube. x705

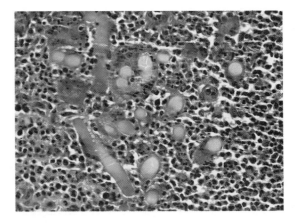

Figure 1.63
Unidentified nonparasitic structures in dermal abscess, possibly synthetic fibers. x240

Figure 1.64
Unidentified nonparasitic structures in stomach wall. x240

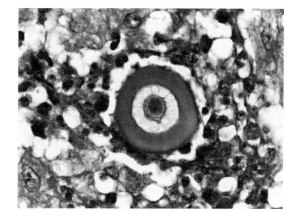

Figure 1.65
Unidentified nonparasitic structure within tissue of tonsil. x620

References

1. Andersen K. Studies of the helminth fauna of Norway. 34. The morphological stability of Diphyllobothrium Cobbold. A comparison of D. dendriticum (Nitzsch), D. latum (L.) and D. ditremum (Creplin) developed in different hosts. *Norw J Zool* 1975;23:45–53.

2. Brooks DR, Bandoni SM, MacDonald CA, O'Grady RT. Aspects of the phylogeny of the Trematoda Rudolphi,1808 (Platyhelminthes: Cercomeria). *Can J Zool* 1989;67:2609–2624.

3. Brooks DR, McLennan DA. *Phylogeny, Ecology, and Behavior: A Research Program in Comparative Biology.* Chicago, Ill: University of Chicago Press; 1991:1–434.

4. Camus D, Zalis MG, Vannier-Santos MA, Banic DM. The art of parasite survival. *Braz J Med Biol Res* 1995;28:399–413.

5. Fitch DH, Bugaj-Gaweda B, Emmons SW. 18S ribosomal RNA gene phylogeny for some Rhabditidae related to Caenorhabditis. *Mol Biol Evol* 1995;12:346–358.

6. Flisser A. Advances in parasitism by larval cestodes [in French]. *Ann Parasitol Hum Comp* 1991;66:32–36.

7. Gardner SL, Stock SP, Kaya HK. A new species of Heterorhabditis from the Hawaiian Islands. *J Parasitol* 1994;80:100–106.

8. Garey JR, Near TJ, Nonnemacher MR, Nadler SA. Molecular evidence for Acanthocephala as a subtaxon of Rotifera. *J Mol Evol* 1996;43:287–292.

9. Gibson DI, Bray RA. The evolutionary expansion and host-parasite relationships of the Digenea. *Int J Parasitol* 1994;24:1213–1226.

10. La Volpe A. A repetitive DNA family, conserved throughout the evolution of free-living nematodes. *J Mol Evol* 1994;39:473–477.

11. Little MD. Laboratory diagnosis of worms and miscellaneous specimens. *Clin Lab Med* 1991;11:1041–1050.

12. Marty AM, Chester AJ. Distinguishing lipid pseudomembranes from larval cestodes by morphologic and histochemical means. *Arch Pathol Lab Med* 1997;121:900–907.

13. McManus DP, Bryant C. Biochemistry, physiology, and molecular biology of Echinococcus. In: Thompson RC, Lymbery AJ, eds. *Echinococcus and Hydatid Disease.* Wallingford, England: CAB Int; 1995:135–181.

14. Poinar, GO Jr. *Nematodes for Biological Control of Insects.* Boca Raton, Fla: CRC Press, Inc; 1979:225–231.

15. Rausch RL, D'Alessandro A, Rausch VR. Characteristics of the larval Echinococcus vogeli Rausch and Bernstein, 1972 in the natural intermediate host, the paca, Cuniculus paca L. (Rodentia: Dasyproctidae). *Am J Trop Med Hyg* 1981;30:1043–1052.

16. Sprent JF. Ascaridoid nematodes of South American mammals, with a definition of a new genus. *J Helminthol* 1982;56:275–295.

17. Yamaguti S. *Synopsis of Digenetic Trematodes of Vertebrates.* 2 vols. Tokyo, Japan: Keigaku Pub Co; 1971.

2

Schistosomiasis

Allen W. Cheever *and*
Ronald C. Neafie

Introduction

Definition

Schistosomiasis is infection by flukes (trematodes) of the genus *Schistosoma*.

Three species of *Schistosoma* cause most of the serious disease in humans: *Schistosoma haematobium* (urinary schistosomiasis); *Schistosoma mansoni* (intestinal schistosomiasis); and *Schistosoma japonicum* (intestinal schistosomiasis in Asia). Additional species that sometimes infect humans are *Schistosoma mekongi*, *Schistosoma intercalatum,* and *Schistosoma mattheei,* the latter being a common schistosome of domestic and wild animals of some African countries. A few infections of humans by other schistosomes have been reported. Certain species parasitic in other mammals and in birds produce cercarial dermatitis in humans.

Synonyms

Schistosomiasis is often called bilharziasis, after Bilharz, who discovered schistosomes.

General Considerations

A disease retrospectively identified as schistosomiasis was described in Egyptian records dating from about 1900 BC. Schistosomiasis is of comparable antiquity in China. Fujii described oriental schistosomiasis in 1847. Bilharz discovered adult schistosomes in 1851 in a cadaver and later established that the parasite and its eggs cause hematuria. Fujinami and Katsurada described *S. japonicum* and its eggs from studies of human and animal infections in 1904. In 1907, Sambon distinguished 2 species in Africa: *S. haematobium* and *S. mansoni*. In 1913 and 1914, Miyairi and Suzuki hatched eggs of *S. japonicum*, saw the miracidia penetrate a snail, and described development of sporocysts and cercariae in the snail. Between 1915 and 1918, after a visit to Japan, Leiper proved experimentally that both *S. haematobium* and *S. mansoni* required snails as intermediate hosts, and that the urinary and intestinal forms of schistosomiasis were produced by different species of schistosomes.[12,24]

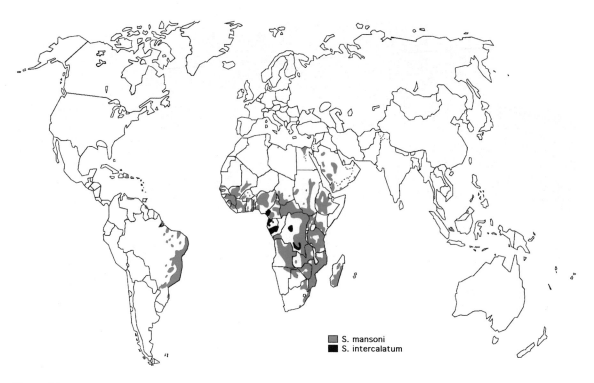

Figure 2.1
Distribution of *S. mansoni* and *S. intercalatum*, schistosomes that infect humans.*

Epidemiology

The geographic distribution of schistosomiasis in humans depends upon the distribution of snail hosts and opportunities for infection of both the snails and humans (Figs 2.1 and 2.2).

Schistosomes infect approximately 10% of the world population and are a major cause of morbidity and mortality.[1,2] Schistosomiasis is spreading geographically and increasing in prevalence. Water conservation and irrigation projects provide additional habitats for freshwater snails that serve as intermediate hosts (Figs 2.3 to 2.5). Widespread treatment programs reduce the intensity of infection and decrease morbidity, but seldom affect transmission for extended periods. To be effective, control must be sustained over years; however, in affected countries resources for prolonged control are seldom available. General improvement in socioeconomic conditions has often resulted in decreased transmission and morbidity. There is hope that vaccines may eventually reduce the intensity of infection and morbidity.

Transmission is characteristically focal: one water source in an endemic area may be highly infectious and a nearby source noninfectious.

Infectious Agent

Morphologic Description

Schistosomes have several unique features distinguishing them from other flukes that infect humans: they inhabit blood vessels, the sexes are separate, the eggs are nonoperculated, and their metacercariae are not encysted.

Schistosomes are slender, elongated flukes 0.6 to 2.6 cm long by 0.01 to 0.1 cm wide. Male worms are shorter and stouter than females of the same species. Males have multiple testes situated posterior to the ventral sucker and a gynecophoral canal in which the female worm lies (Fig 2.6). Females have a single ovary near the midbody. Vitellaria are located in the lateral fields in the posterior half of the body.

Schistosomes have an oral and a ventral sucker at the anterior end (Fig 2.6). Suckers vary considerably in size, ranging from 50 to 400 μm. The intestinal cecum bifurcates anteriorly but unites

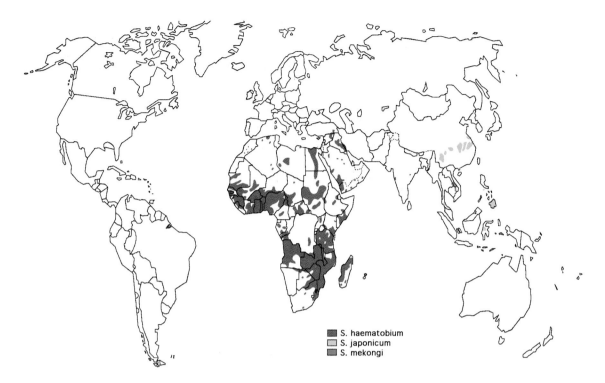

Figure 2.2
Distribution of *S. haematobium*, *S. japonicum*, and *S. mekongi*, schistosomes that infect humans.*

again posteriorly to form a single blind tube. The cecum usually contains brown granular material (schistosomal pigment). The tegument of the worm is a syncytium of living cells rather than a cuticle. There is no body cavity, and the internal organs are suspended in a spongy parenchyma. Schistosome eggs have no opercula and are immature when laid. *Schistosoma mansoni* lays about 400 eggs per day, *S. haematobium* perhaps 200, and *S. japonicum* about 2000. Optimal egg production and passage of eggs in the feces of nude and severe combined immunodeficient (SCID) mice infected with *S. mansoni* indicate the requirement of host immune reaction, including production of tumor necrosis factor.[2] Host immunity may eventually decrease egg production and egg passage in immunologically normal animals. Comparison of circulating worm antigen levels to egg passage suggests that the fecundity of *S. mansoni* does not change with duration of infection in humans, while fewer *S. haematobium* eggs per worm pair are passed in chronic infections.[1]

The tegument of *S. mansoni* males contains numerous prominent tuberculations (Figs 2.6 to 2.8), while the tegument of females of this and other species is nontuberculated. The uterus is short and usually contains a single egg. *Schistosoma mansoni* eggs are 114 to 175 μm by 45 to 68 μm and have a prominent lateral spine (Figs 2.9 and 2.10). Lateral spines are 5 to 10 μm wide at the base, sharply pointed at the tip, and 15 to 25 μm long.

Tuberculations of *S. haematobium* males (Fig 2.11) are fewer and less pronounced than those of *S. mansoni*. The uterus in the female is long and usually contains 20 to 30 eggs. *Schistosoma haematobium* eggs are 112 to 170 μm by 40 to 70 μm and have a small terminal spine (Figs 2.12 and 2.13). Terminal spines are 2 to 3 μm wide at the base, pointed or rounded at the tip, and 5 to 10 μm long.

Schistosoma japonicum males have no tegumental tuberculations (Fig 2.14). The uterus in the female is long and usually contains 50 to 100 eggs. *Schistosoma japonicum* eggs are 70 to 100 μm by 50 to 65 μm (Figs 2.15 and 2.16) and have a minute lateral spine. In tissue sections, however, they are rarely more than 80 μm long and the spine is usually unidentifiable. Extensive descriptions of the morphology and biology of the schistosomes are given by Basch.[6]

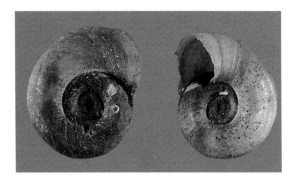

Figure 2.3
Empty shells of typical *Biomphalaria* sp of snails, the intermediate hosts of *S. mansoni*. x3.75

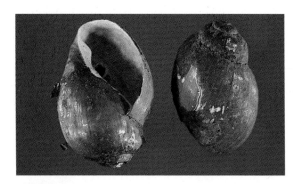

Figure 2.4
Empty shells of *Bulinus* sp of snails, intermediate hosts of *S. haematobium*. x3

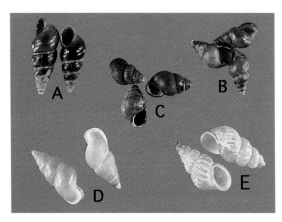

Figure 2.5
Empty shells of *Oncomelania* sp, intermediate hosts of *S. japonicum*. A) *O. hupensis hupensis* from China; B) *O. formosana* from Taiwan, transmits zoophilic strain of *S. japonicum*; C) *O. hupensis chiui* from Taiwan; D) *O. quadrasi* from the Philippines; and E) *O. hupensis nosophora* from Japan.* x3

Figure 2.6
Scanning electron micrograph of adults of *S. mansoni* depicting female worm in gynecophoral canal of male. Note oral and ventral suckers of male worm at top of photograph.*

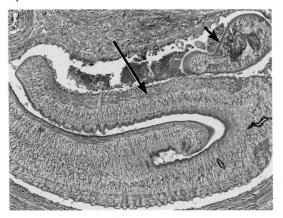

Figure 2.7
Adults of *S. mansoni* in urinary bladder wall. Male (long arrow) has many prominent tegumental tuberculations. Female (short arrow) is nontuberculated. x55

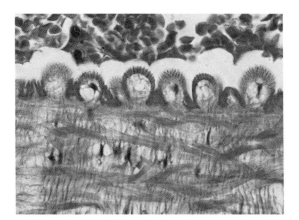

Figure 2.8
Higher magnification of adult *S. mansoni* male in Figure 2.7, depicting prominent tuberculations. x585

SCHISTOSOMIASIS • 2

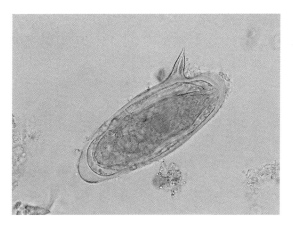

Figure 2.9
Egg of *S. mansoni* from feces. Eggs are 114 to 175 μm by 45 to 68 μm and have a large lateral spine. Unstained x265

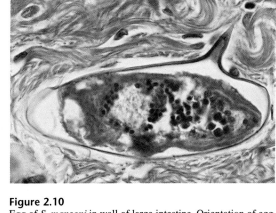

Figure 2.10
Egg of *S. mansoni* in wall of large intestine. Orientation of egg clearly shows large, sharply pointed, yellowish-brown lateral spine. x605

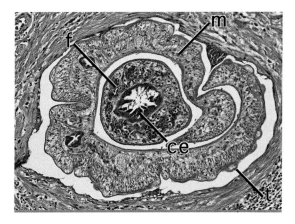

Figure 2.11
Adults of *S. haematobium* in vein of cervix. Male (m) tegument has a few inconspicuous tuberculations (t); female (f) tegument is nontuberculated. Note brown pigment in cecum (ce) of female. x100

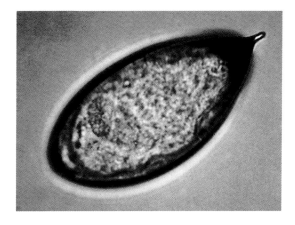

Figure 2.12
Egg of *S. haematobium* from urine. Eggs are similar to those of *S. mansoni* in size, but have a small terminal spine. Unstained x535

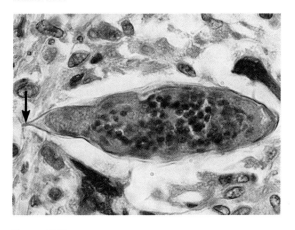

Figure 2.13
Egg of *S. haematobium* in wall of intestine. Orientation of egg shows small terminal spine (arrow). x625

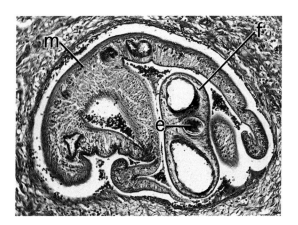

Figure 2.14
Adults of *S. japonicum* in intestinal wall. The tegument of neither male (m) nor female (f) has tuberculations. An egg (e) lies in the uterus of the female. x120

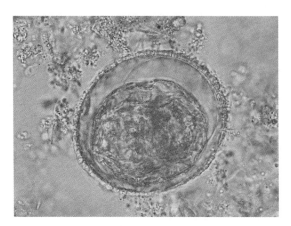

Figure 2.15
Egg of *S. japonicum* from feces. Eggs measure 70 to 100 μm by 50 to 65 μm. The minute lateral spine is not evident here. Unstained x500

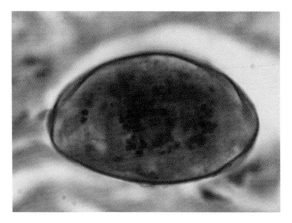

Figure 2.16
Egg of *S. japonicum* in wall of appendix. Minute lateral spine is not evident and is rarely observed in tissue sections. x890

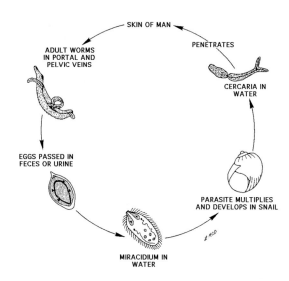

Figure 2.17
Life cycle of *S. haematobium*, *S. mansoni*, and *S. japonicum*.

Life Cycle and Transmission

Humans are the only important definitive host of *S. mansoni* and *S. haematobium*, although nonhuman primates occasionally transmit *S. mansoni* (Fig 2.17). Wild and domestic animals are important reservoirs of *S. japonicum*.

Schistosoma mansoni and *S. haematobium* are transmitted, respectively, by nonoperculated snails of the genera *Biomphalaria* (Fig 2.3) and *Bulinus* (Fig 2.4), and *S. japonicum* by operculated snails of the genus *Oncomelania* (Fig 2.5).

Immature eggs of *S. mansoni* and *S. japonicum* are deposited in small venules of the intestine and rectum; eggs of *S. haematobium* are deposited in venules of the urinary bladder. Embryos within the eggs mature in 1 week. About half the deposited eggs stay in the tissues. The remainder penetrate the venous capillaries, work their way through the wall of the intestine or bladder, and pass in feces or urine (Figs 2.9, 2.12, and 2.15).

Once outside the body and in water, the eggs hatch, liberating miracidia that penetrate a suitable snail (Fig 2.18). Within the snail, a single miracidium multiplies and develops into numerous fork-tailed cercariae in 1 to 2 months (Fig 2.19). Cercariae escape from the snail into fresh water where most die within hours, but some remain viable for up to 2 days. When a cercaria penetrates the skin of a definitive host, the tail drops off. The developing worm, or schistosomule (Fig 2.20), migrates through the tissues, penetrates a vessel, and is carried to the lung (Fig 2.21) where it remains for approximately 1 week. The schistosomule then migrates to the liver, probably via the arterial circulation, and through intestinal capillaries to the portal vein. Schistosomes of humans become sexually mature in hepatic portal venules where they pair (Fig 2.22) and then migrate to veins of the portal system (Fig 2.23) or the urinary tract. *Schistosoma mansoni* and *S. japonicum* mature in 4 to 5 weeks and *S. haematobium* in about 10 weeks.

Clinical Features and Pathogenesis

Cercariae penetrating the skin may cause itching and a maculopapular rash, particularly in previously infected individuals.

Acute Schistosomiasis

Acute schistosomiasis develops after the first exposure of immunologically naive individuals, but seldom in inhabitants of endemic areas. The patient becomes febrile and frequently has a cough, asthma, hives, and dysentery approximately 2 to 4 weeks after exposure. These symptoms are sometimes misdiagnosed as typhoid fever. Eosinophilia is marked. Hepatosplenomegaly and diffuse lymphadenopathy are common. When egg laying begins, these signs are accentuated, then usually subside over a period of weeks or months as immune downregulation develops in the host. Persons with acute schistosomiasis have large, hyperergic circumoval granulomas.

Intermediate Period

Following the acute phase, and in patients who have not experienced the acute phase, the infection is usually silent. Granulomas are smaller and less exudative than those of the acute phase.

Chronic Schistosomiasis

The basic lesions of schistosomiasis are circumscribed granulomas around eggs, or a diffuse cellular infiltrate around eggs. Either reaction may be followed by fibrosis. Eosinophils usually predominate in the diffuse infiltrates (Figs 2.24 and 2.25), but there are also plasma cells, lymphocytes, macrophages, and giant cells. The circumoval granulomas are primarily reactions of cell-mediated immunity in which cytokines play a prominent role in determining both granuloma size and subsequent fibrosis.[2,20,25] Diffuse reactions are probably similarly regulated. Granulomas vary in structure (Figs 2.26 to 2.28). Some eggs are surrounded by a layer of eosinophilic material, the Splendore-Hoeppli phenomenon (Fig 2.29). Immature and dead eggs provoke little or no host response. Many eggs are destroyed in the tissues; others persist, often surrounded by fibrous tissue. Eggs of *S. haematobium* and *S. japonicum*

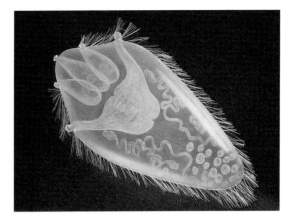

Figure 2.18
Miracidium of *S. japonicum*. Miracidia have only a few hours to find and penetrate a suitable freshwater snail. Darkfield x935

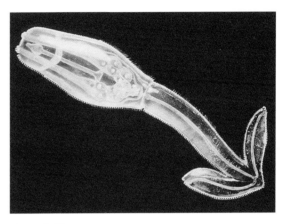

Figure 2.19
Fork-tailed cercaria of *S. japonicum*. Darkfield x240

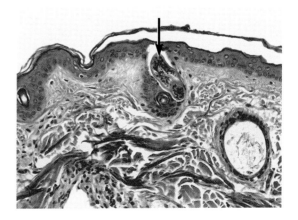

Figure 2.20
Schistosomule in skin (arrow). x250

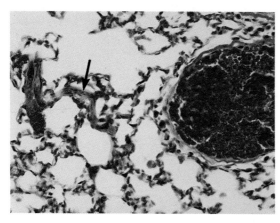

Figure 2.21
Schistosomule of *S. mansoni* (arrow) migrating through lung of mouse. Posterior half of worm is folded upon itself in a terminal arteriole. Schistosomule must navigate capillaries in 2 adjacent alveolar walls to reach venule. x250

Figure 2.22
Adult male *S. mansoni* with female in gynecophoral canal. x20

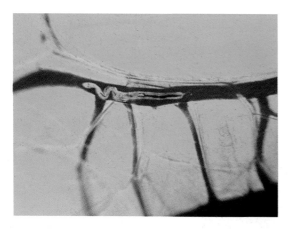

Figure 2.23
Adult male and female *S. mattheei* in mesenteric vein of baboon. Dark streak on right is pigment in female.* x9

frequently become calcified and surrounded by hyalinized scar tissue (Fig 2.30). There are scarred or healed circumoval granulomas in these areas.

Brown, birefringent hematin pigment, produced when adult worms digest erythrocytes, accumulates in Kupffer's cells or in macrophages in granulomas. This pigment causes no reaction and cannot be distinguished from malarial pigment by light microscopy (Fig 2.31).

While alive in the veins, adult schistosomes provoke no host response, but dead worms cause large focal lesions (Fig 2.32). When schistosomicidal drugs kill adult worms, the worms become verminous emboli to the liver or lungs, where they form obstructive intravascular lesions which are gradually resolved and seldom associated with significant clinical signs.

Most patients with chronic schistosomiasis are asymptomatic; microscopically they have only circumoval granulomas and small areas of fibrosis. In some patients, however, hepatosplenic disease or obstructive uropathy develop silently. Severe lesions are more common in patients with heavy infections, but few or no live worms may remain when the patient first seeks attention years after the worms die from old age, host immune response, or treatment. Ectopic localization of worms and eggs in the central nervous system may cause paralyzing or life-threatening focal lesions, which frequently develop in patients with light infections.

Pathologic Features

Intestinal Changes

Most intestinal lesions are caused by *S. mansoni* and *S. japonicum*, but *S. haematobium* may also affect the colon. Petechial hemorrhages and circumoral granulomas in the lamina propria and submucosa are the most common lesions. During the acute stage, diarrhea and abdominal pain are frequently reported. Diarrhea in chronically infected patients has been noted in controlled studies in some, but not all, endemic areas. All 3 species deposit eggs in the appendix (Fig 2.33), but schistosomiasis seldom causes clinical appendicitis.

Inflammatory polyps of the colon (Figs 2.34 to 2.36) are frequently seen in Egyptian patients, but rarely in other endemic areas. Schistosomal colonic polyposis is almost always associated with heavy infection by *S. mansoni*, but less commonly with *S. haematobium* infection. Numerous colonic polyps can cause life-threatening dysentery with loss of fluids, protein, and blood.[13] The mucosa of the polyps is thickened and distorted, without adenomatous change, and the submucosa is inflamed (Fig 2.36). Schistosome eggs are generally numerous in the mucosa, lamina propria, and submucosa. Adult worms often lie in the venules of the stalk. Schistosomal colonic polyps are not precancerous in *S. mansoni* infection, but there is an association between *S. japonicum* infection and colonic cancer.

Colonic fibrosis, usually segmental (Fig 2.37), is a rare complication of schistosomiasis, and segmental fibrosis of the small bowel is even less common.

Bilharziomas

Bilharziomas are localized masses of fibrous and inflammatory tissue containing numerous eggs. They develop most frequently in the intestinal serosa (Fig 2.38) or mesentery. Bilharziomas are usually reactions to numerous eggs produced by 1 or more pairs of worms in a single site, although altered host reactivity may be a factor.[17]

Hepatosplenic Schistosomiasis (Symmers' Fibrosis)

Many eggs laid in the mesenteric venules are carried into the intrahepatic portal circulation (Fig 2.39), where they provoke granulomas (Figs 2.26 to 2.28). In some patients, these reactions result in marked portal fibrosis.

Symmers originally described this fibrosis as follows: "When a portal canal is cut transversely, the mouths of the contained vessels and bile duct are seen embedded in the center of a circular or slightly oval area of white connective tissue, the diameter of the mass being, on the average, from a sixth to a quarter of an inch: whereas longitudinal sections of the canal reveal elongated masses of similar appearance and thickness, so that the cut surface of the liver looks as if a number of white clay pipestems had been thrust at various angles through the organ."[22]

The external surface of the liver may be smooth, macronodular, or micronodular, but more or less broad tracts of portal fibrosis are seen on gross

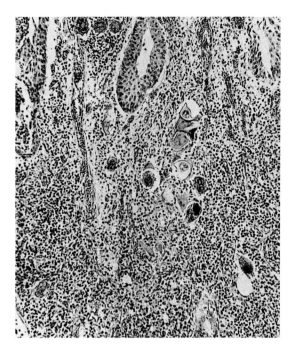

Figure 2.24
Diffuse reaction to eggs of *S. haematobium* in ureter of 15-year-old Egyptian boy. Areas of ureteritis glandularis are at top. x120

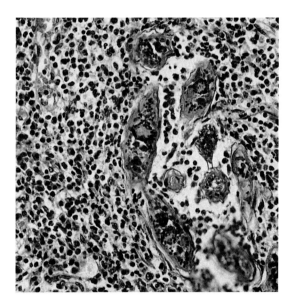

Figure 2.25
Diffuse cellular infiltrate around eggs of *S. haematobium* in urinary bladder of same patient as in Figure 2.24. Embryonated eggs are on right. x315

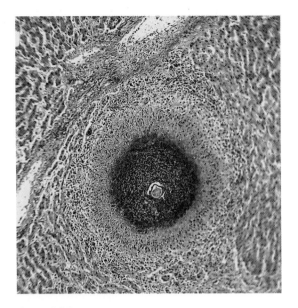

Figure 2.26
Circumoval granuloma with intensely eosinophilic necrotic center, surrounding schistosome egg in liver. x75

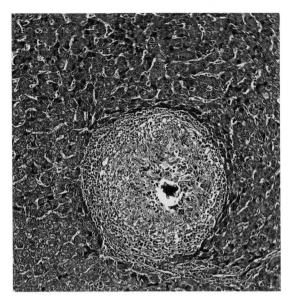

Figure 2.27
Typical advanced circumoval granuloma in liver. Note centrally located, collapsed schistosome egg surrounded by epithelioid cells, and outer zone of fibrous connective tissue infiltrated by lymphocytes. x110

examination, and are pathognomonic of Symmers' pipestem fibrosis (Fig 2.40). Lobular architecture is usually well-preserved, but there may be postnecrotic collapse and regenerative nodules near the liver surface.

Microscopically, dilated capillaries and numerous eggs are usually in the periportal connective tissue (Figs 2.41 to 2.43). Finding remnants of elastic tissue and muscle in the portal fibrous tissue is useful for diagnosis even in the absence of eggs.[5] In some livers, eggs are few or absent. Occlusion of small portal branches is frequent. Areas of dense fibrosis, containing an arteriole and bile ductule but without a venule, are common (Fig 2.42). Parenchymal cells are well-preserved. Arterial branches are increased in size and number, frequently with muscular hypertrophy and intimal sclerosis.

Biopsy specimens of liver taken by needle are not adequate to diagnose Symmers' fibrosis, although they may help to exclude other diseases. Wedge biopsy specimens usually contain characteristic portal areas and frequently show postnecrotic changes. Ultrasonography, the preferred method for diagnosis of Symmers' fibrosis, reveals fibrous bands (Fig 2.44).[10] Many patients diagnosed with portal fibrosis by ultrasonography lack splenomegaly and may escape diagnosis by clinical examination.

Symmers' fibrosis is caused by *S. mansoni* and *S. japonicum*, but is not convincingly associated with *S. haematobium* infection. All portal hypertension caused by schistosomiasis is associated with Symmers' fibrosis. The portal hypertension is presinusoidal, so that wedged hepatic venous pressure is normal or only slightly increased in patients without hepatic failure, in contrast to alcoholic and postnecrotic cirrhosis, in which the wedged pressure is usually elevated. Bogliolo demonstrated pathognomonic splenoportographic findings characterized by cuffing around portal veins caused by filling of the capillaries in the periportal connective tissue. Hepatic function is preserved until late in the disease,[18] and patients usually experience repeated hemorrhage from varices. Hepatic failure develops only in the terminal stage of the disease. Symmers' fibrosis appears to be reversible to a surprising degree following chemotherapy,[11] with morphologically evident collagenolysis.[5]

Symmers' fibrosis is most frequent in heavily infected patients,[7] but these individuals also lack specific inhibitory idiotypic responses that downregulate immune responses in other infected persons.[16]

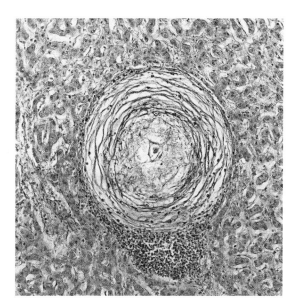

Figure 2.28
Healing circumoval granuloma with debris from necrotic cells and concentric layers of fibrous connective tissue surrounding schistosome egg in liver. x130

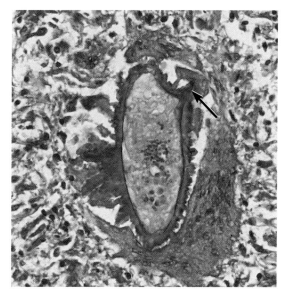

Figure 2.29
Egg of *S. mansoni* in giant cell in liver, surrounded by eosinophilic rays of the Splendore-Hoeppli reaction. Note large, lateral, sharply pointed spine (arrow). x340

Hashem described fine schistosomal periportal fibrosis associated with portal hypertension, but we have been unable to confirm this in many surgical biopsies of liver and livers obtained at autopsy in Brazil and Egypt. Neither cirrhosis nor hepatoma is caused by *S. mansoni* or *S. haematobium,* and both appear with equal frequency in infected and noninfected patients. Hepatomas are associated with *S. japonicum* infection.

In patients with schistosomiasis, salmonella infections often run a prolonged and atypical course, with intermittent daily fever persisting for months, accompanied by marked hepatosplenomegaly.[19] The clinical course resembles kala-azar, but without leukopenia. Patients infected by *S. haematobium* are predisposed to chronic salmonella infections of the urinary tract.[9] Hepatosplenic schistosomiasis is associated with prolonged viremia after hepatitis B infection; hepatic failure in patients with Symmers' fibrosis frequently accompanies chronic active hepatitis B.[15]

Some reactive splenomegaly develops early in infection, but obvious involvement of the spleen is secondary to severe hepatic schistosomiasis. In patients dying with Symmers' fibrosis, spleens are enlarged and average 1000 g, with mixed fibrocongestive and reactive changes. A few patients with hepatosplenic schistosomiasis and marked splenomegaly have retarded growth and delayed sexual maturation. Portal-systemic collateral circulation is similar to that in cirrhosis. Schistosomal glomerulonephritis, probably caused by immune complexes, develops in about 10% of patients with hepatosplenic schistosomiasis caused by *S. mansoni*.[4] Glomerulonephritis has not been convincingly related to *S. haematobium* infection.

Urogenital Schistosomiasis

Although *S. haematobium* is the primary cause of urogenital schistosomiasis, occasionally there are a few eggs of *S. mansoni* in the urinary bladder and other pelvic organs. Kidney, ureter, urinary bladder, urethra, and genital tissue may be involved, but eggs are most numerous in the bladder, ureter, and seminal vesicle. Lung, colon, and appendix also contain eggs of *S. haematobium* in nearly all infections. The most clinically significant lesions of urogenital schistosomiasis are caused by eggs deposited in the wall of the urinary bladder and ureter (Fig 2.45). Adult worms tend to remain fixed for prolonged periods, depositing eggs that accumulate at 1 site and cause focal lesions (Fig 2.46). These gross lesions begin as

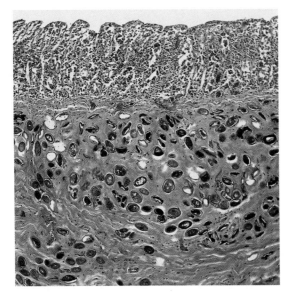

Figure 2.30
Healed lesion in colonic submucosa. Note dense, hyalinized, fibrous connective tissue developed in response to numerous eggs of *S. japonicum*. Most of the eggs are mineralized. x75

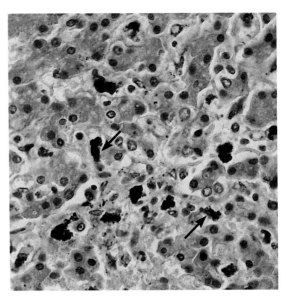

Figure 2.31
Phagocytosed schistomal pigment (arrows) in Kupffer's cells of liver. Giant cells may also engulf this pigment. x320

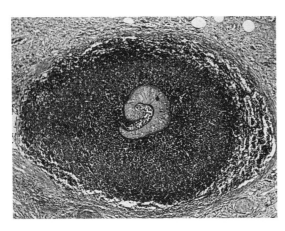

Figure 2.32
Vein in hepatic portal area obstructed and destroyed by intense host reaction to dead adult schistosome. x45

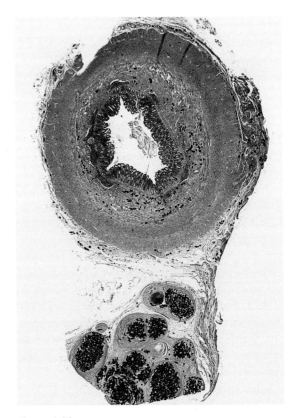

Figure 2.33
Schistosome eggs in submucosa, serosa, and mesentery of appendix. Mesentery contains masses of mineralized eggs in healed granulomas. x10

Figure 2.34
Surgically resected segment of colon from Egyptian patient with severe schistosomal colonic polyposis.* x0.35

Figure 2.35
Air contrast barium enema of 25-year-old Egyptian with diffuse schistosomal colonic polyposis.*

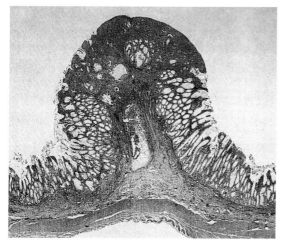

Figure 2.36
Small inflammatory polyp in colon of Egyptian man who had a massive *S. mansoni* infection and died from schistosomal colonic polyposis. Surface of polyp is congested and ulcerated. In other parts of the colon, the mucosa was thickened by proliferating glands. x6.5

Figure 2.37
Diffuse fibrosis of stenotic segment of surgically resected colon from 18-year-old Egyptian man. This tissue contained 112 000 eggs of *S. mansoni* per gram. Infection was almost inactive at surgery and would have been judged light by fecal examination.

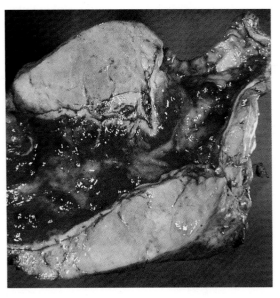

Figure 2.38
Rectosigmoid colon of 40-year-old Egyptian man showing a serosal bilharzioma and several small mucosal polyps. Numerous similar serosal nodules were in remainder of colon and throughout small intestine. Surgery was for ureteral obstruction and hydronephrosis. There were also cirrhosis of liver and splenomegaly. Heavy, active *S. mansoni* and light, inactive *S. haematobium* infections were detected in colon.

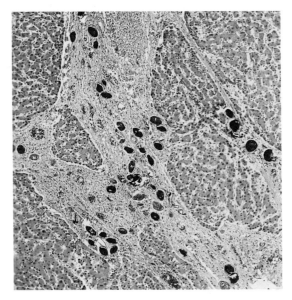

Figure 2.39
Section of liver showing large numbers of eggs of *S. japonicum* embedded in mature fibrous tissue. There is no granulomatous reaction. x65

Figure 2.40
Liver showing marked Symmers' pipestem fibrosis. Lobular architecture of liver is preserved. Portal fibrosis is often less striking than that illustrated here, but is always evident macroscopically.

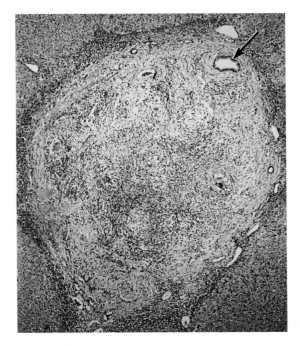

Figure 2.41
Surgical wedge biopsy from liver of Egyptian patient infected with *S. mansoni,* showing Symmers' fibrosis. Portal tract is 2 mm in diameter and contains a bile ductule (arrow) and an arteriole, but portal venule is obstructed. A small granuloma and several eggs, not clearly shown, are in portal granulation tissue. x45

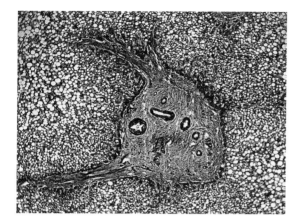

Figure 2.42
Bile ductules and arterioles identify this area of Symmers' fibrosis as a portal tract. The portal venule is obliterated. x45

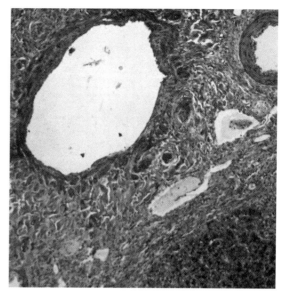

Figure 2.43
This portal fibrosis with telangiectasia of the portal space is diagnostic of Symmers' fibrosis even without eggs. The portal vein is frequently patent, as illustrated here. x60

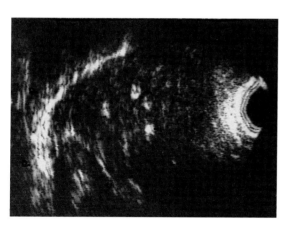

Figure 2.44
Ultrasonographic image of Symmers' fibrosis in patient with chronic *S. mekongi* infection. The highly echogenic portal fibrosis in the liver is to the right. The long curved echogenic area to the left is the diaphragm.

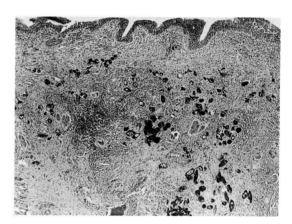

Figure 2.45
Mineralized eggs of *S. haematobium* in wall of bladder. x30

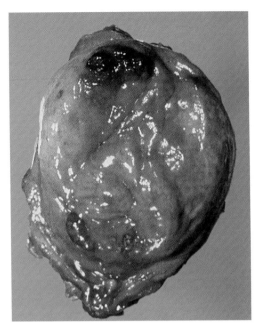

Figure 2.46
Polypoid patches in bladder of 10-year-old Egyptian child infected by *S. haematobium*. Bladder apex is at top of photograph and prostatic urethra at bottom. Sharply localized active lesions are common in *S. haematobium* infection and reflect the sedentary habits of the worms. This was an incidental finding in a patient with medulloblastoma.

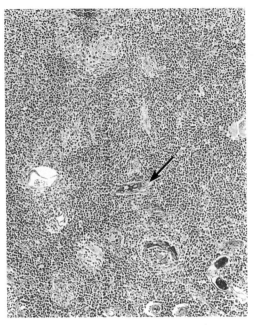

Figure 2.47
Polypoid patch in bladder of an Egyptian. Eggs of *S. haematobium* in lamina propria are surrounded by a diffuse infiltrate composed mainly of eosinophils. One egg contains a mature embryo (arrow). The 2 basophilic embryos in lower right corner are immature miracidia. x100

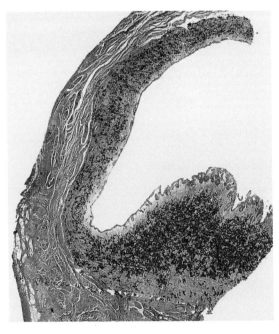

Figure 2.48
Microscopic view of sandy patch in urinary bladder of patient with *S. haematobium* infection. Lamina propria is thickened by sheer number of mineralized eggs. Grossly, sandy patches appear as yellow submucosal lines, which show up as calcified lines on radiologic examination. x2.25

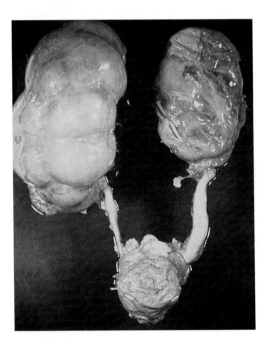

Figure 2.49
Specimen from 35-year-old Egyptian man who died of uremia from schistosomal obstructive uropathy. Kidney at left weighed 3700 g, and kidney at right (collapsed before photography) weighed 1500 g. Thickening of ureters is caused by calcified eggs and fibrosis in some areas, and by hydroureter in other areas. Bladder is calcified. Such extreme hydronephrosis is rare.

red, fleshy masses (polypoid patches) caused by a diffuse exudative and granulomatous reaction to eggs (Fig 2.47). As polypoid patches involute, they become fibrous patches that exhibit more fibrosis and less bulk.[21,26] With time, these may evolve into sandy patches (Fig 2.48) which are yellow-brown accumulations of calcified eggs, with or without associated inflammatory or fibrous reaction. On cross section, sandy patches appear as linear yellow deposits, most commonly submucosal. Bilharziomas, described previously, are exuberant large polypoid and fibrous patches away from a mucosal surface.

Obstructive Uropathy

Obstructed urinary flow may be caused by stricture or distortion of the ureteral orifices, but is more frequently related to egg deposits in the wall of the ureter (Figs 2.49 to 2.51). Polypoid patches, fibrosis, ureteritis cystica, calculi, and sheer accumulation of eggs contribute to obstruction (Figs 2.52 and 2.53). In other patients, particularly those with mild hydroureter, there is no stenosis, and dilatation is presumed to be functional. Obstructive uropathy usually progresses without clinical symptoms or signs other than those related to cystitis or referred pain from the ureters. In young patients, hydronephrosis and hydroureter are often reversible with schistosomicidal treatment. Ultrasonography is useful in assessing urinary tract abnormalities.[8]

Obstruction of the neck of the bladder and fibrotic contraction of the bladder are rare complications in patients with heavy infection by *S. haematobium*.

Cystitis and Metaplasia

Dysuria and hematuria, the predominant manifestations of early *S. haematobium* infection, are caused by cystitis from deposited eggs. Eggs provoke epithelial hyperplasia, cystitis glandularis, and cystitis cystica (Fig 2.54). Urothelium may show colonic (Fig 2.55) or squamous metaplasia (Fig 2.56).

Calcification of Bladder

Confluent sandy patches produce the radiologic picture of the calcified bladder, in which there is a thin band of submucosal calcification (Fig 2.57). Mineralized eggs produce the radiologic image,

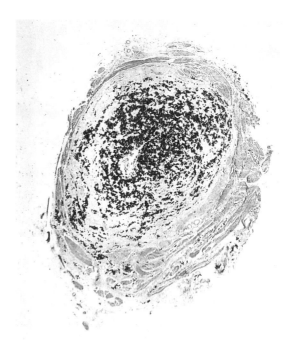

Figure 2.50
Cross section of stenotic ureter shown in Figure 2.49. Note numerous calcified eggs in submucosa. Wall of ureter is fibrotic and lumen obstructed. x7

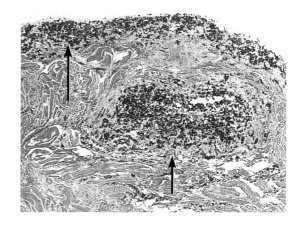

Figure 2.51
Cross section through intravesical portion of ureter shown in Figure 2.49. Numerous calcified eggs outline both submucosa of ureter in bladder wall (short arrow) and submucosa of bladder (long arrow). x6

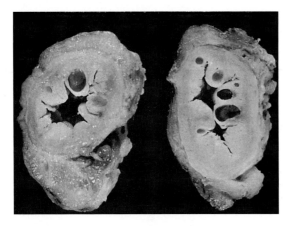

Figure 2.52
Gross cut sections of ureter showing marked fibrous thickening of ureteral wall with ureteritis cystica. x1.15

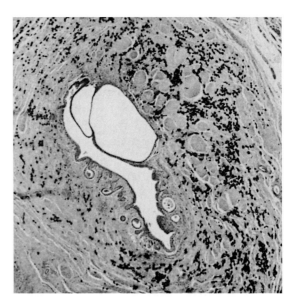

Figure 2.53
Cross section of lower third of ureter with ureteritis glandularis, ureteritis cystica, and numerous schistosome eggs. x11

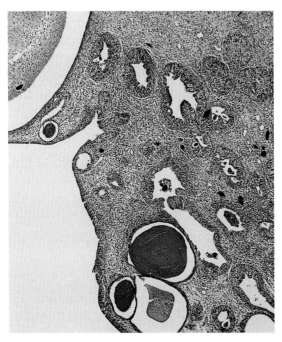

Figure 2.54
Urinary bladder with *S. haematobium* infection showing cystitis cystica and cystitis glandularis. x30

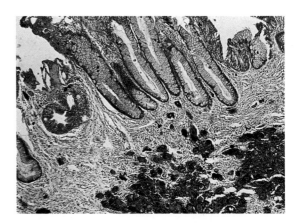

Figure 2.55
Metaplasia of urothelium to intestinal type in this urinary bladder laden with mineralized schistosome eggs. x30

but the tissues of the bladder are not mineralized and the bladder may continue to function normally. Calcification detected radiographically indicates heavy infection; thus, other complications at this stage of schistosomiasis are frequent. Mineralized eggs may be detected radiologically in the ureter, seminal vesicle, and, rarely, in the colon.

Renal Lesions

Pyelonephritis frequently complicates severe obstructive uropathy. In patients with mild ureteral dilatation, or without obstructive uropathy, pyelonephritis appears at the same rate as in noninfected persons. Rapidly progressive uremia in patients with obstructive uropathy is often caused by bacterial infection and chronic salmonellosis. These conditions may be associated with massive proteinuria and are responsive to antibiotic treatment.[9,14]

Cancer of Bladder

The high incidence of carcinoma of the urinary bladder in patients with urinary schistosomiasis is well-recognized. A causal relationship is accepted but unproved. In Egypt over 90% of bladder cancers associated with schistosomiasis develop outside the trigone (Figs 2.58 and 2.59), while elsewhere in the world most carcinomas of the bladder develop in the trigone. Approximately 50% of tumors associated with schistosomiasis are squamous cell carcinomas (Figs 2.58 to 2.60), but prognosis is not worse than with infiltrating transitional cell cancers. Approximately 10% of bladder cancers associated with schistosomiasis are adenocarcinomas, and 40% are transitional cell carcinomas.

Ulcers of Bladder

In addition to numerous small areas of ulceration with schistosomal cystitis, larger isolated, nonhealing ulcers may develop and cause extreme pain, particularly on urination. Usually, these ulcers develop in the most severely infected patients and commonly overlie dense sandy patches. Sometimes there are associated metaplastic and hyperplastic mucosal changes.

Genital Lesions

Deposition of many eggs in the seminal vesicles is accompanied by fibrosis and muscular hypertrophy (Fig 2.61). Clinically evident epididymal and testicular lesions are rare. Variable numbers of

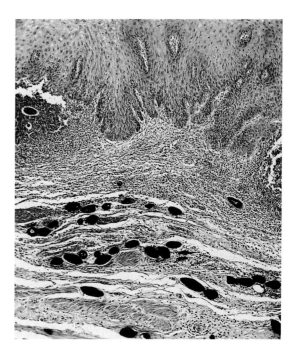

Figure 2.56
Schistosoma haematobium infection of urinary bladder showing marked squamous metaplasia with keratinization in bladder. x60

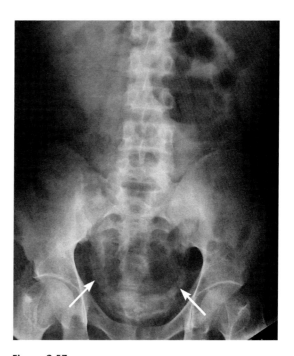

Figure 2.57
Radiograph of pelvis of 24-year-old Egyptian with schistosomal obstructive uropathy. Note linear calcification in base of bladder and along dilated lower segments of ureters (arrows).*

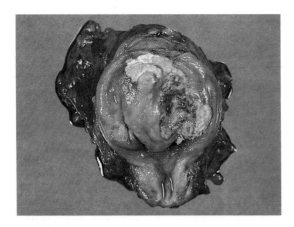

Figure 2.58
Ulcerating squamous cell carcinoma occupies left posterior wall of bladder of 26-year-old Egyptian male. Note leukoplakia in apex and left lateral wall in this specimen with an intense *S. haematobium* infection.

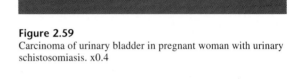

Figure 2.59
Carcinoma of urinary bladder in pregnant woman with urinary schistosomiasis. x0.4

eggs are common in the prostate, vas deferens, ovary, fallopian tube, uterus, and vagina. Clinically significant lesions are rare, but can cause infertility or sterility.

Cardiopulmonary Schistosomiasis

Cardiopulmonary schistosomiasis develops in patients with hepatosplenic schistosomiasis when eggs are shunted from the mesenteric veins to the lungs through portal-systemic collateral veins. Schistosomal cor pulmonale is rare in *S. haematobium* infection. As a rule, cor pulmonale in patients with schistosomiasis without Symmers' fibrosis is caused by pulmonary disease other than schistosomiasis. At autopsy, about 15% of patients with Symmers' fibrosis have schistosomal pulmonary arteritis and cor pulmonale.

Eggs lodge in pulmonary arterioles that are 50 to 100 μm in diameter and produce granulomatous pulmonary endarteritis (Figs 2.62 and 2.63). Angiomatoid and plexiform lesions are common; diffuse endothelial thickening, arteriolar hypertrophy, and cor pulmonale result from pulmonary hypertension.[3] In endemic areas, the majority of human lungs at autopsy contain eggs, but these are seldom of clinical importance.

Cerebrospinal Schistosomiasis

Any of the 3 common species of schistosomiasis may involve the meninges (Figs 2.64 and 2.65), brain (Figs 2.66 and 2.67), and spinal cord (Fig 2.68). Eggs may be transported to the central nervous system from distant sites, but are generally deposited by females living in cerebrospinal venules. *Schistosoma japonicum* frequently involves the brain (Figs 2.64 to 2.67). Some American soldiers infected by *S. japonicum* during World War II developed cerebral lesions. Many of them had jacksonian epilepsy even before they passed eggs in the feces. *Schistosoma mansoni* and *S. haematobium* more frequently involve the spinal cord than the brain. Clinically, severe involvement of the cord usually presents as transverse myelitis. Eggs of *Paragonimus* sp may also produce lesions of the brain and spinal cord; however, *Paragonimus* eggs are brilliantly birefringent, while the slight birefringence of schistosome eggs is usually not evident with unsophisticated polarization.

Schistosomal Cercarial Dermatitis

Numerous species of schistosomes which mature in birds and various mammals cause cercarial dermatitis (swimmer's or clam-digger's itch) in humans. Some schistosome species elicit a mild and transitory dermatitis and then migrate to the lung or beyond, but do not mature and lay eggs. On reexposure, the schistosomules are destroyed in the skin and provoke a maculopapular pruritic rash. Cercarial dermatitis may be contracted in fresh, brackish, or sea water.

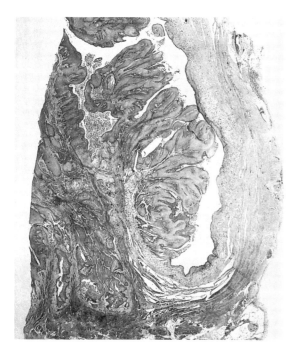

Figure 2.60
Section of urinary bladder from patient with schistosomiasis. An exophytic squamous cell carcinoma occupies central portion; infiltrating squamous cancer extends to left. x4.3

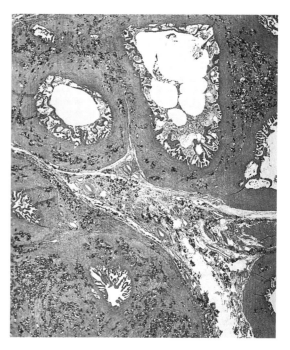

Figure 2.61
Numerous calcified eggs of *S. haematobium* in seminal vesicles of Egyptian patient. Marked involvement of seminal vesicles is common. x7.5

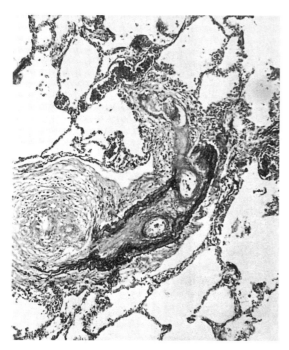

Figure 2.62
Granulomatous pulmonary arteritis in a 37-year-old Brazilian woman with hepatosplenic schistosomiasis caused by *S. mansoni*. Patient had moderately severe cor pulmonale, but died of massive gastrointestinal hemorrhage. x135

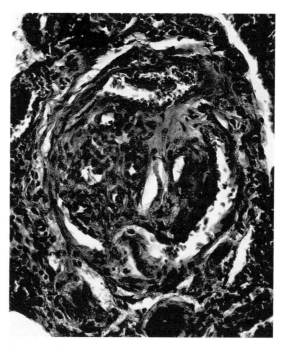

Figure 2.63
Angiomatoid and plexiform alterations of small branches of pulmonary artery, frequent in schistosomal cor pulmonale. x250

Figure 2.64
Dura mater from patient with meningoencephalitic schistosomiasis caused by *S. japonicum*. Note yellowish, slightly elevated, rounded masses of granulomas. x0.75

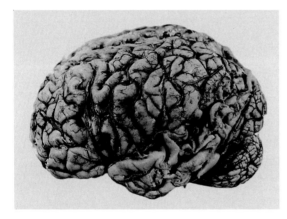

Figure 2.66
Brain of patient who died of severe cerebral involvement by *S. japonicum*. Note hyperemia of meninges overlying granulomas in the brain. x0.43

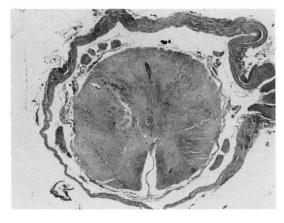

Figure 2.68
Spinal cord of patient with transverse myelitis caused by *S. mansoni*. x6.2

Figure 2.65
Multiple granulomas in meninges and brain of World War II veteran with fatal *S. japonicum* infection acquired in Leyte, Philippines. x13

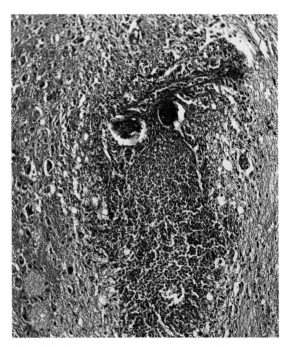

Figure 2.67
Localized abscess of brain around eggs of *S. japonicum*. x210

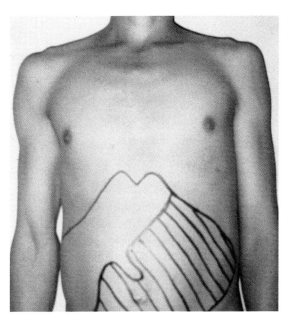

Figure 2.69
Marked splenomegaly in young Filipino male with hepatosplenic schistosomiasis japonica.*

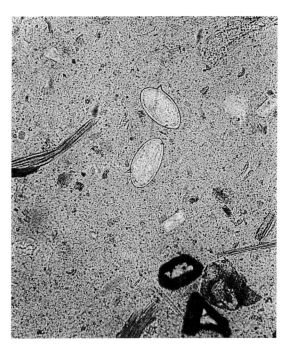

Figure 2.70
Eggs of *S. mansoni* in feces. Kato preparation x100

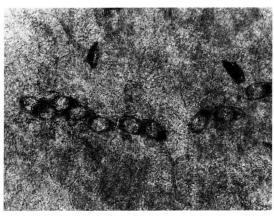

Figure 2.71
Eggs of *S. mansoni* in crushed rectal biopsy from Puerto Rican patient. The more transparent centers of eggs are immature embryos, indicating active infection. Unstained x80

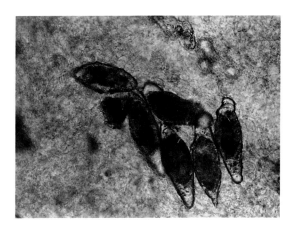

Figure 2.72
Calcified eggs of *S. haematobium* in crushed fragment of fixed tissue from urinary bladder of an Egyptian. Unstained x220

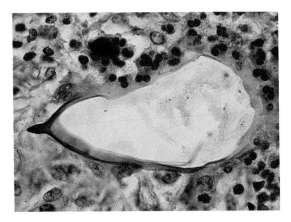

Figure 2.73
Egg of *S. intercalatum* in wall of appendix from 18-year-old Congolese female. Egg is 145 by 50 µm and has a large terminal spine. B&H x600

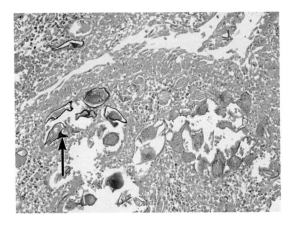

Figure 2.74
Dual infection with eggs of *S. mansoni* and *S. haematobium* in wall of urinary bladder. Eggshells and lateral spine (arrow) of *S. mansoni* are acid-fast; nonacid-fast eggshells are those of *S. haematobium*. Modified ZN x110

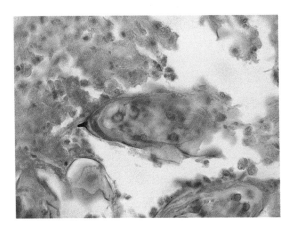

Figure 2.75
Egg of *S. haematobium* in wall of urinary bladder, showing acid-fast terminal spine. Modified ZN x400

Diagnosis

The diagnosis of schistosomiasis in endemic areas is frequently made clinically (Fig 2.69), usually by finding schistosome eggs in concentrated urine or feces (Figs 2.9, 2.12, 2.15, and 2.70). Excreta may contain eggs as early as 5 to 6 weeks after infection with *S. japonicum* or *S. mansoni* and 11 to 13 weeks with *S. haematobium*, but longer delays between exposure and patency are common. In epidemiologic studies of *S. haematobium* infection, hematuria, proteinuria, or eosinophiluria are useful indications of infection and are often related to intensity of infection.

The Kato method is a simple and sensitive technique for quantitating eggs in feces. A 25- to 50-mg sample of feces is placed on a glass slide and covered with a square of semipermeable cellophane (such as dialysis tubing) soaked in 50% glycerol. The slide is inverted on an absorbent surface and pressed until the feces are spread in a thin layer about 2 cm in diameter. After 2 to 4 hours clearing, the feces become semitransparent, and the eggs in the specimen are counted at x50 to x100 (Fig 2.70).

In patients with light or inactive infections, rectal biopsy is particularly effective in detecting eggs, which are best seen by crushing the specimen, preferably unfixed, between glass slides (Fig 2.71). This is more sensitive than histologic examination and allows more accurate assessment of the species and the viability of the eggs. This technique is also effective in demonstrating eggs in fixed or unfixed surgical or autopsy specimens (Fig 2.72). Eggs may also be recovered by hydrolysis of tissues in 4% KOH for 18 hours at 37°C for fresh tissues, and 56°C for fixed tissues.

Eggs of *S. haematobium* are readily demonstrated in bladder specimens taken through the cystoscope, but cystoscopy is rarely justified for diagnosis. Rectal biopsies contain eggs in over 70% of infected individuals. A variety of sensitive and specific immunodiagnostic tests are available, including a test for serum antigen levels.[1,23]

In tissue sections, *S. haematobium* eggs can easily be confused with those of *S. intercalatum*, because both have a terminal spine and are similar in size. *Schistosoma intercalatum* eggs, however, are usually longer (140 to 240 µm) and have a larger, more conspicuous spine (Fig 2.73). Terminal spines of *S. intercalatum* are 3 to 4 µm wide at the base, pointed at the tip, and 10 to 20 µm long. Eggs of *S. mekongi* are similar to those of *S. japonicum* in that they also have a tiny lateral spine, but they are smaller, measuring 51 to 78 µm by 39 to 66 µm. Obviously, finding schistosomal eggs in tissues does not prove the illness is caused by schistosomiasis.

Both the standard and modified ZN stain can be useful in identifying schistosome eggs in tissue sections (Figs 2.74 and 2.75). Most often the shells and spines of *S. mansoni* and *S. japonicum* are acid-fast, whereas only the spine of *S. haematobium* is acid-fast. Use caution, however, because the shells and spines sometimes fail to stain as expected.

Treatment

Praziquantel is effective against all schistosome species. Oxamniquine is effective only against *S. mansoni*, and metrifonate only against *S. haematobium*. Prevention and control measures are discussed under Epidemiology in this chapter. Vaccine strategies are under study.

References

1. Agnew A, Fulford AJ, Mwanje MT, et al. Age-dependent reduction of schistosome fecundity in Schistosoma haematobium but not Schistosoma mansoni infection in humans. *Am J Trop Med Hyg* 1996;55:338–343.
2. Amiri P, Locksley RM, Parslow TG, et al. Tumour necrosis factor alpha restores granulomas and induces parasite egg-laying in schistosome-infected SCID mice. *Nature* 1992;356:604–607.
3. Andrade ZA, Andrade SG. Pathogenesis of schistosomal pulmonary arteritis. *Am J Trop Med Hyg* 1970;19:305–310.
4. Andrade ZA, Andrade SG, Sadigursky M. Renal changes in patients with hepatosplenic schistosomiasis. *Am J Trop Med Hyg* 1971;20:77–83.
5. Andrade ZA, Peixoto E, Guerret S, Grimaud J-A. Hepatic connective tissue changes in hepatosplenic schistosomiasis. *Hum Pathol* 1992;23:566–573.
6. Basch PF. *Schistosomes: Development, Reproduction and Host Relations.* New York, NY: Oxford University Press; 1991.
7. Cheever AW. A quantitative post-mortem study of schistosomiasis mansoni in man. *Am J Trop Med Hyg* 1968;17:38–64.
8. Dittrich M, Doehring E. Ultrasonographical aspects of urinary schistosomiasis: assessment of morphological lesions in the upper and lower urinary tract. *Pediatr Radiol* 1986;16:225–230.
9. Farid Z, Higashi GI, Bassily S, Young SW, Sparks HA. Chronic salmonellosis, urinary schistosomiasis, and massive proteinuria. *Am J Trop Med Hyg* 1972;21:578–581.
10. Homeida M, Abdel-Gadir AF, Cheever AW, et al. Diagnosis of pathologically confirmed Symmers' periportal fibrosis by ultrasonography: a prospective blinded study. *Am J Trop Med Hyg* 1988;38:86–91.
11. Homeida MA, el Tom I, Nash T, Bennett JL. Association of the therapeutic activity of praziquantel with the reversal of Symmers' fibrosis induced by Schistosoma mansoni. *Am J Trop Med Hyg* 1991;45:360–365.
12. Jordan P, Webbe G. *Schistosomiasis: Epidemiology, Treatment and Control.* London, England: William Heinemann Medical Books Ltd; 1982.
13. Lehman JS Jr, Farid Z, Bassily S, Haxton J, Wahab MF, Kent DC. Intestinal protein loss in schistosomal polyposis of the colon. *Gastroenterology* 1970;59:433–436.
14. Lehman JS Jr, Smith JH, Bassily S, el-Masry NA. Urinary schistosomiasis in Egypt: clinical, radiological, bacteriological and parasitological correlations. *Trans R Soc Trop Med Hyg* 1973;67:384–399.
15. Lyra LG, Rebouças G, Andrade ZA. Hepatitis B surface antigen carrier state in hepatosplenic schistosomiasis. *Gastroenterology* 1976;71:641–645.
16. Montesano MA, Freeman GL Jr, Gazzinelli G, Colley DG. Immune responses during human Schistosoma mansoni. XVII. Recognition by monoclonal anti-idiotypic antibodies of several idiotopes on a monoclonal anti-soluble schistosomal egg antigen antibody and anti-soluble schistosomal egg antigen antibodies from patients with different clinical forms of infection. *J Immunol* 1990;145:3095–3099.

17. Mostofi FK. *Bilharziasis*. International Academy of Pathology Special Monograph. New York, NY: Springer-Verlag; 1967.

18. Nash TE, Cheever AW, Ottesen EA, Cook JA. Schistosome infections in humans: perspectives and recent findings. *Ann Intern Med* 1982;97:740–754.

19. Neves J, Martins NR. Long duration of septicaemic salmonellosis: 35 cases with 12 implicated species of salmonella. *Trans R Soc Trop Med Hyg* 1967;61:541–552.

20. Prakash S, Postlethwaite AE, Wyler DJ. Alterations in influence of granuloma-derived cytokines on fibrogenesis in the course of murine Schistosoma mansoni infection. *Hepatology* 1991;13:970–976.

21. Rollinson D, Simpson AJG. *The Biology of Schistosomes. From Genes to Latrines*. London, England: Academic Press; 1987.

22. Symmers D. Note on a new form of liver cirrhosis due to the presence of the ova of Bilharzia haematobia. *J Path Bact* 1904;9:237–239.

23. Tsang VC, Wilkins PP. Immunodiagnosis of schistosomiasis. *Immunol Invest* 1997;26:175–188.

24. Warren, KS. *Schistosomiasis: The Evolution of a Medical Literature. Selected Abstracts and Citations, 1852–1972*. Cambridge, Mass: The MIT Press; 1973.

25. Wynn TA, Cheever AW. Cytokine regulation of granuloma formation in schistosomiasis. *Curr Opin Immunol* 1995;7:505–511.

26. Von Lichtenberg F, Edington GM, Nwabuebo L, Taylor JR, Smith JH. Pathologic effects of schistosomiasis in Ibadan, Western State of Nigeria. II. Pathogenesis of lesions of the bladder and ureters. *Am J Trop Med Hyg* 1971;20:244–254.

Paragonimiasis

Aileen M. Marty *and*
Ronald C. Neafie

Introduction

Definition

Paragonimiasis is infection by trematodes of the genus *Paragonimus*. *Paragonimus westermani* (Kerbert, 1878) Braun, 1899, the most prominent species infecting humans, causes pulmonary, pleuropulmonary, and cerebral paragonimiasis. Adult *P. westermani* preferentially invade the lungs, but adult or immature flukes may lodge in a wide variety of tissues.[45] *Paragonimus* sp are members of the superfamily Troglotrematoidea[74] and are digenetic. Sexual reproduction by adult worms and asexual multiplication of larvae take place within intermediate hosts.

Paragonimus sp other than *P. westermani* that parasitize the lungs of animals may also infect human lungs or cause ectopic lesions as either mature adults or wandering immature forms. The number of *Paragonimus* sp infecting humans is controversial,[28,34] but significant species include *Paragonimus skrjabini* Chen, 1959,[8] *Paragonimus miyazakii* Kamo et al, 1961,[12,20,23] *Paragonimus heterotremus* Chen and Hsia, 1964, *Paragonimus hueitungensis* Chung et al, 1977,[18] *Paragonimus pulmonalis* (Baelz, 1880) Miyazaki, 1978,[41] *Paragonimus mexicanus* Miyazaki and Ishii, 1968,[43] *Paragonimus ecuadoriensis* Voelker and Arzube, 1979,[3,68] *Paragonimus kellicotti* Ward, 1908,[72] *Paragonimus africanus* Voelker and Vogel, 1965,[70] and *Paragonimus uterobilateralis* Voelker and Vogel, 1965.[53,69]

Related trematodes parasitic in the nasal sinuses of animals (*Achillurbainia nouveli* Dollfus, 1939, *Achillurbainia recondita* Travassos, 1942,[67] and *Poikilorchis congolensis* Fain and Vandepitte, 1957) infect humans only rarely (see chapter 6).[4]

Synonyms

Synonyms for paragonimiasis include pulmonary distomiasis, endemic hemoptysis, lung fluke disease, parasitic hemoptysis, *tojil* ("earth-borne disease") in Korea,[7] and Gregarinosis pulmonum. *Paragonimus westermani*, the Oriental lung fluke, has many synonyms, including *Distoma ringeri*, *Distoma westermani*, *Paragonimus edwardsi*, *Paragonimus macacae*, *Paragonimus philippinensis*, and *Polysarcus westermani*. The related species *P. pulmonalis*, the triploid *P. westermani* (sometimes considered a parthenogenic variant and named *P. westermani* var *pulmonalis*) is synonymous with *Distoma pulmonale*.[41] *Para-*

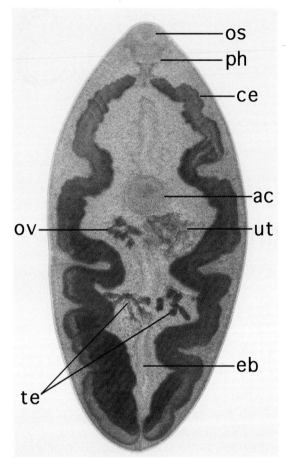

Figure 3.1
Ventral view of stained gross specimen of adult *P. westermani*. Note oral sucker (os), pharynx (ph), cecum (ce), acetabulum (ac), ovary (ov), uterus (ut), testes (te), and excretory bladder (eb). x13

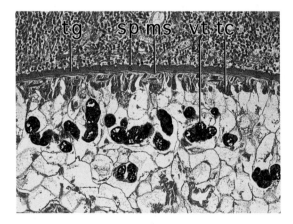

Figure 3.2
Body wall of adult *P. westermani* showing tegument (tg), spine (sp), muscle (ms), tegumental cell (tc), and vitellaria (vt). x110

gonimus mexicanus is identical to *Paragonimus peruvianus* and *Paragonimus caliensis,* and *P. ecuadoriensis* is probably a variant of *P. mexicanus.*[73] *Paragonimus skrjabini* is identical to *Paragonimus szechuanensis,* and *Paragonimus heterotremus* is identical to *Paragonimus tuanshanensis.*

General Considerations

In 1877 an autopsy of a Bengal tiger at the Amsterdam Zoo revealed pulmonary nodules containing pairs of flukes. Westerman, the zoo director, sent the flukes to Kerbert, who compared them to *Distoma compactum* Cobbold, 1859 and *Distoma rude* Diesing, 1850. Kerbert's study of the fluke revealed that it was a new species, which he named *Distoma westermani*.[25] Two years later, Ringer discovered a gravid fluke in the lung of a Portuguese man in Formosa. In 1880 Manson compared the eggs in Ringer's specimen with those he had just recovered from the bloody sputum of a 35-year-old Chinese man. He established that Ringer's specimen was a distomate fluke, and that the eggs in both patients were identical. Manson sent the specimens to Cobbold, who named the fluke *Distoma ringeri*.[2,35] In comparing specimens from Japan with those described by Kerbert, Leuckart and Nakahama concluded that the flukes were identical.[31] In 1899 Braun erected the genus *Paragonimus* from the Greek words *para* (on the side) and *gonimos* (genitalia) for the mammalian lung flukes; that same year, Looss provided the most accurate and complete description of the fluke.

Manson placed unembryonated *Paragonimus* eggs from human sputum in fresh water and observed the release of miracidia after several weeks. This led him, in 1883, to propose that a snail could be a first intermediate host.[35] Hungerford, in 1881, identified *Melania libertina* (now *Semisulcospira libertina*) as the most likely snail host, and in 1914 Nakagawa determined that crabs served as a second intermediate host.[48] Anado, in 1920, established the complete life cycle by exposing crabs to infected snails, then feeding dogs the infected crabs.[1] Yoshida, in 1916, showed that crabs release infectious metacestodes into fresh water through their gills.[79] Japanese researchers were also the first to recognize that migrating flukes occasionally take aberrant paths, sometimes encysting in tissues where they cannot complete their development.[48]

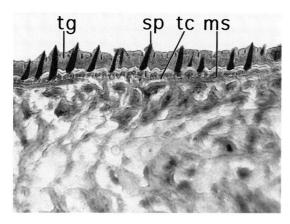

Figure 3.3
Body wall of adult *P. westermani* demonstrating tegument (tg), prominent scale-like spines (sp), double layer of smooth muscle (ms), and tegumental cells (tc). Movat x585

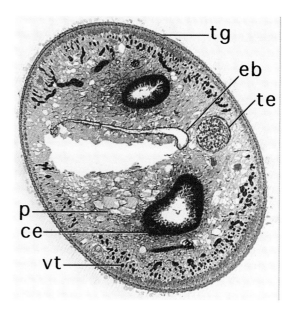

Figure 3.4
Transverse section through posterior region of adult *P. westermani* showing tegument (tg), excretory bladder (eb), testis (te), ceca (ce), vitellaria (vt), and parenchyma (p). x27

Morphologic differences between *P. pulmonalis* and *P. westermani* were established in 1978 by Miyazaki. The studies of Hirai et al show that *P. pulmonalis* is allotriploid, probably induced by hybridization between the diploid *P. westermani* and an unknown species.[17]

In 1894, Ward described the first *Paragonimus* sp in North America, found in a cat in Michigan; in 1915, Ward and Hirsch concluded that this was not *P. westermani*, but a new species that they named *P. kellicotti*. Grove offers a comprehensive historical outline of the discovery of other species of *Paragonimus* that infect humans.[13]

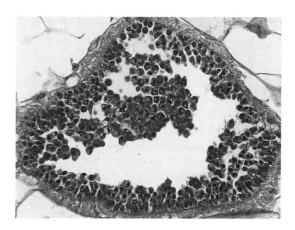

Figure 3.5
Transverse section through ovary of adult *P. westermani* depicting female reproductive cells. x230

Epidemiology

Paragonimiasis is distributed globally. Approximately 20.7 million people are infected with the disease worldwide, and another 195 million are at risk.[73] Chronic pulmonary disease is the most common manifestation, but ectopic (especially cerebral) paragonimiasis is a serious and often fatal disease that mainly afflicts juvenile and young adult males. Approximately 25% of all Chinese paragonimiasis patients have cerebral involvement.[73] The average age of onset of cerebral paragonimiasis is 15 years; clinicians diagnose 75% of all cases before age 20 and 90% before age 30.[29]

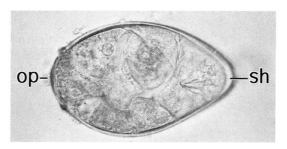

Figure 3.6
Egg of *P. westermani* from feces depicting operculum (op) and thickened shell (sh) at abopercular end. Unstained x475

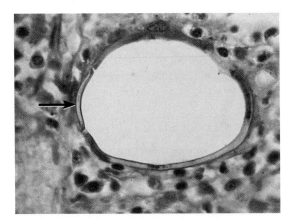

Figure 3.7
Egg of *P. westermani* in exudate of lung, demonstrating operculum (arrow). x635

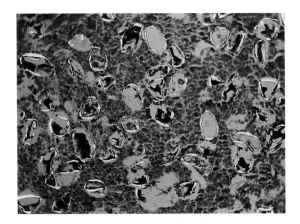

Figure 3.8
Eggs of *P. westermani* in nodule within thoracic wall, demonstrating birefringence under polarized light. x95

Human paragonimiasis is distributed focally in endemic countries. The major endemic region is Asia (Cambodia, China, the Commonwealth of Independent States,[73] India, Indonesia, Japan, Korea, Laos, Myanmar, Nepal, the Philippines, Taiwan, Thailand, Vietnam),[73] but the disease is also reported in Oceania (New Guinea, Samoa, Solomon Islands), Africa (Burkina Faso, Cameroon, Central African Republic, Congo, Côte d'Ivoire, Equatorial Guinea, Gabon,[71] Guinea, Liberia, Nigeria, Sierra Leone, South Africa,[38] Zambia[59]), and the Americas (Canada, Colombia, Costa Rica, Cuba, Ecuador, El Salvador, Guatemala, Honduras, Mexico, Nicaragua, Panama, Peru, the United States, Venezuela).

Paragonimus westermani is the most common species infecting Asians, though *P. pulmonalis* is probably a more significant cause of human infection than *P. westermani* in Japan, Taiwan, and Korea (Tables 3.1 and 3.2).[40] Early reports of *P. westermani* in the Americas and Africa were erroneous. In the Americas, the most common species infecting humans is *P. mexicanus*, but its range is limited to Mexico and Central and South America.[73] In North America, *P. kellicotti* is commonly found in mink, cats, muskrats, and other mammals in the Midwest, Louisiana, Mississippi, South Carolina, and Canada.[52] *Paragonimus kellicotti* sometimes causes pulmonary infection in humans.[36,55] *Paragonimus uterobilateralis*, which causes pulmonary disease and subcutaneous infections, is the dominant species in Nigeria and is also found in Cameroon, Côte d'Ivoire, Liberia, Sudan, and southern Africa.[38] *Paragonimus africanus* is endemic in Cameroon, Congo, Côte d'Ivoire, and Nigeria.[49]

The taxonomy of *Paragonimus* sp has been continuously revised. Recent study has shown that the Philippine strain[5] is a subspecies: *P. westermani filipinus*.[64]

Infectious Agent

Morphologic Description

An adult *Paragonimus* is a plump, oval, hermaphroditic fluke (Fig 3.1). *Paragonimus westermani* is the type species. Living *P. westermani* are pink to red-brown and vary in shape depending on the extent of expansion or contraction. Preserved specimens are gray-brown, rounded anteriorly, and slightly tapered posteriorly. They are 7.5 to 12 mm long, 4 to 6 mm wide and 3.5 to 5 mm thick (Table 3.1). Oral and ventral suckers are about 800 μm in diameter, with the oral sucker slightly smaller than the ventral sucker. The ventral sucker (acetabulum) is slightly anterior to the middle of the fluke (Fig 3.1). The body wall is 50 to 100 μm thick and composed of a tegument, 2 layers of smooth muscle, and tegumental cells (Figs 3.2 and 3.3). The tegument is a syncytium of living cells covered with spines. Tegumental spines are arranged singularly and vary in size and number, depending on their location. The parenchyma is a loose stroma containing fluid-filled spaces and parenchymal cells (Fig 3.4).

The digestive tract consists of a pharynx and a short esophagus that bifurcates into large, unbranched ceca that reach to the posterior end of the worm (Figs 3.1 and 3.4). There is a long excretory bladder extending from the posterior end of the worm to the level of the pharynx. Paired testes are deeply lobed, irregular, and situated side by side in the posterior third of the body. A single lobed ovary (Figs 3.1 and 3.5) lies slightly posterior to the ventral sucker, to the right or left of midline. The uterus is short, tightly coiled, and lies on the side of the body opposite the ovary (Figs 3.1 and 3.16). Vitellaria are well-developed, situated in the lateral fields (Fig 3.4), and extend from the pharynx to the posterior end of the worm (Fig 3.2).

Eggs of *P. westermani* are 80 to 120 µm by 45 to 65 µm and unembryonated when laid (Fig 3.6). In tissue sections, eggs rarely exceed 80 µm (Fig 3.7). Eggs are ovoid, yellow-brown, or red-brown.[35] The double-layered shell is 2 µm thick, except at the abopercular end where it is about 4 µm thick (Fig 3.6). The shell is birefringent (Fig 3.8). The operculum lies at the broader end and is distinctly flattened, seated, and rimmed. Eggs are filled with a delicate, jelly-like material containing 3 or 5 aggregations of smaller, spherical, colorless bodies about twice the size of a white blood cell.[2]

Small morphologic differences exist between *P. westermani* and other *Paragonimus* sp (Figs 3.9 and 3.10). Examining gross intact specimens rather than fragments in tissue sections is the best way to appreciate these differences (Table 3.1).[18,52,65] Morphologic criteria that differentiate species include the shape of the entire body, the arrangement of tegumental spines, the transverse diameter of oral and ventral suckers, and the shape and size of the ovary and testes. Differences in spines are observable in tissue sections, but spines alone do not permit species identification. The arrangement of tegumental spines into groups helps to distinguish the *P. westermani-kellicotti* group, with single spines, from the *P. africanus* group, with clustered spines (Table 3.1).[76] Determining *Paragonimus* sp based on egg morphology is difficult. When many eggs are available for study, however, distinguishing features such as shape, size, and eggshell characteristics become more apparent (Table 3.1).

Life Cycle and Transmission

The complex life cycle of *Paragonimus* sp

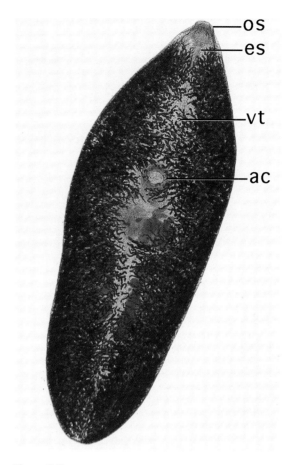

Figure 3.9
Ventral view of stained gross specimen of adult *P. skrjabini* recovered from dog liver, showing oral sucker (os), esophagus (es), acetabulum (ac), and vitellaria (vt). x11

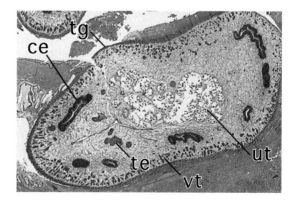

Figure 3.10
Adult *P. kellicotti* in dog lung demonstrating tegument (tg), vitellaria (vt), ceca (ce), testis (te), and uterus (ut). x11.5

Species	Range	Distinguishing features: adults	Distinguishing features: eggs, larvae
P. westermani	India, Sri Lanka, Nepal, Myanmar, Thailand, Malaysia, Indonesia, New Guinea, the Philippines, Taiwan, China, Korea (Cheju Island), Japan, former Soviet Union (Amur River basin). Suspected: Vietnam, Cambodia, Laos	Shape: stout; avg. 8.4 X 4.2 mm Suckers: oral slightly smaller than ventral Tegumental spines: singly spaced, may split as worm ages Ovary: branched into 4-6 lobes Testes: usually divided into 6 lobes; about same size as ovary Vitellaria: limited to lateral fields	Eggs: asymmetrical; 100 X 56 μm (varies with host), elongate, oval with barrel-shaped center. Operculum with shoulders. Abopercular end is 4 μm at thickest point. Shell is about 2 μm thick except at abopercular end. Larvae: metacercaria with 60 flame cells, enveloped with 2 cysts; inner spherical, thick-walled cyst avg. 283 μm, inner cyst wall avg. 18 μm, outer cyst wall avg. 3 μm. Oral sucker smaller than ventral and has stylet. Pale pink granules difficult to discern
P. skrjabini	China (mountains), Southeast Asia	Shape: lanceolate, crescent, or spindled; 12-15 X 3.1-5 mm Suckers: oral smaller than ventral Tegumental spines: singly spaced, rectangular; a few in groups of 2-6 in tail end Ovary: many fine branches Testes: moderately branched; gonads small compared to body size Vitellaria: extensive; project to midline	Eggs: avg. 80.4 X 47.7 μm, only slightly asymmetrical, irregularly thick shell, oval with small knob-like process at abopercular end Larvae: metacercaria with 72 flame cells, spherical, enveloped with 2 kinds of cysts; inner cyst avg. 436-427 μm in diameter; cyst wall is 14 μm thick
P. miyazakii	Japan (southern half of Honshu, Shikoku, and Kyushu)	Shape: elongate; 9.2-12.5 X 5-6 mm Suckers: oral slightly smaller than ventral Tegumental spines: singly spaced throughout body; some split into 2 or 3 at distal ends Ovary: only primary and secondary branches Testes: moderately lobed; usually larger than ovary	Eggs: vary in size but avg. 70.5 X 43.3 μm; shell very thin (avg. 1.2 μm on lateral sides); thinner at abopercular end (avg. 1.06 μm) Larvae: metacercaria spherical, enveloped by 2 cysts; inner cyst avg. 445 μm, outer cyst wall 2-12 μm and enveloped with membranous substance. Granules not visible.
P. heterotremus	Thailand, China, Laos	Suckers: oral almost twice as large as ventral Tegumental spines: singly spaced Ovary: profusely, delicately branched Testes: profusely lobed; larger than ovary	Eggs: avg. 86.4 X 47.4 μm; symmetrical Larvae: metacercaria ovoid, thin outer cyst and thick inner cyst which avg. 292 X 237 μm; inner cyst wall gradually thickens at poles. Oral sucker has stylet. Granules not visible.
P. hueitungensis	China	Shape: elongate; 8.4 mm long, 3 mm wide, 0.3 mm thick Suckers: oral slightly smaller than ventral Tegumental spines: arranged singly; long, slender, "pine needle" appearance over most of worm Ovary: 3-6 short, stout primary branches, each giving rise to 2-3 thick secondary branches; occasional secondary branch gives rise to short tertiary branching Testes: each with elongate central mass that gives rise to 5-6 lumpy primary branches; 2-3 short secondary branches; markedly larger than ovary	Eggs: 73.1 X 45.4 μm; widest part slightly anterior to middle. Shell thin, avg. 1.6 μm. Operculum with visible rim and ridges. Minute knob at abopercular end. Larvae: metacercaria, usually spherical; 3 layers of cyst wall, middle and inner layers in close contact, outer layer very thin (1-2 μm); easily ruptures. Small stylet on dorsal aspect of oral sucker.
P. pulmonalis	Japan, Korea, Taiwan, China	Shape: stout; avg. 13.2 X 7.1 mm Tegumental spines: singly spaced Suckers: oral slightly smaller than ventral Ovary: branches into 4-6 lobes; differs in size from P. westermani; seminal receptacle contains some egg cells and vitelline cells instead of sperm	Eggs: 92.1 X 51.2 μm; elongate Larvae: metacercaria spherical, enveloped by 2 cysts; inner cyst avg. 415 μm. Pink granules easy to recognize.

Table 3.1
Species, range, and distinguishing morphologic features of *Paragonimus* sp.

Species	Range	Distinguishing features: adults	Distinguishing features: eggs, larvae
P. mexicanus	Mexico, Central and South America	Suckers: oral slightly larger than ventral Tegumental spines: singly spaced, but terminal ends split into 2-3 Ovary: multiple fine branches Testes: moderately lobed; usually larger than ovary	Eggs: 79 X 48 µm; eggshell thin and irregularly undulated Larvae: metacercaria with no cyst wall; avg. 1.3-0.56 mm. Contain brilliant red granules on ventral side of body; yellow intestine.
P. ecuadoriensis	Ecuador	Suckers: oral slightly larger than ventral Tegumental spines: singly spaced Ovary: 5-6 short branches that divide into short, broad lobes Testes: 5-6 simple or polymorphic lobes; larger than ovary	Eggs: avg. 91.4 X 50.8 µm; elongate, ovoid; shell has thin, undulated surface Larvae: metacercaria with no cyst wall
P. kellicotti	North and South America	Shape: oval; 7-12 mm long, 4-6 mm wide, 3-5 mm thick Suckers: oral smaller than ventral Tegumental spines: singly spaced except on ventral surface lateral to ventral sucker, where spines may be paired and show splitting or grooving at the tip, which is typically truncate Ovary: 5 or 6 branches arising from a stem of various lengths; branching more complex than in P. westermani	Eggs: 75-118 X 48-68 µm; oval. Shell 2-3 µm thick. Slight thickening at abopercular end; tapers more sharply than in P. westermani Larvae: metacercaria enveloped with 2 cysts; inner cyst membrane homogeneous, avg. 3 µm; outer membrane avg. 2 µm
P. africanus	Cameroon, Congo, Nigeria, and Côte d'Ivoire	Suckers: oral larger than ventral Tegumental spines: mixed growth pattern: spines in groups except at dorsal median line, where they are arranged singly Ovary: basic part cone-shaped; 5 short primary branches and delicate, irregular secondary branches Uterus: unilateral Testes: profusely branched; often antler-shaped; much larger than ovary	Eggs: avg. 96.5 X 49.6 µm Larvae: metacercaria spherical, enveloped with 2 cysts; inner cyst avg. 449 X 416 µm
P. uterobilateralis	Nigeria, Liberia, Cameroon, Côte d'Ivoire, Gabon, Sudan, and southern Africa	Suckers: oral and ventral same size Tegumental spines: singly spaced Ovary: basic part digitate; delicate irregular branches Uterus: bilateral and unilateral Testes: divided into polymorphic lobes	Eggs: oval; avg. 71 X 45.4 µm. Shell 1 µm thick except at abopercular end, where it is 2 µm thick. Larvae: metacercaria enveloped with single cyst, avg. 702 µm in diameter; cyst wall very thin and fragile

Table 3.1 (continued)
Species, range, and distinguishing morphologic features of *Paragonimus* sp.

involves 7 distinct phases: egg, miracidium, sporocyst, redia, cercaria, metacercaria, and adult. *Paragonimus* needs 3 hosts to complete its life cycle (Fig 3.11). The first intermediate hosts are mollusks, usually snails of the families Pleuroceridae and Thiaridae. Fifty-three known species from 21 genera of freshwater crabs and crayfish serve as secondary intermediate hosts for *Paragonimus* sp throughout the world. Carnivorous mammals are the definitive hosts. They vary considerably with the species of *Paragonimus* (Table 3.2), but include humans, dogs and other canids, wild and domestic cats, pigs, beavers, mongooses, drills, and mink. Animal hosts such as dogs and cats contaminate fresh water more than humans do.

Unembryonated eggs exit the definitive host in sputum or feces. Eggs require 2 to 3 weeks in fresh water to embryonate, hatch, and release free-swimming miracidia. Miracidia penetrate the soft tissues of a molluscan host, where they multiply and develop over a period of several weeks into sporocysts, rediae, and finally, cercariae. Cercariae emerge from the mollusk and swim freely until they invade the tissues of a suitable crustacean, usually crab or crayfish, where they encyst as metacercariae.[48] Some mature metacercariae escape from the crustacean's gills and enter fresh water.[79] A definitive host is infected by ingesting metacercariae from a second intermediate host, paratenic host, or contaminated water, depending on the species of *Paragonimus* (Table 3.2).[26]

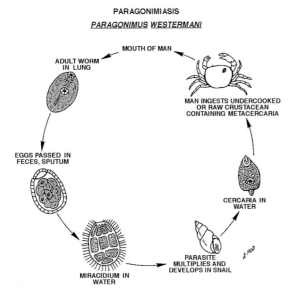

Figure 3.11
Simplified life cycle of *P. westermani*, depicting one of the most common sources of infection.

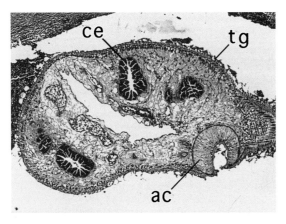

Figure 3.12
Adult *P. africanus* in choroid of 12-year-old Congolese who presented with red eye and swollen eyelid. Note tegument (tg), acetabulum (ac), and ceca (ce). x23

Larvae enter the duodenum of the definitive host and excyst. They penetrate the wall of the small intestine and enter the peritoneal cavity 30 minutes[76] to 48 hours after excysting, depending on the host species. After 5 to 7 days, larvae in the abdominal cavity go through the diaphragm. Most of these immature flukes float awhile in the serous excretion of the thoracic cavity, then attach by their suckers to the pleura. They may remain attached to the pleura, invade the pleura, or enter the parenchyma of the lung and lodge near bronchioles. The total incubation period of the disease is about 70 days. Adult worms may live for 20 years or more in humans.[76]

Most human infections are acquired by ingesting metacercariae from contaminated food such as raw or partially cooked crabs or crayfish (Fig 3.11). Using meat or fluids from parasitized crustacea for medicinal purposes is a significant source of infection in Korea.[7] Handlers of shellfish are at risk when metacercariae contaminate hands or cooking utensils. Contaminated drinking water can transmit paragonimiasis,[79] as can ingesting metacercariae in the flesh of a paratenic host (Table 3.2).[42,47,73] Young, active flukes can penetrate a flesh wound and cause infection.

Clinical Features and Pathogenesis

During the early stages of paragonimiasis, patients are usually asymptomatic, but some develop dull abdominal pain and diarrhea soon after initial infection. Presumably, these symptoms result from the tissue damage and inflammation provoked by the worm as it traverses the intestinal wall and then the diaphragm. If the fluke remains in the abdomen, patients may have palpable intra-abdominal masses, tenderness, nausea, vomiting, and diarrhea. Heavily infected individuals may present with fever, fatigue, generalized myalgia, abdominal pain, and eosinophilia.[73] The eosinophilia is usually 10% to 30% but can be as high as 91%.[24] In humans, only a small percentage of worms in the thoracic cavity successfully penetrate the pleura. Paragonimiasis may have cutaneous and lymphatic manifestations.

Species	First intermediate hosts (snails)	Second intermediate (crabs) and paratenic hosts	Definitive hosts
P. westermani	Semisulcospira sp Brotia sp	Potamon sp Eriocheir sp Procamburus sp (Paratenic hosts: wild boar, pig, some monkeys, white mice and other small mammals, chickens)	Human, felines (tiger most suitable), canines, Taiwanese monkey, other mammals
P. skrjabini	Assiminea lutea Tricula gregoriana	Potamon sp	Human, cat, dog, civet
P. miyazakii	Bythinella nipponica akiyoshiensis Bythinella nipponica nipponica Saganoa sp	Sinopotamon sp	Human, Japanese and Korean weasels, marten, dog, Japanese wild boar, domestic cat, badger (experimentally: monkey, rat, rabbit, guinea pig)
P. heterotremus	Tricula gregoriana	Potamon sp Parathelphusa dugasti	Human, rat, dog, cat, leopard (experimentally: monkey)
P. hueitungensis	Tricula cristella	Sinopotamon denticulatum denticulatum Sinopotamon joshueiense Isolapotamon sinense Isolapotamon papilionaceus	Dog, cat (experimentally: albino rats); immature adults in human subcutaneous tissue
P. pulmonalis	Semisulcospira sp	Eriocheir japonicus Cambaroides similis Procambarus clarkii Palaemon nipponensis (Paratenic hosts: Japanese wild boar, Japanese monkey, pig, rat, rabbit, guinea pig, hamster, mouse)	Human, dog, cat, tiger
P. mexicanus	Aroapyrgus costaricensis Aroapyrgus colombiensis	Hipolobocera bouvieri Other Hipolobocera sp (Paratenic host: rat)	Human, opossum, cat, jaguar, dog, fox, raccoon, coati, skunk
P. ecuadoriensis	Unknown	Hipolobocera aequatorialis	Human, coati, opossum
P. kellicotti	Pomatiopsis lapidaria	Cambarus sp	Human, mink, cat, dog, pig
P. africanus	Potadoma freethii	Sudanonautes africanus Sudanonautes pelli	Human, dog, mongoose, swamp mongoose, African civet, drill, potto (experimentally: cat, monkey)
P. uterobilateralis	Unknown Suspected: Afropomus balanoides Potadoma sanctipauli	Sudanonautes africanus Sudanonautes pelli Liberonautes latidactylus	Human, dog, mongoose, swamp mongoose, African civet, rat (Malacomysll edwardsi), shrew (Crocidura flavescens)

Table 3.2
First intermediate, second intermediate, and definitive hosts of *Paragonimus* sp.

Pulmonary and pleuropulmonary paragonimiasis

Baelz, in 1878, was the first to recognize that, unlike those with tuberculosis, patients with pulmonary paragonimiasis are in apparent good health on physical examination, with no symptoms other than coughing.[33]

Patients usually present with a cough, typically productive of rusty sputum with a foul, fishy odor.[73] Some have chest pain and night sweats. Most have eosinophilia and an incidental lung lesion on a routine chest film. Some patients experience symptoms continuously or have episodes of hemoptysis interrupted by several months without symptoms.[2,35]

In heavy infections, patients experience dyspnea and chest pain secondary to pneumothorax, pleural effusion, or hydropneumothorax.[22,73]

Patients often have fever, fatigue, and myalgia, and eventually develop a chronic cough similar to that of chronic bronchitis, bronchial asthma, or bronchiectasis. Frank hemoptysis is uncommon.[73] Some patients later develop clubbing of the fingers and persistent rales. More than 50% of Korean patients have pleural lesions.[73] Peripheral blood eosinophilia and leukocytosis are moderately high, and serum IgG and IgE levels are usually elevated.

The radiographic appearance of pulmonary lesions varies with the stage of infection and surrounding tissue reaction. Initially there is patchy consolidation representing hemorrhagic pneumonia caused by migrating flukes. Pleural effusions or pneumothorax are frequent during this stage. Cysts form around adult worms and may contain bloody fluid visible on a radiograph or computed tomography (CT) scan.[46] Most cysts are near the lung periphery; apices are frequently spared. Sometimes there are conglomerates of cysts or nodules. A pericystic airspace encases cysts, producing a localized consolidation. Cysts may communicate with an adjacent bronchiole or bronchus and appear as air cysts within the consolidated lung or as ring shadows. Radiographic changes may be transient, appearing and disappearing over short intervals. Secondary bacterial infections are common. Classically, paragonimiasis both masks and confuses the clinical and radiologic diagnosis of tuberculosis.[56] Although tuberculosis and paragonimiasis may coexist, no causal relationship is known. Lung lesions of paragonimiasis may resolve spontaneously within 10 years.

Paragonimus miyazakii infections tend to cause pleural effusion, pleural exudates, and pneumothorax with marked eosinophilia. Intrathoracic chest-wall lesions contain eggs that may produce an abnormal chest x-ray, clinically mimicking an intrathoracic tumor.[60]

Cerebral paragonimiasis

Cerebral paragonimiasis is the most common symptomatic, extrapulmonary form of *Paragonimus* infection. Incidence is highest in men.[7] Most lesions are in areas of the brain supplied by the middle and posterior cerebral arteries, especially in the temporal, occipital, and parietal lobes.[29] Pulmonary symptoms usually precede central nervous system (CNS) symptoms. Yokogawa

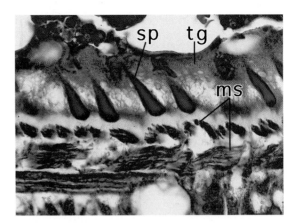

Figure 3.13
Body wall of adult *P. africanus* described in Figure 3.12. Spines (sp) are 30 µm long, pointed, and arranged singly. Note also tegument (tg) and smooth muscle (ms). Movat x595

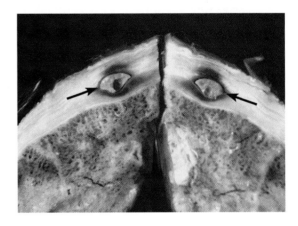

Figure 3.14
Bisected adult *P. westermani* (arrows) in thickened pleura. x3.3

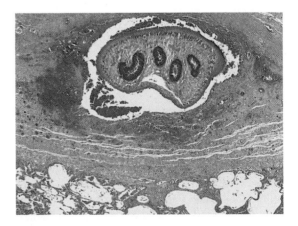

Figure 3.15
Section through posterior end of adult *P. westermani* in thickened, inflamed, hemorrhagic pleura. Note multiple sections of ceca. x23

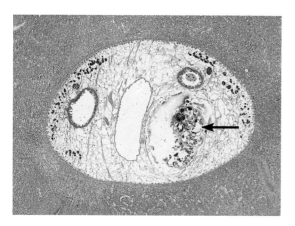

Figure 3.16
Adult *P. westermani* in lung eliciting chronic inflammatory infiltrate of lymphocytes, plasma cells, and giant cells. Note egg-filled uterus (arrow). x15.5

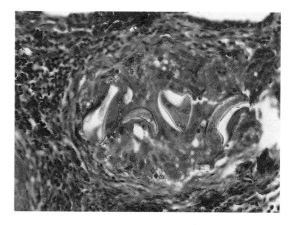

Figure 3.17
Collapsed and distorted eggs of *P. westermani* causing granulomatous inflammation in lung. Note giant cells engulfing eggs. x240

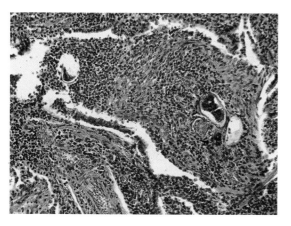

Figure 3.18
Eggs of *P. westermani* provoking granuloma in bronchiole. x110

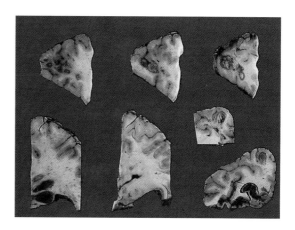

Figure 3.19
Multiple sections of a brain containing parenchymal cysts caused by eggs of *P. westermani*. x0.33

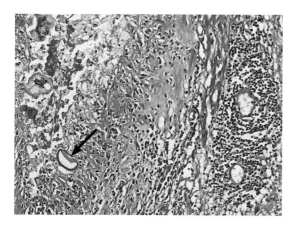

Figure 3.20
Eggs of *P. westermani* in brain provoking cystic, necrotizing granuloma. Note egg (arrow) at edge of necrosis and perivascular cuffing. x120

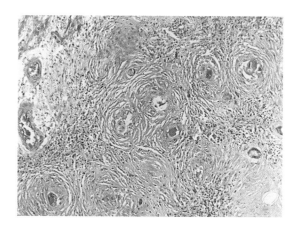

Figure 3.21
Fibrotic granulomas in brain caused by *P. westermani* eggs. x60

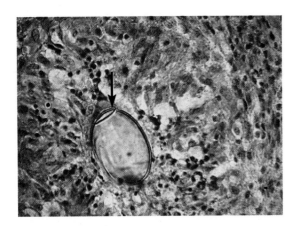

Figure 3.22
Egg of *P. westermani* within necrotizing granuloma in brain, demonstrating operculum (arrow). x210

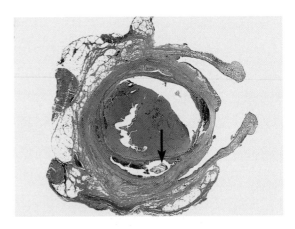

Figure 3.23
Adult *P. africanus* (arrow) in exenteration specimen from eye of 12-year-old African boy. Vitreous cavity is filled with exudate. x1.4

believes that the lung is the only site where *Paragonimus* sp can reach sexual maturity, and that adult flukes accidentally wander out of the lung to aberrant locations such as the brain.[76]

A fluke migrating through the brain can cause cerebral hemorrhage, edema, and meningitis. A minority of patients present with an acute syndrome, usually caused by parasite-induced meningitis or arachnoiditis. These patients have fever, headache, vomiting, and sometimes are markedly disoriented. The major signs are nuchal rigidity, Kernig's sign, and, in about 15% of patients, coma. Cerebrospinal fluid (CSF) has an elevated cell count, elevated protein, increased pressure, and decreased glucose level.[29] Acute meningeal paragonimiasis resembles tuberculous meningitis, but onset is more acute and neurologic deficits are less severe.

Most patients with cerebral paragonimiasis experience an insidious onset of symptoms including fever, severe headache, weakness, nausea, vomiting, and seizures, with seizures being most common. Initially, these are focal motor (jacksonian) seizures that progress to generalized tonic-clonic seizures.[29]

About 70% of patients undergo personality changes, disorientation, and a gradual decline in cognitive function. Sensory disturbances such as hypoesthesia and hemihypoesthesia are common. Some patients have hemiparesis, paresis, aphasia, and meningismus.[73] Anemia and leukocytosis develop in 10% to 30% of patients with cerebral paragonimiasis. Peripheral eosinophilia is variable.

Spinal cord lesions may cause paraplegia, sensory loss, and impaired sphincter control. Blurred vision is common, and some patients develop associated diplopia. Involvement of the eye and periorbital tissue may produce a variety of ophthalmological signs, such as moderate to severe loss of visual acuity, visual field defects, fundal changes (optic atrophy, papilledema, macular degeneration, Foster-Kennedy syndrome), palsy (extraocular muscle, gaze, or convergency), nystagmus, and papillary abnormalities (Figs 3.12 and 3.13).[50]

Other sites

Wherever the fluke or its eggs lodge, host tissue reaction forms a nodule. In skin or subcutaneous tissue, *P. westermani* causes abscesses or ulcers. Subcutaneous paragonimiasis may be concomitant with pulmonary paragonimiasis.[6] More often, subcutaneous manifestations, especially migrating nodules, represent immature migrating *Paragonimus* sp for which humans are aberrant final hosts, and are probably not associated with pulmonary lesions. *Paragonimus skrjabini*, *P. heterotremus*, *P. hueitungensis*,[18] and *P. mexicanus* usually localize ectopically. Migratory nodules are common on the chest, abdominal wall, and extremities. Cysts in the mastoid area are common

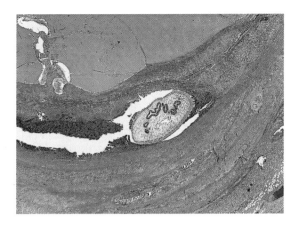

Figure 3.24
Higher magnification of Figure 3.23 showing adult *P. africanus* in thickened, inflamed, hemorrhagic choroid. x5.9

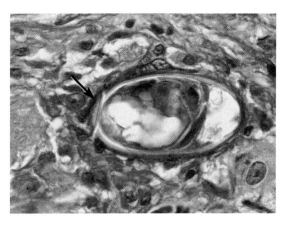

Figure 3.25
Egg of *P. uterobilateralis* in eye of 3-year-old from Sudan. Note giant cell engulfing egg with operculum (arrow). x595

in *P. africanus* infections. There is a report of 3 mastoid abscesses in Nigerian patients that contained operculated eggs consistent with those of *P. uterobilateralis*, but which could also represent those of either *P. congolensis* or *A. nouveli* (see chapter 6).[54]

Paragonimiasis may present as a benign cystic breast mass,[10] where it may mimic a variety of lesions. Adrenal involvement causes granulomatous inflammation with extensive fibrosis and necrosis that may clinically and radiographically resemble a neoplasm.[14]

Adult *Paragonimus* can infect lymph nodes, heart and pericardium,[58] mediastinum, breast, kidney, adrenal gland, omentum, bone marrow, stomach wall, urinary bladder, and male and female reproductive organs.[76] Adult flukes may also parasitize liver, spleen, and pancreas, but clinical symptoms related to these organs are rare. Flukes in bile ducts can damage the mucosa and lead to extrahepatic obstructive jaundice and biliary cirrhosis.[51] Not all ectopic lesions result from adult worm migration. *Paragonimus* eggs may disseminate and embolize to any organ, including the brain.

Patients with *P. hueitungensis* usually have particularly high leukocytosis, marked eosinophilia, increased blood sedimentation rate, and symptoms such as fever, lassitude, night sweats, impaired appetite, and general weakness. Some individuals may merely have high eosinophilia.[18] The differential clinical diagnosis includes various forms of eosinophilic granulomas, tropical eosinophilia of parasitic or nonparasitic origins, especially larva migrans,[44] and so-called hypereosinophilia of obscure causes (eg, allergic reactions to food, chemicals, and pollen, and Weber-Christian syndrome).

Pathologic Features

Larvae penetrate tissues by enzymatic digestion of host tissue as they migrate through the intestinal wall, diaphragm, and other sites. These enzymes have hyaluronidase, collagenase, and protease activities. Some cysteine proteases of *P. westermani* cleave human IgG to evade immune reaction,[9] and can degrade collagen and hemoglobin.[63] Tissue damage from collagen degradation leads to edema, hemorrhage, and inflammation. Lymphoid cells, macrophages, and some neutrophils accumulate around larvae as they penetrate intramuscular connective tissue or serous membranes. There is severe segmentation and fibrin-like degeneration of muscle and connective tissue. Periodic acid-Schiff (PAS) technique reveals partial or total loss of polysaccharide, including glycogen; tropeolin staining reveals a marked decrease of basic protein.[76]

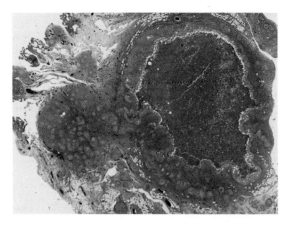

Figure 3.26
Omentum of 38-year-old Guatemalan showing extensive hemorrhagic necrosis and granulomatous inflammation caused by eggs of *P. mexicanus*. x5.8

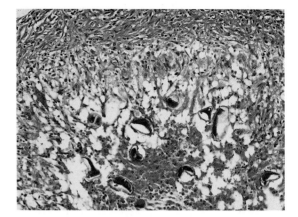

Figure 3.27
Higher magnification of Figure 3.26 depicting collapsed and distorted eggs of *P. mexicanus* in necrotizing granuloma of omentum. x115

Pleural paragonimiasis

Cysts form around immature flukes lodged in the pleura. The wall of a newly formed cyst is composed of young connective tissue with a marked neutrophilic infiltration. Older cysts have a thick, fibrous wall of connective tissue (Figs 3.14 and 3.15).

Pulmonary paragonimiasis

Worms usually reside near large bronchioles or bronchi, eliciting an acute exudate of eosinophils and neutrophils which rapidly becomes chronic; the exudate contains lymphocytes, plasma cells, and giant cells (Fig 3.16). Flukes discharge eggs into the lung parenchyma, exacerbating the hemorrhage, inflammatory infiltrate, and granulomatous reaction (Figs 3.7 and 3.17). A fibrous cyst wall that may be very thin or up to several millimeters thick soon forms around the area. Cysts range up to 1.5 cm in diameter and may develop at any depth in the parenchyma of the lung. The center of the cyst may become necrotic, filling with a fluid made brown by hematin. Many of these cysts penetrate the bronchiolar lumen and discharge eggs (Fig 3.18), cystic contents, and blood into the respiratory tract; patients may cough up this material. Damage to the lung parenchyma by the cystic contents encourages secondary pulmonary infection. Eggs that enter the lung parenchyma directly or through the bronchioles provoke granulomas (Fig 3.18). Older lesions may be mineralized.

Cerebral paragonimiasis

How flukes reach the brain is unknown. They may migrate from the lungs through the jugular foramen to the CNS, but this has not been confirmed in experimental models.[29] Once in the brain, flukes cause necrosis of the parenchyma, producing cavities that range up to 10 cm and have a thin outer membrane (Fig 3.19).

Three characteristic findings distinguish cerebral paragonimiasis: arachnoiditis, encapsulated abscesses, and granulomas that may calcify and become encapsulated.[7]

In acute meningeal paragonimiasis, the arachnoid infiltrate contains neutrophils, eosinophils, and lymphocytes. Later, chronic arachnoiditis develops, characterized by scattered lymphocytic infiltrates with intervening connective tissue. Arachnoiditis is most prominent on the basilar surface of the brain, accounting for the high frequency of optic nerve involvement.[29]

Abscesses take the form of cystic lesions up to 10 cm in diameter, and have a thin connective tissue capsule (Figs 3.19 and 3.20). In the center of the lesion there is thick, yellow pus composed of necrotic cellular debris, including lymphocytes, plasma cells, eosinophils, and Charcot-Leyden crystals. *Paragonimus* eggs usually lie along the inner aspect of the capsule, often in great numbers. Gliosis may be marked. Later, granulomas form around eggs (Figs 3.21 and 3.22). Several granulomatous nodules may be connected by fibrous

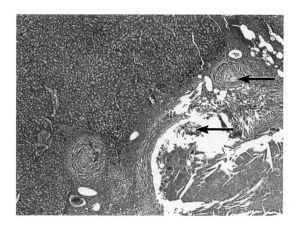

Figure 3.28
Caseation necrosis in liver caused by eggs (arrows) of *P. westermani*. Note focal, mild perivascular cuffing. x27

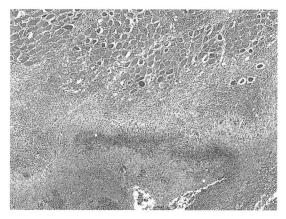

Figure 3.29
Extensive eosinophilic necrosis and cyst formation caused by eggs of *Paragonimus* sp in external oblique muscle of 19-year-old Iranian man. Patient presented with right upper quadrant pain and mass in right flank. x23

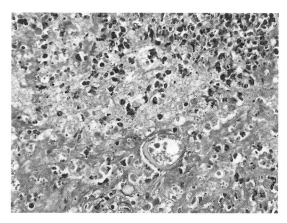

Figure 3.30
Higher magnification of Figure 3.29 depicting fragmented egg of *Paragonimus* sp causing edema and eosinophilia. x230

gliotic tissue. Generally, egg granulomas consist of 3 layers. The center is a mass of caseous necrosis with *Paragonimus* eggs and giant cells. In the middle there is connective tissue with eggs lying along the inner aspect. The outer layer contains lymphocytes, plasma cells, eosinophils, glial cells, and eggs.[15] Perivascular cuffing is readily apparent (Fig 3.20). Calcification may be prominent in older lesions.

Lesions in the spinal cord resemble those in the brain and rarely contain adult worms. Flukes in the eye may cause a necrotizing abscess (Figs 3.23 and 3.24). Older lesions may contain only eggs and be chronically inflamed (Fig 3.25).

Other sites

Immature *Paragonimus* must migrate through various tissues to arrive at their preferred location, the lung, and may produce lesions along the way. Subcutaneous paragonimiasis is one of the more common ectopic sites. Typically, a subcutaneous nodule contains an immature *Paragonimus* within an intense infiltrate of eosinophils and abundant Charcot-Leyden crystals. There are usually no eggs in the nodules or in the fluke's uterus,[18] but in the rare instances that eggs are present, they incite granulomas. Immature or mature flukes that lodge in the omentum (Figs 3.26 and 3.27),[30] in an abdominal organ (liver (Fig 3.28), adrenal, pancreas), or skeletal muscle (Figs 3.29 and 3.30) can produce suppurative inflammation that may become granulomatous.

Cystic granulomas that contain only eggs may result from egg emboli, though some investigators suspect that, in most such instances, migrating adult worms deposited eggs at the site of the granuloma.[66] These lesions are composed of a cyst with a necrotic center containing cellular debris and many *Paragonimus* eggs lining the inner surface of the fibrous wall.

A 60-year-old Chinese man developed severe obstructive jaundice caused by *Paragonimus* sp nodules that formed along his biliary tract. He presented with advanced biliary cirrhosis, many bile thrombi in the hepatic parenchyma, and acute pancreatitis. Autopsy revealed that liver cells were interspersed with fibrous tissue, and the lobular pattern was greatly distorted. There was a periductal infiltrate of lymphocytes and plasma cells with proliferation of bile ducts and deposits of calcified

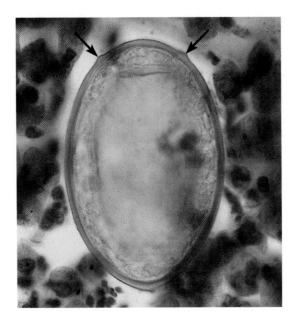

Figure 3.31
Egg of *P. kellicotti* in sputum from patient who never traveled outside the United States. Note opercular shoulders (arrows) and thick shell. Papanicolaou x845

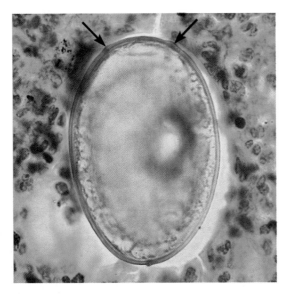

Figure 3.32
Egg of *P. uterobilateralis* in sputum of patient from Liberia showing opercular shoulders (arrows). ZN x1020

eggs. Parasitic granuloma had caused ulcers in the stomach and gallbladder, bilateral adhesions, thickening of the pleura, nephrotic changes of the renal tubules, and an abscess in the mesentery.[51]

Diagnosis

In an endemic area, an otherwise healthy individual who presents with eosinophilia and a cough productive of rusty sputum is likely to have paragonimiasis. The diagnostic triad of pulmonary paragonimiasis (cough, hemoptysis, and eggs in sputum or feces) does not usually develop in patients with ectopic or pleural paragonimiasis.[39]

Microscopic examination of sputum (Figs 3.31 and 3.32) remains the diagnostic method of choice for symptomatic patients. However, eggs that have been swallowed in sputum are sometimes found only in feces (Fig 3.6). In such cases, fecal examination can provide a definitive diagnosis. Histopathologic study of surgical or autopsy specimens is particularly useful for diagnosing nonpulmonary paragonimiasis (Figs 3.22 and 3.33). Occasionally, cytologic examination of fluid obtained by thoracentesis, paracentesis, or fine-needle aspiration reveals eggs.[10,32]

Eggs of *Paragonimus* sp may be mistaken for those of other operculate trematodes, including *Clonorchis sinensis,* which are much smaller (26 to 35 µm), and those of *Fasciola hepatica* and *Fasciolopsis buski,* which are considerably larger (130 to 140 µm). Eggs of *Diphyllobothrium* sp, the fish tapeworms, are also operculated and resemble eggs of *Paragonimus* sp. But *Paragonimus* sp eggs are larger and, except for eggs of *P. hueitungensis*, do not have a knob on the abopercular end. Eggs of *Schistosoma japonicum* resemble *Paragonimus* sp eggs and are about the same size and shape, but schistosome eggs lack an operculum and are not birefringent (Fig 3.8). Eggs of *Nanophyetus salmincola* also resemble those of *Paragonimus*.[73]

Patients with symptoms of cerebral paragonimiasis require radiographic studies of the chest and head, as well as analyses of blood, sputum, feces, and CSF (where there may be specific antibodies to *Paragonimus*). Biopsy specimens of accessible lesions may be diagnostic. Electroencephalograms

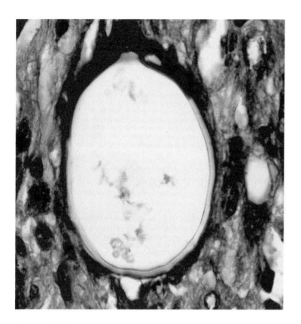

Figure 3.33
Egg of *P. africanus* in eye. Note variation in color of thick shell. x875

are usually abnormal, revealing focal epileptogenic activity and suppression of normal background rhythm. In about 70% of patients, plane films of the skull show aggregate round or oval cystic calcifications with increased peripheral density. Multiple cystic calcifications that increase with duration of symptoms and look like soap bubbles on skull films suggest cerebral paragonimiasis. In recently diagnosed patients, CT scans display multilocular cysts with surrounding edema, which show ring enhancement after contrast infusion. Magnetic resonance imaging (MRI) of chronic cerebral paragonimiasis may show peripheral low-intensity areas surrounding a central hypointense and isointense region on T_1-weighted images, corresponding to foci of granulomatous inflammation. On T_2-weighted images, the central region has a high signal intensity and surrounding low-intensity, reflecting the widespread inflammatory changes in the tissues surrounding the lesions of cerebral paragonimiasis.[21]

Intradermal testing is useful for epidemiologic investigations (a weal of 5 mm or greater is positive), but is not clinically useful since results may be positive for years after the fluke dies.[62,73] Complement-fixing antibodies are usually found in sera during active disease, and become negative within 3 to 9 months after complete recovery.[76] An enzyme-linked immunsorbent assay (ELISA) using fluke cysteine proteinases of *P. westermani* has adequate sensitivity and minimal cross-reactivity with sera from patients with fasciolopsiasis, onchocerciasis, or clonorchiasis, but is not species-specific.[19] Enzyme immunoassay (EIA) for IgG antibodies indicates infection with *Paragonimus*. Simultaneous application of different antigens to test sera can help distinguish between *Paragonimus* sp.[27]

Treatment and Prevention

Praziquantel, the drug of choice for treating paragonimiasis,[57] is not recommended for pregnant women. For such patients, it is preferable to defer treatment until after delivery unless immediate intervention is called for.[73] Bithionol, introduced in 1961, is efficacious, has minimal side effects, and was the drug of choice for 2 decades.[77] It is the logical alternative when praziquantel cannot be used.[75]

Before the advent of praziquantel, cerebral paragonimiasis often required surgical intervention, especially for localized lesions. Today, in-patient drug therapy is recommended, especially for patients with multiple lesions. Along with praziquantel, corticosteroids may be needed to lessen the local inflammation that normally intensifies after the worm dies. Surgical excision is still the treatment of choice for uncomplicated subcutaneous and abdominal lesions.

Eradicating the primary or secondary intermediate hosts of *Paragonimus* sp is neither practical nor ecologically sound. Therefore, eating raw or poorly cooked crabs, or using raw crab products for medicinal purposes, is strongly discouraged. Irradiation of fish and shellfish effectively eliminates metacercariae[73]; education and mass treatment are effective preventive measures.[37] Early diagnosis and treatment of pulmonary paragonimiasis in children will likely prevent progression to more severe cerebral involvement. Populations in endemic areas with no access to protected water should drink only boiled water.

References

1. Ando R. The first intermediate host of Paragonimus westermani [in Japanese; English abstract]. *Trop Dis Bull* 1921;17:51.
2. Anonymous. Parasitical haemoptysis. *Lancet* 1880;2:548–549.
3. Arzube ME, Voelker J. On the occurrence of human paragonimiasis in Ecuador (1972–76) [in German]. *Tropenmed Parasitol* 1978;29:275–277.
4. Beaver PC, Jung RC, Cupp EW. *Clinical Parasitology*. 9th ed. Philadelphia, Pa: Lea & Febiger; 1984:470–471.
5. Cabrera BD. Paragonimiasis in the Philippines: current status. *Arzneimittelforschung* 1984;34:1188–1192.
6. Carrera-Cobos T. Pulmonary distomatosis with subcutaneous dissemination [in Spanish]. *Rev Ecuat Hig Med Trop* 1967;24:267–268.
7. Choi DW. Paragonimus and paragonimiasis in Korea. *Korean J Parasitol* 1990;28:79–102.
8. Chung HL, Tsao WC. Paragonimus westermani (Szechuan variety) and a new species of lung fluke—Paragonimus szechuanensis. 1. Studies on morphology and life history of Paragonimus szechuanensis. *Chin Med J* 1962;81:354–378.
9. Chung YB, Yang HJ, Kang SY, Kong Y, Cho SY. Activities of different cysteine proteases of Paragonimus westermani in cleaving human IgG. *Korean J Parasitol* 1997;35:139–142.
10. Fogel SP, Chandrasoma PT. Paragonimiasis in a cystic breast mass: case report and implications for examination of aspirated cyst fluids. *Diagn Cytopathol* 1994;10:229–231.
11. Fujino T, Ishii Y. Ultrastructural studies on spermatogenesis in a parthenogenetic type of Paragonimus westermani (Kerbert, 1878) proposed as P. pulmonalis (Baelz 1880). *J Parasitol* 1982;68:433–441.
12. Fukunaga H, Baba M. A case of Paragonimus miyazakii infection in man showing bilateral pleural, interlobar and pericardial effusion with spontaneous pneumothorax. *Nippon Kyobu Shikkan Gakkai Zasshi* 1977;15:205–209.
13. Grove DI. *A History of Human Helminthology*. Wallingford, England: CAB Int; 1990:159–185.
14. Hahn ST, Park SH, Kim CY, Shinn KS. Adrenal paragonimiasis simulating adrenal tumor—a case report. *J Korean Med Sci* 1996;11:275–277.
15. Higashi K, Aoki H, Tatebayashi K, Morioka H, Sakata Y. Cerebral paragonimiasis. *J Neurosurg* 1971;34:515–527.
16. Higo H, Ishii Y. Comparative studies on surface ultrastructure of newly excysted metacercariae of Japanese lung flukes. *Parasitol Res* 1987;73:541–549.
17. Hirai H, Sakaguchi Y, Habe S, Imai HT. C-banding analysis of six species of lung flukes, Paragonimus spp. (Trematoda: Platyhelminthes), from Japan and Korea. *Z Parasitenkd* 1985;71:617–629.
18. Huei-Lan C, Chih-piao H, Lien-yin H, P'ei-chih K, Lan S, Fu-hsi C. Studies on a new pathogenic lung fluke—Paragonimus hueit'ungensis sp. nov. *Sci Sin* 1975;18:785–814.
19. Ikeda T, Oikawa Y, Nishiyama T. Enzyme-linked immunosorbent assay using cysteine proteinase antigens for immunodiagnosis of human paragonimiasis. *Am J Trop Med Hyg* 1996;55:435–437.
20. Itakura M, Shinozaki T, Shingyouji M. A case of Paragonimus miyazakii with pleuritis and meningoencephalitis [in Japanese]. *Nippon Kyobu Shikkan Gakkai Zasshi* 1997;35:980–984.
21. Kadota T, Ishikura R, Tabuchi Y, et al. MR imaging of chronic cerebral paragonimiasis. *AJNR Am J Neuroradiol* 1989;10:S21–S22.
22. Kagawa FT. Pulmonary paragonimiasis. *Semin Respir Infect* 1997;12:149–158.
23. Kamo H, Nishida H, Hatsushika R, Tomimura T. On the occurrence of a new lung fluke, Paragonimus miyazakii n. sp. in Japan. (Trematoda: Troglotrematidae). *Yonago Acta Med* 1961;5:43–52.
24. Kan H, Ogata T, Taniyama A, Migita M, Matsuda I, Nawa Y. Extraordinarily high eosinophilia and elevated serum interleukin-5 level observed in a patient infected with Paragonimus westermani. *Pediatrics* 1995;96:351–354.
25. Kerbert C. Zur Trematoden-Kenntnis. *Zoologischer Anzeiger* 1878;1:271–273.
26. Kim DC. Paragonimus westermani: life cycle, intermediate hosts, transmission to man and geographical distribution in Korea. *Arzneimittelforschung* 1984;34:1180–1183.
27. Knobloch J. Application of different Paragonimus antigens to immmunodiagnosis of human lung fluke infection. *Arzneimittelforschung* 1984;34:1208–1210.
28. Kurochkin IuV, Sukhanova GI. Species makeup of the genus Paragonimus and the causative agents of human paragonimiasis [in Russian]. *Med Parazitol (Mosk)* 1978;47:36–39.
29. Kusner DJ, King CH. Cerebral paragonimiasis. *Semin Neurol* 1993;13:201–208.
30. Lee SC, Jwo SC, Hwang KP, Lee N, Shieh WB. Discovery of encysted Paragonimus westermani eggs in the omentum of an asymptomatic elderly woman. *Am J Trop Med Hyg* 1997;57:615–618.
31. Leuckart R. *Die parasiten des Menschen und die von ihnen herrhührenden Krankheiten. Ein Hand - und Lehrbuch für Naturforscher und Aertze*. Vol 2. Leipzig, Germany: CF Winter'sche Verlagshandlung; 1901:404–437.
32. Linford RA, Nguyen GK. Paragonimus westermani ova in pleural effusion. *Diagn Cytopathol* 1994;11:95–96.
33. Lü CH. Clinical observations of 195 cases of Paragonimus in children [in Chinese; English abstract]. *Chin J Pediatr* 1957;8:143–146.
34. Mamiya N. An autopsy case of heterotopic parasitism with Paragonimus in human paraproctal region [in Japanese]. *Jpn J Parasitol* 1961;10:557–562.
35. Manson P. On endemic haemoptysis. *Lancet* 1883;1:532–534.
36. Mariano EG, Borja SR, Vruno MJ. A human infection with Paragonimus kellicotti (lung fluke) in the United States. *Am J Clin Pathol* 1986;86:685–687.
37. Marty AM, Andersen EM. Helminthology. In: Doerr W, Seifert G, eds. *Tropical Pathology*. Berlin, Germany: Springer-Verlag; 1995:801–982.
38. McCallum SM. Ova of the lung fluke Paragonimus kellicotti in fluid from a cyst. *Acta Cytol* 1975;19:279–280.
39. Minh VD, Engle P, Greenwood JR, Prendergast TJ, Salness K, St Clair R. Pleural paragonimiasis in a Southeast Asian refugee. *Am Rev Respir Dis* 1981;124:186–188.
40. Miyazaki I. Paragonimiasis. In: Hillyer GV, Hopla CS, eds. *CRC Handbook Series in Zoonoses. Section C: Parasitic Zoonoses*. Vol 3. Boca Raton, Fla: CRC Press Inc; 1982:143–164.
41. Miyazaki I. Two types of the lung fluke which has been called Paragonimus westermani (Kerbert, 1878). *Med Bull Fukuoka Univ* 1978;5:251–263.
42. Miyazaki I, Habe S. A newly recognized mode of human infection with the lung fluke Paragonimus westermani (Kerbert, 1878). *J Parasitol* 1976;62:646–648.
43. Miyazaki I, Ishii Y. Comparative study of the Mexican lung flukes with Paragonimus kellicotti Ward, 1908. *J Parasitol* 1968;54:845–846.
44. Miyazato T, Ando S, Nakagawa O. One case of dermal heterotopic Paragonimus parasitism which induced creeping disease-like symptoms [in Japanese]. *Jpn J Parasitol* 1961;10:55–60.
45. Mizuki M, Mitoh K, Miyazaki E, Tsuda T. A case of paragonimiasis westermani with pleural effusion eight months after migrating subcutaneous induration of the abdominal wall [in Japanese]. *Nippon Kyobu Shikkan Gakkai Zasshi* 1992;30:1125–1130.

46. Moon WK, Kim WS, Im JG, Kim IO, Yeon KM, Han MC. Pulmonary paragonimiasis simulating lung abscess in a 9-year-old: CT findings. *Pediatr Radiol* 1993;23:626–627.

47. Motoyama S, Suzuki T, Sobue F, et al. Four cases of paragonimiasis due to consumption of wild boar meat [in Japanese]. *Nippon Naika Gakkai Zasshi* 1996;85:1145–1146.

48. Nakagawa K. The mode of infection in pulmonary distomiasis. Certain freshwater crabs as intermediate hosts of Paragonimus westermani. *J Infect Dis* 1916;18:131–142.

49. Nozais JP, Doucet J, Dunan J, N'Dri G. Paragonimiasis in Black Africa. A recent infection focus in Ivory Coast [in French]. *Bull Soc Pathol Exot* 1980;73:155–163.

50. Oh SJ. Ophthalmological signs in cerebral paragonimiasis. *Trop Geogr Med* 1968;20:13–20.

51. Okuda K, Kuratomi S, Moriyama M, Mae A. Biliary cirrhosis secondary to extrapulmonary paragonimiasis. *Digestion* 1969;2:347–353.

52. Olsen OW. *Animal Parasites. Their Life Cycles and Ecology*. 3rd ed. Baltimore, Md: University Park Press; 1974:315–318.

53. Onuigbo WI, Nwako FA. Discovery of adult parasites of Paragonimus uterobilateralis in human tissue in Nigeria. *Tropenmed Parasitol* 1974;25:433–436.

54. Oyediran AB, Fajemisin AA, Abioye AA, Lagundoye SB, Olugbile AO. Infection of the mastoid bone with a Paragonimus-like trematode. *Am J Trop Med Hyg* 1975;24:268–273.

55. Pachucki CT, Levandowski RA, Brown VA, Sonnenkalb BH, Vruno MJ. American paragonimiasis treated wtih praziquantel. *N Engl J Med* 1984;311:582–583.

56. Peroff RP. Paragonimiasis in Hawaii. *Hawaii Med J* 1974;33:329–330.

57. Rim HJ, Chang YS, Lee JS, Joo KH, Suh WH. Clinical evaluation of praziquantel (Embay 8440; Biltricide®) in the treatment of Paragonimus westermani. *Korean J Parasitol* 1981;19:27–37.

58. Saborio P, Lanzas R, Arrieta G, Arguedas A. Paragonimus mexicanus pericarditis: report of two cases and review of the literature. *J Trop Med Hyg* 1995;98:316–318.

59. Sachs R, Cumberlidge N. Distribution of metacercariae in freshwater crabs in relation to Paragonimus infection of children in Liberia, West Africa. *Ann Trop Med Parasitol* 1990;84:277–280.

60. Sawamura T, Takiya H, Yamada T, Sugimoto H, Kawai H, Watanabe H. A case of paragonimiasis with a tumor of the intrathoracic chest wall [in Japanese; English abstract]. *Kyobu Geka* 1994;46:937–939.

61. Shin DH, Joo CY. Prevalence of Paragonimus westermani in some Ulchin school children. *Acta Paediatr Jpn* 1990 32:269–274.

62. Singh TS, Mutum S, Razaque MA, Singh YI, Singh EY. Paragonimiasis in Manipur. *Indian J Med Res* 1993;97:247–252.

63. Song CY, Kim TS. Characterization of a cysteine proteinase from adult worms of Paragonimus westermani. *Korean J Parasitol* 1994;32:231–241.

64. Terasaki K. Karyotope of a lung fluke, Paragonimus westermani filipinus Miyazaki, 1978. *Nippon Juigaku Zasshi* 1983;45:9–14.

65. Terauchi J, Takenori O, Tomimura T. A case of spontaneous infection with Paragonimus miyazakii in a dog. *Jpn J Parasitol* 1961;10:394–397.

66. Thamprasert K. Subcutaneous abscess of neck, a granulomatous reaction to eggs of Paragonimus: a case report from northern Thailand. *Southeast Asian J Trop Med Public Health* 1993;24:609–611.

67. Travassos L. Sobre um interessante trematodeo parasito dos seios maxilares de gamba (Didelphis marsupialis). *Rev Bras Biol* 1942;2:213–218.

68. Voelker J, Arzube M. A new lung fluke from the coastal range of Ecuador: Paragonimus ecuadoriensis n. sp. *Tropenmed Parasitol* 1979;30:249–263.

69. Voelker J, Sachs R. Morphology of the lung fluke Paragonimus uterobilateralis occurring in Gabon, West Africa. *Trop Med Parasitol* 1985;36:210–212.

70. Voelker J, Vogel H. Two new Paragonimus species from West Africa: Paragonimus africanus and Paragonimus uterobilateralis [in German]. *Z Tropenmed Parasitol* 1965;16:125–148.

71. Vuong PN, Bayssade-Dufour C, Mabika B, Ogoula-Gerbeix S, Kombila M. Paragonimus westermani pulmonary distomatosis in Gabon. First case [letter]. *Presse Med* 1996;25:1084–1085.

72. Ward HB, Hirsch EF. The species of Paragonimus and their differentiation. *Ann Trop Med Parasitol* 1915;9:109–163.

73. World Health Organization. Control of foodborne trematode infections. Report of a WHO Study Group. *World Health Organ Tech Rep Ser* 1995;849:1–157.

74. Yamaguti S. *Synopsis of Digenetic Trematodes of Vertebrates*. Vol 1. Tokyo, Japan: Keigaku Pub Co; 1971:787–791.

75. Yokogawa M. Experimental chemotherapy of paragonimiasis. A review. *Arzneimittelforschung* 1984;34:1193–1196.

76. Yokogawa M. Paragonimus and paragonimiasis. *Adv Parasitol* 1965;3:99–158.

77. Yokogawa M, Yoshimura H, Okura T, et al. Chemotherapy of paragonimiasis with bithionol. 2. Clinical observations on the treatment with bithionol. *Jpn J Parasitol* 1961;10:317–327.

78. Yokogawa S, Suyemori S. Observations of abnormal courses of infection of Paragonimus ringeri [in Japanese; English abstract]. *Trop Dis Bull* 1921;17:50.

79. Yoshida S. Some notes on the encysted larva of the lung distome. *J Parasitol* 1916;2:175–180.

Hepatic Trematodiases

Santiago Mas-Coma,
Maria Dolores Bargues Castello,
Aileen M. Marty, *and* Ronald C. Neafie

Introduction

Digenetic flukes from 3 families of trematodes infect the human liver, bile ducts, and gallbladder, and may localize ectopically.

1. **Opisthorchiidae**
   - *Clonorchis sinensis* (Cobbold, 1875) Looss, 1907
   - *Opisthorchis viverrini* (Poirier, 1886) Stiles and Hassall, 1896
   - *Opisthorchis felineus* (Rivolta, 1884) Blanchard, 1895
   - *Metorchis conjunctus* (Cobbold, 1860) Looss, 1899
   - *Pseudamphistomum truncatum* (Rudolphi, 1819) Lühe, 1908
   - *Amphimerus pseudofelineus* (Ward, 1901) Barker, 1911
   - *Amphimerus noverca* (Braun, 1902) Barker, 1911
2. **Fasciolidae**
   - *Fasciola hepatica* (Linnaeus, 1758)
   - *Fasciola gigantica* Cobbold, 1855
3. **Dicrocoeliidae**
   - *Dicrocoelium dendriticum* (Rudolphi, 1819) Looss, 1899
   - *Dicrocoelium hospes* Looss, 1907

Four of these species are the most common cause of human trematodiasis: *C. sinensis*, *O. viverrini*, *O. felineus*, and *F. hepatica*.[69] *Fasciola gigantica*, *D. dendriticum*, and *D. hospes* cause numerous, but usually isolated, human infections. *Pseudamphistomum truncatum*, *A. pseudofelineus*, *A. noverca* and *M. conjunctus* infect humans only sporadically. *Pseudamphistomum aethiopicum* Pierantoni, 1942[89] produces cyst-like nodules within the wall of the small intestine of humans in Ethiopia, and some liver flukes of animals, such as *Metorchis albidus* (Braun, 1893) Looss, 1899, are potential parasites of humans.[77]

Clonorchiasis and Opisthorchiasis
Definition

Clonorchiasis and opisthorchiasis are infections of extrahepatic and intrahepatic biliary tracts by trematodes of the genera *Clonorchis* and *Opisthorchis*.

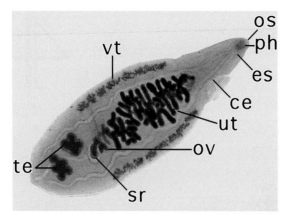

Figure 4.1
Whole mount of adult *O. viverrini* showing lobed testes (te) in posterior end of worm. Note oral sucker (os), pharynx (ph), esophagus (es), ceca (ce), lobed ovary (ov), uterus (ut), seminal receptacle (sr), and vitellaria (vt). Carmine x35

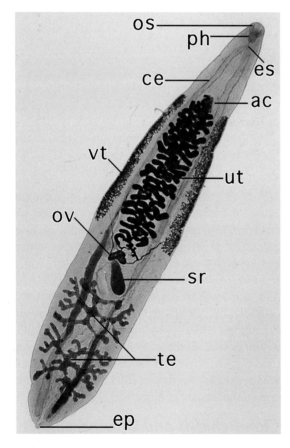

Figure 4.2
Whole mount of adult *C. sinensis* (13 mm by 2.6 mm) from Chinese patient. Note oral sucker (os), acetabulum (ac), pharynx (ph), esophagus (es), ceca (ce), branched testes (te), ovary (ov), uterus (ut), seminal receptacle (sr), vitellaria (vt), and excretory pore (ep). Carmine x8.4

Synonyms

Other names for clonorchiasis and opisthorchiasis include Chinese or Oriental liver fluke infection, biliary distomiasis, and biliary trematodiasis. Older names for *C. sinensis* include *Distoma sinense*, *Distoma spathulatum*, *Distoma endemicum*, and *Opisthorchis sinensis*. Older names for *O. felineus* include *Distoma conus*, *Distoma lanceolatum felis cati*, *Distoma felineum*, *Distoma lanceolatum canis familiaris*, *Distoma sibiricum*, *Dicrocoelium felineus*, and *Opisthorchis wardi*. Synonyms for *O. viverrini* include *Opisthorchis tenuicollis*, *Distoma tenuicollis*, and *Distoma viverrini*.

General Considerations

In 1874 McConnell discovered flukes in the biliary tree of a Chinese carpenter[53] and recognized them as a new species. Looss named the flukes *Clonorchis*, meaning branched testis.[51] A few years later Kobayashi identified fish as the second intermediate host. In 1918 Muto identified the snail *Parafossarulus manchouricus* (*Bithynia striatual* var *japonica*) as the first intermediate host. Mebius presented the first detailed study of the pathology of clonorchiasis in 1920, noting especially the epithelial proliferation of bile ducts. *Opisthorchis felineus* was first described in cats in 1884. Winogradoff reported the first human infection with *O. felineus* in Siberia in 1892. Poirier described *O. viverrini* in 1886.

Epidemiology

Clonorchiasis is prevalent in areas bordering the Sea of Japan, the East China Sea, and the South China Sea, corresponding to the distribution of the first intermediate snail host, *P. manchouricus*. This snail inhabits ditches, streams, ponds, and reservoirs in low-lying areas. An estimated 7 million people have clonorchiasis and another 290 million are at risk for infection.[69] Importation of infected fish and travel in endemic areas account for infections reported in Hong Kong, Thailand, Malaysia, Singapore, the Philippines, Hawaii, continental United States,[23] Canada,[55] Panama,[19] Surinam,[60] Brazil,[48] France,[52] Saudi Arabia,[4] and Australia.[8] In recent decades, 2 other first intermediate snail hosts (*Thiara granifera* and *Melanoides tuberculatus*) have spread widely and successfully colonized new regions on several continents.[56,86]

Opisthorchis viverrini infection is prevalent in Southeast Asia, affecting nearly 68% of inhabitants of some villages in northeastern Thailand. In Vietnam, some reported *O. felineus* infections may actually be *O. viverrini* infections, a confusion arising from the similarity of these flukes' eggs.

Opisthorchis felineus infects 1.5 million people in Russia and parts of the Commonwealth of Independent States.[69] Of 9 endemic regions, 4 have the highest prevalence: western Siberia (90% of the population),[92] Ukraine (up to 82%),[93] central Ural Mountains (average 14.5%; foci 68%),[26] and the Kama River basin (58.3%; eggs in 68% of soil samples).[17] Isolated human infections have also been reported in Central Europe.[12]

Humans are the most common definitive hosts, but reservoir hosts (cats and dogs) also have high infection rates. Because human infection is related to eating habits,[39,68] prevalence and intensity of infection in humans and animals are usually not the same.

Infectious Agent

Morphologic Description

Adult *Opisthorchis* sp that infect humans are distinguished from adult *C. sinensis* by their lobed testes (Fig 4.1). Differentiating *O. viverrini* from *O. felineus* is not as simple. Most authorities[28,40,72] consider the adult worms and eggs of these 2 *Opisthorchis* sp morphologically indistinguishable due to their overlapping variability. Fortunately, there is little overlap in the distribution of *O. viverrini* and *O. felineus*, permitting distinction on a geographic basis. Eggs of all 3 flukes are nearly impossible to distinguish morphologically. The similar distribution of *C. sinensis* and *O. viverrini* across Southeast Asia adds to the problem of species identification.[28,68]

Clonorchis sinensis

Adult: *Clonorchis sinensis* is a flat, transparent, aspinous fluke, 8 to 25 mm by 1.5 to 5 mm. It is tapered anteriorly and somewhat rounded posteriorly (Fig 4.2). The oral sucker (approximately 0.5 mm in diameter) is slightly larger than the ventral

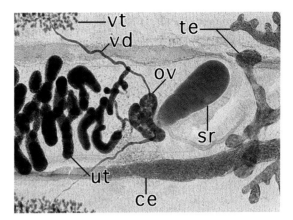

Figure 4.3
Higher magnification of adult *C. sinensis* shown in Figure 4.2. Note branched testes (te), ovary (ov), uterus (ut), seminal receptacle (sr), ceca (ce), vitellaria (vt), and vitelline ducts (vd). x23

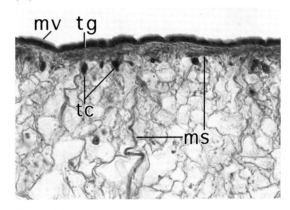

Figure 4.4
Body wall of adult *C. sinensis* in liver, demonstrating aspinous eosinophilic tegument (tg), fine, hair-like microvilli (mv), smooth muscle (ms), tegumental cells (tc), and parenchyma. Movat x800

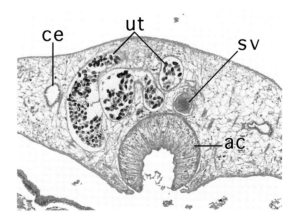

Figure 4.5
Section of adult *C. sinensis* in liver at level of acetabulum revealing muscular wall of acetabulum (ac), eggs within coiled sections of uterus (ut), seminal vesicle (sv), and paired ceca (ce). Movat x55

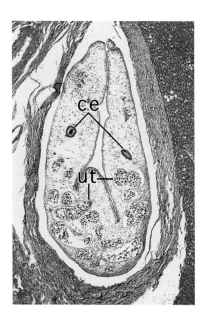

Figure 4.6
Adult *C. sinensis* in bile duct. Section is at midbody, showing 2 ceca (ce) and many sections of coiled, egg-filled uterus (ut). x30

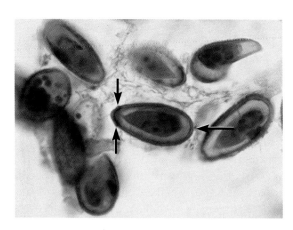

Figure 4.7
Eggs of *C. sinensis* in uterus of adult fluke in liver. Minute shoulders (short arrows) are at rim of operculum. Abopercular knob (long arrow) is barely perceptible. x765

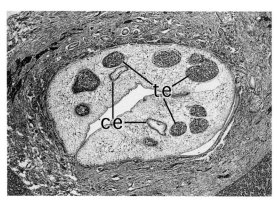

Figure 4.8
Adult *C. sinensis* in bile duct. Section is through posterior region and shows 2 ceca (ce) and several sections of branching testes (te). x30

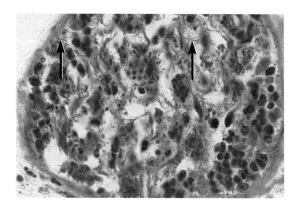

Figure 4.9
Section through testis of adult *C. sinensis* in liver. Note layers of cells consisting of packets of germinal cells in various stages of division, and mature, thin, filamentous sperm (arrows) in ducts. x570

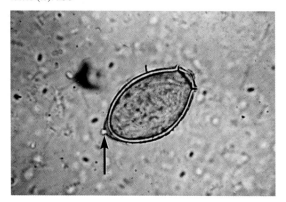

Figure 4.10
Clonorchis sinensis egg in feces. Egg is broadly ovoid, thick-shelled, light yellow-brown, and has large convex operculum that fits into rimmed extension of eggshell. Note small, comma-shaped protuberance (arrow) at abopercular end. Unstained x760

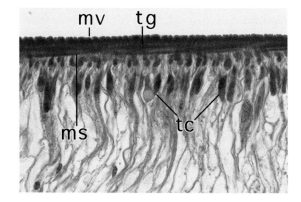

Figure 4.11
Body wall of adult *O. viverrini* in liver depicting thin, aspinous tegument (tg), fine, hair-like microvilli (mv), somatic muscle (ms), tegumental cells (tc), and parenchyma. x565

sucker (acetabulum). The acetabulum is situated at the end of the first anterior quarter of the fluke. A prominent pharynx leads into a short esophagus that divides into 2 unbranched ceca. The ceca extend laterally and end blindly at the posterior end of the body. The bladder terminates as a long, sigmoid excretory vesicle at the posterior end. There are 2 large, branched testes in the posterior third of the body, extending bilaterally beyond the ceca (Figs 4.2 and 4.3). The small, oval or slightly lobed ovary is median, lying anterior to the seminal receptacle (Fig 4.3). The large seminal receptacle is anterior to the testis (Fig 4.3). Vitellaria are small, dense, bilateral, and confined to the middle third of the worm (Fig 4.2). The coiled uterus is directed anteriorly between the ceca and terminates at the genital atrium near the acetabulum.

Morphologic features of *C. sinensis* are readily identified in tissue sections. The thin body wall (15 to 30 μm) (Fig 4.4) is composed of an aspinous, 5-μm-thick tegument lined with microvilli, 2 or more layers of smooth muscle, and irregularly spaced tegumental cells. A loose stroma of parenchymal cells and fibrous interstitial tissue comprise the parenchyma. Sections at the level of the acetabulum reveal the 400-μm-thick muscular sucker, golden-brown eggs within sections of coiled uterus, the seminal vesicle, and lateral intestinal ceca (Fig 4.5). Most prominent in the middle of the fluke are bifurcated, unbranched ceca and sections of coiled uterus (Fig 4.6) containing golden-brown eggs (Fig 4.7). The posterior third of the worm contains paired ceca made up of ciliated columnar epithelial cells, multiple sections of branched testes (Fig 4.8), and the dorsally located excretory bladder. Testes consist of packets of germinal cells in various stages of division, with mature filamentous sperm in ducts (Fig 4.9).

Eggs: Eggs of *C. sinensis* are fully embryonated when laid, 26 to 35 μm by 12 to 19 μm (average 29 μm by 16 μm) (Fig 4.7). They are broadly ovoid, thick-shelled, light yellow-brown, with a large convex operculum that fits into a rimmed extension of the eggshell, producing minute shoulders. Frequently, there is a small tubercular or comma-shaped protuberance at the abopercular end (Fig 4.10).

Opisthorchis viverrini
Adult: Adult *O. viverrini* (Fig 4.1) is a flat,

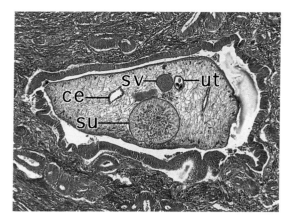

Figure 4.12
Section of adult *O. viverrini* in liver at level of acetabulum demonstrating muscular wall of sucker (su), eggs in uterus (ut), seminal vesicle (sv), and paired ceca (ce). x55

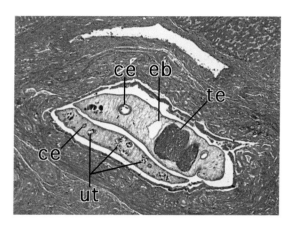

Figure 4.13
Section through posterior end of adult *O. viverrini* in liver demonstrating 1 of 2 lobed testes (te), excretory bladder (eb), and paired ceca (ce). Adjacent thinner section is at midbody and illustrates paired ceca (ce) and multiple sections of coiled uterus (ut) with eggs. x30

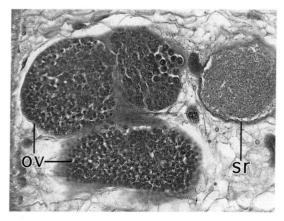

Figure 4.14
Section of adult *O. viverrini* in liver at level of lobed ovary (ov) and seminal receptacle (sr). x230

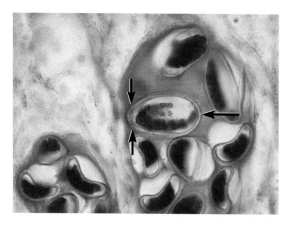

Figure 4.15
Eggs of *O. viverrini* in uterus of adult fluke in liver. Eggs are ovoid, thick-shelled, light yellow-brown, and have convex operculum with minute shoulders (short arrows) and small knob at aboperculum end (long arrow). x755

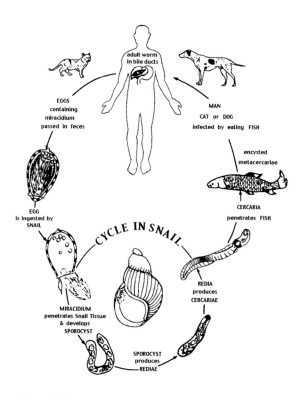

Figure 4.16
Life cycle of *C. sinensis*. Snails serve as first intermediate hosts and freshwater fish as second intermediate hosts.

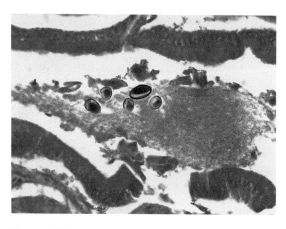

Figure 4.17
Several eggs of *O. viverrini* in bile duct. x230

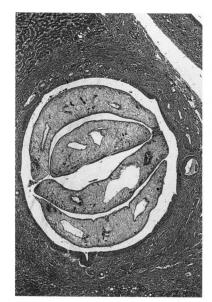

Figure 4.19
Sections through adult *O. viverrini* within dilated bile duct. Epithelial lining of bile duct is intact, but there is marked periductal fibrosis. x33

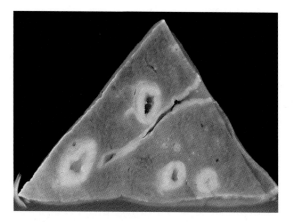

Figure 4.18
Cut surface of liver from patient with clonorchiasis. Note normal parenchyma punctuated by ectatic bile ducts, with walls 2 to 3 times thicker than normal.

elongate, aspinous, lanceolate fluke. It is thin, transparent, rounded posteriorly and tapered anteriorly, 5.5 to 9.5 mm by 0.77 to 1.65 mm (average 7.4 mm by 1.47 mm). Oral and ventral suckers are generally equal. The acetabulum is about one fourth the length of the worm from the anterior end. The pharynx is small, and the esophagus bifurcates into ceca that extend blindly to the posterior end. The long, pouched excretory bladder and 2 lobed testes (Fig 4.1) are in the posterior end of the body. A long, slightly coiled seminal vesicle terminates in a weakly muscular ejaculatory duct, and opens through the male genital pore immediately in front of the acetabulum. The ovary is oval or slightly lobed, median, and pretesticular. The seminal receptacle is immediately anterior to the testes. A few clusters of vitellaria aggregate in 2 lateral fields in the middle third of the body. The coiled uterus (Fig 4.1) extends between the ceca, terminating at the female genital pore near the acetabulum.

Histologic sections of *O. viverrini* reveal a thin body wall (about 25 µm) (Fig 4.11) composed of a thin tegument (3 to 4 µm) lined with microvilli, smooth muscle fibers, and tegumental cells. The loose parenchyma contains parenchymal cells and fibrous interstitial tissue. Sections at the level of the acetabulum reveal the 250-µm-thick muscular sucker, golden-brown eggs within sections of uterus, the seminal vesicle, and lateral intestinal ceca (Fig 4.12). The middle of the fluke contains bifurcated, unbranched ceca lateral to sections of coiled uterus (Fig 4.13). The lobed ovary and seminal receptacle (Fig 4.14) are just anterior to the testes. The posterior end contains 2 lobed testes and the excretory bladder, bordered laterally by paired ceca (Fig 4.13).

Eggs: *Opisthorchis viverrini* eggs are fully embryonated when laid, 22 to 32 µm by 11 to 22 µm (average 28 by 16 µm) (Fig 4.15). They are elongate, thick-shelled, ovoid, and light yellow-brown. The operculum fits into a thickened rim of the shell proper, producing minute shoulders. There is a slight terminal thickening at the abopercular end (Fig 4.15), that may not be visible in all eggs.

Opisthorchis felineus

Adult: *Opisthorchis felineus* is morphologically similar to *O. viverrini*. Some investigators note that *O. felineus* adults are usually larger than *O. viverrini* adults (8 to 18 mm by 1.2 to 2.5 mm), and have a shorter esophagus. Ovary and testes are less lobulated. The anteriormost testis is more posterior from the ovary, and the posterior testis is more anterior from the end of the cecum. The seminal vesicle is longer and more winding, and vitelleria are not aggregated into clusters. Wykoff et al[88] believe these differences are neither consistent nor specific enough to permit a clear differentiation of adult *O. viverrini* and adult *O. felineus*. Currently, the best indication of species is the geographic origin of the infection. DNA sequencing and isoenzyme electrophoresis may eventually make identification more precise.

Eggs: *Opisthorchis felineus* eggs are 21 to 36 µm by 11 to 17 µm (average 30 µm by 14 µm) and fully embryonated when laid. The length/width ratio is 2.15 for *O. felineus* and 1.75 for *O. viverrini*.[68,72] The operculum, shoulders, and abopercular thickening are similar in eggs of both species. Electron microscopy shows that the eggshell of both species has a net-like ultrastructure.[11]

Life Cycle and Transmission

Clonorchis sinensis, *O. viverrini*, and *O. felineus* have a triheteroxenous life cycle (Fig 4.16). Humans and other fish-eating mammals are definitive hosts, snails are first intermediate hosts, and fish are second intermediate hosts.

Adult flukes deposit fully embryonated eggs in mammalian biliary ducts (Fig 4.17). Eggs then pass into the intestinal tract and exit in feces. Eggs must be ingested by an appropriate species of freshwater snail to hatch and release miracidia. Miracidia penetrate the digestive tract wall and become sporocysts in the peridigestive tract regions. Sporocysts give rise to rediae, which produce numerous cercariae. Cercariae escape from rediae and from the snail hosts, and are briefly free-living. They are positively phototactic and geotropic. Cercariae penetrate beneath the scales of freshwater fish, shed their tails, and encyst as ovoid metacercariae, chiefly in muscle and subcutaneous tissue, less often on scales, fins, and gills.[68] In a few weeks, 2 membranes encase the parasite: a hyaline wall secreted by the encysted metaceraria, and an outer capsule formed by the fish host. Humans are infected by eating fish contaminated with metacercariae. In a human host, metacercariae excyst in the duodenum, migrate through

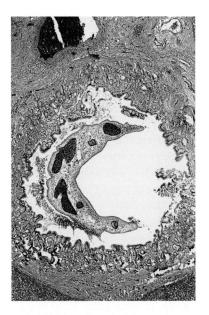

Figure 4.20 Section through posterior end of adult *C. sinensis* in intrahepatic bile duct. Note marked ectasia of bile duct, with mucinous metaplasia and adenomatous hyperplasia. Note also absence of inflammatory infiltrate. Compare with Figure 4.19. x17

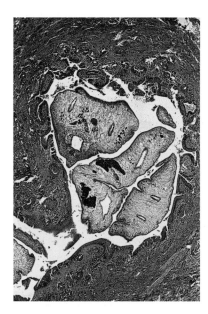

Figure 4.21 Several adult *O. viverrini* flukes clustered together causing periductal fibrosis. Dilation of bile duct may account for mulberry appearance in percutaneous transhepatic cholangiograms. x32

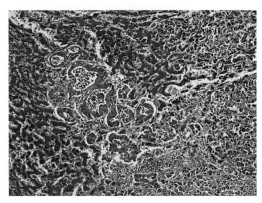

Figure 4.22 Cholangiocarcinoma in patient with clonorchiasis. Note cuboid to columnar malignant cells attempting to form clearly defined glandular and tubular structures. Flukes are not present in this section. x55

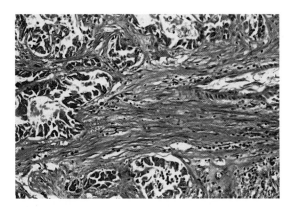

Figure 4.23 Same patient as in Figure 4.22. Note mucus vacuoles in some malignant cells, and desmoplastic change evidenced as dense collagenous stroma separating parenchymal elements. x110

the ampulla of Vater and common bile ducts, and enter distal biliary ducts where they mature to adults. Eggs appear in feces within a few weeks.

Clonorchis sinensis: Humans, domestic dogs and cats, wild cats, pigs, Norway rats, martens, badgers, mink, weasels, and camels all serve as definitive hosts for *C. sinensis*. Self-fertilization is common, although cross-fertilization also takes place. The prepatent period is about 4 weeks in humans. The life cycle is completed in about 3 months.[68] Generally, adult flukes survive 15 to 25 years in humans, depending on host-parasite compatibility and host tolerance.[8,19] Egg production varies with the definitive host: 2400/worm/day in cats, 1600 in guinea pigs, 1125 to 2000 in dogs, and 4000 in rabbits.

Nine species from 4 families of freshwater snails serve as first intermediate hosts for *C. sinensis*. *Parafossarulus manchouricus* is the principal snail host in all regions endemic for *C. sinensis*.[68] Two other snail hosts, *Thiara granifera* and *Melanoides tuberculatus*, have shown a remarkable capacity for colonization, having spread to North, Central, and South America, Europe, Africa, Australia, the Pacific islands, and areas of Asia beyond those traditionally endemic for clonorchiasis.[56] *Clonorchis sinensis* infects 25.8% of *M. tuberculatus* in Vietnam.[45]

At least 113 species of freshwater fish from several families, mostly Cyprinidae,[69] serve as second intermediate hosts for encysted metacercariae of *C. sinensis*.[90] In some parts of China, certain freshwater crayfish are a source of infection.[84]

Opisthorchis viverrini: Humans and other fish-eating mammals are definitive hosts for *O. viverrini*. Metacercariae in consumed fish travel to biliary epithelium, where they attach and mature to adult flukes within 3 to 4 weeks. Average egg production is 2000 to 4000/worm/day in humans (average 180/worm/g of feces),[29] but is lower in heavy infections.[66] The entire life cycle requires approximately 4 months. Miracidia of *O. viverrini* have a marked specificity for aquatic snails of the genus *Bithynia* as first intermediate hosts.[40] The most important second intermediate fish hosts in Southeast Asia are *Cyclocheilichthys* sp, *Hampala* sp, *Puntius* sp, and *Barbodes gonionotus*.[28,35,88] Adult *O. viverrini* can survive for many years.

Opisthorchis felineus: Definitive hosts for *O. felineus* include humans, dogs, foxes, cats, hogs, wolverine, martens, beavers, otters, European polecats, Siberian weasels, sables, Norway rats, water vole, rabbits, seals, and lions.

Three freshwater snail species of the genus *Cordiella* serve as first intermediate hosts: *Cordiella inflata* (*Bithynia inflata*), *Cordiella troscheli*, and *Cordiella leachi*. *Cordiella inflata* is the most common host, but in some areas, certain strains of *C. inflata* are incompatible with *O. felineus*.[10] Nearly 2 dozen species from 17 genera of freshwater cyprinoid fish serve as second intermediate hosts for *O. felineus*.

Clinical Features and Pathogenesis

Clinical manifestations of clonorchiasis and opisthorchiasis become more severe as worm burden gradually increases and infection persists. Most patients with clonorchiasis are asymptomatic and harbor few parasites. Eggs may not appear in feces until 3 or 4 weeks after the onset of symptoms, so diagnosis in the acute phase is often missed. Clonorchiasis has a variable onset; fever may begin abruptly or insidiously. Symptoms and signs in the first few months may include malaise, low-grade fever, hepatomegaly, hepatic or epigastric tenderness, high leukocytosis, slight jaundice, splenomegaly, and eosinophilia. In moderate or progressive infections, patients experience loss of appetite, indigestion, abdominal fullness, epigastric distress unrelated to eating, discomfort in the right upper quadrant, diarrhea, edema, and hepatomegaly. In heavy infections, patients experience weakness and lassitude, epigastric discomfort, paresthesia, weight loss, palpitations and tachycardia, diarrhea, vertigo, tetanic cramps, and tremors. In late stages of heavy infection, toxemia due to liver impairment progresses to portal cirrhosis, splenomegaly, ascites, and edema.

Clonorchiasis may persist for 20 years or longer. Patients commonly develop intrahepatic gallstones. Secondary bacterial infection may produce cholangitis and cholecystitis. Patients who develop hepatic cirrhosis usually develop cholangiocarcinoma as well. There is a close epidemiologic association between *C. sinensis* and cholangiocarcinoma. Metabolic products of the fluke may be carcinogenic.

Most patients with *O. viverrini* infection are also asymptomatic. When symptoms do appear, they usually develop in patients over 30 years of age.[68] Mild infections (less than 100 worms) are usually asymptomatic with normal liver function. Flatulent dyspepsia and dull pain in the right hypochondrium may last for several days or weeks, and may recur. As disease progresses, pain becomes persistent.[65] Moderate infections produce diarrhea, dyspeptic flatulence, pain over the liver, moderate fever (37.5° to 38.5°C), mild to moderate jaundice, hepatomegaly, palpable gallbladder, poor appetite, lassitude, weight loss, and warm skin over the liver and back, or sometimes over the whole abdomen.[68] Eighty percent of patients have normal ultrasonograms.[64] Patients with heavy, long-standing infections have hepatic cirrhosis, ascites, pedal edema, acute abdominal pain, and sometimes carcinoma. There is no correlation between severity of disease and fecal egg output among patients with heavy *O. viverrini* infection. Patients with severe infections may have intense jaundice, secondary infection of the biliary system, cholangitis, intra-abdominal mass, hepatomegaly, high bilirubin transaminase, and low serum albumin. Mortality is low. Some patients develop cholangiocarcinoma but are asymptomatic until the neoplasm is advanced. Incidence of cholangiocarcinoma is significantly higher in Thailand, where *O. viverrini* infection is prevalent,

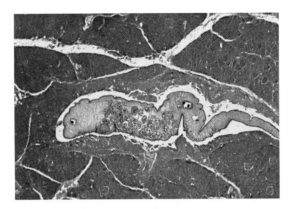

Figure 4.24
Section of adult *C. sinensis* within pancreatic duct causing damage to duct epithelium. x12

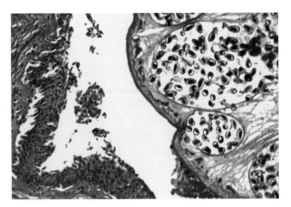

Figure 4.25
Higher magnification of Figure 4.24, showing adult *C. sinensis* in pancreatic duct causing squamous metaplasia. x110

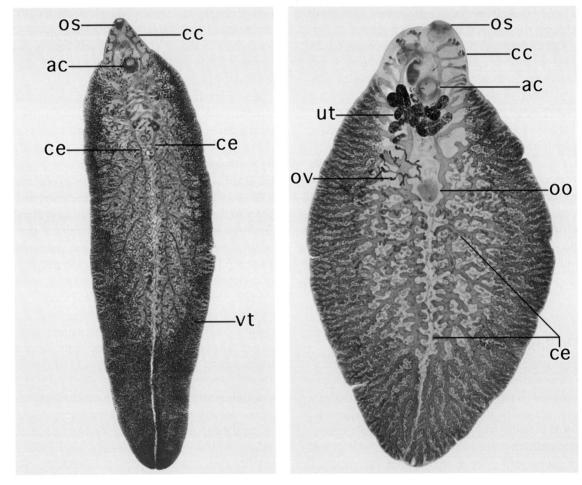

Figure 4.26
Two ventral views of adult *F. hepatica*. Specimen at *left*, fixed in extended position, is from a human and demonstrates oral sucker (os), cephalic cone (cc), acetabulum (ac), branched ceca (ce), and vitellaria (vt).* x2.8 Specimen at *right*, fixed in contracted position, is from bile duct of a sheep and demonstrates oral sucker (os), cephalic cone (cc), acetabulum (ac), uterus (ut), ovary (ov), ootype (oo), and branched ceca (ce). Carmine x4.6

than in areas where this fluke is not endemic.[79]

Signs and symptoms of *O. felineus* infection are similar to those of *O. viverrini*, but morbidity and mortality are very different. Some clinicians recognize a number of clinical stages from acute to late chronic, with accompanying symptoms and pathologic consequences.[16,79] The incubation period is usually 2 to 4 weeks, but may be as short as 7 days. Acute symptoms include fever, abdominal pain, dizziness, and urticaria. There is no relationship between acute symptoms and the number of *O. felineus* eggs in feces. Unlike *O. viverrini* and *C. sinensis* infections, there is no known correlation with cholangiocarcinoma.[18]

Intrahepatic gallstones are a characteristic complication of infection with any of these 3 flukes.[5,85] Death is uncommon, except in heavy, long-standing infections causing serious impairment of liver function, such as chronic obstruction of bile ducts leading to relapsing pyogenic cholangitis or cholecystitis, or cholangiocarcinoma with *C. sinensis* or *O. viverrini* infection.

Pathologic Features

In massive *C. sinensis* and *O. viverrini* infections, or malignancies caused by these infections, the liver may be greatly enlarged. There are often pale cystic areas and white streaks (corresponding to dilated bile ducts or subcapsular bile-lake dilatation) on the external surface.[67] Liver parenchyma is normal on cut surface, though punctuated by the dilated bile ducts with walls 2 to 3 times thicker than normal (Fig 4.18). Flukes lodge and mature in large and medium-sized intrahepatic ducts, especially those of the left lobe. There are no gross hepatic changes in light or early *O. felineus* infections other than liver enlargement. Heavy infections correspond to an enlarged, nonfunctional gallbladder.

Flukes migrate by means of their suckers from small to larger bile ducts, including the common duct. They seldom enter pancreatic ducts or the gallbladder; when they do, they usually die there. Migration of flukes causes mechanical damage to the biliary epithelium (Fig 4.19). In clonorchiasis there is excessive mucin formation, desquamation, and adenomatous hyperplasia of ductal epi-

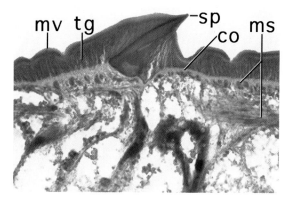

Figure 4.27
Body wall of adult *F. hepatica* removed from bile duct, demonstrating tegument (tg) with microvilli (mv), and 50-μm-long pointed spine (sp). Note layer of collagen (co) and reticulin fibers between the tegument and smooth muscle (ms). Thin cytoplasmic membrane surrounds intracellular spine. Movat x560

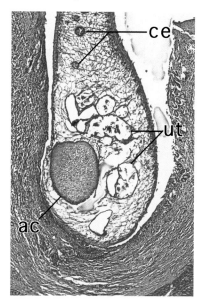

Figure 4.28
Adult *F. hepatica* in human liver. Cross section is at level of acetabulum (ac), revealing egg-filled uterus (ut) and branches of ceca (ce). x24

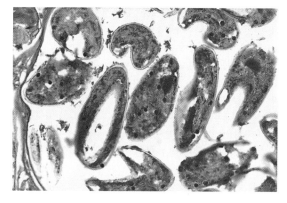

Figure 4.29
Eggs in uterus of adult *F. hepatica*. Eggs are ovoid, yellow, thick-shelled, and have opercula that are difficult to discern in tissue sections. x220

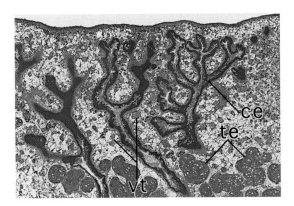

Figure 4.30
Section from midbody of adult *F. hepatica* removed from bile duct, demonstrating highly branched ceca (ce), vitellaria (vt), and several sections of highly branched testis (te).* x22

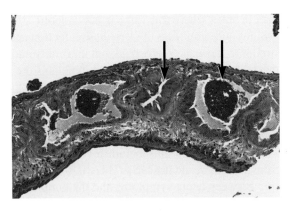

Figure 4.31
Transverse section at midbody of immature *F. hepatica* expressed from skin lesion on chest of 26-year-old Turkish woman, showing several branches of ceca (arrows). Ceca have thick, ciliated columnar epithelium; some contain blood.* x105

thelium with goblet-cell metaplasia (Fig 4.20). In chronic stages, bile ducts become progressively thickened, tortuous, and dilated, with ductal and periductal fibrosis.[79] Portal fibrosis and portal cirrhosis are rare in clonorchiasis. In early *O. viverrini* infections there is no hyperplasia of epithelium or proliferation of fibrous connective tissues, and ducts are lined by a single layer of columnar epithelium. In heavy *O. viverrini* infections, however, there is marked desquamation and inflammatory cell infiltration of secondary bile duct epithelial cells, leading to epithelial hyperplasia, goblet-cell metaplasia and gland formation, followed by adenomatous hyperplasia and periductal fibrosis (Fig 4.21).[33] Necrosis and atrophy of hepatic cells commonly leads to extensive portal fibrosis. Localized dilatation of bile ducts, usually near the free edge of the liver, especially the left lobe, is common, but generalized dilatation of bile ducts is uncommon. Biliary stasis results in secondary infection leading to pericholangitis, cholangiohepatitis, pylephlebitis, and multiple abscesses. *Clonorchis sinensis* does not invade tissue, so there is little or no inflammatory reaction in hepatic parenchyma (Fig 4.20). Pyogenic cholangitis is characterized by dilatation and stricture of second-order biliary radicals and the presence of intrahepatic pigmented biliary stones or sludge. There is a marked increase in mucus production.

Cholangiocarcinoma is so common in areas endemic for *C. sinensis* or *O. viverrini* that presymptomatic cases are detectable. The metabolic products of these flukes cause chemical irritation; secondary bacterial infection damages the biliary epithelium, leading to inflammation and metaplasia (Fig 4.21).[49] The mechanical injury caused by these flukes outlasts the active infection; malignancy may not become apparent until 10 to 20 years after the infection has been cured. Most cholangiocarcinomas are well-differentiated sclerosing adenocarcinomas, with clearly defined glandular and tubular structures lined by cuboid to columnar epithelial cells (Fig 4.22). They are often desmoplastic, with dense collagenous stroma separating the parenchymal elements (Fig 4.23). There may be focal areas of necrosis. Malignant cells are frequently mucous and resemble adenocarcinomas in needle biopsy samples. At autopsy, up to 50% of patients have metastases to regional lymph nodes and/or hematogenous metastasis to lung, bone, adrenal, brain, or other parts of the body.

Opisthorchis felineus often cause a marked disruption of external pancreatic secretion.[43] Occasionally *C. sinensis* fill pancreatic ducts, which then thicken, dilate, and exhibit mucinous and squamous metaplasia (Figs 4.24 and 4.25). Pancreatitis and pancreatic adenocarcinoma are rarely associated with *C. sinensis* infection.

Figure 4.32
Egg of *F. hepatica* in feces. Egg is operculate, ovoid, yellow, and nonembryonated. Unstained x225

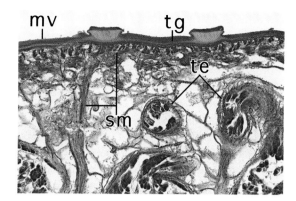

Figure 4.33
Section through body wall of adult *F. gigantica* removed from liver, showing tegument (tg) with microvilli (mv) and 2 flattened intracellular spines. Note layer of collagen and reticulin fibers between tegument and smooth muscle (sm). Note also cross sections of branching testis (te). Movat x220

Diagnosis

Diagnosis depends on finding eggs in feces or biliary tracts. Though some authorities suggest that eggs of *O. viverrini* and *C. sinensis* can be distinguished from eggs of *O. felineus* by their smaller length/width ratio, eggs of these 3 flukes are nearly indistinguishable.[68] Concentration techniques, such as formalin-ether sedimentation, are helpful for detecting eggs in feces. The number of eggs indicates clinical severity only for *C. sinensis*. Stoll's dilution egg-counting method recognizes 4 levels of infection: light (1 to 999 eggs/g of feces), medium (1000 to 9999 eggs/g), heavy (10 000 to 29 999 eggs/g), and very heavy (30 000 or more eggs/g). Patients with biliary obstruction do not pass eggs in feces; percutaneous needle aspiration of bile ducts or examination of biopsy material from the biliary system or pancreatic duct is needed for diagnosis.[68] In biopsy specimens, cross sections of *C. sinensis* differ from *Opisthorchis* sp in having branched rather than lobed testes that extend lateral to the ceca. ELISA is a useful screening test, but there are cross-reactions with other trematodes (*Fasciola, Paragonimus, Schistosoma*). Transhepatic cholangiograms of large and medium-sized intrahepatic ducts reveal curved filling defects within dilated bile ducts or mounds attached to the duct walls, indicating the presence of flukes.[49,50] Ultrasonography reveals flukes in bile ducts and the extent of disease in the hepatobiliary tract, and can be used to evaluate the effectiveness of treatment.[50] Computed tomography (CT) may suggest *C. sinensis* by demonstrating associated liver abscesses, gallstones, and diffuse, uniform dilatation of intrahepatic bile ducts in infected patients.[25] In nonendemic areas, such as North America, most patients present with light infections, are asymptomatic at the time of presentation, and have normal abdominal ultrasound studies and normal liver function tests.[40] Imaging techniques are most useful for evaluating patients with recurrent pyogenic cholangitis, a common complication of clonorchiasis. Endoscopic and percutaneous cholangiography are helpful, but CT appears to be the most sensitive diagnostic tool.[50]

Fascioliasis
Definition
Fascioliasis is infection by the liver fluke *F. hepatica* or, less often, *F. gigantica*.

Synonyms
Fascioliasis is also called fascioliasis hepatica, febrile fasciolitic eosinophilic syndrome, and liver rot (in sheep). Older names for *F. hepatica*, the sheep liver fluke, include *Distoma hepaticum, Distomum hepaticum, Fasciola californica*, and *Fasciola halli*. A synonym for the giant liver fluke *F. gigantica* is *F. indica*.

General Considerations

The first mention of *Fasciola* was by de Brie in 1379; he believed eating a local leaf produced a worm that led to sheep rot.[37] Fitzherbert referred to *Fasciola* as "flokes," from the Anglo-Saxon "floc" for flatfish. Pallas, in 1760, first described a human infection after finding flukes in the bile ducts of a German woman. *Fasciola* was variously interpreted as tapeworm proglottids, leeches, and even slugs. Linnaeus finally recognized the fluke's true nature, creating the genus *Fasciola* and the name *F. hepatica* in 1758.[37] The fluke's anatomy was further delineated in the 19th century. In 1882 Leuckart described the life cycle of *F. hepatica*.

In 1855, Cobbold described *F. gigantica* found in a giraffe. Codvelle et al reported the first indisputable human infection in 1928.[27] Evidence suggests that various strains of *F. gigantica*, such as *F. indica*, evolved through geographic isolation. Recently developed molecular techniques for distinguishing *F. hepatica* from *F. gigantica* should be helpful, especially in Asia, where several morphologic types of *Fasciola* exist.[3,13,57]

Epidemiology

Worldwide, there are approximately 2.4 million people with fascioliasis.[69] Fascioliasis in humans and animals is prevalent in Europe, the Americas, northern Asia, Oceania, and Africa. It is also found in New Zealand, Tasmania, Great Britain, Iceland, Cyprus, Corsica, Sardinia, Sicily, Japan, Papua New Guinea, the Philippines, and the Caribbean. Slight biologic differences between *F. hepatica* and *F. gigantica* influence their prevalence in specific regions. Developmental stages of *F. gigantica* grow more slowly, survive longer at high temperatures, and are more susceptible to desiccation. Intermediate snail hosts of *F. gigantica* share these characteristics. Thus, *F. hepatica* prevails in temperate zones and is dominant in Europe, the Americas, and Oceania,[61] while *F. gigantica* is adapted to tropical and aquatic environments and prevails in Africa.[15,24,31]

Snail hosts of *Fasciola* require moderate temperatures and high humidity. In Europe, a long wet summer often leads to an outbreak of fascioliasis.[24] In the Americas the main endemic area is the Altiplano region of Bolivia, where 53% of Aymara Indians are positive serologically[42] and up to 68% of schoolchildren have positive fecal samples. These are the highest prevalences of human fascioliasis in the world.

The ranges of *F. hepatica* and *F. gigantica* sometimes overlap in Asia, Africa, and North America. In tropical regions where both species exist, *F. gigantica* favors lower elevations and *F. hepatica* the highlands.

Infectious Agent

Morphologic Description

Fasciola hepatica

Adult: *Fasciola hepatica* are large, broad flukes 20 to 50 mm by 6 to 13 mm. The anterior end is cone-shaped and the posterior end is bluntly rounded (Fig 4.26). The tegument contains spines that vary in size and distribution. Oral and ventral suckers are adjacent in the short, cone-shaped area. The oral sucker is about 1 mm in diameter; the ventral sucker is about 1.75 mm in diameter. There is a prominent pharynx, a short esophagus, and 2 highly branched ceca (Fig 4.26) that extend to the posterior end of the fluke. Two large, highly branched testes lie in the second and third quarters of the body. The single branched ovary is on the right side and pretesticular. The short coiled uterus lies at midbody below the acetabulum (Fig 4.26). Vitellaria are in the lateral fields, dorsal and ventral to the ceca and extending from the ventral sucker to the posterior end of the worm. The cirrus sac, containing a protrusible spined cirrus, is easily visible, preacetabular, and opens in a genital pore.

In tissue sections, the body wall is 30 to 50 μm thick and consists of a spinous tegument and several layers of muscle. The tegument is 1 to 5 μm thick with minute (0.5 μm) microvilli and irregularly distributed spines (Fig 4.27). There are 2 layers of smooth muscle immediately beneath the tegument, separated by a layer of collagen and reticulin fibers (1 to 5 μm thick). The outermost layer of muscle consists of well-organized, thick longitudinal fibers; the inner layer begins as a thin layer of longitudinal fibers and intermingles with a layer of circular muscle fibers. The base of the triangular spine extends to the layer of collagen and reticulin fibers (Fig 4.27). A thin cytoplasmic membrane surrounds the intracellular spines. Below the smooth muscles are the cell nuclei, with thin extensions forming the syncytial tegument. Sections at the level of the acetabulum reveal the coiled, egg-filled uterus and branches of the ceca (Figs 4.28 and 4.29). Sections at midbody reveal vitellaria, multiple sections of the highly branched ceca, and numerous branches of testes (Figs 4.30 and 4.31).

Eggs: Eggs of *F. hepatica* (130 to 150 µm by 63 to 90 µm) are operculate, ovoid, yellow, and nonembryonated when laid (Figs 4.29 and 4.32). They are practically identical to eggs of *F. buski*, and differ from those of *F. gigantica* only by their slightly smaller size.

Fasciola gigantica

Adult: Adult *F. gigantica* are larger than adult *F. hepatica* (24 to 76 mm by 5 to 13 mm). The average length/width ratio of *F. gigantica* is 4.39 to 5.20, while that of *F. hepatica* is 1.88 to 2.32.[73] *Fasciola gigantica* also differs from *F. hepatica* in having a shorter cephalic cone, a larger acetabulum, larger eggs, and more anteriorly oriented testes. The average distance between the posterior border of the body and the posterior testis is longer in *F. gigantica* (14.9 mm; range: 6 to 19 mm) than in *F. hepatica* (7.78 mm; range: 3 to 13 mm),[73] and the ovary and ceca are more branched.

In tissue sections, the body wall is about 50 µm thick and consists of a spinous tegument, tegumental cells, and somatic muscle (Fig 4.33). The tegument is 10 to 15 µm thick, covered with microvilli, and has irregularly distributed, flattened intracellular spines (Fig 4.33). Below the tegument is a layer of collagen and reticulin fibers. The extensive branching of ceca, testes, and ovary is prominent. Ceca have a ciliated columnar epithelium.

Eggs: Eggs of *F. gigantica* are nonembryonated when laid and are larger (150 to 196 µm by 70 to 100 µm) than eggs of *F. hepatica* (Fig 4.34).

Life Cycle and Transmission

Fasciola hepatica and *F. gigantica* have similar aquatic, diheteroxenous life cycles. Humans are infected by ingesting encysted metacercariae on raw vegetation, and probably also floating infec-

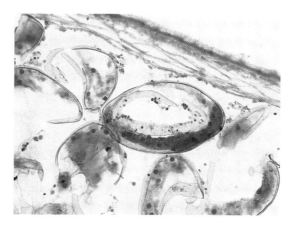

Figure 4.34
Egg of *F. gigantica* in uterus of adult fluke removed from liver. Operculum is inconspicuous. Egg is 150 by 85 µm. x255

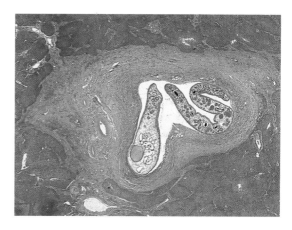

Figure 4.35
Sections of adult *F. hepatica* in intrahepatic bile duct in patient from China. Bile duct is enlarged and dilated; the wall is thickened.* x6

Figure 4.36
Skin and subcutaneous tissue from patient with cutaneous fascioliasis, showing a panniculitis with necrosis and massive eosinophilia. Immature *F. hepatica* pictured in Figure 4.31 was expressed from this lesion. x6

tive metacercariae in contaminated water.[80] Sheep, goats, cattle, horses, donkeys, mules, camels, buffalo, deer, wild sheep, wild pigs, domestic pigs, marsupials, rabbit, hare, nutria[15] and other rodents, and monkeys serve as definitive hosts for both *F. hepatica* and *F. gigantica*.

Adult *Fasciola* sp deposit immature eggs that pass in feces. Egg production is generally inversely proportional to worm burden. Eggs develop in water and hatch in 9 to 21 days in suitable climatic conditions (15° to 25°C).[14,46] In unfavorable conditions, eggs do not hatch and miracidia do not mature, but may remain viable for several months.[46] Light stimulates eggs of *F. hepatica* to release miracidia (approximately 130 μm by 28 μm), which then swim in search of a snail host of the family Lymnaeidae. Miracidia that fail to penetrate a snail die within 24 hours. Miracidia of *F. gigantica* move into deeper waters that harbor *Lymnaea natalensis* in Africa and varieties of *Lymnaea auricularia* in Asia. After penetrating a snail host, miracidia transform into elliptical saccular sporocysts 150 to 500 μm long. Sporocysts produce mother rediae, which produce daughter rediae, and finally cercariae, in about 6 to 7 weeks. Snails release motile cercariae into fresh water, where some find water plants to land on. These cercariae shed their tails, encyst, transform into metacercariae, and become infective within 24 hours. Some infective metacercarial cysts float on the surface of the water. Metacercarial cysts remain viable for long periods, but are susceptible to heat and dryness. Most metacercariae ingested by a definitive host die in the gastrointestinal tract. Surviving metacercariae excyst in the small intestine, penetrate the host's intestinal wall, and appear as immature flukes in the abdominal cavity about 2 hours after ingestion. Immature flukes may enter the bloodstream and travel to various parts of the body, or may reach the liver by wandering up the bile duct.[24] Most immature flukes reach the liver within 6 days, then migrate through the liver for 5 to 6 weeks. They eventually penetrate bile ducts and become sexually mature. The prepatent period is at least 3 to 4 months and varies with the number of adult flukes in the liver. The entire life cycle takes 14 to 23 weeks.[46] *Fasciola hepatica* can survive up to 13 years in humans.

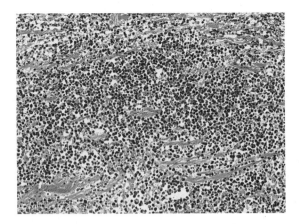

Figure 4.37
Higher magnification of Figure 4.36 demonstrating massive eosinophilia in subcutaneous tissue. x15

Clinical Features and Pathogenesis

The incubation period for fascioliasis infections varies from a few days to a few months.[38] Acute symptoms result mainly from destruction of abdominal peritoneum and liver tissue by migrating larvae. Larvae also cause localized or generalized toxic and allergic reactions lasting 2 to 4 months. Major symptoms are fever, abdominal pain, gastrointestinal disturbances, and urticaria. Fever may be remittent, intermittent, or irregular, with higher temperatures in the evening (up to 42°C). Some patients have a low recurrent fever lasting 4 to 18 months. Mild to excruciating abdominal pain is generalized at the outset, but usually localizes to the right hypochondrium or epigastrium. Loss of appetite, flatulence, nausea, and diarrhea are common; vomiting and constipation are infrequent. Urticaria with dermatographism is a distinctive early feature, sometimes accompanied by bronchial asthma. Nonproductive cough is common. Patients may have hepatomegaly, splenomegaly, ascites, anemia, chest signs, and jaundice.[24]

The latent phase of disease can last for months or years. Eosinophilia during this interval suggests helminthic infection. Patients may have gastrointestinal complaints or recurring acute symptoms.[31]

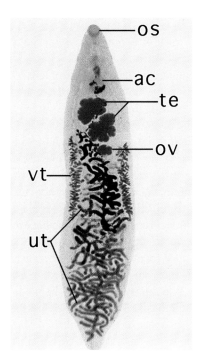

Figure 4.38 Whole mount of adult *D. dendriticum* displaying oral sucker (os), acetabulum (ac), lobed testes (te), lobed ovary (ov), vitellaria (vt), and coiled uterus (ut). Carmine x8.5

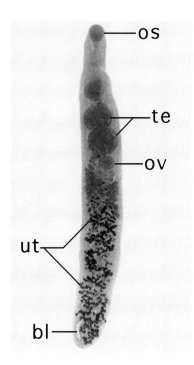

Figure 4.39 Whole mount of adult *D. hospes* showing oral sucker (os), testes (te), ovary (ov), coiled uterus (ut), and bladder (bl). Carmine x7

The chronic or obstructive phase develops after months or years of infection. Adult flukes in bile ducts cause inflammation of ducts and gallbladder, promoting mechanical obstruction, cholangitis, and cholecystitis (Fig 4.35). Clinical manifestations include biliary colic, epigastric pain, fat intolerance, nausea, jaundice, pruritus, tenderness in the right upper quadrant, hepatomegaly, splenomegaly, ascites, enlarged edematous gallbladder, lithiasis in bile ducts or gallbladder, and hemobilia. Death is uncommon.[24]

Eosinophilia is present in all phases of *F. hepatica* infection (from 5% to 83%), accompanied by leukocytosis (10×10^3 to $43 \times 10^3/\mu l$). Anemia is common (usually 7 to 11 g/dl). Erythrocyte sedimentation rate may be high in the acute phase (up to 165 mm/hour). High serum bilirubin is associated with the obstructive phase.[24] IgG, IgM, and IgE are usually elevated.[75] Up to 48% of patients have specific IgE antibodies. Total and specific IgE levels positively correlate with egg burden, age, clinical features, and degree of eosinophilia.

Flukes can localize ectopically and produce painful, wandering, subcutaneous swellings (Figs 4.31 and 4.36).[32] Pharyngeal fascioliasis, an acute dysphagia and laryngeal obstruction resulting from ingestion of raw sheep or goat liver, can be caused by immature *F. hepatica*[74] or by other etiologic agents, including pentastomids.[76]

Pathologic Features

The liver is enlarged with a smooth or uneven surface.[24] The capsule has white or yellow striae. There are often soft, yellow or gray-white nodules 2 to 30 mm in diameter,[83] with hemorrhagic stippling at the margins. Near the nodules there are ribbed or vermiform areas. Microscopically, these nodules display eosinophilic abscesses. The capsule is focally or entirely thickened. Subcapsular lymphatic vessels are dilated. Lymph nodes near the porta hepatis are enlarged. Marked involvement of the peritoneum and liver surfaces produces yellow, opalescent ascites. Splenomegaly is rare.[59]

Bile ducts are dilated and have thickened walls (Fig 4.35). Muscular hypertrophy and perimuscular fibrosis cause thickening of the gallbladder wall. The gallbladder is also edematous, has adhesions to adjacent structures, multiple gray-white nodules in the subserosa, and prominent mucosal folds. There is glandular epithelial hyperplasia. Lymphocytes, plasma cells, and eosinophils

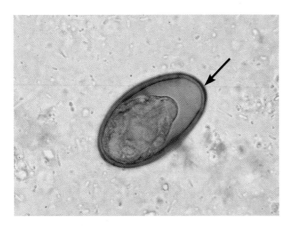

Figure 4.40
Egg of *D. dendriticum* in feces. Operculum (arrow) is not prominent. Unstained x540

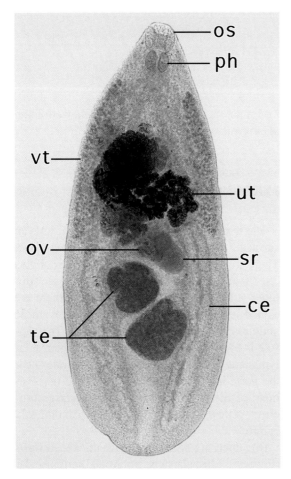

Figure 4.41
Whole mount of adult *M. conjunctus* demonstrating oral sucker (os), pharynx (ph), vitellaria (vt), uterus (ut), ovary (ov), seminal receptacle (sr), testes (te), and ceca (ce). Carmine x45

focally infiltrate all layers of the gallbladder wall. Gallstones often form in common bile ducts and gallbladder.[24]

Flukes digest hepatic tissue and cause extensive parenchymal destruction with hemorrhage and inflammation. Generally, migration tracks are visible in the liver and other organs. The walls of liver tracks contain Charcot-Leyden crystals and eosinophils. Necrotic cellular debris fills the cavities, and eosinophilic infiltrates surround the tracks. Older lesions contain macrophages, lymphocytes, eosinophils, and fibrous tissue. Dead flukes leave cavities filled with necrotic debris, which heal by scarring.[80] Focal calcification at the margin of the necrotic debris may form the outline of a dead fluke.[2,24,36] Extensive hyperplasia results in enlargement of the bile ducts, which may be mediated by proline synthesized and released by the parasites.[44] Eggs can cause granulomas.[2] Portal triads may be edematous and dilated. Frequently there is bile duct proliferation, periductal fibrosis, necrotizing arterial vasculitis, and portal venous thrombosis. Unlike clonorchiasis and opisthorchiasis, fascioliasis is not associated with biliary carcinoma.[24]

The most frequent ectopic lesions in humans appear in the gastrointestinal tract.[63] Other locations include the abdominal wall, pancreas, spleen, subcutaneous tissue (Figs 4.31, 4.36, and 4.37), heart, blood vessels, lung and pleural cavity, brain, orbit, skeletal muscle, appendix, and epididymis.[2,34] Immature flukes migrating through tracts cause inflammation and fibrosis. Ectopic flukes do not mature, but calcify or become incorporated in a granuloma.[31]

Cell- and/or antibody-mediated response vary with the phase of infection.[80] Spontaneous recovery is rare.

Diagnosis

Diagnosis depends on finding eggs in feces by direct smears or various concentration methods,[24] including flotation, sedimentation,[83] and the cellophane fecal thick-smear technique. To avoid a spurious diagnosis, repeat stool analysis after the patient has been on a liver-free diet for a few days. Diagnosing flukes in ectopic locations requires

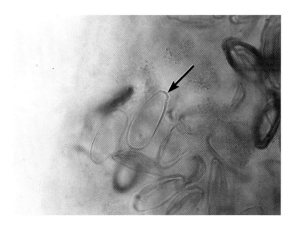

Figure 4.42
Eggs in uterus of adult *M. conjunctus*. Eggs are yellow-brown, slender, and slightly tapered at opercular end (arrow). Carmine x605

biopsy specimens.

Immunologic tests can confirm the diagnosis even before flukes release eggs. Such tests are useful for epidemiologic studies and for monitoring patients after treatment. Skin tests using an antigen from adult flukes[80] are simple and sufficiently sensitive to suggest a diagnosis,[22] but are not specific.[83]

There is no consensus on the optimal serologic method for diagnosis. Almost all serologic tests are highly sensitive, including complement fixation, immunofluorescence assay (IFA), counterelectrophoresis (CEP), ELISA, kinetic-dependent ELISA, double diffusion, indirect hemagglutination, enzyme-linked immunoelectrotransfer blot (EITB), Falcon® assay screening test—enzyme-linked immunosorbent assay (FAST-ELISA), automated assay of anti-P_1 antibodies, circulating antigen, and circulating immune complex. An antigen derived from purified somatic or excretory-secretory products of adult *F. hepatica* used in ELISA, IFA, or CEP assays has the highest sensitivity and specificity.[41] ELISA is rapid, specific, and very useful in areas endemic for either *F. gigantica* or *F. hepatica* infection.[91] IFA has a 92% to 96% sensitivity in the acute phase of *F. hepatica* infection, but in the chronic phase, IFA and CEP assays may not be positive.[22] Most of the assays cross-react with schistosomes and other trematodes, ascarids, and filarids, but specificity can be improved by absorption techniques. *Fasciola*

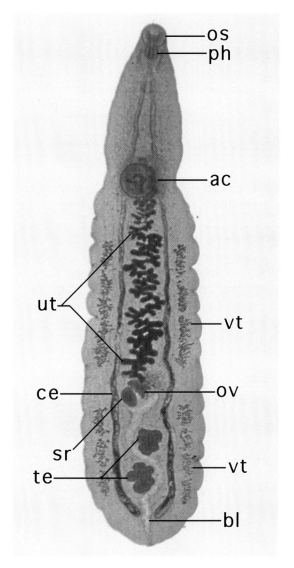

Figure 4.43
Whole mount of adult *A. pseudofelineus* removed from bile duct, showing oral sucker (os), pharynx (ph), unbranched ceca (ce), acetabulum (ac), vitellaria (vt), coiled uterus (ut), ovary (ov), seminal receptacle (sr), lobed testes (te), and bladder (bl). x19

hepatica antibodies are undetectable after therapy.

Abdominal and chest x-rays, oral, percutaneous, and intravenous cholangiography, radioisotope liver scan, and endoscopic retrograde cholangiopancreatography (ERCP) are useful but not diagnostic. CT reveals multiple hypodense areas in the liver that diminish with therapy.[62]

Dicrocoeliasis

Definition

Dicrocoelium dendriticum and *D. hospes* cause dicrocoeliasis, a hepatitis-like syndrome.

Synonyms

Other names for *D. dendriticum* include lancet fluke, lanceolate fluke, little liver fluke, *Fasciola lanceolata*, and *Dicrocoelium lanceatum*. Comparisons of the range and prevalence of various types of *Dicrocoelium* sp suggest that *D. hospes* could be a polymorph of *D. dendriticum*.[54]

Epidemiology

Dicrocoeliasis is most common in herbivorous mammals, especially sheep and cattle. *Dicrocoelium dendriticum* has a cosmopolitan distribution, while *D. hospes* is mainly confined to Africa. Human infections are sporadic. There are several hundred reports of human dicrocoeliasis worldwide (based on identifying eggs in stool[71]) and nearly 30 reports of human *D. hospes* infections from Ghana, Sierra Leone, Nigeria, and Democratic Republic of Congo.

Infectious Agent

Morphologic Description

Adult *D. dendriticum* are 5 to 15 by 2 to 2.5 mm, elongate, and tapered anteriorly and posteriorly (Fig 4.38). They have an aspinous tegument. The acetabulum (500 to 600 µm) is in the second fifth of the body and is slightly larger than the oral sucker. The pharynx is small, the esophagus short, and the unbranched ceca extend only to the posterior fifth of the body. Unlike many other flukes, the 2 large (usually lobed) testes of *D. dendriticum* are immediately posterior to the acetabulum. The slightly lobed ovary lies immediately posterior to the testes. Vitellaria are in 2 fields lateral to the ceca at midbody. The coiled uterus occupies the posterior two thirds of the fluke, terminating at the genital pore just anterior to the acetabulum.

Adult *D. hospes* are similar to *D. dendriticum* but narrower (12 mm by 1.3 mm) (Fig 4.39). The body has a uniform width except for the tapered anterior end. Oral and ventral suckers are nearly equal. The acetabulum is situated far anteriorly. Two large, slightly lobed to unlobed testes lie immediately posterior to the acetabulum (Fig 4.39). Vitellaria are situated between the unbranched ceca, in a compact area posterior to the ovary. The prominent uterus has a descending branch on the ovarian side of the body and an ascending branch on the opposite side; the 2 branches do not intersect. The ascending branch of the uterus terminates at the genital pore near the acetabulum.

Eggs of *D. dendriticum* and *D. hospes* are 35 to 45 µm by 22 to 30 µm, dark brown, thick-shelled, and operculate (Fig 4.40). They are fully embryonated when passed and are morphologically indistinguishable.

Clinical and Pathologic Features, Pathogenesis, and Diagnosis

Patients with dicrocoeliasis are usually asymptomatic, and infections often go undiagnosed. Clinical manifestations are neither uniform nor specific, but suggest a hepatitis-like infection. Patients may have prolonged nausea, vomiting, and constipation or diarrhea. Some experience lassitude, headache, dizziness, and epigastric or abdominal pain that intensifies at night. Patients may exhibit hepatosplenomegaly, mild jaundice, leukocytosis, eosinophilia (8% to 25%), traces of bile in urine, liver dysfunction, and gallbladder or biliary disease. In later stages of disease, leukocytosis disappears, eosinophilia diminishes to 5% to 7%,[71] and there may be slight anemia. Because of the small size and aspinous tegument of *Dicrocoelium* sp, mechanical and toxic damage to bile ducts is less severe than in other liver fluke infections.[71] There is no known correlation with bile duct carcinoma. Cerebral involvement in dicrocoeliasis is rare.[78] In genuine human infection there is prolonged egg shedding. Eggs or flukes may be recovered in surgical specimens or duodenal aspirates. Serologic tests cross-react with *F. gigantica* and *Schistosoma bovis*.

Infection by Minor Opisthorchiidae

Definition

Flukes of lesser medical importance (*M. conjunctus*, *P. truncatum*, *A. pseudofelineus*, and *A. noverca*) sometimes parasitize bile ducts of humans and animals.

Synonyms

Synonyms for *P. truncatum* include *Amphistoma truncatum*, *Distoma lanceolatum*, *Opisthorchis truncatus*, and *Metorchis truncatus*. An older name for *M. conjunctus* is *Distoma conjunctum*. *Amphimerus pseudofelineus* is synonymous with *Opisthorchis guayaquilensis* and *Amphimerus guayaquilensis*. *Amphimerus noverca* has been called *Opisthorchis noverca*, *Distoma caninus*, *Distoma indicus*, and has been misclassified as *D. conjunctum*.[21]

General Considerations and Epidemiology

Cobbold first described *M. conjunctus* (naming it *D. conjunctum*) in an American fox at the London Zoo in 1859 and expanded the description in 1860. Looss established the genus *Metorchis* in 1899.[21] Cameron reported *M. conjunctus* in feces of a Canadian patient,[20] and Babbott et al identifed eggs of *M. conjunctus* in the stools of an indigenous patient from Greenland.[9] Reports of human infection by *P. truncatum*, a parasite of bile ducts of cats, dogs, and other mammals, come from Europe and North America.[89] In 1949 Rodriguez et al described the first human infection with *A. pseudofelineus*, a parasite of dogs, coyotes, domestic cats, and marsupials of the Americas, in an Ecuadorian. There are foci in Ecuador and Panama where opisthorchiid eggs in human feces are presumed to be *A. pseudofelineus* eggs, since this species is found in dogs[70] and cats[81] in these areas. In 1872, Lewis and Cunningham described *A. noverca*, a common parasite of dogs, wolverines, and pigs, as *D. conjunctum*; Blanchard described the same fluke as *O. noverca* in 1895. There are 2 questionable reports of *A. noverca* infection of human gallbladders.[47]

Infectious Agent

Morphologic Description

Adult *M. conjunctus* taper slightly at both ends and vary greatly in size (1 to 6.6 mm by 0.6 to 2.6 mm). Small, thin, regularly spaced spines cover the tegument. Oral and ventral suckers are nearly equal in diameter (175 µm). Ceca are unbranched and extend posteriorly to the end of the body. Vitellaria are prominent in 2 fields lateral to the ceca and extend from the acetabulum to the ovary. Two large, slightly lobed testes are in the posterior end of the worm.[87] The spherical to slightly lobed ovary is near midbody, immediately anterior to the testes. The prominent, coiled, egg-filled uterus lies between the acetabulum and the ovary (Fig 4.41). The slender, thick-shelled eggs of *M. conjunctus* (22 to 32 µm by 11 to 18 µm) are yellow-brown, operculate, and may be slightly constricted anteriorly (Fig 4.42).

Adult *A. pseudofelineus* are about 7 by 2 mm and taper anteriorly (Fig 4.43). The acetabulum is larger than the oral sucker and is about a quarter of the way from the anterior end of the fluke. Two unbranched ceca extend to the posterior end. Vitellaria are lateral to the ceca and extend, with a characteristic interruption,[47] from near the acetabulum to the posterior end. The 2 lobed testes are far posterior. There is a slightly lobed ovary just anterior to the testes. The coiled uterus extends anteriorly, between the ceca, from the ovary to the acetabulum. There is a prominent excretory bladder at the posterior end (Fig 4.43). Eggs of *A. pseudofelineus* (27 to 35 µm by 11 to 17 µm; average 31.5 µm by 13.5 µm) are operculate and have an abopercular knob.

Adult *A. noverca* are 9.5 to 12.7 mm by 2.5 mm, and their eggs are 34 µm by 19 µm. *Pseudamphistomum truncatum* are 1.64 to 2.5 mm by 0.6 to 1 mm, and their eggs are 27 to 35 µm by 12 to 16 µm.

Life Cycle and Transmission

The life cycle of *M. conjunctus* is similar to that of *Clonorchis* sp and *Opisthorchis* sp. Aquatic snails serve as first intermediate hosts; the common sucker fish (*Catostomus commersonii*) is the second intermediate host. Various fish serve as paratenic hosts. Infective metacercariae encyst in the muscles of fish.[21] Eating raw fish is the major cause of *M. conjunctus* infection in humans.[89] An adult *M. conjunctus* may survive for 5 or more years.[21] Life cycles of other minor opisthorchiidae are still under investigation.

Clinical and Pathologic Features

In humans, *M. conjunctus* may cause bile duct lesions similar to those caused by *C. sinensis* and *O. felineus,* including proliferation of biliary epithelium, biliary congestion, and some degree of cirrhosis.[58] Other miscellaneous flukes may cause significant disease in animals, but human infections are generally incidental.

Treatment and Prevention

Praziquantel is the drug of choice for clonorchiasis, opisthorchiasis, and dicrocoeliasis.[6,30] However, it has no therapeutic effect on *Fasciola* sp, even at high doses.[24] Triclabendazole is becoming the drug of choice for fascioliasis.[1,7]

Measures aimed at reducing or eliminating transmission of trematodiases include 1) snail host abatement, 2) health and sanitary education, 3) treatment of infected persons and domestic animals, and 4) preventing contamination of small, stagnant bodies of water. Control programs should take into account the life cycle of the endemic fluke, patterns of disease transmission, and the customs and resources of indigenous peoples.[68]

Acknowledgments:

Information in this chapter concerning human fascioliasis in the Bolivian Altiplano was obtained through research funded by STD-3 Project No. TS3-CT94-0294 of the Commission of the European Communities (DG XII). Research into human fascioliasis in Corsica was funded by PDP Project No. B2/181/125 of the World Health Organization and DGICYT Project No. PB87-0623 of the Spanish Ministry.

References

1. Abdul-Hadi S, Contreras R, Tombazzi C, Alvarez M, Melendez M. Hepatic fascioliasis: case report and review. *Rev Inst Med Trop Sao Paulo* 1996;38:69–73.
2. Acosta-Ferreira W, Vercelli-Retta J, Falconi LM. Fasciola hepatica human infection. Histopathological study of sixteen cases. *Virchows Arch A Pathol Pathol Anat* 1979;383:319–327.
3. Agatsuma T, Terasaki K, Yang L, Blair D. Genetic variation in the triploids of Japanese Fasciola species, and relationships with other species in the genus. *J Helminthol* 1994;68:181–186.
4. Al-Karawi MA, Qattan N. Clonorchis sinensis: a case report. *Ann Saudi Med* 1992;12:93–95.
5. Al'perovich BI, Rodicheva NS, Mitasov VIa. Opisthorchiasis of the liver [in Russian]. *Khirurgiia (Mosk)* 1991;10:96–99.
6. Ambroise-Thomas P, Goullier A, Wegner DG. Praziquantel in the treatment of Far Eastern hepatic distomiasis caused by Clonorchis sinensis or Opisthorchis viverrini [in French]. *Bull Soc Pathol Exot* 1981;74:426–433.
7. Apt W, Aguilera X, Vega F, et al. Treatment of human chronic fascioliasis with triclabendazole: drug efficacy and serologic response. *Am J Trop Med Hyg* 1995;52:532–535.
8. Attwood HD, Chou ST. The longevity of Clonorchis sinensis. *Pathology* 1978;10:153–156.
9. Babbott FL Jr, Frye WW, Gordon JE. Intestinal parasites of man in Arctic Greenland. *Am J Trop Med Hyg* 1961;10:185–190.
10. Beer SA, German SM. Susceptibility of Bithynia inflata mollusks from discrete populations for Opisthorchis felineus infestations from different foci of opisthorchiasis [in Russian]. *Parazitologiia* 1987;21:585–588.
11. Beer SA, Giboda M, Ditrich O. The differentiation of opisthorchis eggs by the ultrastructure of their outer membranes [in Russian]. *Med Parazitol (Mosk)* 1990;5:48–51.
12. Bernhard K. Detection of special kinds of helminths in East German citizens [in German]. *Angew Parasitol* 1985;26:223–224.
13. Blair D. Molecular variation in fasciolids and Paragonimus. *Acta Trop* 1993;53:277–289.
14. Boray JC. Experimental fascioliasis in Australia. *Adv Parasitol* 1969;7:95–210.
15. Boray JC. Fascioliasis. In: Hillyer GV, Hopla CE, eds. *Parasitic Zoonoses*. Boca Raton, Fla: CRC Press Inc; 1982:71–88. CRC Handbook Series in Zoonoses, Section C, Vol 3.
16. Bronshtein AM. 2. Opisthorchiasis and diphyllobothriasis morbidity in the native population of the Kyshik settlement, Khanty-Mansi Autonomous Okrug [in Russian]. *Med Parazitol (Mosk)* 1986;3:44–48.
17. Bronshtein AM, Uchuatkinn EA, Romanenko NA, Kantsan SN, Veretennikova NL. Comprehensive assessment of an opisthorchiasis focus in the Komi-Permiak Autonomous Okrug [in Russian]. *Med Parazitol (Mosk)* 1989;4:66–72.
18. Bychkov VG, Iarotskii LS. The problem of the carcinogenicity of parasites [in Russian]. *Med Parazitol (Mosk)* 1990;3:46–49.
19. Calero C. Clonorchiasis in Chinese residents of Panama. *J Parasitol* 1967;53:1150.
20. Cameron TW. Fish-carried parasites in Canada. *Can J Comp Med* 1945;9:245–254; 283–286; 302–311.
21. Cameron TW. The morphology, taxonomy, and life history of Metorchis conjunctus (Cobbold, 1860). *Can J Res* 1944;22:6–16.
22. Capron A, Wattre P, Capron M, Lefebvre MN. Diagnostic immunologique des parasitoses. *Gaz Med France* 1973;80:273–279.
23. Catanzaro A, Moser RJ. Health status of refugees from Vietnam, Laos, and Cambodia. *JAMA* 1982;247:1303–1308.
24. Chen MG, Mott KE. Progress in assessment of morbidity due to Fasciola hepatica infection: a review of recent literature. Schistosomiasis and other trematode infections. Division of Control of Tropical Diseases, World Health Organization. *Trop Dis Bull* 1990;87:R1–R38.
25. Choi BI, Kim HJ, Han MC, Do YS, Han MH, Lee SH. CT findings of clonorchiasis. *AJR Am J Roentgenol* 1989;152:281–284.
26. Churina NV. Occurrence and epidemiology of opisthorchiasis in the central Urals [in Russian]. *Med Parazitol (Mosk)* 1973;42:149–153.
27. Codvelle, Grandclaude, Vanlande. Un cas de distomatose humaine à Fasciola gigantica (Cholécystite aiguë distomienne avec lésions particulières de la paroi vésiculaire). *Bull Mem Soc Med Hosp Paris* 1928;52:1180–1185.
28. Ditrich O, Giboda M, Sterba J. Species determination of eggs of opisthorchiid and heterophyid flukes using scanning electron microscopy. *Angew Parasitol* 1990;31:3–9.
29. Elkins DB, Sithithaworn S, Haswell-Elkins M, Kaewkes S, Awacharagan P, Wongratanacheewin S. Opisthorchis viverrini: relationships between egg counts, worms recovered and antibody levels within an endemic community in northeast Thailand. *Parasitology* 1991;102:283–288.
30. el-Shiekh Mohamed AR, Mummery V. Human dicrocoeliasis. Report on 208 cases from Saudi Arabia. *Trop Geogr Med* 1990;42:1–7.
31. Facey RV, Marsden PD. Fascioliasis in man: an outbreak in Hampshire. *Br Med J* 1960;ii:619–625.
32. Fain A, Delville J, Jacquerye L. A case of human Fasciola gigantica douve. Simultaneous hepatic and subcutaneous infestation [in French]. *Bull Soc Pathol Exot* 1973;66:400–405.
33. Flavell DJ. Liver-fluke infection as an aetiological factor in bile-duct carcinoma of man. *Trans R Soc Trop Med Hyg* 1981;75:814–824.
34. Garcia-Rodriguez JA, Martin Sanchez AM, Fernandez Gorostarzu JM, Garcia Luis EJ. Fascioliasis in Spain: a review of the literature and personal observations. *Eur J Epidemiol* 1985;1:121–126.
35. Giboda M, Ditrich O, Scholz T, Viengsay T, Bouaphanh S. Human Opisthorchis and Haplorchis infections in Laos. *Trans R Soc Trop Med Hyg* 1991;85:538–540.
36. Grange D, Dhumeaux D, Couzineau P, Bismuth H, Bader JP. Hepatic calcification due to Fasciola gigantica. *Arch Surg* 1974;108:113–115.
37. Grove DI. *A History of Human Helminthology*. Wallingford, England: CAB Int; 1990:106.
38. Hardman EW, Jones RL, Davies AH. Fascioliasis—a large outbreak. *Br Med J* 1970;3:502–505.
39. Harinasuta C, Harinasuta T. Opisthorchis viverrini: life cycle, intermediate hosts, transmission to man and geographical distribution in Thailand. *Arzneimittelforschung* 1984;34:1164–1167.
40. Harinasuta T, Pungpak S, Keystone JS. Trematode infections. Opisthorchiasis, clonorchiasis, fascioliasis, and paragonimiasis. *Infect Dis Clin North Am* 1993;7:699–716.
41. Hillyer GV. Fascioliasis in Puerto Rico: a review. *Bol Asoc Med P R* 1981;73:94–101.
42. Hillyer GV, Soler de Galanes M, Rodriguez-Perez J, et al. Use of the Falcon assay screening test—enzyme-linked immunosorbent assay (FAST-ELISA) and the enzyme-linked immunoelectrotransfer blot (EITB) to determine the prevalence of human fascioliasis in the Bolivian Altiplano. *Am J Trop Med Hyg* 1992;46:603–609.
43. Imamkuliev KD. External secretion of the pancreas in opisthorchiasis [in Russian]. *Med Parazitol (Mosk)* 1971;40:663–667.
44. Isseroff H, Sawma JT, Reino D. Fascioliasis: role of proline in bile duct hyperplasia. *Science* 1977;198:1157–1159.
45. Kieu TL, Bronshtein AM, Fan TI. Clinico-parasitological research in a mixed focus of clonorchiasis and intestinal nematodiasis in Hanamnin Province [in Russian]. *Med Parazitol (Mosk)* 1990;2:24–26.
46. Lapage G, ed. *Veterinary Parasitology*. 2nd ed. London, England: Oliver & Boyd; 1968:329–369.
47. Leiper RT. Observations on certain helminths of man. *Trans Soc Trop Med Hyg* 1913;6:265–297.
48. Leite OH, Higaki Y, Serpentini SL, et al. Infection by Clonorchis sinensis in Asian immigrants in Brazil. Treatment with praziquantel [in Portuguese]. *Rev Inst Med Trop Sao Paulo* 1989;31:416–422.

49. Leung JW, Sung JY, Chung SC, Metreweli C. Hepatic clonorchiasis—a study by endoscopic retrograde cholangiopancreatography. *Gastrointest Endosc* 1989;35:226–231.

50. Lim JH. Radiological findings of clonorchiasis. *AJR Am J Roentgenol* 1990;155:1001–1008.

51. Looss A. On some parasites in the Museum of the School of Tropical Medicine, Liverpool. *Ann Trop Med Parasitol* 1907;1:121–154.

52. Luong Dinh Giap G, Lam Tan B, Faucher P, Roche MC, Ripert C. Hepatic distomatosis caused by Clonorchis/Opisthorchis spp. in refugees from Southeast Asia. Effects of treatment with praziquantel [in French]. *Med Trop (Mars)* 1983;43:325–330.

53. McConnell JF. Remarks on the anatomy and pathological relations of a new species of liver-fluke. *Lancet* 1875;2:271–274.

54. Macko JK, Pacenovsky J. On the variability of Dicrocoelium dendriticum (Rudolphi, 1819) in domestic and free-living animals. 2. On individual variability of the cattle (Bos taurus-race locale) dicrocoeliids in Algeria. *Helminthologia* 1987;24:111–118.

55. McSherry JA. Clonorchis sinensis infestation: a case report. *Can Fam Physician* 1981;27:861–864.

56. Madsen H, Frandsen F. The spread of freshwater snails including those of medical and veterinary importance. *Acta Trop* 1989;46:139–146.

57. Marquez FJ, Mas-Coma S. Caracterización de la secuencia del gen 18S ribosomal de Fasciola gigantica Cobbold, 1855 (Trematoda: Fasciolidae). *Acta Parasitol Port* 1993;1:341.

58. Mills JH, Hirth RS. Lesions caused by the hepatic trematode, Metorchis conjunctus, Cobbold, 1860. A comparative study in carnivora. *J Small Anim Pract* 1968;9:1–6.

59. Moreto M, Barron J. The laparoscopic diagnosis of the liver fascioliasis. *Gastrointest Endosc* 1980;26:147–149.

60. Oostburg BF, Smith SJ. Clonorchiasis in Surinam. *Trop Geogr Med* 1981;33:287–289.

61. Over HJ. Ecological basis of parasite control: trematodes with special reference to fascioliasis. *Vet Parasitol* 1982;11:85–97.

62. Pagola Serrano MA, Vega A, Ortega E, Gonzalez A. Computed tomography of hepatic fascioliasis. *J Comput Assist Tomogr* 1987;11:269–272.

63. Park CI, Kim H, Ro JY, Gutierrez Y. Human ectopic fascioliasis in the cecum. *Am J Surg Pathol* 1984;8:73–77.

64. Pungpak S, Sornmani S, Suntharasamai P, Vivatanasesth P. Ultrasonographic study of the biliary system in opisthorchiasis patients after treatment with praziquantel. *Southeast Asian J Trop Med Public Health* 1989;20:157–162.

65. Pungpak S, Viravan C, Radomyos B, et al. Opisthorchis viverrini infection in Thailand: studies on the morbidity of the infection and resolution following praziquantel treatment. *Am J Trop Med Hyg* 1997;56:311–314.

66. Ramsay RJ, Sithithaworn P, Prociv P, Moorhouse DE, Methaphat C. Density-dependent fecundity of Opisthorchis viverrini in humans, based on faecal recovery of flukes. *Trans R Soc Trop Med Hyg* 1989;83:241–242.

67. Riganti M, Pungpak S, Punpoowong B, Bunnag D, Harinasuta T. Human pathology of Opisthorchis viverrini infection: a comparison of adults and children. *Southeast Asian J Trop Med Public Health* 1989;20:95–100.

68. Rim HJ. Clonorchiasis. In: Hillyer GV, Hopla CE, eds. *Parasitic Zoonoses*. Boca Raton, Fla: CRC Press Inc; 1982:17–32. CRC Handbook Series in Zoonoses, Section C, Vol 3.

69. Rim HJ, Farag HF, Sornmani S, Cross JH. Food-borne trematodes: ignored or emerging? *Parasitol Today* 1994;10:207–209.

70. Rodriguez JD, Gomez LF, Montalvan JA. El Opisthorchis guayaquilensis una nueva especie de Opisthorchis encontrada en el Ecuador. *Rev Ecuat Hig Med Trop* 1949;6:11–24.

71. Rosicky B, Groschaft J. Dicrocoeliosis. In: Hillyer GV, Hopla CE, eds. *Parasitic Zoonoses*. Boca Raton, Fla: CRC Press Inc; 1982;33–52. CRC Handbook Series in Zoonoses, Section C, Vol 3.

72. Sadun EH. Studies on Opisthorchis viverrini in Thailand. *Am J Hyg* 1955;62:81–115.

73. Sahba GH, Arfaa F, Farahmandian I, Jalali H. Animal fascioliasis in Khuzestan, southwestern Iran. *J Parasitol* 1972;58:712–716.

74. Saleha AA. Liver fluke disease (fascioliasis): epidemiology, economic impact and public health significance. *Southeast Asian J Trop Med Public Health* 1991;22:361–364.

75. Salem AI, Abou Basha LM, Farag HF. Immunoglobulin levels and intensity of infection in patients with fascioliasis, single or combined with schistosomiasis. *J Egypt Soc Parasitol* 1987;17:33–40.

76. Self JT, Hopps HC, Williams AO. Pentastomiasis in Africans. *Trop Geogr Med* 1975;27:1–13.

77. Sidorov EG, Belyakova YV. Natural focus of Metorchis and its biology [in Russian]. *Voprosy Prirodnoi Ochagovosti Boleznei* 1972;5:133–150.

78. Siguier F, Feld PM, Welti JJ, Lumbroso P. Tribulations neurologiques d'un jeune berger atteint de distomatose cérébrale à Dicrocoelium lanceolatum. *Bull Mem Soc Med Hosp Paris* 1952;68:353–359.

79. Sithithaworn P, Haswell-Elkins MR, Mairiang P, et al. Parasite-associated morbidity: liver fluke infection and bile duct cancer in northeast Thailand. *Int J Parasitol* 1994;24:833–843.

80. Smithers SR. Fascioliasis and other trematode infections. In: Cohen S, Warren KS, eds. *Immunology of Parasitic Infections*. 2nd ed. Oxford, England: Blackwell Scientific Publications; 1982:608–621.

81. Smrkovski LL, Hendricks LD. A first report: praziquantel treatment of suspected feline fluke (Amphimerus guayaquilensis) infection in a human in Panama. *Program and Abstracts*, American Society of Tropical Medicine and Hygiene, 1992. Abstract 123.

82. Stemmermann G. Human infestation with Fasciola gigantica. *Am J Pathol* 1953;29:731–759.

83. Stork MG, Venables GS, Jennings SM, Beesley JR, Bendezu P, Capron A. An investigation of endemic fascioliasis in Peruvian village children. *J Trop Med Hyg* 1973;76:231–235.

84. Tang CC, Lin YK, Wang PC, et al. Clonorchiasis in South Fukien with special reference to the discovery of crayfishes as second intermediate host. *Chin Med J* 1963;82:545–562.

85. Tun MA, Beloborodova EI, Iushkova GI, Soldatova LP. The functional-morphological state of the liver in chronic opisthorchiasis [in Russian]. *Ter Arkh* 1991;63:63–66.

86. Vaz JF, Teles HM, Correa MA, Leite SP. Occurrence in Brazil of Thiara (Melanoides) tuberculata (O.F. Muller, 1774) (Gastropoda, Prosobranchia), the first intermediate host of Clonorchis sinensis (Cobbold, 1875) (Trematoda, Platyhelminthes) [in Portuguese]. *Rev Saude Publica* 1986;20:318–322.

87. Watson TG. Evaluation of actual and relative measurements used in the description of Metorchis conjunctus (Cobbold, 1860) Looss, 1899 (Trematoda: Opisthorchiidae). *Proc Helminth Soc Wash* 1981;48:172–176.

88. Wykoff DE, Harinasuta C, Juttijudata P, Winn MM. Opisthorchis viverrini in Thailand—the life cycle and comparison with O. felineus. *J Parasitol* 1965;51:207–214.

89. Yamaguti S. *Synopsis of Digenetic Trematodes of Vertebrates*. Vols 1 and 2. Tokyo, Japan: Keigaku Publishing Co; 1971.

90. Yoshimura H. The life cycle of Clonorchis sinensis: a comment on the presentation in the seventh edition of Craig and Faust's *Clinical Parasitology*. *J Parsitol* 1965;51:961–966.

91. Youssef FG, Mansour NS. A purified Fasciola gigantica worm antigen for the serodiagnosis of human fascioliasis. *Trans R Soc Trop Med Hyg* 1991;85:535–537.

92. Zavoikin VD. The structure of the geographic range of opisthorchiasis and the tasks in controlling this invasion [in Russian]. *Med Parazitol (Mosk)* 1991;3:26–30.

93. Zavoikin VD, Beer SA, Pliushcheva GL, Sholokhova SE, Nikiforova TF. Opisthorchiasis at left Dnieper watersheds [in Russian]. *Med Parazitol (Mosk)* 1989;2:9–14.

5

Fasciolopsiasis
and Other Intestinal Trematodiases

Aileen M. Marty *and*
Ellen M. Andersen

Introduction

Definition

Numerous genera of digenetic trematodes (flukes), comprising over 70 species, cause intestinal infection in humans.[26] The most familiar is *Fasciolopsis buski* (family Fasciolidae), the etiologic agent of fasciolopsiasis. Swine are the natural reservoir hosts of adult *F. buski*. Agents of the intestinal trematodiases may cause abdominal, cardiac, or neurologic manifestations. These diseases are most often asymptomatic, but heavy infections may be fatal.

Most intestinal trematodes infecting humans belong to the families Heterophyidae and Echinostomatidae.[5,18] Other pathologic intestinal flukes include members of the families Paramphistomatidae, Gastrodiscidae, Nanophyetidae Lecithodendriidae, Gymnophallidae, Microphallidae, Brachylaimidae, Diplostomidae, Strigeidae, and Plagiorchiidae (see Table 5.1).[6,20] Members of 2 families suspected of infecting humans, but currently categorized as infecting only fish, are Isoparorchiidae[15] and Didymozoidae.[12,25]

Synonyms

The common name for *F. buski* is the giant intestinal fluke. Synonyms for other intestinal trematodes are included in Table 5.1.

General Considerations

Busk, an English surgeon, found 14 large flukes in the duodenum of an Indian sailor who died in England in 1843.[13] The fluke remained unnamed until 1857, when Lankester suggested *Distoma buskii*. In 1860, Cobbold named the worm *Distoma crassum*. In Copenhagen in 1902, Odhner improved the taxonomic description, calling the trematode *Fasciolopsis buski*.

In 1891, Walker identified eggs in the stool of a patient as those of *F. buski* by studying an adult fluke from the same individual.[23] Nakagawa elucidated the life cycle by observing that pigs became infected only by ingesting metacercariae on raw water plants or in water containing contaminated plants.[16] Barlow,[1] an American missionary in China, swallowed eggs, cysts, and adults and described the resulting clinical features. Aware of Nakagawa's studies in swine, Barlow determined that humans acquire fasciolopsiasis by eating

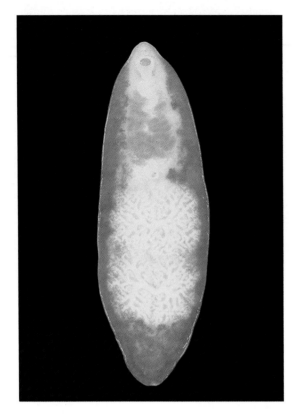

Figure 5.1
Fasciolopsis buski from pig, 7 cm long. Note tiny oral sucker, large acetabulum, highly branched, golden-yellow uterus in anterior portion of body, and 2 white, branched testes in posterior portion. Unstained x1.4

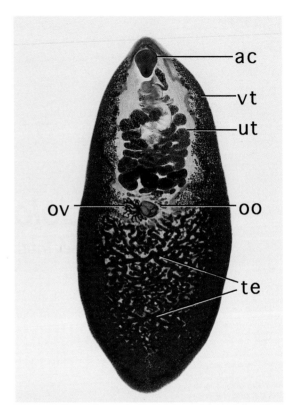

Figure 5.2
Whole mount of *F. buski* from pig, displaying acetabulum (ac), uterus (ut), ootype (oo), ovary (ov), testes (te), and vitellaria (vt). There is no cephalic cone. Compare with whole mount of *F. hepatica* (see chapter 4). Carmine x2.8

water plants contaminated by encysted metacercariae.

Descriptions of other intestinal flukes that infect humans date to the late 19th century. These flukes are less common and not as thoroughly studied as *F. buski*.[22]

Epidemiology

Because of their wide geographic distribution, high prevalence, and severe morbidity, intestinal trematodiases are significant public health problems. *Fasciolopsis buski*, the best known and largest intestinal fluke infecting humans, usually lives in the duodenum and jejunum (Fig 5.1). *Fasciolopsis buski* prevails in Southeast Asia and the southwest Pacific. Infections are most common in children 4 to 13 years old. Individuals acquire *F. buski* by ingesting metacercariae encysted on water plants, especially lotus, water caltrop, water chestnut, and watercress.

Intestinal fluke infections are most common in Asian countries.[4] Table 5.1 lists the geographic areas and sources of infection for most flukes reported in humans.

At least 20 species of echinostomes cause human infections. Asia and regions in the western Pacific have reported most echinostomiasis in humans, but some species infect humans in Europe, North America, and possibly East Africa.[17]

Potentially, all species of the family Heterophyidae may be parasitic for humans. Thus far, 33 species have been recovered from patients with intestinal trematodiasis; of these, *Heterophyes heterophyes*, in parts of Asia, and *Metagonimus yokogawai*, in the Far East, are the most common.

Two genera of intestinal flukes that infect humans and other mammals are important as carriers

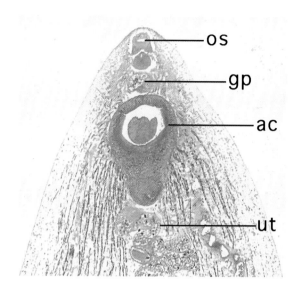

Figure 5.3
Section of anterior end of *F. buski* with small oral sucker (os), large acetabulum (ac), and genital pore (gp). Uterus (ut) is filled with darkly stained eggs. Unstained x6

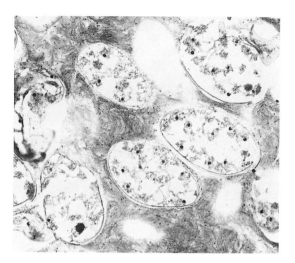

Figure 5.4
Eggs of *F. buski* in uterus of adult fluke. H&E x180

Figure 5.5
F. buski egg from feces showing small operculum (op) at one end. Unstained x475

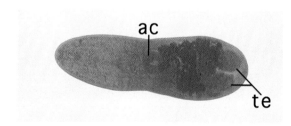

Figure 5.6
Heterophyes heterophyes adult with attenuated anterior end and broadly rounded posterior end. Acetabulum (ac) is midline in center of worm. The 2 testes (te) are in lower fifth of worm. Carmine x70

of ehrlichial infections to dogs, other mammals, and possibly humans: *Stellantchasmus* (transmits the SF ehrlichial agent) and *Nanophyetus* (transmits *Neorickettsia helmintheca*).[24] Asians are the principal victims of *Stellantchasmus falcatus*, a heterophyid fluke of gray mullet fish. *Nanophyetus salmincola* is endemic along the Pacific coast of North America and Siberia. Two subspecies of *N. salmincola* infect humans: *N. s. schikhobalowi*[10] commonly infects indigenous inhabitants of eastern Siberia, where some villages have infection rates of up to 98%; *N. s. salmincola* prevails in the coastal areas of the Pacific Northwest of North America. The significant distinction between the 2 subspecies is that *N. s. schikhobalowi* apparently does not transmit *Ehrlichia* sp.

Infectious Agents

Morphologic Description

Measurements of adults and eggs of flukes discussed below are summarized in Table 5.2.

Fasciolopsis buski is a flat, fleshy, elongate to oval trematode (Fig 5.1). There is no cephalic cone (Fig 5.2), the oral sucker is approximately one fourth the diameter of the ventral sucker (acetabulum) (Fig 5.3), and the tegument is covered with small spines. A pair of unbranched ceca extends to

Genus and species	Disease and geographic area	Source of metacercariae
Family: Fasciolidae	Fasciolopsiasis	Water plants
Fasciolopsis buski (Distomum buskii, D. crassum, D. rathouisi, F. rathouisi, F. fülleborni, and F. goddardi)	India, Bangladesh, Pakistan, Myanmar, Korea, China, Taiwan, Vietnam, Cambodia (Kampuchea), Thailand, Indonesia, Laos	Water chestnut, water caltrop, roots of lotus, water bamboo, other aquatic vegetation
Family: Echinostomatidae	Echinostomiasis	Fish, mollusks, amphibians (raw, salted, or poorly pickled)
Echinostoma hortense	Japan, China, Korea	Fish: loach
Echinostoma cinetorchis	Japan, Taiwan, Korea, Indonesia	Fish: loach Amphibians: tadpoles, frogs, salamanders
Echinoparyphium paraulum	CIS*, China	Fish
Episthmium caninum	Thailand	Fish: freshwater
Echinochasmus perfoliatus	Hungary, Italy, Romania, CIS, Japan, Taiwan, China	Fish: freshwater
Echinochasmus japonicus	China, Korea	Fish
Echinochasmus liliputanus	China	Fish
Echinochasmus jiufoensis	China	Unknown, fish suspected
Echinochasmus fujianensis	China	Fish
Echinostoma angustitestis	China	Fish: freshwater
Echinostoma ilocanum (Fascioletra ilocanum, Euparyphium ilocanum)	Philippines, Indonesia, China, Thailand	Mollusks: snails
Paryphostomum sufrartyfex, (Echinostoma sufrartyfex, Euparyphium malayanum, Artyfechinostomum mehrai)	India	Mollusks: snails
Echinostoma malayanum	Singapore, Malaysia, Thailand, Indonesia, Philippines	Mollusks: snails
Echinostoma macrorchis	Japan	Mollusks: snails
Echinostoma revolutum (E. echinatum, E. acuticauda)	Taiwan, Thailand, China, Indonesia	Mollusks: clams Amphibians: tadpoles
Hypoderaeum conoideum, (Echinostoma conoideum)	Thailand	Mollusks: snails
Echinostoma lindoense	Indonesia	Mollusks: mussels
Himasthla muehlensi	The Americas	Mollusks: bivalve mollusks (clams)
Echinoparyphium recurvatum (Echinoparyphium koidzumii)	Taiwan, Indonesia, Egypt, but cosmopolitan in animals	Tadpoles, frogs, snails
Euparyphium melis (Euparyphium jassyense, Euparyphium beaveri, Isthmiophora melis)	Romania, China, North America	Amphibians: tadpoles Fish: loach
Family: Heterophyidae	Metagonimiasis or Heterophyiasis (names vary with genera)	Fish (raw, salted, or poorly pickled)
Metagonimus yokogawai (Losotrema ovatum, Metagonimus ovatus, Loossia romanica)	Metagonimiasis: Asia (China, Japan, Korea, Indonesia, CIS), Europe (Spain), Middle East (Israel), Balkans, Siberia	Fish: freshwater
Metagonimus takahashii	Korea, Japan	Fish: carp
Metagonimus minutus	Taiwan	Fish: mullet
Heterophyes heterophyes (Distoma heterophyes, Heterophyes aegyptiaca)	Heterophyiasis: Egypt, Sudan, Iran, Tunisia, Turkey, Israel, India, Japan, China, Taiwan, Philippines, Indonesia	Fish: freshwater or brackish-water, mullet
Heterophyes heterophyes nocens	Subspecies of *H. heterophyes* in Korea	Fish: brackish-water, mullet
Heterophyes dispar	Human: Korea, Thailand; animal: Egypt, North Africa, eastern Mediterranean	Fish: mullet
Heterophyes katsuradai	Japan	Fish: mullet
Centrocestus armatus	Korea	Fish: cyprinoid
Centrocestus formosanus	China, Taiwan, Philippines	Fish, frogs
Centrocestus cuspidatus	Egypt, Taiwan	Fish: freshwater
Centrocestus caninus	Taiwan	Fish, frogs
Centrocestus kurokawai	Japan	Fish: freshwater

* CIS: Commonwealth of Independent States

Table 5.1
Epidemiology of intestinal trematodiases.

Genus and species	Disease and geographic area	Source of metacercariae
Centrocestus longus	Taiwan	Fish: freshwater
Haplorchis pumilio (Haplorchis taihokui, Monorchotrema taihokui)	Taiwan, China, Philippines, Egypt, Thailand	Fish: mullet
Haplorchis yokogawai	Philippines, Indonesia, Thailand, China, Taiwan	Crustacean, shrimp, fish: mullet
Haplorchis taichui	Philippines, Thailand, Laos, Taiwan, Bangladesh	Fish
Haplorchis pleurolophocerca	Egypt	Gambosia sp
Haplorchis vanissimus	Philippines	Fish: freshwater
Haplorchis microrchis	Japan	Fish: mullet
Diorchitrema formossanum	Taiwan	Fish: mullet
Diorchitrema amplicaecale (Stellantchasmus amplicaecale)	Taiwan	Fish: mullet
Diorchitrema pseudocirratum	Hawaii, Philippines	Fish: mullet
Heterophyopsis continua (Heterophyopsis expectans)	Japan, Korea, China	Fish: mullet, perch
Stellantchasmus pseudocirratus	Philippines	Fish
Stellantchasmus falcatus	Japan, Philippines, Hawaii, Thailand, Korea	Fish: gray mullet
Stellantchasmus formosanus	Taiwan	Fish
Stichtodora fuscata	Korea, Japan	Fish: mullet, goby
Procercovum calderoni	Philippines, China, Africa	Fish: mullet
Procercovum varium	Japan	Fish: mullet
Pygidiopsis summa	Japan, Korea	Fish: mullet, goby
Phagicola sp	Human: Brazil, USA; animal: Europe, Asia, Africa, the Americas	Fish: mullet
Appophalus donicus	USA	Fish
Cryptocotyle lingua	Greenland	Gobius sp, Labrus sp
Family: Nanophyetidae	Nanophyetiasis	Fish (raw, salted, or poorly pickled)
Nanophyetus salmincola salmincola (Troglotrema salmincola, Nanophyes salmincola)	CIS (Khabarovsk territory), USA	Fish: salmon, trout
Nanophyetus salmincola schikhobalowi	CIS (far eastern Siberia)	Fish: salmon and other fish
Family: Microphallidae	Microphallidiasis	Crustaceans
Spelotrema brevicaeca (Carneophallus brevicaeca, Heterophyes brevicaeca)	Philippines	Crabs, shrimp
Family: Gymnophallidae, Paramphistomatidae	Gymnophallidiasis, Paramphistomatidiasis	Mollusks, vegetation
Gymnophalloides seoi	Korea	Oysters
Watsonius watsoni	Human: Africa; animal: eastern Asia, Africa	Vegetation (suspected)
Fischoederius elongatus	China	Aquatic plants
Family: Gastrodiscidae	Gastrodiscoidiasis	Multiple sources
Gastrodiscoides hominis (Amphistoma hominis, Gastrodiscus hominis)	India, Vietnam, Philippines, Burma, Thailand, China, Kazakhstan	Aquatic plants; crustaceans: crayfish; Amphibians: frogs, tadpoles
Family: Diplostomidae	Diplostomidiasis	Amphibians and reptiles
Neodiplostomum seoulense (Fibricola seoulensis, Neodiplostomum seoulensis)	Korea	Amphibians: frogs Reptiles: terrestrial snakes
Family: Strigeidae	Strigeidiasis	Unknown
Cotylurus japonicus	China	Unknown
Family: Plagiorchiidae	Plagiorhiidiasis	Arthropods
Plagiorchis philippinensis	Philippines	Insect larvae
Plagiorchis javensis	Indonesia	Insect larvae
Plagiorchis muris	Japan	Arthropods: insects; mollusks: snails
Plagiorchis harinasutai	Thailand	Unknown
Family: Lecithodendriidae	Lecithodendriidiasis	Arthropods
Prosthodendrium molenkampi	Human: Indonesia, Thailand; animal: Laos	Odonate insects: dragonflies, damselflies
Phaneropsolus bonnei	Human: Indonesia, Thailand; animal: Laos	Insects suspected
Phaneropsolus spinicirrus	Thailand	Insects suspected
Paralecithodendrium obtusum	Thailand	Unknown
Paralecithodendrium glandulosum	Thailand	Unknown
Family: Brachylaimidae	Brachylaimidiasis	Mollusks
Brachylaima sp	Australia	Snails

Trematode	Location in intestine	Egg size	Adult size (length, width, thickness)
Fasciolopsis buski	Upper small intestine	130–140 μm by 80–85 μm	2–7.5 cm by 8–20 mm by 0.5–3 mm
Heterophyes heterophyes	Small intestine	28–30 μm by 15–17 μm	1–2 mm by 0.3–0.4 mm
Metagonimus yokogawai	Small intestine	26–28 μm by 15–17 μm	1–2.5 mm by 0.4–0.8 mm
Pygidiopsis summa	Intestine	22 μm by 12 μm	0.7 mm by 0.44 mm
Gastrodiscus hominis	Cecum, colon	150 μm by 60–70 μm	5–8 mm by 5–14 mm
Watsonius watsoni	Small intestine	120–130 μm by 75–80 μm	8–10 mm by 4–5 mm
Echinostoma ilocanum	Small intestine	83–116 μm by 58–69 μm	2.5–6.5 mm by 0.5–0.6 mm
Echinostoma lindoense	Small intestine	97–107 μm by 65–73 μm	13–15 mm by 2.5–3 mm
Echinostoma revolutum	Small intestine	90–126 μm by 59–71 μm	1.7–3 mm by 0.25–0.5 mm
Euparyphium melis	Small intestine	132–154 μm by 79–85 μm	3.5–11.2 mm by 1.3–1.6 mm
Echinoparyphium recurvatum	Small intestine	108–110 μm by 81–84 μm	3.7 mm by 0.6 mm
Nanophyetus salmincola	Small intestine	60–80 μm by 34–50 μm	0.8–1.1 mm by 0.3–0.5 mm
Neodiplostomum seoulense	Small intestine	95 μm by 60 μm	1.5 mm by 0.7 mm
Prosthodendrium molenkampi	Small intestine	24–26 μm by 8–10 μm	0.4–0.8 mm by 0.37–0.58 mm
Phaneropsolus bonnei	Small intestine	22–33 μm by 13–18 μm	0.48–0.78 mm by 0.22–0.35 mm
Spelotrema brevicaeca	Small intestine	15 μm by 10 μm	0.5–0.7 mm by 0.3–0.4 mm
Dicrocoelium dendriticum	Bile duct, intestine	38–45 μm by 22–30 μm	5–15 mm by 1.5–2.5 mm

Table 5.2
Host location, and egg and adult sizes of representative intestinal trematodes.

the posterior end. The highly branched uterus lies in the anterior portion, an ootype and a branched ovary in the middle portion, and 2 branched testes in the posterior portion of the body. The lateral fields contain the vitellaria that extend from the ventral sucker to the posterior end of the worm (Fig 5.2). Eggs of *F. buski* (130 to 140 μm by 80 to 85 μm) are operculate and unembryonated when passed, and are morphologically indistinguishable from those of *F. hepatica* (Figs 5.4 and 5.5).

Heterophyes heterophyes adults are ovoid, with attenuated anterior and broadly rounded posterior ends (Fig 5.6). The acetabulum is midline in the center of the worm and 3 times larger than the oral sucker. The ovary lies midline in the anterior part of the posterior third of the body, and the 2 testes are in the posterior fifth. Approximately 14 to 15 polygonal glands (vitellaria) are in the lateral fields of the posterior portion. Eggs are ovoid, embryonated, operculate, and light brown.

In *M. yokogawai* adults (Fig 5.7) the acetabulum is right of the midline and fused to the genital sucker. Eggs are difficult to distinguish from those of other heterophyids (Fig 5.8). *Pygidiopsis summa* adults are piriform (Fig 5.9); eggs are gold-tipped (Fig 5.10).

Gastrodiscus hominis adults are piriform with a conical anterior portion and a broadly discoid posterior portion (Fig 5.11). The acetabulum is conspicuous and lies posteriorly, with a deep posterior cleft measuring 2.5 to 4 mm. Eggs are operculate and unembryonated.

Watsonius watsoni adults are piriform and red-yellow. They have a large posterior acetabulum measuring 1 mm in diameter and a subterminal oral sucker. Eggs are unembryonated, operculate, and light yellow.

Echinostomes have a distinctive collar of spines surrounding the oral sucker (Figs 5.12 to 5.14). *Echinostoma ilocanum* is one of the most common echinostomes infecting humans (Figs 5.13 and 5.14). Adults have 49 to 51 collar spines, a spinous tegument, and an acetabulum close to the oral sucker (Fig 5.14). Testes are deeply lobed and in tandem (Fig 5.13). The uterus fills the anterior portion between the acetabulum and ovary, and the vitellaria lie in the lateral fields of the posterior three fourths of the body. Eggs are unembryonated, operculate, and yellow (Fig 5.15). Species of other echinostomes are similar to *E. ilocanum* except in number and clustering of collar spines. *Echinostoma lindoense* and *E. revolutum* have 37 circumoral collar spines; *E. malayanum* has 43 (Figs 5.16 and 5.17). *Echinoparyphium recurvatum* usually has 45, and *Euparyphium melis* has 27.

Nanophyetus salmincola adults are piriform,

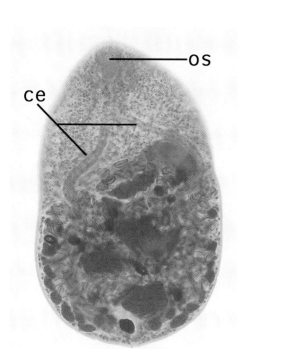

Figure 5.7
Metagonimus yokogawai adult from a cat, showing oral sucker (os) leading into long esophagus that bifurcates into ceca (ce) near acetabulum in center. Reproductive organs are in posterior half of worm. Carmine x135

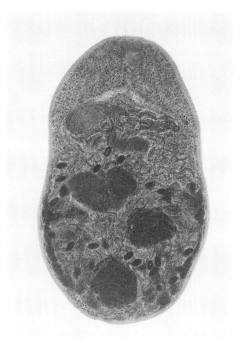

Figure 5.8
Many golden eggs fill coiled uterus of adult *M. yokogawai* from a cat. Numerous minute spines decorate tegument. Carmine x150

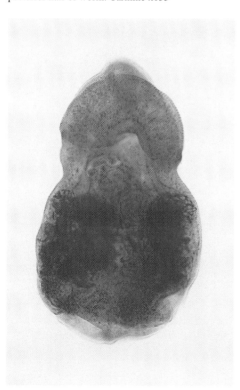

Figure 5.9
Piriform adult *Pygidiopsis summa*. Carmine x145

Figure 5.10
Adult *P. summa* with numerous gold-tipped eggs in uterus. Carmine x215

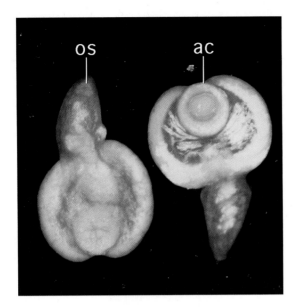

Figure 5.11
Two adult *Gastrodiscus hominis*, 0.75 cm long. Conspicuous acetabulum (ac) is best seen on ventral aspect; oral sucker (os) is best displayed on dorsal side. Specimen obtained from patient in Thailand. Unstained x6

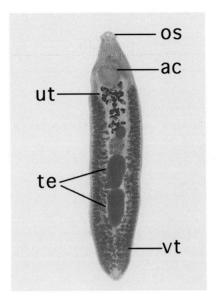

Figure 5.12
Echinostomatidae sp from a rat in the Philippines displaying testes (te) in tandem and darkly stained uterus (ut). Note also oral sucker (os), acetabulum (ac), and vitellaria (vt). Carmine x12

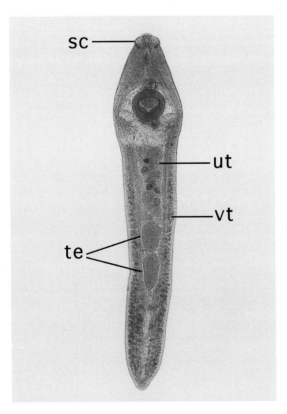

Figure 5.13
Mounted *Echinostoma ilocanum* from Indonesian patient. Note spiny collar (sc) and deeply lobed testes (te) lying in tandem. Uterus (ut) fills area between acetabulum and ovary. Vitellaria (vt) lie in lateral fields. Unstained x29

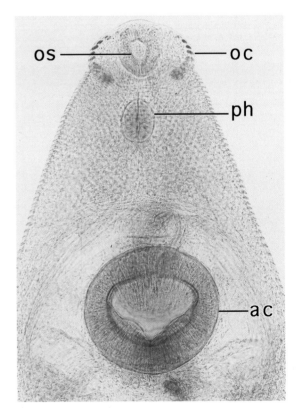

Figure 5.14
Spinous oral collar (oc) surrounds the oral sucker (os) in this whole mount of *E. ilocanum*. Pharynx (ph), acetabulum (ac), and spinous tegument are readily visible. Specimen obtained from patient in Indonesia. Masson x105

and the oral sucker is larger than the midventral acetabulum. Two large ovoid posterior testes lie laterally to each other. Eggs are operculate, thick-shelled, and yellow.

Neodiplostomum seoulense adults are conspicuously bisegmented and spoon-shaped. The acetabulum is anterior, and the uterus usually contains 1 to 10 eggs (Fig 5.18). Eggs are operculate, unembryonated, and brown to golden-yellow (Fig 5.18).

Phaneropsolus bonnei is ovoid and has a large oral sucker. It has a spinous tegument with longer spines posteriorly, an acetabulum at midbody, and prominent vitellaria (Fig 5.19). Eggs are dark, operculate, and unembryonated.

Prosthodendrium molenkampi is ovoid, with an oral sucker slightly larger than the acetabulum lying at midbody (Fig 5.20). The tegument is covered with spines and some dark dots. Eggs are operculate and unembryonated. Eggs of these members of the Lecithodenriidae family can be difficult to distinguish from those of *Opisthorchis viverrini* and the heterophyids.

Spelotrema brevicaeca adults are piriform to triangular. Intestinal ceca are short and reach to the midbody. Testes are globular and posterior to the midventral acetabulum. Eggs are operculate and yellow.

Members of the genus *Plagiorchis* are elongate, piriform, or ovoid. Ovaries are near the acetabulum and anterior to the testes. Eggs are small and embryonated.

Brachylaima sp are elongate with a smooth or spinous tegument. Oral sucker and pharynx are well-developed; ceca terminate at the posterior extremity. Eggs are smooth-shelled, with an inconspicuous operculum and an abopercular knob.

Dicrocoelium dendriticum most commonly causes bile duct infections and is discussed in chapter 4. Eggs, however, can appear in stool and must be differentiated from intestinal trematodes.[9]

There are reports of humans passing eggs, but not adult flukes, of the Didymozoidae family of trematodes.[7] Eggs resemble those of *Dicrocoelium* sp and vary in size. These trematodes are parasites of marine fish, and because no clinical disease has been described in humans, the eggs are probably spurious.

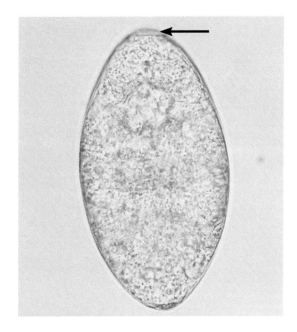

Figure 5.15
Unembryonated egg of *E. ilocanum* in feces. Note operculum (arrow). Unstained x500

Figure 5.16
Adult *Echinostoma malayanum*, 1 cm long. The large acetabulum is prominently located in center of anterior region. Specimen obtained from patient in Thailand. Unstained x9

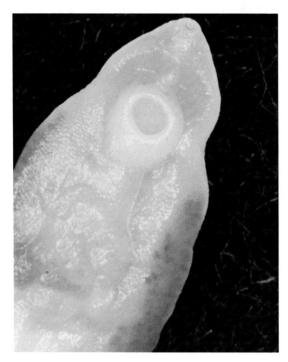

Figure 5.17
Higher magnification of Figure 5.12 demonstrating oral and ventral suckers of *E. malayanum*. Collar spines are undiscernible in this unstained specimen. x12

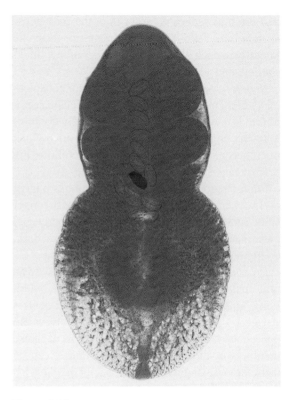

Figure 5.18
Conspicuously bisegmented, rice ladle-shaped adult *Neodiplostomum seoulense*. Operculate, unembryonated eggs line up in uterus. Carmine x85

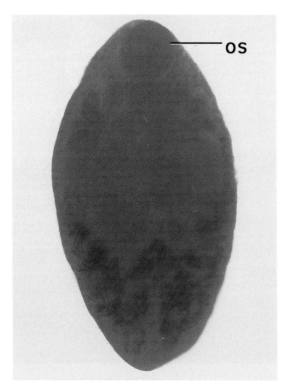

Figure 5.19
Ovoid adult *Phaneropsolus bonnei* with large oral sucker (os). Acetabulum is undiscernible in this photograph. Carmine x160

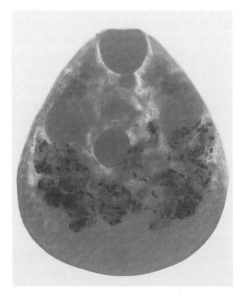

Figure 5.20
Adult piriform *Prosthodendrium molenkampi* from patient in Thailand. Oral sucker is slightly larger than acetabulum, which lies at midbody. Carmine x85

Life Cycle and Transmission

Eggs (embryonated or unembryonated, depending on species) pass in feces of the definitive host, contaminate water, and eventually hatch and release free-swimming miracidia. The miracidia penetrate the soft tissues of appropriate molluscan hosts. In *Metagonimus* and *Plagiorchis* sp, the appropriate snail must ingest the eggs before the eggs will hatch. Within the snail, parasites develop and multiply over a period of several weeks into sporocysts, rediae, and eventually cercariae. Cercariae emerge from snails and attach themselves to water plants or invade a second intermediate host, encyst, and develop into metacercariae. Humans acquire infection by ingesting metacercariae in food (see Table 5.1). Freshwater fish are the major source of intestinal trematodiasis. Humans may also acquire intestinal trematodes from plants, snails, crustaceans, amphibians, reptiles, and insects. Metacercariae excyst in the intestine, attach themselves to the mucosa, and develop into adult flukes in several months.

Figure 5.21
Adult *Haplorchis taichui* obtained from patient in Thailand. Note numerous light yellow eggs in posterior half of fluke.*
Carmine x135

Clinical Features and Pathogenesis

In light infections, adult *F. buski* inhabit the duodenum and jejunum; in heavy infections they may occupy the stomach. Heavy infections cause nausea, an intense gripping abdominal pain relieved by food (reminiscent of peptic ulcer disease), and sometimes hyperperistalsis, flatulence, minimal intestinal bleeding, or ileus. Heavy infections of *F. buski* may produce intestinal obstruction. Slight macrocytic anemia with moderate leukocytosis and eosinophilia are common, but occasionally there is neutropenia. Absorption of toxic or allergic metabolites may cause generalized edema, most pronounced in the face. Ascites and extreme prostration may develop. In a fatal infection of *F. buski* in Thailand, the patient vomited several adult worms and passed several in stools.[19] Egg count was 88 000/ml of stool, and there were 466 adult worms in the small intestine at autopsy.

Heterophyid flukes can penetrate the wall of the small intestine, causing superficial necrosis and producing colicky preprandial pain and mucoid diarrhea. These tiny worms and their eggs

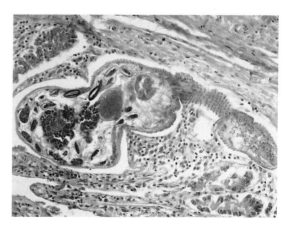

Figure 5.22
Adult heterophyid in intestinal tissue of patient from the Philippines causing mild inflammation. Note spinous tegument and small yellow eggs in uterus. x155

(Fig 5.21) can enter blood vessels and go to the heart, central nervous system, and probably elsewhere. If blood vessels of the heart are involved, myocarditis can lead to cardiac failure. In the brain, the worms or eggs are encapsulated within the neural parenchyma. *Haplorchis pumilio* that lodge in spinal cord vessels can produce motor and sensory abnormalities.[2,26] *Metagonimus yokogawai* infection causes symptoms similar to those of other heterophyids in the gastrointestinal tract: most frequently abdominal pain, diarrhea, and lethargy. The extent of the symptoms varies with worm burden, depth of penetration of the wall of the small intestine, and susceptibility of the patient. Eggs may infiltrate intestinal capillaries and lymphatics and lodge in the myocardium, brain, spinal cord, lung, and other tissue, causing embolic symptoms.

Stellantchasmus falcatus transmits the SF ehrlichial agent to dogs, mice, and possibly humans. The SF agent causes severe splenomegaly and lymphadenopathy in mice, and mild clinical signs in dogs.

Members of echinostomatidae may cause eosinophilia, intestinal colic, and diarrhea from inflammation induced by attachment to the gut wall.

Heavy infections with Gastrodiscidae (*G. hominis*), Paramphistomatidae (*W. watsoni*), and Diplostomidae (*N. seoulense*) can produce headache, epigastric pain, and diarrhea. The diarrhea may be a reaction to metabolic products released by the worms.

Both subspecies of *N. salmincola* cause clinical disease in humans.[8,14] Clinical features of *N. s. schikhobalowi* and *N. s. salmincola* include diarrhea, peripheral blood eosinophilia, abdominal discomfort, nausea, vomiting, weight loss, fatigue, flatulence, and back pain. Most importantly, *N. s. salmincola* is a carrier of *N. helmintheca*, the cause of salmon poisoning in canids.[11]

Brachylaima sp in human infants have produced prolonged foul-smelling diarrhea, mild abdominal pain, and anorexia.[3]

Pathologic Features

In light infections, the larger adult trematodes inhabit the duodenum and jejunum, but in heavy infections they may occupy the stomach, ileum, and colon. Pathophysiologic disturbances are the result of trauma, obstruction, and toxicity. There is inflammation at the site of attachment, usually followed by ulceration, with occasional submucosal and intraluminal hemorrhage, and abscess formation in heavy infections. Hypersecretion of the intestinal mucosa around attachment sites may cause intestinal obstruction.

The tiny heterophyids may provoke a mild inflammatory reaction at the site of attachment of the fluke and where they have burrowed into the mucosa (Fig 5.22). They may penetrate deeply into the small intestine and produce shallow ulcers and mild irritation or superficial necrosis. Worms lodging in blood vessels of the heart may provoke myocarditis; in the brain, they may encapsulate within the neural parenchyma. Eggs may infiltrate through the intestinal wall, enter mesenteric lymphatics, and penetrate cardiac valves and myocardium or brain tissue. *Metagonimus yokogawai* parasitize the middle section of the small intestine, first within the crypts of Lieberkühn and later between the villi, provoking villus atrophy, hyperplasia of the crypt, and varying degrees of inflammation. Adult *M. yokogawai* do not penetrate the submucosa of immunocompetent hosts, but their eggs may enter intestinal capillaries and lymphatics and lodge in vital tissues, where they provoke granulomas or microemboli.

Diagnosis

In endemic areas, clinical features suggest the diagnosis. Identifying eggs in stool or tissue provides direct evidence and confirms the diagnosis. Immunodiagnostic tests, including intradermal and serologic tests, can support the diagnosis if eggs or adults are not found.[21]

Treatment and Prevention

Hexylresorcinol removes adult worms, has low toxicity, and remains the drug of choice against *F. buski*. Patients with toxic symptoms must be treated cautiously because of the risk of cardiac failure. Tetrachloroethylene can be used for *F. buski* and is also effective for heterophyiasis, though niclosamide is the preferred drug against heterophyids. Praziquantel is the drug of choice for echinostomiasis.[21] Control of intestinal trematodiases includes identification and treatment of infected persons, disinfection and sanitary disposal of excreta, and application of molluscacides to the environment. The key to reducing the number of infections, however, is health education, including the proper preparation of edible snails, plants, amphibians, crustaceans, fish, insects, and reptiles.

References

1. Barlow CH. The life cycle of the human intestinal fluke Fasciolopsis buski (Lankester). *Monogr Ser Am J Hyg* 1925;4:99.

2. Beaver PC, Jung RC, Cupp EW. *Clinical Parasitology*. 9th ed. Philadelphia, Pa: Lea & Febiger; 1984:481.

3. Butcher AR, Talbot GA, Norton RE, et al. Locally acquired Brachylaima sp. (Digenea: Brachylaimidae) intestinal fluke infection in two South Australian infants. *Med J Aust* 1996;164:475–478.

4. Chai JY, Lee SH. Intestinal trematodes infecting humans in Korea. *Southeast Asian J Trop Med Public Health* 1991;22:163–170.

5. Chai JY, Lee SH. Intestinal trematodes of humans in Korea: Metagonimus, heterophyids and echinostomes. *Korean J Parasitol* 1990;28:103–122.

6. Chung PR, Jung Y, Kim DS. Segmentina (Polypylis) hemisphaerula (Gastropoda: Planorbidae): a new molluscan intermediate host of a human intestinal fluke Neodiplostomum seoulensis (Trematoda: Diplostomatidae) in Korea. *J Parasitol* 1996;82:336–338.

7. Cross JH. Diagnostic methods in intestinal fluke infections: a review. *SEAMEO Trop Med Tech Mtg: Diagnostic Methods for Important Helminthiasis and Amoebiasis in Southeast Asia and the Far East*. Tokyo, Japan; February 1974:87–108.

8. Eastburn RL, Fritsche TR, Terhune CA Jr. Human intestinal infection with Nanophyetus salmincola from salmonid fishes. *Am J Trop Med Hyg* 1987;36:586–591.

9. el-Shiekh Mohamed AR, Mummery V. Human dicrocoeliasis. Report on 208 cases from Saudi Arabia. *Trop Geogr Med* 1990;42:1–7.

10. Filimonova LV. Biological cycle of the trematode Nanophyetus schikhobalowi. *Tr Gelmintol Lab* 1963;13:347–357.

11. Foreyt WJ, Thorson S, Gorham JR. Experimental salmon poisoning disease in juvenile coyotes (Canis latrans). *J Wildl Dis* 1982;18:159–162.

12. Gibson DI, Bray RA. The evolutionary expansion and host-parasite relationships of the Digenea. *Int J Parasitol* 1994;24:1213–1226.

13. Grove DI. *A History of Human Helminthology*. Wallingford, England: CAB Int; 1990:127-140.

14. Harrell LW, Deardorff TL. Human nanophyetiasis: transmission by handling naturally infected coho salmon (Oncorhynchus kisutch). *J Infect Dis* 1990;161:146–148.

15. Manning GS, Nganpanya B. Recovery of Isoparorchis hypselobargri (Billet, 1898) from the catfish Wallago attu in a tributary of the Mekong in northeastern Thailand. *Southeast Asian J Trop Med Public Health* 1971;2:412–413.

16. Nakagawa K. Development of Fasciolopsis buski (Lankester). *J Parasitol* 1922;8:161–166.

17. Poland GA, Navin TR, Sarosi GA. Outbreak of parasitic gastroenteritis among travelers returning from Africa. *Arch Intern Med* 1985;145:2220–2221.

18. Radomyos R, Bunnag D, Harinasuta T. Haplorchis pumilio (Looss) infection in man in northeastern Thailand. *Southeast Asian J Trop Med Public Health* 1983;14:223–227.

19. Sadun EH, Maiphoom C. Studies on the epidemiology of the human intestinal fluke, Fasciolopsis buski (Lankester), in central Thailand. *Am J Trop Med Hyg* 1953;2:1070–1084.

20. Seo BS. Fibricola seoulensis Seo, Rim and Lee, 1964 (Trematoda) and fibricoliasis in man. *Seoul J Med* 1990;31:61–96.

21. Singh S. Trematode infections in India: a review. *Trop Gastroenterol* 1991;12:119–132.

22. Waikagul J. Intestinal fluke infections in Southeast Asia. *Southeast Asian J Trop Med Public Health* 1991;22:158–162.

23. Walker JH. Two cases of beri-beri associated with Distoma crassum, Anchylostoma duodenale, and other parasites. *Br Med J* 1891;2:1205.

24. Wen B, Rikihisa Y, Yamamoto S, Kawabata N, Fuerst PA. Characterization of the SF agent, an Ehrlichia sp. isolated from the fluke Stellantchasmus falcatus, by 16S rRNA base sequence, serological, and morphological analyses. *Int J Syst Bacteriol* 1996;46:149–154.

25. Yamaguti S. *Synopsis of Digenetic Trematodes of Vertebrates*. Tokyo, Japan: Keigaku Pub Co; 1971:247–822.

26. Yu SH, Mott KE. Epidemiology and morbidity of food-borne intestinal trematode infections. World Health Education Document (WHO/SHISTO/94.108), Geneva, Switzerland, 1995;C-562:1–17.

6

Miscellaneous Trematodiases

Aileen M. Marty,
Santiago Mas-Coma, *and*
Maria Dolores Bargues Castello

Introduction

Definition

Certain zoonotic trematodes cause infection outside the intestinal tract in humans. These parasites are medical curiosities, not public health problems. Members of the genus *Alaria* produce a visceral larva migrans syndrome. *Eurytrema pancreaticum* is a rare parasite of pancreatic ducts. Three species from the family Achillurbainiidae cause subcutaneous lesions: *Achillurbainia nouveli*, *Achillurbainia recondita*, and *Poikilorchis congolensis*. *Clinostomum complanatum* parasitizes upper respiratory passages, and *Philophthalmus* sp invade the conjunctiva.

Synonyms

The genus *Alaria* is also known as *Conchosomum*. Synonyms for *E. pancreaticum* include *Eurytrema dajii*, *Eurytrema medium*, and *Eurytrema ovis*. *Eurytrema parvum* is an immature form of *E. pancreaticum*.

Life Cycle and Transmission

All trematodes have complex life cycles involving 1 or more intermediate hosts, and sometimes paratenic hosts. In most instances, the first host is a mollusk. The second host may be 1 or more invertebrate or vertebrate animals, or larvae may attach to a plant or other aquatic object.

■ *Alaria* sp

In 1782, Goeze identified the type species, *Alaria alata*.[58] Shea et al, in 1973, identified a living mesocercaria of *Alaria* sp in the retina of a Canadian woman.[48,55] Fernandes et al first described systemic alariasis.[22] *Alaria* sp are found in the Americas (Canada, the United States, Mexico, Paraguay) and Europe (the Netherlands, Germany). Of the 19 known species of *Alaria*,[58] at least 4 are endemic in North America: *Alaria americana*, *Alaria marcianae*, *Alaria arisaemoides*, and *Alaria mustelae*. There are 5 reports of *Alaria* infection in humans, 2 in Canada,[22,23,48] 1 in Louisiana,[7] and 2 in San Francisco, all due to mesocercariae.[38]

Infectious Agent

Mesocercariae of *Alaria* sp are approximately 0.5 mm long by 80 μm at mid-hindbody and 120 μm at the level of the acetabulum (Fig 6.1). The tegument of *A. americana* has spines on both the ventral and dorsal surfaces. The acetabulum is approximately 50 μm long by 40 μm wide. The acetabula of *A. americana* and *A. marcianae* have several rows of spines. There are 2 pairs of large, elongate, unicellular penetration glands. The size, shape, and distribution of penetration glands provide a means of species identification. In *A. marcianae*, both pairs of penetration glands are preacetabular. In *A. mustelae*, both pairs are lateral in positon and mostly anterior to the level of the acetabulum. In *A. arisaemoides* and *A. americana*, 1 pair is preacetabular and median; the other is lateral to the acetabulum.

Adult *Alaria* sp inhabit the small intestine of carnivorous mammals. Their life cycle involves at least 3 hosts: a snail as intermediate host, a tadpole or frog as second intermediate host, and a carnivore as definitive host. The metacercaria is essentially a young adult, ready for growth and full maturation on reaching the intestine of its final host. The unencysted, migratory larva of *Alaria*, the mesocercaria, is a stage between the cercaria and the metacercaria. The mesocercaria sheds its tail and penetrates the skin of an intermediate host, where it grows considerably and develops a more complex excretory system. When ingested by a paratenic host, the mesocercaria undergoes little, if any, further development or growth and is not destroyed by tissue reaction. By predation, a mesocercaria may pass through a series of paratenic hosts, including alligators, bullfrogs, snakes, birds, and mammals such as rodents, raccoons, opossum, and humans. A human infant can become a paratenic host by ingesting a mesocercaria from breast milk.

Figure 6.1
Mesocercaria of *Alaria* sp found in intradermal mass on thigh of 43-year-old man from Louisiana. Biopsy revealed 100-by-160-μm section of worm (arrow) in mid-dermis, surrounded by inflammatory cells.* x30

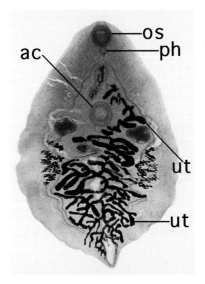

Figure 6.2
Adult *E. pancreaticum*. Note oral sucker (os), pharynx (ph), acetabulum (ac), and uterus (ut).

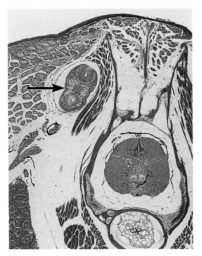

Figure 6.3
Mesocercaria (arrow) of *C. complanatum* in skeletal muscle of fish (*Poecilia reticulata*). x12

Clinical and Pathologic Features

Clinical presentation of *Alaria* differs with worm burden and location of parasites. Patients with ocular mesocercariae present with symptoms of ocular larva migrans, typically experiencing blurred vision and decreased visual acuity, but no other symptoms.[38] A 29-year-old woman who presented with *Alaria* sp infection reported having had intermittent visual problems for 6 years. The larval fluke slowly migrated from the conjunctiva along the superior retinal field to the macula, after which visual symptoms intensified. The infected eye was photophobic, and the patient described seeing a shadowy spot that undulated with worm-like movements.[48] A massive infection in a 24-year-old Canadian man began with tightness of the chest and abdominal pain. Within 2 days he developed headache and cough, then severe dyspnea, hemoptysis, and fever of 39°C. Over the next 2 days he developed petechiae, became comatose, and died. A patient with subcutaneous mesocercariae complained of small intradermal masses.[7]

Mesocercariae of *Alaria* sp secrete enzymes from their penetration glands that permit migration. A stringy, mucinous material containing inflammatory cells and eosinophils surrounds the mesocercariae as they travel through the dermis. In the surrounding dermis, scattered collections of lymphocytes, plasma cells, and eosinophils extend from vascular channels into subcutaneous fat. Vessel walls are thickened and endothelium is swollen. Vessel changes may progress to fibrinoid necrosis and thrombosis. A dense infiltrate of chronic inflammatory cells surrounds vessels. In the dermis and subcutaneous fat, focal areas of necrosis are filled with mucinous material mixed with inflammatory cells or dense infiltrates of lymphoctyes and eosinophils, and surrounded by palisades of histiocytes, apparent tracks left by the migrating parasite.[7] Overwhelming systemic infection elicits inflammatory cells only along tracts in the stomach, the presumed site of entry. Elsewhere, extensive hemorrhage accompanies the mesocercariae.[23]

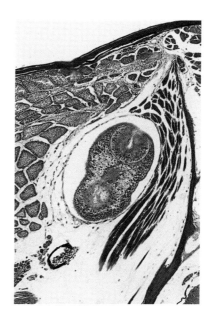

Figure 6.4
Higher magnification of mesocercaria of *C. complanatum* shown in Figure 6.3. x60

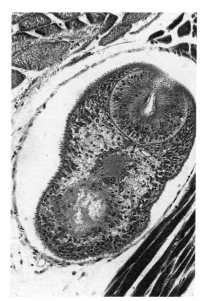

Figure 6.5
Mesocercaria of *C. complanatum* described in Figures 6.3 and 6.4, showing prominent oral sucker at anterior end and genital primordia at midbody. x120

Treatment

Argon laser therapy (photocoagulation) eliminates *Alaria* sp[48] on or below the retina, especially those located away from the macula.[38] Vitrectomy to remove intravitreal or subretinal worms has

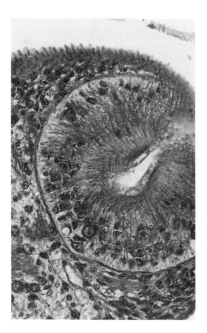

Figure 6.6
Higher magnification of oral sucker of mesocercaria shown in Figure 6.5. Note prominent tegumental spines. x230

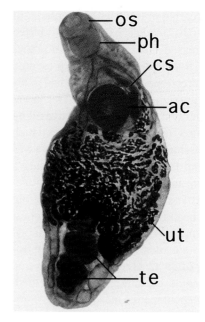

Figure 6.7
Adult *Philophthalmus* sp extracted from palpebral conjunctiva of 66 year-old man who had traveled to Galapagos Islands, Ecuador, and Colombia. Note oral sucker (os), pharynx (ph), acetabulum (ac), cirrus sac (cs), uterus (ut), and testes (te).*

successfully treated similar parasitic infections.[38] Anthelmintics such as praziquantel and albendazole should be used in conjunction with corticosteroids, especially for patients with ocular or systemic infection, to lower the risk of inflammation.[38]

■ *Eurytrema pancreaticum*

Janson introduced *E. pancreaticum* to Railliet and other European scientists during the Paris Exhibition of 1889.[25] Castellani and Chambers later discovered *E. pancreaticum* eggs in a Chinese patient.[9] In 1964, Chang and Li described adult flukes in the pancreatic ducts of a Chinese man.[10] Basch determined the fluke's life cycle in 1965 by feeding sporocysts from a snail to grasshoppers, and the infected grasshoppers to goats.[4]

Eurytrema pancreaticum is endemic in Asia, especially in the Commonwealth of Independent States (Kazakhstan,[32] Kirgizia,[36] Amur region,[19] and Primorsk Territory[43]). It is also endemic in Korea,[29] Japan,[13,14,50,53] Mongolia,[26] mainland China, Hong Kong,[51,52] Vietnam,[44] Malaysia,[4] the Philippines,[20] and Mauritius.[34] Reports of human eurytremiasis come from Japan,[28,41,50,53] Hong Kong, and mainland China. Climatic conditions affect the development of *E. pancreaticum* in the snail host. In Russia, eggs remain viable from April to November but do not survive the cold winter.[19] In Korea, snails consume eggs in autumn. Miracidia become mother sporocysts in winter and form daughter sporocysts in spring and summer.[29]

Infectious Agent

Eurytrema pancreaticum is a thick fluke (8 to 16 mm by 5.5 to 8 mm), oval to fusiform or piriform, and sometimes has a tail cone (Fig 6.2). The tegument has caducous spines. The oral sucker (2 to 2.1 mm by 1.9 to 2 mm) is subterminal and larger than the acetabulum (1.4 to 1.6 mm by 1.5 to 1.6 mm). It has a small pharynx followed by a short esophagus, and ceca that do not reach the posterior extremity. *Eurytrema pancreaticum* is a parasite of cattle, sheep, goats, rabbits, hogs, water buffalo, camels, monkeys, and sometimes humans.

Eggs of *E. pancreaticum* are 40 to 51 μm by 23 to 33 μm and embryonated. They are morphologically indistinguishable from eggs of *Dicrocoelium dendriticum* (see chapter 4). Eggs exit in feces of a definitive host and hatch within a snail. Daughter sporocysts develop within mother sporocysts into large, complicated sacs containing cercariae, then exit the snail host. Cercariae, which have short, stumpy tails,[52] penetrate the gut and encyst in the hemocoelom of grasshoppers (*Conocephalus*,[4] *Oecanthus*, or *Epocromia*[48]). In 9 days, metacercarial cysts grow to full size and mature in 2 more weeks. Humans consume metacercarial cysts in or from infected grasshoppers. Young flukes develop within the duodenum and then migrate to pancreatic ducts.[4] About 45 days after reaching the pancreas, *E. pancreaticum* matures to an adult; approximately 30 days later, adults begin to lay eggs.[47]

Clinical and Pathologic Features

Clinical presentation with *E. pancreaticum* is variable, but patients typically complain of prolonged hypochondrodynia. *Eurytrema pancreaticum* causes dilatation of pancreatic ducts without peripheral eosinophilia.[28,50] At autopsy, a 70-year-old Japanese woman with gastric cancer harbored 15 adult *E. pancreaticum* in dilated pancreatic ducts and had no peripheral eosinophilia.[28] A 57-year-old woman from Japan who complained of a 3-month history of hypochondrodynia had 3 *E. pancreaticum* in dilated pancreatic ducts at pancreatectomy.[50]

Animals with light *E. pancreaticum* infections have few changes in the pancreas. In moderate to heavy infections, there may be catarrh, ectasis, hyperplasia, or fibrosis of pancreatic ducts.[49] Eggs may penetrate duct walls, causing inflammatory foci and granulomas consisting mostly of plasma cells and eosinophils.[5]

Diagnosis of human infection with *E. pancreaticum* is based on histopathologic examination after pancreatectomy[50] or at autopsy.[28] Eggs in stool are not diagnostic since *E. pancreaticum* eggs are indistinguishable from *D. dendriticum* eggs. They are also indistinguishable from *Dicrocoelium hospes* eggs, but here a different geographic distribution is helpful, *D. hospes* being confined to the African continent.

Treatment

Many anthelmintics have been used to treat animals with *E. pancreaticum* infection without success. However, oral doses of praziquantel (50 to 70 mg/kg body weight) appear to be effective in treating heavy infections in sheep.[35] There is no standard treatment for *E. pancreaticum* infection in humans.

■ Achillurbainiidae

In 1939, Dollfus identified the genus *Achillurbainia* and described the type species, *A. nouveli*, in an abscess of the frontal sinuses of a black panther (*Felis pardus*).[17] In 1964 Chen et al reported excising an adult *A. nouveli* from a postauricular nodule in a 10-year-old Chinese girl.[11,12] Travassos, in 1942, described *A. recondita* in the maxillary sinuses of a Brazilian opossum. In 1965, Duron reported unusual omental lesions in a young Honduran man.[18] Beaver et al later identified the cause of these lesions as eggs of *A. recondita*.[6]

Yarwood and Elmes recovered eggs later identified as *P. congolensis* from the postauricular cyst of a patient in 1943, but did not recover an adult worm.[60] In 1957, Fain and Vandepitte found a mature fluke and its eggs in the postauricular cyst of a Central African patient and named it *P. congolensis*.[21] That same year, Vandepitte reported 4 more cysts of this type containing similar eggs.[54]

Achillurbainia nouveli and *A. recondita* are parasitic in a variety of mammals worldwide.[58] *Poikilorchis congolensis* is a parasite of the subcutaneous tissues of humans[45] in Africa[46,58] and Asia.[57]

Infectious Agent

Adult *A. nouveli* are 9.5 to 11 mm by 4.5 to 6 mm. They have somewhat foliate bodies, broadest behind the middle, and an aspinous tegument. The oral sucker is subventral, followed directly by a globular pharynx, a very short esophagus, and narrow, serpentine ceca that reach to the posterior extremity. The acetabulum is smaller than the oral sucker and lies about one third the body length from the anterior extremity. Testes are in lateral fields and the postovarian median field. The seminal vesicle winds forward from the anterodorsal side of the acetabulum. The ejaculatory duct is narrow. The genital pore is median and a short distance posterior to the intestinal bifurcation. The ovary is pre-equatorial, a little left of the median line. The seminal receptacle is postovarian. The uterus is convoluted in the acetabulovarian zone, and is intercecal. Small vitelline follicles extend nearly the length of the body.[18] *Achillurbainia recondita* differs from *A. nouveli* in that its testicular follicles are smaller than its vitelline follicles and its intestinal ceca are attached to the distal end of the excretory bladder by ligamentous bands.

Eggs of *A. nouveli* (55 to 64 μm by 32 to 36 μm) are unembryonated, have a distinctly flattened operculum, and may be surrounded by a low rim of outwardly thickened shell. Eggs of *A. recondita* (64 to 75 μm by 38 to 45 μm) are amber-colored and smooth, except at the abopercular end where the shell is slightly irregular. The operculum follows the regular contour of the egg and is neither flattened nor seated. With no raised border, the operculum is difficult to distinguish from the rest of the egg unless it dislodges. The egg tapers toward the abopercular end. *Poikilorchis congolensis* eggs in utero are operculated, measuring 52 to 62 μm by 33 to 38 μm; eggs in cysts are 60 to 68 μm by 38 to 41 μm. There is no abopercular knob.

Life cycles of *A. nouveli*, *A. recondita*, and *P. congolensis* are not fully understood. Epidemiologic studies suggest they are transmitted similarly or identically to *Paragonimus* sp. Freshwater crabs are suspected vectors.[11,46]

Clinical and Pathologic Features

Two reports describe a girl with *A. nouveli* infection who had a hard, painless, postauricular cyst.[11] Pathologic examination revealed a thick, 2-layered cyst wall. The inner layer consisted of foam cells, a few giant cells, and fibrous tissue; the outer layer had dense fibrous tissue infiltrated by lymphocytes and foam cells. The cyst was filled with an odorless yellow fluid containing the fluke, numerous eggs, a few lymphocytes, and many eosinophils and giant cells. Some of the giant cells contained phagocytized eggs.

A Honduran man with an infection suspected to be *A. recondita* was asymptomatic.[6,18] Numerous granulomas scattered over the peritoneal cavity and viscera were incidentally discovered during a hernia repair with subsequent exploratory laparotomy. Tissue reaction had damaged many eggs, apparently beginning at the abopercular end.

Infections with *P. congolensis* present as subcutaneous egg-filled cysts near the ear. Such lesions are tender and cause earache. There are also reports of suppurative granulomatous lesions in subcutaneous tissue.[21,54,60]

Treatment

Though praziquantel may prove to be therapeutic, surgical drainage and extraction is currently the only treatment for *P. congolensis* infection.

■ *Clinostomum complanatum*

In 1899, Braun redescribed an avian parasite discovered by Rudolphi and placed it in the genus *Clinostomum*.[58] In 1938, Yamashita reported the first human infection with *C. complanatum* in a Japanese woman with laryngitis.[59] At least 16 reports of infection with *C. complanatum* come from Japan; others come from Korea,[15] Israel,[56] and India.[8]

Infectious Agent

Clinostomum complanatum (4.5 mm by 1.62 mm) has a stout, linguiform body that is convex dorsally and concave ventrally, and a spinous tegument. A collar-like fold surrounds the oral sucker when it is retracted. The posterior end of the esophagus is bulbous. Ceca have a sinuous wall and open into an excretory vesicle by a narrow passage. The acetabulum is usually in the anterior third of the body. Testes are in the mid-hindbody or near the posterior extremity. The ovary lies to the right of the median line, between the 2 testes. The cirrus sac is anterior to the ovary or right half of the anterior testis, and contains a seminal vesicle and ejaculatory duct. There is no visible prostatic complex.

Clinostomum complanatum is primarily a parasite of the esophagus and pharynx of herons, egrets, and gulls.[3] Humans are infected with *C. complanatum* by eating raw or improperly cooked freshwater fish harboring mesocercariae (Figs 6.3 to 6.6).

Clinical Features

Clinostomum complanatum produces edema at its attachment site on the pharyngeal wall.[15] Patients with *C. complanatum* infection have laryngitis resembling halzoun.

■ *Philophthalmus* sp

Looss identified *Philophthalmus palpebrarum* in the eyelids of birds in Egypt in 1899.[37] In 1939, Markovic reported the first human infection with *Philophthalmus* sp, identifying *Philophthalmus lacrymosus* in a 28-year-old Yugoslavian man with follicular conjunctivitis.[39] Altogether there are 6 reports of human infection with *Philophthalmus* sp from Yugoslavia,[39] Japan,[40] the United States,[27] Israel,[33] and Sri Lanka.[16,30,31] Based on epidemiologic and morphologic evidence, 3 species are implicated in these infections: *P. lacrymosus*,[39] *Philophthalmus gralli*,[27] and *P. palpebrarum*.[24]

Infectious Agent

Philophthalmus sp have elongate or piriform bodies (Fig 6.7). The acetabulum lies about one third the body length from the anterior extremity. The oral sucker is followed by a short prepharynx, then a large pharynx and short esophagus. Ceca terminate at the posterior extremity. Testes are tandem, near the posterior extremity. The cirrus sac is long and encloses an elliptical seminal vesicle, a prostatic complex, and an ejaculatory duct. The genital pore is median. The ovary is in the posterior third of the body. The uterus winds between the testes and acetabulum. Vitellaria are arranged in symmetrical longitudinal series outside the ceca and anterior to the testes. The excretory vesicle has a distinct constriction just before it divides into 2 lateral arms.

Eggs of *Philophthalmus* are 68 to 79 µm by 29 to 36 µm (average 73 by 33 µm), yellow, and operculated. The anterior coils contain a developed miracidium with eyespots at the broader anterior end.[16,27]

Philophthalmus sp lay eggs in the eyes. Eggs pass into water, where miracidia hatch and find a snail intermediate host. Cercariae exit the snail and encyst on solid objects in water. They invade a definitive host by ingestion or direct contact with the eyes.[1,2] Most reports attribute human infections to swimming in or washing with contaminated water. It is less likely, but possible, that humans are infected by ingesting metacercariae in intermediate or paratenic hosts.[40]

Clinical and Pathologic Features

There are 2 forms of ocular philophthalmosis: 1) an external conjunctival form with follicular conjunctivitis and superficial keratitis, and 2) a subconjunctival form producing only mild edema and minimal cellular reaction.[31] Patients with follicular conjunctivitis have a watery discharge involving the palpebral conjunctiva of the upper and lower eyelids. The bulbar conjunctiva may also be infected. In these instances the fluke is generally on the surface of the palpebral conjunctiva.

Treatment

Extracting the parasite immediately alleviates symptoms.

References

1. Alicata JE. Life cycle and developmental stages of Philophthalmus gralli in the intermediate and final hosts. *J Parasitol* 1962;48:47–54.
2. Alicata JE, Ching HL. On the infection of birds and mammals with the cercariae and metcercariae of the eye fluke, Philophthalmus. *J Parasitol* 1960;46:16.
3. Aohagi Y, Shibahara T, Machida N, Yamaga Y, Kagota K, Hayashi T. Natural infections of Clinostomum complanatum (Trematoda: Clinostomatidae) in wild herons and egrets, Tottori Prefecture, Japan. *J Wildl Dis* 1992;28:470–471.
4. Basch PF. Completion of the life cycle of Eurytrema pancreaticum (Trematoda: Dicrocoeliidae). *J Parasitol* 1965;51:350–355.
5. Basch PF. Patterns of transmission of the trematode Eurytrema pancreaticum in Malaysia. *Zahnarztl Prax* 1966;17:234–240.
6. Beaver PC, Duron RA, Little MD. Trematode eggs in the peritoneal cavity of man in Honduras. *Am J Trop Med Hyg* 1977;26:684–687.
7. Beaver PC, Little MD, Tucker CF, Reed RJ. Mesocercaria in the skin of man in Louisiana. *Am J Trop Med Hyg* 1977;26:422–426.
8. Cameron TW. Fish-carried parsites in Canada. 1. Parasites carried by fresh water fish. *Can J Comp Med* 1945;9:245–254; 283–286; 302–311.
9. Castellani A, Chalmers AJ. *Manual of Tropical Medicine*. 3rd ed. London, England: Bailliére, Tindall & Cox; 1919:436.
10. Chang Y, Li K. Cited by: Grove DI. *A History of Human Helminthology*. Wallingford, England: CAB Int; 1990.
11. Chen HT. Paragonimus, Pagumogonimus and a Paragonimus-like trematode in man. *Chin Med J* 1965;84:781–791.
12. Chen HT, et al. A newly discovered human trematode, Achillurbainia nouveli Dollfus, 1939. *Zhong Yixue Z* 1964;50:164.
13. Chinone S, Itagaki H. Development of Eurytrema pancreaticum (Trematoda). 2. Development in definitive hosts. *Bull Azabu Vet Coll* 1976;1:73–81.
14. Chinone S, Maruyama K, Itagaki H. Development of Eurytrema pancreaticum (Trematoda). 1. Development in the first intermediate snail host. *Bull Azabu Vet Coll* 1976;1:15–22.
15. Chung DI, Moon CH, Kong HH, Choi DW, Lim DK. The first human case of Clinostomum complanatum (Trematoda: Clinostomidae) infection in Korea. *Korean J Parasitol* 1995;33:219–223.
16. Dissanaike AS, Bilimoria DP. On an infection of a human eye with Philophthalmus sp. in Ceylon. *J Helminthol* 1958;32:115–118.
17. Dollfus R. Distome d'un abcès palpébro-orbitaire chez une panthère. Possibilité d'affinités lointaines entre ce distome et les Paragonimidae. *Ann Parasitol Hum Comp* 1939;17:209–235.
18. Duron RA. Granulomatosis omento-mesentérico parasitaria. Reporte de un caso. *Rev Med Hondur* 1965;33:3–6.
19. Dvoryadkin VA. Biology of Eurytrema pancreaticum in the Amur region [in Russian]. *Parazitologiya* 1969;3:431–435.
20. Eduardo SL, Manuel MF, Tongson MS. Eurytrema escuderoi, a new species, and two other previously known species of the genus Eurytrema Looss, 1907 (Digenea: Dicrocoeliidae) in Philippine cattle and carabao. *Philippine J Vet Med* 1976;15:104–116.
21. Fain A, Vandepitte J. Description du nouveau distome vivant dans des kystes ou abcès rétroauricularis chez l'homme au Congo Belge. *Ann Soc Belg Med Trop* 1957;37:251–258.
22. Fernandes BJ, Cooper JD, Cullen JB, et al. Systemic infection with Alaria americana (Trematoda). *Can Med Assoc J* 1976;115:1111–1114.
23. Freeman RS, Stuart PF, Cullen JB, et al. Fatal human infection with mesocercariae of the trematode Alaria americana. *Am J Trop Med Hyg* 1976;25:803–807.
24. Gold D, Lang Y, Lengy J. Philophthalmus species, probably P. palpebrarum, in Israel: description of the eye fluke from experimental infection. *Parasitol Res* 1993;79:372–377.
25. Grove, DI. *A History of Human Helminthology*. Wallingford, England: CAB Int; 1990:32.
26. Gu JT, Liu RK, Li QF, et al. Epidemiological survey on Eurytrema pancreaticum and Dicrocoelium chinensis in sheep in the southern area of Daxinganling Mountain of Inner Mongolia [in Chinese]. *Chin J Vet Sci Tech* 1990;3:15–16.
27. Gutierrez Y, Grossniklaus HE, Annable WL. Human conjunctivitis caused by the bird parasite Philophthalmus. *Am J Ophthalmol* 1987;104:417–419.
28. Ishii Y, Koga M, Fujino T, et al. Human infection with the pancreas fluke, Eurytrema pancreaticum. *Am J Trop Med Hyg* 1983;32:1019–1022.
29. Jang DH. Study on Eurytrema pancreaticum. 2. Life cycle [in Korean]. *Korean J Parasitol* 1969;7:178–200.
30. Kalthoff H, Janitschke K, Mravak S, Schopp W, Werner H. Mature avian fluke (Philophthalmus sp.) under the human conjunctiva [in German]. *Klin Monatsbl Augenheilkd* 1981;179:373–375.
31. Kalthoff H, Mravak S. To the editor [letter]. *Am J Trop Med Hyg* 1983;32:436.
32. Ksembaeva GK. The first and second intermediate hosts of Eurytrema pancreaticum and Dicrocoelium dendriticum in south-eastern Kazakh SSR [in Russian]. *Izv Akad Nauk Kazakh SSR* 1967;5:51–56.
33. Lang Y, Weiss Y, Garzozi H, Gold D, Lengy J. A first instance of human philophthalmosis in Israel. *J Helminthol* 1993;67:107–111.
34. Le Roux PL, Darne A. The probable intermediate hosts of the pancreatic fluke (Eurytrema pancreaticum) in Mauritius. *Trans R Soc Trop Med Hyg* 1955;49:292.
35. Li JY, Wang FY, Huo XC, Chen YJ. Treatment of pancreatic flukes in sheep with praziquantel [in Chinese]. *Chin J Vet Med* 1983;9:15.
36. Logacheva LS. The biology of Eurytrema pancreaticum (Janson, 1889) in Kirgizia [in Russian]. *Sbornik Nauchnykh Trudov Kirgizskogo Gosudarstvennogo Meditsinskogo Instituta* 1974;95:31–34.
37. Looss A. Weitere Beiträge zur Kenntnis der Trematoden-Fauna Aegyptens, zugleich Versuch einer natürlichen Gliederung des Genus Distomum, Retzius. *Zool Jahrb Syst* 1899;12:521–784.
38. McDonald HR, Kazacos KR, Schatz H, Johnson RN. Two cases of intraocular infection with Alaria mesocercaria (Trematoda). *Am J Ophthalmol* 1994;117:447–455.
39. Markovic A. Der erste Fall von Philophthalmose beim Menschen. *Arch Ophthalmol (von Graefes)* 1939;140:515–526.
40. Mimori T, Hirai H, Kifune T, Inada K. Philophthalmus sp. (Trematoda) in a human eye. *Am J Trop Med Hyg* 1982;31:859–861.
41. Moriyama N. Taxonomic studies of Japanese bovine pancreatic flukes (Eurytrema sp.) [in Japanese]. *Japan J Parasitol* 1982;31:67–79.
42. Nadikto MV. Development of Eurytrema pancreaticum (Janson, 1889) (Trematoda: Dicrocoeliidae) in the Primorsk Territory [in Russian]. *Parazitologiya* 1973;7:408–417.
43. Nadikto MV, Romanenko PT. On the second intermediate host of the trematode Eurytrema pancreaticum (Janson, 1889) in the Far East [in Russian]. *Mater Nauch Konf Vses Obshch Gel'mint* 1969;Part I:182–186.
44. Nguen Tki Le, Matekin PV. The helminth fauna of domestic pigs in North Vietnam [in Russian]. *Vestnik Moskovskogo Universiteta, Biologiya* 1978;3:15–19.
45. Nieuwenhuyse E van, Gatti F. A case of chronic mastoid infestation by a rare parasite (Poikilorchis congolensis). *Acta Otolaryngol (Stockh)* 1968;66:444–448.
46. Oyediran AB, Fajemisin AA, Abioye AA, Lagundoye SB, Olugbile AO. Infection of the mastoid bone with a Paragonimus-like trematode. *Am J Trop Med Hyg* 1975;24:268–273.

47. Panin VY, Ksembaeva GK. Migration and morphogenesis of the marita of Eurytrema pancreaticum (Trematoda: Dicrocoeliidae) in the definitive host [in Russian]. *Parazitologiya* 1971;5:330–334.

48. Shea M, Maberley AL, Walters J, Freeman RS, Fallis AM. Intraretinal larval trematode. *Trans Am Acad Ophthalmol Otolaryngol* 1973;77:OP784–OP791.

49. Shien YS, Yang PC, Liu JJ, Huang SW. Studies on eurytremiasis. 2. Pathological study of the pancreas of cattle and goats naturally infected with Eurytrema pancreaticum [in Chinese]. *J Chin Soc Vet Sci* 1979;5:133–138.

50. Takaoka H, Mochizuki Y, Hirao E, Iyota N, Matsunaga K, Fujioka T. A human case of eurytremiasis: demonstration of adult pancreatic fluke, Eurytrema pancreaticum (Janson, 1889) in resected pancreas. *Japan J Parasitol* 1983;32:501–508.

51. Tang C, Cui G, Dong Y, et al. Studies on the biology and epidemiology of Eurytrema pancreaticum (Janson, 1889) in Heilungkiang Province [in Chinese]. *Acta Zool Sin* 1979;25:234–242.

52. Tang CC. Studies on the life history of Eurytrema pancreaticum Janson, 1889. *J Parasitol* 1950;36:559–573.

53. Tang Z, Tang C. The biology and epidemiology of Eurytrema coelomaticum (Giard and Billet, 1892) and Eurytrema pancreaticum (Janson, 1889) in cattle and sheep in China [in Chinese]. *Acta Zool Sin* 1977;23:267–282.

54. Vandepitte J, Job A, Delaisse J, Tabary MJ. Quatre cas d'abcès rétro-auriculaires chez des Congolais, produits par un nouveau distome. *Ann Soc Belg Med Trop* 1957;37:309–315.

55. Walters JC, Freeman RS, Shea M, Fallis AM. Penetration and survival of mesocercariae (Alaria spp.) in the mammalian eye. *Can J Ophthalmol* 1975;10:101–106.

56. Witenberg G. What is the cause of the parasitic laryngo-pharyngitis in the Near East ("Halzoun")? *Acta Med Orient (Jerusalem)* 1944;3:191–192.

57. Wong Soon Kai, Lie Kian Joe. Another periauricular abscess from Sarawak, probably caused by a trematode of the genus Poikilorchis, Fain and Vandepitte. *Med J Malaya* 1965;19:229–230.

58. Yamaguti S. *Synopsis of Digenetic Trematodes of Vertebrates.* Vol 1. Tokyo, Japan: Keigaku Pub Co; 1971:638–639; 813; 818–819.

59. Yamashita J. Clinostomum complanatum, a trematode parasite new to man. *Annot Zool Japon* 1938;17:563–566.

60. Yarwood GR, Elmes BG. Paragonimus cyst in a West African native. *Trans R Soc Trop Med Hyg* 1943;36:347–351.

7

Taeniasis and Cysticercosis

Ronald C. Neafie,
Aileen M. Marty, and
Linda K. Johnson

Introduction

Definition

Human taeniasis is infection by an adult tapeworm of the genus *Taenia*. Three organisms cause taeniasis in humans: *Taenia saginata* Goeze, 1782; *Taenia solium* Linnaeus, 1758; and *Taenia saginata*-Taiwanese variant. Cysticercosis is infection by *Cysticercus cellulosae*, the metacestode larva of *T. solium*.

Synonyms

The term tapeworm applies to all members of the Eucestoda (Cestoda) subclass of Cestoidea. Older scientific names for *T. solium*, the pork tapeworm, include *Taenia cucubitina*, *Taenia pellucida*, *Taenia vulgaris*, *Taenia armata umana*, and *Taenia africana*. *Taeniarhynchus saginata*, and *Taeniarhynchus mediocanellata* are older scientific names for *T. saginata*, the beef tapeworm. Indonesians call human taeniasis *adeon*.[7] Synonyms for *T. saginata*-Taiwanese variant include Taiwan Taenia, Asian Taenia, *Taenia asiatica*, and the Asian *Taenia saginata*-like tapeworm.[10]

Synonyms for cysticercosis include larval taeniasis and bladder worm infection. The metacestode of *T. saginata* is *Cysticercus bovis*, and of *T. saginata*-Taiwanese, *Cysticercus viscerotropica*.

General Considerations

Adult worms can be many meters long, and individual proglottids are obvious to the naked eye. Recognition of tapeworms or segments of tapeworms undoubtedly dates to prehistoric times. Nearly all ancient medical writers of Western and Eastern cultures mention large tapeworms.[9,19] Hippocrates, Aristotle, and Galen recognized tapeworms as animals, but others believed they were transformed strips of intestinal lining.[17] Serapion, an Arab author, mistook individual proglottids for complete worms. Still, for over 2000 years discussions of the large tapeworms that commonly infect humans, *T. saginata*, *T. solium*, and *Diphyllobothrium latum*, were hopelessly confused. In 1683, Tyson made the sentinel discovery that tapeworms have heads and necks, and that the heads attach themselves to the upper small intestine.[42] This finding indicated that there are different species of tapeworms. Rumler, in 1558, described a cystic infection of humans that was most likely cysticercosis, and in 1675, Wepfer identified *C. bovis* cysts in cattle. In 1688, Hartmann correctly surmised that these cysts were worms. Almost a century later, in 1784, Goeze perceived the similarities between the heads of tapeworms found in the human intestinal tract and the invaginated heads of *C. cellulosae* in pigs. In 1854, van Beneden fed eggs of *T. solium* to a pig and found *C. cellulosae* in the animal's muscles. In 1855, Küchenmeister fed *C. cellulosae* to a convict and recovered an adult *T. solium*. Virchow, in 1860, described a peculiar form of cysticercus in the brain, that Zenker later designated *Cysticercus racemosus*.

Using only morphologic criteria, *T. saginata* was originally thought to cause taeniasis in aborigines of East Asia. Huang et al questioned this when they noted that aborigines in mountainous areas of Taiwan do not raise cattle and seldom eat beef. In addition, they knew from Park and Cheu's work that Koreans from Cheju Island frequently eat uncooked pig liver.[20,31] Cross et al noted that aboriginal Taiwanese also eat raw meat and viscera of many mammals, especially favoring uncooked livers of pigs and goats.[4] While infection with the newly recognized *T. saginata*-Taiwanese variant was first found in aboriginal Taiwanese, the disease prevails in many other Asian countries. Molecular biologic studies indicate that this variant is genetically distinct but closely related to *T. saginata*, and that classification as a subspecies or strain of *T. saginata* is probably more appropriate than designation as a new species.[2] In this chapter, we refer to this organism as *T. saginata*-Taiwanese.

Epidemiology

Taenia saginata is the longest tapeworm that infects humans (Fig 7.1). Both *T. saginata* and *T. solium* are cosmopolitan, but the former is less widely distributed. Asian taeniasis (*T. saginata*-Taiwanese) is prevalent in Indonesia (up to 21% of the population), Taiwan (12%), Korea (7%), Thailand (2.5%), and the Philippines (1%).[11,13]

Cysticercosis, caused by the metacestode of *T. solium*, is a global health concern. Neurocysticercosis afflicts over 2.5 million people worldwide and is a common cause of seizure disorders. Clinical prevalence of neurocysticercosis is 7% in Mexico and 18% among the Ekari population of New Guinea.[14,16] Cysticercosis prevails in impoverished regions where personal hygiene tends to be poor. The most highly endemic regions are Africa, Asia, Central America, Eastern Europe, Mexico, Portugal, and Spain.

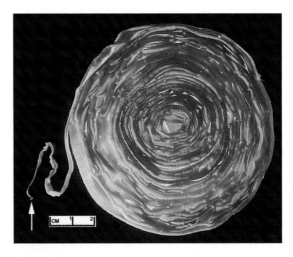

Figure 7.1
Coiled adult *T. saginata*. Thin anterior end contains tiny scolex (arrow).

Infectious Agent

Morphologic Description
Adult *Taenia* sp

The 3 *Taenia* infecting humans, *T. saginata*, *T. saginata*-Taiwanese, and *T. solium*, inhabit the small intestine, attaching themselves to the mucosa by the scolex while the strobila folds itself back and forth in the intestinal lumen. Characteristically, adult *Taenia* are long (Fig 7.2), have a scolex, narrow neck, and proglottids. Immature proglottids are broader than long, mature proglottids are nearly square, and gravid proglottids are longer than broad. The term strobila refers only to the chain of proglottids, but is sometimes applied to the entire worm distal to the scolex. *Taenia saginata*, a member of the subgenus *Taenia*-

Figure 7.2
Adult *T. saginata* demonstrating proglottids at various stages of maturity.

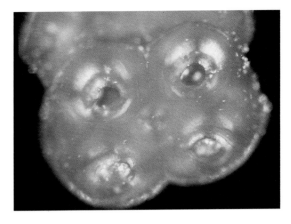

Figure 7.3
Scolex of *T. saginata* showing 4 suckers, but no rostellum or hooklets.

rhynchus, is the only one of the 3 having a scolex without a rostellum or hooklets.

Taenia saginata

Most *T. saginata* are 3 to 8 m long, but range up to 25 m. There are often over 1000 proglottids. The scolex is quadrate, 1 to 2 mm in diameter, and has 4 suckers (Fig 7.3). The apex of the scolex is concave and superficially pigmented, and has no rostellum or hooklets. Gravid proglottids are longer than broad (18 to 20 mm by 5 to 7 mm) and have 15 to 20 lateral uterine branches per side (Fig 7.4). The single genital pore is on a lateral side (Fig 7.4). Each proglottid has a thick tegument overlying well-organized smooth muscle, and a spongy parenchyma containing calcareous corpuscles (Figs 7.5 and 7.6). Terminal proglottids are motile and usually appear singly in feces. A gravid proglottid may contain over 100 000 eggs (Fig 7.7).

Taenia saginata-Taiwanese

Taenia saginata-Taiwanese is similar morphologically to *T. saginata*. Adults of the Taiwanese variant, however, have a conspicuous unarmed rostellum on the scolex, and mature proglottids have uterine buds (short uterine branches without subbranching).

Taenia solium

Adult *T. solium* are 2 to 7 m long. The scolex is globular, quadrangular, and about 1 mm in diameter (Fig 7.8). Crowned by a rostellum armed with a double row of large and small hooklets, the scolex resembles a traditional portrayal of the sun, hence the species name *solium* (Fig 7.9). There are 22 to 36 hooklets in the rostellum, with small hooklets measuring 100 to 150 μm and large hooklets 140 to 200 μm. Four large (0.5 mm), deeply cupped suckers decorate the sides of the scolex (Fig 7.9). A short, narrow neck precedes the chain of proglottids that usually numbers less than 1000. Gravid proglottids are longer than broad (11 by 5 mm) and have 7 to 13 lateral uterine branches per side (Fig 7.10). Each gravid proglottid has a single genital pore located on a lateral side. Gravid terminal proglottids are quite muscular, enabling them to migrate out of the anus and onto adjacent skin. They usually appear in feces in a chain of 5 to 6 segments. Each gravid proglottid may contain 30 000 to 50 000 eggs.

Eggs of *Taenia* sp

The eggs of all 3 *Taenia* sp infecting humans are morphologically indistinguishable. Eggs are spherical, yellow-brown, 31 to 43 μm in diameter

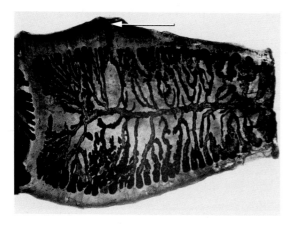

Figure 7.4
Gravid proglottid of *T. saginata*. There are numerous uterine branches per side, and a single genital pore (arrow) on a lateral side.*

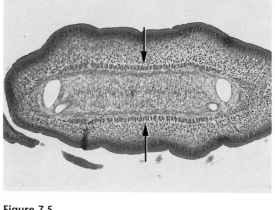

Figure 7.5
Section of immature proglottid of *T. saginata*. Proglottid has a thick tegument and an inner layer of muscle (arrows) separating outer cortical zone from inner medullary zone. x50

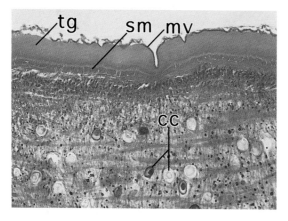

Figure 7.6
Section of proglottid of tapeworm. Note thin microvillar layer (mv) and thick tegument (tg) overlying 2 layers of smooth muscle (sm). Numerous calcareous corpuscles (cc) lie in the parenchyma. x225

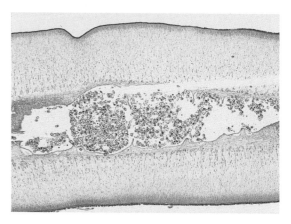

Figure 7.7
Section of gravid proglottid of *T. saginata* filled with eggs. x25

with a thick, radially striated shell, and contain an oncosphere with 6 hooklets (Fig 7.11). These eggs are also morphologically indistinguishable from those of *Echinococcus* sp.

The Metacestode (Cysticercus) Larva of *Taenia solium*

Cysticerci are milky-white, spherical to oval cysts, and are usually about 1 cm in diameter when fully developed (Fig 7.12). Each cyst contains fluid and a single invaginated protoscolex. The protoscolex has 4 large spherical suckers and a rostellum armed with a double row of 22 to 36 large and small hooklets (Figs 7.13 and 7.14). Suckers are approximately 300 μm in diameter. Hooklets are large (100 to 130 μm), readily visible on hematoxylin-eosin-stained sections (Figs 7.14 and 7.15), birefringent, and may be acid-fast.

The wall of the bladder is 100 to 200 μm thick (Figs 7.16 and 7.17) and consists of a tegument, muscle cells, tegumental cells, and parenchyma. Tegument in the region of the bladder is thin, usually less than 5 μm, bears microvilli (microtriches) on the outer surface, and is frequently raised into projections that are 10 to 25 μm wide (Fig 7.17). Numerous haphazardly arranged

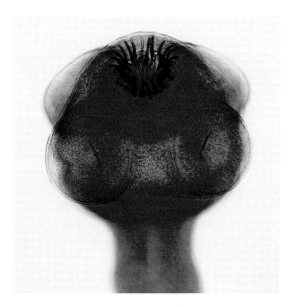

Figure 7.8
Scolex of *T. solium* with prominent armed rostellum. Two of the 4 suckers are clearly visible.* Carmine

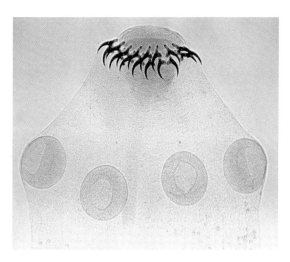

Figure 7.9
Cleared scolex of *T. solium* demonstrating double row of hooklets and 4 suckers.*

smooth muscle fibers lie immediately beneath the tegument and extend into the parenchyma. A row of tegumental cells lies below the muscle fibers. The parenchyma consists of loose stroma, fluid-filled spaces, calcareous bodies, mesenchymal fibers, and smooth muscle. Adjacent to the invagination of the neck of the protoscolex, the bladder wall may contain a thin layer of nucleated cells.

The tegument of the neck region is thicker than that of the bladder wall, measuring 10 to 20 μm (Fig 7.18). In the neck region uniform layers of smooth muscle fibers lie above a row of tegumental cells, and the parenchyma contains numerous calcareous corpuscles that vary from 10 to 20 μm in diameter and contain layered organic and inorganic material (Figs 7.18 and 7.19). The fine morphologic details of these larvae vary; special stains such as the Movat may be necessary for optimal visualization.

Racemose Cysticercus

Cysticercus racemosus consists of several bladders of various sizes forming an irregular mass, with channels or stalks connecting the different parts and proliferates, usually without a protoscolex, in a racemose pattern (resembling a compact cluster of grapes) (Figs 7.20 and 7.21). The proliferating bladder wall of the racemose cysticercus has the same morphology as the bladder wall of cysticercus cellulose (Figs 7.22 and 7.23). Occasionally there are portions of a rudimentary protoscolex (Figs 7.24 and 7.25). The origin of a racemose cysticercus is unknown. One possible explanation is that the scolex of a cysticercus disintegrates soon after it begins to develop in brain tissue, while the bladder wall continues to grow and proliferate. Some authorities believe that *C. racemosus* is an aberrant *T. solium* cysticercus, but it may represent metacestodes of other species of cestodes (eg, *Taenia serialis* and *Taenia multiceps*).[6,15,21,29]

Life Cycle and Transmission

These tapeworms have both definitive and intermediate hosts. Humans are the natural definitive hosts of *T. saginata* and *T. solium*, and probably also for *T. saginata*-Taiwanese. The natural intermediate hosts are cattle (*T. saginata*) and pigs (*T. solium* and *T. saginata*-Taiwanese). Human consumption of a metacestode (cysticercus) in raw or undercooked tissues of an infected intermediate host produces infection with the adult tapeworm.

Adult *Taenia* live in the small intestine of humans, their definitive hosts. Infected humans

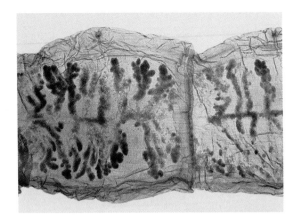

Figure 7.10
Gravid proglottids of *T. solium*. There are only a few branches of uterus on each side.*

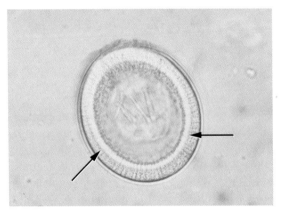

Figure 7.11
Spherical, yellow-brown *T. saginata* egg from feces. Note thick, radially striated shell (arrows). Five of 6 hooklets of the oncosphere are clearly discernible. Unstained x1140

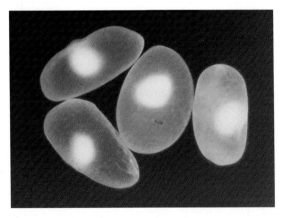

Figure 7.12
Four cysticerci extracted from a patient. The single invaginated protoscolex forms a white opacity within the bladder of each cysticercus. Unstained

usually have a single adult worm; however, because tapeworms are hermaphroditic, they release gravid proglottids into the stools. Eggs from the proglottids contaminate the environment, usually soil, where they may be consumed by the usual intermediate hosts, cattle (*T. saginata*) or pigs (*T. solium* or *T. saginata*-Taiwanese). Eggs hatch in the small intestine of the intermediate host, releasing oncospheres that enter the bloodstream and encyst in muscle, viscera, and other tissues of the intermediate host. The encysted form is called a cysticercus. The life cycle is completed when a human ingests raw or inadequately cooked beef or pork containing cysticerci. Upon reaching the small intestine, the scolex everts, attaches to the mucosa, and develops into an adult tapeworm.

The life cycle of *T. solium* can vary from that just described. Unlike those of *T. saginata* and *T. saginata*-Taiwanese, the metacestodes of *T. solium* accept humans as intermediate hosts (Fig 7.26). Eggs of *T. solium* are transmitted from contaminated fingers, food, or water. When swallowed by humans, the eggs release oncospheres in the intestines. The oncospheres penetrate the mucosa, enter lymphatics and blood vessels, and travel to various organs and tissues, where they develop into cysticerci. Cysticercosis may also result from internal autoinfection by regurgitation of gravid proglottids into the stomach. Cysticercosis can develop in any human organ, but is more common in skin or subcutaneous tissue (Fig 7.27), the brain, or the eye.

Clinical Features and Pathogenesis
Adult *Taenia*

Symptoms from adult tapeworms are numerous and nonspecific, but are more common in *T. solium* than *T. saginata* infections. The only pathognomonic sign is the spontaneous release of proglottids per anum, noted by over 90% of patients. Symptoms are more common and often more serious in undernourished, ill, elderly, or infant patients. The prepatent period is 3 to 5 months.

Common gastrointestinal symptoms include hunger pains, altered appetite, loss of body weight, nausea that disappears after eating, vomiting, constipation or diarrhea, and itching in the anal

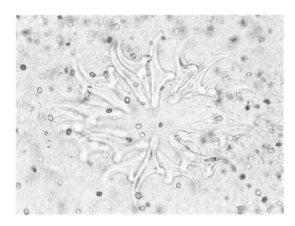

Figure 7.13
Protoscolex of *T. solium* cysticercus in unstained section of brain. Note double row of hooklets and numerous small, spherical calcareous corpuscles. x125

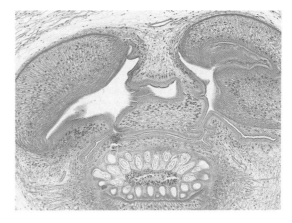

Figure 7.14
Section of protoscolex of *T. solium* cysticercus from the subcutis. Note double row of hooklets on rostellum and 2 of 4 suckers. x105

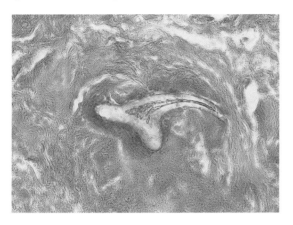

Figure 7.15
Section of cysticercus showing a hooklet 115 μm in largest dimension. x120

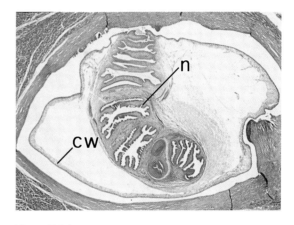

Figure 7.16
Section of heart containing a cysticercus, showing thin cyst wall (cw) and neck (n) with protoscolex. Movat x17

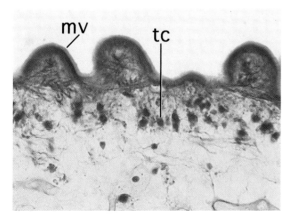

Figure 7.17
Section of cysticercus from heart, displaying several projections of bladder wall. Note thin layer of microvilli (mv), tegument, haphazardly arranged muscle fibers extending into the parenchyma, and tegumental cells (tc). Movat x580

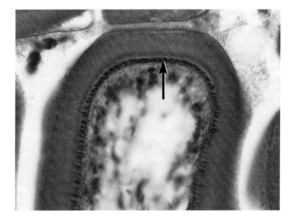

Figure 7.18
Section of cysticercus from heart showing thick tegument of the neck region. Smooth muscle fibers (arrow) lie in organized layers above a row of tegumental cells and parenchyma. x585

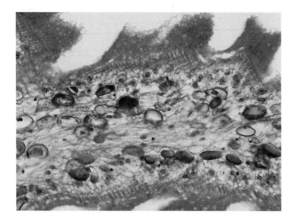

Figure 7.19
Section of cysticercus containing numerous calcareous corpuscles in invaginated neck region. x410

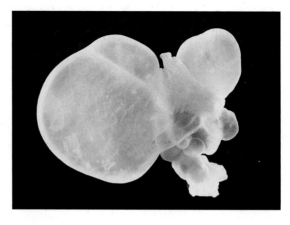

Figure 7.20
Cysticercus racemosus removed from fourth ventricle of the brain. Several interconnected bladders of various sizes make up the structure.

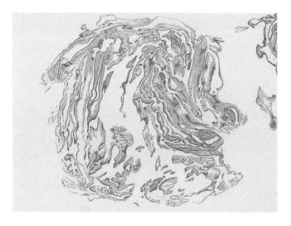

Figure 7.21
Section of proliferating bladder wall of racemose cysticercus with no identifiable protoscolex. x4

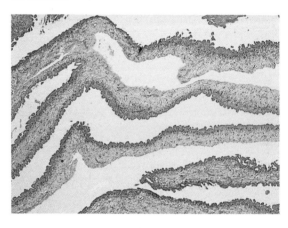

Figure 7.22
Section of proliferating bladder wall of racemose cysticercus, demonstrating multiple layers and numerous projections. x55

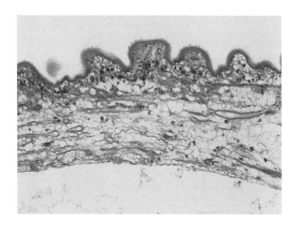

Figure 7.23
Section of proliferating bladder wall of racemose cysticercus, demonstrating several projections with microvillar layer overlying thin tegument. x230

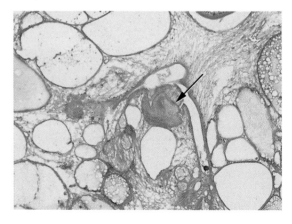

Figure 7.24
Section of racemose cysticercus from brain, showing a portion of a rudimentary protoscolex (arrow). x55

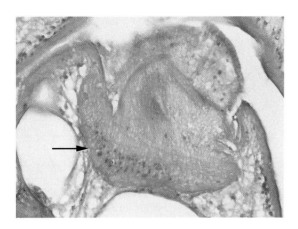

Figure 7.25
Higher magnification of racemose cysticercus in Figure 7.24, demonstrating sucker (arrow) of rudimentary protoscolex. x225

region.[33] Gastric secretions may increase or decrease.[43] Occasionally there is dysentery or intestinal obstruction. Proglottids of *T. saginata* in the appendix may produce appendicitis. Intestinal perforation may lead to granulomatous peritonitis around *Taenia* eggs.[39]

Toxic products of the parasite may produce neurologic signs: general weakness, dizziness, headache, restlessness, insomnia, and occasionally epileptic-like seizures and temporary paresis.

Hypersensitive individuals may develop allergic disorders of the skin, such as urticaria, prurigo nodularis, or prurigo ani. Chronic infections produce anemia and cachexia. Eosinophilia is most pronounced at the onset of infection. Symptoms disappear with the expulsion of the tapeworm.

Rarely, migrating *T. saginata* produce ectopic parasitism of the common bile duct, respiratory tract, uterine cavity, or nasopharynx.[34] Some patients develop cholangitis or cholecystitis.

Over 90% of patients with *T. saginata*-Taiwanese pass proglottids per anum and 71% have anal pruritus. Other symptoms include nausea (46%), dizziness (43%), increased appetite (25%), abdominal pain (23%), headache (19%), diarrhea and hunger (17%), and vomiting (11%).[12]

Metacestode (Cysticercus) of *Taenia solium*

Cysticerci frequently cause cerebral, ocular (Fig 7.28), and subcutaneous cysticercosis (Figs 7.27 and 7.29). They may lodge in other organs, including heart, lung, liver, kidney, breast, soft tissue, tongue, and skeletal muscle.

Cerebral cysticercosis is usually symptomatic, and can be fatal. Type and severity of symptoms depend on size, number, and location of parasites.

Figure 7.26
Life cycle of *T. solium*.

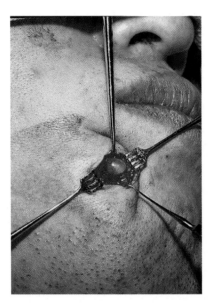

Figure 7.27
Extraction of a cysticercus from the subcutaneous facial tissue of a Korean.

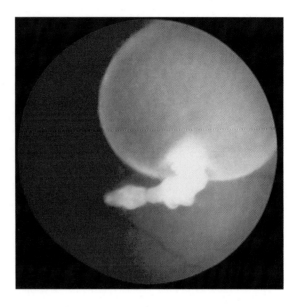

Figure 7.28
Evaginated protoscolex of cysticercus in vitreous body of the eye, visible by ophthalmoscopy.

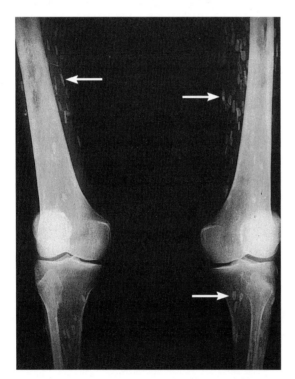

Figure 7.29
Radiograph of lower extremities of patient with multiple calcified cysticerci (arrows) in subcutaneous tissues of thigh and leg.

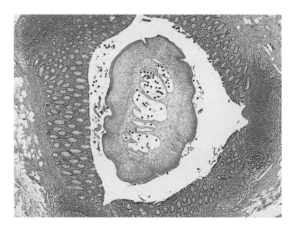

Figure 7.30
Section of appendix with intact proglottid of adult *Taenia* in lumen. x16

Symptoms and signs include meningitis, hydrocephalus, jacksonian epilepsy, hemiparesis, motor or sensory disturbances, or those associated with increased intracranial pressure (headache, nausea, vomiting). Involvement of the optic nerve or chiasm can produce papilledema and decreased visual acuity. Other cranial nerves may sustain damage. Spinal cord involvement is rare, and most lesions are in the intramedullary space. Lesions in the peripheral nervous system are uncommon, but may cause localized symptoms over the area subserved by the nerve.[30]

Ocular cysticercosis is the most common helminthic infection of the eye. In the eye, cysticerci may cause periorbital pain, flashes of light, grotesque shapes in the visual fields, and blurring or loss of vision. Examination often reveals a mobile cysticercus within the vitreous (Fig 7.28),[25] which may be removed without loss of the eye.

In locations where they are palpable, such as skin, subcutaneous tissue, breast, soft tissue, tongue, and skeletal muscle, cysticerci can cause single or multiple painless masses or lumps that gradually grow to 1 cm in diameter or greater.

Lesions of other organs, for example the heart, deep skeletal muscle, lung, liver, and kidney, are usually asymptomatic, even in heavy infections.

Pathologic Features

Human Taeniasis

Humans usually harbor only a single adult *Taenia* in the small intestine, but there are reports of as many as 25. The adult tapeworm attaches to the intestinal mucosa by its scolex. A portion of the mucosa is sucked into the central concave surface of the sucker and firmly fixed by the musculature of the tapeworm. This results in mild local inflammation of the mucosa at the site of attachment. The large mobile strobila may virtually fill the small intestine, impairing function and sometimes causing obstruction. A very active parasite may cause intestinal strangulation. Enteric intussusception, volvulus, and even rupture, especially with *T. saginata*, develop sporadically.[23] Dying proglottids release toxic substances that may affect the gastrointestinal tract, endocrine glands, nervous system, and reproductive system.[40] Consumption of nutrients by the parasite may lead to symptoms

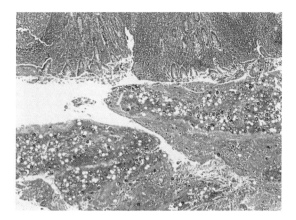

Figure 7.31
Section of appendix showing contents of a ruptured *Taenia* proglottid filling the lumen. x25

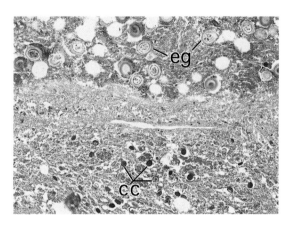

Figure 7.32
Multiple eggs (eg) and calcareous corpuscles (cc) from the ruptured *Taenia* proglottid in Figure 7.31. x115

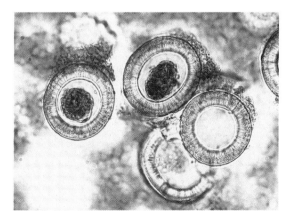

Figure 7.33
Taenia eggs from the ruptured proglottid in Figure 7.31. Eggs are spherical and have a thick, radially striated shell. x55

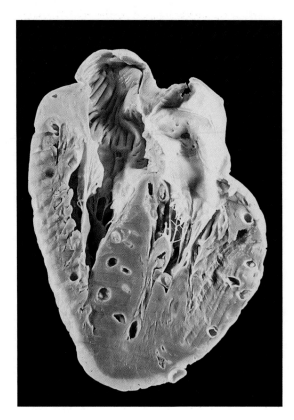

Figure 7.34
Heart with multiple cysticerci in myocardium. x0.6

of starvation, even though the patient is consuming a normal quantity of food.[35] Khechenov describes anemia in a patient who had 10 *T. saginata* worms at autopsy.[23] The mucosa of the small intestine is pallid and edematous, and Peyer's patches are swollen. Adult *Taenia* in the gastrointestinal tract may adversely affect fetal development. Chaban noted several changes in the placentas of infected patients: increased permeability of capillary walls, infarcts, necrosis, hemorrhage, fibrinoid necrosis of villi, intervilliosis, and perivascular reactions, resulting in dystrophic and proliferative changes and thromboses of the umbilical cord vessels.[3]

Intact or degenerated proglottids may lodge in the appendix and present incidental findings or initiate appendicitis. Eggs and calcareous corpuscles may be the only evidence of ruptured proglottids in the appendix (Figs 7.30 to 7.33).

Cysticercosis

Cysticerci can invade virtually any organ: brain, eye (Fig 7.28), subcutaneous tissue, heart (Figs 7.16 and 7.34), lung, liver, kidney, breast (Figs 7.35 and 7.36), soft tissue, tongue (Fig 7.37), and skeletal muscle (Fig 7.38). The pattern of reaction is similar in all organs involved. Viable cysticerci compress adjacent structures, inducing mild fibrosis while causing little or no other signs of inflammation (Fig 7.16). As cysticerci degenerate, they provoke an intense inflammatory response, initially consisting of neutrophils (Figs 7.39 and 7.40), then macrophages, fibrin, and sometimes eosinophils (Figs 7.41 and 7.42). If the bladder wall degenerates before the protoscolex, the neutrophils may invade the bladder (Fig 7.39). Eventually, a granulomatous reaction develops that includes macrophages, epithelioid cells, and foreign body giant cells (Fig 7.43), and leads to fibrosis and calcification. Often, plump macrophages line the cavity containing the cysticercus.

Cerebral cysticercosis may be meningeal, ventricular, parenchymatous, or mixed (Figs 7.44 and 7.45). Meningeal involvement produces arachnoiditis or leptomeningitis, sometimes with inflammatory adhesions that block circulation of cerebrospinal fluid (CSF). Invasion of brain parenchyma most commonly involves the gray matter (Fig 7.46). While viable cysticerci may provoke little or no host response (Fig 7.46), nonviable cysticerci incite an intense lymphocytic and monocytic infiltrate in the brain (Figs 7.47 to 7.49). In the brain parenchyma, there is usually a thin, fibrous capsule around the parasite, which in turn is surrounded by a rim of gliosis. In the absence of other significant parasitic structures, finding calcareous corpuscles in necrotic brain tissue permits a presumptive diagnosis of cysticercosis (Figs 7.50 and 7.51).

Cysticercus racemosus is an uncommon entity that arises at the base of the brain and sometimes extends to the spinal cord. Most commonly, the racemose cysticercus is in the fourth ventricle and subarachnoid space (Figs 7.20 to 7.25), where it induces considerable inflammation that may obstruct circulation of CSF.

Obstructive effects of a cysticercus in the eye can produce glaucoma and atrophy of the retina (Figs 7.28, 7.52, and 7.53). A cyst in the eye causes iridocyclitis, retinitis, conjunctivitis (Fig 7.54),

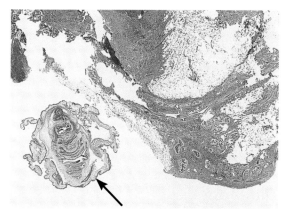

Figure 7.35
Section of palpable breast mass demonstrating a viable cysticercus (arrow) with surrounding mild inflammation. x7.5

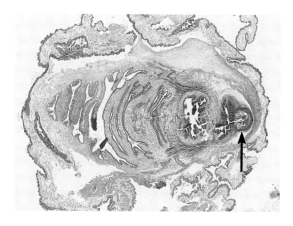

Figure 7.36
Higher magnification of cysticercus in Figure 7.35, showing armed rostellum (arrow) of protoscolex and prominent neck region within bladder. x22

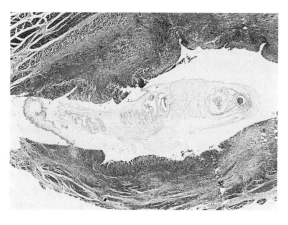

Figure 7.37
Section of tongue containing a cysticercus in muscle. Note surrounding inflammation. Movat x10.4

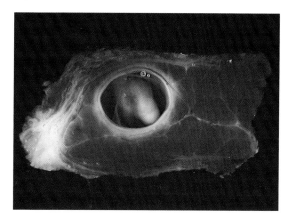

Figure 7.38
Cut surface of gross specimen of latissimus dorsi muscle containing a cysticercus. x3.4

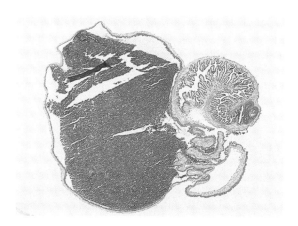

Figure 7.39
Section of cysticercus showing intense inflammatory exudate adjacent to bladder (left). Neck region and protoscolex of the cyst (right) are not inflamed. x11.7

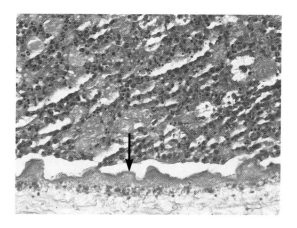

Figure 7.40
Higher magnification of cysticercus in Figure 7.39, showing interface of the cyst tegument (arrow) and the bladder filled with an exudate of neutrophils, macrophages, fibrin, and erythrocytes. x250

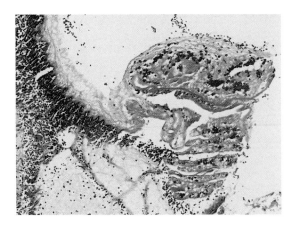

Figure 7.41
Section of degenerated and partially calcified protoscolex of a cysticercus and the mixed inflammatory response, including eosinophils. x60

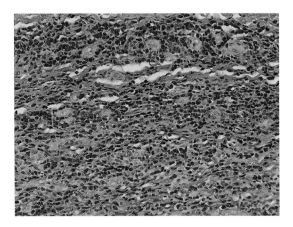

Figure 7.42
Higher magnification of inflammatory infiltrate in Figure 7.41, composed mostly of eosinophils. x155

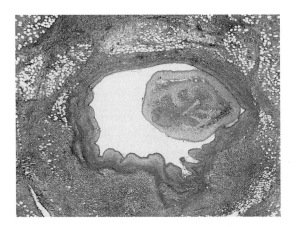

Figure 7.43
Section of subcutaneous tissue containing a degenerated cysticercus provoking a granulomatous reaction. x10.3

and vitritis. Removing the parasite may lead to complications such as sectorial iris atrophy, cataract, or chorioretinal scarring.

Rarely, the protoscolex of a cysticercus evaginates in an extraintestinal site and appears as a juvenile tapeworm (Figs 7.55 to 7.61). In these instances, there can be limited strobilization. How or why this happens is not known.

Diagnosis

Identifying eggs (Figs 7.11 and 7.33) or gravid proglottids (Figs 7.4, 7.7, and 7.10) in stools or tissues is diagnostic of taeniasis. Identifying the species of *Taenia* based on eggs alone is impossible because eggs of all species of *Taenia*, and the related genus *Echinococcus*, are morphologically indistinguishable. Eggs recently released from a ruptured proglottid rarely retain the thin, outer primary membrane. Eggs are spherical (31 to 43 μm in diameter), have a thick, yellow-brown, radially striated shell, and contain a 6-hooked embryo (Fig 7.11). Prestaining a tissue section with hematoxylin, then staining overnight in dilute polychrome blue, followed by differentiating with tartrazine in Cellosolve® helps demonstrate cestode hooklets of eggs in tissue section, even in granulomas and scars. Hooklets stain bright blue, nuclei brown to brown-green, and cytoplasm yellow.[38] Proglottids of *T. saginata*, *T. saginata*-Taiwanese, and *T. solium* passed in feces are longer than broad, distinguishing them from proglottids of *D. latum*, which normally pass in feces as short chains of segments that are broader than long. The number of uterine branches or buds within proglottids, as viewed when pressed between 2 glass slides, distinguishes between the 3 *Taenia* sp (Table 7.1). In the laboratory, eggs and proglottids should be handled carefully because eggs of *T. solium* are potentially infectious to humans.

A very effective test for diagnosing human cysticercosis detects antibodies in an immunoblot assay incorporating a fraction of lentil-lectin affinity purified cysticercus glycoprotein antigens.[41] There are also enzyme-linked immunosorbent assay (ELISA) techniques that detect antigens of or antibodies to *T. solium* in stool samples of tape-

	Adult tapeworm		
Characteristic	*Taenia solium*	*Taenia saginata*	*Taenia saginata*-Taiwanese
Host (definitive)	Human	Human	Human
Mode of infection	Ingestion of cysticercus in infected pork (muscle, brain, or viscera)	Ingestion of cysticercus in infected beef (muscle or viscera)	Ingestion of cysticercus in infected liver of pig
Prepatent period	3–5 months	3–5 months	22 days–1 month
Normal life span	Up to 25 years	Up to 25 years	N/A
Scolex	Quadrate, 1 mm in diameter, rostellum with hooklets (22–36 hooklets: smaller, 100–150 μm; larger, 140–200 μm), and 4 suckers	Quadrate, 1–2 mm in diameter, lacks rostellum, has 4 suckers, apex concave and superficially pigmented	Quadrate, 1.4–1.7 mm in diameter with 4 suckers and a conspicuous unarmed rostellum
Length	2–7 meters	3–8 meters; can be 25 meters	4–8 meters
Proglottids	≤ 1000 (700–1000)	Usually 1000–2000	260–1016
Uterine branches	7–13, dendritic	15–20, dichotomous	11–32 short uterine buds (not branches)
Gravid proglottids	Longer than broad (11 by 5 mm), usually appear in feces in chain of 5–6 segments, eggs 31–43 μm	Longer than broad (18–20 by 5–7 mm), usually appear singly in feces, eggs 31–43 μm	Longer than broad, usually appear singly in feces, eggs 31–43 μm

	Metacestode		
Characteristic	*Cysticercus cellulosae*	*Cysticercus bovis*	*Cysticercus viscerotropica*
Host (intermediate)	Human, nonhuman primate, sheep, pig, dog, cat, some marine mammals	Cattle, buffalo, giraffe, llama, reindeer (goat & sheep experimentally)	Pig (liver)
Prepatent period	7–9 weeks but variable (from days to greater than 10 years)	10–12 weeks not applicable to humans	4 weeks not applicable to humans
Mode of infection	Ingestion of eggs or gravid proglottids: 1) *heteroinfection*, contaminated food or drink; 2) *self-contamination* by patient harboring adult; 3) possible internal *autoinfection*, eggs carried to duodenum or stomach by reversed peristalsis	No unequivocal human infection. Three questionable infections	No unequivocal human infection
Hooklets	100–130 μm long	No hooklets	2 rows of rudimentary hooklets
Size of metacestode	5 mm long by 8–10 mm wide, except in brain or submeningeal spaces where they can have volume of ≥ 60 ml	Ovoid, milky-white, opalescent, 7.5–10 mm wide by 4–6 mm long	Ovoid, milky-white, 5–18 mm wide by 4.5–20 mm long

Table 7.1
Characteristics of the adult tapeworm and the metacestode.

worm carriers, even when proglottids are not found.[22] ELISA tests for antigens in CSF and in serum may give false negative results in HIV-positive patients.[24] Enzyme-linked immuno-electrotransfer blot (EITB) has higher specificity and sensitivity than ELISA.[8] DNA probes can differentiate *T. saginata* and *T. solium*.[18]

Though the definitive diagnosis of cysticercosis depends on the gross and microscopic identification of the tapeworm, computed tomography, magnetic resonance, and other radiologic modalities aid in the diagnosis of cysticercosis, especially neurocysticercosis and ocular cysticercosis.[27] It is important to differentiate neurocysticercosis from tuberculous lesions as seen radiographically.[28]

Treatment and Prevention

Human taeniasis requires treatment with albendazole or low-dose praziquantel (Biltricide®). Other effective drugs include bithionol (Bitin®), mebendazole, quinacrine (Atabrine®), and

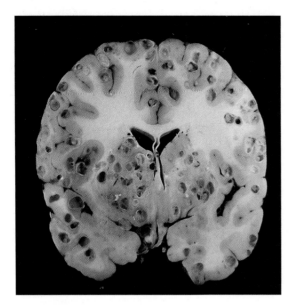

Figure 7.44
Coronal section of brain infiltrated by multiple cysticerci.

niclosamide (Niclocide® or Yomesan®). Although praziquantel can be effective, albendazole is the drug of choice for neurocysticercosis.[5,37,44] Dead cysticerci release antigens that enhance the inflammatory reaction; thus, concomitant use of steroids with chemotherapy is recommended to minimize complications, especially for patients with neurocysticercosis. Because of potential complications, surgical intervention may be preferable to chemotherapy in some patients with neurocysticercosis or optic lesions.[1,26]

Proper cooking of all meats is essential to preventing infection with an adult tapeworm. Pigs and cattle acquire cysticercosis by contact with environments contaminated by human feces. Proper disposal of human feces and careful animal husbandry are mandatory in the control of cysticercosis in animals. Public health education on transmission, combined with rapid diagnosis and immediate treatment of patients with intestinal taeniasis, is crucial.[32] Vaccines against cestoid infections are currently under study.[36]

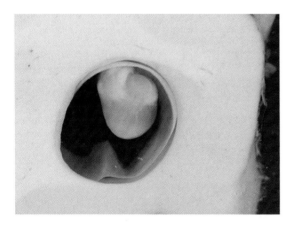

Figure 7.45
Cut surface of brain showing a cysticercus with the neck and protoscolex invaginated into its bladder.

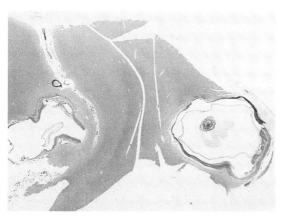

Figure 7.46
Section of cerebral gray matter containing a viable cysticercus and demonstrating minimal host response. x3.3

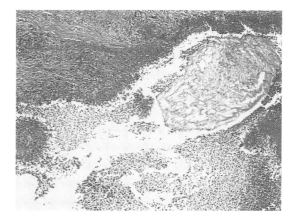

Figure 7.47
Section of brain demonstrating a degenerated cysticercus provoking an intense inflammatory reaction. x30

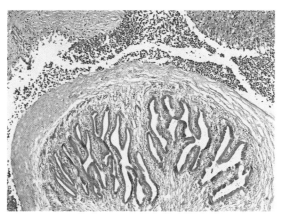

Figure 7.48
Higher magnification of Figure 7.47 displaying interface of neck region of parasite and an infiltrate of neutrophils. x70

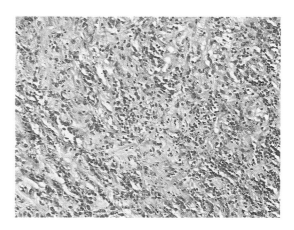

Figure 7.49
Section of brain with gliosis, lymphocytes, and monocytes in chronic abscess provoked by the degenerated cysticercus seen in Figures 7.47 and 7.48. x120

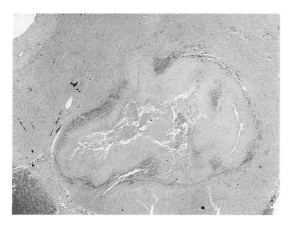

Figure 7.50
Section of brain with caseating granuloma and fibrosis. There is no visible intact parasite at this magnification (see Figure 7.51). x11.5

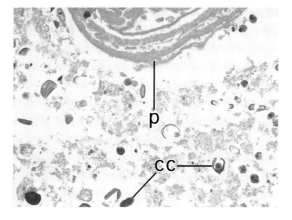

Figure 7.51
Higher magnification of Figure 7.50, revealing a degenerated fragment of the parasite (p) and numerous calcareous corpuscles (cc) in a necrotic lesion. These findings permit a presumptive diagnosis of cysticercosis. x310

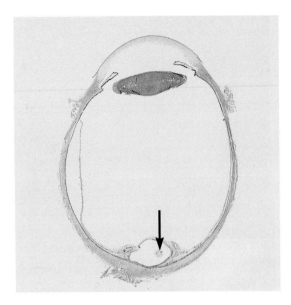

Figure 7.52
Section of eye containing a cysticercus with its protoscolex (arrow) between the retina and vitreous near the macula. x2.4

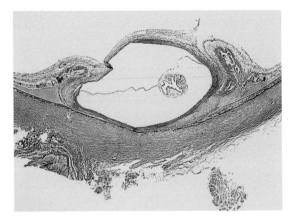

Figure 7.53
Higher magnification of cysticercus in Figure 7.52 showing slight chronic inflammatory reaction around the parasite. x8

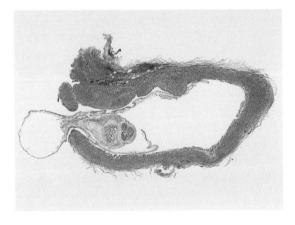

Figure 7.54
Section of conjunctiva containing a cysticercus provoking necrotizing and granulomatous inflammation. x5.8

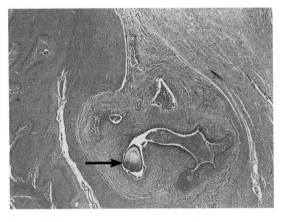

Figure 7.55
Section of brain abscess from a 50-year-old man from Ohio who traveled extensively in South America. Patient presented with irritability, aphasia, and other neurologic changes, then went into a coma and, despite treatment, died 3 months after presentation. Note evaginated protoscolex (arrow) in the abscess. x7.8

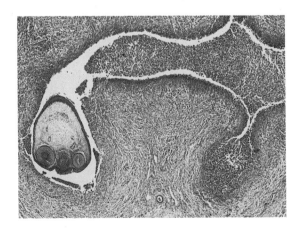

Figure 7.56
Higher magnification of abscess in Figure 7.55, depicting evaginated protoscolex surrounded by acute inflammatory response. x23

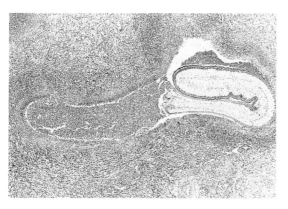

Figure 7.57
Abscess in brain described in Figures 7.55 and 7.56 illustrating elongated portion of juvenile tapeworm. Giemsa x10

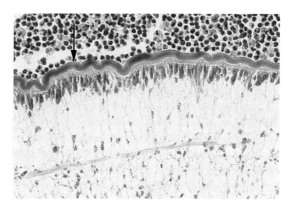

Figure 7.58
Higher magnification of juvenile tapeworm in Figure 7.57, showing interface of parasite and exudate. Parasite has thick tegument (arrow) with subjacent smooth muscle fibers in organized layers, and a row of tegumental cells. Body of parasite is solid and composed of spongy parenchyma. Giemsa x215

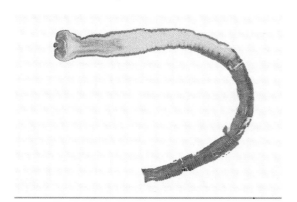

Figure 7.59
This 6-cm-long juvenile strobilate tapeworm was still alive when removed from left triceps muscle of a 22-year-old man from Laos. In addition to this worm, the patient had a typical cysticercus in subcutaneous tissue of his abdomen.* x8.6

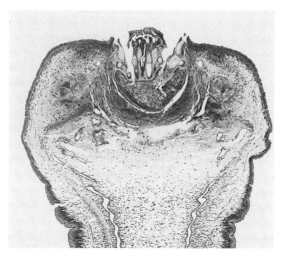

Figure 7.60
Higher magnification of juvenile strobilate tapeworm from Figure 7.59, demonstrating scolex with armed rostellum. Movat x70

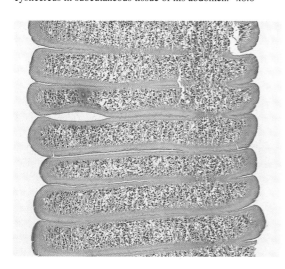

Figure 7.61
Higher magnification of juvenile strobilate tapeworm described in Figures 7.59 and 7.60, depicting immature proglottids. x140

References

1. Bandres JC, White AC Jr, Samo T, Murphy EC, Harris RL. Extraparenchymal neurocysticercosis: report of five cases and review of management. *Clin Infect Dis* 1992;15:799–811.

2. Bowles J, McManus DP. Genetic characterization of the Asian Taenia, a newly described taeniid cestode of humans. *Am J Trop Med Hyg* 1994;50:33–44.

3. Chaban AS. The effect of taeniarhynchosis on the course of pregnancy [in Russian]. *Med Parazitol (Mosk)* 1972;41:605–607.

4. Cross JH, Murrell KD, Cates MD. Survey for intestinal parasites in aborigines in Nantou County, Central Taiwan with a report of two spurious infections of Macracanthorhynchus hirudinaceus. *Chin J Microbiol* 1971;4:116–122.

5. Cruz M, Cruz I, Horton J. Clinical evaluation of albendazole and praziquantel in the treatment of cerebral cysticercosis. *Southeast Asian J Trop Med Public Health* 1991;22:279–283.

6. de Rivas D. Cited by: Beaver PC, Jung RC, Cupp EW. *Clinical Parasitology*. 9th ed. Philadelphia, Pa: Lea & Febiger; 1984:505–543.

7. Depary AA, Kosman ML. Taeniasis in Indonesia with special reference to Samosir Island, north Sumatra. *Southeast Asian J Trop Med Public Health* 1991;22:239–241.

8. Diaz JF, Verastegui M, Gilman RH, et al. Immunodiagnosis of human cysticercosis (Taenia solium): a field comparison of an antibody-enzyme-linked immunosorbent assay (ELISA), an antigen-ELISA, and an enzyme-linked immunoelectrotransfer blot (EITB) assay in Peru. The Cysticercosis Working Group in Peru (CWG). *Am J Trop Med Hyg* 1992;46:610–615.

9. Ebers G, Bryan CP, trans. *The Papyrus Ebers*. London, England: G Bles; 1930:167.

10. Eom KS, Rim HJ. Morphologic descriptions of Taenia asiatica sp. n. *Korean J Parasitol* 1993;31:1–6.

11. Fan PC. Asian Taenia saginata: species or strain? *Southeast Asian J Trop Med Public Health* 1991;22:245–250.

12. Fan PC. Taiwan Taenia and taeniasis. *Parasitol Today* 1988;4:86–88.

13. Fan PC, Chung WC, Soh CT, Kosman ML. Eating habits of east Asian people and transmission of taeniasis. *Acta Trop* 1992;50:305–315.

14. Flisser A. Taeniasis-cysticercosis: an introduction. *Southeast Asian J Trop Med Public Health* 1991;22:233–235.

15. Fontan C. Cited by: Beaver PC, Jung RC, Cupp EW. *Clinical Parasitology*. 9th ed. Philadelphia, Pa: Lea & Febiger; 1984:505–543.

16. Gajdusek DC. Introduction of Taenia solium into west New Guinea with a note on an epidemic of burns from cysticercus epilepsy in the Ekari people of the Wissel Lakes area. *P N G Med J* 1978;21:329–342.

17. Grove DI. Taenia solium and Taeniasis solium and cysticercosis. In: Grove DI, ed. *A History of Human Helminthology*. Wallingford, England: CAB Int; 1990:355–383.

18. Harrison LJ, Delgado J, Parkhouse RM. Differential diagnosis of Taenia saginata and Taenia solium with DNA probes. *Parasitology* 1990;100:459–461.

19. Hoeppli R. *Parasites and Parasitic Infections in Early Medicine and Science*. Singapore: University of Malaya Press; 1959:526.

20. Huang SW, Lin CY, Khaw OK. Cited by: Fan PC, Chung WC, Soh CT, Kosman ML. Eating habits of east Asian people and transmission of taeniasis. *Acta Trop* 1992;50:305–315.

21. Lachberg S, Thompson RC, Lymbery AJ. A contribution to the etiology of racemose cysticercosis. *J Parasitol* 1990;76:592–594.

22. Maass M, Delgado E, Knobloch J. Isolation of an immunodiagnostic Taenia solium coproantigen. *Trop Med Parasitol* 1992;43:201–202.

23. Magdiyev R (Magdiev R). Problems of pathology of taeniasis. In: Prokopic J, ed. *The First International Symposium on Human Taeniasis and Cattle Cysticercosis*. Praha, Czechoslovakia: Academia; September 1982(1983):196–208.

24. Mason P, Houston S, Gwanzura L. Neurocysticercosis: experience with diagnosis by ELISA serology and computerised tomography in Zimbabwe. *Cent Afr J Med* 1992;38:149–154.

25. Messner KH, Kammerer WS. Intraocular cysticercosis. *Arch Ophthalmol* 1979;97:1103–1105.

26. Mitchell WG, Snodgrass SR. Intraparenchymal cerebral cysticercosis in children: a benign prognosis. *Pediatr Neurol* 1985;1:151–156.

27. Monteiro L, Coelho T, Stocker A. Neurocysticercosis, a review of 231 cases. *Infection* 1992;20:61–65.

28. Moses PD, Kirubakaran C, Chacko DS. Disappearing CT lesions in focal seizures. *Ann Trop Paediatr* 1992;12:331–333.

29. Niño FL. Tratamiento de la teniasis por Taenia saginata con sales de acridina. *Prensa Med Argent* 1950;39:547-549.

30. Nosanchuk JS, Agostini JC, Georgi M, Posso M. Pork tapeworm of cysticercus involving peripheral nerve. *JAMA* 1980;244:2191–2192.

31. Park HJ, Chyu I. Cited by: Fan PC, Chung WC, Soh CT, Kosman ML. Eating habits of east Asian people and transmission of taeniasis. *Acta Trop* 1992;50:305–315.

32. Pawlowski ZS. Control of Taenia solium taeniasis and cysticercosis by focus-oriented chemotherapy of taeniasis. *Southeast Asian J Trop Med Public Health* 1991;22:284–286.

33. Pawlowski ZS. Clinical expression of Taenia saginata infection in man. In: Prokopic J, ed. *The First International Symposium on Human Taeniasis and Cattle Cysticercosis*. Praha, Czechoslovakia: Academia; September 1982(1983):138–144.

34. Pawlowski Z, Schultz MD. Taeniasis and cysticercosis (Taenia saginata). *Adv Parasitol* 1972;10:269–343.

35. Podyapolskaya VP. Cited by: Magdiyev R (Magdiev R). Problems of pathology of taeniasis in man and cysticercosis in cattle. In: Prokopic J, ed. *The First International Symposium on Human Taeniasis and Cattle Cysticercosis*. Praha, Czechoslovakia: Academia; September 1982 (1983):196–208.

36. Rickard MD. Cestode vaccines. *Southeast Asian J Trop Med Public Health* 1991;22:287–290.

37. Sotelo J, del Brutto OH, Penagos P, et al. Comparison of therapeutic regimen of anticysticercal drugs for parenchymal brain cysticercosis. *J Neurol* 1990;237:69–72.

38. Sterba J, Milacek P, Vitovec J. A new method for selective diagnostic staining of hooks of echinococci, cysticerci and tapeworms in histological sections. *Folia Parasitol (Praha)* 1989;36:341–344.

39. Sterba J. Granulomatous peritonitis induced by T. saginata eggs after perforative appendicitis. In: Prokopic J, ed. *The First International Symposium on Human Taeniasis and Cattle Cysticercosis*. Praha, Czechoslovakia: Academia; September 1982(1983):209–213.

40. Talyzin FF. Cited by: Magdiyev R (Magdiev R). Problems of pathology of taeniasis in man and cysticercosis in cattle. In: Prokopic J, ed. *The First International Symposium on Human Taeniasis and Cattle Cysticercosis*. Praha, Czechoslovakia: Academia; September 1982(1983):196–208.

41. Tsang VC, Brand JA, Boyer AE. An enzyme-linked immunoelectrotransfer blot assay and glycoprotein antigens for diagnosing human cysticercosis (Taenia solium). *J Infect Dis* 1989;159:50–59.

42. Tyson E. *Lumbricus latus*, or a discourse read before the Royal Society, of the joynted worm, wherein a great many mistakes of former writers concerning it, are remarked; its natural history from more exact observations is attempted; and the whole urged, as a difficulty against the doctrine of univocal generation. *Philos Trans R Soc London* 1683;12:113–144.

43. Valent M, Sobota K, Sterba J. Clinical aspects of taeniasis. In: Prokopic J, ed. *The First International Symposium on Human Taeniasis and Cattle Cysticercosis*. Praha, Czechoslovakia: Academia; September 1982(1983):127–137.

44. Vanijanonta S, Bunnag D, Riganti M. The treatment of neurocysticercosis with praziquantel. *Southeast Asian J Trop Med Public Health* 1991;22:275–278.

Dipylidiasis

Aileen M. Marty *and*
Ronald C. Neafie

Introduction

Definition

Dipylidiasis is infection by *Dipylidium caninum* (Linnaeus, 1758) Leuckart, 1863; Railliet, 1892. This zoonotic intestinal tapeworm infection is primarily found in children.

Synonyms

Common names for *D. caninum* include dog tapeworm, double-pored tapeworm, cucumber seed tapeworm (*Gurkenkernbandwurm*), and pumpkin seed tapeworm (*Kürbiskernbandwurm*). Older scientific names include *Taenia canina*, *Taenia cucumerina*, *Taenia elliptica*, *Dipylidium cucumerinum*, and *Dipylidium sexcoronatum*.

General Considerations

Dipylidium caninum and *Hymenolepis diminuta* cause the 2 tapeworm infections customarily acquired by ingesting an infected flea. Linnaeus[20] first described *D. caninum* in 1748 and revised his description in 1758. Leuckart[18] placed it in the genus *Dipylidium* in 1863, from the Greek word *dipylos,* meaning 2 gates. In 1751, Dubois, a pupil of Linnaeus, confirmed the first recorded human infection. In 1869, Melnikov established that arthropods serve as intermediate hosts by feeding eggs of *D. caninum* to larvae of the dog louse *Trichodectes canis,* then tracing their transformation into cysticercoids in the body cavity of the louse. In 1888, Grassi and Rovelli showed that *D. caninum* eggs can also hatch in the intestine of larvae of the dog flea *Ctenocephalides canis*, and larvae of the human flea *Pulex irritans*, and develop into cysticercoid larvae. In the United States, the first known case of dipylidiasis in humans was in a 16-month-old infant. Duffield recovered the specimen, and Stiles[30] identified the parasite and reported it in 1903. In 1963, Belmar, a Chilean physician, first employed niclosamide, the first consistently safe and effective medication for dipylidiasis.[3]

Epidemiology

Dipylidium caninum is a common parasite of dogs and cats worldwide and is one of the most common tapeworms in dogs and cats in the United States.[11] There are no apparent geographic limitations to this infection. Reports of human infection come from Argentina, Australia, Brazil, the Caribbean, Chile, Europe (Austria, Great Britain, Italy, Portugal, and Romania), India, Mexico, the Philippines, South Africa, Sri Lanka, the United States, and Zimbabwe.[5,10,14,19,21,23,24,27,29,32] Nearly one third of the patients are 6 months of age or younger, and over 80% are under 8 years of age, but infection can also appear in adults.[1,4,21]

Infectious Agent

Morphologic Description
Adults

An adult *D. caninum* is a white to pale pink, flat, cyclophyllidean tapeworm measuring 10 to 70 cm (usually 20 to 40 cm) long.[28] The scolex is conical with a transverse width of 250 to 500 µm (Fig 8.1). It has 4 deeply cupped oval suckers and a retractile rostellum armed with 1 to 7 rows of small spines. The rostellum is capable of protruding to 185 µm and can completely invaginate into the scolex.[2] The spines have a short curved arm and a large rounded base. The neck of the tapeworm is short and slender. Young immature proglottids are broader than long, but as they develop they become trapezoidal, with the dorsal width nearly equal to the length of each proglottid (Fig 8.2). Mature proglottids are longer than broad, with 2 genital pores, 1 on each lateral margin (Fig 8.3), and a double set of reproductive organs, including bilobed, lobulate ovaries (Fig 8.4). Gravid proglottids are longer than broad and vary considerably in size, ranging from less than 8 mm long by 1.5 mm wide, to 23 mm long by 8 mm wide (Figs 8.5 and 8.10). The gravid proglottid has a distinctive shape: the ends tend to narrow, giving it the appearance of a cucumber seed or grain of rice (Fig 8.5). Clusters of eggs become encased in uterine tissue and form discrete egg-filled packets. The uterus of a fully gravid proglottid is compartmentalized into packets, with each packet containing clusters of eggs (Figs 8.6 and 8.7).

Eggs

Each *D. caninum* egg packet contains 5 to 20 eggs (usually 8 to 15) (Figs 8.7 and 12.29). Eggs

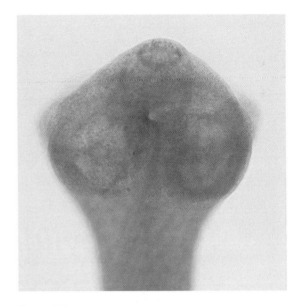

Figure 8.1
Scolex of *D. caninum*. Note conical shape, cupped oval suckers, and armed retractile rostellum. Carmine x300

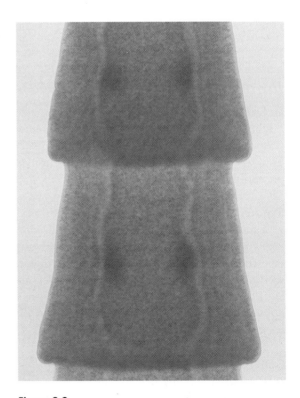

Figure 8.2
Older immature proglottids of *D. caninum* are trapezoidal, with maximum width and length of each proglottid nearly equal. There are no reproductive organs. Carmine x150

are spherical, 25 to 40 µm in diameter, and contain an oncosphere that has 6 delicate hooklets. Four envelopes or membranes surround each egg: a thin inner envelope adjacent to the onchosphere, a thicker shell-like embryophore envelope, a thick outer eosinophilic envelope, and an outermost, basophilic envelope (Fig 8.8). It is not always possible to observe all 4 envelopes. Unstained eggs sometimes have a light red tinge.

Life Cycle and Transmission

Dipylidium caninum requires 2 hosts to complete its life cycle (Fig 8.9).[8] Domestic and feral dogs, other canids, and cats are optimal, but humans can also serve as definitive hosts.[7,15] Fleas and lice are intermediate hosts, especially the dog flea *(C. canis)*, dog louse *(T. canis)*, cat flea *(Ctenocephalides felis)*, and human flea *(P. irritans)*.[9] Male fleas are more often, but less heavily, infected than female fleas.[13]

Gravid proglottids are expelled in feces or emerge spontaneously from the anus of the definitive host (Fig 8.10). After exiting the definitive host, the proglottids disintegrate and release egg packets that larval fleas or lice ingest. Each egg packet contains 6-hooked embryos capable of developing into a cysticercoid in the body cavity (hemocoelom) of the larval insect. Within the hemocoelom, the embryo matures into a procercoid larva and then a cysticercoid metacestode larva, and remains as such until the insect larva metamorphoses into an adult. When the definitive host eats the infected flea or louse, the cysticercoid develops into an adult tapeworm in the small intestine. Maturation to adult tapeworm takes 3 to 4 weeks.

Human infection most often takes place in either of the following ways: 1) a child may ingest an infected flea or louse, 2) a cysticercoid on the tongue of a pet that has recently nipped an infected flea may enter the human mouth by the pet's licking.

Clinical Features and Pathogenesis

Dipylidium caninum is mildly to moderately pathogenic depending on the worm burden. Infection with multiple adult tapeworms is not uncom-

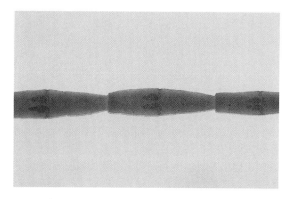

Figure 8.3
Mature proglottids of *D. caninum* are longer than broad and have paired genital pores, 1 on each lateral margin. Carmine x7

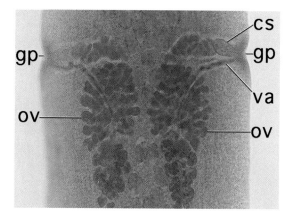

Figure 8.4
Mature proglottid of *D. caninum* demonstrating paired genital pores (gp) and paired reproductive organs, including bilobed, lobulate ovaries (ov), cirrus sac (cs), and vagina (va). Carmine x60

Figure 8.5
Gravid proglottids of *D. caninum* are longer than broad and narrow toward the ends. Uterus is compartmentalized, encasing discrete packets of eggs. Carmine x15

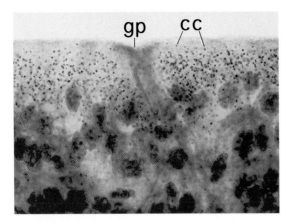

Figure 8.6
Section of gravid proglottid of *D. caninum* demonstrating eggs in packets, genital pore (gp), and calcareous corpuscles (cc). x40

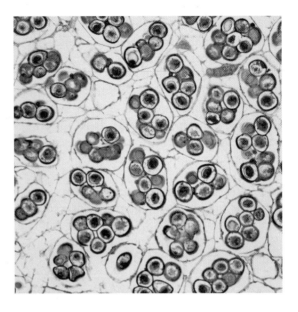

Figure 8.7
Section of gravid proglottid of *D. caninum* showing uterus compartmentalized into packets of eggs. x145

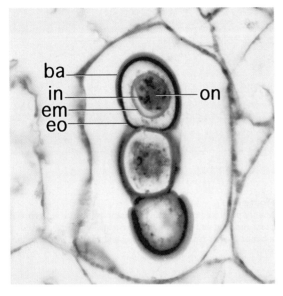

Figure 8.8
Section of gravid proglottid of *D. caninum* depicting 3 eggs in a packet. Note oncosphere (on) and 4 envelopes: inner (in), embryophore (em), eosinophilic (eo), and basophilic (ba). x555

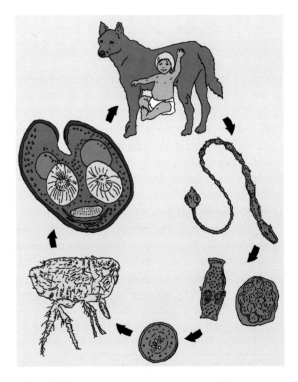

Figure 8.9
Life cycle of *D. caninum*. Eggs or proglottids shed in feces of definitive host are ingested by appropriate larval flea. Humans, usually children, are infected by ingesting cysticercoids in contaminated fleas.

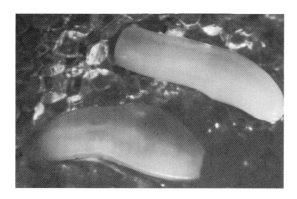

Figure 8.10
Gravid proglottids of *D. caninum* that emerged spontaneously from the anus of a dog. x12

mon, because there are many eggs in each egg packet and each egg can potentially mature to an adult worm. Infected infants may show nocturnal irritability,[32] anorexia and weight loss, or failure to thrive.[31] These symptoms frequently go unnoticed until a parent observes motile gravid proglottids in the child's diaper. Parents often describe the proglottids as moving grains of rice, or maggots.

Patients may complain of abdominal pain, or a baby may appear colicky or have episodic vomiting,[26] all of which may be mistaken for an allergy to milk. Some children have diarrhea, and anal pruritus and rash.[24] Infrequently, patients develop urticaria and/or eosinophilia. Infection may produce changes in the intestinal flora or exacerbate intercurrent illness.

The scarcity of reports of dipylidiasis in adults suggests that previous exposure confers a degree of immunity.

Pathologic Features

Tissues demonstrating pathologic changes are not readily available because the symptoms of dipylidiasis are minor, rarely prompting surgical intervention. Some degree of tissue damage at the point of attachment would be expected, because adult tapeworms inhabiting the small intestine attach to the mucosa by an armed scolex.[22] There is a report of *D. caninum* with its scolex embedded in the intestinal mucosa, making a channel through which proglottids may pass.[30]

Clinicians should submit whole proglottids to the laboratory. Experienced pathologists or parasitologists can usually identify gross *D. caninum* proglottids. Processing and embedding a proglottid as a routine biopsy specimen often makes diagnosis difficult. The morphologic features of proglottids in sections vary depending upon the stage of the proglottid. Most commonly, the specimen is a late gravid proglottid filled with egg packets (Figs 8.11 and 8.12). The body wall of a late gravid proglottid is thin, but consists of tegument, smooth muscle fibers and calcareous corpuscles (Fig 8.12). Sometimes, however, the specimen is an early gravid proglottid that contains only a few eggs (Figs 8.13 to 8.15). Early gravid proglottids have a thick body wall and include well-defined tegument, tegumental cells, smooth muscle fibers, and parenchyma (Fig 8.15). The parenchyma of an early gravid proglottid consists predominantly of uterine ducts, mesenchymal fibers, smooth muscle fibers, and variable numbers of calcareous corpuscles. The multiple envelopes of *D. caninum* eggs are characteristic and permit a definitive diagnosis even when eggs are scarce (Fig 8.14).

Diagnosis

Diagnosing dipylidiasis depends on finding characteristic proglottids (Fig 8.5) or egg packets (Fig 8.7). Egg packets disintegrate rapidly, making it essential to examine fresh stool to find the occasional packet. Microscopic study of unstained proglottids compressed between glass slides reveals the characteristic double genital pores and eggs in packets (Fig 8.16).

Routine examination of feces is unreliable because proglottids do not usually release eggs within the intestine. If dipylidiasis is suspected, but the initial exam fails to yield proglottids or egg packets, it is advisable to give the patient's parent a preserving fluid, such as 70% alcohol, to salvage any suspicious objects found in the patient's stool or diaper, or on the perineum. The adhesive-tape method used for pinworm diagnosis can also reveal proglottids of *D. caninum*.[24]

The clinical presentation of dipylidiasis often resembles pinworm infection.[6,25] The differential

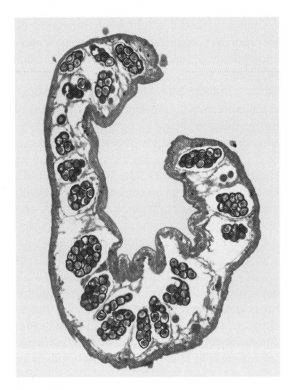

Figure 8.11
Section of late gravid proglottid of *D. caninum* filled with egg packets. x80

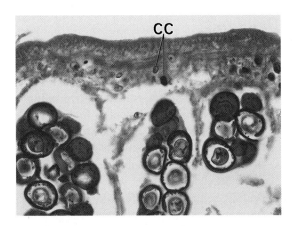

Figure 8.12
Higher magnification of Figure 8.11 illustrating thin body wall, calcareous corpuscles (cc), and several egg packets. x250

clinical diagnosis also includes *Hymenolepis nana* infection, but *H. nana* proglottids are not double-pored and the eggs are not in packets.[29] Eggs of *Inermicapsifer madagascariensis* and *Raillietina* sp are in packets like those of *D. caninum*; however, the proglottids of *I. madagascariensis* and *Raillietina* sp are single-pored (Fig 12.29).

Treatment and Prevention

Treatment with niclosamide (Yomesan® or Nicloside®) is highly effective but can cause nausea and abdominal discomfort.[16,17] Praziquantel (Droncit® or Biltricide®) is the drug of choice for dipylidiasis in animals,[25] and Schenone et al[28] successfully treated an infant with praziquantel, which may prove to be the drug of choice for human dipylidiasis. Quinacrine (Atabrine®) is no longer recommended,[24] and mebendazole is ineffective.[12]

Diagnosis of dipylidiasis in a patient should prompt the examination and treatment of household pets. Preventing infection in humans calls for routine examination of pets for worms, periodic treatments with a taeniafuge, and control of fleas. Kissing canines is discouraged.[26]

Figure 8.13
Section of early gravid proglottid of *D. caninum* with thick body wall, parenchyma, and only a few eggs in uterine ducts. x35

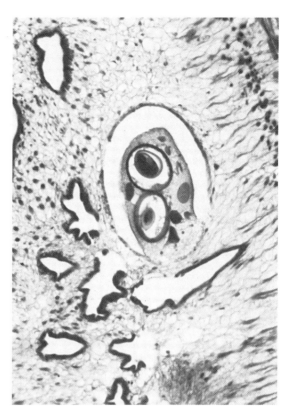

Figure 8.14
Early gravid proglottid of *D. caninum* demonstrating development of egg packets in uterine duct. Note that eggs have multiple envelopes. x1065

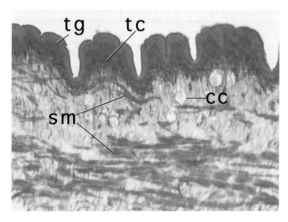

Figure 8.15
Thick body wall of early gravid proglottid demonstrating well-defined tegument (tg), tegumental cells (tc), smooth muscle fibers (sm), and parenchyma containing scattered calcareous corpuscles (cc). x235

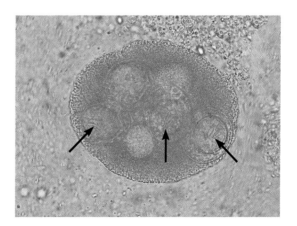

Figure 8.16
Packet of eggs of *D. caninum* from a compressed proglottid. Eggs are spherical and contain an oncosphere with 6 delicate hooklets (arrows). Unstained x305

References

1. Anderson OW. Dipylidium caninum infestation. *Am J Dis Child* 1968;116:328–330.
2. Beaver PC, Jung RC, Cupp EW. *Clinical Parasitology*. 9th ed. Philadelphia, Pa: Lea & Febiger; 1984:508–509.
3. Belmar R. Dipylidium caninum en niños. Comunicación de 13 casos y tratamiento con un derivado de la Salicilamida. *Bol Chil Parasitol* 1963;18:63–67.
4. Blanchard R, Leroux C, Labbé R. Encore un cas de Dipylidium caninum à Paris. *Arch Parasitol* 1913–1914;16:438–448.
5. Brandstetter W, Auer H. Dipylidium caninum, a rare parasite in man [in German]. *Wien Klin Wochenschr* 1994;106:115–116.
6. Chappell CL, Enos JP, Penn HM. Dipylidium caninum, an under-recognized infection in infants and children. *Pediatr Infect Dis J* 1990;9:745–747.
7. Coman BJ, Jones EH, Driesen MA. Helminth parasites and arthropods of feral cats. *Aust Vet J* 1981;57:324–327.
8. Currier RW 2d, Kinzer GM, DeShields E. Dipylidium caninum infection in a 14-month-old child. *South Med J* 1973;66:1060–1062.
9. Engbaek K, Madsen H, Larsen SO. A survey of helminths in stray cats from Copenhagen with ecological aspects. *Z Parasitenkd* 1984;70:87–94.
10. Gadre DV, Kumar A, Mathur M. Infection by Dipylidium caninum through pet cats [letter]. *Indian J Pediatr* 1993;60:151–152.
11. Georgi JR. Tapeworms. *Vet Clin North Am Small Anim Pract* 1987;17:1285–1305.
12. Guerrero J, Pancari G, Michael B. Comparative anthelmintic efficacy of two schedules of mebendazole treatment in dogs. *Am J Vet Res* 1981;42:425–427.
13. Hinaidy HK. The biology of Dipylidium caninum. Part 2 [in German]. *Zentralbl Veterinarmed [B]* 1991;38:329–336.
14. Jackson D, Crozier WJ, Andersen SE, Giles W, Bowen TE. Dipylidiasis in a 57-year-old woman. *Med J Aust* 1977;2:740–741.
15. Jones A, Walters TM. The cestodes of foxhounds and foxes in Powys, mid-Wales. *Ann Trop Med Parasitol* 1992;86:143–150.
16. Jones WE. Niclosamide as a treatment for Hymenolepis diminuta and Dipylidium caninum infection in man. *Am J Trop Med Hyg* 1979;28:300–302.
17. Katz M. Anthelmintics. *Drugs* 1977;13:124–136.
18. Leuckart R. *The Parasites of Man and the Diseases Which Proceed from Them*. Hoyle, trans. Edinburgh, Scotland: Young J Pentland; 1886.
19. Link A, Cassorla E. Dipylidium caninum in an infant [in Spanish]. *Rev Chil Pediatr* 1966;37:33–34.
20. Linne C (Linnaeus). *Systema naturae per regna tria naturae, secundum classes, ordines, genera, species cum characteribus differentiis, synonymis, locis*. 10th ed. Laurentii Salvii: Holmiae; 1758;1:823.
21. Marinho RP, Neves DP. Dipylidium caninum (Dilepididae-Cestoda). Report of 2 human cases [in Portuguese]. *Rev Inst Med Trop Sao Paulo* 1979;21:266–268.
22. Neafie RC, Marty AM. Unusual infections in humans. *Clin Microbiol Rev* 1993;6:34–56.
23. Nitzulescu V, Feldioreanu T, Purcherea A. A further case of human dipylidiasis with Dipylidium caninum [in Romanian]. *Pediatria (Bucur)* 1966;15:357–358.
24. Oberle MW, Knight WB, Hernandez L. Dipylidium caninum in Puerto Rico: report of a human case. *Bol Asoc Med P R* 1979;71:258–260.
25. Raitiere CR. Dog tapeworm (Dipylidium caninum) infestation in a 6-month-old infant. *J Fam Pract* 1992;34:101–102.
26. Reddy SB. Infestation of a 5-month-old infant with Dipylidium caninum. *Del Med J* 1982;54:455–456.
27. Reid CJ, Perry FM, Evans N. Dipylidium caninum in an infant. *Eur J Pediatr* 1992;151:502–503.
28. Schenone H, Thompson L, Quero MS. Infection by Dipylidium caninum in a young girl treated with praziquantel [in Spanish]. *Bol Chil Parasitol* 1987;42:74–75.
29. Shane SM, Adams RC, Miller JE, Smith RE, Thompson AK. A case of Dipylidium caninum in Baton Rouge, Louisiana. *Int J Zoonoses* 1986;13:59–62.
30. Stiles CW. A case of infection with the double-pored dog tapeworm (Dipylidium caninum) in an American child. *Am Med* 1903;5:65–66.
31. Turner JA. Human dipylidiasis (dog tapeworm infection) in the United States. *J Pediatr* 1962;61:763–768.
32. Wijesundera MD, Ranaweera RL. Case reports of Dipylidium caninum: a pet-associated infection. *Ceylon Med J* 1989;34:27–30.

Hydatidosis
(Echinococcosis)

Aileen M. Marty,
Linda K. Johnson, *and*
Ronald C. Neafie

Introduction
Definition
Hydatidosis is infection by larval tapeworms of the genus *Echinococcus*. In endemic areas, infections are easily acquired and often fatal. At least 4 species cause infection in humans: *Echinococcus granulosus* (Batsch, 1786) Rudolphi, 1805 (cystic hydatid disease); *Echinococcus multilocularis* (Leuckart, 1863) Vogel, 1955 (alveolar hydatid disease); *Echinococcus vogeli* Rausch and Bernstein, 1972 (polycystic hydatid disease); and *Echinococcus oligarthrus* (Diesing, 1863) Luke, 1910 (polycystic hydatid disease).

Synonyms
Other names for hydatidosis are hydatid disease and echinococcosis. In their strictest sense, hydatid disease and hydatidosis refer to infection in humans by metacestodes, and echinococcosis refers to infection in nonhuman carnivores by the adult tapeworm. The term echinococcosis is frequently misapplied to infection in humans by metacestodes; however, no attempt will be made to correct this terminology. The common name for *E. granulosus* is hydatid tapeworm. *Taenia echinococcus* was formerly a synonym for various species of *Echinococcus*.

General Considerations
Ancient writers spoke of hydatid cysts in both animals and humans, but their parasitic origin was unknown. The Talmud mentions cystic lesions in the viscera of sacrificial animals, and Hippocrates[29] alluded to water-filled tumors in the lungs of cattle, sheep, and pigs, and in human livers. Galen[16] noted that the liver is a common site for hydatid cysts in slaughtered animals. In 1684, Redi[56] deduced from their spontaneous movements that the cystic swellings in humans and animals were animals themselves. In 1695, Hartmannus[25] first studied adult worms in the dog intestine, but the relationship between adult and larval forms was not recognized. Morgagni,[45] in 1760, suggested there were numerous kinds of bladderworms, but confused the various cysticerci with echinococcal cysts. In 1782, Goeze[22] reported living cysts containing scolices resembling heads of tapeworms in a mutilated sheep liver. Küchenmeister,[38] in 1851, fed verminous cysts to carnivores and observed that

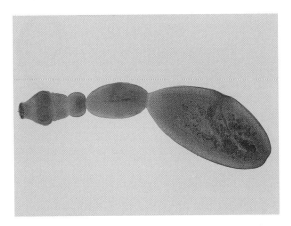

Figure 9.1
Adult *E. granulosus* is a minute tapeworm that has a scolex with an armed rostellum at the anterior end, followed by 4 suckers and a neck. Behind the neck are 3 proglottids: the first is immature, wide and short; the middle is mature and elongated; the last is gravid and long.*

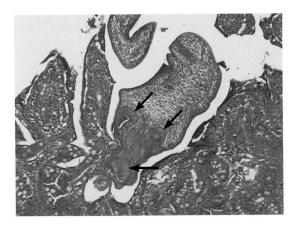

Figure 9.2
Adult *E. granulosus* attached to intestinal mucosa of a lion. Note scolex with suckers (arrows). x110

they were probably larval tapeworms. Von Siebold,[67] in 1852, fed the hydatid sand from hydatid cysts of herbivores to domestic dogs and observed echinococcal tapeworms in the intestines of these dogs. Several other investigators successfully repeated Von Siebold's experiments using hydatid cysts from humans to produce the adult tapeworm in dogs.

Many investigators, including Virchow in 1855, recognized but could not explain the differences in the echinococcal cysts. In 1863, Leuckart designated the invasive variant as *Taenia echinococcus multilocularis*. Morin,[46] in 1875, proposed that there were 2 distinct species, and this was substantiated by Vogel[66] in 1955. In 1972, Rausch and Bernstein[53] examined proglottids from a bush dog (*Speothus venaticus*) from Ecuador in the Los Angeles Zoo, and established a new species of *Echinococcus*, *E. vogeli*. D'Alessandro et al examined Colombian patients with polycystic hydatid disease and fed hydatid sand from these patients to dogs. The dogs developed adult *E. vogeli*, establishing the larval stage of *E. vogeli* as a polycystic hydatid.[15]

In 1906, Ghedini[19] developed a complement fixation test, and in 1911, Cassoni devised a skin test for hydatidosis.[11] Heath and Chevis,[26] in 1974, reported that mebendazole was effective against daughter cysts of *E. granulosus* in mice, and in 1975, together with Christie,[27] they demonstrated that mebendazole had antihydatid activity.

Epidemiology

Echinococcosis is widely distributed and has been a serious problem in sheep- and cattle-raising regions, especially Australia, Iceland, Peru, and New Zealand; however, eradication campaigns have greatly reduced incidence in many of these areas. Globally, transmission to humans appears to be rising, but improved diagnostic methods and surveillance have increased awareness of hydatid disease.[33] Infections are more common in women.[6,36]

Dogs are the most common definitive hosts of *E. granulosus*, although many other animals are found infected in nature, including wolves, cats, jaguars, arctic foxes, pumas, jackals, dingoes, and hyenas. Endemic areas of human infection by *E. granulosus* larvae include: in Eurasia: France, Germany, Great Britain, Greece, India, Iran, Iraq, Italy, Lebanon, Norway, Russia, Spain, and Turkey; most countries of Africa, including Madagascar; in Oceania: Australia and New Zealand; in North America: Canada and the United States, particularly Alaska, Arizona, California, Minnesota, New Mexico,

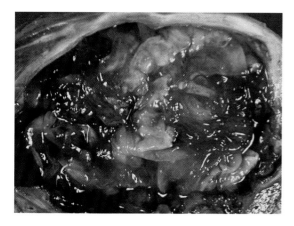

Figure 9.3
Unilocular metacestode of *E. granulosus* in liver. Laminated wall of cyst has collapsed and is infolded. x0.8

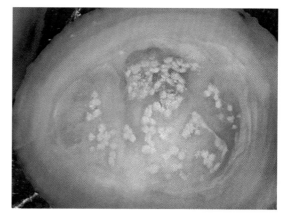

Figure 9.4
Cut section of cyst showing numerous brood capsules attached to inner germinal membrane of metacestode of *E. granulosus* from pelvic area of 24-year-old Greek man living in New Jersey. A thick fibrous capsule of host origin surrounds the cyst. Unstained

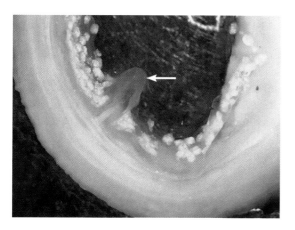

Figure 9.5
Cut section from cyst in Figure 9.4, demonstrating brood capsules attached to germinal membrane, laminated membrane, and fibrous capsule. A small piece of laminated membrane (arrow) projects into the cyst cavity. Unstained

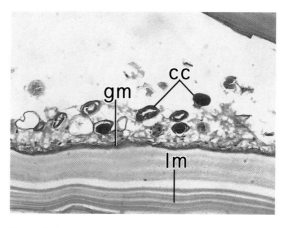

Figure 9.6
Section of metacestode of *E. granulosus* showing the 2 layers of cyst wall: inner germinal membrane (gm) with calcareous corpuscles (cc), and outer laminated membrane (lm). x600

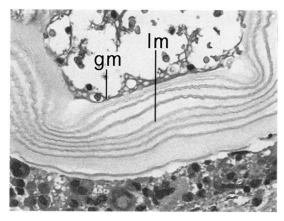

Figure 9.7
Metacestode of *E. granulosus* in spleen. Note thin germinal membrane (gm) and thick outer laminated membrane (lm). x380

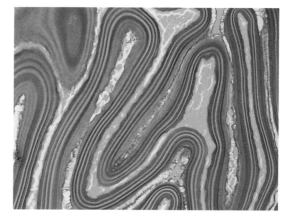

Figure 9.8
Metacestode of *E. granulosus* showing accentuation of laminations in wall of cyst. GMS x165

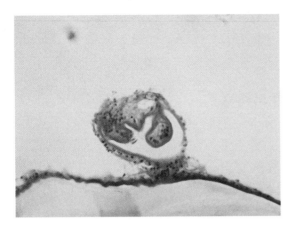

Figure 9.9
Primary cyst metacestode of *E. granulosus* with protoscolex arising from germinal membrane. x265

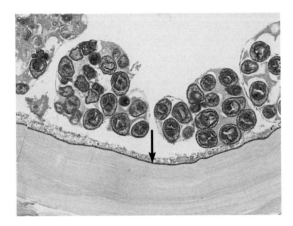

Figure 9.10
Echinococcus granulosus metacestode in liver demonstrating several brood capsules arising from thin germinal membrane (arrow) overlying laminated wall. x60

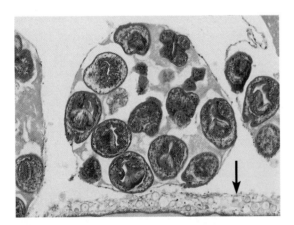

Figure 9.11
Brood capsule of *E. granulosus* attached to germinal membrane (arrow). There are at least a dozen protoscolices within this brood capsule. x120

North and South Dakota, and Utah[51]; and in South America: Argentina, Chile, and Peru. There are 2 major strains. The northern strain has a Holarctic distribution, and the European strain is cosmopolitan.[73]

Echinococcus multilocularis, the second most common echinococcal larval tapeworm infecting humans, is focally distributed throughout much of the world. In the United States, Alaska, Illinois, Indiana, Minnesota, Montana, North Dakota, and Ohio are known endemic areas.[17,40,62,74] Other countries with *E. multilocularis* infections are Argentina, Australia, Austria, Canada, France, Germany, Italy, Japan, New Zealand, Russia (Siberia), Switzerland, and Uruguay.[9,60] The animal cycle typically involves foxes or dogs and arvicoline rodents, but cats and house mice in North America may also serve as hosts.[40]

Echinococcus vogeli is distributed widely in Central and South America, bush dogs being the natural definitive hosts. Domestic dogs are also definitive hosts and are probably the main source of larval infection in humans. There are *E. vogeli* infections in humans in rural areas of Argentina, Bolivia, Brazil, Colombia, Costa Rica, Ecuador, Panama, and Venezuela.[15,18,44,53]

Felids are the natural definitive hosts for *E. oligarthrus*, while rodents serve as intermediate hosts. The geographic range of *E. oligarthrus* is Central America (Costa Rica and Panama) and South America (Argentina, Brazil, and Venezuela). There are 2 reports of human infections: 1 in Venezuela and 1 in Brazil.[14,42]

Intraspecies variants exist, but their taxonomic status remains uncertain, and some authorities consider many subspecies as taxonomically invalid. Variants of *Echinococcus* are perhaps best classified as strains.[39] Characterization of strains is based on their distinct morphology, rate of maturation, and DNA hybridization profiles.

Infectious Agent

Morphologic Description
Adult Tapeworms

The anterior end of adult *Echinococcus* sp has a piriform scolex containing 4 cup-like suckers that lie behind a ring (rostellum) composed of an inner row of small hooklets and an outer row of larger hooklets. The scolex is followed by a short neck and a strobila of 3 or 4 proglottids (Figs 9.1 and 9.2). The testes and single uterus, ovary, ootype, seminal receptacle, vagina, and genital pore are in the centrally located, mature proglottid. As in all cestodes, *Echinococcus* has no digestive tract; all metabolic interchange takes place across the tegument.[64] The size of rostellar hooklets in both adult and larval stages is a distinguishing characteristic of *Echinococcus* sp (Tables 9.1 and 9.2).

The strobila of *E. granulosus* is 2 to 11 mm long (Table 9.1 and Fig 9.1). The scolex has 4 suckers and an armed rostellum with 2 rows of 28 to 50 hooklets (usually 30 to 36). Behind the narrow piriform neck there is usually 1 immature proglottid, 1 mature proglottid, and 1 gravid proglottid. The mature proglottid contains 25 to 80 (usually 32 to 68) testes, a uterus that usually has lateral outpockets, and a genital pore that opens near (most often posterior to) the middle. The gravid proglottid comprises about half the length of the adult *E. granulosus*.

The strobila of *E. multilocularis* is 1.2 to 4.5 mm long (Table 9.1). The mature proglottid contains 16 to 35 (usually 18 to 26) testes, a uterus that lacks lateral outpockets, and a genital pore that opens anterior to the middle of the proglottid.

The strobila of *E. vogeli* is 3.9 to 5.5 mm long (Table 9.1). The scolex is crowned by a rostellum armed with 28 to 36 hooklets in 2 rows. Mature proglottids have 50 to 67 (average 56) testes, a long tubular uterus that lacks outpockets, and a slightly lobulated ovary. The genital pore is lateral and opens slightly posterior to the middle of the proglottid.

The strobila of *E. oligarthrus* is 2.2 to 2.9 mm long (Table 9.1). Mature proglottids contain 15 to 46 (average 29) testes, and a uterus that lacks lateral outpockets. The genital pore opens anterior to the middle of the proglottid.

The Metacestode (Larva)
Echinococcus granulosus

The metacestode of *E. granulosus* is a white, spherical, fluid-filled cyst that varies from a few millimeters to many centimeters in diameter (Fig 9.3). A fertile (mature) cyst contains brood capsules with protoscolices (Figs 9.4 and 9.5). The cyst wall is composed of 2 layers (Figs 9.6 and 9.7). There is a thin (10 to 25 μm) inner layer

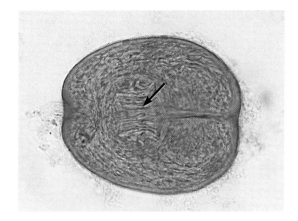

Figure 9.12
Invaginated protoscolex of *E. granulosus* from hydatid sand in lung. Hooklets (arrow) and 2 of 4 suckers are visible. Unstained x400

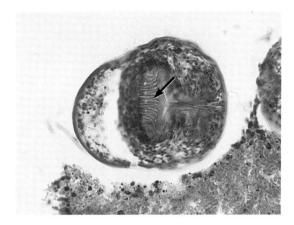

Figure 9.13
Section of ruptured hydatid cyst in lung showing invaginated protoscolex of *E. granulosus*. Two of 4 suckers, numerous calcareous corpuscles, and hooklets (arrow) are visible. x310

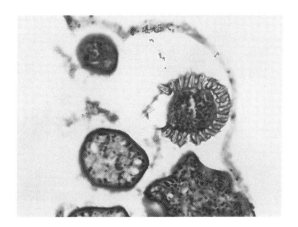

Figure 9.14
Brood capsule of *E. granulosus* demonstrating protoscolex with 2 rows of hooklets. x345

Figure 9.15
Echinococcus granulosus cyst of orbit containing a large (30 μm) loose hooklet. x750

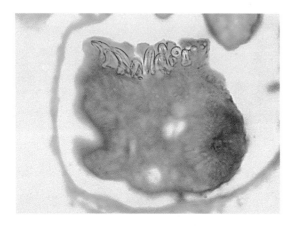

Figure 9.16
Rostellum of *E. granulosus* protoscolex demonstrating highlighting of hooklets by GMS. x600

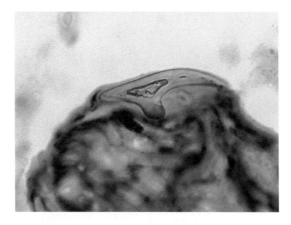

Figure 9.17
Protoscolex of *E. granulosus* with large hooklet. GMS x1200

(germinal membrane) with nuclei and occasionally calcareous corpuscles (Figs 9.6 and 9.7). The outer layer (laminated membrane) varies in thickness but is approximately 1 mm thick (Figs 9.6 to 9.8). The germinal membrane probably secretes the laminated membrane, which stains well with H&E, but the layering is more distinct with the Gomori methenamine-silver (GMS) stain (Fig 9.8).[43] Brood capsules, and sometimes individual protoscolices, arise from the germinal layer (Fig 9.9). A mature cyst may contain thousands of brood capsules (Fig 9.4). Brood capsules measure up to 700 μm in diameter and may contain more than a dozen protoscolices (Figs 9.10 and 9.11). Protoscolices are ovoid and approximately 100 μm across, and each contains 4 cup-like suckers (up to 50 μm across), numerous calcareous corpuscles, a row of large hooklets, and a row of small hooklets (Figs 9.12 to 9.14). Hooklets of metacestodes are smaller than those of adult tapeworms. Reported lengths of hooklets in metacestodes vary. The large hooklets range from 19.4 to 44 μm and the small hooklets from 17 to 31 μm (see Table 9.2). In our experience, loose hooklets of *E. granulosus* in paraffin-embedded tissues measure between 16 and 31 μm (Fig 9.15). Hooklets are highlighted by the GMS stain (Figs 9.16 and 9.17), are birefringent (Fig 9.18), and focally acid-fast by the Ziehl-Neelsen (ZN) technique (Figs 9.19 to 9.21). Most protoscolices are invaginated but some evaginate (Fig 9.22). Brood capsules may rupture, releasing protoscolices and hooklets into the fluid of the cyst. Formation of daughter cysts (Figs 9.23 and 9.24) is poorly understood, though free protoscolices injected into appropriate hosts develop into daughter cysts. Daughter cysts are morphologically identical to the parent cyst, having both laminated and germinal membranes, and may develop brood capsules (Fig 9.25). Daughter cysts may exit a damaged parent cyst, generating exogenous daughter cysts. Hydatid sand is the collective term for all of the particulate matter within a hydatid cyst. It may include brood capsules, free protoscolices, hooklets, calcareous corpuscles, and daughter cysts. Some *E. granulosus* cysts in humans are sterile; although they have a germinal membrane, they do not form brood capsules, daughter cysts, or protoscolices.

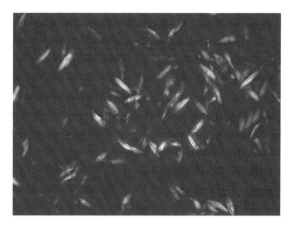

Figure 9.18
Necrotic cyst of *E. granulosus* in orbit of eye, containing numerous birefringent hooklets. x245

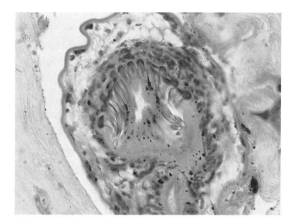

Figure 9.19
Armed rostellum of invaginated protoscolex of *E. granulosus*, demonstrating focally acid-fast hooklets. ZN x600

Echinococcus multilocularis

Oncospheres infecting abnormal hosts commonly develop into sterile cysts. Typically, the metacestode of *E. multilocularis* in humans is a sterile mass of proliferative tissue of the cestode cyst wall (Figs 9.26 to 9.28). Protoscolices are usually absent, but when present their large and small rostellar hooklets have average lengths of 27 and 24 μm respectively (Figs 9.29 and 9.30).

Grossly, these alveolar hydatids are amorphous and solid rather than cystic. Although the cut surfaces reveal irregular spaces and collapsed membranes, they contain no fluid (Figs 9.31 and 9.32). Though usually sterile, the metacestode is nonetheless alive. If, for example, an alveolar hydatid cyst of *E. multilocularis* from a human is experimentally fed to a more suitable intermediate host, it will become fertile. In normal hosts, protoscolices and calcareous corpuscles are plentiful (Figs 9.33 and 9.34). Alveolar hydatid cysts form multiple exogenous daughter cysts that invade surrounding host tissue. They may be focally or diffusely necrotic and mineralized. Some invade blood vessels and produce metastatic lesions.

Echinococcus vogeli

The polycystic hydatid of *E. vogeli* in its natural intermediate host, the paca, proliferates only by endogenous budding of the germinal membrane forming folds and pockets within the primary cyst. In humans, these cysts not only grow by endogenous budding, but by exogenous proliferation that produces an invasive growth pattern.[54] Polycystic hydatids are characterized by internal division of fluid-filled cysts to form multichambered growths. Aggregates of cysts can be 10 mm to several centimeters in diameter. Grossly, they resemble irregular clusters of variously sized grapes. Cysts are translucent, revealing brood capsules containing infective protoscolices. The cyst has a laminated membrane varying from 8 to 65 μm thick in the same vesicle (Fig 9.35). Germinal membrane is uniformly thin (3 to 13 μm) and has calcareous corpuscles (Fig 9.35). Brood capsules have numerous protoscolices (up to 480) (Fig 9.36).[54] The large and small hooklets on the rostellum of *E. vogeli* have average lengths of 41 and 33 μm respectively (Fig 9.37).

Echinococcus oligarthrus

The metacestode of *E. oligarthrus* is polycystic and has an internal division similar to that of *E. vogeli*. The laminated membrane of the metacestode larva *E. oligarthrus* is thin and covers a thick germinal membrane; the former is 5 to 33 μm thick (average 17.8 μm), and the latter 10.5 to 105 μm thick (average 44 μm).[14] The germinal membrane contains numerous calcareous corpuscles. Brood capsules are small and contain 7 to 15 (usually 13) small protoscolices per capsule. The large and small hooklets on the rostellum of *E. oligarthrus* are 30.5 to 33.4 μm long (average 32 μm), and 22.6 to 29.2 μm long (average 26.4 μm) respectively.

Morphology	E. granulosus	E. multilocularis	E. vogeli	E. oligarthrus
Large hooklets	Mean length: 32–42 µm Range: 25–49 µm	Mean length: 31 µm Range: 24.9–34 µm	Mean length: 53 µm Range: 49–57 µm	Mean length: 52 µm Range: 43–60 µm
Small hooklets	Mean length: 22.6–27.8 µm Range: 17–31 µm	Mean length: 27 µm Range: 20.4–31 µm	Mean length: 42.6 µm Range: 30–47 µm	Mean length: 39 µm Range: 28–45 µm
Segments	Mean: 3 Range: 2–7	Mean: 5 Range: 2–6	Mean: 3	Mean: 3
Strobila length	2–11 mm	1.2–4.5 mm	3.9–5.5 mm	2.2–2.9 mm

Table 9.1
Morphologic features of adult *Echinococcus* sp.[56,64]

Life Cycle and Transmission

All *Echinococcus* sp are minute tapeworms with an indirect life cycle requiring 2 hosts. Adults are hermaphroditic and larvae proliferate asexually (Fig 9.38). Canidae (bush dog, dingo, dog, fox, hyena, jackal, and wolf) and Felidae (cat and lion) families are hosts for adult worms. Intermediate hosts are usually sheep or pigs, but also include antelope, bison, deer, elk, goats, kangaroos, moose, pacas, wallabies, and humans. The larval form in intermediate hosts is a metacestode.

Gravid proglottids passed in feces of definitive hosts rupture and release eggs. Eggs may contaminate fingers, water, fruits, or vegetables and be eaten by humans or other intermediate hosts. After passing through the stomach and into the small intestine, each egg releases an oncosphere (hexacanth embryo).

The oncosphere penetrates the intestinal mucosa, including the lamina propria, where it causes necrosis, loses its 6 hooklets, and enters a vascular channel. At this stage the parasite is approximately 25 µm in diameter, making it susceptible to entrapment in the microcirculation of the liver. If the oncosphere passes through the liver and reaches the lung, it will often lodge in pulmonary tissue, but may continue to the left side of the heart and be carried to other foci.

Once enmeshed in host tissue, the oncosphere develops into a metacestode (Fig 9.3). Development of the metacestode in a suitable host is similar in all 4 species, but in humans the cysts of *E. multilocularis*, unlike those of the other species, do not usually form fertile brood capsules. The cyst membrane of *E. multilocularis* propagates extensively both into the cyst and the surrounding tissue, creating a proliferative mass resembling a clump of grapes. Growth rate of a metacestode varies with the tissue site where the cyst localizes, species of *Echinococcus*, and host. Growth is particularly rapid in well-vascularized organs or tissues where a fibrous capsule does not form. Echinococcal cysts have a thin inner germinal membrane that may be nonfertile or studded by budding brood capsules and protoscolices. Most protoscolices develop from the germinal layer within each brood capsule. Many brood capsules detach from the germinal membrane and float in the cyst fluid. Some brood capsules disintegrate and liberate protoscolices. Primary cysts of *E. granulosus* have survived for up to 53 years in humans.

Daughter cysts may form within the primary cyst. Daughter cysts are structurally identical to the primary cyst; each has its own laminated layer and is capable of producing protoscolices and brood capsules from its germinal layer. The origin of daughter cysts remains speculative, but protoscolices released from the germinal membrane or from ruptured brood capsules can produce daughter cysts. Daughter cysts that escape from the primary hydatid cyst can implant and grow exogenously.

Fertile cysts may contain thousands of protoscolices. Each protoscolex is infective for the definitive host and can develop into an adult tapeworm; thus a suitable carnivore that consumes infected flesh may develop a massive intestinal infection. In the small intestine of the definitive host the protoscolex evaginates, attaches to the mucosa (Fig 9.2), and develops into an adult tapeworm within 7 weeks. Adult *E. granulosus* can survive up to 2.5 years in canines. Despite massive intestinal infections by adult tapeworms, dogs are not usually symptomatic.

Morphology	E. granulosus	E. multilocularis	E. vogeli	E. oligarthrus
Large hooklets	Mean length: 25.9–35 µm Range: 19.4–44 µm	Mean length: 26.7–28.5 µm Range: 25–29.7 µm	Mean length: 39.3–41.6 µm Range: 38.2–45.6 µm	Mean length: 30.5–33.4 µm Range: 29.1–37.9 µm
Small hooklets	Mean length: 22.6–27.8 µm Range: 17–31 µm	Mean length: 23.1–25.4 µm Range: 21.8–27 µm	Mean length: 32.5–34 µm Range: 30.4–36.9 µm	Mean length: 25.4–27.3 µm Range: 22.6–29.2 µm
Nature of cyst	Unilocular	Multivesicular	Polycystic	Polycystic
Growth of cyst	Endogenous	Endogenous and exogenous	Definitive host: endogenous Humans: endogenous and exogenous	Endogenous
Laminated (LM) and germinal (GM) membrane	LM: size varies, generally 1 mm thick GM: 10–25 µm thick, has nuclei and occasional calcareous corpuscles	LM: thin	LM: thick, 8–65 µm GM: delicate and thin, 3–13 µm	LM: thin and covers a thick germinal membrane bearing numerous calcareous corpuscles

Table 9.2
Morphologic features of larval stage (metacestode) of *Echinococcus* sp.[56,64]

The life cycle of all echinococcal species is similar, but the definitive and intermediate hosts vary. The natural cycle of the northern strain of *E. granulosus* involves wolves (*Rangier* sp) and large deer (*Alces* sp), and that of the European strain mainly domestic dogs and domestic ungulates. Definitive hosts for *E. multilocularis* are dogs, foxes, and cats, and intermediate hosts are usually rodents. In the natural cycle of infection of *E. vogeli*, bush dogs and domestic dogs serve as definitive hosts and the paca (*Cuniculus paca*) as the intermediate host. Members of the cat family, including pumas (*Felis concolor*) and jaguarundi (*Felis yagouaroundi*), are definitive hosts for *E. oligarthrus*, and rodents serve as intermediate hosts. There are 2 accepted and several disputed reports of *E. oligarthrus* infections in humans from South America and Panama. After *E. vogeli* was recognized, several infections mistakenly classified as *E. oligarthrus* proved to be *E. vogeli* infections.

There is a report of a calcified cyst of *E. granulosus* in a transplanted liver.[4]

Clinical Features and Pathogenesis

Patients with cystic hydatid disease have 1 or more unilocular, slow-growing cysts in tissues. Generally, an intact *E. granulosus* cyst is life-threatening only if it grows within or compresses a vital organ. Metacestodes lodge most commonly in the liver (Fig 9.39) or lung, but can localize to any tissue (bone marrow, brain, endocrine organs, eye, heart, kidney, muscle, pelvic organs, spleen, or subcutaneous tissue). Hydatids grow most rapidly in compressible tissues such as brain or lung, and slowest in bone. Cysts grow about 1 mm in diameter a month in the liver, but clinical symptoms do not appear until cysts reach about 10 cm in diameter. Cysts are not usually palpable until they approach 20 cm in diameter. A cyst in the liver may produce no clinical symptoms for 65 years or longer,[70] and most asymptomatic cysts are incidental radiographic findings. Patients may complain of ill-defined hepatic or biliary disturbances, or a noticeable mass. Enlarging cysts may cause jaundice or portal hypertension from compression of parenchyma or the porta hepatis. Complications include cholangitis and rupture. Rupture into the gallbladder may cause acalculous cholecystitis. Ruptured cysts in the peritoneal, pericardial, or pleural space may seed adjacent areas causing new cysts, usually with severe complications.

Pulmonary cysts may be silent for years and detected only by routine radiographs (Figs 9.40 to 9.42). Symptomatic cysts in lungs may cause coughing, hemoptysis, dyspnea, and chest pain (Fig 9.43). When a cyst ruptures into a bronchus,

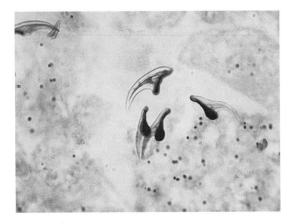

Figure 9.20
Disintegrated metacestode of *E. granulosus* in liver, containing 4 focally acid-fast hooklets lying free in necrotic debris. ZN x685

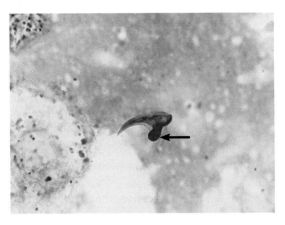

Figure 9.21
Necrotic metacestode of *E. granulosus* showing a single 27-μm-long, partially acid-fast hooklet. The guard (arrow) of this hooklet is especially large and stains deeply acid-fast. ZN x280

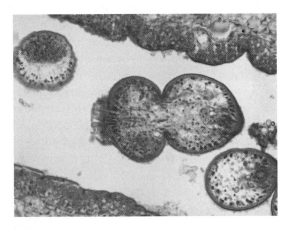

Figure 9.22
Evaginated protoscolex of *E. granulosus* in brood capsule. x235

the patient coughs up portions of the cyst wall and contents. Suppuration in the lung persists for months. Rupture into the pleural space causes pneumothorax, empyema, and development of secondary cysts.[48,68]

Neuroechinococcosis affects children most frequently, usually causing symptoms of increased intracranial pressure. When adults develop cysts in the brain, the presentation is often of focal neurologic symptoms (hemiparesis, jacksonian epilepsy, speech disorders, hemianopia, meningeal irritation).[13] Eye involvement (Fig 9.44) is typically orbital. Patients have proptosis, and can develop exposure keratitis, phthisis bulbi, optic neuritis, and optic atrophy.

A hydatid in the heart most often involves the left ventricle. Symptoms include precordial pain, dyspnea, palpitations, congestive heart failure or sudden death. Depending on the location, rupture of a cyst in the heart can produce pulmonary[2] or abdominal aortic embolism.[58] Serologic examination of patients with myocardial echinococcosis may be negative.[28]

Involvement of bone is infrequent (2% to 3% of all cases). The most common site of bone lesions is the lower vertebrae, often causing low back pain (Fig 9.45). This condition is called hydatid Pott's disease and can lead to paraplegia. Bone involvement can produce erosion of bone and pathologic fractures.[23]

Cysts in the spleen and kidneys may cause pain and, on rupture, peritonitis. Renal hydatid cysts compress parenchyma. Patients present clinically with a flank mass (84% of cases), pain (73% of cases), and sometimes hydatiduria (29% of cases), often accompanied by hematuria and albuminuria.[5,69] Retrovesical cysts may produce cystovesicular fistulas resulting in hematuria and hydatiduria.[49] Uncommonly, a retroperitoneal or hepatic hydatid can form a cystogastric or cystointestinal fistula, causing abdominal pain, hydatidemesis or hydatidenteria.[50,63] A hydatid cyst in the fallopian tube can contribute to infertility. Pregnant patients with hydatid disease present diagnostic and therapeutic dilemmas.[32]

Immunologic response to the initial growth of the cyst varies with the host and can be mild, moderate, or strong. If strong, the host may destroy the cyst; if moderate or mild, the host forms a protective fibrous capsule.

Release of cyst contents can initiate systemic allergic reactions. Uncommonly, some patients have anaphylaxis to hydatid disease without obvious rupture of the cyst.[8] Allergic symptoms from leakage of cysts include itching, pain, vomiting, urticaria, edema, diarrhea, colicky pain, and in pulmonary infection can include dyspnea and other asthmatic symptoms. Sudden rupture of hydatid cysts can cause anaphylactic shock and death.[37] Secondary bacterial infection can produce complications and present clinical symptoms.[12]

Echinococcus multilocularis causes a slowly progressive but often fatal alveolar hydatid disease (Fig 9.46), one of the most routinely fatal helminthic infections of humans.[3] A patient who acquires this metacestode in childhood, however, may live 30 to 40 years before dying from the infection.

Echinococcus vogeli and *E. oligarthrus* cause polycystic hydatid disease. The clinical characteristics of these infections are intermediate between those of cystic and alveolar hydatid disease, but polycystic hydatid disease progresses more rapidly than either cystic or alveolar hydatid disease.

Pathologic Features

In early unilocular cystic disease of *E. granulosus* infection, the tissues around the oncosphere develop an initial inflammatory response that slowly subsides (Fig 9.47). Eventually a wall of fibrous host tissue, up to 3 mm thick, surrounds these cysts (Fig 9.48). Subsequent changes result from compression of adjacent tissues. Collapsed and degenerating cysts of *E. granulosus* that release daughter cysts into adjacent tissues may resemble the alveolar cyst of *E. multilocularis*. Rincon et al associate hydatids in the kidney with mesangiocapillary glomerulonephritis.[57]

When larvae die and rupture, fragments of the laminated membrane and other debris of the parasite may initiate severe granulomatous reactions (Figs 9.49 and 9.50). Host reaction can destroy the parasite, leaving only fragments in necrotic debris (Fig 9.51). In such lesions, a GMS stain will often demonstrate remnants of laminated membrane (Fig 9.8)[43] and hooklets (Figs 9.16 and 9.17). Hooklets are also birefringent (Fig 9.18) and par-

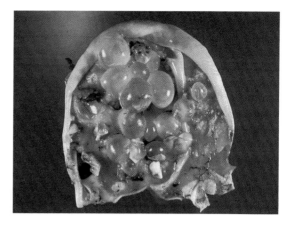

Figure 9.23
Multiple daughter cysts within a primary cyst of *E. granulosus*. x0.4

Figure 9.24
Multiple daughter cysts extracted from a primary cyst of *E. granulosus*. x0.95

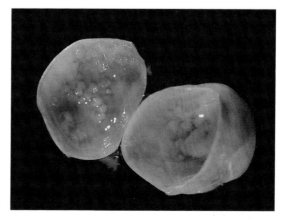

Figure 9.25
Daughter cysts of *E. granulosus* opened to reveal numerous brood capsules on inner surface of germinal membrane.

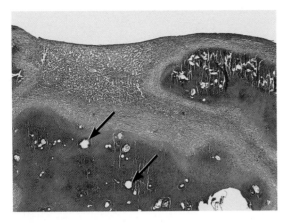

Figure 9.26
Cyst of *E. multilocularis* infection in liver, demonstrating multiple infiltrating portions (arrows) of sterile cyst wall. x17

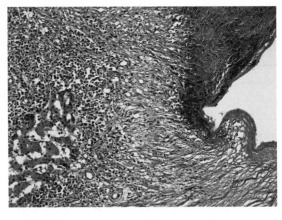

Figure 9.27
Infection of *E. multilocularis* showing a portion of sterile cyst wall infiltrating the liver. Note extensive chronic inflammation and fibrosis around the parasite. x105

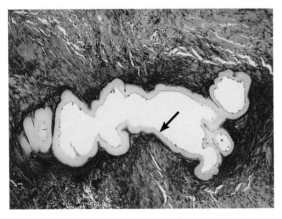

Figure 9.28
Cyst of *E. multilocularis* in liver. Stain shows sterile membrane (arrow) and collagenous tissue surrounding worm. Movat x120

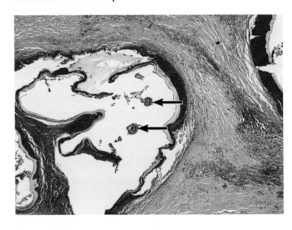

Figure 9.29
Fertile cyst of *E. multilocularis* in left lobe of liver of 34-year-old Chinese man, containing 2 protoscolices (arrows). Casoni test was 4 plus.* x45

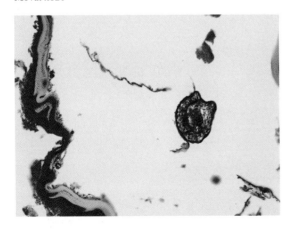

Figure 9.30
Higher magnification of protoscolex of *E. multilocularis* shown in Figure 9.29.* x165

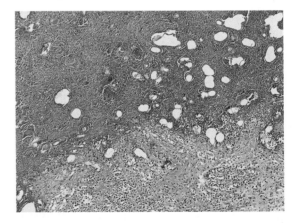

Figure 9.31
Sterile alveolar cyst of *E. multilocularis* in brain. Note many small cystic fragments of parasite in necrotic upper portion of specimen and extensive gliosis in lower portion. x80

tially acid-fast when stained by ZN technique (Figs 9.19 to 9.21). Cysts eventually calcify. There are usually focal deposits of calcium at sites of degenerating or dead cysts (Figs 9.52 and 9.53). Secondary bacterial infection and suppuration may accompany degeneration of cysts.

Pulmonary changes following rupture of a metacestode of *Echinococcus* into a bronchus are mainly those of persistent suppuration with abscess formation. Individual protoscolices can disperse throughout the lung tissue and provoke inflammation (Fig 9.54). Fragments of cyst walls that lodge in the lung cause necrotizing granulomas. A hydatid cyst in bone generally causes erosion, with multiple fractures or disintegration of the bony architecture.

The larval mass of *E. multilocularis* propagates indefinitely by external budding of the germinal membrane, and produces an alveolar pattern of microvesicles (Fig 9.32). Multilocular cysts have poorly defined borders that encroach on and invade surrounding host tissue in the manner of malignant neoplasms (Fig 9.28). The histologic appearance can suggest a benign cystic tumor. Numerous small vesicles embed themselves into a dense stroma of connective tissue, but there is no limiting host-tissue barrier (Fig 9.32). There is usually necrotic debris around the cyst (Fig 9.26), sometimes accompanied by granulomatous tissue reaction and, in advanced disease, a central necrotic cavity. In zones of cellular reaction there are lymphocytes, plasma cells, and numerous eosinophils. Calcareous corpuscles are rare or absent, and protoscolices are rare in human lesions. Experimental evidence suggests that cell-mediated immunity is crucial to suppression of larval growth.[52]

With the polycystic hydatid of *E. vogeli,* pathologic changes are intermediate between those of unilocular hydatid cysts and alveolar hydatid disease. In human liver the polycystic mass is usually visible at the hepatic surface, but extends down into the hepatic parenchyma and may involve bile ducts. Polycystic hydatids can also lodge in other host tissues, including the mesentery, omentum, pericardium, right auricle, lung, pleura, and superior vena cava. Similar to patients with unilocular cysts of *E. granulosus*, the host forms a 2-to-3-mm wall of fibrous tissue around the polycystic hydatid of *E. vogeli*.

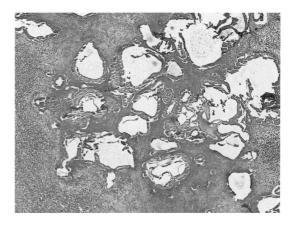

Figure 9.32
Same specimen as in Figure 9.31. At some sites the infiltrating mass of sterile alveolar cysts forms large cystic areas. x25

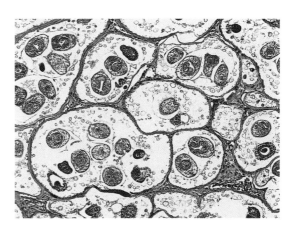

Figure 9.33
Numerous vesicles of *E. multilocularis* in a vole. Protoscolices and calcareous corpuscles are plentiful. x60

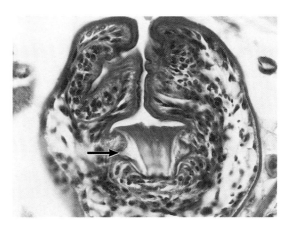

Figure 9.34
Armed rostellum of *E. multilocularis* in a vole. Hooklet (arrow) is 27 μm long. x595

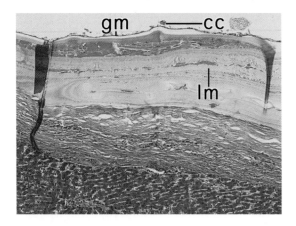

Figure 9.35
Metacestode of *E. vogeli* in liver of a paca. Note uniformly thin, delicate germinal membrane (gm) with calcareous corpuscles (cc) and thick laminated membrane (lm). x95

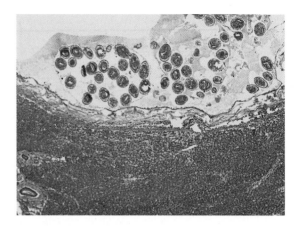

Figure 9.36
Brood capsules of *E. vogeli* in liver of a paca, demonstrating numerous protoscolices. x35

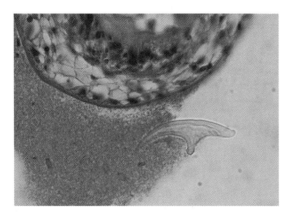

Figure 9.37
Polycystic hydatid of *E. vogeli* in liver of a paca. Note long hooklet (41 μm). x750

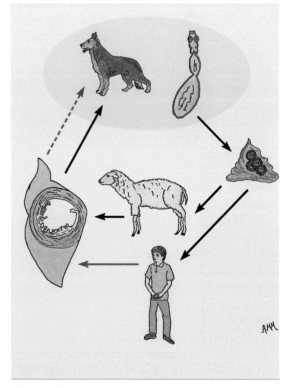

Figure 9.38
Life cycle of domestic strain of *E. granulosus*. Asexual reproduction takes place in the intermediate host. Sexual reproduction requires that the definitive host consume larvae from the intermediate host. Humans are suitable intermediate hosts but are usually dead-end hosts. Life cycles of other species differ only in preferred definitive and intermediate hosts.

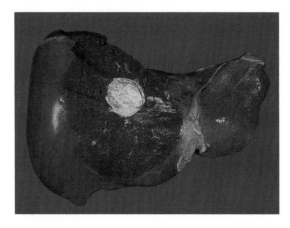

Figure 9.39
Hydatid cyst of *E. granulosus* in liver of 65-year-old woman from Arkansas. Patient had 4 well-circumscribed unilocular cysts containing yellow-green pasty material. Cysts were found incidentally at autopsy. x2.7

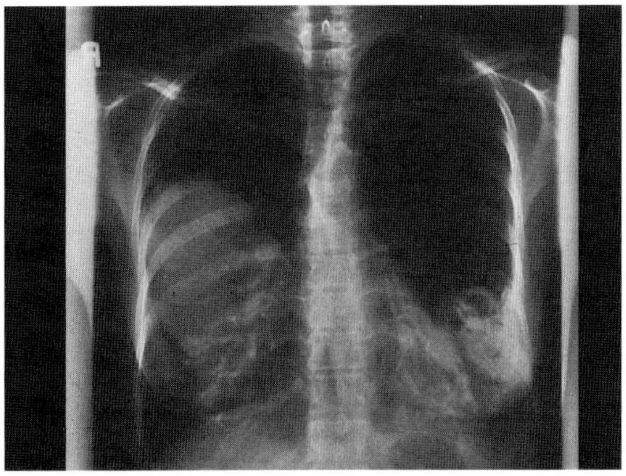

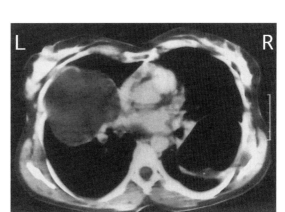

Figure 9.40
Radiograph of 37-year-old man with multiple pulmonary echinococcal cysts. Note large, circumscribed orb of an unruptured cyst in right lower lobe, and ruptured cyst with daughter cysts in left lower lobe. Daughter cysts within a large sphere are diagnostic.*

Figure 9.41
Computed tomography from same patient as in Figure 9.40, illustrating fluid-filled cyst of right lung and collapsed cyst in left lung.*

Diagnosis

Histopathologic demonstration of hydatid structures from any site is diagnostic. The cyst wall is finely laminated and amphophilic, an appearance that, when recognized, is pathognomonic. The wall is usually identifiable even in old calcified cysts.[43] Examination of cysts may reveal daughter cysts, brood capsules, protoscolices, and hooklets.

Imaging of cystic hydatid disease reveals sharply circumscribed orbs; daughter cysts within the larger spheres are diagnostic (Fig 9.40).[24] Roentgenograms are particularly helpful in diagnosing calcified cysts. Radiological findings most suggestive of hydatid disease of bone are translucent lesions with the "water lily sign" (Fig 9.42) associated with bone expansion, thinning of the cortex, and calcification of surrounding soft tissue.[7] Angiography of the liver reveals a "rim sign." Magnetic resonance, ultrasound, and computed tomography (CT) aid in diagnosis.[61]

There are a variety of serologic tests for echinococcosis. Indirect hemagglutination (IHA) is sensitive but nonspecific. Successfully treated patients become antibody-negative by IHA (<1:128) within a year after beginning treatment.[59] ELISA yields results similar to IHA.[30] Monoclonal antibody detection of antigen arc-5 or antigen B is more specific, but has low sensitivity.[41] Western blot for detection of antibody to Em18 (a protein of *E. multilocularis*) seems to be a reliable and useful method to differentiate active from inactive alveolar hydatid disease.[31] Serology of echinococcosis is an active area of investigation.

In the Casoni skin test, 0.2 ml of hydatid fluid is inoculated intradermally. In patients with live hydatids, positive reactions develop at the inoculation site within 20 minutes.

Examination of vomitus, stool, or sputum can reveal protoscolices, hooklets, or pieces of cyst wall, any of which confirms the diagnosis.[1]

Identification of specific DNA sequences of *Echinococcus* sp in stool are used epidemiologically to identify infected carnivores.[10]

Treatment and Prevention

Cystic hydatids should be removed surgically, taking every precaution to prevent rupture that could lead to seeding and recurrence. Alveolar hydatids may require lobectomy of the lung.[72] Pulmonary cystic hydatids tend to heal spontaneously and need not be surgically excised, but chemotherapy is recommended.[21] Those that become symptomatic can be removed before serious

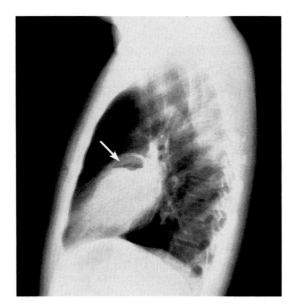

Figure 9.42
Radiograph of chest of 33-year-old man with a pulmonary hydatid cyst. Collapsed membrane forms the "water lily sign" (arrow) that is diagnostic of echinococcosis.*

Figure 9.43
Well-circumscribed, unilocular hydatid cyst in lung.

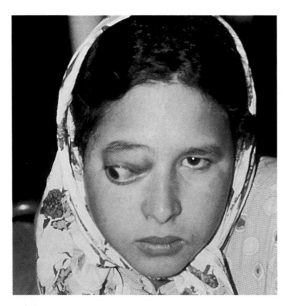

Figure 9.44
Hydatid cyst of right orbit.

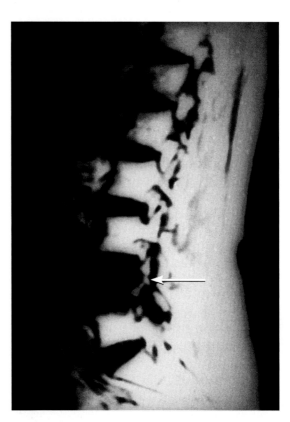

Figure 9.45
Computed tomography of lower vertebrae, showing a unilocular cyst destroying a lumbar vertebra (arrow).

complications develop. Cysts may be excised by laparoscopy,[34] and should be sterilized before drainage or extirpation to prevent anaphylaxis. Chemotherapy with albendazole, while not completely effective, is useful in treating both cystic and alveolar echinococcosis.[71] Chemotherapy (albendazole[65] or, alternatively, mebendazole combined with praziquantel) is used when lesions are surgically inaccessible, patients are unfit for surgery or have recurrent disease, surgical facilities are inadequate, or there is a risk of rupture during surgery.[47] After a week of chemotherapy, hydatid cysts in the liver can be treated with double fine-needle aspiration and injection of 95% ethanol, then continued chemotherapy for 3 more weeks.[20,35] This procedure requires caution because penetrating a living cyst can cause metastatic proliferation and anaphylactic shock. Surgery is the treatment of choice for cerebral hydatid cysts. Surgery of alveolar hydatids can provide relief, but rarely cures the disease.

Egg-contaminated fingers, food, and water are the source of infection, so strict personal hygiene and cleaning of food are necessary. Handling infected dogs is dangerous. The best prevention is to eliminate infection in definitive hosts, particularly domestic dogs and cats. Pets should be dewormed semiannually and prevented from consuming raw, contaminated flesh.

Figure 9.46
Alveolar hydatid cyst diffusely infiltrated the liver and caused the death of this Chinese patient.

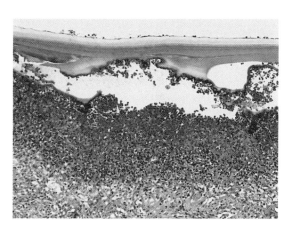

Figure 9.47
Cystic hydatid in lung showing adjacent acute inflammatory reaction with eosinophilia. x24

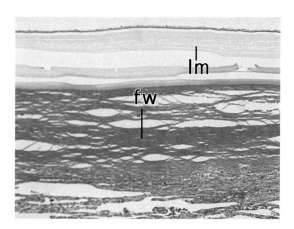

Figure 9.48
Metacestode of *E. granulosus* in lung with thick fibrous wall (fw) surrounding laminated membrane (lm). x65

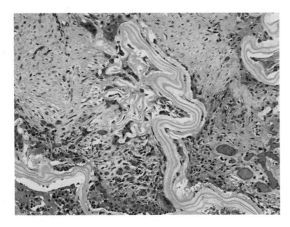

Figure 9.49
Granulomatous reaction to fragments of laminated membrane of metacestode of *E. granulosus* in spleen. x120

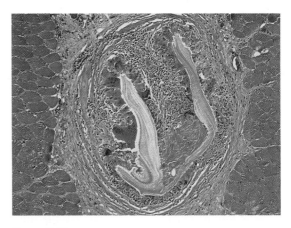

Figure 9.50
Granulomatous reaction with giant cells in response to fragments of laminated membrane of *E. granulosus* in paraspinal mass. x60

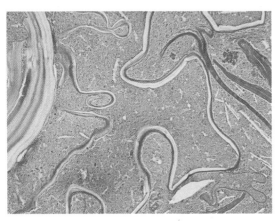

Figure 9.51
Necrotic laminated membrane of *E. granulosus*. x60

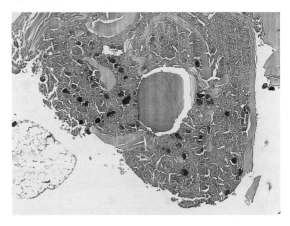

Figure 9.52
Necrotic orbital lesion with numerous calcified protoscolices of *E. granulosus*. x25

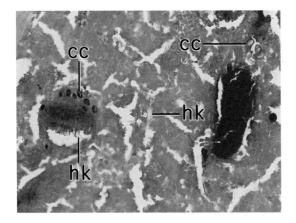

Figure 9.53
Two partially calcified protoscolices of *E. granulosus* in necrotic orbital lesion, demonstrating calcareous corpuscles (cc) and hooklets (hk). x230

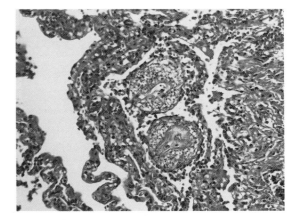

Figure 9.54
Two protoscolices of *E. granulosus* provoking pulmonary inflammation. x170

References

1. Allen AR, Fullmer CD. Primary diagnosis of pulmonary echinococcosis by the cytologic technique. *Acta Cytol* 1972;16:212–216.

2. Bayezid O, Ocal A, Isik O, Okay T, Yakut C. A case of cardiac hydatid cyst localized on the interventricular septum and causing pulmonary emboli. *J Cardiovasc Surg (Torino)* 1991;32:324–326.

3. Beaver PC, Jung RC, Cupp EW. *Clinical Parasitology*. 9th ed. Philadelphia, Pa: Lea & Febiger; 1984:536.

4. Bein T, Haerty W, Haller M, Forst H, Pratschke E. Organ selection in intensive care: transplantation of a liver allograft, including calcified cyst of Echinococcus granularis [letter]. *Intensive Care Med* 1993;19:182.

5. Benjelloun S, Elmrini M. Hydatid cyst of the kidney [in French]. *Prog Urol* 1993;3:209–215.

6. Berrada S, Essadki B, Zerouali NO. Hydatid cyst of the liver. Treatment by resection of the cyst wall. Our experience apropos of a series of 495 cases [in French]. *Ann Chir* 1993;47:510–512.

7. Booz MY. The value of plain film findings in hydatid disease of bone. *Clin Radiol* 1993;47:265–268.

8. Boyano T, Moldenhauer F, Mira J, Joral A, Saiz F. Systemic anaphylaxis due to hepatic hydatid disease. *J Investig Allergol Clin Immunol* 1994;4:158–159.

9. Bresson-Hadni S, Laplante JJ, Lenys D, et al. Seroepidemiologic screening of Echinococcus multilocularis infection in a European area endemic for alveolar echinococcosis. *Am J Trop Med Hyg* 1994;51:837–846.

10. Bretagne S, Guillou JP, Morand M, Houin R. Detection of Echinococcus multilocularis DNA in fox faeces using DNA amplification. *Parasitology* 1993;106:193–199.

11. Cassoni T. La diagnosa biolgica del'echinococcosi umana mediante l'intradermoreazione. *Folia Clinica Chemica e Microscopia* 1911;4:5–16.

12. Chang R, Higgins M, DiLisio R, Hawasli A, Camaro LG, Khatib R. Infected hepatic Echinococcus cyst presenting as recurrent Escherichia coli empyema. *Ann Thorac Surg* 1993;55:774–775.

13. Dietrichs E, van Knapen F, Bakke SJ. Meningeal irritation: possible manifestation of cerebral Echinococcus infestation. *Scand J Infect Dis* 1994;26:631–634.

14. D'Alessandro A, Ramirez LE, Chapadeiro E, Lopes ER, de Mesquita PM. Second recorded case of human infection by Echinococcus oligarthrus. *Am J Trop Med Hyg* 1995;52:29–33.

15. D'Alessandro A, Rausch RL, Cuello C, Aristizabal N. Echinococcus vogeli in man, with a review on polycystic hydatid disease in Colombia and neighboring countries. *Am J Trop Med Hyg* 1979;28:303–317.

16. Galeni CC. *Medicorum graecorum opera quae extant*. Kühn CG, trans. Leipzig, Germany: Georg Olms Verlagsbuchhandlung; 1827:781–782;1829:165–166.

17. Gamble WG, Segal M, Schantz PM, Rausch RL. Alveolar hydatid disease in Minnesota. First human case acquired in the contiguous United States. *JAMA* 1979;241:904–907.

18. Gardner SL, Rausch RL, Camacho OC. Echinococcus vogeli Rausch and Bernstein, 1972, from the paca, Cuniculus paca L. (Rodentia: Dasyproctidae), in the Departamento de Santa Cruz, Bolivia. *J Parasitol* 1988;74:399–402.

19. Ghedini G. Ricerche sui siero di sangue di individuo affectta da cisti da echinococco e sui liguido in essa contenuto. *Gazzetta degli Ospedali e delle Cliniche* 1906;27:1616–1617.

20. Giorgio A, Tarantino L, Francica G, et al. Unilocular hydatid liver cysts: treatment with US-guided, double percutaneous aspiration and alcohol injection. *Radiology* 1992;184:705–710.

21. Gocmen A, Toppare MF, Kiper N. Treatment of hydatid disease in childhood with mebendazole. *Eur Respir J* 1993;6:253–257.

22. Goeze JA. Cited by: Kean BH, Mott KE, Russell AJ. *Tropical Medicine and Parasitology: Classic Investigations*. Vol 2. Ithaca, NY: Cornell University Press; 1978:636–652.

23. Gorun N. Necessary hip disarticulation in extended echinococcosis of the femur [in French]. *Rev Chir Orthop Reparatrice Appar Mot* 1992;78:255–257.

24. Gouliamos AD, Kalovidouris A, Papailiou J, Vlahos L, Papavasiliou C. CT appearance of pulmonary hydatid disease. *Chest* 1991;100:1578–1581.

25. Hartmannus PJ. Cited by: Kean BH, Mott KE, Russell AJ. *Tropical Medicine and Parasitology: Classic Investigations*. Vol 2. Ithaca, NY: Cornell University Press; 1978:636–652.

26. Heath DD, Chevis RA. Mebendazole and hydatid cysts [letter]. *Lancet* 1974;2:218–219.

27. Heath DD, Christie MJ, Chevis RA. The lethal effect of mebendazole on secondary Echinococcus granulosus, cysticerci of Taenia pisiformis, and tetrathyridia of Mesocestoides corti. *Parasitology* 1975;70:273–285.

28. Hindricks G, Bocker D, Konertz W, Bongartz G, Borggrefe M. Polycystic disease of the kidneys complicating the diagnosis of myocardial echinococcosis. *Eur Heart J* 1993;14:141–142.

29. Hippocrates. Aphorisms (VII, 55) in Hippocratic writings. In: *The Great Books*. Adams F, trans. 23rd ed. Chicago, Ill: Encyclopedia Britannica Inc;1980;10:143.

30. Iacona A, Pini C, Vicari G. Enzyme-linked immunosorbent assay (ELISA) in the serodiagnosis of hydatid disease. *Am J Trop Med Hyg* 1980;29:95–102.

31. Ito A, Schantz PM, Wilson JF. Em18, a new serodiagnostic marker for differentiation of active and inactive cases of alveolar hydatid disease. *Am J Trop Med Hyg* 1995;52:41–44.

32. Jackisch C, Schwenkhagen A, Louwen F, Raber G, Oehme A, Schneider HP. Echinococcosis in pregnancy. A rare differential diagnosis in cystic intra-abdominal tumors in pregnancy [in German]. *Zentralbl Gynakol* 1993;115:263–272.

33. Kammerer WS, Schantz PM. Echinococcal disease. *Infect Dis Clin North Am* 1993;7:605–618.

34. Khoury G, Geagea T, Hajj A, Jabbour-Khoury S, Baraka A, Nabbout G. Laparoscopic treatment of hydatid cysts of the liver. *Surg Endosc* 1994;8:1103–1104.

35. Khuroo MS, Dar MY, Yattoo GN, et al. Percutaneous drainage versus albendazole therapy in hepatic hydatidosis: a prospective, randomized study. *Gastroenterology* 1993;104:1452–1459.

36. Klungsoyr P, Courtright P, Hendrikson TH. Hydatid disease in the Hamar of Ethiopia: a public health problem for women. *Trans R Soc Trop Med Hyg* 1993;87:254–255.

37. Kok AN, Yurtman T, Aydin NE. Sudden death due to ruptured hydatid cyst of the liver. *J Forensic Sci* 1993;38:978–980.

38. Küchenmeister F. Experimente über die Entstehung der Cestoden Zweiter Stufe zunächst des Coenurus cerebralis, Unter Mitwirkung des Herrn Professor Haubner auf Befehl und Kosten des hohen Königliche sächsischen. Staatsministerium des Innern. *Zeitschrift für klinische Medizin XL* 1853;4:448–451.

39. Kumaratilake LM, Thompson RC. A review of the taxonomy and speciation of the genus Echinococcus Rudolphi 1801. *Z Parasitenkd* 1982;68:121–146.

40. Leiby PD, Kritsky DC. Echinococcus multilocularis: a possible domestic life cycle in central North America and its public health implications. *J Parasitol* 1972;58:1213–1215.

41. Liu D, Rickard MD, Lightowlers MW. Assessment of monoclonal antibodies to Echinococcus granulosus antigen 5 and antigen B for detection of human hydatid circulating antigens. *Parasitology* 1993;106:75–81.

42. Lopera RD, Melendez RD, Fernandez I, Sirit J, Perera MP. Orbital hydatid cyst of Echinococcus oligarthrus in a human in Venezuela. *J Parasitol* 1989;75:467–470.

43. Marty AM, Hess SJ. Tegumental laminations in Echinococci using Gomori's methenamine silver stain. *Lab Med* 1991;22:419–420.
44. Meneghelli UG, Martinelli AL, Llorach Velludo MA, Bellucci AD, Magro JE, Barbo ML. Polycystic hydatid disease (Echinococcus vogeli). Clinical, laboratory and morphological findings in nine Brazilian patients. *J Hepatol* 1992;14:203–210.
45. Morgagni JB. *De sedibus et causus morborum per anatomen indagatis i quinque etc*. Venetiis, Italia: Remondini-ana; 1760–1761.
46. Morin A. Cited by: Grove DI. Echinococcus granulosus and echinococosis or hydatid disease. In: *A History of Human Helminthology*. Wallingford, England: CAB Int; 1990:319–353.
47. Morris DL, Richards KS. *Hydatid Disease: Current Medical and Surgical Management*. Oxford, England: Butterworth-Heinemann; 1992:1–150.
48. Munzer D. New perspectives in the diagnosis of Echinococcus disease. *J Clin Gastroenterol* 1991;13:415–423.
49. Njeh M, Hajri M, Chebil M, el Ouakdi M, Ayed M. Retrovesical hydatid cyst. Apropos of 2 cases [in French]. *Ann Urol (Paris)* 1993;27:97–100.
50. Noguera M, Alvarez-Castells A, Castella E, Gifre L, Andreu J, Quiroga S. Spontaneous duodenal fistula due to hepatic hydatid cyst. *Abdom Imaging* 1993;18:234–236.
51. Pappaioanou M, Schwabe CW, Sard DM. An evolving pattern of human hydatid disease transmission in the United States. *Am J Trop Med Hyg* 1977;26:732–742.
52. Playford MC, Kamiya M. Immune response to Echinococcus multilocularis infection in the mouse model: a review. *Jpn J Vet Res* 1992;40:113–130.
53. Rausch RL, Bernstein JJ. Echinococcus vogeli sp. n. (Cestoda: Taeniidae) from the bush dog, Speothos venaticus (Lund). *Z Prakt Anasth* 1972;23:25–34.
54. Rausch RL, D'Alessandro A, Rausch VR. Characteristics of the larval Echinococcus vogeli Rausch and Bernstein, 1972 in the natural intermediate host, the paca, Cuniculus paca L. (Rodentia: Dasyproctidae). *Am J Trop Med Hyg* 1981;30:1043–1052.
55. Rausch RL, Rausch VR, D'Alessandro A. Discrimination of the larval stages of Echinococcus oligarthrus (Diesing, 1863) and E. vogeli Rausch and Bernstein, 1972 (Cestoda: Taeniidae). *Am J Trop Med Hyg* 1978;27:1195–1202.
56. Redi F. *Osservazioni intorno agli animali viventi che si trovano negli animali viventi*. Firenze, Italia: Piero Matini; 1684:253.
57. Rincon B, Bernis C, Garcia A, Traver JA. Mesangiocapillary glomerulonephritis associated with hydatid disease [letter]. *Nephrol Dial Transplant* 1993;8:783–784.
58. Rosenberg T, Panayiotopoulos YP, Bastounis E, Papalambros E, Balas P. Acute abdominal aorta embolism caused by primary cardiac echinococcus cyst. *Eur J Vasc Surg* 1993;7:582–585.
59. Schantz PM, Wilson JF, Wahlquist SP, Boss LP, Rausch RL. Serologic tests for diagnosis and post-treatment evaluation of patients with alveolar hydatid disease (Echinococcus multilocularis). *Am J Trop Med Hyg* 1983;32:1381–1386.
60. Seiferth T, Endsberger G, Stolte M. Echinococcosis—current status of diagnosis and therapy [in German]. *Leber Magen Darm* 1993;23:161–164.
61. Shamsi K, Deckers F, De Schepper A. Unusual cystic liver lesions: a pictorial essay. *Eur J Radiol* 1993;16:79–84.
62. Storandt ST, Kazacos KR. Echinococcus multilocularis identified in Indiana, Ohio, and east-central Illinois. *J Parasitol* 1993;79:301–305.
63. Thomas S, Mishra MC, Kriplani AK, Kapur BM. Hydatidemesis: a bizarre presentation of abdominal hydatidosis. *Aust N Z J Surg* 1993;63:496–498.
64. Thompson RC. Biology and systematics of Echinococcus. In: Thompson RC, ed. *The Biology of the Echinococcosis and Hydatid Disease*. London, England: George Allen & Unwin; 1986:5–43.
65. Todorov T, Vutova K, Mechkov G, et al. Chemotherapy of human cystic echinococcosis: comparative efficacy of mebendazole and albendazole. *Ann Trop Med Parasitol* 1992;86:59–66.
66. Vogel H. About the species and the life cycle of the European alveolar Echinococcus [in German]. *Deutsche medizinische Wochenschrift* 1955;8:931–932.
67. Von Siebold CT. Über die Verwandlung der Echinococcus-brut in Taenien. *Zeitschrift für wissenschaftliche Zoolgoie* 1853;4:409–424.
68. Von Sinner W. Pleural complications of hydatid disease (Echinococcus granulosus). *Rofo Fortschr Geb Rontgenstr Neuen Bildgeb Verfahr* 1990;152:718–722.
69. Von Sinner WN, Hellstrom M, Kagevi I, Norlen BJ. Hydatid disease of the urinary tract. *J Urol* 1993;149:577–580.
70. Weirich WL. Hydatid disease of the liver. *Am J Surg* 1979;138:805–808.
71. Wen H, Zou PF, Yang WG, et al. Albendazole chemotherapy for human cystic and alveolar echinococcosis in north-western China. *Trans R Soc Trop Med Hyg* 1994;88:340–343.
72. Wen H, New RR, Craig PS. Diagnosis and treatment of human hydatidosis. *Br J Clin Pharmacol* 1993;35:565–574.
73. Wilson JF, Diddams AC, Rausch RL. Cystic hydatid disease in Alaska. A review of 101 autochthonous cases of Echinococcus granulosus infection. *Am Rev Respir Dis* 1968;98:1–15.
74. Wilson JF, Rausch RL. Alveolar hydatid disease. A review of clinical features of 33 indigenous cases of Echinococcus multilocularis infection in Alaskan Eskimos. *Am J Trop Med Hyg* 1980;29:1340–1355.

10

Diphyllobothriasis and Sparganosis

Aileen M. Marty *and*
Ronald C. Neafie

Introduction

Definition

Diphyllobothriasis is infection of the intestine by adult tapeworms of the genus *Diphyllobothrium*. *Diphyllobothrium latum* (Linnaeus, 1758) Cobbold, 1858,[62] and *Diphyllobothrium nihonkaiense* Yamane et al, 1986,[77] are the principle agents, but adult worms of other species of *Diphyllobothrium* and of other genera (*Diplogonoporus, Ligula, Digramma,* and *Braunia*) can infect the human intestine.

Sparganosis is infection by spargana, a common term for plerocercoid (third-stage) metacestode larvae of pseudophyllidean tapeworms. Plerocercoid larvae of at least 4 species can produce sparganosis in humans: *Spirometra mansoni* (Cobbold, 1883) Mueller, 1937; *Spirometra ranarum* (Mueller, 1937); *Spirometra mansonoides* (Mueller, 1935) Mueller, 1937; and *Spirometra erinacei* (Rudolphi, 1819, Diesing, 1854) Mueller, 1937. A fifth entity, *Sparganum proliferum,* may be an aberrant form of either *S. mansonoides* or *S. erinacei,* or of both, or may represent a separate species.

Synonyms

Older synonyms for *D. latum,* the broad or fish tapeworm, include *Diphyllobothrium americanus, Dibothrium serratum, Dibothrium fuscum, Dibothrium luxi, Taenia lata, Bothriocephalus latus, Bothriocephalus taenioides, Dibothriocephalus latus,* and *Dibothriocephalus minor. Diphyllobothrium parvum,*[36] *Diphyllobothrium strictum,* and *Diphyllobothrium tungussicum* are probably all variants of *D. latum.*[7] *Diphyllobothrium nihonkaiense,* the broad tapeworm of Japan, previously accepted as *D. latum,* is a distinct species.[19,77]

Other members of the genus *Diphyllobothrium* that cause diphyllobothriasis in humans include *Diphyllobothrium pacificum* (Nybelin, 1931) Margolis, 1956; *Diphyllobothrium cordatum* (Leuckart, 1863) Faust, 1929; *Diphyllobothrium alascense* Rausch and Williamson, 1958; *Diphyllobothrium klebanovskii*[48]; *Diphyllobothrium ursi* Rausch, 1954[39]; *Diphyllobothrium dendriticum* (Nitzsch, 1824)[76]; *Diphyllobothrium lanceolatum* (Krabbe, 1865) Linstow, 1905; *Diphyllobothrium dalliae* Rausch, 1956[58]; and *Diphyllobothrium yonagoensis* Yamane et al, 1981.[78] There are rare reports of infections in Japan with other species:

Diphyllobothrium cameroni Raush, 1969[24]; *Diphyllobothrium hians* Diesing, 1850[28]; *Diphyllobothrium scoticum* (Rennie and Reid, 1912)[18]; and *Diphyllobothrium orcini* (Hatsushika and Shirouzu, 1990).[50]

Diplogonoporus balaenopterae (Lönnberg, 1892) is probably the only bona fide species of *Diplogonoporus*.[15,75] Synonyms for this worm include *Diplogonoporus grandis* and *Diplogonoporus fukuokaensis*.

Schmidt[62] considers the generic name *Spirometra* a synonym for *Diphyllobothrium*. In 1929, Faust et al[17] demonstrated that worms causing sparganosis and those that cause diphyllobothriasis should be placed in 2 separate subgenera, and Mueller et al argued that they belong to a separate genus. Thus, we use the generic name *Spirometra* for these worms.[17,44] In 1908, Stiles put the proliferating sparganum in the genus *Gatesius*, after Gates, the physician who drew his attention to the matter.[23,67] Consequently, if *S. proliferum* is a mutant or aberrant form of the *S. mansoni* complex, the name *Gatesius* should have priority over *Spirometra*.[67] *Ligula mansoni*, *Bothriocephalus liguloides*, *Bothriocephalus mansoni*, *Diphyllobothrium houghtoni*, *Spirometra houghtoni*, and *Sparganum mansoni* are all synonyms for *S. mansoni*.

General Considerations

Diphyllobothriasis

Dunus and Wolphius gave recognizable descriptions of *D. latum* in 1592, but did not realize that tapeworms have scolices and did not separate them from other tapeworms. Platter,[71] in 1602, distinguished between 2 types of tapeworms, *species prima* (broad tapeworm) and *species secunda* (*Taenia*). In 1777, Bonnet published the first accurate description of the scolex of *D. latum*. Formulation of the life cycle took many years. In 1747, Spöring disagreed with the prevailing theory that the infection developed spontaneously and related the broad tapeworm disease to eating improperly cooked or raw fish. More than a century elapsed before Braun proved Spöring correct when, in the 1880s, he demonstrated that pike and burbot in lakes of Estonia harbored plerocercoid larvae of the broad tapeworm. How fish acquire the infection was established in 1917 by Janicki and Rosen.

Beginning in 1877, Reyher in Estonia studied patients with pernicious anemia who were infected with *D. latum*. In 1886, he discovered that ridding patients of this tapeworm cured their anemia.[60]

Scandinavia is the dominant historical endemic focus of *D. latum*. Immigrant Scandinavian fishermen probably transmitted *D. latum* infection to fish of the lake regions of Minnesota and Wisconsin in the 19th century. Sampling of these undercooked fish led to numerous infections in midwestern Jewish housewives preparing gefilte fish.

Diplogonoporus balaenopterae, a parasite of whales, is a rare intestinal tapeworm infection of humans. In 1894, Blanchard recovered the parasite from a whale and named it *Krabbea grandis*. In the same year Ijima and Kurimoto described the first known infected patient. Lühe renamed the worm *Diplogonoporus* in 1899. Because of their fondness for raw seafood, most patients have been Japanese.[25,26,27]

Digramma brauni (Leon, 1907) Joyeux and Baer, 1929 and *Ligula intestinalis* (Goeze, 1782) Gmelin, 1790, have infected patients in Romania. *Braunia jasseyensis* Leon, 1908, a pseudophyllidean tapeworm found in French patients, may be an immature *L. intestinalis*.

Sparganosis

According to Zhong et al, Chinese physicians documented sparganosis infections as early as 1596.[80] In 1881, Manson recovered numerous spargana from tissues of a 34-year-old Chinese man and sent specimens to Cobbold, who reported them as a new species in *Lancet* in 1882/1883. Manson and others recognized that these were immature tapeworms and suspected they were related to *Diphyllobothrium*. In 1907, Verdum suggested that until the adult was identified, these worms be given the provisional generic name *Sparganum*. Based on observations in 1927 on a patient who applied poultices containing dismembered frogs to her eyes, Casaux established the transmission of sparganosis by direct penetration of plerocercoids. Many investigators confirmed this observation.[22]

The first patient known to be infected with *S. proliferum* was a Japanese woman who, in 1904, presented with subcutaneous nodules in the thigh and trunk. Ijima found proliferating larval cestodes in these nodules and named them *Plerocercoides prolifer*.[23] In 1908, Stiles[67] reported the first patient with sparganosis in the United States. This patient, who was born in Minnesota and had lived in Florida since 1872, had multiple nodules of a proliferating form of sparganosis. Mueller identified *Spirometra* in cats over a wide area of the eastern United States and correctly predicted that as physicians became aware of this organism, they would find sporadic autochthonous infections in humans.[45]

Epidemiology

Diphyllobothrium latum is a common parasite of humans in Baltic and Scandinavian countries and some other temperate or colder regions. Estonia, Finland, Lithuania, northern Russia, and Sweden have the highest incidence of infection. Recent efforts have reduced the incidence in Finland to under 1% of the population and have limited the distribution of the broad tapeworm in Europe.[34] In the past, *D. latum* infections prevailed widely in Europe.[8] Among patients with bothriocephalus anemia, over 70% are from Finland. Finno-Ugric ethnic groups seem more susceptible.[7] Once prevalent in North America, especially around the Great Lakes, eradication programs have successfully rid northern Michigan and many other sites of *D. latum*[51] and *D. pacificum*.[4] In Chile and Argentina there are autochthonous infections of other *Diphyllobothrium* sp.[59] In Japan the dominant species is *D. nihonkaiense*,[37,55] but there are other species of *Diphyllobothrium* in northern Japan. *Diphyllobothrium dendriticum* is probably the third most frequent cause of diphyllobothriasis in humans.[71] Gulls are the definitive hosts of this parasite that has a more northerly range than *D. latum*.

Sparganosis in humans is widely distributed,[14] but China, Japan, and southeast Asian countries report most infections. Sparganosis in animals is endemic throughout the United States, but humans rarely acquire the disease. Most patients in the United States are males from the Eastern Seaboard and Gulf Coast, from Texas to New York. Asian patients usually have eaten frogs or snakes, or used poultices made from frogs. In the Western Hemisphere, patients often have eaten snakes or drunk from freshwater ponds. Current data suggest that third-stage larvae of *S. mansoni*, *S. ranarum*, or *S. erinacei* cause sparganosis in humans in Asia. The common species in animals in the United States is *S. mansonoides*, while a second rare species in animals resembles a pantropical form generally named *S. erinacei*. Presumably 1 or both of these species cause autochthonous sparganosis in humans in the United States. Investigators in Argentina report both *S. mansonoides* and *S. erinacei* in animals.[14] *Spirometra erinacei* is also found in animals in Australia (especially New South Wales) and Egypt, where they are potential pathogens for humans. Although uncommon, cerebral and visceral sparganosis can be serious diseases.

Sparganum proliferum is a rare and unusual sparganosis-like infection. There are reports of autochthonous infections of *S. proliferum* from Japan, Paraguay, Taiwan, the United States, and Venezuela, and of an unconfirmed infection in the Philippines.[6,35,49]

Infectious Agent

Morphologic Description

Adult members of the Diphyllobothriidae family have a scolex with 2 slit-like bothria or sucking grooves (Fig 10.1). Grooves extend down the ventral and dorsal areas of the scolex, dividing the scolex into 2 lips or leaves, hence the family name Diphyllobothriidae. Proglottids throughout most of the length of the strobila discharge eggs through their uterine pores until the supply is exhausted. Thus terminal proglottids of diphyllobothriid tapeworms are senile rather than gravid, and usually detach in chains instead of individually.[2,21] Diphyllobothriid tapeworms have 5 distinct morphologic stages: embryo in egg, coracidium, procercoid larva, plerocercoid larva (sparganum), and adult (Fig 10.2).

Adults

Diphyllobothrium latum

The adult *D. latum* is an ivory-white tapeworm measuring up to 10 meters long with 3000 or more proglottids (Fig 10.2).[33,38] The scolex is small, spatulate, about 1 mm in cross section and 2.5 mm in length (Fig 10.1). The long, slender, unsegmented neck gives rise to proglottids. Proglottids are typically broader (10 to 12 mm) than long (2 to 4 mm) (Fig 10.3). About 80% of the worm consists of mature and gravid proglottids (Figs 10.4 to 10.7).

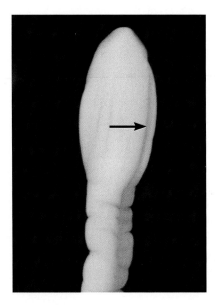

Figure 10.1
Scolex of *D. latum* demonstrating 1 of 2 sucking grooves (arrow).* x20

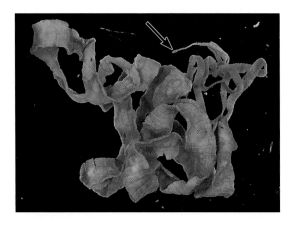

Figure 10.2
Portions of adult *D. latum* demonstrating the variation in size of the tapeworm, from the narrow anterior end with its scolex (arrow) to the broad proglottids of the posterior end.* x0.8

Figure 10.3
Proglottids of an adult *D. latum*. Proglottids are usually broader than long.* Unstained x6

The vagina and the cirrus open medially in a common genital sucker (common genital atrium) on the ventral surface, anterior and medial to the uterine pore (Fig 10.4). Testes are multiple minute follicles located in the dorsal plane on both sides of each proglottid (Fig 10.7). The vasa efferentia merge just in front of the ootype to form the vas deferens, a convoluted tubule that travels anteriorly from the region of the ootype, and leads into a muscular cirrus at the upper border of the common genital atrium. The bilobed ovary is in the posterior third of the proglottid. The ootype is between the 2 lobes. The ootype receives eggs from the ovary and secretions from the vitellaria and Mehlis' gland that surrounds the ootype. The seminal receptacle lies adjacent to the ootype, and the vagina extends anteriorly away from the ootype toward the common genital pore. The uterus also extends anteriorly away from the ootype, and its outer coils form a rosette that leads to the uterine pore. Eggs exit through uterine pores. A single worm may release a million eggs per day. There is no clear distinction between mature and gravid proglottids. As proglottids cease to function, they slowly disintegrate and detach in strands of variable lengths.

Clinicians sometimes submit proglottids of *D. latum* to the laboratory where the specimen is embedded and processed routinely. Stained sections reveal slices of the internal organs described in the preceding paragraph (Figs 10.8 to 10.17). The body wall of the proglottid varies in thickness (Figs 10.10 and 10.11) and is composed of microvilli, tegument, tegumental cells, and smooth muscle fibers. Variable numbers of calcareous corpuscles are scattered throughout the parenchyma of the proglottid (Fig 10.18). In tissue section, the ovary contains multiple polygonal cells (Fig 10.16).

Diphyllobothrium nihonkaiense

Adult *D. nihonkaiense* have a maximum length of 7.4 meters and up to 665 proglottids. The scolex averages 2.6 mm long and 1.4 mm wide and tends to be spatulate with numerous pits at the crown. Bothria are deep and may extend the length of the scolex. The neck is well-developed and averages 15.6 mm long by 1.2 mm wide. All proglottids are broader than long, but their length increases as they approach the posterior end of the strobila. In

Figure 10.4
Diagram of mature proglottid of *D. latum* demonstrating ovary, ootype, and vagina (blue), seminal receptacle, uterus, and uterine pore (red), testes and vasa efferentia (brown), vas deferens and cirrus (green), vitelline glands and ducts (yellow), common genital sucker (black), and Mehlis' gland (brown). This diagram depicts only a small percentage of the numerous testes and vitelline glands that fill the lateral portion of each proglottid.*

mature proglottids the common genital sucker (cirrus sac) is oval and large (420 to 480 μm long by 390 to 400 μm in diameter), and opens into the genital atrium, sharing an opening (the genital pore) with the vagina ventrally in the midline. The thick-walled seminal vesicle is large and elliptical (approximately 250 μm long by 100 μm in diameter), and turns sharply as it connects to the back of the cirrus sac. A transverse muscle layer 5 to 10 μm thick covers the testes, which are spherical and lie in a single layer in the medullary parenchyma. The ovary is kidney-shaped, with neither anterior nor posterior horns, and is situated in the posterior third of the mature proglottid. Uterine loops (usually 6 to 7 per side) extend laterally, with peripheral loops opening into the uterine pore. The uterine pore opens posterior to the genital pore.[77] Gravid proglottids are much broader (6.7 to 6.9 mm) than long (1.9 to 2.1 mm).

Spirometra

Adult tapeworms of the genus *Spirometra* are small (85 to 100 cm long) and more delicate than those of *D. latum*. The scolex is small and spatulate; the neck is long and slender. Terminal proglottids are rectangular and slightly broader than long. Tegument is composed of an outer noncellular layer and an inner cellular layer (the subtegument). Immediately beneath the nuclei of the tegumental cells are clusters of vitelline glands. A thick, ill-defined band of longitudinal muscle fibers lies beneath the vitelline glands and divides the cortical and medullary layers of the parenchyma. The uterus is spiral with 2 to 7 closely packed coils. The cirrus and vagina open separately. The uterine pore is separate and posterior.[44]

Eggs

Eggs of diphyllobothriid tapeworms are not immediately infective for the copepod intermediate host. They resemble eggs of trematodes by having an operculum at 1 pole.

Eggs of *D. latum* are 58 to 76 μm by 40 to 51 μm, broadly ovoid, and have an operculum at 1 end and a knob at the abopercular end (Figs 10.17 and 10.19). The shell is moderately thick and light golden-yellow. The egg contains an immature embryo.

Eggs of *S. mansonoides* are 57 to 66 μm by 33 to 37 μm, ellipsoid, and have a conical, prominent operculum and a small knob-like thickening at the abopercular end.[3] The shell is thick and yellow-brown. The egg contains an immature embryo.

Larvae

Coracidia of *Diphyllobothrium* and *Spirometra* are morphologically similar, and are covered with cilia of equal length. Coracidia of *D. latum* swim with a slow rolling motion, while those of *Spirometra* swim rapidly with their hooks to the rear.

Coracidia and procercoid larvae of *Diphyllobothrium* sp do not infect humans. Plerocercoid larvae of *Diphyllobothrium* mature to adult tapeworms in human intestines.[33,77]

Figure 10.5
Mature proglottids of *D. latum*. Carmine x13

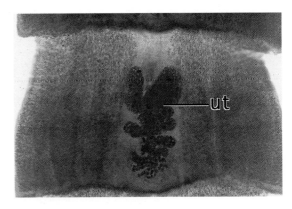

Figure 10.6
Mature proglottid of *D. latum* demonstrating uterus (ut) with eggs. Carmine x20

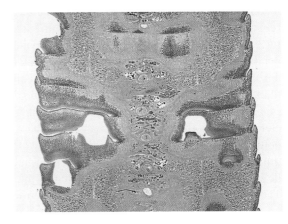

Figure 10.8
Low magnification of sections through several proglottids of *D. latum*. x8

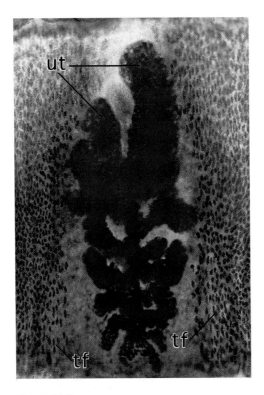

Figure 10.7
Mature proglottid of *D. latum* demonstrating uterus (ut) with eggs and multiple testicular follicles (tf) on both sides of the uterus. Carmine x25

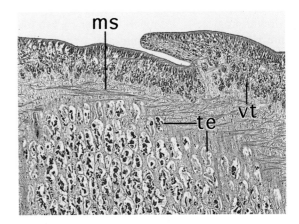

Figure 10.9
Section of proglottid of *D. latum* depicting body wall and parenchyma, containing smooth muscle fibers (ms), vitellaria (vt), and testes (te). x60

Plerocercoid larvae (spargana) of *Spirometra* are ivory-white, flat, ribbon-like worms that vary from a few millimeters to 50 cm or more in length (Fig 10.20). Larvae vary considerably in width; some areas are broad (up to 3 mm), while others are narrow and string-like. The immature scolex at the anterior end of the worm is flattened and has a cleft-like invagination called the rudimentary ventral groove (bothrium) (Fig 10.21). Grossly, the body wall is irregular, producing a pseudo-segmented appearance that in sections presents as ridges and inward folds (Fig 10.22). The body wall varies in thickness and is composed of a layer of microvilli, tegument (5 to 15 μm thick), 2 layers of smooth muscle, and a row of tegumental cells (Fig 10.23). The parenchyma contains irregularly scattered bundles of longitudinal muscle fibers, branching excretory channels, mesenchymal fibers, and calcareous corpuscles in a loose stroma (Fig 10.24). There are no reproductive organs.

Sparganum proliferum

Sparganum proliferum is unusual among the metacestodes of pseudophyllidean tapeworms in having continuous branches and buds (Figs 10.25 and 10.26). These larvae present in 2 morphologic forms: some are large, motile, vermiform structures with irregular branching[23,67]; others are small with vesicular budding.[1,49] Both forms have histologic features of a typical sparganum, but organization and symmetry are lost.[53] Vesicular forms appear more primitive and have poorly developed muscle fibers. There is no definitive anterior end or scolex.

Microscopic cross sections of *S. proliferum* tend to be spherical with poorly developed transverse and dorsoventral muscles in the parenchyma. Parenchymal cells are distributed in an abundant granular and fibrillated extracellular matrix containing many calcareous bodies. There are abundant excretory channels lacking specific arrangement. A proliferating sparganum (Fig 10.27) has a layered tegument: tegumental syncytium (approximately 9.5 μm thick), basal lamina, tegumental muscle, and tegumental cells.[53] Separation of the tegumental syncytium from the basal membrane produces tegumental vesicles.

Sparganum proliferum can be maintained in vitro and in vivo. Inoculation of a single larval fragment appears as effective as that of complete larvae. A minimal essential medium allows *S. proliferum* to survive as long as 14 weeks, but growth takes place only during the first 4 weeks of culture. Despite initial in vitro growth of larvae, they neither differentiate into a more developed stage nor multiply.[1]

Life Cycle and Transmission

The life cycle of pseudophyllidean tapeworms involves 5 distinct phases: 1) embryonated egg; 2) free-swimming first-stage larva (coracidium); 3) second-stage larva (procercoid); 4) third-stage larva (plerocercoid or sparganum); and 5) adult tapeworm in definitive host (Figs 10.28 and 10.29). Three hosts participate in the life cycle. The first intermediate host for *Diphyllobothrium* and *Spirometra* is a copepod (water flea) crustacean. Selection of an appropriate species of copepod host is ecologically significant. *Diaptomus* (first intermediate host for *D. latum*) lives in open lakes, where it is ingested by fish (second intermediate hosts for *D. latum*). *Cyclops* (first intermediate host for *Spirometra*) thrives in stagnant pools along shorelines, where it is ingested by secondary intermediate hosts of *Spirometra* (snakes, frogs, and mammals). The definitive hosts for these genera are carnivorous or omnivorous mammals.

Diphyllobothrium latum

Humans are the optimal definitive host for *D. latum,* but many fish-eating mammals (bears, canines, felines, seals, pigs, and other carnivores) can harbor adult *D. latum* (Fig 10.28).[57]

Feces of infected definitive hosts contaminate water with eggs of *D. latum*, exposing copepods of the genus *Diaptomus* (the first intermediate host) to infection. *Cyclops* sp are only accidental or abnormal hosts.[44] Infected copepods swim sluggishly, making them easy prey for fish.[43,56] The second intermediate hosts and paratenic hosts are freshwater fish or fish that spawn in fresh water, and include pike, perch, lawyer, salmon, trout, whitefish, grayling, ruff, eel, and barbel. Humans are not paratenic hosts for *D. latum*; ingested plerocercoid larvae develop into adult tapeworms in human intestines, where adult worms can live for at least 29 years.

Other species of *Diphyllobothrium*

Humans are the natural definitive host of

D. nihonkaiense; the usual second intermediate and paratenic hosts are Masu salmon and pink salmon. *Diphyllobothrium ursi, D. dendriticum,* and *D. dalliae* also use freshwater fish as second intermediate hosts. Bears are the optimal definitive host for *D. ursi* and sockeye salmon the usual second intermediate host. Some authorities believe that *D. ursi* is an infrequent cause of human infection, mainly because the plerocercoid larvae of *D. ursi* tend to develop on the serosal surface of the stomach of the salmon and not in the muscles.[7] *Diphyllobothrium* sp whose second intermediate or paratenic hosts are saltwater fish include *D. cordatum, D. alascense, D. lanceolatum, D. pacificum, D. yonagoensis,* and *D. cameroni*. Seals, walrus, bears, and dogs are definitive hosts for *D. cordatum*, and fur seals the optimal definitive host for *D. pacificum*.

Spirometra

The life cycle of *Spirometra* resembles that of *Diphyllobothrium*, although the intermediate and paratenic hosts and the definitive hosts are different (Fig 10.29). Second intermediate and paratenic hosts may overlap; optimal hosts of each worm differ.

Adult *Spirometra* tapeworms live in the intestines of dogs and cats. Dogs are the optimal definitive hosts for *S. mansoni*, and cats, especially bobcats, for *S. mansonoides*,[43] *S. ranarum, S. erinacei,* and *Spirometra decipiens*.

Humans are not definitive hosts for *Spirometra*, but serve as second intermediate and paratenic hosts, and thus develop sparganosis.

Infected dogs and cats pass eggs in feces, which hatch in fresh water and release coracidia. The first intermediate host, a copepod of the genus *Cyclops*, consumes a rapidly swimming coracidium. Second intermediate hosts (fish, snakes, amphibians, birds, or mammals) ingest infected *Cyclops* and acquire procercoid larvae. Within the tissues of the second intermediate host the procercoid larva develops into a plerocercoid larva–the sparganum. When a dog or cat eats the flesh of an intermediate host, the sparganum develops in the gut into an adult tapeworm and begins laying eggs within 15 to 18 days.

Humans acquire sparganosis in 3 distinct ways: 1) by drinking water containing an infected *Cyclops* (the procercoid larva produces proteases that enable the parasite to penetrate the intestinal wall and migrate to various organs and tissues)[12,31,61]; 2) by ingesting a sparganum in improperly cooked flesh of an infected second intermediate or paratenic host (amphibian, reptile, bird, or mammal)[72]; or, 3) by applying the flesh of an infected second intermediate or paratenic host to an open wound. Spargana live up to 20 years in humans.

Clinical Features and Pathogenesis

Diphyllobothriasis

Clinical symptoms vary with the number of adult tapeworms, length of each worm, amount of essential nutrients each worm absorbs, and host reaction to products of the tapeworm. Most patients have a single worm, but some may harbor over 100 worms. One patient had 201.[69]

Patients may present with diarrhea, abdominal pain, or anemia. Occasionally, portions of worm are vomited, some as long as several meters. Massive worm burdens can obstruct the bowel. Often patients have a slight leukocytosis with eosinophilia.

About 2% of patients suffer from bothriocephalus anemia, a nonlethal form of pernicious anemia. Patients with this complication complain of fatigue, weakness, sore tongue, paresthesias of hands and feet, and diarrhea. Some present with weight loss and difficulty in walking. Serum vitamin B_{12} levels are low, but serum iron levels are normal or high. Patients infected by *Diphyllobothrium* who produce a limited amount of gastric intrinsic factor are at increased risk for bothriocephalus anemia. *Diphyllobothrium latum* competes with the host for available vitamin B_{12}. Anemia results when the supply of vitamin B_{12} is low or when the host's ability to assimilate the vitamin is poor. Unlike classic pernicious anemia, patients with bothriocephalus anemia often are not achlorhydric.

Rarely, an adult worm migrates into the bile duct and produces biliary obstructive symptoms.

Sparganosis

Most patients with sparganosis harbor a single plerocercoid larva (sparganum). This sparganum can migrate, causing various symptoms depend-

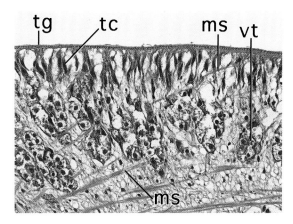

Figure 10.10
Section of body wall of proglottid of *D. latum,* composed of thin tegument (tg) and tegumental cells (tc). Vitellaria (vt) and smooth muscle fibers (ms) lie beneath the tegumental cells. x230

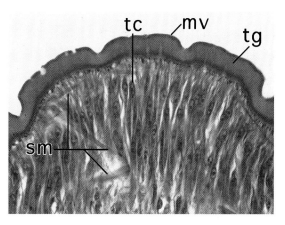

Figure 10.11
Section of body wall of proglottid of *D. latum,* demonstrating microvilli (mv), thick tegument (tg), smooth muscle fibers (sm), and tegumental cells (tc). x580

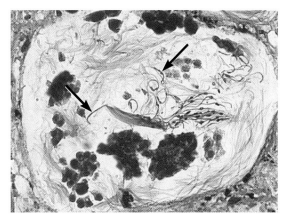

Figure 10.12
Section of proglottid of *D. latum* depicting a testicular follicle. Note that mature sperm are filamentous (arrows). x580

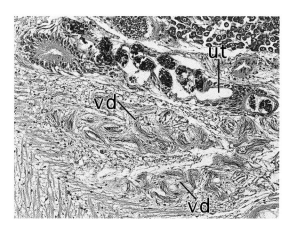

Figure 10.13
Section of proglottid of *D. latum* demonstrating portions of the uterus (ut) with eggs, and sinuous vas deferens (vd) containing sperm. x120

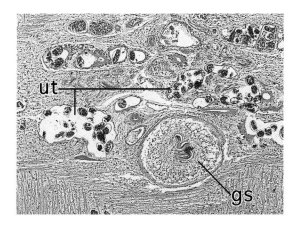

Figure 10.14
Section from medial section of proglottid of *D. latum,* demonstrating common genital sucker (gs) and uterus (ut) with eggs. x60

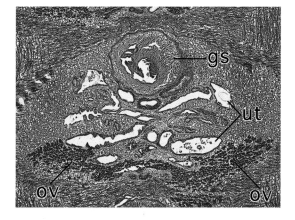

Figure 10.15
Section through proglottid of *D. latum* depicting genital sucker (gs), uterus (ut), and ovary (ov). The bilobate ovary (ov) extends toward both sides of the proglottid. x60

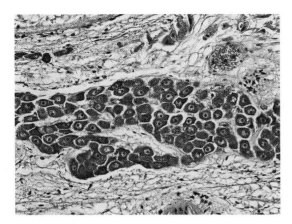

Figure 10.16
Section through ovary of proglottid of *D. latum*. Note multiple polygonal ovarian cells. x230

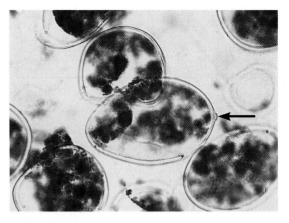

Figure 10.17
Section through uterus of gravid proglottid of *D. latum* showing characteristic broadly ovoid eggs. The abopercular knob (arrow) is readily visible, but it is difficult to see the operculum in sections. x610

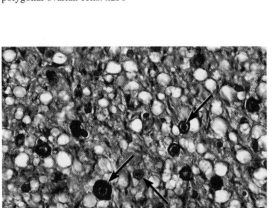

Figure 10.18
Section of proglottid of *D. latum* demonstrating several calcareous corpuscles (arrows) in parenchyma. x595

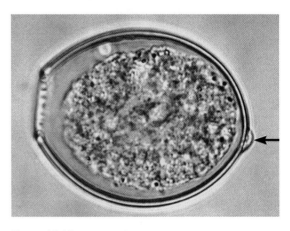

Figure 10.19
Egg of *D. latum* in feces demonstrating operculum at 1 end and knob (arrow) at abopercular end. Egg has moderately thick, light golden-yellow shell, and is broadly ovoid. Unstained x1800

ing on its final location. Most larvae lodge and die in subcutaneous tissue or muscle, but a sparganum may locate anywhere, including breast (Fig 10.30), orbit, urinary tract, pleural cavity, pericardium, abdominal cavity, abdominal viscera, brain, and spinal cord. Symptoms do not usually develop until the sparganum lodges or dies in tissue.

Migration of the sparganum through subcutaneous tissue is often painless and goes unnoticed. Some patients, however, especially those from the Western Hemisphere, have slowly growing, painful, sometimes migratory subcutaneous nodules.[16] Lesions in the breast may produce a palpable nodule visible on mammography. Surgically re-

moved nodules may contain a motile metacestode. Some sparganaare incidental surgical findings.

Ocular sparganosis is common in Asia and is usually attributed to application of infected frog flesh to eyes. The sparganum often invades only the conjunctiva, but can enter periorbital tissues as well. In the subconjunctival tissues sparganaprovoke pruritus, pain, edema, and lacrimation.[5] Rarely, a sparganum may enter the globe and cause endophthalmitis.[64]

The major presenting symptoms of cerebral sparganosis are headache, seizure, and progressive weakness.[11] Patients may develop hemiparesis or hemianesthesia,[47] or a variety of other focal

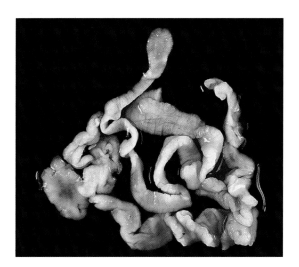

Figure 10.20
Sparganum from subcutaneous nodule in inguinal region. Anterior end is flattened and grooved vertically. Body width varies from broad to narrow and string-like. x1.6

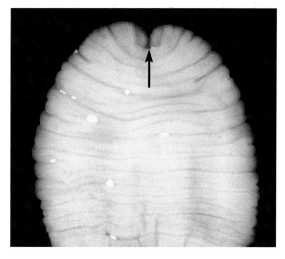

Figure 10.21
Enlarged anterior end of sparganum in Figure 10.20. Note flattened, immature, wrinkled anterior end with cleft-like invagination, and rudimentary bothrium (arrow). x25

neurologic deficits.[9] A dead or degenerated sparganum may appear as a calcified mass in the brain, and can produce intraventricular hemorrhage and obstructive hydrocephalus.[73]

Intraspinal sparganosis can induce paraparesis and autonomic functional deficits,[13,29] and inner ear involvement can provoke vertigo or deafness.[30] Spargana may perforate the intestine, causing peritonitis. Sparganosis of the urinary bladder sometimes leads to cystitis.[54] Other complications include pulmonary and renal infarction, or involvement of the perirenal fat, intestinal wall, peritoneal cavity, or pleural cavity.

Sparganum proliferum usually forms gradually, expanding and ulcerating cutaneous and subcutaneous papules and nodules.[42] Elephantiasis-like thickening of soft tissue and pruritic vermiculous skin nodules are common. If the infection becomes systemic, various symptoms may arise, including hemoptysis, enlargement of the liver and spleen, vocal disturbances, seizures, and motor and sensory deficits.[42] Electron microscopy has revealed virus-like particles, suggesting that a viral infection may genetically alter a sparganum to produce *S. proliferum*.[46]

Pathologic Features

Diphyllobothriasis

The adult tapeworm lies folded in loops in the small intestine. Attachment to the intestinal wall is usually at the level of the ileum, and less commonly in the jejunum or other levels. Rarely, the worm attaches in a bile duct. There are no known specific histologic changes in the intestine related to the adult tapeworm. In patients with bothriocephalus anemia, the peripheral blood reveals abnormal red and white cells and platelets. Mean corpuscular volume is increased (110 to 150 μm),[3] and there are many macro-ovalocytes. Anisopoikilocytosis is prominent. Red blood cells may contain inclusions, Howell-Jolly bodies, basophilic stippling, and Cabot rings. Multilobular neutrophils are prominent, with at least 3% of the neutrophils having 5 or more lobes. Platelets are decreased and often in bizarre forms. Bone marrow biopsy reveals a hypercellular, megaloblastic marrow, with a decreased myeloid:erythroid ratio (1:1 or less). Elements of the granulocytic series have giant metamyelocytes that contain large, C-shaped nuclei with moderately fine chromatin,

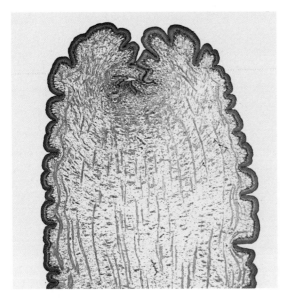

Figure 10.22
Stained section of anterior end of sparganum demonstrating irregular body wall with ridges and inward folds that produce the pseudosegmentation visible in Figure 10.21. Note also the cleft-like rudimentary bothrium. x52

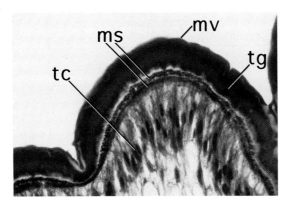

Figure 10.23
Section of body wall of sparganum depicting the thin outer layer of microvilli (mv), thick tegument (tg), 2 layers of smooth muscle (ms), and a row of tegumental cells (tc). Movat x560

and there may be neutrophils with multilobular nuclei. Megakaryocytes may be reduced in numbers.[63]

Sparganosis

A sparganum may be extracted directly from an abscess or cyst. If viable, the larva has a slow, undulating movement. A single lesion may sometimes contain more than 1 sparganum.[29]

Nodules containing living spargana may have little inflammatory reaction. To facilitate migration, spargana secrete proteases that degrade extracellular matrices of collagen, fibronectin, and myosin.[66]

Living spargana escape immunologic detection, probably by cleavage of immunoglobulin by the product of a cathepsin S-like, allergenic protease,[32] or by selective release of prostaglandin E_2.[20]

A dead or dying sparganum elicits a marked inflammatory reaction (Fig 10.31). These relatively long larval cestodes may degenerate in a particular section and remain viable at other levels. Tissue reaction near the more viable portion of sparganum may be minimal or consist of acute inflammation (Figs 10.32 and 10.33), whereas a more chronic inflammatory process surrounds the degenerated portion (Figs 10.34 and 10.35).

In subcutaneous tissue, an abscess may surround sections of viable or dying sparganum. The abscess may include granulation tissue, fibrous tissue, and variable numbers of inflammatory cells, especially neutrophils, but also lymphocytes, macrophages, and plasma cells (Figs 10.32 and 10.33). Eosinophils may be absent or numerous (Fig 10.33); when numerous, the reaction may contain Charcot-Leyden crystals. Perivascular chronic inflammation may be prominent and usually includes lymphoid follicles (Figs 10.31 and 10.32). Granulomatous reactions with palisading epithelioid cells usually surround degenerated worms (Fig 10.34). Long-standing granulomas around degenerated spargana may contain caseation necrosis (Fig 10.35).

In cerebral sparganosis, the metacestode initially causes an abscess, and there is perivascular cuffing by lymphocytes and plasma cells (Figs 10.36 and 10.37). There may be extensive destruction of white matter and granuloma formation with giant cells and fibrosis if the sparganum is degenerated (Figs 10. 38 to 10.40). Necrotic areas of the brain containing an almost completely destroyed sparganum may contain only calcareous corpuscles (Fig 10.41).

Nodules of *S. proliferum* reveal multiple, variably sized cystic spaces, with surrounding chronic inflammation and mild fibrosis (Figs 10.25 and 10.26). The cystic spaces contain sections of proliferating larval tissue.

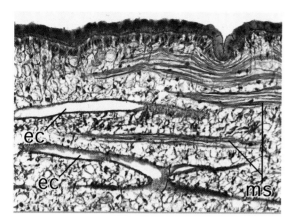

Figure 10.24
Section of sparganum demonstrating irregularly scattered bundles of longitudinal muscle fibers (ms), branching excretory channels (ec), and mesenchymal fibers in a loose stroma. Movat x220

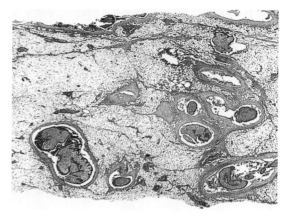

Figure 10.25
Section of subcutaneous tissue from thigh of Japanese woman. Panniculitis and multiple sections of branching, proliferating sparganum are evident.* Movat x7.8

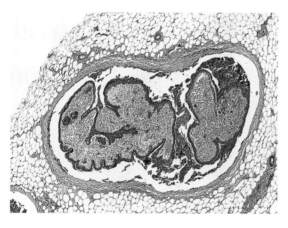

Figure 10.26
Higher magnification of branching, proliferating sparganum from Figure 10.25. Note band of fibrous tissue surrounding this pair of sections of different branches of the sparganum. Movat x25

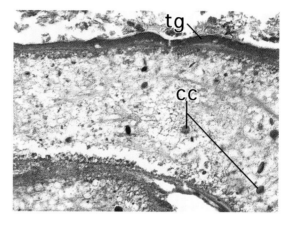

Figure 10.27
Higher magnification of proliferating sparganum in Figure 10.25. Note tegument (tg) and calcareous corpuscles (cc) in loose stroma. Movat x240

Diagnosis

Diagnosis of diphyllobothriasis is usually based on recovery and identification of characteristic proglottids or eggs in feces or, less often, in vomitus. Direct fecal smears readily demonstrate eggs. Eggs of *D. latum* are difficult to distinguish from those of other *Diphyllobothrium* sp and from those of *D. grandis*. Eggs of *Paragonimus* sp and other trematodes are sometimes confused with those of *Diphyllobothrium*. *Paragonimus* eggs are larger (usually greater than 80 μm) and do not have a knob at the abopercular end. Eggs of the fluke *Nanophyetus salmincola* strongly resemble those of *Diphyllobothrium* sp and should be ruled out in areas where *Nanophyetus* is endemic.[74]

Sparganosis in humans has varying clinical presentations, but a presumptive diagnosis can be rendered in a patient with a migrating subcutaneous crepitant mass and a history of consuming raw snakes or frogs. Since the landmark article by Kim et al,[29] serologic study combined with radiologic evidence is the standard approach to diagnosing sparganosis. There are several serologic tests for sparganosis.[79] Chemiluminescence ELISA appears

Figure 10.28
Life cycle of *D. latum* has 5 phases: 1) embryo in egg; 2) free-swimming first-stage larva (coracidium); 3) second-stage larva (procercoid) in *Diaptomus*, the copepod host; 4) third-stage larva (plerocercoid or sparganum) in fish; and 5) adult tapeworm in definitive host.

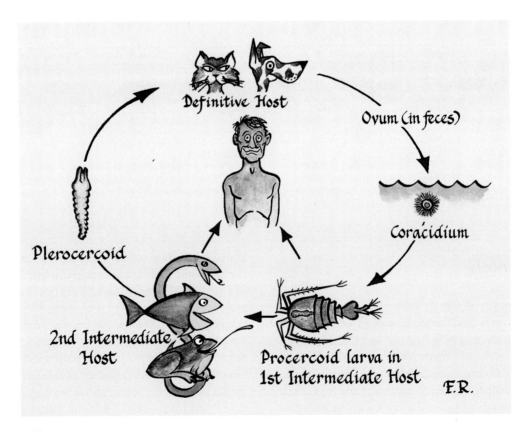

Figure 10.29
Life cycle of *S. mansoni*. Intermediate and definitive hosts differ from those of *D. latum*. Humans may serve as second intermediate or paratenic hosts, but not as definitive hosts.

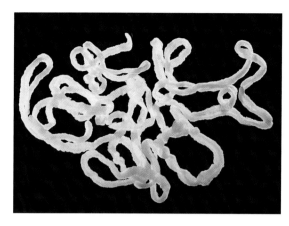

Figure 10.30
Sparganum removed from breast of 49-year-old Cambodian woman with a subcutaneous nodule. Worm was 50 cm long. x1.8

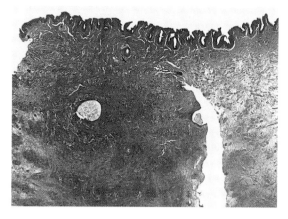

Figure 10.31
Section of scrotal skin depicting a sparganum provoking marked inflammatory reaction. Note multiple lymphoid follicles and abscess around sparganum. x9

Figure 10.32
Sparganum in same specimen shown in Figure 10.31. Note lymphoid follicles and acute inflammation surrounding worm. x25

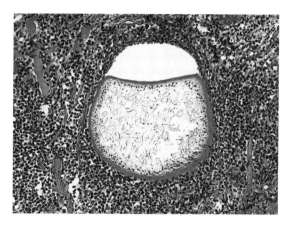

Figure 10.33
Numerous eosinophils surround this viable section of sparganum in scrotal skin. x135

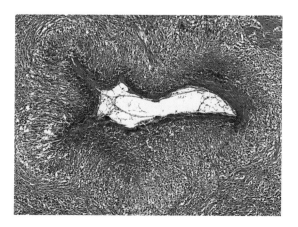

Figure 10.34
Degenerated fragment of sparganum in scrotal skin. Granulomatous reaction with palisading epithelioid cells surround the sparganum. x65

Figure 10.35
Section of scrotal skin showing area where sparganum has completely degenerated and left only a granuloma with central caseation necrosis. x11

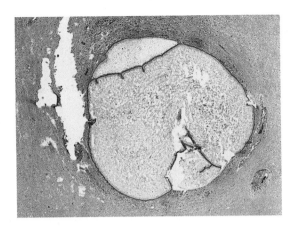

Figure 10.36
Cerebral cortex showing viable segment of sparganum, with chronic inflammation surrounding the worm and adjacent perivascular cuffing of blood vessels. x25

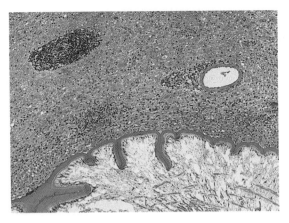

Figure 10.37
Cerebral cortex showing interface of chronic inflammation between a viable portion of sparganum and brain tissue. Note also perivascular cuffing by lymphocytes and plasma cells of blood vessels. x60

highly sensitive and specific for *S. mansoni*.[52] At present, the only method for determining the species of sparganum is to place the living specimen in an appropriate definitive host and identify the resultant adult. Thus, the diagnosis of sparganosis in humans is usually not carried to species level.

In cerebral sparganosis, magnetic resonance imaging gives the best contrast between the normal and degenerated brain tissues and defines lesions more precisely than computed tomography (CT).[41] CT, however, is more helpful for detecting small foci.[10]

Gross or microscopic identification of the sparganum provides a definitive diagnosis. The elongated, wormlike form usually distinguishes a sparganum from the cystic or bladder structure of a cysticercus or a coenurus. Cysticerci and coenuri have protoscolices containing suckers and hooklets, whereas spargana do not. If only the bladder wall of a cyclophyllidean metacestode (cysticercus or coenurus) is visible, it is usually thinner than that of a sparganum and lacks the inwardly branching tegument.[40] A small brain biopsy may be sufficient to diagnose sparganosis.[70]

Treatment and Prevention

Praziquantel is the current drug of choice for diphyllobothriasis.[55,68] Other useful drugs include niclosamide, bithionol, paromomycin sulfate, and quinacrine hydrochloride. Preventing the contamination of lakes and rivers by sewage, and thorough cooking of fish before consumption, control diphyllobothriasis.[69] Freezing infected fish or roe at -18 °C for 1 or 2 days (depending on the size of the fish) kills the metacestode.

Spargana are resistant to praziquantel and gamma-irradiation.[65] Currently, drug therapy is not satisfactory for sparganosis and surgical excision is recommended. Data suggest that the vitality center of the sparganum seems to be near the anterior end of its scolex. Thus, it is important to excise the entire sparganum, taking special care to remove the anterior end of the larva. Effective preventive measures for sparganosis include avoiding unfiltered drinking water, cleaning vegetables and fruits with filtered water, thoroughly cooking all meats, and not using animal flesh as a poultice.

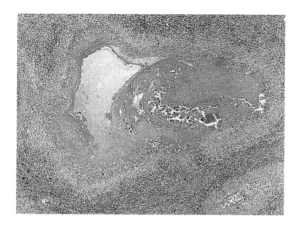

Figure 10.38
Granuloma in brain around degenerated segment of sparganum. Granuloma is centered around calcareous corpuscles and other remnants of degenerated worm. Epithelioid cells palisade around periphery of granuloma. Note also cerebral edema. x25

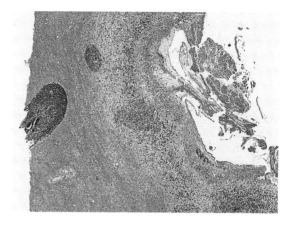

Figure 10.39
Lesion in brain around degenerated segment of sparganum. Note chronic inflammation, white-matter degeneration, and prominent perivascular cuffing. x25

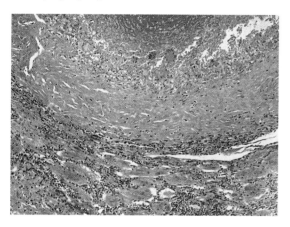

Figure 10.40
Edge of granuloma around degenerating segment of sparganum in cerebral cortex. Note fibrosis, giant cells, and necrosis. x60

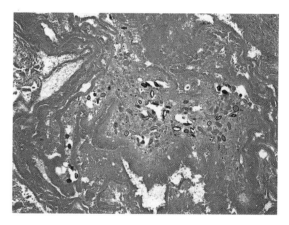

Figure 10.41
Necrotic area of brain previously occupied by a sparganum. Only calcareous corpuscles remain. x115

References

1. Alarcon de Noya B, Torres JR, Noya O. Maintenance of Sparganum proliferum in vitro and in experimental animals. *Int J Parasitol* 1992;22:835–838.
2. Andersen K. Cited by: Yamane Y, Kamo H, Bylund G, Wikgren BP. Diphyllobothrium nihonkaiense sp. nov. (Cestode: Diphyllobothriidae)– revised identification of Japanese broad tapeworm. *Shimane J Med Sci* 1986;10:29–48.
3. Ash LR, Orihel TC. *Atlas of Human Parasitology*. 3rd ed. Chicago, Ill: ASCP Press; 1990:216–219.
4. Atías A, Cattan PE. First case of human infection with Diphyllobothrium pacificum in Chile [in Spanish]. *Rev Med Chil* 1976;104:216–217.
5. Ausayakhun S, Siriprasert V, Morakote N, Taweesap K. Ocular sparganosis in Thailand. *Southeast Asian J Trop Med Public Health* 1993;24:603–606.
6. Beaver PC, Rolon FA. Proliferating larval cestode in a man in Paraguay. A case report and review. *Am J Trop Med Hyg* 1981;30:625–637.
7. Beaver PC, Jung RC, Cupp EW. *Clinical Parasitology*. 9th ed. Philadelphia, Pa: Lea & Febiger; 1984:494–504.
8. Catár G, Sobota K, Kvasz L, Hruzik J. The first nonimported case of diphyllobothriasis in Czechoslovakia [English abstract]. *Bratisl Lek Listy* 1967;47:241–244.
9. Chamadol W, Tangdumrongkul S, Thanaphaisal C, Sithithaworn P, Chamadol N. Intracerebral hematoma caused by sparganum: a case report. *J Med Assoc Thai* 1992;75:602–605.
10. Chang KH. Cited by: Kradel J, Drolshagen LF, MacDade A. MR and CT findings in cerebral sparganosis. *J Comput Assist Tomogr* 1993;17:989–990.
11. Chang KH, Chi JG, Cho SY, Han MH, Han DH, Han MC. Cerebral sparganosis: analysis of 34 cases with emphasis on CT features. *Neuroradiology* 1992;34:1–8.
12. Cho SY, Chung YB, Kong Y. Component proteins and protease activities in excretory-secretory product of sparganum. *Korean J Parasitol* 1992;30:227–230.
13. Cho YD, Huh JD, Hwang YS, Kim HK. Sparganosis in the spinal canal with partial block: an uncommon infection. *Neuroradiology* 1992;34:241–244.
14. de Roodt AR, Suarez G, Ruzic A, Bellegarde E, Braun M, Blanco CM. A case of human sparganosis in Argentina [in Spanish]. *Medicina (B Aires)* 1993;53:235–238.
15. Deliamure SL, Skriabin AS. Identification of species of Diplogonoporus balaenopterae and D. grandis [in Russian]. *Parazitologiia* 1986;20:69–72.
16. Deshpande VV. Subcutaneous sparganosis–a rare infection in man. *J Assoc Physicians India* 1993;41:685.
17. Faust EC, Campbell H, Kellogg CR. Morphological and biological studies on the species of Diphyllobothrium in China. *Am J Hyg* 1929;9:560–583.
18. Fukumoto S (1988a). Cited by: Ohnishi K, Murata M. Single dose treatment with praziquantel for human Diphyllobothrium nihonkaiense infections. *Trans R Soc Trop Med Hyg* 1993;87:482–483.
19. Fukumoto S (1988b). Cited by: Ohnishi K, Murata M. Single dose treatment with praziquantel for human Diphyllobothrium nihonkaiense infections. *Trans R Soc Trop Med Hyg* 1993;87:482–483.
20. Fukushima T, Isobe A, Hojo N, Shiwaku K, Yamane Y, Torii M. The metabolism of arachidonic acid to prostaglandin E2 in plerocercoids of Spirometra erinacei. *Parasitol Res* 1993;79:634–638.
21. Georgi JR. Tapeworms. *Vet Clin North Am Small Anim Pract* 1987;17:1285–1305.
22. Grove DI. *A History of Human Helminthology*. Wallingford, England: CAB Int; 1990:428–431.
23. Ijima I. Cited by: Nakamura T, Hara M, Matsuoka M, Kawabata M, Tsuji M. Human proliferative sparganosis. A new Japanese case. *Am J Clin Pathol* 1990;94:224–228.
24. Kamo H. Cited by: Ohnishi K, Murata M. Single dose treatment with praziquantel for human Diphyllobothrium nihonkaiense infections. *Trans R Soc Trop Med Hyg* 1993;87:482–483.
25. Kamo H, Hatsushika R, Yamane Y, Ishihara K, Tanaka M. Vital strobilae of Diplogonoporus grandis evacuated spontaneously from man. *Yonago Acta Med* 1969;13:31–36.
26. Kamo H, Hatsushika R, Yamane Y, Nishida H. Diplogonoporus grandis from man in the coastal area of the Japan Sea. *Yonago Acta Med* 1968;12:183–191.
27. Kamo H, Miyazaki I. A case of human infection with unknown species of Diplogonoporus in Japan. *Yonago Acta Med* 1971;15:55–60.
28. Kamo H, Yazaki S, Fukumoto S, et al. The first human case infected with Diphyllobothrium hians (Diesing, 1850). *Jpn J Parasitol* 1988;37:29–35.
29. Kim H, Kim SI, Cho SY. Serological diagnosis of human sparganosis by means of micro-ELISA. *Korean J Parasitol* 1984;22:222–228.
30. Kittiponghansa S, Tesana S, Ritch R. Ocular sparganosis: a cause of subconjunctival tumor and deafness. *Trop Med Parasitol* 1988;39:247–248.
31. Kong Y, Chung YB, Cho SY, Choi SH, Kang SY. Characterization of three neutral proteases of Spirometra mansoni plerocercoid. *Parasitology* 1994;108:359–368.
32. Kong Y, Chung YB, Cho SY, Kang SY. Cleavage of immunoglobulin G by excretory-secretory cathepsin S-like protease of Spirometra mansoni plerocercoid. *Parasitology* 1994;109:611–621.
33. Kuhlow F. Bau und Differentialdiagnose heimischer Diphylloboth-rium-Plerocercoide. *Z Tropenmed Parasitol* 1953;4:186–202.
34. Kyrönseppä H. The occurrence of human intestinal parasites in Finland. *Scand J Infect Dis* 1993;25:671–673.
35. LaChance MA, Clark RM, Connor DH. Proliferating larval cestodiasis: report of a case. *Acta Trop* 1983;40:391–397.
36. Lee SH, Chai JY, Seo M, et al. Two rare cases of Diphyllobothrium latum parvum type infection in Korea. *Korean J Parasitol* 1994;32:117–120.
37. Lou YS, Koga M, Higo H, et al. A human infection of the cestode, Diphyllobothrium nihonkaiense Yamane et al, 1986 [in Japanese]. *Fukuoka Igaku Zasshi* 1989;80:446–450.
38. Magath TB. Experimental studies on Diphyllobothrium latum. *Am J Trop Med* 1929;9:17–48.
39. Margolis L, Rausch RL, Robertson E. Diphyllobothrium ursi from man in British Columbia–first report of this tapeworm in Canada. *Can J Public Health* 1973;64:588–589.
40. Marty AM, Chester AJ. Distinguishing lipid pseudomembranes from larval cestodes by morphologic and histochemical means. *Arch Pathol Lab Med* 1997;121:900–907.
41. Moon WK, Chang KH, Cho SY, et al. Cerebral sparganosis: MR imaging versus CT features. *Radiology* 1993;188:751–757.
42. Moulinier R, Martinez E, Torres J, Noya O, de Noya BA, Reyes O. Human proliferative sparganosis in Venezuela: report of a case. *Am J Trop Med Hyg* 1982;31:358–363.
43. Mueller JF. The life history of Diphyllobothrium mansonoides Mueller, 1935, and some considerations with regard to sparganosis in the United States. *Am J Trop Med* 1938;18:41–66.
44. Mueller JF. A repartition of the genus Diphyllobothrium. *J Parasitol* 1937;23:308–310.
45. Mueller JF, Hart EP, Walsh WP. Human sparganosis in the United States. *J Parasitol* 1963;49:294–296.

46. Mueller JF, Strano AJ. Sparganum proliferum, a sparganum infected with a virus? *J Parasitol* 1974;60:15–19.
47. Munckhof WJ, Grayson ML, Susil BJ, Pullar MJ, Turnidge J. Cerebral sparganosis in an East Timorese refugee. *Med J Aust* 1994;161:263–264.
48. Muratov IV, Posokhov PS, Romanenko NA, Kozyreva TG, Skulkina AI. The infectivity of the population with the tapeworm Diphyllobothrium klebanovskii in an area of infection drift in the Khabarovsk Territory [in Russian]. *Med Parazitol (Mosk)* 1992;2:30–32.
49. Nakamura T, Hara M, Matsuoka M, Kawabata M, Tsuji M. Human proliferative sparganosis. A new Japanese case. *Am J Clin Pathol* 1990;94:224–228.
50. Nakazawa M, Amano T, Oshima T. The first record of human infection with Diphyllobothrium orcini Hatsushika and Shirouzu, 1990 [in Japanese; English abstract]. *Jpn J Parasitol* 1992;41:306–313.
51. Neghme A, Donckaster R, Silva R. Diphyllobothrium latum in Chile [in Spanish]. *Rev Med Chil* 1950;78:410–411.
52. Nishiyama T, Ide T, Himes SR Jr, Ishizaka S, Araki T. Immuno-diagnosis of human Sparganosis mansoni by microchemiluminescence enzyme-linked immunosorbent assay. *Trans R Soc Trop Med Hyg* 1994;88:663–665.
53. Noya O, Alarcon de Noya B, Arrechedera H, Torres J, Arguello C. Sparganum proliferum: an overview of its structure and ultrastructure. *Int J Parasitol* 1992;22:631–640.
54. Oh SJ, Chi JG, Lee SE. Eosinophilic cystitis caused by vesical sparganosis: a case report. *J Urol* 1993;149:581–583.
55. Ohnishi K, Murata M. Single dose treatment with praziquantel for human Diphyllobothrium nihonkaiense infections. *Trans R Soc Trop Med Hyg* 1993;87:482–483.
56. Pasternak AF, Huntingford FA, Crompton DW. Changes in metabolism and behavior of the freshwater copepod Cyclops strenuus abyssorum infected with Diphyllobothrium spp. *Parasitology* 1995;110:395–399.
57. Peters L, Cavis D, Robertson J. Is Diphyllobothrium latum currently present in northern Michigan? *J Parasitol* 1978;64:947–949.
58. Rausch RL, Hilliard DK. Studies on the helminth fauna of Alaska. 49. The occurrence of Diphyllobothrium latum (Linnaeus, 1758) (Cestoda: Diphyllobothriidae) in Alaska, with notes on other species. *Can J Zool* 1970;48:1201–1219.
59. Revenga JE. Diphyllobothrium dendriticum and Diphyllobothrium latum in fishes from southern Argentina: association, abundance, distribution, pathological effects, and risk of human infection. *J Parasitol* 1993;79:379–383.
60. Reyher G. Beiträge zur Aetiologie und Heilbarkeit der perniciösen Anämie. *Dtsch Arch Klin Med* 1886;39:31–69.
61. Romanenko NA, Novosil'tsev GI, Skripova LV, Muratov IV, Glazyrina GF, Pogorel'chuk TI. The sanitary parasitological characteristics of different sources of drinking water supply [in Russian]. *Med Parazitol (Mosk)* 1993;5:56–59.
62. Schmidt GD. *CRC Handbook of Tapeworm Identification*. Boca Raton, Fla: CRC Press; 1986.
63. Schumacher HR, Garvin DF, Triplett DA. *Introduction to Laboratory Hematology and Hematopathology*. New York, NY: Alan R Liss Inc; 1984:287–303.
64. Sen DK, Muller R, Gupta VP, Chilana JS. Cestode larva (Sparganum) in the anterior chamber of the eye. *Trop Geogr Med* 1989;41:270–273.
65. Sohn WM, Hong ST, Chai JY, Lee SH. Infectivity of the sparganum treated by praziquantel, gamma-irradiation and mechanical cutting [in Korean; English abstract]. *Korean J Parasitol* 1993;31:135–139.
66. Song CY, Chappell CL. Purification and partial characterization of cysteine proteinase from Spirometra mansoni plerocercoids. *J Parasitol* 1993;79:517–524.
67. Stiles CW. The occurrence of a proliferating cestode larva (Sparganum proliferum) in man in Florida. *Bulletin of the Hygienic Lab, Government Printing Office, Treasury Department, Public Health and Marine-Hospital Service of the United States.* 1908;40:7–18.
68. Tanowitz HB, Weiss LM, Wittner M. Diagnosis and treatment of intestinal helminths. 1. Common intestinal cestodes. *Gastroenterologist* 1993;1:265–273.
69. Torres P, Franjola R, Weitz JC, Pena G, Morales E. New records of human diphyllobothriasis in Chile (1981–1992), with a case of multiple Diphyllobothrium latum infection [in Spanish]. *Bol Chil Parasitol* 1993;48:39–43.
70. Tsai MD, Chang CN, Ho YS, Wang AD. Cerebral sparganosis diagnosed and treated with stereotactic techniques. Report of two cases. *J Neurosurg* 1993;78:129–132.
71. von Bonsdorff B. The broad tapeworm story. *Acta Med Scand* 1978;204:241–247.
72. Whittington R, Middleton D, Spratt DM, et al. Sparganosis in the monotremes Tachyglossus aculeatus and Ornithorhynchus anatinus in Australia. *J Wildl Dis* 1992;28:636–640.
73. Wong CW, Ho YS. Intraventricular haemorrhage and hydrocephalus caused by intraventricular parasitic granuloma suggesting cerebral sparganosis. *Acta Neurochir (Wien)* 1994;129:205–208.
74. World Health Organization. Control of foodborne trematode infections. Report of a WHO study group. *World Health Organ Tech Rep Ser* 1995;849:1–157.
75. Yamane Y. An electron-microscope study of the tegument of Diplogonoporus grandis. *Yonago Acta Med* 1969;13:25–29.
76. Yamane Y, Bylund G, Abe K, Osaki Y, Okamoto T. Scanning electron microscopic study of four Diphyllobothrium species. *Parasitol Res* 1989;75:238–244.
77. Yamane Y, Kamo H, Bylund G, Wikgren BP. Diphyllobothrium nihonkaiense sp. nov. (Cestode: Diphyllobothriidae)–revised identification of Japanese broad tapeworm. *Shimane J Med Sci* 1986;10:29–48.
78. Yamane Y, Kamo H, Yazaki S, Fukumoto S, Maejima J. On a new marine species of the genus Diphyllobothrium (Cestoda: Pseudophyllidea) found from a man in Japan. *Jpn J Parasitol* 1981;30:101–111.
79. Yeo IS, Yong TS, Im K. Serodiagnosis of human sparganosis by a monoclonal antibody-based competition ELISA. *Yonsei Med J* 1994;35:43–48.
80. Zhong HL, Shao L, Lian DR, et al. Ocular sparganosis caused blindness. *Chin Med J (Engl)* 1983;96:73–75.

11

Coenurosis

Aileen M. Marty *and*
Ronald C. Neafie

Introduction

Definition

Coenurosis is infection by a coenurus, the metacestode of *Taenia multiceps* (Leske, 1780) or *Taenia serialis* (Gervais, 1847) Baillet, 1863. Coenuri are bladder worms with multiple protoscolices. The metacestode of *T. multiceps* is the most commonly recognized coenurus and the one that most frequently causes serious illness. The coenurus of *T. multiceps* tends to locate in the central nervous system and eye, while that of *T. serialis* more commonly localizes in subcutaneous tissue.[10]

Synonyms

Multiceps multiceps, *Multiceps gaigeri*, and *Multiceps skrjabini* are synonyms for the adult *T. multiceps*. Older names for *T. serialis* are *Taenia packii*, *Taenia antarctica*, *Taenia laruei,* and *Multiceps serialis*. Beaver et al list 2 additional *Taenia* sp that reproduce with a coenurus as a larval stage: *Taenia brauni* (Setti, 1897) Fain et al, 1956, and *Taenia glomerata* (Railliet and Henry, 1915) Brumpt, 1932.[3] Division of these adult tapeworms into multiple species is based on the most common intermediate hosts: *T. multiceps* (brain of sheep and other ungulates), *T. serialis* (intramuscular and subcutaneous tissue of rabbits and rodents), *T. brauni* (monkeys and wild rodents in Africa), and *T. glomerata* (murine rodents in tropical Africa). We accept the view of Graber[15] and others who consider *T. brauni* and *T. glomerata* synonymous with, or as subspecies of, *T. serialis*. *Taenia brauni* and *T. glomerata* are morphologically indistinguishable from *T. serialis*. Larval stages of cestode parasites of carnivorous mammals are far less host-specific than the adult worms. Thus, classifying metacestode larvae according to their intermediate hosts is probably not reliable. The term *M. serialis*, often applied to coenuri found outside the central nervous system (CNS), is a synonym for *T. serialis*. Common names for metacestodes of *T. multiceps* include gid worm and gid tapeworm. Gid, sturdy, and blind staggers are popular names for coenurosis in sheep and calves. *Coenurus cerebralis* is a descriptive term for a coenurus in the CNS and has no taxonomic significance. *Coenurus glomeratus*, *Coenurus glomerulatus*, and *Multiceps glomeratus* are older terms for larval

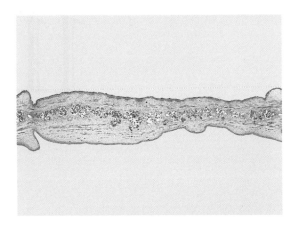

Figure 11.1
Gravid proglottids of *T. serialis* are longer than wide. x15

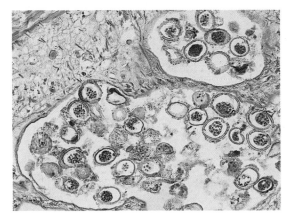

Figure 11.2
Gravid proglottid of *T. serialis* demonstrating many oval eggs with thick striated shells in uterus. x115

and adult *T. glomerata*, which we accept as synonymous with *T. serialis*.

General Considerations

Leske described *T. multiceps* in 1780 and Batsch described *C. cerebralis* in 1786.[10] In 1913, Brumpt recorded the first known case of human coenurosis: a Parisian locksmith with convulsions and aphasia, who at autopsy had 2 coenuri (probably *T. multiceps*) in the brain.[7] In 1919, Turner and Leiper found a coenurus in the intercostal muscle of a Nigerian and reported it as *C. glomerulatus*.[37] In 1933, Bonnal et al confirmed the first infection with *T. serialis* in humans by feeding a living coenurus from a French woman to a dog and recovering 7 adult tapeworms with characteristic scolices.[6] In 1950, Johnstone and Jones described the first known infection acquired in the United States: an infant from California, who had lived in Nevada and Idaho, and had neural coenurosis.[20] Fain et al documented 8 human infections with *T. serialis* (synonym, *T. brauni*) in Rwanda-Urundi in 1956.[14]

Epidemiology

Taenia multiceps and *T. serialis* infections have a wide distribution and prevail in children.[42] Of the several hundred documented patients with coenurosis, the majority are from Africa,[21,36] but the disease occasionally appears in sheep-raising areas of Europe (France, Great Britain, and Sardinia),[11] South America,[9] and the United States.[2,17,27,35] There have also been sporadic reports of human infection in Canada.[4] Identifying a coenurus species from gross or histologic features may be impossible. In tropical Africa, most patients with coenurosis who have a single subcutaneous swelling, visible or palpable, in striated muscle or subcutaneous tissue probably are infected with *T. serialis*.[41] In South Africa, nearly all patients come from sheep-raising areas and have cerebral coenurosis, most likely of *T. multiceps*. The majority of patients in the United States and Great Britain have lesions in the CNS, with *T. multiceps* as the etiologic agent. In France, both subcutaneous and cerebral coenurosis are well-known, implicating both *T. multiceps* and *T. serialis*. Prevalence of coenurosis in rabbits in the central plains of the United States ranges from 4.2% to 19.2%; infection is also common in mink in this area. These data suggest that coenurosis from *T. serialis* in the United States may be more prevalent than is currently suspected.

Infectious Agent

Morphologic Description

Because of their shared morphologic features, it is difficult to separate adult tapeworms reproducing by a coenurus from those reproducing by a

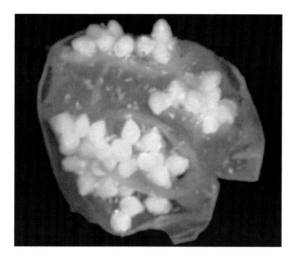

Figure 11.3
Coenurus (1 cm in diameter) from a human. The cyst is polycephalic and contains multiple protoscolices in clumps. ×6.5

Figure 11.4
Coenurus from jackrabbit in the United States. Cyst is polycephalic, oval, and 3.1 cm in diameter. Cyst wall is white-yellow. Protoscolices are in clumps on 1 side of cyst.

cysticercus. Adult tapeworms that produce coenuri differ from those forming cysticerci by having a rostellum armed with sinuous handles on the rostellar hooklets, and by usually having proglottids with a reflex loop of the vagina near the portal osmoregulatory canals.[38]

Adults

Taenia multiceps

Adult *T. multiceps* are 40 to 60 cm long and have a piriform scolex with 4 suckers and an armed rostellum with a double circle of 22 to 33 hooklets. Gravid proglottids are longer than wide and measure 8 to 10 mm by 3 to 4 mm. Gravid proglottids have 18 to 26 uterine branches per side. Vaginal sphincters in the proglottid are absent or poorly developed.

Taenia serialis

Adult *T. serialis* are 20 to 72 cm long. The scolex has 4 suckers and an armed rostellum with a double circle of 26 to 32 hooklets. Proglottids are longer than wide (Fig 11.1). Gravid proglottids have 20 to 25 uterine branches per side. In contrast to *T. multiceps*, gravid proglottids have well-developed vaginal sphincters.

Eggs

Eggs of *T. multiceps* and *T. serialis* are 29 to 37 μm in diameter, spherical, and have a thick, radially striated shell (Fig 11.2). Eggs contain a 6-hooked oncosphere and are morphologically indistinguishable from eggs of other *Taenia* sp.

Metacestode (Coenurus)

Metacestodes (coenuri) are usually thin-walled, white or gray, spherical or ovoid polycephalic cysts, varying from a few millimeters to 6 cm in diameter (Figs 11.3 and 11.4). A coenurus contains gray or white jelly-like fluid and multiple yellow-white protoscolices. The numerous protoscolices (Figs 11.3 and 11.4) distinguish a coenurus from a cysticercus, which contains a single protoscolex (Fig 7.12). There are usually 50 to 100 protoscolices per cyst, but there may be as many as 700. Coenuri vary morphologically depending on the intermediate host and the tissue in which they lodge. Some authorities state that the serial arrangement of protoscolices is diagnostic of *T. serialis*,[23,34] but others believe that characteristics such as the arrangement of protoscolices, number of calcareous corpuscles, and the tendency to bud exogenously are unpredictable and cannot be used to identify the species of a coenurus larva. Coenuri may be multilocular, but most are unilocular. Multilocular coenuri have been interpreted as those of *T. multiceps* in the eye[39] and in the CNS.

The body wall of the coenurus varies from 50 to 150 μm thick and is composed of a tegument,

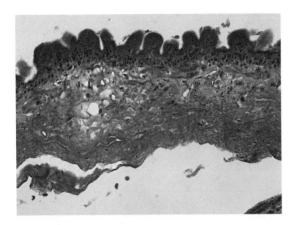

Figure 11.5
Section of cyst wall of coenurus composed of tegument, muscle cells, tegumental cells, and parenchyma. x230

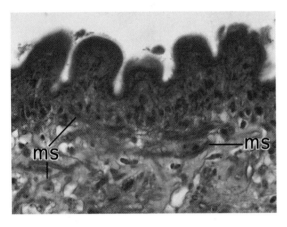

Figure 11.6
Coenurus wall displaying eosinophilic tegument that is raised into irregular projections and contains haphazardly arranged muscle fibers (ms). A layer of tegumental cells separates the tegument from underlying parenchyma. x725

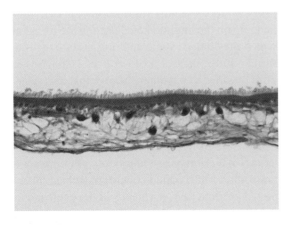

Figure 11.7
Microvilli line outer surface of tegument of a coenurus. x550

muscle cells, and tegumental cells (Fig 11.5). The eosinophilic tegument, about 3 to 5 μm thick, is raised into irregular projections (papillae) up to 25 μm wide, and may contain minute, haphazardly arranged muscle fibers (Fig 11.6). On the surface of the tegument are fine, hair-like structures called microtriches or microvilli, 1 to 2 μm long. The microvilli form a border that completely covers the outer surface of the cyst wall and the invaginated surface of each protoscolex (Fig 11.7). The basement layer, or subcuticle, contains acid mucosaccharides and some reticulin, and separates the tegument from the parenchyma. The parenchyma of the cyst wall contains loose connective tissue with fibrillar ground substance, haphazardly arranged muscle fibers, and few calcareous corpuscles. Within this parenchyma, a system of branched tubules runs parallel to the tegument and radiates vertically towards the surface of the tegument.

Invaginated protoscolices develop from the cyst wall (Figs 11.8 and 11.9) and are attached to the wall by their necks (Fig 11.10). In comparison with other areas of the cyst, the tegument of the neck region is thicker (up to 10 μm) and the parenchymal layer is denser and contains more calcareous corpuscles (Fig 11.11). Multiple invaginated protoscolices usually lie in irregular groups (Figs 11.3 and 11.4). The age of a coenurus can be estimated by the number of protoscolices in each cluster on the coenurus wall.[40] An occasional protoscolex may be everted. Morphologic features of the protoscolex are similar to that of the scolex of the adult tapeworm. A protoscolex is about 500 μm in diameter and has 4 suckers (Fig 11.12) ranging from 130 to 200 μm in diameter. The protoscolex contains reticulin fibers and stains strongly with PAS in the muscular regions, particularly near the suckers. The rostellum of the protoscolex is located at the apex of the neck, and is armed with a double row of hooklets (Fig 11.13). The number of hooklets in each rostellum ranges from 20 to 33. Hooklets range from 46 to 175 μm in length and fall into 3 groups: small, medium, and large (Figs 11.14 and 11.15).[41] The length of large hooklets is considered of greatest diagnostic value.[8,13] Large hooklets of *T. multiceps* range from 120 to 170 μm and those of *T. serialis* from 110 to 175 μm. Hooklets are birefringent and partially acid-fast (Fig 11.16).

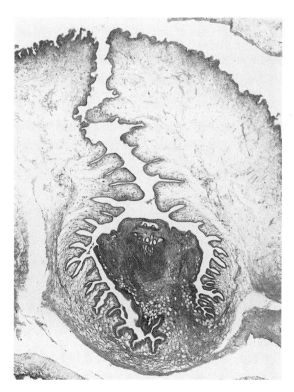

Figure 11.8
Invaginated protoscolex of a coenurus developing from cyst wall. Note invaginated canal. x80

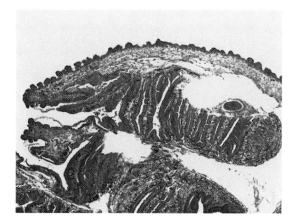

Figure 11.9
Neck of invaginated protoscolex developing from cyst wall. Tegument of neck region is continuous with tegument of cyst wall. x55

Life Cycle and Transmission

Taenia sp require 2 hosts to complete their life cycle (Fig 11.17). The definitive host is a carnivore. Many canids serve as definitive hosts for *T. multiceps*, while *T. serialis* is a tapeworm of dogs and foxes only.

Numerous animal species serve as intermediate hosts; most are herbivores, commonly sheep and rabbits. Other herbivore intermediate hosts include antelope (especially gazelles and gemsboks), chamois, cows, goats, horses, wild rodents, and yaks.[5,31,32] Animals that serve as accidental intermediate hosts include nonhuman primates[24] and cats.[18,34] Humans, especially young children, become inadvertent intermediate hosts by ingesting eggs.

The intermediate host ingests eggs passed in the stool of the definitive host. Eggs hatch in the intestine, releasing oncospheres that actively penetrate the intestinal wall and enter the bloodstream. Oncospheres lodge and develop into coenuri in suitable organs, including the brain, eye, skeletal muscle, and subcutaneous tissue. About 3 months after ingestion, oncospheres develop into coenuri. Coenuri of *T. multiceps* have a predilection for the CNS and those of *T. serialis* for subcutaneous tissue. In sheep, migrating oncospheres of *T. multiceps* may lodge in the liver, heart, kidney, diaphragm, or skeletal muscle. Attempts by metacestode larvae of *T. multiceps* to develop in sites other than the CNS are usually aborted by an acute host response; however, in rare instances these coenuri develop in subcutaneous tissue.[16]

The life cycle is completed when a dog or other suitable carnivore eats the carcass of an infected intermediate host and ingests a coenurus. Each protoscolex of the coenurus is capable of developing into an individual adult tapeworm in a definitive host. Thus, a single coenurus may develop into many adult tapeworms in the intestine of the carnivore.

Clinical Features and Pathogenesis

Coenuri in the skin or subcutaneous tissue usually present as solitary painless nodules. The lesion is less than 6 cm in diameter and can be fluctuant and moderately tender. Most nodules develop on the trunk, but the sclera, subconjunctiva, neck, head, or limbs may be involved.[22] A lesion in

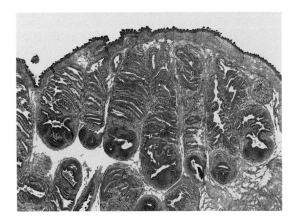

Figure 11.10
Section showing multiple protoscolices, each attached by the neck to cyst wall of coenurus. x23

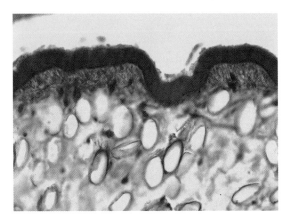

Figure 11.11
Section of neck region of coenurus showing thick tegument, dense parenchyma, and numerous calcareous corpuscles. x570

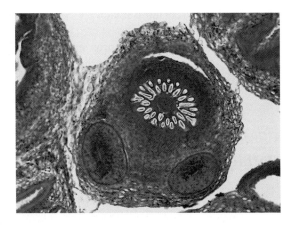

Figure 11.12
Section of protoscolex of coenurus revealing 2 of 4 suckers and armed rostellum. x115

the neck may limit neck movement and swallowing.[41] Coenuri tend to lodge just below the deep fascia in muscle and are often mistaken clinically for lipomas, ganglions, pseudotumors, or neurofibromas. Infection may provoke systemic manifestations. One woman presented with recurrent rash, pyrexia, night sweats, lymphadenopathy, and splenomegaly. Clinically, her symptoms mimicked a lymphoma. Treatment with vinblastine partially controlled her symptoms for 3 years, but removal of a coenurus in her breast cured her.[4]

In the CNS, infections may develop in the brain and spinal cord or in the eye.[25] In the spinal cord, symptoms are mostly related to the space-occupying lesion and pressure. In the brain, coenuri tend to develop in the subarachnoid space and cause basal arachnoiditis or ependymitis, but may also invade the brain parenchyma. Common symptoms include headache (mainly suboccipital), fever, and vomiting. In most instances, these symptoms result from increased intracranial pressure as a result of the arachnoiditis. Localizing neurologic symptoms may develop. Some patients may have cranial nerve palsies, most commonly of the abducens nerve, or signs of pyramidal tract involvement. Other complications include jacksonian epilepsy, pachymeningitis, obstructive or communicating hydrocephalus, or intracranial arteritis with transient hemiparesis.[26] Cerebrospinal fluid may reveal many mononuclear cells.[2]

Patients with eye involvement present with varying degrees of visual impairment. There are reports of both intraocular and orbital infections. Orbital lesions include those in the subconjunctival space, eyelid, or extraocular muscle. Intraocular lesions follow invasion of the eye by the oncosphere via a posterior ciliary artery. A coenurus develops in the subretinal space and sometimes perforates the retina to reach the vitreous body. A coenurus growing within the vitreous is therefore usually attached to the retina and the choroid (Fig 11.18). In subretinal areas, the coenurus may clinically resemble a neoplasm or a granuloma. If not removed, the coenurus may provoke a painful inflammatory reaction, glaucoma, and blindness. Some patients present with panophthalmitis.

Pathologic Features

In soft tissue, the gross appearance of a coenurus is that of an encapsulated, lilac-colored, nodular tumor. This tumor is composed of an inner parasitic cyst (Figs 11.19 and 11.20) and an outer, host-tissue-generated, adventitial capsule that may have attached to patches of skeletal muscle. Sectioning the mass exposes the unilocular, polycephalic, yellow-white larva. The elongated oval nodules representing the protoscolices are arranged along the wall of the cyst in irregular groups.

The viable coenurus usually does not adhere to the adventitial capsule formed by the host (Fig 11.20). This capsule is made up of an outer and an inner zone. The outer zone is composed of dense, collagenous, fibrous tissue and acid mucosaccharides heavily infiltrated by plasma cells, lymphocytes, and occasional macrophages and eosinophils. The inner zone contains acid mucosaccharides and compressed palisading macrophages, fibroblasts, and occasional lymphocytes intermixed with fibrin and compacted reticular fibrils. Necrosis in the capsule may obscure sections of the zones (Fig 11.21).

In the brain, the reaction surrounding a coenurus is often minimal, limited only to a thin rim of fibrous tissue (Fig 11.22). Coenuri detach readily from the brain during dissection of the specimen. The brains of patients with chronic leptomeningitis have variable pathologic changes, the most severe being plastic meningeal fibrosis with a foreign body giant cell reaction around the cyst wall. As the coenurus dies, acute inflammatory cells attack and infiltrate the larva (Fig 11.23). Degeneration of the coenurus leads to necrosis and eventually to a chronic inflammatory reaction composed of lymphocytes, macrophages, and foreign body giant cells (Figs 11.24 and 11.25). There is often perivascular inflammation and occasionally tissue eosinophilia (Fig 11.26). This degeneration leads to fibrosis and calcification (Fig 11.27).

Macroscopic examination of ocular coenurosis readily reveals the metacestode with its multiple protoscolices (Figs 11.28 and 11.29). Microscopically, the larva lies behind the retina, pushing the retina into the vitreous (Figs 11.30 and 11.31). The coenurus provokes an inflammatory reaction that

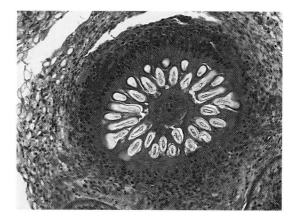

Figure 11.13
Higher magnification of Figure 11.12, showing rostellum of protoscolex of coenurus armed with double row of hooklets. x230

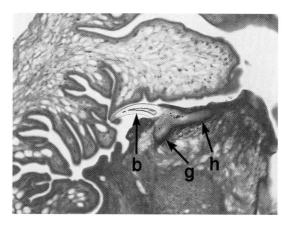

Figure 11.14
Section of coenurus, demonstrating blade (b), handle (h), and guard (g) of a hooklet. x225

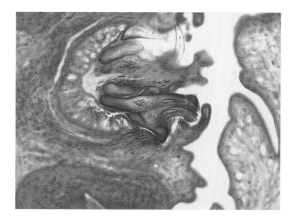

Figure 11.15
Section displaying hooklets. Note range in size of hooklets within a single coenurus. x230

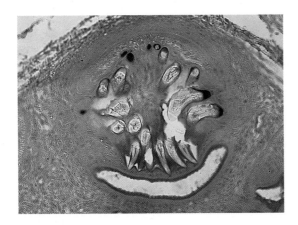

Figure 11.16
Section demonstrating that some hooklets of the protoscolex are acid-fast. ZN x220

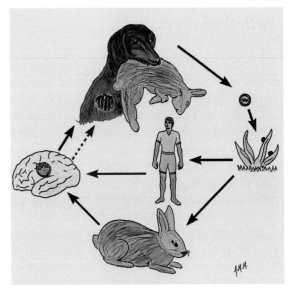

Figure 11.17
Life cycle of *T. multiceps*.

Figure 11.18
Section of eye showing a coenurus attached to the retina and choroid and growing within the vitreous body. Protoscolices (arrows) are dispersed in random clumps. x2.5

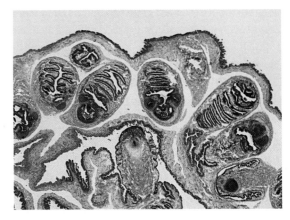

Figure 11.19
Section through a coenurus, extracted from tumor in right temple. Note protoscolices. Movat x23

may cause necrosis and chorioretinitis (Figs 11.32 and 11.33), uveitis, and scleritis.

Patients with coenurosis may develop hypoplasia or hyperplasia of regional lymph nodes, and immunosuppression similar to that sometimes seen in patients with echinococcosis.[4] In vitro studies indicate that these larval cestodes release components that alter macrophage and T-lymphocyte function.[30]

Diagnosis

Diagnosis depends upon identifying a coenurus in surgical or autopsy specimens. The most obvious feature distinguishing coenuri from cysticerci and sparganas is the multiple protoscolices within the coenuri. Metacestodes of *Echinococcus* also have multiple protoscolices, but the protoscolices

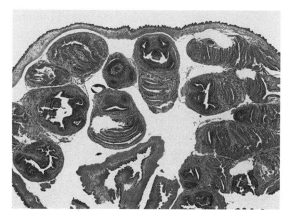

Figure 11.20
Section of viable coenurus extracted from chest wall of patient. Cyst was not adherent to surrounding adventitial capsule. x23

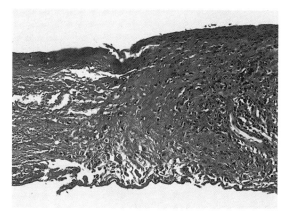

Figure 11.21
Adventitial capsule from coenurus in Figure 11.20. Note necrosis, fibrosis, and chronic inflammation. x110

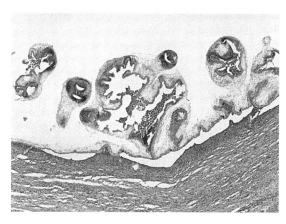

Figure 11.22
Section of viable coenurus in the brain. Note thin rim of host fibrous tissue around parasite. x20

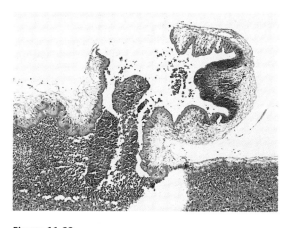

Figure 11.23
Section of brain showing early marked, acute inflammation with eosinophils surrounding a recently degenerated coenurus. x55

Figure 11.24
Degenerating coenurus with several protoscolices provoking extensive necrosis, extravasation of blood and chronic inflammatory infiltrate. x20

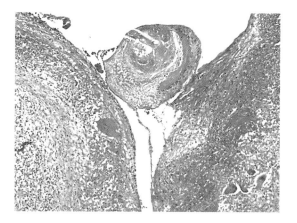

Figure 11.25
Protoscolex from degenerating coenurus causing extensive necrosis and chronic inflammation with giant cells. Mononuclear cells are invading the degenerating protoscolex. x55

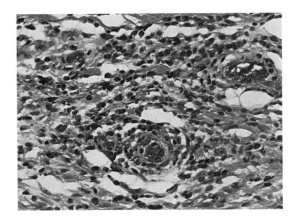

Figure 11.26
Section of brain parenchyma adjacent to a coenurus larva demonstrating tissue eosinophilia and perivascular inflammation. x220

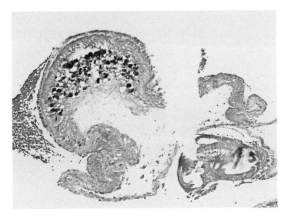

Figure 11.27
Section showing partial calcification of protoscolex in brain of patient with cerebral coenurosis. x75

of *Echinococcus* arise from a germinal membrane that lines a laminated wall, and the protoscolices are much smaller than those of coenuri. The largest hooklets of coenuri are nearly 6 times the length of the largest hooklets of metacestodes of *Echinococcus*.

A metrizamide-enhanced, computed tomography (CT) scan displays a viable coenurus as a thin-walled cyst. If the cyst is degenerating and initiating a host response, the CT scan displays a contrast-enhanced peripheral rim surrounding a translucent lesion.[33] The layer of fibrous tissue surrounding the viable cyst produces the contrast on the CT scan. Using multiple echo sequences, a magnetic resonance image (MRI) displays the cyst contents as a cerebrospinal fluid-like pattern.[29] CT scan and MRI studies allow a precise evaluation of both the size and location of the cysts.

Treatment and Prevention

Neurosurgical extirpation of the cyst effectively treats intracranial coenurosis with an excellent prognosis.[28] Ophthalmic surgeons have also developed techniques for successful surgical removal of viable coenuri from the eye, with recovery of sight.[39] Praziquantel is effective in killing the coenurus; however, caution is advised. A 25-year-old man with intraocular coenurosis, treated with praziquantel, lost his vision through toxic endophthalmitis and retinal detachment, despite the concomitant use of prednisone to reduce inflammation.[19] Thus, surgical removal remains the preferred treatment for patients with a single coenurus.[1] Patients with multiple neural coenuri present a particularly difficult problem; surgical intervention is not always indicated. These patients have a poor prognosis.

Infected dogs are the major source of human infection. Thus, regularly treating dogs for tapeworms reduces the chance of accidental human infection with the metacestodes. Attempts to vaccinate intermediate hosts such as sheep have had equivocal results.[12]

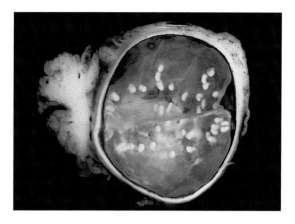

Figure 11.28
Thin, translucent, white cyst attached to retina and bulging into vitreous of eye. Multiple protoscolices are arranged in rows and clumps.

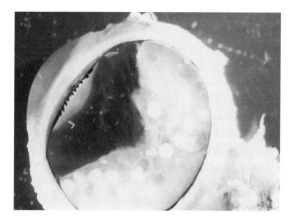

Figure 11.29
Macroscopic section of eye of 6-year-old German containing thickened white cyst of a coenurus. Larva is behind the retina but bulges into the vitreous.

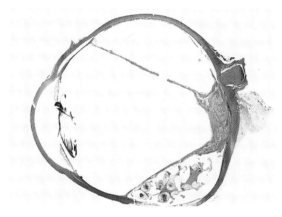

Figure 11.30
Section of eye demonstrating coenurus larva growing behind the retina and pushing into the vitreous. x2.2

Figure 11.31
Higher magnification of coenurus in Figure 11.30, demonstrating multiple protoscolices. x11

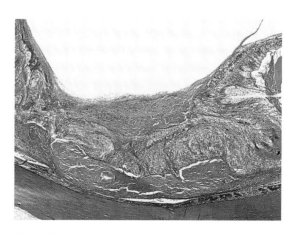

Figure 11.32
Chorioretinitis in eye pictured in Figure 11.29. x11

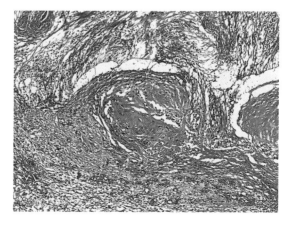

Figure 11.33
Higher magnification of Figure 11.32, illustrating chronic inflammation and fibrosis. x55

References

1. Baily GG. Editorial review of: Pau A, et al. Long-term follow-up of the surgical treatment of intracranial coenurosis. *Br J Neurosurg* 1990;4:41–42.
2. Barlow JF, Church BG. Coenurosis in the brain of a child from South Dakota. *S D J Med* 1969;22:37–44.
3. Beaver PC, Jung RC, Cupp EW. *Clinical Parasitology*. 9th ed. Philadelphia, Pa: Lea & Febiger; 1984:523–527.
4. Benger A, Rennie RP, Roberts JT, Thornley JH, Scholten T. A human coenurus infection in Canada. *Am J Trop Med Hyg* 1981;30:638–644.
5. Bohrmann R. Coenurus in the muscles of a gemsbok (Oryx gazella). *Vet Parasitol* 1990;36:353–356.
6. Bonnal G, Joyeux C, Bosch P. Un cas de cénurose humaine dû à Multiceps serialis (Gervais). *Bull Soc Pathol Exot* 1933;26:1060–1071.
7. Brumpt E. *Précis de parasitologie*. 2nd éd. Paris, France: Libraires de l'académie de médecine; 1913:281–283.
8. Clapham PA. On identifying Multiceps spp. by measurement of the large hook. *J Helminthol* 1942;20:31–40.
9. Corrêa FM, Ferriolli-Filho F, Forjaz S, Martelli N. (Reistra de casos) Cenurose cerebral. A propósito de um caso humano. *Rev Inst Med Trop Sao Paulo* 1962;4:38–45.
10. Debrie JC, Faugere JM, Menard M, Brunetti G, Aubry P. Human coenurosis. Localization to the sternocleidomastoid muscle [in French]. *Ann Otolaryngol Chir Cervicofac* 1982;99:477–481.
11. Deiana S. Current state of the diffusion of cerebral cenuriasis (coenurus of Taenia multiceps multiceps) in sheep and humans in Sardinia [in Italian]. *Parassitologia* 1971;13:173–175.
12. Edwards GT, Herbert IV. Preliminary investigations into the immunization of lambs against infection with Taenia multiceps metacestodes. *Vet Parasitol* 1982;9:193–199.
13. Esch GW, Self JT. A critical study of the taxonomy of Taenia pisiformis Bloch, 1780; Multiceps multiceps (Leske, 1780); and Hydatigera taeniaeformis Batsch, 1786. *J Parasitol* 1965;51:932–937.
14. Fain A, Denisoff N, Homans L, et al. Cénurose chez l'homme et les animaux due à Taenia brauni, Setti au Congo Belge et au Ruanda-Urundi. II. Relation de huit cas humains. *Ann Soc Belg Med Trop* 1956;36:679–696.
15. Graber M. Cenuriasis of small ruminants of central Africa. African cenuriasis of man and animals [in French]. *Rev Elev Med Vet Pays Trop* 1976;29:323–325.
16. Herbert IV, Edwards GT, Willis JM. Some host factors which influence the epidemiology of Taenia multiceps infections in sheep. *Ann Trop Med Parasitol* 1984;78:243–248.
17. Hermos JA, Healy GR, Schultz MG, Barlow J, Church WG. Fatal human cerebral coenurosis. *JAMA* 1970;213:1461–1464.
18. Huss BT, Miller MA, Corwin RM, Hoberg EP, O'Brien DP. Fatal cerebral coenurosis in a cat. *J Am Vet Med Assoc* 1994;205:69–71.
19. Ibechukwu BI, Onwukeme KE. Intraocular coenurosis: a case report. *Br J Ophthalmol* 1991;75:430–431.
20. Johnstone HG, Jones OW Jr. Cerebral coenurosis in an infant. *Am J Trop Med* 1950;30:431–441.
21. Kaminsky RG, Gatei DG, Zimmermann RR. Human coenurosis from Kenya. *East Afr Med J* 1978;55:355–359.
22. Kurtycz DF, Alt B, Mack E. Incidental coenurosis: larval cestode presenting as an axillary mass. *Am J Clin Pathol* 1983;80:735–738.
23. Lapage G. *Veterinary Parasitology*. 2nd ed. Edinburgh, Scotland: Oliver & Boyd; 1968:428–429.
24. Lau D, Casey WJ, Jones MD. Coenurosis in a whitehanded gibbon. *J Am Vet Med Assoc* 1973;163:633–635.
25. Malomo A, Ogunniyi J, Ogunniyi A, Akang EE, Shokunbi MT. Coenurosis of the central nervous system in a Nigerian. *Trop Geogr Med* 1990;42:280–282.
26. Michal A, Regli F, Campiche R, et al. Cerebral coenurosis. Report of a case with arteritis. *J Neurol* 1977;216:265–272.
27. Orihel TC, Gonzalez F, Beaver PC. Coenurus from neck of a Texas woman. *Am J Trop Med Hyg* 1970;19:255–257.
28. Pau A, Perria C, Turtas S, Brambilla M, Viale G. Long-term follow-up of the surgical treatment of intracranial coenurosis. *Br J Neurosurg* 1990;4:39–43.
29. Pau A, Turtas S, Brambilla M, Leoni A, Rosa M, Viale GL. Computed tomography and magnetic resonance imaging of cerebral coenurosis. *Surg Neurol* 1987;27:548–552.
30. Rakha NK, Dixon JB, Skerritt GC, Carter SD, Jenkins P, Marshall-Clarke S. Lymphoreticular responses to metacestodes: Taenia multiceps (Cestoda) can modify interaction between accessory cells and responder cells during lymphocyte activation. *Parasitology* 1991;102:133–140.
31. RangaRao GS, Sharma RL, Hemaprasanth. Parasitic infections of Indian yak Bos (poephagus) grunniens–an overview. *Vet Parasitol* 1994;53:75–82.
32. Razig SA, Magzoub M. Goat infected with Coenurus cerebralis–clinical manifestations. *Trop Anim Health Prod* 1973;5:278–280.
33. Schellhas KP, Norris GA. Disseminated human subarachnoid coenurosis: computed tomographic appearance. *AJNR Am J Neuroradiol* 1985;6:638–640.
34. Smith MC, Bailey CS, Baker N, Kock N. Cerebral coenurosis in a cat. *J Am Vet Med Assoc* 1988;192:82–84.
35. Templeton AC. Anatomical and geographical location of human coenurus infection. *Trop Geogr Med* 1971;23:105–108.
36. Templeton AC. Human coenurus infection. A report of 14 cases from Uganda. *Trans R Soc Trop Med Hyg* 1968;62:251–255.
37. Turner M, Leiper RT. On the occurrence of Coenurus glomerulatus in man in West Africa. *Trans R Soc Trop Med Hyg* 1919;13:23–24.
38. Wardle RA, McLeod JA, Radinovsky S. Order Taeniidea, new order. In: *Advances in the Zoology of Tapeworms, 1950–1970*. Minneapolis: University of Minnesota Press; 1974:139–166.
39. Williams PH, Templeton AC. Infection of the eye by tapeworm Coenurus. *Br J Ophthalmol* 1971;55:766–769.
40. Willis JM, Herbert IV. A method for estimating the age of coenuri of Taenia multiceps recovered from the brains of sheep. *Vet Rec* 1987;121:216–218.
41. Wilson VC, Wayte DM, Addae RO. Human coenurosis-the first reported case from Ghana. *Trans R Soc Trop Med Hyg* 1972;66:611–623.
42. Yoshino T, Momotani E. A case of bovine coenurosis (Coenurus cerebralis) in Japan. *Nippon Juigaku Zasshi* 1988;50:433–438.

12

Hymenolepiasis and Miscellaneous Cyclophyllidiases

Aileen M. Marty *and*
Ronald C. Neafie

Introduction

Definition

Hymenolepis nana (Von Siebold, 1852) Blanchard, 1891, causes hymenolepiasis nana, and is a major cause of intestinal tapeworm infection in humans, especially in developing countries.

Other cyclophyllidean tapeworms that infect humans are also discussed in this chapter: *Hymenolepis diminuta* (Rudolphi, 1819) Blanchard, 1891; *Mesocestoides lineatus* (Goeze, 1782) Railliet, 1893; *Mesocestoides variabilis* Mueller, 1928; *Bertiella studeri* (Blanchard, 1891) Stiles and Hassall, 1902; *Bertiella mucronata* (Meyner, 1895) Stiles and Hassall, 1902; *Inermicapsifer madagascariensis* (Davaine, 1870) Baer, 1956; *Raillietina celebensis* (Janicki, 1902) Fuhrmann, 1920; *Raillietina demerariensis* (Daniels, 1895) Dollfus, 1939 to 1940; and other *Raillietina* sp.

The strobilocercus (larval stage) of *Taenia taeniaeformis* (Batsch, 1786) Wolffügel, 1911, and larvae of *Taenia crassiceps* rarely infect humans.[2,33,78]

Synonyms

Hymenolepis nana is popularly known as the dwarf tapeworm. *Hymenolepis fraterna, Taenia murina, Taenia nana*, and *Vampirolepis nana* are other synonyms.

Older scientific names for *H. diminuta*, the rat tapeworm, include *Taenia diminuta, Taenia flavopunctata*, and *Taenia minima*.

The proper nomenclature for members of the genus *Mesocestoides* is controversial, possibly involving host-induced variations.[68] There may be only 1 species, *M. lineatus*, the Eastern mesocestoides worm. Synonyms for *M. lineatus* include *Taenia lineatus* (Goeze, 1782), *Mesocestoides variabilis*, the Western mesocestoides worm, and the worm common to the red fox, *Mesocestoides leptothylacus*.[57]

Blanchard originally used the generic name *Bertia* for *B. studeri*. Synonyms include *Bertia satyri, Bertiella satyri, Bertiella conferta, Bertiella polyorchis*, and *Bertiella cercopitheci*. *Taenia mucronata* is a synonym for *B. mucronata*.

Synonyms for *Inermicapsifer madagascariensis* include 5 generic and 6 specific names. Baer demonstrated that *Taenia madagascariensis*,

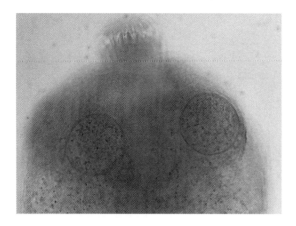

Figure 12.1
Scolex of *H. nana* (225 μm in diameter) demonstrating 2 of 4 suckers and an everted rostellum, 50 μm in diameter, with hooklets. Carmine x220

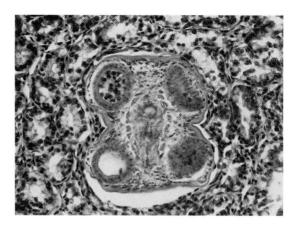

Figure 12.2
The 4 suckers are readily visible in this cross section of an *H. nana* scolex embedded in rat intestine. x235

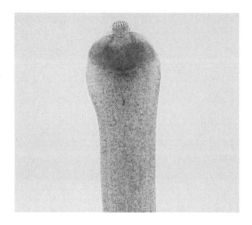

Figure 12.3
Armed rostellum and long neck of *H. nana*. Carmine x75

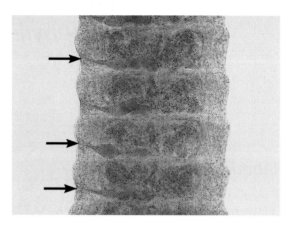

Figure 12.4
Mature proglottids of *H. nana* are broader than long. Note marginal genital pore (arrows) near anterior border on left side of each proglottid. Carmine x130

Davainea madagascariensis, Anoplocephala arvicanthidis, Inermicapsifer arvicanthidis, Raillietina madagascariensis, Raillietina cubensis, Inermicapsifer cubensis, Raillietina davainei, Raillietina loeche-salavezi, Raillietina kouridovali, and *Inermicapsifer cubanensis,* are all synonyms of *I. madagascariensis.*[3,24]

Raillietina celebensis originally had the name *Davainea celebensis,*[48] and *R. demerariensis* the name *Taenia demerariensis.* Both of these *Raillietina* sp infect humans. Synonyms for *R. celebensis* include *Davainea formosana, D. madagascariensis, Raillietina garrisoni, Raillietina sinensis, Raillietina murium, Raillietina siriraji,* and *Meggittia celebensis.*[32]

Synonyms for *T. taeniaeformis* include *Taenia infantis, Hydatigera taeniaeformis, Taenia crassicollis,* and *Cysticercus fasciolaris.*

General Considerations
Hymenolepis nana

Bilharz discovered *H. nana* in 1851 in the small intestine of an Egyptian boy and sent the specimen to Von Siebold. Noting its smallness, Von Siebold named the tapeworm *T. nana.* In 1887, Grassi showed that transmission from rat to rat did not require a vector or intermediate host.[41] In 1920, Saeki proved direct transmission of *H. nana* in humans without an intermediate host.[73] In addition to the direct cycle, Nicholl and Minchin

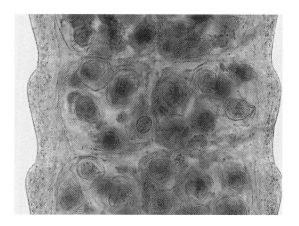

Figure 12.5
Gravid proglottids of *H. nana* filled with eggs. Carmine x215

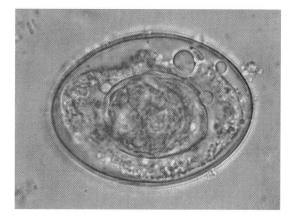

Figure 12.6
Ovoid, thin-shelled egg of *H. nana* in stool. Polar filaments and polar thickenings are not discernible. Unstained x1100

found that fleas harbored the cysticercoids and served as intermediate hosts.[66]

Hymenolepis diminuta

In 1766, Olfers found *H. diminuta* in rats in Rio de Janeiro.[15] The first adequate description was provided by Rudolphi in 1819, who recovered worms from rats and mice and named them *T. diminuta*. Weinland described the worm in humans in 1858. In 1892, Grassi and Rovelli demonstrated that various insects and insect larvae serve as intermediate hosts.

Mesocestoides

Goeze first identified a *Mesocestoides* in 1782 in dogs in Germany and named it *T. lineatus*.[37] Mueller described a similar parasite in mammals in 1928, naming it *M. variabilis*.[64] In 1942, Chandler reported the first human infected with a *Mesocestoides*—an infant in Texas who, after treatment, passed 4 to 5 feet of *Mesocestoides* tapeworm segments in stool, including 4 scolices.[22]

Bertiella

Bertiella studeri and *B. mucronata* are tapeworms of primates. Blanchard's original description in 1891 of *B. studeri* was based on a complete tapeworm that Studer obtained from a chimpanzee. A second specimen, which he called *B. satyri (B. studeri)*, consisted of portions of a tapeworm Blanchard procured from an orangutan.[17] In 1913, Blanchard found the same type of cestode in a child in Mauritius.[18] Cram in 1928 described the first infection with *B. mucronata* in humans.[29]

Inermicapsifer madagascariensis

In 1870, in the Comoro Islands, Grenet recovered an adult tapeworm from a West Indian child, and Davaine identified the worm as *T. madagascariensis*. Baer reclassified this tapeworm and put it in the genus *Inermicapsifer*.[4,5]

Raillietina celebensis

Leuckart probably was the first to report infection by *R. celebensis*, in a Tahitian child in 1891. He originally named this worm *T. madagascariensis*. Janicki, in 1902, named a worm of the same species, that he had obtained from a mouse, *D. celebensis*.

Epidemiology

Hymenolepis nana is a common tapeworm of humans, and the only tapeworm that can pass directly from human to human. Although more common in warm climates, *H. nana* is also observed in cold regions. In India, *H. nana* infects 50% of children, in Mexico 27%, and in Egypt 16%.[27,53,70] *Hymenolepis nana* is widespread in southern and eastern Europe, the Commonwealth of Independent States (CIS), and Africa. *Hymenolepis nana* infection is highly prevalent in Central

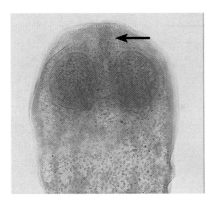

Figure 12.7
Scolex of *H. diminuta* showing 2 of 4 deeply cupped suckers and inverted unarmed rostellum (arrow). Carmine x125

Figure 12.8
Immature proglottids of *H. diminuta* are much broader than long. x55

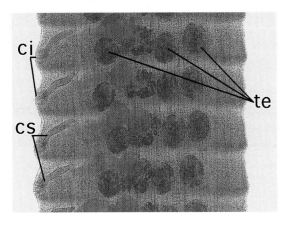

Figure 12.9
Mature proglottids of *H. diminuta* demonstrating the 3 testes (te): 2 testes on the right side, and the third testis, cirrus sac (cs), and cirrus (ci) on the left. Carmine x35

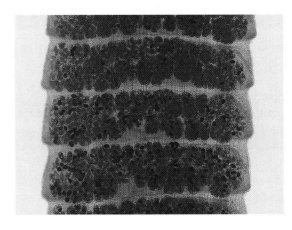

Figure 12.10
Gravid proglottids of *H. diminuta* filled with eggs. Most eggs are eosinophilic but some appear black. Carmine x23

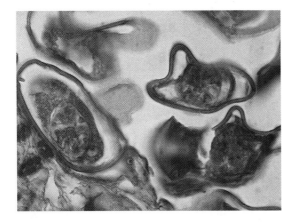

Figure 12.11
Section through gravid proglottid of *H. diminuta* depicting eggs. x590

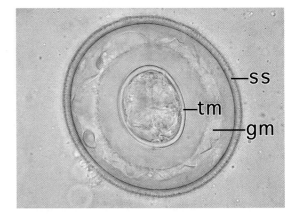

Figure 12.12
Spherical unstained egg of *H. diminuta* in feces. Note thick outer, radially striated shell (ss), colorless, gelatinous matrix (gm), and thick inner membrane (tm). x610

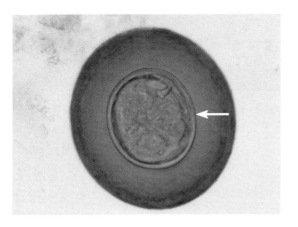

Figure 12.13
Egg of *H. diminuta* preserved in merthiolate-iodine-formaldehyde (MIF) solution, demonstrating thick inner membrane (arrow), thick outer shell, and hooklets in the oncosphere. x675

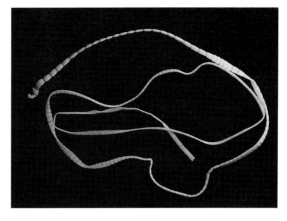

Figure 12.14
Portion of strobila (25 cm long) of *Mesocestoides* sp recovered from stool of a 3-year-old boy from Virginia. The scolex is missing. x1

and South American countries, and is frequently encountered in humans in the southern United States and Hawaii. In North America, infection is most frequent among institutionalized people and malnourished or immunocompromised patients. Incidence is slightly higher in males, but highest in children 4 to 9 years of age.[15,53] The higher prevalence in children suggests specific acquired resistance in adults. Immunodeficiency or internal autoinfection can cause persistent heavy infections.

Hymenolepis diminuta has a cosmopolitan distribution among rodents, but is rare in humans. About 300 human infections are known.[25] There are reports of infections in humans in Africa (Malawi, Zambia, Zimbabwe)[38]; Asia (China, CIS, India, Iran, Japan, the Philippines, Thailand)[81]; the Caribbean (Cuba, Grenada, Jamaica, Martinique); Europe (Belgium, Italy); South and Central America (Argentina, Brazil, Colombia, Ecuador, Nicaragua, Venezuela); and several states of the United States (Alabama, Arkansas, California, Florida, Georgia, Indiana, Louisiana, Minnesota, Nebraska, New York, North Carolina, Tennessee, Texas, Virginia, and Washington, DC).[25,56]

Mesocestoides lineatus is mainly a parasite of dogs, cats, foxes, raccoons, and other mammals.[57] *Mesocestoides* infections are uncommon in humans. There are 24 cases reported in world literature, and we have studied 2 additional cases: a 3-year-old boy from Virginia and a 1-year-old girl from Indiana.[31,74] *Mesocestoides* infections are widely distributed: Greenland, Japan, Rwanda, and the continental United States (California, Mississippi, Missouri, New Jersey, Ohio, and Texas).[10] Thirteen of the human infections were in Japanese adults with a history of eating uncooked snake liver, blood, and gallbladder, or turtle blood.[52] In the United States, most infections have been in children; the source of these infections is not known.[42]

Most human infections with *B. studeri* or *B. mucronata* are in children. Even though parasites are widespread in primates in warmer areas, human infections are rare. There are reports of patients infected in Africa, Argentina, Borneo, Brazil, the Caribbean (Cuba, Saint Kitts), India, Java, Malaysia, Mauritius, Paraguay, the Philippines, Singapore, Sri Lanka, Sumatra, Thailand, the United States, and Yemen.[9,16,19,82] Lower primates are believed to serve as reservoir hosts.[79]

Inermicapsifer madagascariensis is endemic in Cuba,[5] with over 100 human infections recorded. Infections have also been documented in the Comoro Islands, Kenya, Madagascar, Malawi, Mauritius, Rwanda, South Africa, Venezuela, Zambia, and Zimbabwe. The majority of patients, in both Africa and Cuba, are children of European descent.[5,40] Baer suggests that in Cuba and Madagascar, the parasite has adapted to human-to-human transmission.[3] Significantly, all infections recorded outside Africa have been in or near sugar plantations,

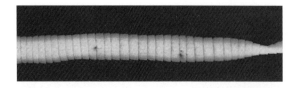

Figure 12.15
Immature proglottids of *Mesocestoides* sp, approximately 3 times broader than long. x4.5

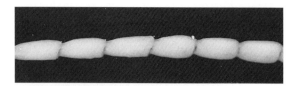

Figure 12.16
Gravid proglottids of *Mesocestoides* sp, twice as long as broad. x4.8

where consumption of raw cane is common.[65]

Raillietina celebensis is a common tapeworm of rats in the Orient and Oceania.[67] There are reports of infections in humans, mostly children, from Australia, Burma, China, French Polynesia, Japan, Malagasy Republic, Taiwan, Thailand, and Vietnam.[4,23,32,71,72] Other *Raillietina* sp infect humans in Cuba, Ecuador, Iran, the Philippines, and Thailand.[15]

Mathevotaenia sp, a member of the family Anoplocephalidae, rarely infects humans.[54]

The adult tapeworm *T. taeniaeformis* is a common parasite of cats, which acquire the infection by eating rodents infected with a strobilocercus. Two human infections are known.[12]

Infectious Agents

Morphologic Description

Hymenolepis nana
Adult

Hymenolepis nana is the smallest adult tapeworm (25 to 45 mm long by up to 1 mm wide) that infects humans. The small rhomboid scolex has 4 suckers and a rostellum with a single ring of 20 to 30 hooklets (Figs 12.1 and 12.2). The neck region is long and narrow (Fig 12.3) and the strobila is composed of 100 to 200 proglottids. All proglottids, from immature to gravid, are broader than long (Fig 12.4). Proglottids gradually enlarge as they mature, the largest measuring 0.15 to 0.3 mm long and 0.8 to 1 mm wide. Mature proglottids have a marginal genital pore near the anterior border (Fig 12.4). Gravid proglottids are filled with eggs (Fig 12.5); the proglottids disintegrate within the intestine and discharge eggs in the feces.

Eggs

Eggs are ovoid, measure 30 to 55 μm in diameter, have an inner hyaline membrane, and a thin shell (Fig 12.6). The egg contains an oncosphere and the inner envelope has 2 polar thickenings. Four to 8 polar filaments arise from each polar thickening and occupy the clear space between the inner membrane and the outer shell. Polar thickenings and filaments are sometimes difficult to discern. Oncospheres have 3 pairs of lanceolate hooklets.

Hymenolepis diminuta
Adult

Hymenolepis diminuta is 20 to 60 cm long. The scolex has 4 deeply cupped suckers and a retractable unarmed rostellum (Fig 12.7). The neck is long and slender, and the strobila consists of 800 to 1000 proglottids which gradually increase in width from 0.5 mm at the neck to about 3.5 mm near the distal end. Proglottids are always broader than long by a ratio of about 5 to 1 (Fig 12.8). Mature proglottids have 3 ovoid testes; 2 testes are on the right side while the third testis, seminal vesicle, cirrus sac, and genital pore are on the left side (Fig 12.9). Gravid proglottids (Figs 12.10 and 12.11) disintegrate within the intestine, discharging eggs into the feces.

Eggs

Eggs of *H. diminuta* are transparent yellow, nearly spherical (up to 86 μm in greatest dimension), and have a thick inner membrane without polar filaments and a thick outer shell with fine radial striations (Fig 12.12). A colorless, gelatinous matrix lies between the membrane and shell. The 6 lanceolate hooklets of the oncosphere are arranged in a fan-shaped pattern (Fig 12.13).

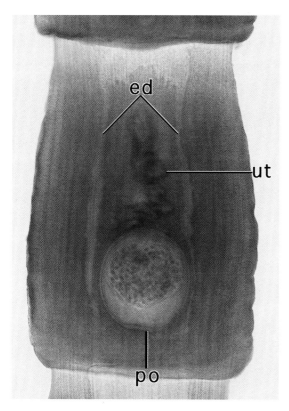

Figure 12.17
Gravid proglottid of *Mesocestoides* sp, demonstrating parauterine organ (po) enclosing a mass of eggs. Uterus (ut) lies in the center and there are 2 lateral excretory ducts (ed). Mayer's paracarmine x55

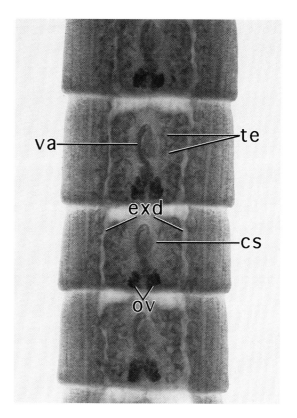

Figure 12.18
Mature proglottids of *Mesocestoides* sp, showing lateral excretory ducts (exd), numerous testes (te), bilobed ovary (ov), vagina (va), and cirrus sac (cs). Mayer's paracarmine x70

Mesocestoides
Adults and Eggs

The scolex of *M. lineatus* measures up to 470 to 600 µm wide by 350 to 400 µm long and has 4 oval suckers (180 to 225 µm long by 195 µm wide), but no rostellum or hooklets.[22,68] The strobila is 40 to 150 cm long and up to 2 mm wide, and contains 400 or more proglottids (Fig 12.14).[22,31] Immature proglottids are about 3 times broader than long (Fig 12.15). Mature proglottids are broader than long to nearly square, while gravid proglottids are longer than broad (1.7 to 2.5 mm long by 1.25 to 1.6 mm wide and 550 to 700 µm thick) (Figs 12.16 and 12.17). Mature proglottids have numerous testes, usually arranged in 1 or 2 layers dorsally on either side of the lateral excretory ducts, forming 2 lateral groups extending the length of the proglottid (Fig 12.18). The ovary is bilobed. The cirrus sac and vaginal duct enter the genital atrium on the ventral surface of the proglottid at midline (Fig 12.19). The vagina is considerably convoluted, passing forward to about the anterior level of the cirrus sac, then turning posteriorly to the genital pore. The uterus develops as a thin-walled, slightly spiral tube, extending forward to about midway between the cirrus sac and the anterior end of the proglottid. Parauterine organs (egg capsules) form from a bowl-like thickening of the posterior portion of the uterus. Proglottids have a body wall composed of a thick, ciliated tegument, well-organized smooth muscle cells, tegumental cells, and a spongy parenchyma containing calcareous corpuscles (Fig 12.20). In the gravid proglottid, the parauterine organ encloses a mass of eggs (Figs 12.17, 12.21, and 12.22). The parauterine organ is conspicuous in stained sections, approximately 500 µm in diameter, and is unique to this genus. Eggs are spherical to oval and up to 40 µm in greatest dimension (Fig 12.23).

Larvae

The first larval form, a cysticercoid, has not been described in detail.[68,74] The second larval form, the tetrathyridium, is cysticercoid-like with an invaginated protoscolex at the anterior end and tapered toward the posterior end. Tetrathyridia are up to several centimeters long in mammals, but only 2 to 5 mm long in reptiles.

Bertiella studeri and Bertiella mucronata
Adult

Bertiella tapeworms are white, but sometimes are darkly pigmented in the most terminal segments. *Bertiella studeri* range from 11 to 30 cm long,[79] but most are 25 to 27 cm long, 6 to 10 mm wide, and 1 mm thick. *Bertiella mucronata* are up to 40 cm long, 14 mm wide, and 3 mm thick. The scolex of both species has a rudimentary unarmed rostellum and 4 oval suckers. Mature proglottids are much broader (usually 6 to 7 mm) than long (0.5 to 0.8 mm) (Fig 12.24).[6,80] The body wall consists of a thick tegument, well-organized muscle cells, and tegumental cells (Fig 12.25). Mature proglottids of *B. studeri* and *B. mucronata* are similar, except for the vagina. In *B. studeri*, the vagina is poorly developed, funnel-shaped, and glandular only in the outer portion; in *B. mucronata*, the vagina is thick-walled and glandular along its full length.[16,29] Gravid proglottids exit the bowel in strands of 10 to 20 segments.

Eggs

Eggs of *Bertiella* have an irregular oval shape: those of *B. studeri* are approximately 45 to 47 μm by 48 to 50 μm, and those of *B. mucronata*, 38 μm by 41 μm.[28] Eggs have an external membrane, a delicate middle envelope composed mainly of mucin, and an inner shell (Fig 12.26). The inner shell, or piriform apparatus, has a distinctive double horn-like protrusion on 1 side (Fig 12.26). Within the piriform apparatus is a round oncosphere (17 to 20 μm in diameter). Hooklets in the oncosphere vary from 6 to 12 and are in pairs.[49,79]

Larvae

Larvae of *Bertiella* develop as spherical, oval, or piriform cysticercoids in mites. These cysticercoids measure 0.1 to 0.15 mm in diameter and contain a long, knob-like appendage called a cercomer,[79] which may be as long as the greatest dimension of the cystic portion of the cysticercoid.

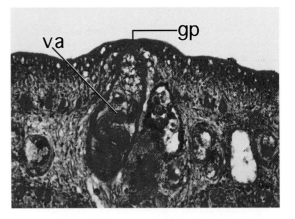

Figure 12.19
Section of mature proglottid of *Mesocestoides* sp, displaying genital pore (gp) and coiled vagina (va) within cirrus sac. x170

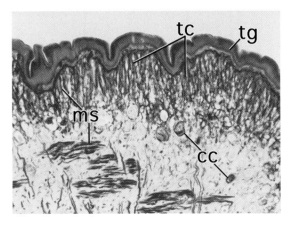

Figure 12.20
Body wall of gravid proglottid of *Mesocestoides* sp. Note tegument with microvilli (tg), well-organized smooth muscle cells (ms), and tegumental cells (tc). The spongy parenchyma contains calcareous corpuscles (cc) and smooth muscle (ms). Movat x230

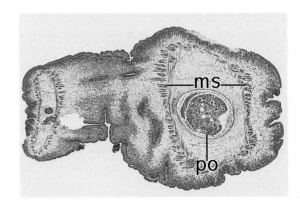

Figure 12.21
Section of gravid proglottid of *Mesocestoides* sp, demonstrating parauterine organ (po) enclosing a mass of eggs, framed by a band of smooth muscle (ms). Movat x35

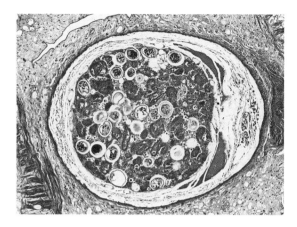

Figure 12.22
Stained section through egg-filled parauterine organ of gravid proglottid of *Mesocestoides* sp. x120

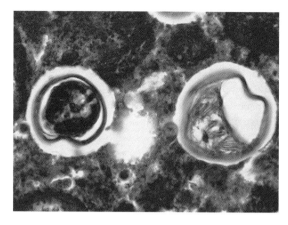

Figure 12.23
Section through parauterine organ of *Mesocestoides* sp, showing 2 eggs containing 6-hooked hexacanth embryos. x765

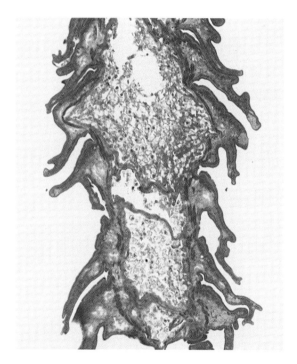

Figure 12.24
Section of 2 gravid proglottids of *Bertiella* sp from a monkey. In this rare specimen the 2 proglottids are abnormally long. Movat x45

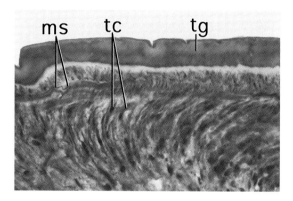

Figure 12.25
Body wall of gravid proglottid of *Bertiella* sp showing thick tegument (tg), well-organized smooth muscle cells (ms), and tegumental cells (tc). Movat x585

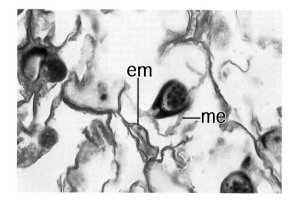

Figure 12.26
Section of proglottid of *Bertiella* sp containing several eggs. Eggs have an oval external membrane (em), a delicate middle envelope (me), and an inner shell (piriform apparatus) with a double horn-like protrusion on 1 side. Movat x590

Inermicapsifer madagascariensis
Adult

Adult *I. madagascariensis* are long, narrow tapeworms, up to 42 cm in length. The scolex is round and flat, with 4 round suckers. There is no rostellum or other hooklet-bearing structure (Fig 12.27). The strobila consists of about 315 to 360 proglottids.[8] Immature proglottids are broader than long. Mature proglottids are nearly square. Gravid proglottids resemble white rice grains, are slightly longer than broad, and contain up to 175 egg capsules (Fig 12.28). The egg capsules usually contain 6 to 10 mature or immature eggs, but may have up to 15.[5,24] Gravid proglottids have a unilateral genital pore in the middle of the lateral border of each margin (Fig 12.29).[46]

Eggs

Mature eggs of *I. madagascariensis* are spherical, measuring 34 to 60 μm. The egg capsules are pressed against one another and are polyhedral (Fig 12.29).[5,15,24] Oncospheres have 6 hooklets, and are surrounded by a vitelline membrane.[8]

Raillietina sp
Adult

Raillietina celebensis tapeworms are flat and up to 60 cm long by approximately 3 mm wide. The scolex has an armed, cushion-shaped rostellum and 4 suckers. There are 80 to 160 hammer-shaped hooklets on the rostellum measuring 18 to 25 μm long.[4,32] Younger proglottids tend to be trapezoidal, while more mature proglottids are nearly square, and older proglottids are barrel-shaped.[35] Mature proglottids have a unilateral genital pore in the anterior third of the lateral border (Fig 12.29).[15] The uterus consists of a median cavity with multiple pouches. Egg packets form around these pinched-off portions. Gravid proglottids are 2 to 3 mm long and 1 to 1.75 mm wide, contain 100 to 400 egg packets,[61] and grossly resemble those of *Dipylidium caninum*. Egg packets range from 130 to 510 μm and contain 1, 2, or 3 eggs (Fig 12.29). There are small calcareous corpuscles within the inner and outer zones of the egg packets. Other species of *Raillietina* infecting humans have similar morphologic characteristics.

Eggs

Eggs of *Raillietina* sp vary widely in size and

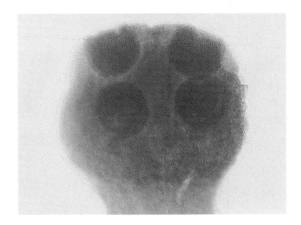

Figure 12.27
The scolex of *I. madagascariensis* is about 500 μm wide and has 4 suckers, but no rostellum. Carmine x125

Figure 12.28
Gravid proglottids of *I. madagascariensis* are about 2 mm wide and trapezoidal. Eggs are in capsules. Carmine x32

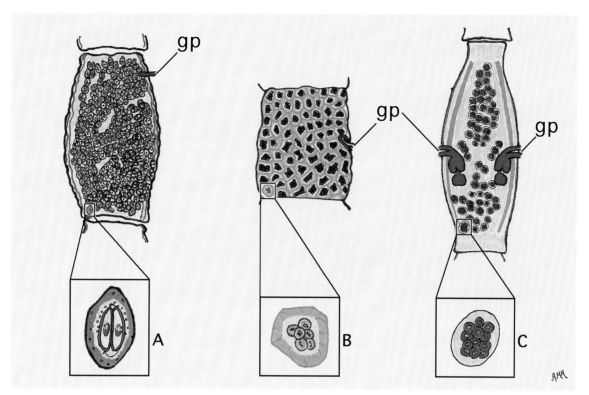

Figure 12.29
Comparison of gravid proglottids and egg packets of (A) *R. celebensis*, (B) *I. madagascariensis*, and (C) *D. caninum*. Note position of genital pores (gp) in the proglottids: *Raillietina,* anterior; *Inermicapsifer,* middle portion; *Dipylidium,* bilateral.

shape, from oval to elongated forms that range from 99 to 185 μm long and 34 to 46 μm wide. Eggs contain an oncosphere (14 to 20 μm) with 6 small hooklets.[61]

Life Cycle and Transmission

Unique among tapeworms in humans, *H. nana* requires only a single host to complete its life cycle. Humans, especially young children, and animals such as mice, can serve as both intermediate and definitive hosts. Humans acquire infections from 3 separate sources (Fig 12.30): most commonly by swallowing infective eggs from human feces, but also from eggs in rodent feces, or by ingesting cysticercoids in an infected flea or beetle.[70] A cysticercoid is a larval form of tapeworm that develops from a hexacanth embryo and has a protoscolex, usually invaginated, but lacks a bladder. The oncosphere exits the egg in the small intestine, penetrates a villus, and develops into an encysted cysticercoid within 96 to 140 hours. The mature cysticercoid reenters the bowel lumen, migrates down the intestine where the scolex evaginates, attaches to the mucosa, and matures in 1 to 2 weeks. Although internal autoinfection sometimes ensues, most eggs pass in feces. Ingestion of eggs from a patient's own feces may lead to autoinfection.[70] Person-to-person fecal-oral transmission is common.

Hymenolepis diminuta requires 2 hosts to complete its life cycle. Intermediate hosts are coprozoic or scavenger insects (fleas, beetles, cockroaches). Rodents are the usual definitive hosts. In insects, ingested eggs release oncospheres that penetrate the hemocoelom and metamorphose into cysticercoid larvae. When rodents, or uncommonly, humans, consume infected insects, the cysticercoid larvae lodge in the intestine and mature into adult tapeworms that can live 14 years or longer.[69] Eggs resist desiccation, putrefaction, and many chemicals, but are killed by temperatures above 60°C. In Africa, cereal grains (such as maizemeal)

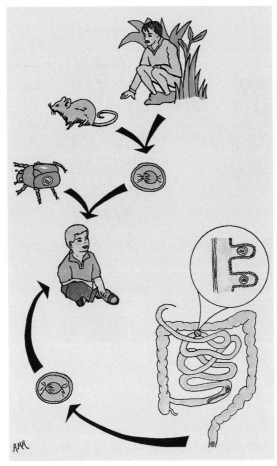

Figure 12.30
Life cycle of *H. nana*. Humans may acquire infection by swallowing infective eggs in human or rodent feces, or by swallowing a cysticercoid in an infected flea or beetle.

contaminated by parasitized insects are a common source of infection.[39]

The life cycle of *Mesocestoides* is not well-understood, but may involve 4 stages: egg, cysticercoid-like larva, tetrathyridean larva, and adult. The cysticercoid-like metacestode larva may use an arthropod vector.[74] A tetrathyridium is a large cysticercoid-like larval tapeworm with an invaginated protoscolex, common to the genus *Mesocestoides*. Tetrathyridean metacestode larvae develop in secondary intermediate hosts, including multiple species of reptiles, birds, and mammals.[31] The tetrathyridean larva may lie free or encyst in any part of the secondary intermediate or paratenic host.[22,51,83] To complete its life cycle, natural definitive hosts such as dogs, cats, or rodents must eventually eat the flesh of a second intermediate or paratenic host containing a tetrathyridean larva.[75]

Life cycles of *Bertiella* species are not fully known. Mites are intermediate hosts for *B. studeri*, and mammals, especially nonhuman primates, the definitive hosts.[16] Humans infected with *Bertiella* frequently have had contact with monkeys.[49,82]

The life cycle of *I. madagascariensis* is unknown, but insects or mites are probably intermediate hosts.[40,65] In Cuba, and in the endemic islands of the Indian Ocean, there are no known nonhuman definitive hosts. In continental Africa, rodents are definitive hosts. Human infections may depend on transmission by intermediate insect hosts.

Raillietina celebensis has 2 hosts, an insect intermediate host and a mammalian definitive host. Adult ants carry gravid proglottids into the colony and feed them to larval ants. The eggs pass into the gut of larval ants and develop into cysticercoids. Humans, rodents, chickens, and quail acquire the infection by consuming infected ants. Other *Raillietina* sp have an insect as the intermediate host and birds as definitive hosts.

Clinical Features and Pathogenesis

The clinical features produced by these tapeworms resemble those of human taeniasis (see chapter 7).

Hymenolepis nana usually produces symptoms only in massive infections. Most symptoms probably represent allergic reactions.[11] Abdominal pain and anorexia are the most common complaints. Patients may have a history of irritability, weight loss, abdominal gas (tympanites), flatulence, headache, nausea, diarrhea, anal itching, nasal itching, nocturnal enuresis, vomiting, and, uncommonly, constipation.[53,70] Children may also have epileptiform convulsions and impaired growth.[53] Circulation of antigens or excretory products of worms may cause id-like papular eruptions in the skin.[21] Even in relatively asymptomatic patients, immunosuppression is not uncommon (decreased number and functional activity of T- and B-lymphocytes, decreased IgA titer, and increased IgE titer). Immunosuppression can

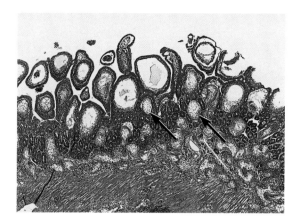

Figure 12.31
Multiple cysticercoids of *H. nana* (arrows) in small intestinal villi of experimentally infected mouse. x80

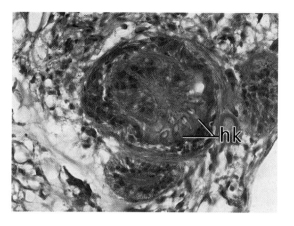

Figure 12.32
Cysticercoid of *H. nana* in pancreas of experimentally infected mouse, demonstrating invaginated protoscolex with hooklets (hk). x585

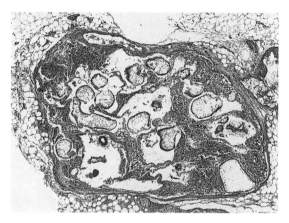

Figure 12.33
Lymph node from patient treated with immunosuppressive drugs for Hodgkin's disease. Note multiple cystic fragments of an aberrant larval cyclophyllidean tapeworm. x25

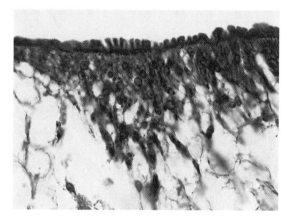

Figure 12.34
Body wall of large cystic fragment of larval cyclophyllidean tapeworm, shown in Figure 12.33, in a lymph node, demonstrating microvilli, thin tegument, and tegumental cells. x580

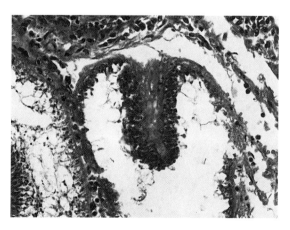

Figure 12.35
Cyst in lymph node shown in Figures 12.33 and 12.34. Cyst wall of larva is invaginated. x237

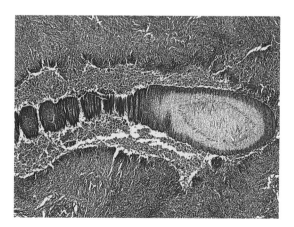

Figure 12.36
Brain abscess of 24-year-old Filipino woman containing solid-bodied cyclophyllidean larva, probably a tetrathyridium of *Mesocestoides* sp. x25

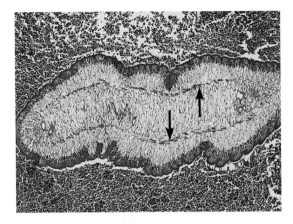

Figure 12.37
Solid-bodied cyclophyllidean larva in brain, demonstrating band of muscle (arrows) dividing cortical from medullary region of larva. x55

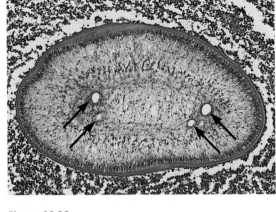

Figure 12.38
Solid-bodied cyclophyllidean larva depicting the 2 pairs of lateral excretory canals (arrows). x85

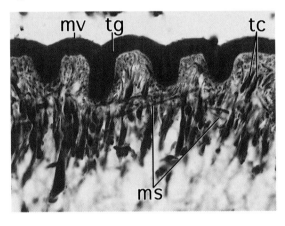

Figure 12.39
Body wall of solid-bodied cyclophyllidean larva, demonstrating microvilli (mv), tegument (tg), tegumental cells (tc), and smooth muscle fibers (ms). x600

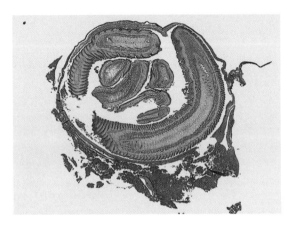

Figure 12.40
Strobilocercus of *T. taeniaeformis* in liver of a rat. x7

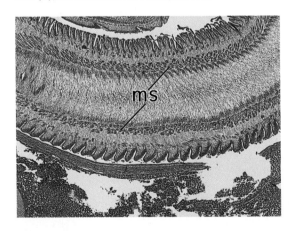

Figure 12.41
Higher magnification of strobilocercus larva of *T. taeniaeformis* seen in Figure 12.40. Note prominent, organized bands of smooth muscle fibers (ms). x25

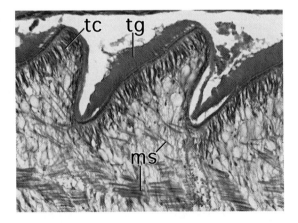

Figure 12.42
Body wall of strobilocercus illustrated in Figures 12.40 and 12.41. Note thick tegument (tg), tegumental cells (tc), and numerous smooth muscle fibers (ms). x230

persist for up to 6 months after effective treatment for hymenolepiasis.[60] Some individuals show impaired d-xylose absorption, decreased serum lysozyme activity, and increased serum complement activity. Patients may have a moderate (4% to 16%) peripheral eosinophilia and anemia.[15] Immunocompromised patients are at risk for extraintestinal infection with cysticercoids of *H. nana*, but this seems to be rare.[34]

Infection acquired by consuming an egg or a cysticercoid induces immunity to reinfection.[44,47] Infections with *H. nana* in humans are often associated with other intestinal parasitic diseases, especially giardiasis.

Light infections with *H. diminuta* generally have minimal symptoms, but worm burdens above 10 provoke symptoms similar to heavy infections of *H. nana*. Infected children are irritable, fail to thrive, and have abdominal pain, anorexia, vomiting, headache, diaphoresis, and black stools.[43] Concomitant intestinal parasitic infections are common.[56]

Mesocestoides infections are generally light. Patients are asymptomatic, or present with anorexia, abdominal colic, anemia, and irritability.[22]

Bertiella sp infections are often asymptomatic. Patients sometimes have abdominal discomfort, anorexia, nausea, vomiting, or loose or fatty stools.[6]

Infections by *I. madagascariensis, R. celebensis, R. demerariensis* and other *Raillietina* sp are most frequent in children. They commonly present as wriggling, rice-like grains (gravid proglottids) migrating out the anus or found in the diaper.[61] Symptoms are often mild or absent, but some patients lose weight and are irritable.[24,45] Infection with *R. celebensis* is generally mild, but may include diarrhea, malaise, abdominal distention, and loss of appetite.[61,72]

Patients infected with an adult *T. taeniaeformis* or its larva (strobilocercus) have minimal symptoms.[2,78]

Pathologic Features

All adult tapeworms discussed in this chapter inhabit the upper ileum and usually incite minimal histopathologic changes in the intestines. In heavy infections the mucosa may be damaged, causing catarrhal enteritis. Jejunal biopsies may show stunted villi, with normal enterocytes but increased numbers of inflammatory cells.[27] One infection of metastatic adult *H. nana* is known: a tumor in the chest wall of a Japanese woman.[63]

Cysticercoid larvae of *H. nana* may invade villi of the small intestine or other parts of the gastrointestinal system of infected humans or rodents (Figs 12.31 and 12.32). The peripheral blood and sputum of immunodeficient patients may contain cysticercoids of *H. nana*, suggesting wide dissemination.[34,76] Connor et al[26] reported aberrant larval tapeworms in the blood vessels and deep organs of a man from Pennsylvania with Hodgkin's disease who had been treated with immunosuppressive drugs and radiotherapy. Many organisms were in the blood vessels and parenchyma of all deep organs, and were especially numerous in lymph nodes (Figs 12.33 and 12.34). The organisms were spherical or elongated. Some were fluid-filled cysts and some had invaginated walls, but without protoscolices (Fig 12.35). Connor et al believed the organism to be a larval cestode, but did not classify the parasite further. Others have speculated that the parasites were abnormal cysticercoids of *H. nana*.[13,58,59] Classification of this aberrant cestode larva remains undetermined.

Solid-bodied cyclophyllidean cestode larvae sometimes infect humans. A biopsy of the brain of a Filipino who presented with convulsions and other neurologic symptoms revealed a solid-bodied cyclophyllidean larva in an abscess in the right cerebral cortex (Fig 12.36). The organism is most likely a tetrathyridium of *Mesocestoides* sp, one of the solid-bodied cyclophyllidean larvae. Solid-bodied cyclophyllidean larvae have well-organized inner muscle fibers arranged in a band that separates the cortical and medullary regions (Fig 12.37).[1] There are 2 pairs of longitudinal excretory canals, a dorsal and a ventral pair, located at each lateral area of the medullary region (Fig 12.38). The body wall of a cyclophyllidean larva is composed of microvilli, tegument, tegumental cells, and smooth muscle fibers (Fig 12.39). Solid-bodied cyclophyllidean larvae are easily confused with pseudophyllidean larvae (spargana). Spargana, also solid-bodied, have muscle fibers and numerous excretory canals scattered randomly throughout the parenchyma.

Adult tapeworms of *H. diminuta, Mesocestoides*

sp, *Bertiella* sp, *I. madagascariensis*, and *Raillietina* sp cause minimal pathologic changes.

The larval stage of *T. taeniaeformis*, a strobilocercus, has infected human liver in 2 reported instances. A strobilocercus is a cysticercus which bears an immature strobila anterior to the bladder, and is described in Figs 12.40 to 12.42.

Diagnosis

Diagnosis of *H. nana* depends on finding characteristic eggs and/or proglottids in feces. Eggs of *H. nana* are sometimes confused with those of *H. diminuta*.[25]

Diagnosis of *Mesocestoides* sp is based on finding gravid proglottids with the characteristic parauterine organ. Eggs are not usually found in stool. Diagnosis of *Bertiella* sp, *I. madagascariensis,* and *Raillietina* sp is based on recovering and identifying proglottids from stool. The proglottids of these 3 genera of tapeworms commonly present as mobile bodies resembling grains of rice and are similar to those of *D. caninum* (Fig 12.29).[24] *Bertiella* sp differ in that the eggs are not in packets. *Inermicapsifer madagascariensis, Raillietina* sp, and *D. caninum* differ in the number of egg capsules per proglottid, the shape and size of the eggs, the length and position of the cirrus sac, and the position of the genital pore or pores (Figs 12.29 and 8.4).[71]

Treatment and Prevention

People acquire these tapeworm infections either by the fecal-oral route or by consuming infected arthropod intermediate hosts. Prevention consists of improved personal and environmental hygiene, along with mass treatment and education of endemic populations. Eliminating rats and mice from housing areas helps reduce exposure to *H. nana*, *H. diminuta*, *Raillietina* sp, and, to a lesser extent, to *Mesocestoides* sp and *I. madagascariensis*.

Praziquantel is the drug of choice for adult tapeworm infections[20] and is superior to niclosamide for *H. nana* infections.[7,14] Niclosamide, a poorly absorbed anthelmintic, is effective against many adult tapeworms and has minimal side effects.[50] Garlic (*Allium sativum*) may be a useful therapeutic alternative for *H. nana* infections.[62,77] Treating *H. diminuta* involves the same drugs, but does not require prolonged administration. Niclosamide has long been the drug of choice for *Mesocestoides* sp, *Bertiella* sp, *I. madagascariensis*, and *R. celebensis* infections.[30,36,72]

References

1. Andersen KI. Description of musculature differences in spargana of Spirometra (Cestoda; Pseudophyllidea) and tetrathyridia of Mesocestoides (Cestoda; Cyclophyllidea) and their value in identification. *J Helminthol* 1983;57:331–334.

2. Bacigalupo J. Sobre una nueva especie de Taenia, Taenia infantis. *Semana Med* 1922;2:302–305.

3. Baer JG, Kourí P, Sotolongo F. Anatomie, position systématique et épidémiologie de Inermicapsifer cubensis (Kourí, 1938) Kourí, 1940, cestode parasite de l'Homme à Cuba. *Acta Trop* 1949;6:120–129.

4. Baer JG, Sandars DF. The first record of Raillietina (R.) celebensis (Janicki, 1902) (Cestoda) in man from Australia, with a critical survey of previous cases. *J Helminthol* 1956;30:173–182.

5. Baer JG. The taxonomic position of Taenia madagascariensis Davaine, 1870, a tapeworm parasite of man and rodents. *Ann Trop Med Parasitol* 1956;50:152–156.

6. Bandyopadhyay AK, Manna B. The pathogenic and zoonotic potentiality of Bertiella studeri. *Ann Trop Med Parasitol* 1987;81:465–466.

7. Baranski MC, Gomes NR, de Godoy OF, et al. Treatment of taeniasis and hymenolepiasis nana with a single oral dose of praziquantel. Study of therapeutic efficacy, tolerance and safety. *Mater Med Pol* 1984;16:129–133.

8. Baylis HA. A new human cestode in Kenya: Inermicapsifer arvicanthidis, a parasite of rats. *Trans R Soc Trop Med Hyg* 1949;42:531–542.

9. Beaver PC, Jung RC, Cupp EW. *Clinical Parasitology*. 9th ed. Philadelphia, Pa: Lea & Febiger; 1984:506.

10. Beaver PC, Jung RC, Cupp EW. *Clinical Parasitology*. 9th ed. Philadelphia, Pa: Lea & Febiger; 1984:507–508.

11. Beaver PC, Jung RC, Cupp EW. *Clinical Parasitology*. 9th ed. Philadelphia, Pa: Lea & Febiger; 1984:509–511.

12. Beaver PC, Jung RC, Cupp EW. *Clinical Parasitology*. 9th ed. Philadelphia, Pa: Lea & Febiger; 1984:522.

13. Beaver PC, Rolon FA. Proliferating larval cestode in a man in Paraguay. A case report and review. *Am J Trop Med Hyg* 1981;30:625–637.

14. Becker B, Mehlhorn H, Andrews P, Thomas H. Scanning and transmission electron microscope studies on the efficacy of praziquantel on Hymenolepis nana (Cestoda) in vitro. *Z Parasitenkd* 1980;61:121–133.

15. Belding DL. *Textbook of Clinical Parasitology*. 3rd ed. New York, NY: D Appleton-Century Co; 1965:387–440.

16. Bhaibulaya M. Human infection with Bertiella studeri in Thailand. *Southeast Asian J Trop Med Public Health* 1985;16:505–507.

17. Blanchard R. Cited by: Grove DI. *A History of Human Helminthology*. Wallingford, England: CAB Int; 1990:421.

18. Blanchard R. Cited by: Cram EB. A species of the cestode genus Bertiella in man and the chimpanzee in Cuba. *Am J Trop Med Hyg* 1928;8:339–344.

19. Bolbol AS. Bertiella sp. infection in man in Saudi Arabia. *Ann Trop Med Parasitol* 1985;79:643–644.

20. Bouree P. Successful treatment of Taenia saginata and Hymenolepis nana by single oral dose of praziquantel. *J Egypt Soc Parasitol* 1991;21:303–307.

21. Buslau M, Marsch WC. Papular eruption in helminth infestation—a hypersensitivity phenomenon? Report of four cases. *Acta Derm Venereol* 1990;70:526–529.

22. Chandler AC. First record of a case of human infection with tapeworms of the genus Mesocestoides. *Am J Trop Med* 1942;22:493–497.

23. Charoenlarp P, Radomyos P. Treatment of Raillietina siriraji with atabrine. *Southeast Asian J Trop Med Public Health* 1973;4:288.

24. Chunge RN, Kabiru EW, Mugo BM. A human case of infection with a rodent cestode (Inermicapsifer) in Kenya. *East Afr Med J* 1987;64:424–427.

25. Cohen IP. A case report of Hymenolepis diminuta infection in a child in St. James Parish, Jamaica. *J La State Med Soc* 1989;141:23–24.

26. Connor DH, Sparks AK, Strano AJ, Neafie RC, Juvelier B. Disseminated parasitosis in an immunosuppressed patient. Possibly a mutated sparganum. *Arch Pathol Lab Med* 1976;100:65–68.

27. Cooper BT, Hodgson HJ, Chadwick VS. Hymenolepiasis: an unusual cause of diarrhoea in Western Europe. *Digestion* 1981;21:115–116.

28. Costa HM de, Corrêa L, Brener Z. Nôvo caso humano de parasitismo por Bertiella mucronata (Meyner, 1895) Stiles and Hassall, 1902 (Cestoda-anoplocephalidae). *Rev Inst Med Trop Sao Paulo* 1967;9:95–97.

29. Cram EB. A species of the cestode genus Bertiella in man and the chimpanzee in Cuba. *Am J Trop Med Hyg* 1928;8:339–344.

30. Elowni EE, Nurelhuda IE, Hassan T. The effect of niclosamide on Raillietina tetragona. *Vet Res Commun* 1989;13:451–453.

31. Eom KS, Kim SH, Rim HJ. Second case of human infection with Mesocestoides lineatus in Korea. *Korean J Parasitol* 1992;30:147–150.

32. Fain A, Limbos P, Van Ros G, De Mulder P, Herin A. Présence du cestode Raillietina (R) celebensis (Janicki, 1902), chez un enfant originaire de Tahiti. *Ann Soc Belg Med Trop* 1977;57:137–142.

33. Freeman RS, Fallis AM, Shea M, Maberley AL, Walters J. Intraocular Taenia crassiceps (Cestoda). II. The parasite. *Am J Trop Med Hyg* 1973;22:493–495.

34. Gamal-Eddin FM, Aboul-Atta AM, Hassounah OA. Extra-intestinal nana cysticercoidiasis in asthmatic and filarised Egyptian patients. *J Egypt Soc Parasitol* 1986;16:517–520.

35. Garrison PE. Davainea madagascariensis (Davaine) in the Philippine Islands. *Phil J Sci* 1911;6:165–175.

36. Gleason NN, Kornblum R, Walzer P. Mesocestoides (cestoda) in a child in New Jersey treated with niclosamide (Yomesan®). *Am J Trop Med Hyg* 1973;22:757–760.

37. Goeze JA. Cited by: Grove DI. *A History of Human Helminthology*. Wallingford, England: CAB Int; 1990:426.

38. Goldsmid JM. A note on the occurrence of Hymenolepis diminuta (Rudolphi, 1819) Blanchard, 1891 (Cestoda) in Rhodesia. *Cent Afr J Med* 1973;19:51–52.

39. Goldsmid JM, Fleming F. The tapeworm infections of children in Rhodesia. *Cent Afr J Med* 1977;23:7–10.

40. Goldsmid JM, Muir M. Inermicapsifer madagascariensis (Davaine, 1870), Baer, 1956 (platyhelminthes: cestoda) as a parasite of man in Rhodesia. *Cent Afr J Med* 1972;18:205–207.

41. Grassi B. Entwicklungscyclus der Taenia nana. Dritte Präliminarnote. *Centralblatt für Bakteriologie und Parasitenkunde* 1887;2:305–312.

42. Gutierrez Y, Buchino JJ, Schubert WK. Mesocestoides (Cestoda) infection in children in the United States. *J Pediatr* 1978;93:245–247.

43. Hamrick HJ, Bowdre JH, Church SM. Rat tapeworm (Hymenolepis diminuta) infection in a child. *Pediatr Infect Dis J* 1990;9:216–219.

44. Hearin JT. Studies on the acquired immunity to the dwarf tapeworm Hymenolepis nana var fraterna in the mouse host. *Am J Hyg* 1941;33:71–87.

45. Hira PR. Human and rodent infection with the cestode Inermicapsifer madagascariensis (Davaine, 1870), Baer, 1956 in Zambia. *Ann Soc Belg Med Trop* 1975;55:321–326.

46. Hira PR. Some helminthozoonotic infections in Zambia. *Afr J Med Med Sci* 1978;7:1–7.

47. Ito A. Hymenolepis nana: protective immunity against mouse-derived cysticercoids induced by initial inoculation with eggs. *Exp Parasitol* 1978;46:12–19.

48. Janicki C. Cited by: Fain A, Limbos P, Van Ros G, De Mulder P, Herin A. Présence du cestode Raillietina (R) celebensis (Janicki, 1902), chez un enfant originaire de Tahiti. *Ann Soc Belg Med Trop* 1977;57:137–142.

49. Jones R, Hunter H, Van Rooyen CE. Bertiella infestation in a Nova Scotia child formerly resident in Africa. *Can Med Assoc J* 1971;104:612.

50. Jones WE. Niclosamide as a treatment for Hymenolepis diminuta and Dipylidium caninum infection in man. *Am J Trop Med Hyg* 1979;28:300–302.

51. Joyeux C, Baer JG. Recherches sur les cestodes appartenant au genre Mesocestoides Vaillant. *Bull Soc Pathol Exot* 1932;25:993–1010.

52. Kawamoto F, Fujioka H, Mizuno S, Kumada N, Voge M. Studies on the post-larval development of cestodes of the genus Mesocestoides: shedding and further development of M. lineatus and M. corti tetrathyridia in vivo. *Int J Parasitol* 1986;16:323–331.

53. Khalil HM, el Shimi S, Sarwat MA, Fawzy AF, Sorougy AO. Recent study of Hymenolepis nana infection in Egyptian children. *J Egypt Soc Parasitol* 1991;21:293–300.

54. Lamon C, Greer GJ. Human infection with an anoplocephalid tapeworm of the genus Mathevotaenia. *Am J Trop Med Hyg* 1986;35:824–826.

55. Leuckart R. Cited by: Fain A, Limbos P, Van Ros G, De Mulder P, Herin A. Présence du cestode Raillietina (R) celebensis (Janicki, 1902), chez un enfant originaire de Tahiti. *Ann Soc Belg Med Trop* 1977;57:137–142.

56. Levi MH, Raucher BG, Teicher E, Sheehan DJ, McKitrick JC. Hymenolepis diminuta: one of three enteric pathogens isolated from a child. *Diagn Microbiol Infect Dis* 1987;7:255–259.

57. Loos-Frank B. Mesocestoides leptothylacus n. sp. and the problem of nomenclature in the genus Mesocestoides Vaillant, 1863 [in German] [author's translations]. *Tropenmed Parasitol* 1980;31:2–14.

58. Lucas SB, Hassounah OA, Muller R, Doenhoff MJ. Abnormal development of Hymenolepis nana larvae in immunosuppressed mice. *J Helminthol* 1980;54:75–82.

59. Lucas SB, Hassounah OA, Doenhoff MJ, Muller R. Aberrant form of Hymenolepis nana: possible opportunistic infection in immunosuppressed patients [letter]. *Lancet* 1979;2:1372–1373.

60. Makarova IA, Astaf'ev BA. The clinico-immunological characteristics of hymenolepiasis [in Russian]. *Med Parazitol (Mosk)* 1992;3:40–43.

61. Margono SS, Handojo I, Hadidjaja P, Mahfudin H. Raillietina infection in children in Indonesia. *Southeast Asian J Trop Med Public Health* 1977;8:195–199.

62. Mikhailitsyn FS, Kovalenko FP, Sergovskaia NL, et al. The search for new antiparasitic agents. 11. The acute toxicity and anticestodal activity of the new anthelmintic Tizanox compared with Azinox [in Russian]. *Med Parazitol (Mosk)* 1992;2:32–34.

63. Mori Y, Shirayama T, Agui Y, Fujiwara A, Nishiura T. A case of chest wall tumor brought on by Hymenolepis nana. *Bull Osaka Med Sch* 1967;13:52–54.

64. Mueller JF. Cited by: Grove DI. *A History of Human Helminthology*. Wallingford, England: CAB Int; 1990:426–427.

65. Nelson GS, Pester FR, Rickman R. The significance of wild animals in the transmission of cestodes of medical importance in Kenya. *Trans R Soc Trop Med Hyg* 1965;59:507–524.

66. Nicholl W, Minchin EA. Cited by: Grove DI. *A History of Human Helminthology*. Wallingford, England: CAB Int; 1990:425.

67. Nowak RM. *Walker's Mammals of the World*. 5th ed. Vol 2. Baltimore, Md: The Johns Hopkins University Press; 1991.

68. Rausch RL. Family Mesocestoididae Fuhrmann, 1907. In: Khalil LF, Jones A, Bray RA, eds. *Keys to the Cestode Parasites of Vertebrates*. International Institute of Parasitology. Wallingford, England: CAB Int; 1994:309–314.

69. Read CP. Longevity of the tapeworm, Hymenolepis diminuta. *J Parasitol* 1967;53:1055–1056.

70. Romero-Cabello R, Godínez-Hana L, Gutiérrez-Quiroz M. Clinical aspects of hymenolepiasis in pediatrics [in Spanish]. *Bol Med Hosp Infant Mex* 1991;48:101–105.

71. Rougier Y, Legros F, Durand JP, Cordoliani Y. A propos de trois cas d'une cestodose rare en polynésie Française. *Bull Soc Pathol Exot* 1980;73:86–89.

72. Rougier Y, Legros F, Durand JP, Cordoliani Y. Four cases of parasitic infection by Raillietina (R.) celebensis (Janicki, 1902) in French Polynesia. *Trans R Soc Trop Med Hyg* 1981;75:121.

73. Saeki Y. Experimental studies on the development of Hymenolepis nana [in Japanese]. *Jika Zasshi* 1920;238:203–244. Taken from: *Trop Dis Bull* 1921;18:112. [English abstract.]

74. Schultz LV, Roberto RR, Rutherford GW 3d, Hummert B, Lubell I. Mesocestoides (Cestoda) infection in a California child. *Pediatr Infect Dis J* 1992;11:332–334.

75. Schwartz B. The life history of tapeworms of the genus Mesocestoides. *Science* 1927;66:17–18.

76. Sidky HA, Hassan ZA, Hassan RR, Gaafar SA, el-Zahraa F, Awadallah M. Disseminated Hymenolepis nana in blood of a filarial patient. *J Egypt Soc Parasitol* 1987;17:155–159.

77. Soffar SA, Mokhtar GM. Evaluation of the antiparasitic effect of aqueous garlic (Allium sativum) extract in hymenolepiasis nana and giardiasis. *J Egypt Soc Parasitol* 1991;21:497–502.

78. Sterba J, Barus V. First record of Strobilocercus fasciolaris (Taeniidae-larvae) in man. *Folia Parasitol (Praha)* 1976;23:221–226.

79. Stunkard HW. The morphology and life history of the cestode, Bertiella studeri. *Am J Trop Med* 1940;20:305–333.

80. Subbannayya K, Achyutha Rao KN, Shivananda PG, Kundaje GN, Adams LJ, Healy GR. Bertiella infection in an adult male in Karnataka. A case report. *Indian J Pathol Microbiol* 1984;27:269–271.

81. Tesjaroen S, Chareonlarp K, Yoolek A, Mai-iam W, Lertlaituan P. Fifth and sixth discoveries of Hymenolepis diminuta infections in Thai people. *J Med Assoc Thai* 1987;70:49–50.

82. Thompson CD, Jellard CH, Buckley JJ. Human infection with a tapeworm, Bertiella sp, probably of African origin. *Br Med J* 1967;3:659–660.

83. Widmer EA, Engen PC, Bradley GL. Intracapsular asexual proliferation of Mesocestoides sp tetrathyridia in the gastrointestinal tract and mesenteries of the prairie rattlesnake (Crotalus viridis viridis). *J Parasitol* 1995;81:493–496.

13

Lymphatic Filariasis

Eric A. Ottesen, Wayne M. Meyers,
Ronald C. Neafie, *and*
Aileen M. Marty

Introduction

Definition

Lymphatic filariasis is infection by the adult filarial nematodes *Wuchereria bancrofti* (Cobbold, 1877) Seurat, 1921, *Brugia malayi* (Brug, 1927) Buckley, 1958, and *Brugia timori* Partono, Purnomo, Dennis, Atmosoedjono, Oemijati, and Cross, 1977. The infection initially presents as lymphangiectasis; secondary infections or host immune reactions may further damage lymphatic vessels. Long-standing alterations in lymphatic circulation can lead to elephantiasis (Fig 13.1), chyluria, and many other clinical complications. Severe host response to the microfilariae of these parasites causes tropical eosinophilic syndrome.

Synonyms

Other names for infection by *W. bancrofti* include Bancroft's filariasis, bancroftian filariasis, and wuchereriasis. Infection by *B. malayi* is also called brugian filariasis or Malayan filariasis. Timorian filariasis is another name for infection by *B. timori*. Lymphatic filariasis is also known as elephantiasis, elephantiasis arabum, curse of St. Thomas, and Barbados leg. Tropical eosinophilic syndrome is synonymous with Meyers and Kouwenaar's syndrome, Weingarten's syndrome, tropical eosinophilia, occult filariasis, and filarial hypereosinophilia. The common pulmonary form of tropical eosinophilia is known as eosinophilic lung and tropical pulmonary eosinophilia.

Synonyms for *W. bancrofti* include *Filaria bancrofti*, *Filaria sanguinis hominis*,[7] *Filaria nocturna*, *Filaria wuchereria*, *Wuchereria filaria*, *Wuchereria pacifica*, *Wuchereria bancrofti* var *pacifica*, and Wucherer's filaria. Synonyms for *B. malayi* include *Filaria malayi*, *Filaria bancrofti*, *Wuchereria malayi*, and microfilaria malayi. The synonym for microfilariae of *B. timori* is *Timor microfilariae*.

General Considerations

Writers of antiquity recorded numerous instances of the condition now known as elephantiasis. Hindu,[39] Persian,[12] and Roman[35] physicians of the pre-Christian era described the clinical manifestations, and works of art unearthed in Egypt,[35] Nigeria, and the Yucatan[33] depict the characteristic swellings and deformities of elephantiasis.

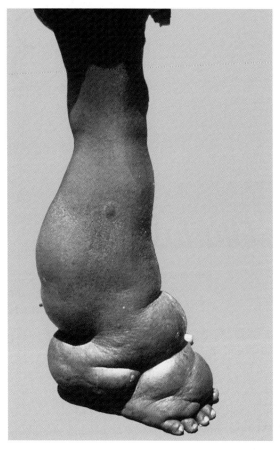

Figure 13.1
Elephantiasis in right leg of Asian man.*

In the 1860s Demarquay found larval nematodes in the hydrocele fluid of a patient,[22] and Wucherer discovered microfilariae in a patient with chyluria.[38] In the 1870s, Lewis found microfilariae in chylous urine and blood,[3,4,6,7] and Cobbold published J. Bancroft's discovery of 5 adult worms in a single patient,[17,19] as well as Manson's observation that larvae developed within mosquitoes that had fed on the blood of infected patients.[18] Within a few years, the mosquitoes *Culex quinquefasciatus* and *Culex pipiens* were identified as intermediate hosts. At the turn of the century, the work of T. Bancroft and Low confirmed that *C. quinquefasciatus* was also a vector for filariae.[5]

In 1927 Lichtenstein reported filarial infections in Indonesia from microfilariae morphologically different from *W. bancrofti*. He and Brug named the new microfilariae *Filaria malayi*. In 1960 Buckley described the adult form of these Pacific worms and renamed them *B. malayi*. A few years later *B. timori* was identified on Timor Island, and *Brugia pahangi* was discovered to be a potential human parasite.[29]

Wise and Minett, in 1912, confirmed that bacterial infections aggravate filarial disease. In 1932, O'Connor suggested that living worms produce no

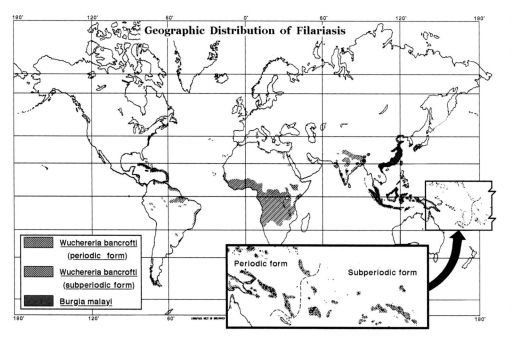

Figure 13.2
Distribution of *W. bancrofti* and *B. malayi*.

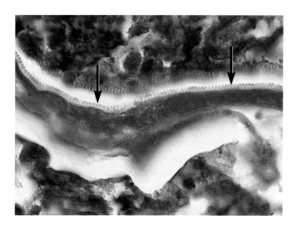

Figure 13.3
Longitudinal section of adult male *W. bancrofti* in lymph node showing fine cuticular striations approximately 1 μm apart (arrows). Movat x760

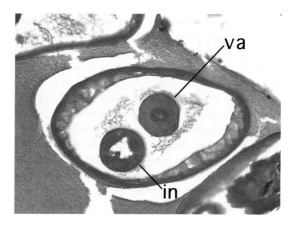

Figure 13.4
Transverse section through anterior region of adult female *W. bancrofti* (180 μm in diameter) within dilated lymphatic in human testis. Note intestine (in) and thick muscular vagina (va). x235

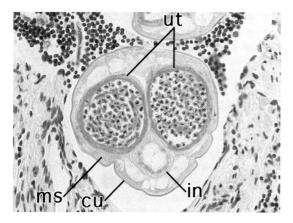

Figure 13.5
Transverse section of gravid *W. bancrofti* (175 μm in diameter) from lymph node, depicting cuticle (cu), muscle (ms), intestine (in), and uteri (ut) filled with mature microfilariae. x230

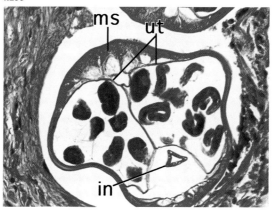

Figure 13.6
Transverse section of gravid *W. bancrofti* (125 μm in diameter) from lymph node, depicting intestine (in), muscle (ms), and uteri (ut) containing developing microfilariae. x420

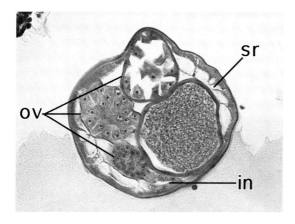

Figure 13.7
Transverse section of posterior region of gravid *W. bancrofti* (140 μm in diameter) from lymph node, showing 3 sections of ovary (ov), seminal receptacle (sr) filled with spermatozoa, and intestine (in). x330

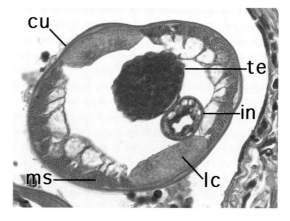

Figure 13.8
Transverse section through anterior region of adult male *W. bancrofti* (90 μm in diameter) in lymph node, demonstrating thin cuticle (cu), prominent lateral cords (lc), muscle (ms), intestine (in), and testis (te). x600

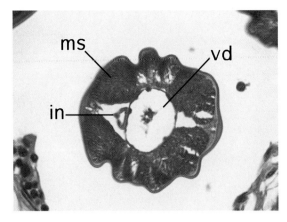

Figure 13.9
Transverse section through posterior region of adult male *W. bancrofti* (90 µm in diameter) in lymph node, depicting intestine (in) and vas deferens (vd). Somatic muscle cells (ms) are prominent at this level. x500

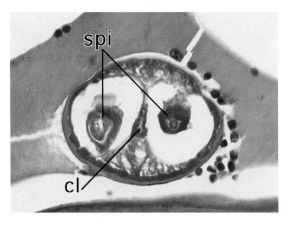

Figure 13.10
Transverse section through posterior region of adult male *W. bancrofti* (85 µm in diameter) in lymph node, depicting cloaca (cl) and 2 spicules (spi). x500

Figure 13.11
Three adult female *B. malayi* recovered from experimentally infected gerbils. Worms are white, thread-like, and 160 µm in maximum diameter.*

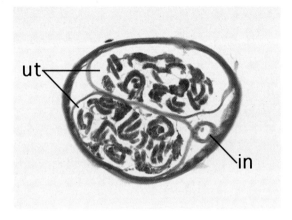

Figure 13.12
Transverse section through gravid *B. malayi* (135 µm in diameter) recovered from experimentally infected gerbil. Note 2 uteri (ut) filled with microfilariae, and intestine (in). Lateral cords and somatic muscle cells are inconspicuous. x315

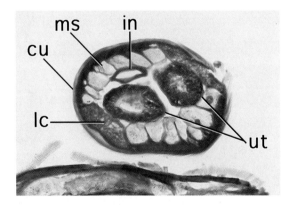

Figure 13.13
Transverse section through adult female *B. malayi* (125 µm in diameter) in lung, demonstrating cuticle (cu), lateral cords (lc), muscles (ms), intestine (in), and 2 uteri (ut). Lateral cords and somatic muscle cells are prominent. Movat x370

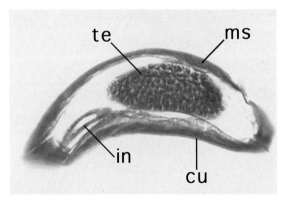

Figure 13.14
Oblique section through anterior region of adult male *B. malayi* (65 µm in diameter) recovered from experimentally infected gerbil, demonstrating thin cuticle (cu), muscle (ms), testis (te), and intestine (in). x350

serious pathology, but that dead or degenerating worms can provoke an immune response that leads to elephantiasis independent of secondary bacterial infection.[46] In 1996 Dreyer et al proved that living worms do cause lymphangiectasis, confirming that damage by living adult worms can produce clinical symptoms, especially when accompanied by secondary bacterial or fungal infection, or by immunologic response to dead or dying worms.

In 1939 Meyers and Kouwenaar first described tropical eosinophilic syndrome in Javanese patients who presented with hypereosinophilia and eosinophilic lesions caused by microfilariae in lymph nodes. They suggested that these symptoms were an allergic reaction to the microfilariae.[46] Later discoveries by investigators in Indonesia and India[13,31,61] established the various clinical manifestations of the syndrome and confirmed that filariasis was a major cause of tropical eosinophilia.[59] In 1979 Ottesen et al reported that patients with tropical eosinophilia had extraordinarily vigorous, immediate hypersensitivity reactions to filarial antigens.[50]

Epidemiology

In tropical and subtropical countries where it is endemic, lymphatic filariasis is a major cause of disease, and the number of cases is increasing as a result of rapid, haphazard urbanization.[22] An estimated 120 million people worldwide have lymphatic filariasis, and an additional billion are at risk for infection.[44,48] Approximately 88% of cases are bancroftian filariasis; the remainder are predominantly brugian filariasis. All patients have either overt symptoms (lymphedema, elephantiasis, hydrocele, recurrent infections, tropical eosinophilia), or subclinical abnormalities in lymphatic and renal function.

Bancroftian filariasis is distributed widely in tropical and subtropical areas of South America, the Caribbean, Africa, Asia, and the Pacific (Fig 13.2). The 2 forms of *W. bancrofti* are distinguished by the periodicity of their circulating microfilariae. Nocturnally periodic microfilariae are primarily detectable in peripheral blood at night. Subperiodic microfilariae are detectable at all hours, but are more concentrated in blood drawn in the late afternoon. Generally, subperiodic *W. bancrofti* are found only in Pacific islands east of Australia and the Solomon Islands (New Caledonia, Fiji, Samoa, Ellice Islands, Cook Islands, Society Islands, and the Marquesas); elsewhere, *W. bancrofti* are generally nocturnally periodic.

The distribution of brugian filariasis is much more restricted. Both the subperiodic and the more common nocturnally periodic forms of *B. malayi* are found primarily in parts of Malaysia, Indonesia, India, China, Korea, the Philippines, and Japan (Fig 13.2). Timorian filariasis is limited to several islands of the Indonesian Archipelago.

The distribution of filariasis is directly related to human migration and changes in living standards. Improved living conditions and intensive mosquito control have eradicated filariasis in some previously endemic regions.[35,57]

Infectious Agents

Morphologic Description

Tables 13.1 and 13.2 outline the distinguishing features of the thin, hair-like filarial nematodes that commonly infect humans.

Adult *W. bancrofti*

In both sexes, the cuticle of adult *W. bancrofti* is thin (1 to 2 μm), finely striated (Fig 13.3), and thicker in the lateral cord regions, forming low rounded ridges. Trichrome stains highlight cuticular and muscular features. Scanning electron microscopy reveals transverse striations with deep periodic annulations, and many small spherical bosses on the ventral and dorsal cuticle.[8]

Adult female *W. bancrofti* are 60 to 100 mm by 150 to 250 μm. The vulva lies in the anterior end of the worm before the esophageal-intestinal junction, and leads into the thick, muscular vagina (Fig 13.4). The intestine and 2 uteri occupy the body cavity throughout most of the length of the worm (Figs 13.5 and 13.6). Female worms vary considerably in appearance, depending on the size of the intestine and reproductive tubes, and the amount of somatic muscle cells visible. The ovaries, oviducts, and seminal receptacles (Fig 13.7) are confined to the posterior region, extending to within 1

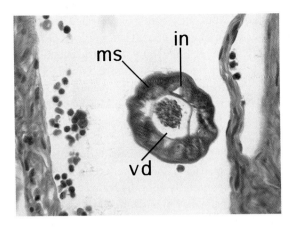

Figure 13.15
Transverse section through posterior region of adult male *B. malayi* (30 μm in diameter) in lymph node of Korean patient, showing vas deferens (vd) containing spermatozoa, and intestine (in). Somatic muscle cells (ms) are well-developed at this level. x415

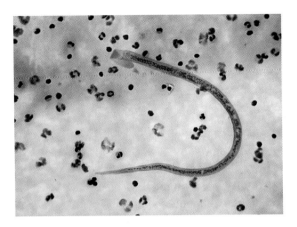

Figure 13.16
Sheathed microfilaria of *W. bancrofti* (290 μm long) in thick blood film. Delafield's hematoxylin x240

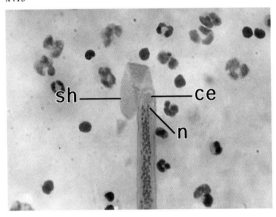

Figure 13.17
Anterior end of microfilaria of *W. bancrofti* in thick blood film demonstrating sheath (sh), cephalic space (ce), and anteriormost nuclei (n). Delafield's hematoxylin x600

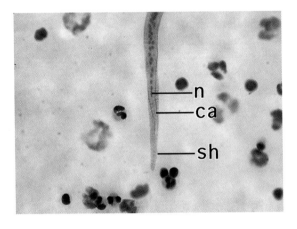

Figure 13.18
Posterior end of microfilaria of *W. bancrofti* in thick blood film depicting sheath (sh), long caudal space (ca), and terminal nucleus (n). Delafield's hematoxylin x625

mm of the tip of the tail. Somatic muscle cells are slightly to moderately developed, with few muscle cells per quadrant. Lateral cords are usually inconspicuous.

Adult male *W. bancrofti* are about 40 mm by 100 to 150 μm. The testis (Fig 13.8) begins in the anterior region, slightly posterior to the esophageal-intestinal junction. The testis is followed by the long vas deferens that occupies most of the length of the worm. In much of the posterior half of the worm, somatic muscle cells are very prominent (Fig 13.9). The vas deferens unites with the intestine to form the cloaca, which terminates at the anus near the tip of the tail. Male worms have 2 unequal and dissimilar copulatory spicules (Fig 13.10).

Adult *B. malayi*

Adults are slender, white, and hair-like (Fig 13.11). They are very similar to adult *W. bancrofti*, but are considerably shorter and slightly more slender. The cuticle of adult *B. malayi* of both sexes is thin (1 to 2 μm), finely striated, and thicker in the lateral cord regions, forming low rounded ridges.

Adult female *B. malayi* are 50 to 60 mm by 130 to 170 μm. The intestine and 2 uteri occupy most of the body cavity (Figs 13.12 and 13.13). Lateral cords are inconspicuous to prominent, and somatic muscle shows slight to moderate development.

Adult male *B. malayi* are 22 to 25 mm by 90 μm in maximum diameter. In the anterior region, containing the intestine and testis (Fig 13.14), lateral cords may be prominent and somatic muscle moderately developed. Most of the posterior region contains the intestine and the long vas deferens (Fig 13.15). In this region lateral cords are inconspicuous, but somatic muscle is well-developed and occupies most of the body cavity.

Adult *B. timori*

Adult female *B. timori* are 21 to 39 mm by 140 μm in maximum diameter; males are 13 to 23 mm by 80 μm in maximum diameter in experimentally infected gerbils. There is no description of adult *B. timori* in humans.

Microfilariae

Table 13.3 lists the morphologic features of the most common microfilariae in humans. Microfilariae of *W. bancrofti*, *B. malayi*, *B. timori*, and *Loa loa* are sheathed, whereas microfilariae of *Onchocerca volvulus* and *Mansonella* sp are unsheathed. Any sheathed microfilaria may lose its sheath and appear sheathless. Table 13.4 lists features that differentiate microfilariae of *W. bancrofti*, *B. malayi*, and *B. timori* in blood films.

Microfilariae of *W. bancrofti* are 230 to 300 μm by 7 to 10 μm (Fig 13.16). The sheath stains poorly or not at all with Giemsa, but is easily seen with hematoxylin. The short cephalic space is 5 to 7 μm long and the anteriormost nuclei are side by side (Fig 13.17). The tapered tail has a caudal space 5 to 15 μm long, and the terminal nuclei are elongated (Fig 13.18). A conspicuous *Innenkörper* (inner body) lies in the posterior half of the microfilariae and stains bright pink with Giemsa (Fig 13.19).

Microfilariae of *B. malayi* are 175 to 260 μm by 4 to 6 μm. The sheath typically stains pink with Giemsa (Figs 13.20 and 13.21) and pale blue with hematoxylin. The long cephalic space measures 6 to 11 μm, and the anteriormost nuclei are side by side. The posterior end tapers to a point and contains minute subterminal and terminal nuclei which sometimes cannot be identified. A constriction separates the subterminal and terminal nuclei (Figs 13.20 and 13.21). The *Innenkörper* does not stain,

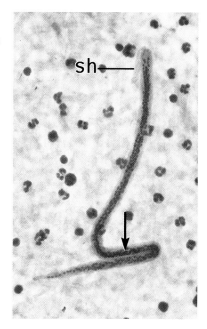

Figure 13.19
Microfilaria of *W. bancrofti* in thick blood film demonstrating faintly stained sheath (sh) at anterior end, and red-stained *Innenkörper* (arrow) in posterior half. Giemsa x450

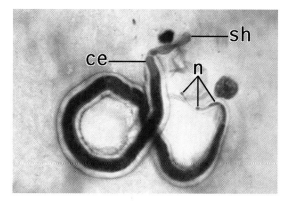

Figure 13.20
Microfilaria of *B. malayi* in thick blood film depicting pink-stained sheath (sh), cephalic space (ce), and terminal nuclei (n). Column of nuclei stains intense blue. Giemsa x725

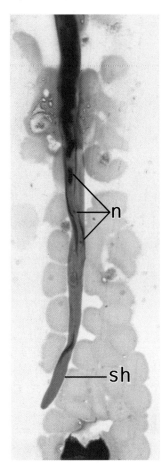 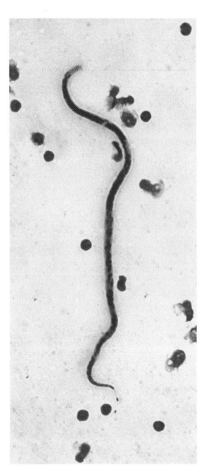 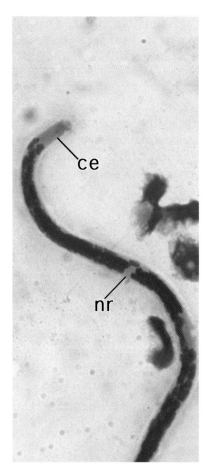

Figure 13.21
Posterior end of microfilaria of *B. malayi* in thin blood film depicting pink-stained sheath (sh) and terminal nuclei (n). Giemsa x985

Figure 13.22
Microfilaria of *B. timori* (310 μm long) in thick blood film. Sheath does not stain with Giemsa. Giemsa x350

Figure 13.23
Anterior end of microfilaria of *B. timori* in thick blood film demonstrating long cephalic space (ce) and nerve ring (nr). Giemsa x845

but is sometimes marked by a large area where the nuclear column is sparse.

Microfilariae of *B. timori* are 290 to 340 μm by 6 to 7 μm (Fig 13.22). The sheath stains pale blue with hematoxylin, but unlike microfilariae of *B. malayi*, the sheath of *B. timori* does not stain with Giemsa (Figs 13.22 to 13.24). Microfilariae of *B. timori* have a long cephalic space (Fig 13.23), a densely stained nuclear column, and subterminal and terminal nuclei with no constriction between them (Fig 13.24). Hematoxylin reveals the sparse nuclei in the region of the large *Innenkörper*.

Spores of helicosporous fungi contaminating blood films may resemble microfilariae (Fig 13.25).

Life Cycle and Transmission

Cats and humans can be natural definitive hosts for subperiodic forms of *B. malayi*, but humans are the only natural definitive and reservoir hosts for *W. bancrofti* and nocturnally periodic *B. malayi*. Mosquitoes of the genera *Culex*, *Anopheles*, *Aedes*, and *Mansonia* serve as intermediate hosts and vectors of *W. bancrofti*.[34] *Culex quinquefasciatus* is the common vector in urban areas; anopheline or aedean mosquitoes are common vectors in rural areas. Natural vectors for the nocturnal form of *B. malayi* are mansonian and anopheline mosquitoes, but aedean mosquitoes can harbor the infective stage. Mansonian mosquitoes are the major

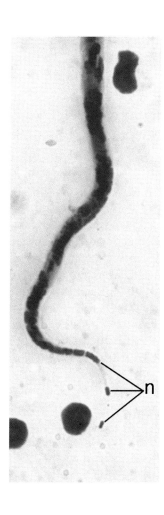

Figure 13.24
Posterior end of microfilaria of *B. timori* in thick blood film demonstrating terminal nuclei (n). Giemsa x850

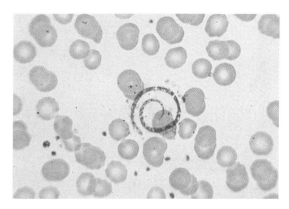

Figure 13.25
Spore of helicosporous fungus in thin blood film. Such spores are sometimes mistaken for microfilariae. Giemsa x735

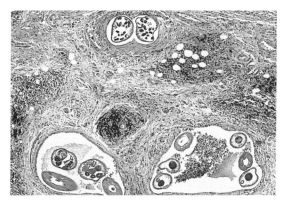

Figure 13.26
Viable adult female and male *W. bancrofti* lying within dilated lymphatics in lymph node. x55

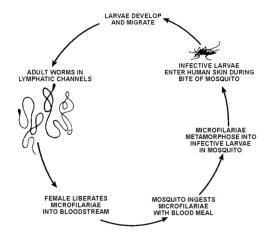

Figure 13.27
Life cycle of the lymphatic-dwelling filariae. Humans serve as definitive hosts and mosquitoes as intermediate hosts.

vectors for the subperiodic form of *B. malayi*. Anopheline mosquitoes are the vector for *B. timori*.[52]

Adult worms live in lymphatic vessels (Fig 13.26). Female worms discharge microfilariae that circulate in the blood until they are ingested by the appropriate mosquito, or die. When ingested by a mosquito, microfilariae exsheath, penetrate the stomach wall, and migrate to the thoracic muscles within 24 hours. In 6 to 12 days, larvae molt twice and metamorphose into infective filariform (L3) larvae within the thoracic muscles. Later, L3 larvae migrate to the mosquito's proboscis, exit to human skin when the mosquito feeds, and penetrate the skin at the puncture site. Within a human host, larvae migrate to a suitable site within lymphatic vessels, where they develop into adult worms. Maturation is a slow process; it may be 6 to 12 months before microfilariae appear in

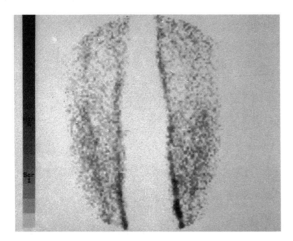

Figure 13.28
Lymphoscintigram, using [99]Technetium-labeled dextran, depicting normal lower limb lymphatics. Contrast with Figure 13.29.*

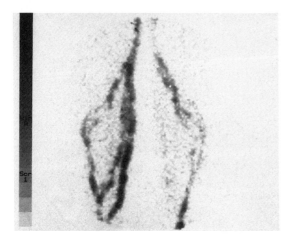

Figure 13.29
Lymphoscintigram, using [99]Technetium-labeled dextran, depicting dilated major lymphatic and collateral trunks in microfilaremic patient with no symptoms or clinical evidence of lymphedema. Contrast with Figure 13.28.*

peripheral blood (Fig 13.27). Adult *W. bancrofti* may be fertile for as long as 40 years (generally 5 years).[16,58]

Clinical Features and Pathogenesis

The clinical symptoms of lymphatic filariasis in inhabitants of endemic regions, with their lifelong exposure to these infectious agents, differ from those in expatriates encountering infection for the first time. Long-term visitors, military personnel, and settlers usually have localized inflammatory reactions, especially adenolymphangitis, and signs of immediate hypersensitivity such as urticaria, eosinophilia, and elevated IgE. These patients rarely have the characteristic clinical findings of indigenous patients, such as asymptomatic microfilaremia.

Nevertheless, patients with asymptomatic microfilaremia do have significantly abnormal lymphatic function, visible on lymphoscintigraphy (Figs 13.28 and 13.29), and may have renal injuries leading to hematuria and proteinuria. For these patients, microfilariae in peripheral blood smears are usually incidental findings in diagnostic evaluations or surveys of endemic regions. Ultrasonography (Fig 13.30) reveals living adult worms even in indigenous patients with no symptoms of filariasis and no detectable microfilariae. Ultrasound also reveals lymphangiectasis at the sites of adult worms.[28]

The major symptoms of brancroftian and brugian filariasis relate to structural and functional abnormalities of lymphatic channels, changes that develop even in patients with asymptomatic microfilaremia (Fig 13.30). Adult worms induce local reactions that cause dilatation and torsion of lymphatic vessels, hypertrophy of vessel walls, loss of valvular function, and backflow of lymph. Compromised lymphatic function leads to lymphedema that is initially reversible. By a variety of pathogenic mechanisms,[25] continued assault by

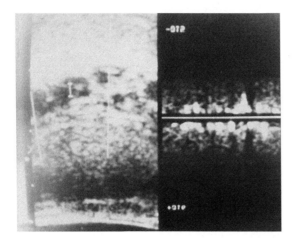

Figure 13.30
Ultrasound of right hemiscrotum of patient with asymptomatic microfilaremia. The random, nonregular pattern of sound in left panel reflects rapid, irregular movements of parasites within the dilated lymphatic, visible in right panel (marked by bar).

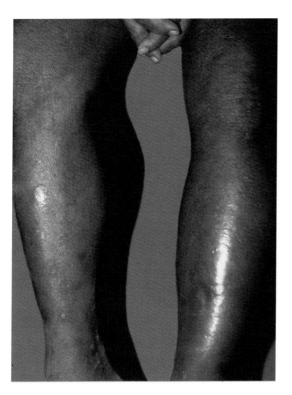

Figure 13.31
Adenolymphangitis (ADL) of leg on left, with acute inflammation caused by bacterial superinfection of lymphedematous limb. This form of ADL is more common than inflammation originating in nodes and spreading along lymphatics (filarial fever).

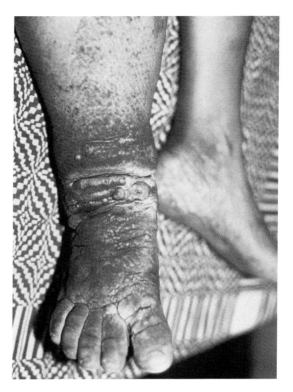

Figure 13.32
Hyperkeratosis and lesions in lymphedematous right leg and foot that serve as points of entry for bacterial or fungal superinfection, leading to inflammatory ADL (see Figure 13.31).

persistent parasites, frequently complicated by localized bacterial (Fig 13.31) and fungal superinfection of the skin, leads to increasing fibrosis and chronic elephantiasis of limbs (Figs 13.32 and 13.33), breasts, and genitalia (Fig 13.34), as well as to hydrocele (Fig 13.35), and/or chyluria.[24] Patients may also develop abscesses (Fig 13.36), ulcers (Fig 13.37), breast masses (Fig 13.38), pleural effusions, constrictive pericarditis, and fibrosing mediastinitis.[32] The site of lymphatic damage determines the site and type of clinical symptoms. Dead worms can cause inflammatory obstruction of lymphatic channels. Collateral lymphatic channels form and some obstructed vessels recanalize, but lymphatic function remains compromised.

Filarial fever is an acute febrile episode of adenolymphangitis accompanied by painful lymphatic inflammation. Patients have lymphadenitis and/or lymphangitis (Fig 13.39) and transient local edema, but are frequently not microfilaremic.

Filarial fever can recur up to 10 times a year, usually lasting 3 to 7 days before subsiding spontaneously. There are 2 distinct causes of filarial fever. The most common is bacterial superinfection of tissue whose lymphatic function is already compromised. A second cause of filarial fever is immune response to lymphatic-dwelling parasites. When bacteria are the cause, patients present with cellulitis and a warm, edematous extremity. If immune response is the cause, patients have a retrograde lymphangitis and cold edema of the affected limb. In both bancroftian and brugian filariasis, these reactions are most common in the lower or upper extremities. Involvement of the genital lymphatics is almost exclusively a feature of *W. bancrofti* infection; acute bancroftian filariasis may include symptoms of funiculitis, epididymitis, scrotal pain, and tenderness that result from local lymphangitis.

As lymphatic damage progresses, the initially transient edema and anatomic distortion become

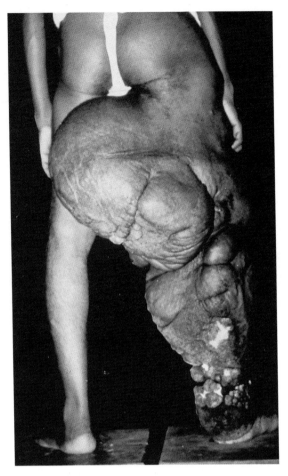

Figure 13.33
Gross deformity resulting from chronic elephantiasis.*

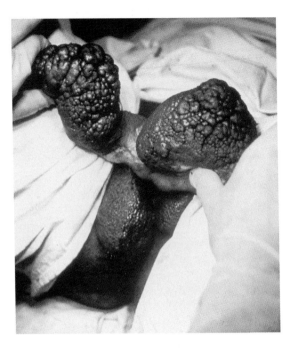

Figure 13.34
Pendulous clitoral tumor in 40-year-old Indian woman, attributed to lymphatic filariasis.*

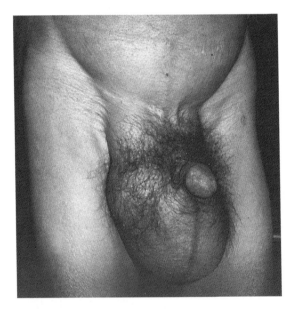

Figure 13.35
Large bilateral hydroceles in patient with bancroftian filariasis. There is no evidence of scrotal or penile lymphedema/elephantiasis.

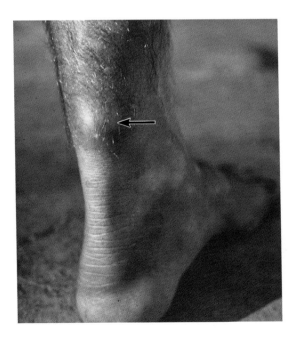

Figure 13.36
Nodular swelling (arrow) on lower calf caused by *B. malayi.**

the permanent changes of elephantiasis. Pitting edema turns into a brawny edema, with hyperkeratosis and thickening of subcutaneous tissue. Fissuring of the skin follows, along with nodular and papillomatous hyperplastic changes (Fig 13.32). Superinfection with bacteria and/or dermatophytes becomes an increasing problem. Patients with bancroftian filariasis may develop scrotal lymphedema, hydrocele (Fig 13.35), and chyluria. Characteristically, the chyluria is intermittent, sometimes lasting days or weeks before abating spontaneously and then recurring sometime later. Some patients with elephantiasis develop a crusty, verrucous skin change that resembles mossy foot (lymphostatic verrucosa) (Figs 13.40 and 13.41). Hematuria and proteinuria are probably the result of circulating immune complexes common in infected individuals.

Tropical eosinophilia commonly has an insidious, subacute onset, then progresses slowly. It is a form of occult filariasis in which the immunologic hyperresponsiveness of the host results in such rapid clearance of microfilariae from the blood that detecting microfilariae in peripheral blood smears is almost impossible.

Generally, microfilarial clearance takes place in the lungs and produces tropical pulmonary eosinophilia (TPE), an acute and chronic lung disease. Symptoms initially result from allergic and inflammatory reactions triggered by the cleared parasites. Patients with TPE usually have a primarily nocturnal paroxysmal cough, with some sputum production and wheezing. Most patients report a low-grade fever and myalgia 1 to 2 weeks prior to the onset of respiratory complaints. Patients have adenopathy and occasional weight loss and gastrointestinal symptoms (nausea, vomiting, diarrhea). Extreme blood eosinophilia (>3000/µl) is the rule; leukocytosis is common (70 to 90 x 10^3/µl, with 70% to 90% eosinophils), as are elevated levels of circulating immune complexes.[45] Chest roentgenograms may be normal, but generally show increased bronchovascular markings, diffuse interstitial lesions (Fig 13.42), or mottled opacities primarily involving the mid and lower lung fields. They may show patchy areas of alveolar filling disease. Pulmonary function tests usually reveal significant restrictive abnormalities, often with superimposed mild to moderate obstructive defects. Without appropriate treatment,

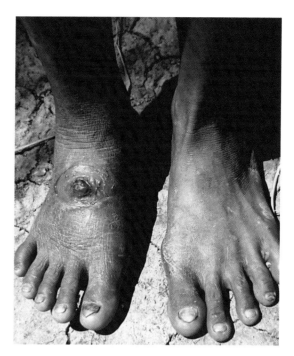

Figure 13.37
Ulcer on foot caused by *B. malayi*.*

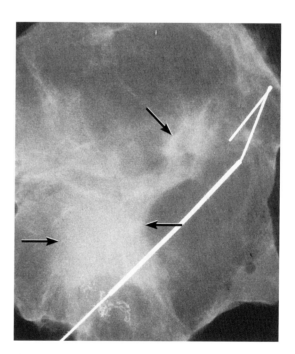

Figure 13.38
Mammogram of breast biopsy specimen. Spiculated margins (arrows) suggest tumor; however, lesion contained adult *W. bancrofti*.

Figure 13.39
Lymphangitis caused by *B. malayi* in man from Borneo.*

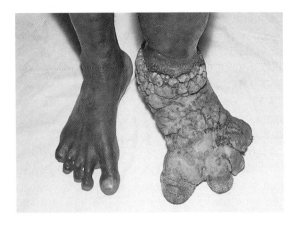

Figure 13.40
Mossy foot (lymphostatic verrucosa) of unknown etiology in an African. Similar warty lesions develop in patients with filarial elephantiasis. x25

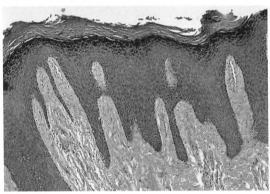

Figure 13.41
Biopsy of mossy foot (lymphostatic verrucosa) demonstrating acanthosis, elongation of rete ridges, edema, and chronic inflammation of dermis with perivascular lymphocytes and plasma cells typical of this lesion. x55

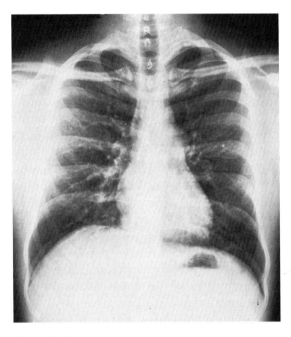

Figure 13.42
Chest radiograph of patient with tropical pulmonary eosinophilia showing diffuse mottling throughout parenchyma. Many variations of this mottling are possible.

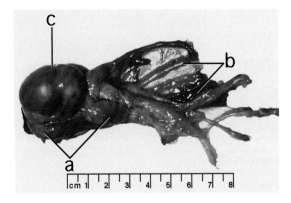

Figure 13.43
Surgically dissected specimen from patient with filariasis demonstrating (a) thickened epididymis and cord, (b) dilated veins and lymphatics, and (c) normal testis.

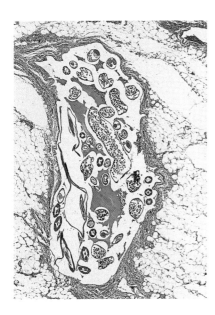

Figure 13.44
Viable adult male and female *W. bancrofti* within dilated lymphatic of lymph node. x2.5

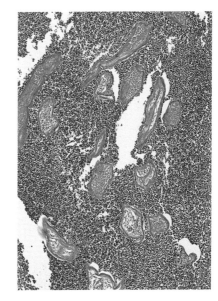

Figure 13.45
Degenerated coiled female *W. bancrofti* centered in abscess of lymph node. x60

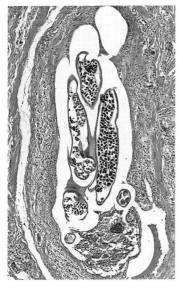

Figure 13.46
Viable adult female *W. bancrofti* within dilated lymphatic of testis. Note thickening of vessel wall and granulomatous reaction. x50

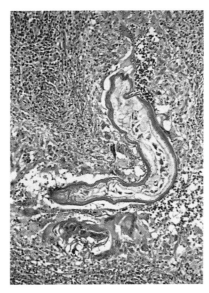

Figure 13.47
Degenerated adult female *W. bancrofti* provoking granulomatous reaction in inguinal lymph node. x95

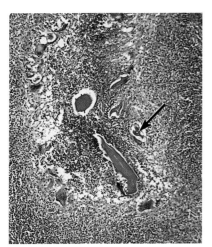

Figure 13.48
Degenerated adult male *W. bancrofti* causing granuloma in inguinal lymph node. One spicule is visible (arrow). x55

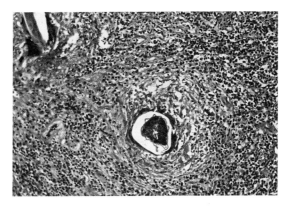

Figure 13.49
Giant cell engulfing degenerated adult male *W. bancrofti* in lymph node. Note massive tissue eosinophilia. x110

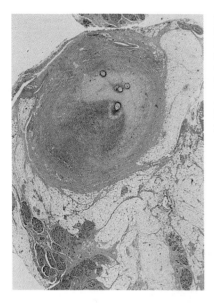

Figure 13.50
Coagulated lymph surrounding degenerated coiled female *W. bancrofti* within necrotic breast lesion. x6.2

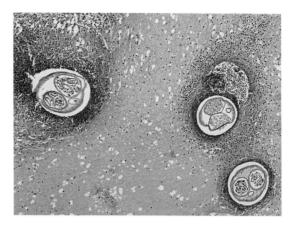

Figure 13.51
Higher magnification of female *W. bancrofti* shown in Figure 13.50. x60

Figure 13.52
Fragments of degenerated *W. bancrofti* provoking caseation necrosis in epididymis. x70

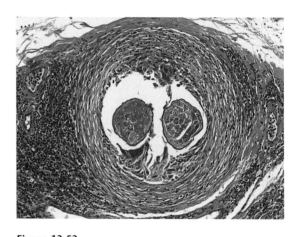

Figure 13.53
Transverse sections through dead gravid *W. bancrofti* provoking concentric fibrosis in lymph node. x115

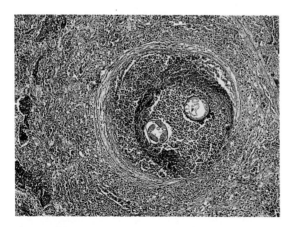

Figure 13.54
Fibrinous thrombus surrounding degenerated adult male *W. bancrofti* and obstructing pulmonary vessel in patient from New Guinea. Movat x55

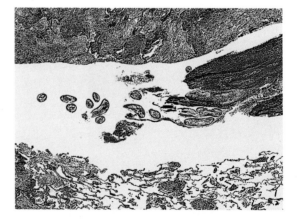

Figure 13.55
Coiled gravid *B. malayi* (125 µm in diameter) within pulmonary lesion in patient from India. Movat x25

patients progress to a debilitating chronic interstitial pulmonary fibrosis. Late-stage interstitial pulmonary fibrosis can potentially cause cor pulmonale.[54] Rarely, patients develop eosinophilic pleural effusions.[14] Even more uncommon is a filarial pleural effusion in patients with no signs or symptoms of TPE.[1]

A related hyperresponsive syndrome develops in patients in whom microfilariae lodge in the liver, spleen, and/or lymph nodes. In these patients, the major clinical findings are hepatomegaly, splenomegaly, or lymphadenopathy.

Figure 13.56
Higher magnification of worm depicted in Figure 13.55. Movat x120

Pathologic Features

Adult filariae live in lymphatic vessels, often those afferent to lymph nodes. In bancroftian filariasis, adult worms also inhabit the lymphatics of the scrotum, testis, epididymis, and spermatic cord (Fig 13.43).

Initially, the only damage is dilatation of the lymphatic vessels (Fig 13.29). Histologic studies of adult filarial worms typically reveal no inflammatory reaction around living worms in their natural state of parasitism. Serious damage begins as the worms degenerate, or when a bacterial or fungal superinfection of the tissue subserved by the abnormally functioning lymphatics provokes an inflammatory response.

A poorly understood interplay develops between dead or dying worms and proinflammatory elements of the host immune system, which leads to local inflammatory and granulomatous reactions. Degenerating worms provoke an acute inflammation that may extend to areas where the worm is still viable (Fig 13.44). Characteristically, necrosis with neutrophils and macrophages surrounds the worm (Fig 13.45). Acute inflammatory cells infiltrate the lymphatic vessel wall and obliterate the lumen. In time, the endothelial lining of the vessel thickens, and a chronic inflammatory infiltrate (lymphocytes, macrophages, plasma cells, and eosinophils) develops in tissue surrounding the dying worm. The tissue may develop a granulomatous reaction consisting of epithelioid and giant cells (Figs 13.46 to 13.49), which may include coagulated lymph (Figs 13.50 and 13.51) or central caseation necrosis around the worm (Fig

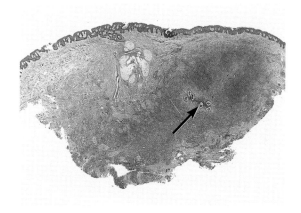

Figure 13.57
Gravid *W. bancrofti* in skin of breast, provoking necrotizing granuloma (arrow) and massive chronic inflammation. x7.8

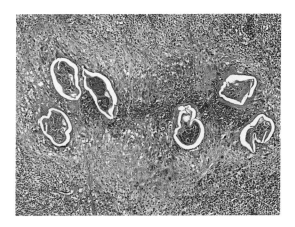

Figure 13.58
Higher magnification of worm depicted in Figure 13.57 revealing multiple transverse sections of coiled, gravid *W. bancrofti* in center of necrosis. x60

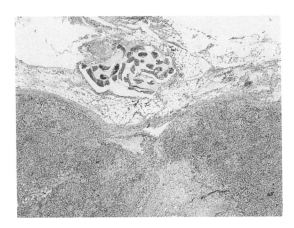

Figure 13.59
Coiled, degenerated gravid *W. bancrofti* in lymphatic vessel adjacent to adrenal gland. x11

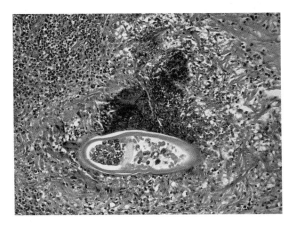

Figure 13.60
Degenerated gravid *W. bancrofti* in conjunctival lesion. Note granulomatous reaction and eosinophilic coagulum. x105

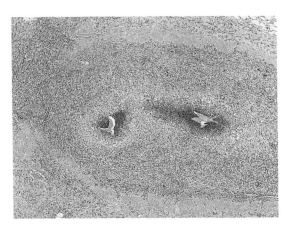

Figure 13.61
Two sections of degenerated male *W. bancrofti* within necrotic granuloma in mediastinal mass. x25

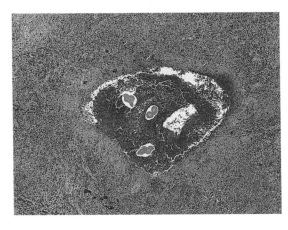

Figure 13.62
Degenerated gravid *B. malayi* within soft tissue of scrotum in patient from India. Note massive tissue eosinophilia and worm in center of eosinophilic coagulum. x30

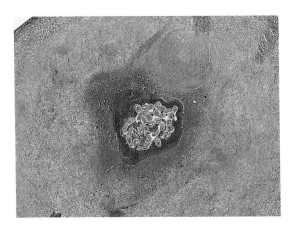

Figure 13.63
Degenerated adult male *B. malayi* coiled in epididymal granuloma. x30

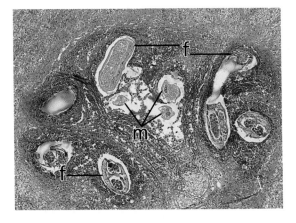

Figure 13.64
Degenerated adult female and male worms in epididymal granuloma from same patient as in Figure 13.63. Female (f) is 110 µm in diameter; male (m) is 80 µm in diameter. x55

13.52). Tissue containing the dead worm may develop concentric fibrosis (Fig 13.53). Dead worms either calcify or are resorbed by surrounding tissue.

Adult worms are sometimes found outside lymphatic vessels (Fig 13.54) or in coagulated lymph (Fig 13.50). In rare instances, adult worms localize in the lung (Figs 13.54 to 13.56), skin (Figs 13.57 and 13.58), adrenal glands (Fig 13.59), or conjunctiva (Fig 13.60). They may even cause a mediastinal mass (Fig 13.61). Adult *B. malayi* sometimes lodge in genital tissue (Figs 13.62 to 13.64). Microfilariae have also been found in vaginal and endometrial smears, bone marrow smears,[53] pericardial fluid, synovial fluid, and mamillary discharges. The presence of microfilariae in extravascular spaces such as bone marrow reflects their ability to penetrate intact vessel walls.

Fine-needle aspirates,[10,36] biopsy, and autopsy specimens may reveal microfilariae or adults within blood or lymphatic vessels of any organ. Microfilariae in tissue sections are similar to those found in peripheral blood, except that they are usually no more than 5 μm in diameter, and the sheath is rarely visible (Figs 13.65 to 13.67). Viable microfilariae do not usually provoke an inflammatory response (Figs 13.65 to 13.69), as degenerating microfilariae do. The tissue around degenerating microfilariae (Figs 13.70 to 13.73) becomes hyperemic and edematous as inflammatory cells (especially eosinophils, lymphocytes, and plasma cells) surround the dying parasites. A single degenerating microfilaria usually incites a discrete eosinophilic microabscess (Fig 13.74), while massive numbers of dying microfilariae typically provoke necrotizing granulomas (Figs 13.75 and 13.76). As they invade vessel walls, microfilariae sometimes generate focal vasculitis which can lead to thrombosis and occlusion (Figs 13.77 to 13.81).

Pulmonary biopsy of patients with tropical eosinophilic syndrome reveals a patchy inflammatory reaction of varying size and severity, consisting of cellular and fibrinous exudates. PTAH stain shows well-defined focal fibrinous exudates in alveolar spaces,[20] composed mainly of eosinophils, macrophages, lymphocytes, and plasma cells. An interstitial infiltrate causes an increase in cellularity of the alveolar septa. Eosinophilic abscesses

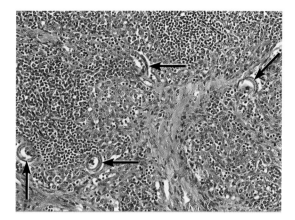

Figure 13.65
Viable coiled microfilariae (arrows) of *W. bancrofti* in lymph node. Microfilariae are only 5 μm in diameter in tissue sections. Sheath is rarely visible in tissue sections. x110

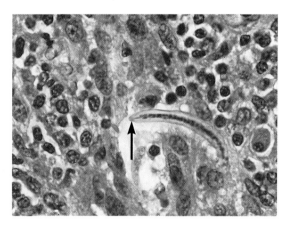

Figure 13.66
Anterior end of viable microfilaria of *W. bancrofti* in sinus of lymph node demonstrating short cephalic space (arrow). x750

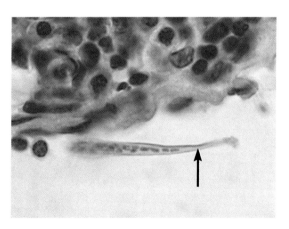

Figure 13.67
Posterior end of viable microfilaria of *W. bancrofti* in sinus of lymph node depicting long caudal space (arrow). x760

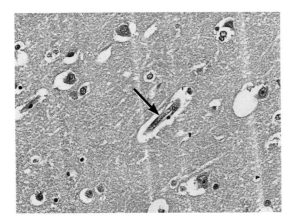

Figure 13.68
Viable microfilaria of *W. bancrofti* (arrow) within cerebral capillary, provoking no inflammation. x230

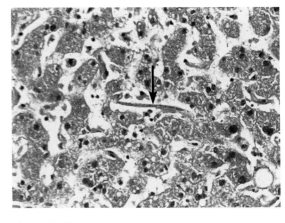

Figure 13.69
Viable microfilaria of *W. bancrofti* (arrow) within sinusoid of liver. x215

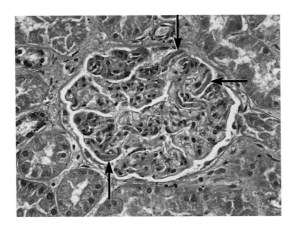

Figure 13.70
Degenerating microfilariae (arrows) of *W. bancrofti* in renal glomerulus. x240

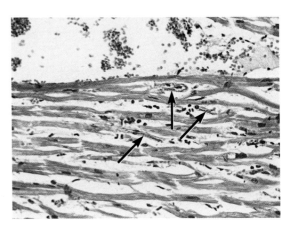

Figure 13.71
Degenerating microfilariae (arrows) of *W. bancrofti* causing myocarditis. x215

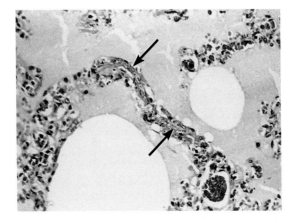

Figure 13.72
Degenerating microfilaria (arrows) of *W. bancrofti* causing inflammation of pulmonary capillary. x220

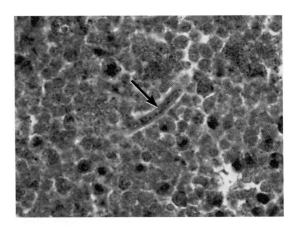

Figure 13.73
Degenerating microfilaria (arrow) of *W. bancrofti* in testicular microabscess. x585

produce gross nodules; microfilariae are sometimes found in the center of such abscesses.[20] Meyers and Kouwenaar (MK) bodies are granulomas up to 500 μm in diameter composed of epithelioid cells surrounding an acidophilic hyaline material that has replaced the eosinophils. The eosinophilic area of necrosis stains pink with PAS and is not birefringent. MK bodies may contain fragments of degenerating microfilariae. Most granulomas contain foreign-body giant cells mingled with thin strands of refractile hyaline material. Older granulomas with fibrosis at the periphery generally have no visible microfilariae.[19] The focal exudative and granulomatous lesions are usually distal to the bronchial tree and close to pulmonary venules. The surrounding lung tissue has areas of compensatory emphysema. Reticulin stain reveals destruction of the alveolar framework in the center of the granulomas, but minimal parenchymal damage. Rarely, lung biopsy reveals microfilariae within areas of eosinophilic inflammation.[42] Despite an intense reaction to microfilariae, patients with tropical eosinophilic syndrome have no apparent reaction to living adult worms.[27] After anthelmintic treatment, however, some patients with TPE develop nodules consisting of eosinophilic granulomas containing dead or degenerating adult *W. bancrofti*.[51]

In patients with occult filariasis where microfilariae lodge mainly in the liver, spleen, and/or lymph nodes, affected tissues are often massively infiltrated with eosinophils. The liver and spleen are enlarged, and their cut surfaces reveal white nodules. Liver biopsy reveals eosinophils around portal tracts and nodules similar to those in pulmonary lesions. Spleen and lymph nodes are soft and fleshy, with irregularly scattered brown or gray circumscribed areas on cut surfaces. Germinal follicles are prominent, and the nodes have a diffuse increase in vascularity with endothelial proliferation. Many eosinophils fill the medullary cords and sinuses, forming small eosinophilic abscesses or MK bodies. A systematic search of liver, spleen, or lymph nodes may reveal fragments of microfilariae in the nodules.[11] Rarely, lymph nodes contain adult worms.[9]

Diagnosis

Traditionally, blood, hydrocele fluid, and chylous urine have been the most common sources of diagnostic specimens. Microfilariae can be isolated by directly examining fluid specimens (20 μl on a slide with or without lysis of red blood cells), processing fluids by centrifuge in 2% formalin (Knott's technique), or filtering fluids through a polycarbonate membrane (3-to-5-μm Nucleopore®). Covering smears with agar gel before staining enhances sensitivity by retaining more microfilariae.[62] To identify species, it is best to stain some slides with hematoxylin and others with Giemsa. The time of blood collection should allow for the parasites' possible nocturnal periodicity. Finding adult worms or microfilariae in fine-needle aspirates,[37] biopsy, or autopsy samples can also be diagnostic.

Diagnostic methods based on serologic techniques (usually detection of antibodies) have not been satisfactory in the case of brugian filariasis. Such tests cannot distinguish between active and inactive infections and cross react with common gastrointestinal parasites and other organisms. Fortunately, these problems have been overcome in serologic tests for *W. bancrofti*, the causative agent of nearly 90% of lymphatic filariasis worldwide.

Circulating filarial antigen (CFA) detection is now regarded as the method of choice for diagnosing *W. bancrofti* infection. This method is highly specific and more sensitive than earlier assays.[60] All microfilaremic individuals have detectable circulating antigen, as do amicrofilaremic individuals with active infections and clinical manifestations of filariasis (eg, lymphedema or elephantiasis). Some asymptomatic individuals also have detectable circulating antigen; the fact that CFA disappears after effective treatment with DEC indicates that these individuals had occult infections. Two forms of the CFA assay are now commercially available. The one based on ELISA yields semiquantitative results. The other, a simple card (immunochromatographic) test, yields only positive/negative results.[60] Together, these assays provide a diagnostic approach to clinical or field evaluation of bancroftian filariasis. Though no such test is currently available for *B. malayi*

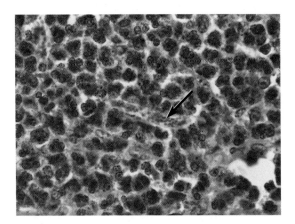

Figure 13.74
Degenerating microfilaria (arrow) of *W. bancrofti* in eosinophilic microabscess of lymph node. x575

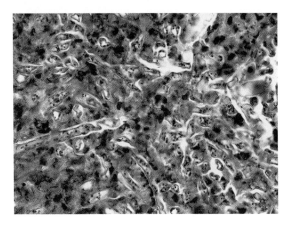

Figure 13.76
Higher magnification of degenerating microfilariae of *W. bancrofti* depicted in Figure 13.75. x475

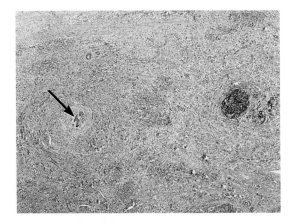

Figure 13.77
Thickened, inflamed capsule of inguinal lymph node in patient treated with DEC. Note granuloma on left side and obstructed lymphatic vessel on right side (dark circular area). Granuloma contains fragment of phagocytosed adult male *W. bancrofti* (arrow). Fibrin, inflammatory cells, and immune reaction to degenerating microfilariae obstruct lymphatic vessel. Movat x20

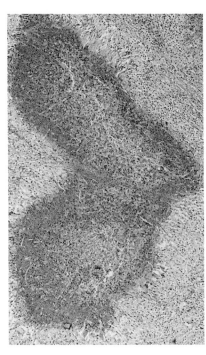

Figure 13.75
Degenerating microfilariae of *W. bancrofti* causing necrotizing granuloma in spermatic cord. x65

infection, the serologic detection of circulating parasite DNA by polymerase chain reaction techniques may prove to be a useful method for diagnosing brugian filariasis.[40]

In 1994, Amaral et al first described finding live adult filarial worms in human tissue through ultrasonography.[2] Ultrasound can show the rapid "filarial dance" characteristic of adult *W. bancrofti* (Fig 13.30), and can even reveal living adult *W. bancrofti* in patients with tropical eosinophilia.[27] Ultrasound studies can also assess the efficacy of various therapies. Computed tomography (CT) and chest roentgenograms are useful in diagnosing TPE. CT clearly demonstrates the reticulonodular pattern, bronchiectasis, air trapping, calcification, and mediastinal adenopathy.[56] Roentgenograms can reveal calcified worms.

Diagnosing amicrofilaremic patients with chronic filariasis or tropical eosinophilia has traditionally depended on observing clinical signs and symptoms. The clinical differential for TPE includes drug reactions, allergic bronchopulmonary aspergillosis, allergic granulomatosis and angiitis, Löffler's syndrome, Churg-Strauss syndrome, hypereosinophilic syndrome, and chronic eosinophilic pneumonia.[30] Löffler's syndrome differs from TPE in that it appears suddenly, disappears within 2 weeks, and leaves no trace on resolution.

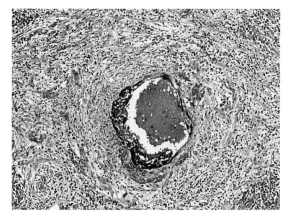

Figure 13.78
Different level of dilated lymphatic described in Figure 13.77. Note organizing fibrin along portion of intima, and fibroinflammatory reaction surrounding lymphatic vessel. Movat x60

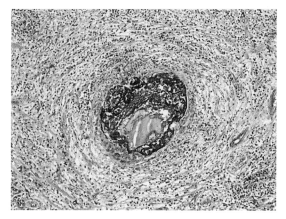

Figure 13.79
Dilated lymphatic described in Figure 13.77 at level where large fibrin thrombus nearly occludes lumen. Movat x60

Lung biopsy is impractical for patients with TPE. Though microfilariae may be present in the lung, thousands of tissue sections may be required to find a single worm. While all inhabitants of endemic areas with lifelong exposure to mosquito-borne filarial larvae develop antifilarial antibodies whether or not they are infected, only TPE patients can be reliably diagnosed by antibody detection, since they have extreme elevations of both IgG and IgE antibody to their filarial infections.

Treatment and Prevention

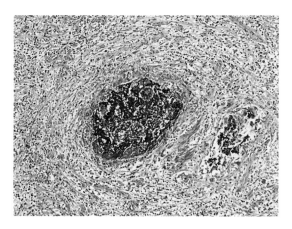

Figure 13.80
Dilated lymphatic described in Figure 13.77 at level where large fibrin thrombus completely occludes lumen. Movat x60

Diethylcarbamazine (DEC) is the drug of choice for treating lymphatic filariasis. A single dose of DEC has the same long-term effectiveness in decreasing microfilaremia, and probably in killing adult worms, as the 1- to 3-week courses previously recommended. This has enormous implications for control programs, but studies should be done to determine if the standard 6- to 12-day regimen for most infections and the 34-week regimen for tropical eosinophilic syndrome should be altered for individual patients. DEC alone can sometimes precipitate acute inflammatory reactions that accelerate the natural course of disease, especially in brugian filariasis.[47]

Ivermectin, another potent microfilaricide,[15] is as effective as DEC in suppressing microfilaremia for prolonged periods; it does not, however, kill adult worms.[23,26] Administering single doses of

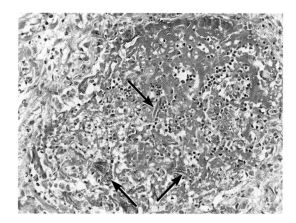

Figure 13.81
Dilated lymphatic described in Figure 13.77 showing degenerating microfilariae (arrows) within fibrinous material and penetrating wall of lymphatic lumen. x155

ivermectin (200 µg/kg body weight) and DEC (6 µg/kg body weight) provides a significant therapeutic synergism that reduces microfilaremia by 90% to 99% for a full year after treatment. A single dose of albendazole (400 mg), in combination with ivermectin or DEC, provides similar long-term microfilarial reduction.[49]

These treatment regimens are so effective, they form the basis of a global effort to eradicate lymphatic filariasis initiated by the World Health Assembly Resolution. The principal strategy calls for a single-dose, 2-drug regimen carried out once yearly for 4 to 6 years, using either albendazole plus invermectin (in Africa, where the use of DEC in contraindicated) or albendazole plus DEC.[48] Mass treatment of at-risk populations is recommended. Each of the 3 drugs used in these regimens is essentially nontoxic in single doses. Any side effects are caused almost exclusively by dying parasites and host response to them. Since albendazole and ivermectin affect a broad spectrum of intestinal parasites and ectoparasites, the public health benefit of their use against lymphatic filariasis goes beyond the treatment of filarial infection alone.[15] The likelihood of success for this public health initiative has been greatly enhanced by the generous donation of albendazole by SmithKline Beecham, and invermectin by Merck and Co., Inc.

Aggressive treatment of chronic lymphedema and elephantiasis can be dramatically effective. It must entail long-term, low-dose DEC to eradicate persistent or new filarial infections and, especially, diligent attention to the hygiene of lymphedematous extremities. Preventing superficial bacterial and fungal infection is essential. Patients should elevate and massage affected limbs, and wear elastic stockings whenever possible.

Patients with severely damaged extremities may benefit from surgical decompression of the lymphatic system through nodovenous shunt, followed by excision of redundant tissue. Surgical correction or repeated drainage is the best treatment for hydroceles. While chyluria is sometimes corrected surgically, it should mainly be controlled by dietary management. Diagnostic lymphangiography itself often stops leakage of chyle into urine, probably by its sclerosing effect on lymphatic vessels that have ruptured into the renal pelvis.

Patients with tropical eosinophilia respond dramatically to DEC. Untreated patients may experience spontaneous resolution of symptoms, but eosinophilia usually persists, and relapses are common. Bronchoalveolar lavage reveals that patients treated with DEC often have persistent mild chronic alveolitis that can cause mild chronic interstitial lung disease.[55] Steroids combined with anthelmintics can mitigate minor allergic reactions.

Molecular engineering technology may someday produce mosquitoes resistant to filarial infections.[41] Until then, mosquito abatement and improved living conditions are necessary ancillary measures.[34] The availability of simple, safe, cost-effective treatments, along with new intervention strategies, have prompted the International Task Force for Disease Eradication to identify lymphatic filariasis as a potentially eradicable infectious disease, and have led directly to the new global program to eliminate lymphatic filariasis initiated by WHO and its partners in the private sector, the academic community, international organizations, and nongovernmental development agencies.

	Length (mm)	Maximum diameter (μm)	Cuticular thickness (μm)	Other cuticular traits	Lateral cords	Somatic muscle	Usual number and aspect of uteri at midbody
Wuchereria bancrofti	60–100	250	1–2 Slightly thicker in lateral cords	Fine transverse striations	Usually inconspicuous	Slightly to moderately developed	2; fill body cavity
Brugia malayi	50–60	170	1–2 Thicker in lateral cords	Fine transverse striations	Inconspicuous to prominent	Slightly to moderately developed	2; fill body cavity
Brugia timori	21–39	140	Not available	Not available	Not available	Not available	2
Loa loa	40–70	600	4–10 Thicker in lateral cords	Irregularly spaced bosses	Conspicuous	Well-developed	3 or 4; fill body cavity
Onchocerca volvulus	230–500	450	4–10	Multilayered; regularly spaced transverse ridges (annular thickenings)	Inconspicuous to prominent	Slightly to well-developed	2; do not fill body cavity
Mansonella perstans	60–80	150	1–2	Very fine transverse striations; thicker in lateral cords	Usually inconspicuous; contain pigmented granules	Slightly to moderately developed	2; fill body cavity
Mansonella ozzardi	65–81 From humans	250	Not available	Not available	Not available	Not available	Not available
Mansonella streptocerca	19–27	85	1–2	Smooth; thicker in lateral cords	Usually inconspicuous; contain pigmented granules	Slightly to moderately developed	2; fill body cavity
Dirofilaria immitis	In dog: mature; 230–310 by 1–2 mm	In humans: always immature; length unknown, 350 μm max. diameter	5–25 Often distorted	Multilayered, transverse striations; internal longitudinal ridges	Usually poorly preserved	Well-developed	2; do not fill body cavity
Dirofilaria tenuis	In raccoon: mature; 80–130 mm by 260–360 μm	In humans: can reach maturity but usually nongravid; length unknown, 360 μm max. diameter	5–15 Often distorted	Multilayered, transverse striations; internal and external longitudinal ridges	Usually prominent	Well-developed	2; do not fill body cavity
Dirofilaria repens	In dog: mature; 100–170 mm by 460–650 μm	In humans: can reach maturity but usually nongravid; length unknown, 650 μm max. diameter	12–20 Often distorted	Multilayered, transverse striations; internal and external longitudinal ridges	Usually prominent	Well-developed	2; do not fill body cavity

Table 13.1
Morphologic features of the most common adult female filariae in humans.

	Length (mm)	Maximum diameter (μm)	Cuticular thickness (μm)	Other cuticular traits	Lateral cords	Somatic muscle
Wuchereria bancrofti	40	150	1–2 Thicker in lateral cords	Fine transverse striations	Inconspicuous to prominent	Moderately to well–developed
Brugia malayi	22–25	90	1–2 Thicker in lateral cords	Fine transverse striations	Inconspicuous to prominent	Moderately to well–developed
Brugia timori	13–23	80	Not available	Not available	Not available	Not available
Loa loa	30–34	400	4–10 Thicker in lateral cords	Irregularly spaced bosses	Usually conspicuous	Well–developed
Onchocerca volvulus	16–42	200	3–5	Deep striations (annulations) spaced 5 μm apart	Inconspicuous to prominent	Moderately developed
Mansonella perstans	35–45	70	1–2 Thicker in lateral cords	Smooth	Usually inconspicuous; contain pigmented granules	Moderately to well–developed
Mansonella ozzardi	24–28 From experimentally infected monkeys	80	Not available	Not available	Not available	Not available
Mansonella streptocerca	13–18	50	1–2 Thicker in lateral cords	Smooth	Usually inconspicuous; contain pigmented granules	Moderately to well–developed
Dirofilaria immitis	In dog: mature; 120–190 by 1–2 mm	In humans: always immature; length unknown, 300 μm max. diameter	5–25 Often distorted	Multilayered, transverse striations; internal longitudinal ridges	Usually poorly preserved	Well–developed
Dirofilaria tenuis	In raccoon: mature; 40–48 mm by 190–260 μm	In humans: can reach maturity; length unknown, 260 μm max. diameter	5–15 Often distorted	Multilayered, transverse striations; internal and external longitudinal ridges	Usually prominent	Well–developed
Dirofilaria repens	In dog: mature; 50–70 mm by 370–450 μm	In humans: can reach maturity; length unknown, 450 μm max. diameter	12–20 Often distorted	Multilayered, transverse striations; internal and external longitudinal ridges	Usually prominent	Well–developed

Table 13.2
Morphologic features of the most common adult male filariae in humans.

	Length (μm)	Diameter (μm)	Sheath	Length of cephalic space (μm)	Anteriormost nuclei	Terminal nucleus	Length of caudal space (μm)	Shape of tail
Wuchereria bancrofti	230–300	7–10	Yes	5–7	Side by side	Elongate	5–15	Pointed
Brugia malayi	175–260	4–6	Yes	6–11	Side by side	Round	0	Terminal and subterminal nuclei with constriction; nuclei not always discernible
Brugia timori	290–340	6–7	Yes	10–14	Side by side	Elongate to oval	0	Terminal and subterminal nuclei without constriction
Loa loa	175–300	5–8	Yes	3–6	Side by side	Elongate	0–1	Pointed
Onchocerca volvulus	220–360	5–9	No	7–13	Side by side	Elongate	9–15	Finely pointed
Mansonella perstans	100–200	3.5–4.5	No	1–3	Side by side	Round	0	Bluntly rounded
Mansonella ozzardi	170–240	3–5	No	2–6	Side by side	Oval	3–8	Pointed
Mansonella streptocerca	180–240	2.5–5	No	3–5	First 4 staggered but not overlapping	Oval or round	1	Tip of tail forms "shepherd's crook"

Table 13.3
Morphologic features of the most common microfilariae in humans.

	General	Cephalic space	Anterior nuclear column	Tail	Sheath on Giemsa
Wuchereria bancrofti	Head and tail usually well-separated; 3 or 4 major curves; graceful	As long as broad	Well-defined and well-spaced; easy to count	Tapers to point; long caudal space (5–15 μm)	Blue
Brugia malayi	Usually closely folded with head close to tail; several major curves; minor angular curves	Twice as long as broad	Blurred and intermingled; difficult to count	Tapers to fine point. Typically 1 terminal nucleus at tip; 2 in terminal thread. Constriction between terminal and subterminal nuclei; nuclei not always discernible	Pink
Brugia timori	Longer than *B. malayi*; usually folded with head close to tail	Three times as long as broad	Blurred and intermingled; difficult to count	Tapers to fine point. Terminal and subterminal nuclei without constriction	Nearly invisible

Table 13.4
Differentiating microfilariae of *Brugia* sp and *W. bancrofti* in blood films.

References

1. Aggarwal J, Kapila K, Gaur A, Wali JP. Bancroftian filarial pleural effusion. *Postgrad Med J* 1993;69:869–870.
2. Amaral F, Dreyer G, Figueredo-Silva J, et al. Live adult worms detected by ultrasonography in human Bancroftian filariasis. *Am J Trop Med Hyg* 1994;50:753–757.
3. Anonymous. Hæmatozoa and chyluria. *Br Med J* 1873;i:147–148.
4. Anonymous. Indian Sanitary Commissioners' Reports. Madras. The hæmatozoon. *Lancet* 1873;i:55–57.
5. Anonymous. Mosquitos and elephantiasis. *Br Med J* 1900;ii:682.
6. Anonymous. The newly-discovered hæmatozoon inhabiting human blood. *Lancet* 1872;ii:889–890.
7. Anonymous. Worms in urine and blood. *Lancet* 1872; ii:310.
8. Araujo AC, Figueredo-Silva J, Souto-Padrón T, Dreyer G, Norões K, De Souza W. Scanning electron microscopy of Wuchereria bancrofti (Nematoda: Filarioidea). *J Parasitol* 1995; 81:468–474.
9. Arora VK, Sen B, Dev G, Bhatia A. Fine needle aspiration identification of the adult worm of Brugia malayi and its ovarian fragment from an epitrochlear lymph node. *Acta Cytol* 1993;37:436–438.
10. Arora VK, Singh N, Bhatia A. Cytomorphologic profile of lymphatic filariasis. *Acta Cytol* 1996;40:948–952.
11. Beaver PC. Filariasis without microfilaremia. *Am J Trop Med Hyg* 1970;19:181–189.
12. Beaver PC, Jung RC, Cupp EW. *Clinical Parasitology*. 9th ed. Philadelphia, Pa: Lea & Febiger; 1984:351–364.
13. Bonne C. Over hypereosinophilie in de milt geocombineerd met een filaria-infectie. *Ned Tijdschr Geneeskd* 1939;79:874–876.
14. Boornazian JS, Fagan MJ. Tropical pulmonary eosinophilia associated with pleural effusions. *Am J Trop Med Hyg* 1985;34:473–475.
15. Cao WC, Van der Ploeg CP, Plaisier AP, van der Sluijs IJ, Habbema JD. Ivermectin for chemotherapy of bancroftian filariasis: a meta-analysis of the effect of single treatment. *Trop Med Int Health* 1997;2:393–403.
16. Carme B, Laigret J. Longevity of Wuchereria bancrofti var. pacifica and mosquito infection acquired from a patient with low level parasitemia. *Am J Trop Med Hyg* 1979;28:53–55.
17. Cobbold TS. Discovery of the adult representative of microscopic filariae. *Lancet* 1877;ii:70–71.
18. Cobbold TS. Discovery of the intermediary host of Filaria sanguinis hominis (F. bancrofti). *Lancet* 1878;i:69.
19. Cobbold TS. On filaria bancrofti. *Lancet* 1877;ii:495–496.
20. Danaraj TJ, Pacheco G, Shanmugaratnam K, Beaver PC. The etiology and pathology of eosinophilic lung (tropical eosinophilia). *Am J Trop Med Hyg* 1966;15:183–189.
21. Demarquay JN. Note sur une tumeur des bourses contenant un liquide laiteux (galactocèle de Vidal) et renfermant des petits êtres vermiformes que l'on peut considérer comme des helminthes hématoïdes a l'état d'embryon. *Gaz Med Paris* 1863;18:665–667.
22. Dhanda V, Das PK, Lal R, Srinivasan R, Ramaiah KD. Spread of lymphatic filariasis, re-emergence of leishmaniasis and threat of babesiosis in India. *Indian J Med Res* 1996;103:46–54.
23. Dreyer G, Addiss D, Norões J, Amaral F, Rocha A, Coutinho A. Ultrasonographic assessment of the adulticidal efficacy of repeat high-dose ivermectin in bancroftian filariasis. *Trop Med Int Health* 1996;1:427–432.
24. Dreyer G, Brandão AC, Amaral F, Medeiros Z, Addiss D. Detection by ultrasound of living adult Wuchereria bancrofti in the female breast. *Mem Inst Oswaldo Cruz* 1996;91:95–96.
25. Dreyer G, Figueredo-Silva J, Neafie RC, Addiss DG. Lymphatic filariasis. In: Horsburgh CR, Nelson AM, eds. *Pathology of Emerging Infections 2*. Washington, DC: ASM Press; 1998:317–342.
26. Dreyer G, Norões J, Amaral F, et al. Direct assessment of the adulticidal efficacy of a single dose of ivermectin in bancroftian filariasis. *Trans R Soc Trop Med Hyg* 1995;89:441–443.
27. Dreyer G, Norões J, Rocha A, Addiss D. Detection of living adult Wuchereria bancrofti in a patient with tropical pulmonary eosinophilia. *Braz J Med Biol Res* 1996;29:1005–1008.
28. Dreyer G, Santos A, Norões J, Rocha A, Addiss D. Amicrofilaraemic carriers of adult Wuchereria bancrofti. *Trans R Soc Trop Med Hyg* 1996;90:288–289.
29. Edeson JFB, Wilson T, Wharton RH, Laing ABG. Experimental transmission of Brugia malayi and B. pahangi to man. *Trans R Soc Trop Med Hyg* 1960;54:229–234.
30. Enright T, Chua S, Lim DT. Pulmonary eosinophilic syndromes. *Ann Allergy* 1989;62:277–285.
31. Frimodt-Möller C, Barton RM. A pseudo-tuberculous condition associated with eosinophilia. *Indian Med Gaz* 1940;75:607–613.
32. Gilbert HM, Hartman BJ. Short report: a case of fibrosing mediastinitis caused by Wuchereria bancrofti. *Am J Trop Med Hyg* 1996;54:596–599.
33. Grove DI. *A History of Human Helminthology*. Wallingford, England: CAB Int; 1990:597–640.
34. Grove DI. Selective primary health care: strategies for the control of disease in the developing world. 7. Filariasis. *Rev Infect Dis* 1983;5:933–944.
35. Hoeppli R. Parasitic diseases in Africa and the Western Hemisphere. Early documentation and transmission by the slave trade. *Acta Trop Suppl* 1969;10:1–240.
36. Jayaram G. Microfilariae in fine needle aspirates from epididymal lesions. *Acta Cytol* 1987;31:59–62.
37. Kapila K, Verma K. Gravid adult female worms of Wuchereria bancrofti in fine needle aspirates of soft tissue swellings. Report of three cases. *Acta Cytol* 1989;33:390–392.
38. Lanceraux. Clinique médicale. De la filariose. *Sem Med* 1888;8:332–333.
39. Laurence BR. Elephantiasis in Greece and Rome and the Queen of Punt. *Trans R Soc Trop Med Hyg* 1967;61:612–613.
40. Lizotte MR, Supali T, Partono F, Williams SA. A polymerase chain reaction assay for the detection of Brugia malayi in blood. *Am J Trop Med Hyg* 1994;51:314–321.
41. Lowenberger CA, Ferdig MT, Bulet P, Khalili S, Hoffmann JA, Christensen BM. Aedes aegyptii induced antibacterial proteins reduce the establishment and development of Brugia malayi. *Exp Parasitol* 1996;83:191–201.
42. Marty AM, Neafie RC. Protozoal and helminthic diseases. In: Saldana MJ, ed. *Pathology of Pulmonary Disease*. Philadelphia, Pa: JB Lippincott; 1994:489–502.
43. Meyers FM, Kouwenaar W. Over hypereosinophilie en over een merkwaardigen vorm van filariasis. *Ned Tijdschr Geneeskd* 1939;79:853–873.
44. Michael E, Bundy DAP, Grenfell BT. Re-assessing the global prevalence and distribution of lymphatic filariasis. *Parasitology* 1996;112:409–428.
45. Nath G, Mohapatra TM, Sen PC. Circulating immune complexes and complement in bancroftian filariasis. *Indian J Pathol Microbiol* 1991;34:92–98.
46. O'Connor FW. The ætiology of the disease syndrome in Wuchereria bancrofti infections. *Trans R Soc Trop Med Hyg* 1932;26:13–33.
47. Ottesen EA. Efficacy of diethylcarbamazine in eradicating infection with lymphatic-dwelling filariae in humans. *Rev Infect Dis* 1985;7:341–356.
48. Ottesen EA, Duke BO, Karam M, Behbehani K. Strategies and tools for the control/elimination of lymphatic filariasis. *Bull World Health Organ* 1997;75:491–503.
49. Ottesen EA, Ismail MM, Horton J. The role of albendazole in programmes to eliminate lymphatic filariasis. *Parasitol Today*. In press.

50. Ottesen EA, Neva FA, Paranjape RS, Tripathy SP, Thiruvengadam KV, Beaven MA. Specific allergic sensitisation to filarial antigens in tropical eosinophilia syndrome. *Lancet* 1979;i:1158–1161.
51. Perera CS, Perera LM, de Silva C, Abeywickreme W, Dissanaike AS, Ismail MM. An eosinophilic granuloma containing an adult female Wuchereria bancrofti in a patient with tropical pulmonary eosinophilia. *Trans R Soc Trop Med Hyg* 1992;86:542.
52. Peterson RK. US Navy Disease Vector Ecology and Control Center. Lymphatic filariasis. Available at: http://www.ianr.unl.edu/ianr/entomol/history_bug/filariasis.htm. Accessed 13 May 98.
53. Pradhan S, Lahiri VL, Elhence BR, Singh KN. Microfilaria of Wuchereria bancrofti in bone marrow smear. *Am J Trop Med Hyg* 1976;25:199–200.
54. Quah BS, Khairul Anuar A, Rowani MR, Pennie RA. Cor pulmonale: an unusual presentation of tropical eosinophilia. *Ann Trop Paediatr* 1997;17:77–81.
55. Rom WN, Vijayan VK, Cornelius MJ, et al. Persistent lower respiratory tract inflammation associated with interstitial lung disease in patients with tropical pulmonary eosinophilia following conventional treatment with diethylcarbamazine. *Am Rev Respir Dis* 1990;142:1088–1092.
56. Sandhu M, Mukhopadhyay S, Sharma SK. Tropical pulmonary eosinophilia: a comparative evaluation of plain chest radiography and computed tomography. *Australas Radiol* 1996;40:32–37.
57. Savitt TL. Filariasis in the United States. *J Hist Med Allied Sci* 1977;32:140–150.
58. Vanamail P, Ramaiah KD, Pani SP, Das PK, Grenfell BT, Bundy DA. Estimation of the fecund life span of Wuchereria bancrofti in an endemic area. *Trans R Soc Trop Med Hyg* 1996;90:119–121.
59. Webb JK, Job CK, Gault EW. Tropical eosinophilia; demonstration of microfilariae in lung, liver, and lymph nodes. *Lancet* 1960;i:835–842.
60. Weil GJ, Lammie PJ, Weiss N. The ICT filariasis test: a rapid-format antigen test for diagnosis of bancroftian filariasis. *Parasitol Today* 1997;13:401–404.
61. Weingarten RJ. Tropical eosinophilia. *Lancet* 1943;1:103–105.
62. Youssef FG, Hassanein SH, Cummings CE. A modified staining method to detect Wuchereria bancrofti microfilariae in thick-smear preparations. *Ann Trop Med Parasitol* 1995; 89:93–94.

14

Mansonelliasis

J. Kevin Baird,
Ronald C. Neafie, *and*
Wayne M. Meyers

Introduction

Mansonelliasis is infection by filarial worms of the genus *Mansonella*. Three species infect humans: *Mansonella perstans, Mansonella streptocerca,* and *Mansonella ozzardi*. Prior to 1982, *M. perstans* and *M. streptocerca* were classified successively in the genera *Tetrapetalonema, Acanthocheilonema,* and, most recently, *Dipetalonema*.[27] Microfilariae of *M. perstans* and *M. ozzardi* ordinarily dwell in blood, and those of *M. streptocerca* in tissue, especially dermal collagen. These 3 filarial infections have some common features, but because they differ in many respects, they will be described separately.

■ Mansonelliasis perstans

Introduction

Definition
Mansonelliasis perstans is infection by the filarial worm *M. perstans*. Adult filariae seem to cause most of the clinical manifestations of infection.

Synonyms
Synonyms include dipetalonemiasis perstans, Uganda eye worm, bulge-eye, bung-eye, and Kampala eye worm (Figs 14.1 and 14.2).

General Considerations
Manson first described the microfilariae of *M. perstans* in 1891 in a Congolese; Daniels reported finding the adult worm in an inhabitant of British Guiana in 1898.[7] The infection is common in Africa, where it affects not only humans but also gorillas and chimpanzees.

Epidemiology

Mansonella perstans is endemic in much of tropical Africa from Senegal east to Uganda and south to Zimbabwe,[11,15,22,33] and is transmitted by biting midges (Fig 14.3). In South America, *M. perstans* is found along the Atlantic Coast, from Panama south to Argentina, including Trinidad.[31] Microfilaremia increases dramatically with the age of the host.

Infectious Agent

Morphologic Description

Adult worms are creamy-white and thread-like (Tables 13.1 and 13.2). Females measure 60 to 80 mm long by 100 to 150 μm in diameter, and the males are 35 to 45 mm by 50 to 70 μm. The posterior end flexes ventrally to a half coil in the female and a full coil in the male. The vulva in the female is situated about 1 mm from the anterior end of the worm. Females have 2 terminal papillae at the posterior tip. At the posterior end, males have 2 unequal, rod-like spicules, and 4 preanal and 1 postanal pair of papillae.

Intact worms are rarely available for study; identification of adults is usually made from tissue sections.[3] Female worms in tissue are 80 to 125 μm in diameter and males are 45 to 60 μm. Adult females contain an intestine and 2 uteri that fill the body cavity at midbody (Fig 14.4); paired ovaries and oviducts are confined to the posterior region (Fig 14.5). Females have coelomyarian musculature with 8 to 12 muscle cells per quadrant. Somatic musculature is tallest in the dorsal and ventral aspects (10 to 20 μm) and tapers to 2 to 5 μm; it is inconspicuous in the lateral cord region. Lateral cords contain fine granules of melanin (Fig 14.6) that persist in degenerated worms.[3] The cuticle is 1 to 2 μm thick, single-layered, very finely striated, and thickened at the lateral cords (Fig 14.7). The anterior region of the male worm contains the intestine and testis (Fig 14.8), and most of the posterior region contains the intestine and vas deferens (Fig 14.9). The lateral cords are inconspicuous at all levels of the male worm, but as in the female they also contain pigment granules (Fig 14.8). Near the posterior end of the worm, the musculature occupies most of the pseudocoelom (Fig 14.9). The cuticle of the male is 1 to 2 μm thick, single-layered, smooth, and thickened at the lateral cords to form internal ridges.

The microfilariae are unsheathed and measure 100 to 200 μm long by 3.5 to 4.5 μm wide (Fig 14.10 and Table 13.3). Cephalic space is 1 to 3 μm. There is no caudal space. The round terminal nucleus extends to the tip of the tail, and the posterior end is characteristically blunt.

Life Cycle

Biting midges of the species *Culicoides austeni*

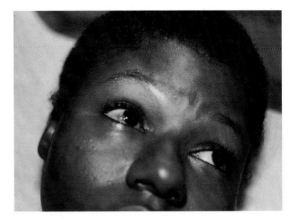

Figure 14.1
Bung-eye in a Ugandan, caused by adult *M. perstans*.

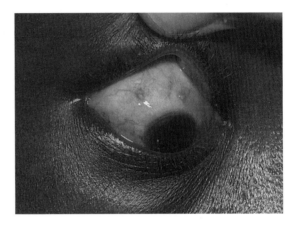

Figure 14.2
Nodule in conjunctiva of a Ugandan, caused by adult *M. perstans*.

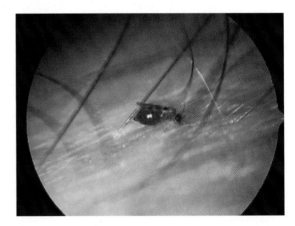

Figure 14.3
Biting midge, *Culicoides grahami*, a vector of *M. perstans*.

and *C. grahami* (Fig 14.3) are the major natural insect vectors of *M. perstans*. These vectors ingest microfilariae while biting a definitive host. The microfilariae traverse the stomach wall and invade the thoracic muscle. Infective larvae develop in the vector 6 to 9 days after feeding on an infected patient. When the vector bites another host, these larvae emerge from the labium and enter the new host through the skin. Less than 10% of *Culicoides* in endemic areas are infected. The prepatent period following infection has not been established, but is estimated at 9 to 12 months.

Clinical Features and Pathogenesis

Most patients are asymptomatic. Expatriates in endemic areas seem to experience more symptoms than local inhabitants. Symptoms include subcutaneous swellings of the arm, shoulder, and face, and abdominal pain, pruritus, pleuritis, arthralgia, and fatigue. Liver disease, gallbladder symptoms, and cardiac involvement have been postulated. One patient who died of fibrinopurulent pericarditis had adult and microfilarial *M. perstans* in the pericardium.[12] Pleural effusion and chronic lymphedema have been reported.[25] Ocular manifestations include periorbital edema and conjunctival irritation, sometimes with granulomatous nodules in the conjunctiva (Figs 14.1 and 14.2).[29] Acute periorbital inflammation is common in Uganda and is called bung-eye or bulge-eye. Similar inflammatory changes around the eye are seen in Nigeria and Sudan.[2]

Pathologic Features

Live adult *M. perstans* most frequently inhabit connective tissues of the abdominal cavity and viscera, and ordinarily provoke little or no inflammation (Figs 14.11 to 14.13). In a study of 12 adult worms (7 female and 5 male) found at autopsy or surgery in 9 patients in the Democratic Republic of Congo, worms were discovered in or around the liver, kidney (Fig 14.12), gallbladder (Fig 14.14), pancreas, rectum, and hernial sacs (Figs 14.15 and

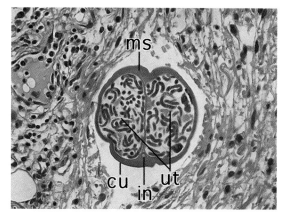

Figure 14.4
Midbody of adult gravid *M. perstans* in spermatic cord of a Congolese. Worm has a thin cuticle (cu), somatic muscle cells (ms), intestine (in), and 2 uteri (ut) filled with microfilariae. x230

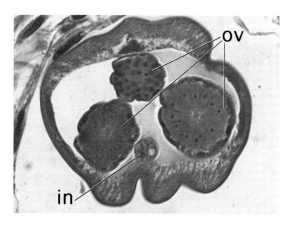

Figure 14.5
Posterior region of adult female *M. perstans* from worm shown in Figure 14.4, demonstrating 3 large sections of ovary (ov) and the intestine (in). x575

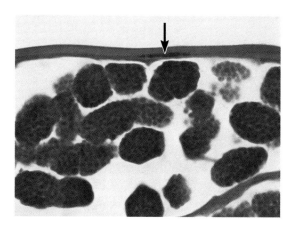

Figure 14.6
Higher magnification of worm shown in Figures 14.4 and 14.5, depicting granules of melanin pigment (arrow) in lateral cord. x770

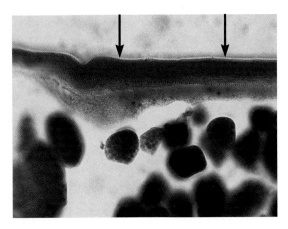

Figure 14.7
Higher magnification of female worm shown in Figures 14.4 to 14.6, illustrating very fine transverse striations of cuticle (arrows). Movat x770

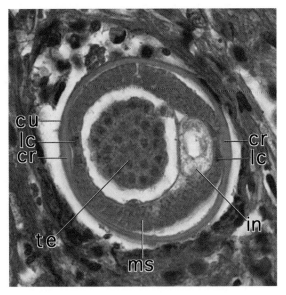

Figure 14.8
Anterior region of male *M. perstans* in hernial sac of a Congolese, demonstrating thin cuticle (cu), internal longitudinal cuticular ridges (cr), pigment in lateral cords (lc), somatic muscle cells (ms), intestine (in), and testis (te). x825

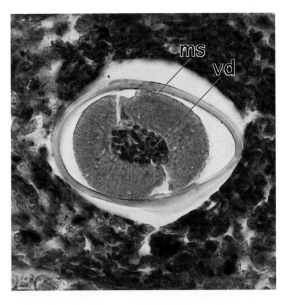

Figure 14.9
Posterior region of same worm depicted in Figure 14.8, showing sperm in vas deferens (vd) and well-developed somatic muscle cells (ms). Movat x800

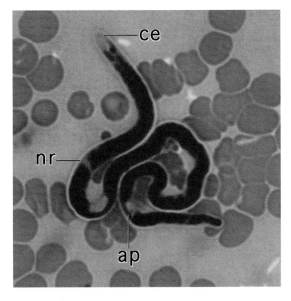

Figure 14.10
Microfilaria of *M. perstans* in peripheral blood film. Microfilaria is unsheathed and terminal round nucleus extends to tip of tail. Note cephalic space (ce), nerve ring (nr), and anal pore (ap). Giemsa x1400

14.16).³ Microfilariae are detected almost exclusively in peripheral blood, but occasionally are found in cerebrospinal fluid and urine. They also appear incidentally in blood in biopsy and autopsy specimens. Peripheral eosinophilia is common.

Dead worms provoke a vigorous immune response. In a patient who had received diethylcarbamazine, there were several dead worms in the wall of a markedly thickened hernial sac (Figs 14.15 and 14.16). The degenerating worms were surrounded by inflammatory exudates made up of eosinophils, neutrophils, plasma cells, and histiocytes. There were focal abscesses, granulomas, and scarring. Presumably, breakdown of the worms following death releases materials that are antigenic or provoke inflammation.

Adult *M. perstans* in periorbital tissue stimulates an inflammatory response similar to that described above, and appears clinically as bungeye. Worms in the conjunctiva are usually surrounded by necrotizing granulomas and palisading epithelioid cells. Remnants of the worm (Fig 14.17) may be seen in the necrobiotic coagulum of degranulating eosinophils,² and identified by persisting melanin granules in the lateral cords.

Diagnosis

Mansonella perstans infections are most frequently diagnosed by finding characteristic microfilariae in peripheral blood films stained by the Giemsa technique (Fig 14.10). The microfilariae of *M. perstans* are more abundant in blood taken at night, but can be found at any hour of the day (subperiodic). Immunologic diagnostic reagents are under development.[32]

Microfilariae of *M. perstans* have distinctive features, both in peripheral blood and tissue sections (Fig 14.18), but must be distinguished from other microfilariae that rarely infect humans: *Meningonema peruzzii*,[26,28] *Microfilaria semiclarum*,[10] *Microfilaria bolivarensis*,[14] and *Microfilaria rodhaini* (see chapter 36).[30]

Identification of intact adult *M. perstans* and fragments of worms in tissue sections can be made from the specific morphologic features described above and listed in Table 13.1.

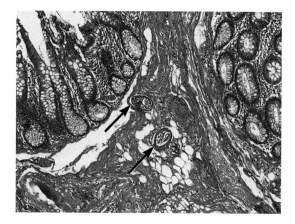

Figure 14.11
Gravid *M. perstans* (arrows) in submucosa of cecum, provoking minimal inflammation. x60

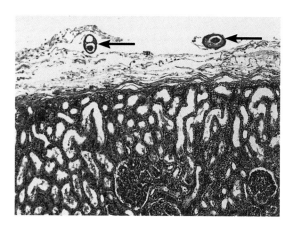

Figure 14.12
Female *M. perstans* (arrows) in capsule of kidney without significant inflammation. x60

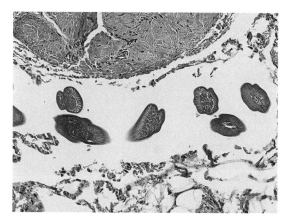

Figure 14.13
Coiled male *M. perstans* in connective tissue of spermatic cord, causing little inflammation. x120

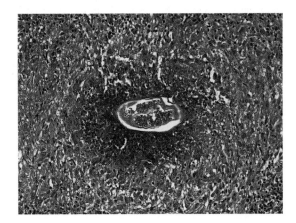

Figure 14.14
Gravid *M. perstans* in necrotic granuloma in gallbladder wall. x125

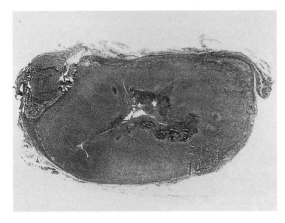

Figure 14.15
Hernial sac from a Congolese containing adult male and female *M. perstans*. Wall of sac is markedly thickened by necrosis and inflammatory exudate. Patient had received DEC several weeks previously. x5.5

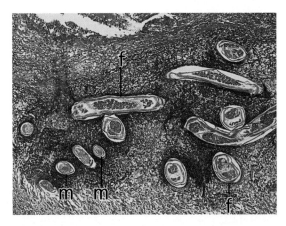

Figure 14.16
Higher magnification of section shown in Figure 14.15, depicting coiled, degenerated adult male (m) and female (f) *M. perstans* in areas of necrosis and fibrosis. x60

Treatment

Therapy with diethylcarbamazine (DEC) eliminates the microfilaremia of *M. perstans* infection, but only after prolonged therapy: 2 mg/kg body weight 4 times daily for 8 periods of 10 days each, with 3-week intervals between each period. The slow elimination of microfilariae from the bloodstream suggests an effect on adults rather than on the microfilariae. Treatment seems to have a salutary clinical effect.[1] The several degenerating adult *M. perstans* seen in random surgical specimens were from patients who had taken DEC. One series of patients responded favorably to mebendazole at 100 mg 2 times daily for 4 to 7 weeks.[17,36] Preliminary studies suggest that a single dose of ivermectin is not effective for *M. perstans* infection, but information on prolonged ivermectin therapy is not available.[34]

■ Streptocerciasis

Introduction

Definition

Streptocerciasis is infection by the filarial nematode *M. streptocerca*.[16] In contrast to other members of the genus *Mansonella*, *M. streptocerca* causes distinctive lesions and is frequently mistaken for leprosy (Fig 14.19).[19] The term streptocerca comes from the Greek for twisted tail, a distinctive feature of these microfilariae.

Synonyms

Previous generic designations of *M. streptocerca* were, successively: *Agamofilaria*, *Tetrapetalonema*, *Acanthocheilonema*, and *Dipetalonema*. This filarid was placed in the genus *Mansonella* in 1982.[27]

General Considerations

While conducting observations on onchocerciasis in 1922 in the Gold Coast (Ghana), Macfie and Corson noted microfilariae in skin snip preparations that were distinctly different from those of *Onchocerca volvulus* in their morphology and movements.[16] Relevant adult filariae were not found at that time, and the microfilariae were appropriately put in the genus *Agamofilaria*. Adult filariae presumed to be *M. streptocerca* were found in chimpanzees in 1946[30] and gorillas in 1964.[35] The discovery of similar adult filarids in wild animals suggests that streptocerciasis may be a zoonosis, but some authorities believe that the agent in chimpanzees and gorillas differs from the species infecting humans. Adult females of the worm were described in human skin in 1972,[19] and males in 1977.[21]

Epidemiology

Streptocerciasis extends throughout all of western and central sub-Saharan Africa, as far east as western Uganda, and south to northern Angola. Prevalence varies widely, but in the densely forested terrain of equatorial Congo it may range up to 90% in focal regions.

Infectious Agent

Morphologic Description

Adult worms of both sexes obtained from humans have been described, both intact specimens extracted from tissue[13] and in tissue sections (Tables 13.1 and 13.2).[19,21,23] In both sexes the cuticle is thin and smooth (Fig 14.20), but thicker in the lateral cord areas. The lateral cords are inconspicuous but contain melanin granules (Fig 14.21 and Table 13.1).

Females are 19 to 27 mm long by 85 μm wide. Most transverse sections of the female contain paired uteri and a much smaller intestine, which fill the body cavity (Fig 14.22). Paired ovaries and oviducts are confined to the posterior portion of the worm (Fig 14.23). Somatic muscle cells in the female are slightly to moderately developed. Males are 13 to 18 mm long by a maximum 50 μm wide. In the male, the anterior region contains the intestine and testis; the posterior portion contains the intestine and vas deferens (Fig 14.24). There are paired spicules at the posterior extremity (Fig 14.25). Somatic muscle is prominent at all levels, especially posteriorly, where muscle may occupy up to two thirds of the diameter of the worm (Fig 14.24).

Features of microfilariae are summarized in

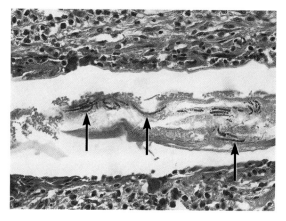

Figure 14.17
Section of degenerated gravid *M. perstans* in conjunctival nodule of a Ugandan. Note microfilariae (arrows) within worm and surrounding cellular exudate. x230

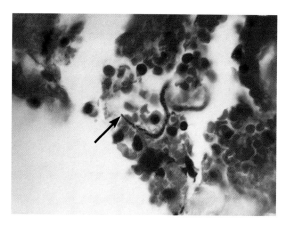

Figure 14.18
Posterior end of microfilaria of *M. perstans* in extravasated blood from skin. Terminal bulbous nucleus (arrow) reaches tip of tail. x780

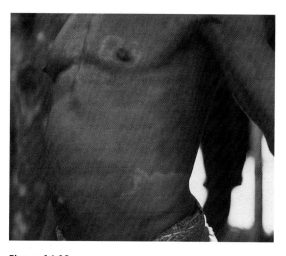

Figure 14.19
Streptocerciasis in a Congolese. The well-defined hypopigmented plaques resemble borderline leprosy, but are not hypoesthetic.

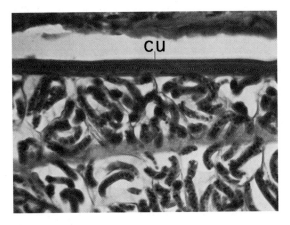

Figure 14.20
Longitudinal section of gravid *M. streptocerca,* showing thin, smooth cuticle (cu). x620

Figure 14.21
Transverse section through lateral cord area of *M. streptocerca* showing pigment granules (arrows). x755

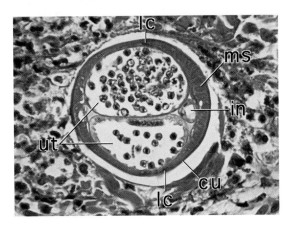

Figure 14.22
Transverse section at midbody of gravid *M. streptocerca* in skin. Note thin cuticle (cu), pigment in lateral cords (lc), somatic muscle cells (ms), intestine (in), and paired uteri (ut) filled with microfilariae. x570

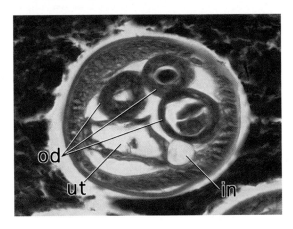

Figure 14.23
Transverse section through posterior region of gravid *M. streptocerca* in skin. Internal organs consist of an intestine (in), uterus (ut), and 3 sections of oviducts (od) containing ova. x605

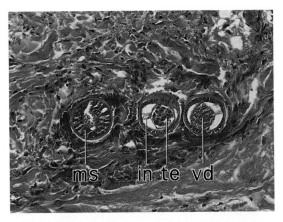

Figure 14.24
Coiled male *M. streptocerca* in skin, showing transverse sections at 3 different levels. Internal organs include testis (te), vas deferens (vd), and intestine (in). Note well-developed somatic musculature (ms) in most posterior section. x265

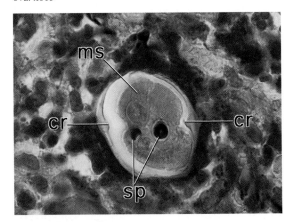

Figure 14.25
Transverse section through posterior extremity of male *M. streptocerca* in dermis. Note internal cuticular ridges (cr), muscle cells (ms), and 2 spicules (sp). Movat x770

Table 13.3. Microfilariae of *M. streptocerca* are unsheathed and measure 180 to 240 μm long by 2.5 to 5 μm wide. The sharply curved tail and straight body of the microfilaria produce the classic shepherd's crook configuration of this species (Fig 14.26). Cephalic space is 3 to 5 μm long, and the first 4 nuclei are oval and staggered but do not overlap (Fig 14.27). The first 4 nuclei are followed by a single row of 7 to 10 smaller nuclei. The terminal nucleus is oval or round and extends to the end of the tail, leaving a 1-μm caudal space (Fig 14.28).

Life Cycle

In 1954, Duke established that the midge *C. grahami* (Fig 14.3) ingests microfilariae when feeding on a patient, and that larvae of *M. streptocerca* develop in this insect.[8] There is general acceptance of transmission of streptocerciasis by this insect vector, and possibly by another species, *Culicoides milnei*.[9]

Clinical Features and Pathogenesis

The incubation period of streptocerciasis following infection in humans is unknown but estimated to be several months. Many patients are asymptomatic. The most consistent symptom is a pruritic dermatitis, most severe over the shoulder girdle and trunk. Physical findings include trauma from scratching, thickening and inflammation of the skin, and occasionally 1 or more papules. Lymphadenopathy is frequently found, and in some patients is probably related to the streptocerciasis. Streptocercal elephantiasis has been postulated, but there is no convincing evidence for this in the authors' experience. The macules and plaques of streptocerciasis in endemic areas of leprosy are frequently mistaken for leprosy (Figs 14.19, 14.29 and 14.30).[19] In a study of a group of 40 Congolese patients known to have streptocerciasis, 62% had hypopigmented dermal lesions that could readily be mistaken for leprosy. In lesions of streptocerciasis, there are no sensory changes as are common in leprosy. However, since early lesions of leprosy often show minimal or no sensory changes, histopathologic evaluation of lesions is strongly advised.

Figure 14.26
Skin snip preparation from patient with streptocerciasis, depicting entire microfilaria of *M. streptocerca*, showing shepherd's crook configuration. Giemsa x450

A response analogous to the Mazzotti reaction in onchocerciasis develops in patients with streptocerciasis following the administration of a single small dose of DEC (50 mg). The skin becomes edematous, itching increases, and papules appear. Often there are showers of 50 or more papules, almost always over the shoulder girdle and trunk (Fig 14.31). Mazzotti-like reactions are often diminished or absent in leprosy patients who also have streptocerciasis.[18]

Pathologic Features

Microfilariae of *M. streptocerca* have not been detected in peripheral blood, but sometimes a microfilaria is seen in lymphatics in biopsy specimens of skin. Peripheral blood may show eosinophilia.

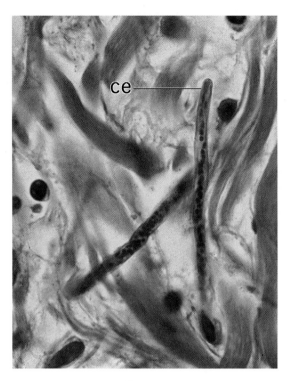

Figure 14.27
Section of skin showing anterior region of microfilaria of *M. streptocerca* in dermal collagen. Cephalic space (ce) is short and anterior nuclei are distributed in a characteristic manner. x1080

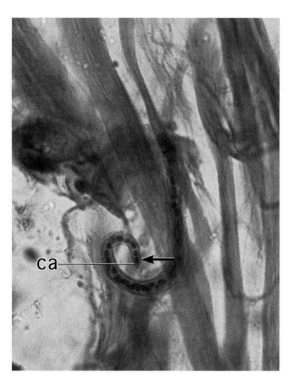

Figure 14.28
Posterior end of *M. streptocerca* in same specimen shown in Figure 14.27, depicting classic shepherd's crook configuration. Terminal nucleus (arrow) is oval and caudal space is 1 µm long (ca). x1500

Microfilariae in varying numbers inhabit the dermal collagen, most abundantly in the upper third of the dermis (Fig 14.32). In untreated patients, histopathologic changes in the skin caused by microfilariae are limited to the dermis. Prominent microscopic features include sclerosis of dermal papillae, edema, incontinence of melanin, fibrosis, and infiltrations of lymphocytes, macrophages, and eosinophils around blood vessels and appendages (Figs 14.33 and 14.34). Lymphatics in the dermis are frequently dilated. In patients who have not received DEC, adult worms are sometimes found in the dermis. They are well-adapted to humans and provoke no inflammatory reaction while alive (Fig 14.35).

Following administration of DEC, when the patient experiences Mazzotti-like reactions, the histopathologic picture abruptly changes within the first 24 hours. Most of the skin changes cited above are enhanced, especially the edema and cellular infiltration, but there are new features (Fig 14.36). The epidermis may contain microabscesses of inflammatory cells and fragments of microfilariae (Fig 14.37). If the specimen comes from a newly formed papule, the dermis will nearly always contain a single coiled male (Fig 14.24) or female (Figs 14.38 and 14.39) *M. streptocerca*, surrounded by acute and chronic inflammatory cells with many eosinophils, and an amorphous precipitate probably composed mostly of immune complexes. Biopsy specimens taken from nodules on day 2 through day 31 following initiation of DEC therapy establish that DEC kills the worms of both sexes and that they completely disintegrate (Fig 14.40). Granulomas develop around the dead and disintegrating worms (Fig 14.41).

In a series of lymph nodes from patients studied at the AFIP, there appears to be a specific lymphadenitis related to streptocerciasis. This lymphadenitis is characterized by sinus histiocytosis, eosinophilia, atrophy of germinal centers, fibrosis, and small numbers of microfilariae of *M. streptocerca* in collagen (Figs 14.42 to 14.44).[24]

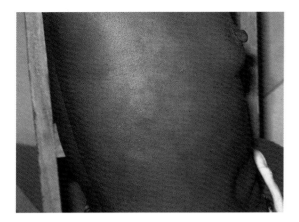

Figure 14.29
Congolese patient with streptocerciasis. Note multiple hypopigmented macules over trunk that clinically resemble lepromatous leprosy. There were no sensory changes. Histopathologic findings were typical of streptocerciasis.

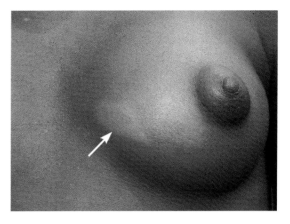

Figure 14.30
Adolescent Congolese female with single plaque (arrow) of streptocerciasis. Lesion was clinically diagnosed as early leprosy.

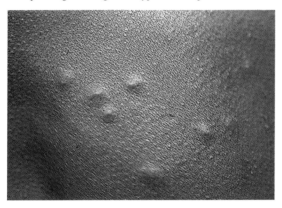

Figure 14.31
Pectoral area of Congolese patient with streptocerciasis 24 hours after ingesting 50 mg of DEC. There were approximately 50 papules over trunk and upper extremities. Papules surround damaged adult *M. streptocerca* in dermis.

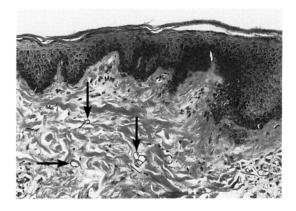

Figure 14.32
Section of skin from Congolese with streptocerciasis. Note multiple fragments of microfilariae (arrows) of *M. streptocerca* in dermal collagen. Giemsa x115

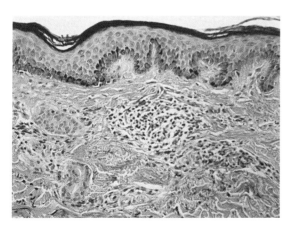

Figure 14.33
Skin of patient with streptocerciasis. Note mild edema, perivascular infiltrates of lymphocytes, and eosinophils. Movat x140

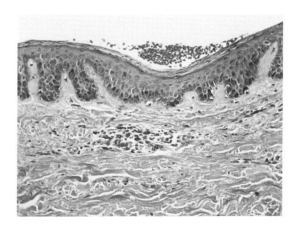

Figure 14.34
Skin from patient with streptocerciasis, showing sclerosis of papillary dermis, cellular exudates, and fibrosis. x140

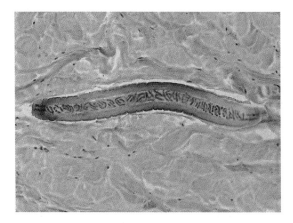

Figure 14.35
Gravid *M. streptocerca* in dermal collagen of untreated Congolese. Worm is uncoiled and provoking no inflammation. x115

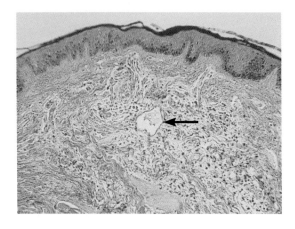

Figure 14.36
Skin of patient with streptocerciasis 24 hours after starting treatment with DEC. Note enhanced edema of dermis and dilatation of lymphatics (arrow). Movat x90

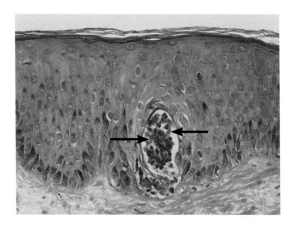

Figure 14.37
Microabscess in epidermis of patient treated with DEC analogous to Mazzotti reaction in onchocerciasis. Abscess contains 2 fragments (arrows) of degenerating microfilariae. x235

Diagnosis

Clinical diagnosis is difficult unless the patient has received DEC or other filaricidal drugs. If streptocerciasis is suspected, skin snips should be taken, preferably from over the scapula, and evaluated as wet mounts in saline. The microfilariae swim sluggishly in comparison to those of *O. volvulus*, and assume a shepherd's crook configuration when immobile. Microfilariae of *M. streptocerca* have distinctive features and are usually readily identifiable in histopathologic sections of skin in the dermal collagen (Figs 14.27 and 14.28). See chapter 13 for comparative features of other microfilariae.

Remember that, because leprosy is found in all endemic areas of streptocerciasis, diagnosis of one does not rule out coexistence of the other.

Treatment

Long-term or pulsed treatment with DEC kills the adult worms (Figs 14.40 and 14.41) and should achieve a radical cure of streptocerciasis. In one study, however, new papules appeared over a 21-day period of therapy, suggesting that early larval forms of *M. streptocerca* are not susceptible to DEC.[20] This aspect of DEC treatment of streptocerciasis requires further study. Take care in administering DEC to patients coinfected with *Loa loa* (see chapter 15). Recent preliminary reports suggest that single-dose ivermectin therapy is effective against *M. streptocerca*, but this too requires further evaluation.

■ Mansonelliasis ozzardi

Introduction

Definition
Mansonelliasis ozzardi is infection by the filaria *M. ozzardi* (Manson, 1897) Faust, 1929. The species was named for Ozzard, who provided the specimens that Manson later studied.

Synonyms
Synonyms include Ozzard's filariasis and

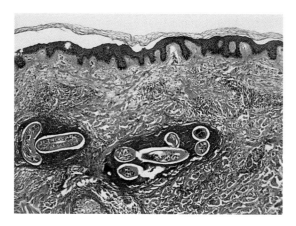

Figure 14.38
Coiled, gravid *M. streptocerca* in dermis 24 hours after starting DEC therapy. Worm is surrounded by eosinophilic coagulum containing inflammatory cells, fibrin, degranulating eosinophils, and antigen-antibody complexes. x45

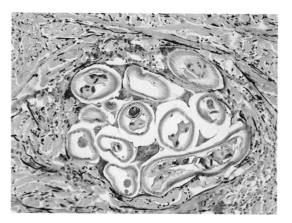

Figure 14.39
Coiled, nongravid female *M. streptocerca* in dermis 5 days after start of DEC therapy. Reaction is more chronic and less intense than that in Figure 14.38. Movat x125

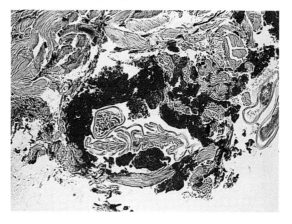

Figure 14.40
Advanced degeneration of female *M. streptocerca* in dermis, 21 days after initiation of DEC therapy. Worm remains surrounded by reaction similar to that described in Figure 14.38. Movat x130

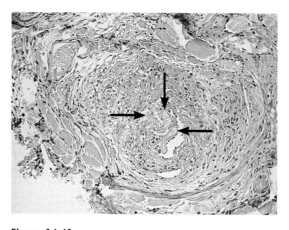

Figure 14.41
Remnants of cuticle (arrows) of disintegrated adult *M. streptocerca* in a granuloma in dermis, 31 days after starting DEC therapy. Movat x100

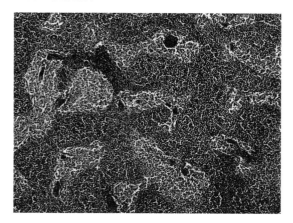

Figure 14.42
Axillary lymph node of patient with streptocerciasis. Note sinus histiocytosis and atrophy of germinal centers. There was also eosinophilia. x60

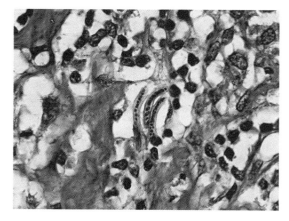

Figure 14.43
Coiled microfilaria of *M. streptocerca* in collagen of lymph node described in Figure 14.42. x575

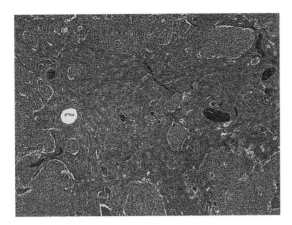

Figure 14.44
Inguinal lymph node from patient with streptocerciasis, showing marked fibrosis. Node contained a few microfilariae of *M. streptocerca*. x25

mansonellosis. Former names for the etiologic agent are *Filaria ozzardi* and *Filaria demarquayi*.

General Considerations

Manson first described microfilariae of *M. ozzardi* in 1897 in Carib Indians in British Guiana; Daniels discovered the adult worms in 1898.[6] In 1929, Faust further described the adult worms and established the new genus *Mansonella*.

Epidemiology

Mansonelliasis ozzardi is limited to the Western Hemisphere and favors remote rural inhabitants, especially the Amerindians of the West Indies, Central America, and South America. In South America, endemic areas are northern Argentina, Bolivia, Brazil, Colombia, Ecuador, and Peru.[4,24]

Infectious Agent

Morphologic Description

Only a few intact adult worms have been recovered from humans. Adult females from humans measure 65 to 81 mm long by 210 to 250 µm wide. The cuticle is smooth (Table 13.1). The single recovered fragment of a male measured 38 mm long by 200 µm wide. The cuticle is smooth (Table 13.2). The morphology of adult *M. ozzardi* relies largely on specimens from experimentally infected monkeys. Female worms from monkeys are 32 to 62 mm long by 130 to 160 µm wide, and males 24 to 28 mm long by 70 to 80 µm wide. The posterior end of the female bears 4 papillae. For a more detailed description of worms from experimental monkeys, consult Orihel and Eberhard.[27]

Microfilariae are unsheathed and 170 to 240 µm long by 3 to 5 µm wide (Fig 14.45) with a 2-to-6-µm cephalic space (Fig 14.46) (see Table 13.3). The anterior 2 or 3 nuclei either overlap or are side by side. Caudal space is 3 to 8 µm long and the terminal nuclei are oval (Fig 14.47). The microfilariae resemble those of *M. perstans* except that the tail of *M. ozzardi* is pointed, slightly flexed, and has a longer caudal space. See chapter 13 for differential diagnostic features of other microfilariae.

Life Cycle

In the West Indies, Surinam, and Argentina, *Culicoides* sp of biting midges are vectors of *M. ozzardi*. In Brazil and Colombia, black flies (*Simulium amazonicum* and *Simulium sanguineum*) are vectors. These vectors take up microfilariae in blood meals and infective larvae develop in the thoracic muscles. Humans are the only known natural definitive hosts.

Clinical Features and Pathogenesis

The incubation period in humans is unknown. Symptoms vary in different populations. In the West Indies, no specific manifestations of the infection are noted. Patients and clinicians in the Amazon region attribute a wide variety of complaints to *M. ozzardi* infection, including lymphadenopathy, varices, pain in the knees and ankles, dermatitis with pruritus, fever, headache, insomnia, and vertigo. There may be peripheral eosinophilia. No consistent clinical picture, however, has emerged, and these complaints and findings may be incidental. Prevalence is equal in the sexes and increases with age, reaching up to 90% of local populations. Infection in children less than 6 years old is uncommon.

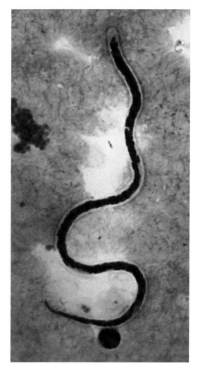 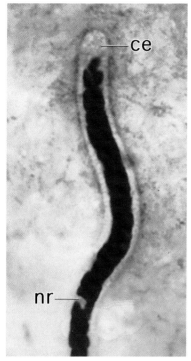 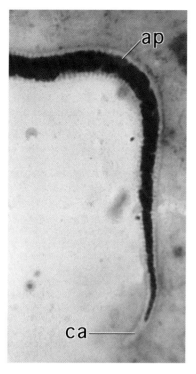

Figure 14.45
Unsheathed microfilaria of *M. ozzardi* in thick peripheral blood film. Giemsa x900

Figure 14.46
Anterior end of microfilaria of *M. ozzardi*, depicting cephalic space (ce) and nerve ring (nr). Giemsa x1900

Figure 14.47
Posterior end of microfilaria of *M. ozzardi*, showing anal pore (ap) and caudal space (ca). Giemsa x2000

Pathologic Features

Adult worms probably are most common in the thoracic and peritoneal cavities, but may invade lymphatic vessels. These worms are apparently well-adapted parasites, usually provoke no inflammation, and are rarely seen grossly or microscopically. Whether or not some of the symptoms (eg, varices and lymphadenopathy) that patients in the Amazon area experience represent responses to adult worms is unknown.

Histologically, in the skin and subcutaneous tissue there are small numbers of microfilariae in blood vessels, and sometimes small perivascular infiltrations of lymphocytes and plasma cells.

Diagnosis

Diagnosis depends on demonstrating microfilariae of *M. ozzardi* in peripheral blood. This may be in thick films of peripheral blood, skin snips, histopathologic study of biopsy specimens of skin, or by Nucleopore® filtration (pore size less than 5 μm) of venous blood. The microfilariae are nonperiodic, so specimens may be taken at any time of day.

Treatment

No effective therapy has been reported. DEC is ineffective.[5,15] Prognosis is excellent.

References

1. Adolph PE, Kagan IG, McQuay RM. Diagnosis and treatment of Acanthocheilonema perstans filariasis. *Am J Trop Med Hyg* 1962;11:76–88.
2. Baird JK, Neafie RC, Connor DH. Nodules in the conjunctiva, bung-eye, and bulge-eye in Africa caused by Mansonella perstans. *Am J Trop Med Hyg* 1988;38:553–557.
3. Baird JK, Neafie RC, Lanoie L, Connor DH. Adult Mansonella perstans in the abdominal cavity in nine Africans. *Am J Trop Med Hyg* 1987;37:578–584.
4. Beaver PC, Jung RC, Cupp EW. *Clinical Parasitology*. 9th ed. Philadelphia, Pa: Lea & Febiger; 1984:380–387.
5. Chadee DD, Tilluckdharry CC, Rawlins SC, Doon R, Nathan MB. Mass chemotherapy with diethylcarbamazine for the control of Bancroftian filariasis: a twelve-year follow-up in northern Trinidad, including observations on Mansonella ozzardi. *Am J Trop Med Hyg* 1995;52:174–176.
6. Daniels CW. The filaria Ozzardi and their adult forms. *Br Guiana Med Ann* 1898;10:1–5.
7. Daniels CW. Discovery of the parental form of a British Guiana blood worm. *Br Med J* 1898;1:1011–1012.
8. Duke BO. The uptake of the microfilariae of Acanthocheilonema streptocerca by Culicoides grahami, and their subsequent development. *Ann Trop Med Parasitol* 1954;48:416–420.
9. Duke BO. The intake of the microfilariae of Acanthocheilonema streptocerca by Culicoides milnei, with some observations in the potentialities of the fly as a vector. *Ann Trop Med Parasitol* 1958;52:123–128.
10. Fain A. Dipetalonema semiclarum sp. n. from the blood of man in the Republic of Zaire (Nematoda: Filarioidea). *Ann Soc Belg Med Trop* 1974;54:195–207.
11. Fischer P, Kilian AH, Bamuhiiga J, Kipp W, Buttner DW. Prevalence of Mansonella perstans in western Uganda and its detection using the QBC-fluorescence method. *Appl Parasitol* 1996;37:32–37.
12. Foster DG. Filariasis-a rare cause of pericarditis. *J Trop Med Hyg* 1956;59:212–214.
13. Gardiner CH, Meyers WM, Lanoie LO. Recovery of intact male and female Dipetalonema streptocerca from man. *Am J Trop Med Hyg* 1979;28:49–52.
14. Godoy GA, Orihel TC, Volcan GS. Microfilaria bolivarensis: a new species of filaria from man in Venezuela. *Am J Trop Med Hyg* 1980;29:545–547.
15. Hawking F. The distribution of human filariasis throughout the world. Part III. Africa. *Trop Dis Bull* 1977;74:649–679.
16. Macfie JW, Corson JF. A new species of filarial larva found in the skin of natives in the Gold Coast. *Ann Trop Med Parasitol* 1922;16:465–471.
17. Maertens K, Wery M. Effect of mebendazole and levamisole on Onchocerca volvulus and Dipetalonema perstans. *Trans R Soc Trop Med Hyg* 1975;69:359–360.
18. Meyers WM, Connor DH. Onchocerciasis and streptocerciasis in patients with leprosy. Altered Mazzotti reactions [letter]. *Trans R Soc Trop Med Hyg* 1975;69:524–525.
19. Meyers WM, Connor DH, Harman LE, Fleshman K, Moris R, Neafie RC. Human streptocerciasis. A clinicopathologic study of 40 Africans (Zairians) including identification of the adult filaria. *Am J Trop Med Hyg* 1972;21:528–545.
20. Meyers WM, Moris R, Neafie RC, Connor DH, Bourland J. Streptocerciasis: degeneration of adult Dipetalonema streptocerca in man following diethylcarbamazine therapy. *Am J Trop Med Hyg* 1978;27:1137–1147.
21. Meyers WM, Neafie RC, Moris R, Bourland J. Streptocerciasis: observation of adult male Dipetalonema streptocerca in man. *Am J Trop Med Hyg* 1977;26:1153–1155.
22. Mommers EC, Dekker HS, Richard P, Garcia A, Chippaux JP. Prevalence of L. loa and M. perstans filariasis in southern Cameroon. *Trop Geogr Med* 1995;47:2–5.
23. Neafie RC, Connor DH, Meyers WM. Dipetalonema streptocerca (Macfie and Corson, 1922): description of the adult female. *Am J Trop Med Hyg* 1975;24:264–267.
24. Neafie RC, Meyers WM, Connor DH. Mansonelliasis. In: Binford CH, Connor DH, eds. *Pathology of Tropical and Extraordinary Diseases*. Vol 2. Washington, DC: Armed Forces Institute of Pathology; 1976:390–391.
25. Olumide YM, Obembe BM. Dipetalonema perstans in a patient with chronic lymphoedema. Case report. *East Afr Med J* 1983;60:186–189.
26. Orihel TC. Cerebral filariasis in Rhodesia-a zoonotic infection? *Am J Trop Med Hyg* 1973;22:596–599.
27. Orihel TC, Eberhard ML. Mansonella ozzardi: a redescription with comments on taxonomic relationships. *Am J Trop Med Hyg* 1982;31:1142–1147.
28. Orihel TC, Esslinger JH. Meningioma peruzzii gen. et sp. n. (Nematoda: Filarioidea) from the central nervous system of African monkeys. *J Parasitol* 1973;59:437–441.
29. Owen HB, Hennessey RS. A note on some ocular manifestations of helminthic origin occurring in natives of Uganda. *Trans R Soc Trop Med Hyg* 1932;25:267–273.
30. Peel E, Chardome M. Sur des filarides de chimpanzés Pan panicus et Pan satyrus au Congo Belge. *Ann Soc Belg Med Trop* 1946;26:117–156.
31. Price DL. Dipetalonemiasis. In: Marcial-Rojas RA, ed. *Pathology of Protozoal and Helminthic Diseases with Clinical Correlations*. Baltimore, Md: Williams & Wilkins; 1971:948–954.
32. Tawill SA, Kipp W, Lucius R, Gallin M, Erttmann KD, Buttner DW. Immunodiagnostic studies on Onchocerca volvulus and Mansonella perstans infections using a recombinant 33 kDa O. volvulus protein (Ov33). *Trans R Soc Trop Med Hyg* 1995;89:51–54.
33. Useh MF, Ejezie GC. The status and consequences of Mansonella perstans infection in Calabar, Nigeria. *East Afr Med J* 1995;72:124–126.
34. Van den Enden E, Van Gompel A, Van der Stuyft P, Vervoort T, Van den Ende J. Treatment failure of a single high dose of ivermectin for Mansonella perstans filariasis. *Trans R Soc Trop Med Hyg* 1993;87:90.
35. Van den Berghe L, Chardome M, Peel E. The filarial parasites of the Eastern gorilla in the Congo. *J Helminthol* 1964;38:349–368.
36. Wahlgren M, Frolov I. Treatment of Dipetalonema perstans infections with mebendazole [letter]. *Trans R Soc Trop Med Hyg* 1983;77:422–423.

15

Loiasis

Aileen M. Marty,
Brian O.L. Duke, and
Ronald C. Neafie

Introduction

Definition

Loiasis is infection by the filarial nematode *Loa loa* (Cobbold, 1864) Castellani and Chalmers, 1913, transmitted by tabanid flies of the genus *Chrysops*.[8] Characteristic clinical manifestations are Calabar swellings and subconjunctival migration of the adult worms, but many patients, even those with high microfilaremia, are asymptomatic.

Synonyms

Common names for *L. loa* include African eye worm, Kampala eye worm, and loa worm. Archaic synonyms include *Dracunculus loa*, *Filaria lacrymalis*, *Filaria loa*, *Filaria oculi humani*, *Filaria subconjunctivalis*, and *Microfilaria diurna*.

General Considerations

Loa loa has been known for many centuries in endemic areas. Possibly the first documentation of loiasis was by Pigafetta in 1589, who recounted a story originated by Lopez about an infected Congolese. Mongin, a French surgeon, published the first conclusive report in 1770, in which he described a parasite found in the eye of a black slavewoman in Hispaniola.[27] In 1777, Bajon reported additional cases in Cayenne,[3] as did Guyot in 1778 in Angola.[2] Guyot noted that the name for the parasite in Angola was *loa*, meaning worm, leading Cobbold in 1864 to designate *loa* as the species. In 1891, Manson described the microfilariae of this worm in the blood of West Africans and named them *Microfilaria diurna,* because the microfilaremia peaked at midday. Argyll-Robertson,[1] a British ophthalmologist, in collaboration with Manson, first described the male and female worms in some detail in 1895. In Germany, in 1904, Looss[24] gave a detailed description of this worm by reviewing the literature and dissecting specimens. In 1905, Stiles and Hassall placed the parasite in the subgenus *Loa* of the genus *Filaria*, and in 1913, Castellani and Chambers[8] established the genus *Loa*. In a telegram to the London School of Tropical Medicine on December 27, 1912, Leiper[23] reported that the vector was a biting fly of the genus *Chrysops,* and later, in 1913, described the larval development of *L. loa* in the fly. Smith and Rivas, in 1914, developed a concentration technique for demonstrating microfilariae in peripheral blood. In the early 1920s Connal dissected larvae out of 2255 specimens of *Chrysops silacea* and 249 specimens of *Chrysops dimidiata*, and described the development of the larvae in these flies.[10] In 1948, Stefanopoulos and Schneider effectively used diethylcarbamazine to treat loiasis. In 1955, Duke demonstrated that smoke from wood fires increases the biting rate of *Chrysops* and may attract the flies from forests into houses. Duke also suggested that movement attracted the flies.[12,13] In 1958, Duke and Wijers showed that the natural simian *Loa*, found in the drill (*Mandrillus leucophaeus*) and some species of *Cercopithecus* monkeys, has a nocturnal microfilarial periodicity. They further observed that nocturnal, canopy-dwelling *Chrysops langi* and *Chrysops centurionis* transmit the simian parasite, and that the human parasite could be transmitted experimentally to monkeys, where it maintained its diurnal periodicity.[17] Duke and Wijers found little evidence of natural infections of human diurnal *L. loa* in monkeys. But later, in 1964, Duke demonstrated that the simian and human strains of *L. loa* hybridized under experimental conditions.[16]

Figure 15.1
Chrysops silacea, a common vector of *L. loa*, feeding on skin of a human volunteer and causing erythema. The head, eyes, and wings of these tabanids are large.*

Epidemiology

Loa loa infects as many as 13 million inhabitants of endemic areas.[35] More than 20% of residents in some endemic areas are microfilaremic, and probably more than twice as many have amicrofilaremic infections.[18] In highly endemic foci more than 90% of inhabitants may be infected. Only flies of the genus *Chrysops* transmit loiasis; *C. silacea* (Fig 15.1) and *C. dimidiata* are the most important vectors. The endemic area extends from latitudes 8° north to 5° south in countries along the Gulf of Guinea, and eastward to southwestern Sudan and Uganda. These regions are primarily rain forests and transitional savanna that support the development of the vector. Endemic countries are Angola, Cameroon, Central African Republic, Chad, Congo, Equatorial Guinea, Gabon, Nigeria, Rwanda, Sudan, Uganda, and Democratic Republic of Congo.

Adults are infected more frequently than children. In some foci, probably related to occupational exposure, men are infected more often than women.

Infectious Agent

Morphologic Description

Loa loa is a filarial nematode of the superfamily Filarioidea. Adult worms are white, cylindrical, and thread-like (Figs 15.2 and 15.3). Understanding of this worm's morphology has not changed much since Looss's description in 1904.[24]

The surface of the cuticle in both sexes contains irregularly spaced bosses (Figs 15.4 and 15.5), which are minute, rounded elevations located throughout most of the length of the worm. The cuticle is 4 to 10 μm thick, but each lateral cord region, where the cuticle forms internal cuticular ridges, is slightly thicker. Cuticular ridges are especially evident in males (Fig 15.6). The hypodermis is a thin layer lying beneath the cuticle, except on the lateral sides of each worm, where the hypodermis projects conspicuously into the body cavity to form lateral cords. Frequently, lateral cords have a band dividing them into sublaterals, especially in the male (Fig 15.7). There are numerous somatic muscle cells that project conspicuously into the body cavity in each quarter of the worm (Fig 15.7). Individual muscle cells are easy to discern, and the contractile and sarcoplasmic portions of each muscle cell are well-developed. The muscular esophagus has a triradiate lumen and is about 1 mm long. The intestine varies in diameter and consists of cuboid cells lined with microvilli.

Adult females are 40 to 70 mm long by 450 to 600 μm wide (Fig 15.2). Reproductive organs extend throughout the length of the worm (Fig 15.8). The vulva is located about 2 mm from the anterior end of the worm, just posterior to the esophageal-intestinal junction. The vagina is about 9 mm long and has no cuticle throughout most of its length. The distal portion of the vagina is thick-walled and muscular (Fig 15.9), whereas the proximal portion is thin-walled. The remainder of the reproductive system consists of paired uteri, seminal receptacles, oviducts, and ovaries (Fig 15.10). Reproductive tubes loop several times throughout the worm, making their combined length about 3 times the length of the worm. Most transverse sections contain 3 to 5 sections of reproductive tube.

Adult males are 30 to 34 mm long by 350 to 400 μm wide (Fig 15.3). The male has a single reproductive tube consisting of a testis (Fig 15.11), vas deferens (Fig 15.12), and ejaculatory duct. The testis begins posterior to the esophageal-intestinal junction, then turns anteriorly toward the nerve ring, where it loops posteriorly. The reproductive tube reflexes 2 more times in the anterior region of the worm before assuming its permanent posterior orientation. The vas deferens comprises most of the length of the reproductive tube. Mature spermatozoa are spherical cells 5 to 6 μm in diameter. The ejaculatory duct unites with the intestine at the posterior end of the worm to form the cloaca that exits at the anus. The anus is about 90 μm from the posterior tip of the worm. There are 2 unequal and dissimilar copulatory spicules (Fig 15.13). The caudal end of the male curves ventrally (Fig 15.3).

Microfilariae of *L. loa* have a sheath (Fig 15.14) which may extend beyond the tip of the tail and the anterior end. Microfilariae, exclusive of the sheath, are 5 to 8 μm in diameter and 175 to 300 μm long (Fig 15.15), but microfilariae under 200 μm in length are uncommon (Fig 15.16). The anterior end is blunted and has a 3-to-6-μm-long cephalic space void of nuclei. The most anterior nuclei are paired (Fig 15.15). The posterior end tapers to a point and the elongated terminal nucleus is at the tip of the

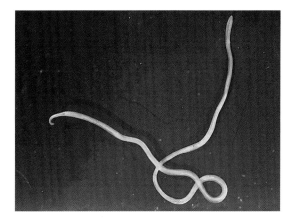

Figure 15.2
Adult female *L. loa* from a Congolese. Worm is 5 cm long by 500 μm in diameter.

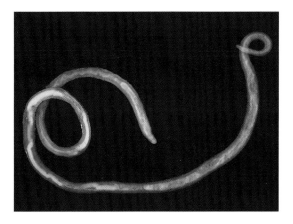

Figure 15.3
Adult male *L. loa* from an African. Worm is 3 cm long by 400 μm in diameter. Note curve at caudal end.*

tail (Fig 15.15).

The insect vector for *L. loa* is the female red fly, also known as the mango (or mangrove) fly, of the family Tabanidae and genus *Chrysops* (Fig 15.1). The 2 most important and most widely distributed vector species in the rain forest are *C. silacea* and *C. dimidiata*. Other vector species include *Chrysops zahrai*, *Chrysops distinctipennis*, and *Chrysops longicornis*. Experimental infections have been established in other members of the genus *Chrysops*, including those native to North America.

These flies breed in the shade of rain forests, depositing their eggs on leaves. The eggs hatch and the larvae drop to the ground or into shallow pools of water, and live in the mud for roughly a

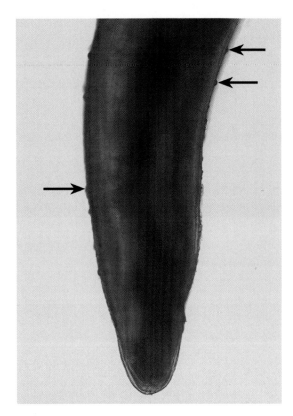

Figure 15.4
Posterior end of female *L. loa* with irregularly spaced bosses (arrows) on cuticle. x150

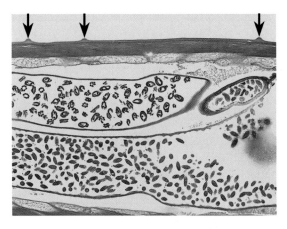

Figure 15.5
Longitudinal section of adult female *L. loa*, showing irregularly spaced bosses (arrows) on cuticle. x95

year before pupating (Fig 15.17). Approximately 1 to 2 weeks after pupation, adult flies emerge and live for about a month.

Biting increases during the rainy season. Female flies are attracted to the movement of humans through the forest, and apparently prefer to bite dark skin. Male flies do not feed on humans and are not involved in transmission.

For comparison with other filariids, see tables in chapter 13.

Life Cycle and Transmission

Chrysops sp are the intermediate hosts and humans the definitive hosts. Monkeys may also serve as definitive hosts, but apparently only to their own species-specific filaria, often referred to as *Loa loa papionis*.

Female *Chrysops* ingest circulating microfilariae with their blood meal (Fig 15.1). Microfilariae traverse the stomach wall and enter the thoracic muscles and fat body of the fly. In the muscles and fat body, the parasites undergo 2 molts to become infective third-stage larvae (L3) in 10 to 12 days. Infective L3 larvae are about 2 mm long. They collect in the hemocoelom and emerge through the proboscis when the fly takes its next blood meal. While feeding, the fly leaves a small pool of hemocoelic fluid on the skin. This fluid may contain 10 to 100 or more infective L3 larvae, which quickly penetrate the host through the wound made by the fly's proboscis. Infected flies often die after feeding on a new host.

Within humans, infective larvae undergo 2 molts before maturing to adults. Juvenile and mature adult worms meander through the subcutaneous and intermuscular connective tissue. Worms mate and gravid females release microfilariae into surrounding tissues. Microfilariae enter the bloodstream via the lymphatic system. They accumulate in pulmonary capillaries and invade the peripheral blood during the day, giving them a diurnal periodicity.

Development from infective larva to mature adult usually takes about 5 months; microfilariae appear in the blood in 6 to 12 months. Adult worms can live for 15 years or longer.[14]

Clinical Features and Pathogenesis

The bite of *Chrysops* causes erythema, swelling, and itching (Fig 15.1). Infective larvae increase the swelling and itching normally experienced from the bite of an uninfected fly. Later, as the infective larvae move under the skin and molt to L4, they often give rise to local and discrete red urticarial papules in the skin. As L4 larvae mature over the next 3 months or more, the patient may experience vague symptoms, such as pain or temporary swelling of a limb, itching, paresthesia, and hives.

Transient migratory angioedema is a common and classic manifestation of loiasis. The high incidence of loiasis around the town of Calabar in southeastern Nigeria gave rise to the term "Calabar swellings" for the angioedema of loiasis. Calabar swellings are itchy, red, swollen areas in the skin, 2 to 10 cm in diameter. They may or may not be painful.[29] These fugitive swellings may develop in any portion of the skin (Fig 15.18), but are most frequent around wrists and ankles, and are more common in expatriate Europeans than in Africans. Onset is sudden, but swellings regress gradually over a period of several hours to several days. They usually disappear within 3 days but can recur at irregular intervals, often at the same sites. Mild local trauma may precipitate the swellings. Calabar swellings are probably local hypersensitivity reactions in an area occupied by or recently traversed by an adult worm. Expatriate patients may have only chronic urticaria and Calabar swellings.[34,38]

Adult worms migrate freely in the connective tissue, often without causing clinically apparent local reaction or other symptoms of infection (Fig 15.19). Nomadic worms that appear beneath the conjunctiva produce the most spectacular manifestation of loiasis (Fig 15.20). The worm usually crosses the lower half of the conjunctiva and may take from 1 to 30 minutes to cross the eye, permitting surgical extraction (Fig 15.21). This migration may produce irritation and itching, and patients may present with unilateral palpebral edema and tearing that lasts several days. Occasionally, adult worms are noted as they move through areas of loose subcutaneous tissue such as the eyelid, breast, and scrotum.

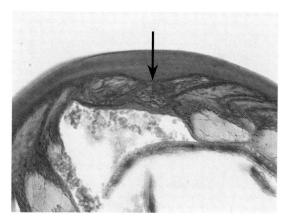

Figure 15.6
Transverse section through male *L. loa* depicting internal longitudinal cuticular ridge (arrow) in lateral cord region. x580

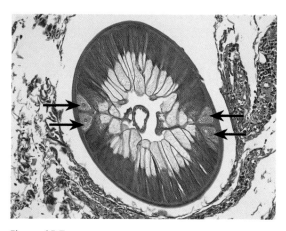

Figure 15.7
Transverse section through male *L. loa* demonstrating lateral cords (arrows) divided into sublaterals and well-developed somatic muscles. Specimen is in spermatic cord of a 55-year-old Congolese. x105

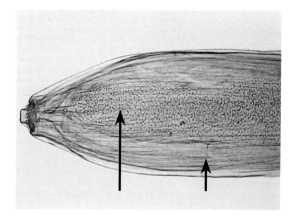

Figure 15.8
Anterior end of gravid *L. loa* with uterus extending to tip of worm. Note that microfilariae-filled uterus (long arrow) is anterior to esophageal-intestinal junction (short arrow). x55

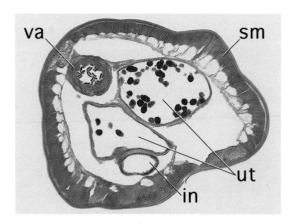

Figure 15.9
Transverse section through adult female *L. loa*, depicting thick-walled, muscular, distal portion of vagina (va), paired uteri (ut), intestine (in), and prominent somatic musculature (sm). x105

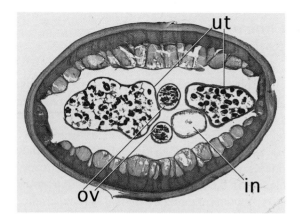

Figure 15.10
Transverse section through adult female *L. loa*. Note paired ovaries (ov), paired uteri containing developing microfilariae (ut), and intestine (in). Contractile and sarcoplasmic portions of somatic muscle cells are equally developed in this worm. x115

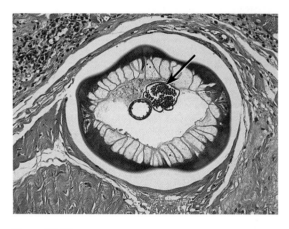

Figure 15.11
Transverse section through anterior region of male *L. loa* in subcutaneous tissue demonstrating testis (arrow) and intestine. x120

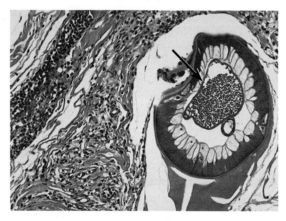

Figure 15.12
Transverse section through male *L. loa* showing sperm-filled vas deferens (arrow). The worm has provoked eosinophilia in surrounding subcutaneous tissue. x120

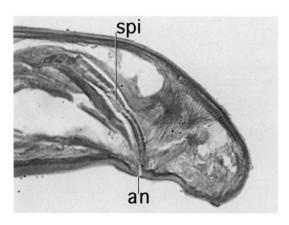

Figure 15.13
Longitudinal section through posterior tip of male *L. loa*, demonstrating anus (an) and 1 of 2 golden-brown spicules (spi). x395

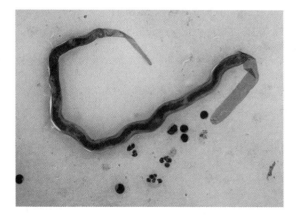

Figure 15.14
Microfilaria of *L. loa* in thick blood film stained by Delafield's hematoxylin, which prominently displays the sheath at both ends of the microfilaria. x425

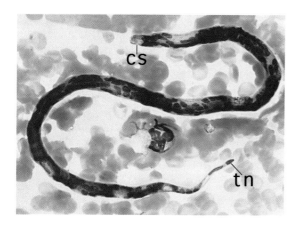

Figure 15.15
Microfilaria of *L. loa* in thin film of peripheral blood. Anterior end is blunted. Note 5-µm-long cephalic space (cs), and elongated terminal nucleus (tn) at tip of tail. Though this stain does not usually demonstrate it, the sheath occupies the apparent void immediately anterior to the microfilaria. Giemsa x695

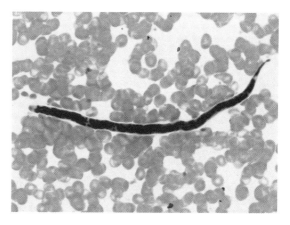

Figure 15.16
Thin blood film containing microfilaria of *L. loa* measuring only 180 µm in length. Microfilariae under 200 µm are uncommon. Giemsa x400

Figure 15.17
Left: Typical forest breeding site of *Chrysops* sp. *Right*: Abundant vegetation and mud necessary for development of fly larvae.

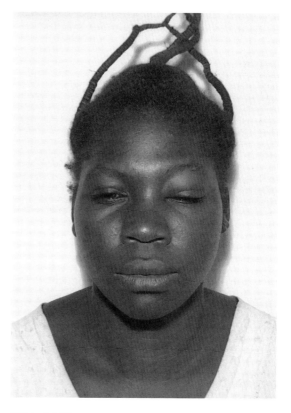

Figure 15.18
Congolese woman with Calabar swelling around left eye.

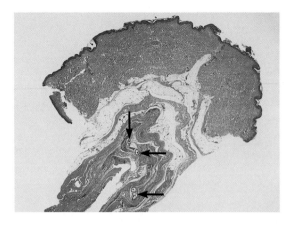

Figure 15.19
Subcutaneous tissue containing multiple sections of adult male *L. loa* (arrows). Worm was an incidental finding in a surgically removed lipoma. Note minimal host reaction, including eosinophils, lymphocytes, plasma cells, and mild edema. x3.7

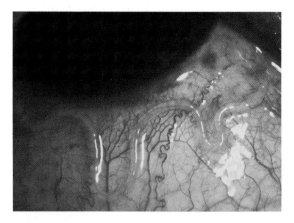

Figure 15.20
Loa loa migrating beneath ocular conjunctiva of a Cameroonian.*

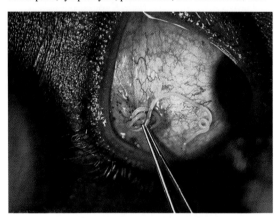

Figure 15.21
Surgical extraction of *L. loa* from subconjunctiva of a Ghanaian.*

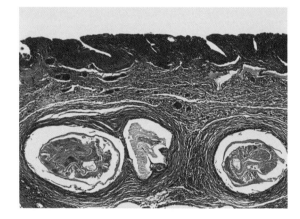

Figure 15.22
Three sections of dead female *L. loa* in subconjunctiva, surrounded by fibrosis. x60

Figure 15.23
Skin and subcutaneous tissue of 49-year-old woman who traveled extensively in Africa. Calabar swellings developed over elbow a few hours after initiating DEC therapy. Biopsy specimen revealed multiple sections of gravid *L. loa* in subcutaneous tissue. Although worm appears viable, DEC led to release of antigens that provoked extensive host reaction. x10.5

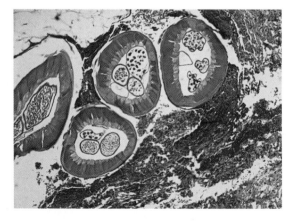

Figure 15.24
Higher magnification of subcutaneous lesion in Figure 15.23 shows gravid *L. loa* in fibrinopurulent hemorrhagic lesion. Inflammatory cells around worm are primarily neutrophils with scattered eosinophils. x40

Cerebral involvement is more frequent in loiasis than in other filarial infections. Microfilariae may appear in cerebrospinal fluid; in heavy infections, microfilariae block cerebral capillaries, causing encephalitis, occlusive cerebrovascular lesions, and even death. Encephalitis and hemiplegia are particularly frequent in heavily infected persons undergoing treatment with diethylcarbamazine (DEC).[6,37] Dead microfilariae in patients undergoing DEC treatment, and the resulting tissue reaction, probably cause obstruction of cerebral capillaries. Dead worms near or within major peripheral nerves can damage the nerves and cause paresthesia and focal paralysis.

Other signs include diffuse edema of the hand or forearm, generalized hives, fever, irritability, confusion, and jacksonian epilepsy. Loiasis may afflict the male genitalia, causing orchitis, scrotitis, funiculitis, and hydrocele. Rarely, patients develop nephropathy, cardiomyopathy, pulmonary infiltrates, and pleural effusion. Acute arthritis with effusion is an uncommon complication, and loiasis may aggravate psoriasis.[11,20,28,31,32,40]

Eosinophilia is common and may reach 50% to 80% ($10 \times 10^3/\mu l$); however, there is no correlation between the number of microfilariae in the blood and the level of eosinophilia.

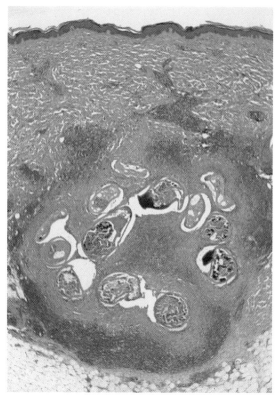

Figure 15.25
Skin from thigh of woman described in Figure 15.23 after a second course of DEC. This second lesion contained a dead gravid *L. loa* in subcutaneous tissue. Worm is partially calcified. Note fibrosis and inflammatory infiltrate made up of lymphocytes, plasma cells, and scattered eosinophils. x25

Pathologic Features

Calabar swellings show edema with lymphocytic and eosinophilic infiltration of the neighboring connective and perivascular tissues, and sometimes leukocytoclastic vasculitis.

Adult *L. loa* inhabit subcutaneous tissue of most parts of the body, including subconjunctiva (Fig 15.22), hernia sac, fingers, and scalp. Uncommonly, they enter the anterior chamber of the eye.[36] Migrating worms usually provoke only slight reactions (Fig 15.19). Dying worms, however, cause significant inflammation, and initial suppuration followed by granulomatous reaction and fibrosis (Figs 15.22 to 15.25). The fibrosis provoked by a dead adult worm in the spermatic cord region can produce a hydrocele (Figs 15.26 to 15.28). *Loa loa* sometimes localize to the peritoneum or, rarely, penetrate the wall of the bowel, causing peritonitis and even bowel obstruction (Fig 15.29).

Microfilariae of *L. loa* circulate in peripheral blood and are found within blood vessels of any tissue. Routine hematoxylin and eosin (H&E) stained sections readily reveal segments of microfilariae either in blood vessels or extravasated blood. In tissue sections they are approximately 5 µm wide. Morphologic features of anterior and posterior ends are diagnostic (Figs 15.30 and 15.31). The sheath is not usually visible on H&E-stained sections, but Warthin-Starry stain may reveal the sheath (Fig 15.32). Sections of skin occasionally reveal microfilariae in dermal vessels, especially in capillaries around sweat glands (Fig 15.33). There is usually a mild focal, chronic inflammatory cell infiltrate in the dermis and fibrosis of dermal papillae. Uncommonly, there is a mild dermal eosinophilia.

The lymph nodes from patients with *L. loa*

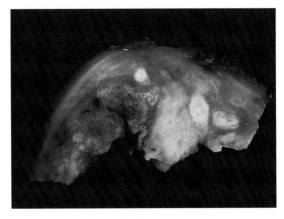

Figure 15.26
Tissue from scrotum of 47-year-old Congolese with bilateral hydroceles caused by loiasis. Dead worm appears grossly as 3 well-circumscribed areas of calcification.

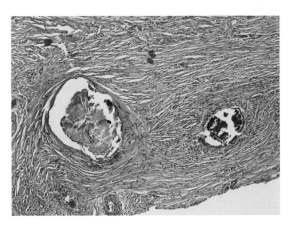

Figure 15.27
Section of calcified adult *L. loa* in tunica vaginalis testis surrounded by extensive fibrosis, from patient described in Figure 15.26. x65

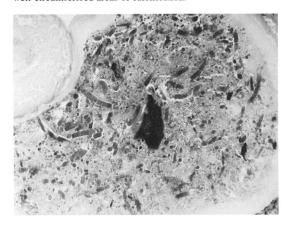

Figure 15.28
Section of calcified fragments of degenerating microfilariae of *L. loa* within dead worm shown in Figures 15.26 and 15.27. Giemsa x175

Figure 15.29
Biopsy specimen of colon from 48-year-old Congolese man with peritonitis and constricting lesion of the colon. Note adult male *L. loa* in thickened fibrotic wall of bowel. Near this section a sinus tract extended from the bowel wall through the peritoneum. x40

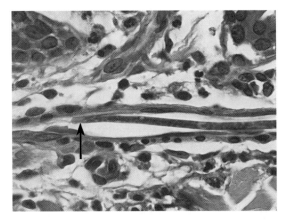

Figure 15.30
Anterior end of *L. loa* microfilaria in dermal capillary. Blunted anterior end is 5 μm wide, with a cephalic space (arrow) followed by paired nuclei. x520

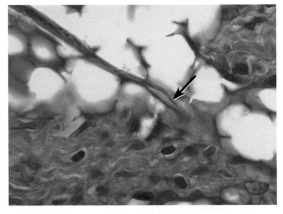

Figure 15.31
Posterior end of *L. loa* microfilaria in lymph node. Pointed tip of tail has elongated terminal nucleus (arrow). x650

lymphadenitis are 2 to 5 cm in largest dimension and soft (Fig 15.34), but are usually not clinically apparent. The 2 cardinal features are distension of subcapsular and medullary sinuses by macrophages (lysozyme positive and KP-1 [P3] positive), eosinophils, and a few lymphocytes, and atrophy of lymph node follicles (Fig 15.35). There may also be fibrous thickening of the capsule and trabeculae, with dilation of lymphatic vessels of the capsule and medulla. Prominent collections of mast cells, plasma cells, and Russell's bodies are also apparent.[33] Microfilariae are in lymphoid sinuses (Fig 15.36) and in blood vessels (Fig 15.37).

Focal inflammatory lesions may be seen in any tissue of the body including the brain (Figs 15.38 and 15.39), retina, spinal cord, heart (Fig 15.40), spleen, kidney (Fig 15.41), lung, and liver (Fig 15.42).[29] These lesions consist of degenerating microfilariae, surrounded by lymphocytes, macrophages, and occasional foreign body giant cells.

Diagnosis

Routine diagnosis of loiasis is made by examining peripheral blood for *L. loa* microfilariae. Because of *L. loa*'s diurnal periodicity, the best time to draw blood for diagnosis is between 10:00 AM and 3:00 PM. The specimen is hemolyzed with saponin, then concentrated by centrifugation or Nucleopore® filtration. Sometimes microfilariae in asymptomatic individuals are noted incidentally in peripheral blood smears, urine, sputum, cerebrospinal fluid, cervicovaginal smears, blood vessels in biopsy or autopsy specimens, and even in uncommon situations such as during oocyte retrieval.[39] The microfilariae are sheathed, but the Giemsa stain will not usually show the sheath (Fig 15.15). Smears stained with Delafield's hematoxylin readily reveal the sheath (Fig 15.14). The elongated terminal nucleus at the tip of the tail is a key diagnostic feature (Fig 15.15).

Occasionally, diagnosis is made by identifying adult worms removed from the eye. Microfilariae or adult worms are frequently identified in tissue sections. Tissue sections of Calabar swellings taken prior to therapy seldom reveal adult worms.

Travelers returning from endemic areas may have typical Calabar swellings and a high degree of

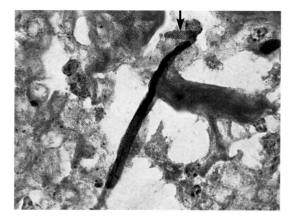

Figure 15.32
Section of lymph node showing blunted anterior end of *L. loa* microfilaria. This stain reveals sheath (arrow) and transverse striations of cuticle. Warthin-Starry x800

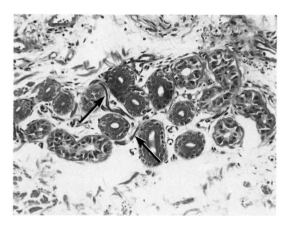

Figure 15.33
Microfilariae (arrows) of *L. loa* occasionally congregate within blood vessels adjacent to eccrine sweat glands. x160

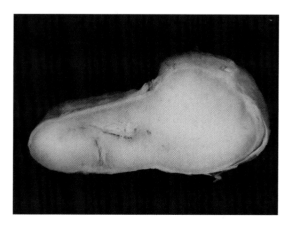

Figure 15.34
Greatly enlarged soft lymph node from patient with *L. loa* lymphadenitis. No grossly apparent discrete lesions are visible in cross section. x2.3

Figure 15.35
Lymph node of patient with *L. loa* lymphadenitis reveals atrophy of follicles, and sinuses dilated by numerous macrophages. x25

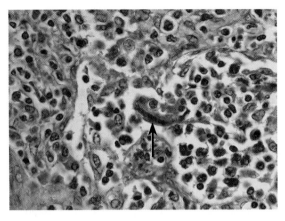

Figure 15.36
Numerous eosinophils and macrophages surround fragment of *L. loa* microfilaria (arrow) in lymph node sinus. x165

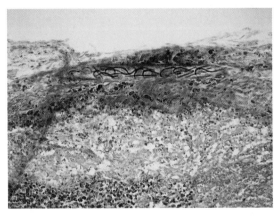

Figure 15.37
Tangled mass of *L. loa* microfilariae within vessel in capsule of lymph node. Giemsa x125

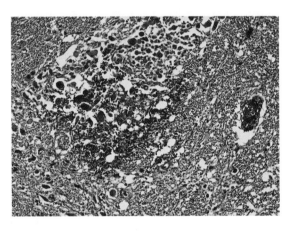

Figure 15.38
Brain of 45-year-old Congolese with loiasis who became comatose and died shortly after use of native medicine (suspected to contain DEC) and morphine. Note microinfarct near obstructed vessel. x115

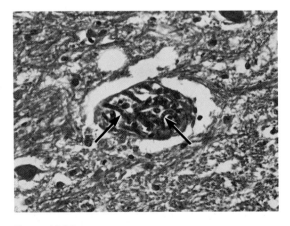

Figure 15.39
Cerebral vessel of patient described in Figure 15.38, showing numerous sections of degenerating microfilariae of *L. loa* (arrows) in a fibrin thrombus, and perivascular edema. x325

Figure 15.40
Inflammatory nodule around dying microfilariae of *L. loa* (arrows) in subendocardial myocardium. x150

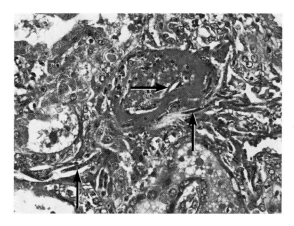

Figure 15.41
Edema and mild inflammation in kidney with degenerating microfilariae of *L. loa* (arrows) in blood vessels. x230

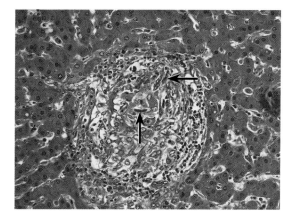

Figure 15.42
Liver with granuloma containing fragments of degenerating *L. loa* microfilaria (arrows). x175

eosinophilia, though peripheral blood examination fails to reveal microfilariae.[38] Skin tests and complement fixation tests using dirofilarial antigens are usually positive in these patients; however, there are frequent cross reactions with other filarial infections.

Molecular diagnostic tests are under development and an *L. loa*-specific repetitive DNA sequence has been identified.

Treatment and Prevention

Since its discovery in 1947, DEC has been the treatment of choice for loiasis. The routine regimen is 2 mg/kg body weight 3 times a day for 21 days. Higher doses (8 to 10 mg/kg/day for 21 days) are well-tolerated by amicrofilaremic persons, which includes most foreigners infected while visiting endemic areas for prolonged periods. Others advocate gradually increasing dosages combined with antihistamines or corticosteroids to reduce the risk of serious side effects. It is imperative that DEC be withdrawn if neurological disorders develop. Filariopheresis to reduce the microfilaremia may decrease the risk of encephalitis during DEC therapy.[4] DEC treatment is curative in about 65% of nonendemic patients, but they may require up to 4 21-day courses. Most relapses develop within the first 12 months after treatment, but patients may relapse as late as 8 years after therapy.[22]

Newer drugs, such as mebendazole and ivermectin, that are effective against other filarial nematodes initially showed little or no efficacy against *L. loa*.[5,7] Recent reports, however, suggest that a single high dose of ivermectin can reduce microfilaremia for at least 3 months, decreasing transmission.[9,25] Ivermectin may also have a macrofilaricidal effect.[26] Hovette et al suggest that ivermectin is superior to DEC in treating patients with loiasis who are amicrofilaremic.[19] In high doses, mebendazole is active against both microfilariae and adult *L. loa*, but intestinal absorption is poor and variable, and adverse effects are frequent. Albendazole, a more reliably absorbed benzimidazole with fewer side effects, appears to kill young adult worms and may have an embryotoxic effect on developing microfilariae.[21] Suramin kills adult *L. loa* but toxicity prohibits its use.

Controlling loiasis depends on chemoprophylaxis with DEC, and eliminating insect vectors by larval insecticides.[15,30] Both methods have limited usefulness. Clearing vegetation around dwellings, screening houses against insects, and wearing protective clothing can create a barrier against the vector.

References

1. Argyll-Robertson D. Case of Filaria loa in which the parasite was removed from under the conjunctiva. *Trans Ophthalmol Soc UK* 1895;15:137–167.

2. Arrachart JN. Mémoires, Dissertations et observations de Chirurgie, Paris. (Presented in 1778.) *Academy of Surgery in Paris* 1805;228–233.

3. Bajon B. *Mémoires pour servir á l'histoire de Cayenne et de la Guiane française*. Vols 1&2. Paris, France; 1777–1778.

4. Bouree P, Duedari N, Bisaro F, Norol F. Value of filariopheresis in the treatment of Loa loa filariasis [in French]. *Pathol Biol (Paris)* 1993;41:410–414.

5. Burchard GD, Kern P. Failure of high dose mebendazole as a microfilaricide in patients with loiasis. *Trans R Soc Trop Med Hyg* 1987;81:420.

6. Carme B, Boulesteix J, Boutes H, Puruehnce MF. Five cases of encephalitis during treatment of loiasis with diethylcarbamazine. *Am J Trop Med Hyg* 1991;44:684–690.

7. Carme B, Ebikili B, Mbitsi A, Copin N. Therapeutic trial with ivermectin in loiasis with medium and high microfilaremia [in French]. *Ann Soc Belg Med Trop* 1991;71:47–50.

8. Castellani A, Chalmers AJ. *Manual of Tropical Medicine*. 2nd ed. New York, NY: William Wood & Co; 1913:1747.

9. Chippaux JP, Ernould JC, Gardon J, Gardon-Wendel N, Chandre F, Barbier N. Ivermectin treatment of loiasis. *Trans R Soc Trop Med Hyg* 1992;86:289.

10. Connal A, Connal SL. The development of Loa loa (Guyot) in Chrysops silacea (Austen) and in Chrysops dimidiata (Van der Wulp). *Trans R Soc Trop Med Hyg* 1922;16:64–89.

11. de la Herran Herrera A, Gonzalez Garrido EA, Pila Perez R, Leon Diaz R. Myocardiopathy in a patient with loiasis. Presentation of a case [in Spanish]. *Rev Cubana Med Trop* 1981;33:201–206.

12. Duke BO. Studies on the biting habits of Chrysops. II. The effect of wood fires on the biting density of Chrysops silacea in the rain-forest at Kumba, British Cameroons. *Ann Trop Med Parasitol* 1955;49:260–272.

13. Duke BO. Studies on the biting habits of Chrysops. III. The effect of groups of persons, stationary and moving, on the biting density of Chrysops silacea at ground level in the rain-forest at Kumba, British Cameroons. *Ann Trop Med Parasitol* 1955;49:362–367.

14. Duke BO. Studies on loiasis in monkeys. II. The population dynamics of the microfilariae of Loa in experimentally infected drills (Mandrillus leucophaeus). *Ann Trop Med Parasitol* 1960;54:15–31.

15. Duke BO. Studies on the chemoprophylaxis of loiasis. II. Observation on diethylcarbamazine citrate (Banocide) as a prophylactic in man. *Ann Trop Med Parasitol* 1963;57:82–96.

16. Duke BO. Studies on loiasis in monkeys. IV. Experimental hybridization of the human and simian strains of Loa. *Ann Trop Med Parasitol* 1964;58:390–408.

17. Duke BO, Wijers DJ. Studies on loiasis in monkeys. I. The relationship between human and simian Loa in the rain-forest zone of the British Cameroons. *Ann Trop Med Parasitol* 1958;52:158–175.

18. Fain A. Epidémiologie et pathologie de la loase. *Ann Soc Belg Med Trop* 1981;61:277–285.

19. Hovette P, Debonne JM, Touze JE, et al. Efficacy of ivermectin treatment of Loa loa filariasis patients without microfilaremias. *Ann Trop Med Parasitol* 1994;88:93–94.

20. Klion AD, Eisenstein EM, Smirniotopoulos TT, Neumann MP, Nutman TB. Pulmonary involvement in loiasis. *Am Rev Respir Dis* 1992;145:961–963.

21. Klion AD, Massougbodji A, Horton J, et al. Albendazole in human loiasis: results of a double-blind, placebo-controlled trial. *J Infect Dis* 1993;168:202–206.

22. Klion AD, Ottesen EA, Nutman TB. Effectiveness of diethylcarbamazine in treating loiasis acquired by expatriate visitors to endemic regions: long-term follow-up. *J Infect Dis* 1994;169:604–610.

23. Leiper RT. Report of the helminthologist, London School of Tropical Medicine, for the half-year ending April 30th 1913. Report to the Advisory Committee of the Tropical Diseases Research Fund. Abstracted in *Trop Dis Bull* 1913;2:195–196.

24. Looss A. Zur Kenntnis des Baues der Filaria loa Guyot. *Zool Jahrb, XX. Abth Syst* 1904;20:549–574.

25. Martin-Prevel Y, Cosnefroy JY, Ngari P, Pinder M. Reduction of microfilaraemia with single high-dose ivermectin in loiasis [letter]. *Lancet* 1993;342:442.

26. Martin-Prevel Y, Cosnefroy JY, Tshipamba P, Ngari P, Chodakewitz JA, Pinder M. Tolerance and efficacy of single high-dose ivermectin for the treatment of loiasis. *Am J Trop Med Hyg* 1993;48:186–192.

27. Mongin. Sur un ver trouvé sous la conjonctive, á Maribarou, isle Saint-Domingue. *Journal de Médecine, Chirurgie, Pharmacie etc.* 1770;32:338–339.

28. Morel L, Delaude A, Girard M, Bouzekri M. Eosinophilic pulmonary infiltrates in filariasis of the Loa loa type [in French]. *Poumon Coeur* 1967;23:685–694.

29. Negesse Y, Lanoie LO, Neafie RC, Connor DH. Loiasis: "Calabar" swellings and involvement of deep organs. *Am J Trop Med Hyg* 1985;34:537–546.

30. Nutman TB, Miller KD, Mulligan M, et al. Diethylcarbamazine prophylaxis for human loiasis. Results of a double-blind study. *N Engl J Med* 1988;319:752–756.

31. Nutman TB, Miller KD, Mulligan M, Ottesen EA. Loa loa infection in temporary residents of endemic regions: recognition of a hyperresponsive syndrome with characteristic clinical manifestations. *J Infect Dis* 1986;154:10–18.

32. Osamulia-Soendjojo N, Prens EP, Sluiters JF, Naafs B. Psoriasis and filariasis. *Br J Dermatol* 1994;131:723–724.

33. Paleologo FP, Neafie RC, Connor DH. Lymphadenitis caused by Loa loa. *Am J Trop Med Hyg* 1984;33:395–402.

34. Rakita RM, White AC Jr, Kielhofner MA. Loa loa infection as a cause of migratory angioedema: report of three cases from the Texas Medical Center. *Clin Infect Dis* 1993;17:691–694.

35. Sasa M. *Human Filariasis: A Global Survey of Epidemiology and Control*. Baltimore, Md: University Park Press; 1976.

36. Satyavani M, Rao KN. Live male adult Loa loa in the anterior chamber of the eye, a case report. *Indian J Pathol Microbiol* 1993;36:154–157.

37. Stanley SL Jr, Kell O. Ascending paralysis associated with diethylcarbamazine treatment of M. loa loa infection. *Trop Doct* 1982;12:16–19.

38. Van Dellen RG, Ottesen EA, Gocke TM, Neafie RC. Loa loa. An unusual case of chronic urticaria and angioedema in the United States. *JAMA* 1985;253:1924–1925.

39. Wisanto A, Laureys M, Camus M, Devroey P, Verheyen G, Van Steirteghem AC. Loa loa microfilariae aspirated during oocyte retrieval. *Hum Reprod* 1993;8:2096–2097.

40. Zuidema PJ. Renal changes in loiasis. *Folia Med Neerl* 1971;14:168–172.

16

Dirofilariasis

Aileen M. Marty *and*
Ronald C. Neafie

Introduction

Definition

Dirofilariasis is infection by a filarial nematode of the genus *Dirofilaria*. Humans are aberrant hosts for *Dirofilaria* sp. In humans the worms usually die before maturing, provoking a focal granulomatous reaction in subcutaneous tissue, or small pulmonary infarcts. There are 2 subgenera, *Dirofilaria Dirofilaria* and *Dirofilaria Nochtiella*, based on the absence (*Dirofilaria*) or presence (*Nochtiella*) of external longitudinal cuticular ridges. *Dirofilaria Dirofilaria immitis* is the only *D. Dirofilaria* sp that infects humans, and hereafter will be referred to as *D. immitis*. *Dirofilaria Nochtiella* sp that infect humans will be designated by the genus *Dirofilaria* and the species.

Approximately 20 species of *Dirofilaria* are in the subgenus *Nochtiella*, but only *Dirofilaria repens*, *Dirofilaria tenuis*, and *Dirofilaria striata* definitely infect humans. *Dirofilaria ursi*-like filariae, representing either *Dirofilaria ursi*, *Dirofilaria subdermata*, or both, also infect humans.

Synonyms

Dirofilaria immitis (Leidy, 1856), the dog heartworm, has been known as *Filaria sanguinis*, *Filaria immitis*, and *Dirofilaria louisianensis*. Older names for *D. repens* (Railliet and Henry, 1911), the subcutaneous filaria of dogs and cats, include *Filaria conjunctivae* and *Filaria acutuscula*. Four species of *D. Nochtiella* were once commonly called *Dirofilaria conjunctivae*: *D. tenuis* (Chandler, 1942), *D. ursi* (Yamaguti, 1941), *D. subdermata* (Mönnig, 1924), and *D. striata* (Molin, 1858). In the United States, *D. tenuis* (and *D. repens* in Europe, Sri Lanka, and Russia) is still frequently referred to as *D. conjunctivae*.[6]

General Considerations

Bebesiu described the first case of dirofilariasis in humans, in the gastrosplenic ligament of a Hungarian woman in 1879.[4] In 1911 Railliet and Henry recognized features that distinguished a group of worms as a new genus, *Dirofilaria*. Most of the species in this new genus were previously in the genus *Filaria*. Faust, in 1937, proposed splitting

Dirofilaria into 2 subgenera. Four years later, Faust et al first diagnosed dirofilariasis in humans in the United States. Their patient was a woman from New Orleans with a male *D. immitis* in the inferior vena cava near the heart.[9] Increasing numbers of *Dirofilaria* infections were reported in the United States in 1957, as the medical community became generally aware of the disease.[7] Dirofilariasis is an emerging zoonosis. Because of the rarity of symptoms, human dirofilariasis is probably more common than now documented; however, advances in early diagnosis and greater awareness in the medical community have increased detection rates.

Epidemiology

Dirofilaria immitis causes pulmonary dirofilariasis in humans. This heartworm commonly infects dogs throughout the world, from the tropics to the temperate regions, but can infect other mammals as well, including cats, foxes, sea lions, wolves, and otters.[2] In the United States, *D. immitis* is prevalent in the southern and southeastern states, but is found as far north as Massachusetts, Michigan, and Minnesota. *Dirofilaria immitis* is rare in northern Europe, but common in Italy and other Mediterranean countries. Other enzootic areas of *D. immitis* infection are Africa, Australia, China, Japan, Southeast Asia, and the Pacific Islands. Humans are most often infected in the United States, Japan, Europe, and Australia.

Dirofilaria tenuis is a subcutaneous parasite of the raccoon in the United States. *Dirofilaria repens* is a subcutaneous parasite of dogs and cats in Europe, Africa, and Asia, and accounts for most human infections in these areas. *Dirofilaria ursi* parasitizes bears in Canada, northern United States, and Japan,[1] and *D. subdermata* infects porcupines in North America. There are 5 published reports of human infection by *D. ursi* along the U.S.-Canadian border.[12] Presumably, both *D. ursi* and *D. subdermata* cause disease in humans in North America. Because of the difficulty in distinguishing female *D. ursi* from female *D. subdermata*, as they present in the immature necrotic state in humans, Beaver et al[3] applied the designation *D. ursi*-like to these worms. *Dirofilaria striata* infects

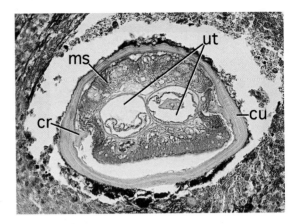

Figure 16.1
Lung specimen with transverse section of immature female *D. immitis,* demonstrating paired uteri (ut), cuticle (cu), muscle (ms), and internal longitudinal cuticular ridges (cr). Movat x200

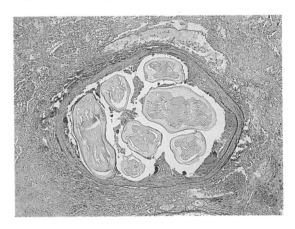

Figure 16.2
Several transverse sections of coiled, immature male *D. immitis* in pulmonary vessel. Worm has prominent musculature. Movat x40

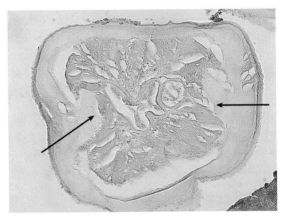

Figure 16.3
Transverse section of degenerating, immature male *D. immitis* in lung, showing single reproductive tube and intestinal duct. Worm has thick, multilayered cuticle, internal longitudinal cuticular ridges (arrows), and prominent somatic musculature. Movat x225

Species	Females length/diameter	Males length/diameter	Cuticle	External longitudinal cuticular ridges*
D. immitis†	230 to 310 mm 350 µm	120 to 190 mm 300 µm	5 to 25 µm thick. Fine transverse striations 2 to 7 µm apart	None
D. tenuis	80 to 130 mm 260 to 360 µm	40 to 48 mm 190 to 260 µm	5 to 15 µm thick. Fine transverse striations 2 to 7 µm apart	Female: 81 to 115 ridges at midbody. Male: 60 to 81 ridges at midbody that are 1 to 4 µm tall and spaced irregularly at 7-to-12-µm intervals
D. repens	100 to 170 mm 460 to 650 µm	50 to 70 mm 370 to 450 µm	12 to 20 µm thick. Course transverse striations	95 to 105 ridges at midbody, 3 to 4 µm tall, spaced at 12-to-20-µm intervals
D. ursi	117 to 224 mm 460 to 700 µm	51 to 86 mm 330 to 480 µm	5 to 10 µm thick. Fine transverse striations	Female: 70 to 72 ridges at midbody, spaced 10 to 25 µm apart. Male: 58 to 62 ridges at midbody, spaced 14 to 30 µm apart
D. subdermata	117 to 185 mm 440 to 660 µm	41 to 66 mm 280 to 420 µm	13 to 16 µm thick. Course transverse striations 7 µm apart	Female: 70 to 92 ridges at midbody Male: 68 to 77 ridges at midbody
D. striata	250 to 360 mm 440 to 500 µm	80 to 120 mm 350 to 380 µm		Small, inconspicuous, except at lateral edge where they mimic lateral alae

* There is great variation within these species. Spacing and number of ridges are approximations.
† Maximum diameter in dog is 1 to 2 mm.

Table 16.1
Morphologic features of *Dirofilaria* sp infecting humans.

wild cats such as bobcats, Florida panthers, ocelots, and margays in North and South America. In humans, infection with *D. striata*, while rare, produces subcutaneous lesions.

Infectious Agent

Morphologic Description

Adult female *D. immitis* in the definitive host, at 230 to 310 mm long and 1 to 2 mm in diameter, are longer than males (120 to 190 mm), but have roughly the same diameter. Females have 2 sets of tubular reproductive organs (Fig 16.1) consisting of ovaries, oviducts, seminal receptacles, and uteri. The uteri join in the anterior portion of the worm to form a single vagina that terminates in the vulva. The male reproductive system consists of a single tube (Figs 16.2 and 16.3), beginning anteriorly with a testis and continuing posteriorly as the seminal vesicle, vas deferens, and ejaculatory duct. The ejaculatory duct joins with the intestine to form the cloaca at the posterior end of the worm. In both sexes, the digestive tube is a single, narrow tube divided into a short anterior esophagus and a long intestine. All *Dirofilaria* sp that infect humans have a multilayered, striated cuticle and prominent lateral cords. Multilayering is most obvious in the lateral cord region. The cuticle of all *Dirofilaria* bulges inward in the area of the lateral cords, forming internal longitudinal ridges that vary considerably in prominence. Many rows of large muscle cells in each quadrant project far into the body cavity. Each muscle cell has contractile fibers on 3 sides. The 2 subgenera of *Dirofilaria* are distinguished by the presence (*Nochtiella*) or absence (*Dirofilaria*) of external longitudinal cuticular ridges (Table 16.1) (see also Tables 13.1 and 13.2).

Only immature worms lodge in human pulmonary vessels. These worms are 100 to 350 µm in diameter and are nearly always partially degenerated in tissue sections.[8] The thick (5 to 25 µm), multilayered cuticle of *D. immitis* has transverse striations 2 to 7 µm apart (Fig 16.4). Lateral cords are usually poorly preserved and difficult to identify. Somatic muscle is abundant and prominent. Immature worms have no ova, spermatozoa, or microfilariae.

The lateral cords of the *Nochtiella* subgenus extend into the body cavity and are divided into sublaterals. In tissue sections, the different

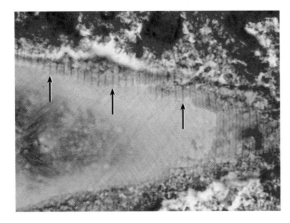

Figure 16.4
Lung specimen containing immature female *D. immitis* with transverse striations on cuticular surface (arrows). Masson x810

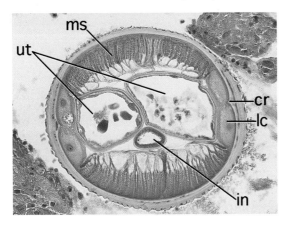

Figure 16.5
Transverse section through midbody of nongravid female *D. tenuis* in subcutaneous tissue. Note internal longitudinal cuticular ridges (cr), lateral cords (lc), prominent somatic musculature (ms), paired uteri (ut), and intestine (in). Movat x200

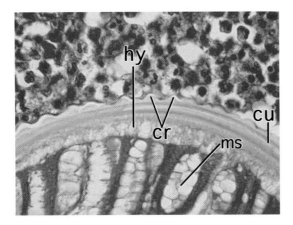

Figure 16.6
Transverse section of nongravid female *D. tenuis* in subcutaneous tissue, illustrating external longitudinal cuticular ridges (cr), laminated cuticle (cu), hypodermis (hy), and somatic musculature (ms). x660

D. Nochtiella sp can sometimes be differentiated by the size, number, and spacing of the external longitudinal cuticular ridges.[12] Intraspecies variation is marked, but usually such worms can be identified only as members of the *Nochtiella* subgenus.

Adult female *D. tenuis* are 80 to 130 mm long and 260 to 360 μm in diameter, and males are 40 to 48 mm long and 190 to 260 μm in diameter. The cuticle in both sexes is 5 to 15 μm thick (Figs 16.5 and 16.6). The ridges of *D. tenuis* are usually distinctive, being low and rounded, with a wavy, broken, branching pattern. The space between each ridge and the height of each ridge varies (Table 16.1). In some transverse sections, especially in degenerated worms, external cuticular ridges are difficult to identify. External ridges may be indiscernible, even in a section of well-preserved *D. tenuis* (Fig 16.7); however, additional sections may reveal ridges. The surface of the cuticle contains transverse striations 2 to 7 μm apart. In both sexes the lateral cords are divided by internal cuticular ridges (Fig 16.8). At midbody the female contains an intestine and 2 uteri that rarely contain microfilariae. One patient, a 39-year-old man from Illinois, had a gravid worm in his right thigh[14]; another patient, a 47-year-old man from Florida, had a gravid worm in his left abdominal wall.[21] In neither case were viable microfilariae found outside the worms.

Adult female *D. repens* are 100 to 170 mm long and 460 to 650 μm in diameter, and males are 50 to 70 mm long and 370 to 450 μm in diameter.[16] The cuticle is 12 to 20 μm thick in both sexes and contains external longitudinal cuticular ridges 3 to 4 μm in height, regularly spaced at approximately 12-to-20-μm intervals (Figs 16.9 and 16.10). At midbody there are 95 to 105 external longitudinal ridges[15] that appear beaded because of the transverse striations of the cuticle.[11] Each sublateral cord has an irregular row of nuclei confined to the basal zone near the cuticle. Somatic muscle cells extend deep into the body cavity. The female worm is usually nongravid, but a unique report describes a gravid *D. repens* recovered from the subcutaneous tissue of a 53-year-old Italian woman.[23]

Adult female *D. ursi* are 117 to 224 mm long and 460 to 700 μm in diameter. Male *D. ursi* are 51 to 86 mm long and 330 to 480 μm in diameter. The

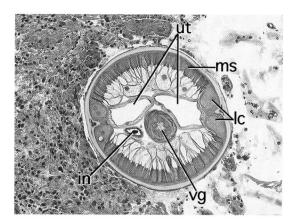

Figure 16.7
Transverse section of nongravid female *D. tenuis* in subcutaneous tissue. External longitudinal cuticular ridges are not apparent at this level. Note distinctive lateral cords (lc), prominent somatic musculature (ms), paired uteri (ut), muscular vagina (vg), and intestine (in). Movat x145

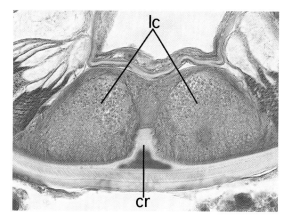

Figure 16.8
Nongravid female *D. tenuis* in subcutaneous tissue. Thick, multilayered cuticle forms internal longitudinal cuticular ridge (cr) that divides broad lateral cord (lc) into sublaterals. x595

cuticle is 5 to 10 µm thick. In the female, there are 70 to 72 external longitudinal cuticular ridges spaced 10 to 25 µm apart (Figs 16.11 and 16.12).[1] In the male there are 58 to 62 external longitudinal cuticular ridges spaced 14 to 30 µm apart. The external longitudinal cuticular ridges of *D. ursi* are usually fewer and farther apart than those of the other *D. Nochtiella* sp.[30]

Adult female *D. subdermata* are 117 to 185 mm long by 440 to 660 µm in diameter, and males are 41 to 66 mm long by 280 to 420 µm in diameter. The cuticle of *D. subdermata* in tissue sections is 13 to 16 µm thick and forms prominent external longitudinal ridges (Figs 16.13 and 16.14). The number of these ridges varies with the level of the worm, but most females have 70 to 92, and males 68 to 77.[13] External longitudinal cuticular ridges are absent in the male worm anterior to the nerve ring, and in the female worm anterior to the vulva. Terminal ends of both sexes have external longitudinal cuticular ridges, and prominent lateral cords in the anterior end. Somatic muscle is prominent at all levels of the worm.

In their natural hosts, adult female *D. striata* are 250 to 360 mm long and 440 to 500 µm in diameter. Adult males are 80 to 120 mm long and 350 to 380 µm in diameter.[19] The cuticle of *D. striata* has weakly developed, irregular, external longitudinal ridges, transverse striations, and minute, rounded lateral alae. A specimen recovered by

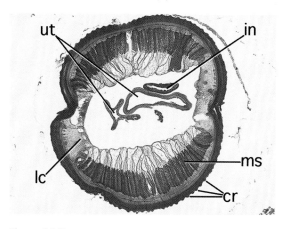

Figure 16.9
Transverse section of nongravid female *D. repens* in subcutaneous tissue. Worm has external longitudinal cuticular ridges (cr), distinctive lateral cords (lc), prominent somatic musculature (ms), paired uteri (ut), and intestine (in). Movat x120

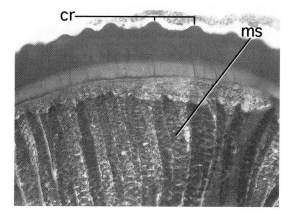

Figure 16.10
Section of nongravid female *D. repens* in subcutaneous tissue. Note prominent external longitudinal cuticular ridges (cr), multilayered cuticle, and somatic musculature (ms). Movat x625

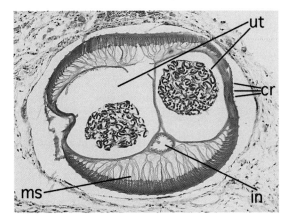

Figure 16.11
Transverse section of gravid *D. ursi* from subcutaneous tissue of black bear from Canada. Note pronounced external longitudinal cuticular ridges (cr), prominent somatic musculature (ms), paired uteri filled with microfilariae (ut), and intestine (in). x95

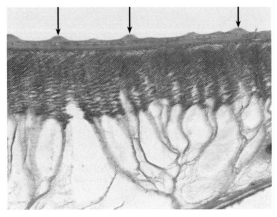

Figure 16.12
Higher magnification of *D. ursi* from Figure 16.11. Note external longitudinal cuticular ridges (arrows) 14 to 30 µm apart. x590

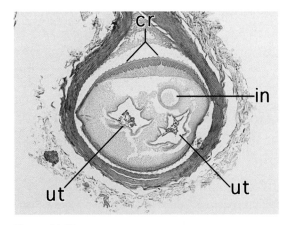

Figure 16.13
Transverse section of gravid *D. subdermata* from subcutaneous tissue of porcupine. Note prominent external longitudinal cuticular ridges (cr), paired uteri (ut), and intestine (in). Movat x60

Orihel and Isbey from the periorbital lesion of a 6-year-old boy measured 220 mm long and 420 µm in diameter. The anterior end of the worm tapered slightly forward of the vulva.[20]

Life Cycle and Transmission

All species of *Dirofilaria* have insects as vectors and intermediate hosts and mammals as definitive hosts. Insects ingest microfilariae from the mammalian host. Within the insect, microfilariae shed their sheaths and become short, thick, sausage-shaped L1 larvae. These larvae mature through 2 molts to the L2 and L3 stages. The L3 stage is the infective filariform larva that enters the definitive mammalian host and undergoes 2 further molts to the L4 stage and, finally, to the adult. *Dirofilaria immitis* adults live in the lumen of the right ventricle of the heart, where gravid females produce microfilariae that circulate in peripheral blood.

In humans, however, immature *D. immitis* usually die on reaching the right ventricle and are swept into the pulmonary artery, eventually lodging in the smaller pulmonary vessels. Rarely, L3- or L4-stage larvae of *D. immitis* die in subcutaneous tissue before they mature, and cause subcutaneous dirofilariasis.

Adult members of the *Nochtiella* subgenus live and reproduce in the subcutaneous tissue of their normal animal hosts, but usually die in human subcutaneous tissue. Though it may be fully mature, there is usually only a single worm, precluding the production of microfilariae. In rare instances, *D. repens* migrate into the circulatory system and reach pulmonary vessels, where they cause pulmonary dirofilariasis.[22]

At least 15 species of mosquitoes serve as vectors for all species of *Dirofilaria* except *D. ursi*.[5,28] Mosquito susceptibility varies within and among species. Temperature and the availability of breeding habitat affect the dynamics of parasite transmission. The most important vectors are: North America: *Aedes sierrensis*, *Aedes sollicitans*, *Aedes taeniorhynchus*, and *Anopheles bradleyi*[24,29]; Oceania: *Aedes polynesiensis* and *Aedes samoanus*[27]; and Australia: *Aedes notoscriptus*.[25,26] Black flies *(Simulium* sp, especially *Simulium venustum)* are vectors for *D. ursi*.

Clinical Features and Pathogenesis

Mosquitoes deposit *D. immitis* larvae in human subcutaneous tissue, which may provoke lesions. Some larvae migrate to the heart and die. Dead worms produce infarcts when they lodge in pulmonary vessels. Small pulmonary infarcts are usually asymptomatic and appear in routine chest roentgenograms as spherical "coin lesions" (Fig 16.15). There is usually only a single lesion. Following embolization, some patients experience chest pain, cough, hemoptysis, fever, chills, malaise, and mild eosinophilia. There are several reports from Europe of *D. repens* causing pulmonary dirofilariasis, wherein the clinical and pathologic changes in the lung are the same as those caused by *D. immitis*.[22]

Patients infected with *D. Nochtiella* sp usually present with inflamed subcutaneous nodules that are painful, erythematous, and sometimes migratory. Worms localize most frequently in the tissue of the orbit, scrotum, breast, arm, or leg.[17] The most common sites of ocular involvement are the subcutaneous tissue of the eyelid and periorbital region. Subconjunctival involvement is uncommon (Fig 16.16).[10] Occasionally the worm or portions thereof can be extracted from these lesions (Fig 16.17).

Pathologic Features

Pulmonary dirofilariasis usually presents as a sharply defined lesion in the periphery of the lung (Figs 16.18 and 16.19). Lesions show a recent, organizing, or fully organized infarct with an obliterated, medium-sized artery within the area of necrosis (Fig 16.20). Typically, there is a central zone of ischemic necrosis.[18] A single necrotic, degenerating, and sometimes partially calcified immature worm (Figs 16.21 and 16.22) is usually found within the lumen of a vessel, surrounded by obliterative thrombosis and fibroblastic proliferation (Fig 16.23). Around the necrosis is a narrow zone of granulomatous reaction composed of epithelioid cells, plasma cells, lymphocytes, and giant cells. A fibrous wall encases this reaction (Fig 16.24).

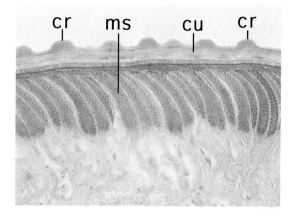

Figure 16.14
Higher magnification of *D. subdermata* in Figure 16.13, demonstrating external longitudinal cuticular ridges (cr), thick multilayered cuticle (cu), and somatic musculature (ms). Movat x560

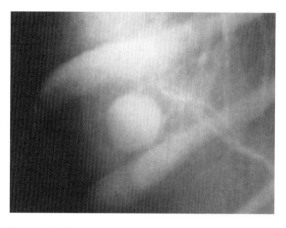

Figure 16.15
Roentgenogram of chest showing spherical coin lesion in right lower lung of patient with pulmonary dirofilariasis.

Figure 16.16
Female *D. tenuis* (arrow) in subconjunctiva of 66-year-old man from North Carolina.

Figure 16.17
Adult nongravid *D. repens* extracted from spermatic cord of 9-month-old boy living in Greece. Worm is 160 mm long and 650 μm in diameter. Patient's right hemiscrotum was swollen, painful, and red.

Figure 16.18
Section of lung with pulmonary dirofilariasis. Note characteristic spherical lesion with necrotic center surrounded by thick fibrous wall. Lesion is 17 mm in diameter.

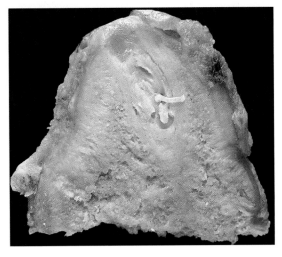

Figure 16.19
Section through necrotic lung from patient with pulmonary dirofilariasis, depicting fragment of immature *D. immitis* protruding from necrotic vessel.

Lesions of subcutaneous dirofilariasis usually contain a single coiled worm (Figs 16.25 and 16.26) that may be viable, degenerated, or dead. In an early lesion, where the worm is still well-preserved, the zone immediately around the worm contains necrotic debris surrounded by an abscess made up largely of neutrophils and eosinophils (Fig 16.26). Worms in subcutaneous fat provoke an eosinophilic panniculitis (Figs 16.27 to 16.29). Older lesions containing a degenerated or partially calcified worm show granulomatous inflammation consisting of epithelioid cells, giant cells, macrophages, lymphocytes, and eosinophils (Figs 16.30 and 16.31). Eosinophils are most numerous in the periphery of the lesion (Fig 16.32).

Diagnosis

Diagnosis of pulmonary dirofilariasis is made by identifying the worm in surgical or autopsy specimens. A degenerating nematode measuring 100 to 350 μm in diameter within an artery in a pulmonary infarct permits a presumptive diagnosis of *D. immitis*. Carcinoma, tuberculosis, fungal infections, and hamartomas should be considered in the differential diagnosis of a coin lesion in the lung. A negative serologic test for filariasis helps rule out a filarial origin for such a lesion, but a positive filarial test does not exclude carcinoma. Diagnosis of subcutaneous dirofilariasis caused by *D. Nochtiella* sp is made by identifying the

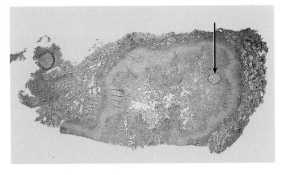

Figure 16.20
Section of lung demonstrating immature male *D. immitis* coiled within pulmonary vessel (arrow) and surrounded by necrosis. Figures 16.2 and 16.3 show higher magnifications of same worm. Movat x2.8

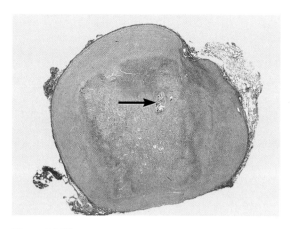

Figure 16.21
Partially calcified, degenerating, immature male *D. immitis* (arrow) within fibronecrotic nodule in lung. x4

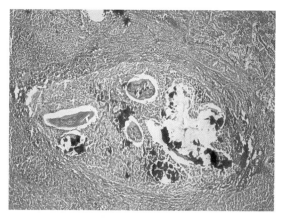

Figure 16.22
Higher magnification of worm in Figure 16.21, demonstrating fragments of partially calcified and degenerated worm. x40

worm within the biopsy specimen or, less commonly, by extracting the worm from a lesion.

Treatment and Prevention

The only treatment for dirofilariasis in humans is surgical removal of the lesion or extraction of the worm. Ivermectin is the treatment of choice for dogs with heartworm. Routine protection against mosquito and fly bites helps prevent infection. Deworming pets, especially dogs, protects against *D. immitis* and *D. repens* infection in humans.

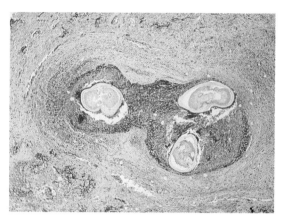

Figure 16.23
Section of lung with pulmonary dirofilariasis. Note 3 transverse sections of immature female *D. immitis* in pulmonary artery, and intense fibroblastic proliferation leading to occlusion of artery. Movat x35

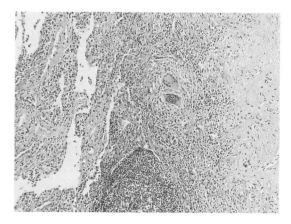

Figure 16.24
Section of lung from patient with pulmonary dirofilariasis, demonstrating zone of granulomatous inflammation containing giant cells and epithelioid cells at periphery of lesion. x40

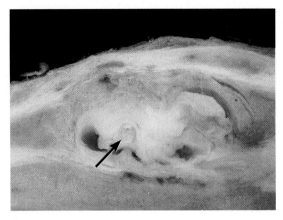

Figure 16.25
Abscess containing single coiled female *D. repens,* from patient with subcutaneous dirofilariasis. Note transverse section of worm (arrow).

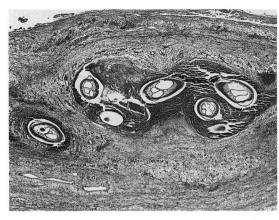

Figure 16.26
Section of lesion illustrated in Figure 16.25. Within abscess are several transverse sections of coiled, viable nongravid female *D. repens.* x12

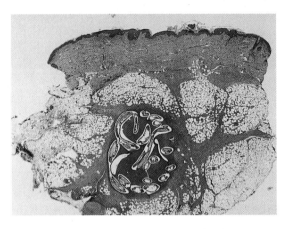

Figure 16.27
Single coiled female *D. tenuis* causing subcutaneous abscess. x6

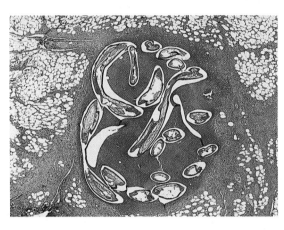

Figure 16.28
Higher magnification of worm in Figure 16.27, demonstrating single, coiled, viable nongravid female *D. tenuis*. Note extensive panniculitis. x11

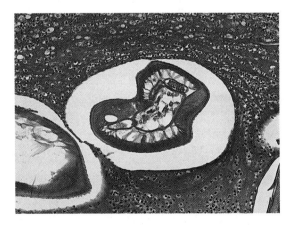

Figure 16.29
Higher magnification of Figure 16.28, showing transverse section of worm surrounded by neutrophils. External longitudinal cuticular ridges are present, but not prominent. x95

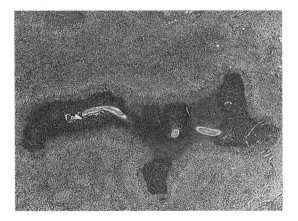

Figure 16.30
Submucosal tissue of lip, demonstrating degenerating, nongravid female *D. tenuis* provoking necrotizing granulomatous reaction. x30

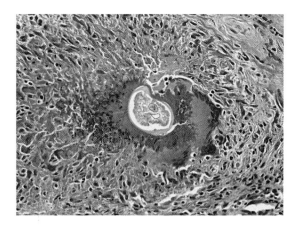

Figure 16.31
Section from lesion illustrated in Figure 16.30. A giant cell is engulfing a fragment of worm. x165

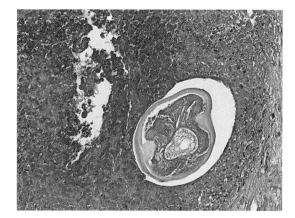

Figure 16.32
Submental mass containing degenerating male *D. tenuis* in necrotizing granuloma. Note extensive tissue eosinophilia. Eosinophils are degranulating adjacent to worm, but are intact and more numerous away from worm. x155

References

1. Anderson RC. Description and relationships of Dirofilaria ursi Yamaguti, 1941, and a review of the genus Dirofilaria Raillet and Henry, 1911. *Trans R Can Inst* 1952;29:35–64.
2. Beaver PC, Orihel TC. Human infection with filariae of animals in the United States. *Am J Trop Med Hyg* 1965;14:1010–1029.
3. Beaver PC, Wolfson JS, Waldron MA, Swartz MN, Evans GW, Adler J. Dirofilaria ursi-like parasites acquired by humans in the northern United States and Canada: report of two cases and brief review. *Am J Trop Med Hyg* 1987;37:357–362.
4. Bruijning CF. Human dirofilariasis. A report of the first case of ocular dirofilariasis in the Netherlands and a review of the literature. *Trop Geogr Med* 1981;33:295–305.
5. Buxton BA, Mullen GR. Field isolations of Dirofilaria from mosquitoes in Alabama. *J Parasitol* 1980;66:140–144.
6. Cancrini G, D'Amelio S, Mattiucci S, Coluzzi M. Identification of Dirofilaria in man by multilocus electrophoretic analysis. *Ann Trop Med Parasitol* 1991;85:529–532.
7. Ciferri F. Human pulmonary dirofilariasis in the United States: a critical review. *Am J Trop Med Hyg* 1982;31:302–308.
8. Dayal Y, Neafie RC. Human pulmonary dirofilariasis. A case report and review of the literature. *Am Rev Respir Dis* 1975;112:437–443.
9. Faust EC, Thomas EP, Jones J. Discovery of human heartworm infection in New Orleans. *J Parasitol* 1941;27:115–122.
10. Font RL, Neafie RC, Perry HD. Subcutaneous dirofilariasis of the eyelid and ocular adnexa. *Arch Ophthalmol* 1980;98:1079–1082.
11. Gardiner CH, Oberdorfer CE, Reyes JE, Pinkus WH. Infection of man by Dirofilaria repens. *Am J Trop Med Hyg* 1978;27:1279–1281.
12. Gutierrez Y. Diagnostic features of zoonotic filariae in tissue sections. *Hum Pathol* 1984;15:514–525.
13. Gutierrez Y. Diagnostic characteristics of Dirofilaria subdermata in cross sections. *Can J Zool* 1983;61:2097–2103.
14. Jung RC, Espenan PH. A case of infection in man with Dirofilaria. *Am J Trop Med Hyg* 1967;16:172–174.
15. Kotlan A. On a new case of human filariidosis in Hungary. *Acta Vet Acad Sci Hung* 1951;1:60–79.
16. Levine ND. *Nematode Parasites of Domestic Animals and of Man*. Minneapolis, Minn: Burgess Publishing Co; 1968:358–359.
17. MacDougall LT, Magoon CC, Fritsche TR. Dirofilaria repens manifesting as a breast nodule. Diagnostic problems and epidemiologic considerations. *Am J Clin Pathol* 1992;97:625–630.
18. Neafie RC, Piggott J. Human pulmonary dirofilariasis. *Arch Pathol* 1971;92:342–349.
19. Orihel TC, Ash LR. Occurrence of Dirofilaria striata in the bobcat (Lynx rufus) in Louisiana with observations on its larval development. *J Parasitol* 1964;50:590–591.
20. Orihel TC, Isbey EK Jr. Dirofilaria striata infection in a North Carolina child. *Am J Trop Med Hyg* 1990;42:124–126.
21. Pacheco G, Schofield HL Jr. Dirofilaria tenuis containing microfilariae in man. *Am J Trop Med Hyg* 1968;17:180–182.
22. Pampiglione S, Rivasi F, Paolino S. Human pulmonary dirofilariasis. *Histopathology* 1996;29:69–72.
23. Pampiglione S, Schmid C, Montaperto C. Human dirofilariasis: discovery of a gravid female of Dirofilaria repens in a subcutaneous nodule [in Italian]. *Pathologica* 1992;84:77–81.
24. Parker BM. Variation of mosquito (Diptera: Culicidae) relative abundance and Dirofilaria immitis (Nematoda: Filarioidea) vector potential in coastal North Carolina. *J Med Entomol* 1993;30:436–442.
25. Russell RC. The relative importance of various mosquitoes for the transmission and control of dog heartworm in southeastern Australia. *Aust Vet J* 1990;67:191–192.
26. Russell RC, Geary MJ. The susceptibility of the mosquitoes Aedes notoscriptus and Culex annulirostris to infection with dog heartworm Dirofilaria immitis and their vector efficiency. *Med Vet Entomol* 1992;6:154–158.
27. Samarawickrema WA, Kimura E, Sones F, Paulson GS, Cummings RF. Natural infections of Dirofilaria immitis in Aedes (Stegomyia) polynesiensis and Aedes (Finlaya) samoanus and their implication in human health in Samoa. *Trans R Soc Trop Med Hyg* 1992;86:187–188.
28. Sauerman DM Jr, Nayar JK. A survey for natural potential vectors of Dirofilaria immitis in Vero Beach, Florida. *Mosq News* 1983;43:222–225.
29. Scoles GA, Dickson SL, Blackmore MS. Assessment of Aedes sierrensis as a vector of canine heartworm in Utah using a new technique for determining the infectivity rate. *J Am Mosq Control Assoc* 1993;9:88–90.
30. Uni S, Kimata I, Takada S. Cross-section morphology of Dirofilaria ursi in comparison with D. immitis. *Jpn J Parasitol* 1980;29:489–497.

17

Onchocerciasis

Ronald C. Neafie,
Aileen M. Marty, and
Brian O.L. Duke

Introduction

Definition

Onchocerciasis is infection by the filarial nematode *Onchocerca volvulus*. The name *Onchocerca* is derived from the Greek words *onkos,* meaning tumor or hook, and *kerkos,* meaning tail. Dipteran black flies of the genus *Simulium* transmit the parasite from person to person. Clinical manifestations vary in different geographic regions and among individuals. Patients with heavy infections usually have 1 or more of the 3 cardinal manifestations: subcutaneous nodules, dermatitis, or ocular lesions.

Synonyms

Regional names for onchocerciasis include river blindness, blinding filarial disease, craw-craw, *gâle filarienne, sowda, enfermedad de Robles, erisípela de la costa, mal morado,* and *ceguera de Los Ríos*.

General Considerations

In 1875, O'Neill[49] found and described the microfilariae of *O. volvulus* while examining West African patients suffering from a papular dermatitis known in the Gold Coast as craw-craw. In 1891, Leuckart[42] described adult male and female filarial nematodes in subcutaneous nodules excised by an anonymous medical missionary, and named them *Filaria volvulus*. In 1904, Brumpt[11] reported, as indigenous Africans had noted, that people living along rivers had more severe infections and suggested that a riverine black fly transmitted the parasite. In 1910, Railliet and Henry[51] recognized a relationship between adult *F. volvulus* worms and *O. volvulus* microfilariae and renamed the worms *O. volvulus*. Robles,[53] who removed nodules from patients in Guatemala, established that onchocerciasis afflicted humans in the Western Hemisphere and first proposed that *O. volvulus* produced ocular lesions. This visual impairment was further characterized by Luna,[41] who noted that the punctate keratitis improved rapidly after removal of the nodules. In 1926, Blacklock[7] reported larval stages of *O. volvulus* within *Simulium damnosum* that pinpointed this black fly as a vector of the parasite. From his observations in Africa in 1931, Hissette[37] confirmed, as Guatemalan physicians had suspected, that onchocerciasis produces ocular lesions and blindness. According to Davies,[19] the first successful vector control was carried out in a region of Kenya in 1943 by removal of vegetation. Five years later, *Simulium neavei* was completely eradicated from the Kodera Valley in Kenya using DDT as a larvicide.[30] In 1948, Mazzotti[44] noted the systemic and cutaneous response to oral diethylcarbamazine (DEC) in Mexican patients with onchocerciasis and recommended this procedure as a diagnostic tool (the Mazzotti reaction). In 1966, Duke et al[23] recognized different strains of *O. volvulus* in different geographic regions and found disparities in their infectivity for black flies. The first detailed description of the histologic anatomy and embryogenesis of *O. volvulus* was published in 1972, based on serial sections of adult worms in nodules.[47] In 1975 the World Health Organization (WHO) initiated its Onchocerciasis Control Program in the Volta River basin. This program calls for the weekly application of temefos to rivers that serve as breeding sites for vectors.[19] Schulz-Key et al[56] greatly simplified the study of the adult worms in 1977 when they isolated living adult *O. volvulus* from nodules by digesting the nodules with collagenase (Fig 17.3). Discovering the efficacy of ivermectin (Aziz et al[3]) in 1982 was a major therapeutic breakthrough.

Epidemiology

Onchocerciasis is indigenous to both the Old and New Worlds, and is limited to those tropical regions where the vectors, *Simulium* sp, are abundant (Fig 17.1). The WHO Expert Committee on Onchocerciasis Control estimates global prevalence at 17.5 million patients, of whom about 270,000 are blind[66] and a similar number have severe visual impairment.[5,59]

Onchocerciasis is a major public health problem in Africa, the home of approximately 95% of all patients. Major foci are along streams and rivers (Fig 17.2) within countries spanning sub-Saharan Africa, particularly Benin, Burkina Faso, Burundi, Cameroon, Central African Republic, Chad, Côte d'Ivoire, Ethiopia, Ghana, Guinea, Guinea-Bissau, Liberia, Mali, Niger, Nigeria, Senegal, Sierra Leone, Sudan, Togo, Uganda, and Democratic Republic of Congo.

Figure 17.1
Female *S. damnosum* (3 mm long), the common vector of *O. volvulus* in Africa. The fly is feeding on human skin. Note erythema around the site.

A less severe form of onchocerciasis prevails in the rain forests of Congo, Equatorial Guinea, Gabon, and Democratic Republic of Congo, and extends into the savannas of Angola, Malawi, and Tanzania.[13] In Latin America there are foci of onchocerciasis in Brazil, Colombia, Ecuador, Guatemala, Mexico, and Venezuela. In Arabia, onchocerciasis is confined to the Asir Province in southwestern Saudi Arabia and to the Republic of Yemen.

Differences in prevalence between men and women are most marked in savanna areas with high transmission rates. In savanna areas, worm burdens and ocular lesions are lower in females from early childhood on. The lower frequency of onchocerciasis in females is probably related more to their greater resistance to infection than to reduced exposure to black flies, as had been postulated earlier.[8,26] In forest areas, infection rates and frequency of ocular involvement are similar in men and women.

In different geographic regions, there are major variations in the clinical manifestations of onchocerciasis: the nature and distribution of the dermatitis, frequency of various ocular lesions, type of lymphadenitis, and distribution of onchocercomata. These variations may be caused by concomitant infections with other filarial

Figure 17.2
Turbulent rivers and streams with abundant overhanging vegetation are breeding sites for *Simulium* sp.*

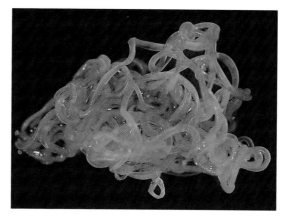

Figure 17.3
Tangled mass of adult *O. volvulus* worms digested out of skin nodule by collagenase. x6.5

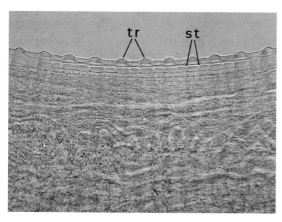

Figure 17.4
Gross longitudinal section of adult female *O. volvulus* showing transverse ridges (tr) of outer layer of cuticle with 2 striae (st) per ridge in inner layer of cuticle. Portion of cuticle between striae is the interstrial region. Unstained x385

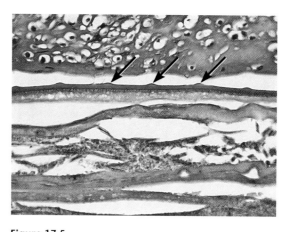

Figure 17.5
Longitudinal section of gravid *O. volvulus* at midbody within onchocercal nodule, demonstrating transverse ridges (arrows) encircling worm. x265

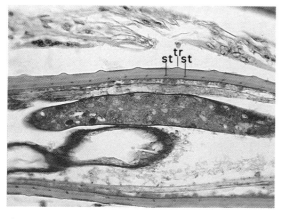

Figure 17.6
Longitudinal section of female *O. volvulus* demonstrating 2 striae (st) in inner layer of cuticle for each transverse ridge (tr) on outer layer. PAS x265

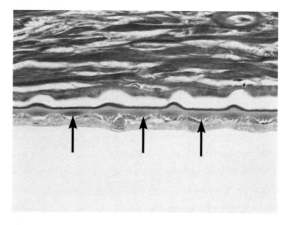

Figure 17.7
Longitudinal section of female *O. volvulus* showing interstrial regions (arrows) in inner layer of cuticle. Masson trichrome x250

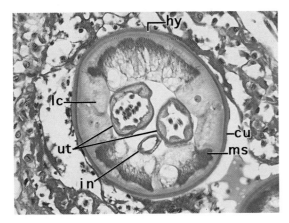

Figure 17.8
Transverse section of anterior region of gravid *O. volvulus* in nodule. Note prominent cuticle (cu), hypodermis (hy), lateral cords (lc), and somatic muscles (ms), and normal-sized intestine (in). Paired uteri (ut) contain cross sections of microfilariae. x265

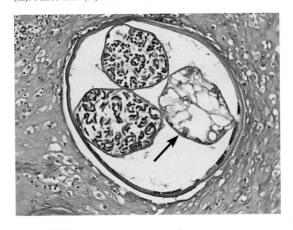

Figure 17.9
Cross section of gravid *O. volvulus* at midbody within nodule. Hypodermis, lateral cords, and somatic muscles are barely perceptible. Intestine is uncommonly large (arrow). Microfilariae fill large paired uteri. x140

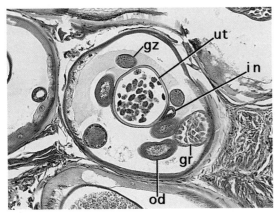

Figure 17.10
Cross section of posterior region of gravid *O. volvulus* in nodule. Note thick-walled oviducts (od), germinal zone of ovary (gz), growth region of ovary (gr), uterus (ut), and intestine (in). x120

parasites, or by the diverse strains of *O. volvulus* that inhabit different parts of the world. Dissimilarities in the innate immunity of local residents in endemic areas may also contribute to the clinical and pathologic variations.[14]

The insect vector, female black flies of the genus *Simulium*, breed in fast-flowing streams and rivers (Figs 17.1 and 17.2). Although black flies can travel long distances with the wind, transmission is most active along the banks of rivers.[4] *Simulium* sp that transmit the parasites to humans vary from region to region. In West Africa, the common vectors are related sibling species belonging to the *S. damnosum* complex. In East Africa, members of the *S. neavei* complex are additional vectors. The major vectors in the Western Hemisphere are *Simulium ochraceum* in Mexico and Guatemala; *Simulium metallicum* in Venezuela; *Simulium oyapockense* and *Simulium guianense* in the border region of Venezuela and Brazil; and *Simulium exiguum*, *Simulium sanguineum*, and *Simulium quadrivittatum* in Ecuador and Colombia.[18]

Infectious Agent

Morphologic Description

The various strains of *O. volvulus* have different biologic properties, vector preferences, and pathogenicity. Although morphologically they are virtually indistinguishable, the strains can be separated by distinctive DNA sequences.[55]

Adult *O. volvulus* are white-yellow, thin, threadlike roundworms (Fig 17.3). The cuticle of each sex has distinguishing characteristics. The gut and reproductive organs lie within the pseudocoelom, with the reproductive organs occupying most of the pseudocoelom throughout the length of the male and female worms.

Adult female worms are 230 to 500 mm by 250 to 450 µm. The cuticle is made up of 3 layers (cortical, median, and basal)[20,28] and varies from 4 to 10 µm thick. Transverse (annular) thickenings or ridges of cuticle encircle the worm at regular intervals and are 20 to 70 µm apart throughout most of the length of the worm (Fig 17.4). Histologically, ridges are most apparent in longitudinal sections of worm fragments, where they appear as slight elevations (up to 4 µm in height) (Fig 17.5).

Sometimes these ridges are barely discernible or even absent. In the inner layer of cuticle, there are 2 striae for each external transverse ridge.[6] Striae are transverse lines or grooves in the cuticle that are usually not visible on H&E-stained sections, but may stain red with PAS (Fig 17.6). The interstrial regions are pink with Masson's trichrome stain (Fig 17.7). Even with special stains, however, striae are not always clearly visible.

In female worms, the hypodermis, lateral cords, and somatic muscle cells vary from prominent to barely perceptible (Figs 17.8 and 17.9). The reproductive system consists of a vulva, vagina, paired uteri, seminal receptacles, oviducts, and ovaries. Ovaries, oviducts, and seminal receptacles are limited to the posterior 35 mm of the worm (Fig 17.10).[47] The paired uteri extend throughout most of the worm's length and unite in the anterior end of the worm to form the vagina. The 2-to-4-mm-long vagina opens exteriorly through the vulva about 0.5 mm behind the anterior end of the worm, near the esophageal-intestinal junction. The intestine is a single tube with few nuclei and is usually much smaller in diameter than the reproductive organs. Reproductive tubes do not extend to the posterior tip of the worm, so sections in this region contain only the intestine (Fig 17.11). A typical transverse section at midbody contains 2 thin-walled uteri and the intestine; these do not fill the body cavity. Occasionally the uteri may form loops, producing many sections of uteri in a cross section of worm (Fig 17.12).

Adult males are 16 to 42 mm by 125 to 200 µm. The cuticle is composed of 2 layers, varies from 3 to 5 µm in thickness, and has deep striae (annulations) that are about 5 µm apart (Fig 17.13). The region between 2 annulations is called an annule. In tissue sections, annulations and annules are best observed in longitudinal sections and stain well with H&E (Fig 17.14). Striae in the inner layer of the cuticle have no diagnostic importance for the male worm. Lateral cords and somatic muscle are usually prominent (Fig 17.15). The single reproductive tube begins slightly posterior to the esophageal-intestinal junction and extends to the posterior end of the worm, where it joins the intestine to form the cloaca. The reproductive tube is composed of a testis, vas deferens, and ejaculatory duct (Fig 17.16). There are 2 copulatory spicules (Fig 17.17).

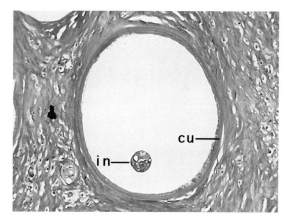

Figure 17.11
Cross section at posterior end of female *O. volvulus* in nodule. At this level, the intestine (in) is the only structure within the pseudocoelom. A thin cuticle (cu) blends in with the surrounding collagen. x120

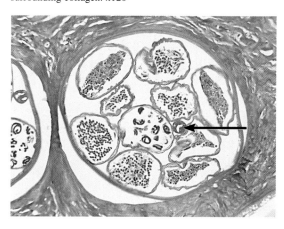

Figure 17.12
Cross section at midbody through gravid *O. volvulus* in nodule. Here uteri are looped, with 10 sections visible. Intestine (arrow) is much smaller in diameter than uteri. x110

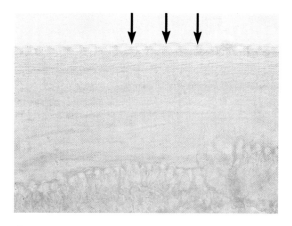

Figure 17.13
Longitudinal section of male *O. volvulus*, demonstrating deep striae (annulations) spaced about 5 µm apart (arrows). The area between 2 annulations is an annule. Unstained x500

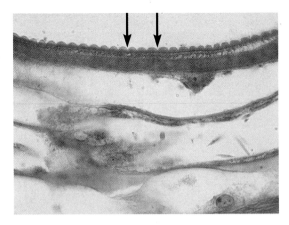

Figure 17.14
Longitudinal section at midbody of male *O. volvulus* in onchocercal nodule, demonstrating annulations (arrows) and annules. x485

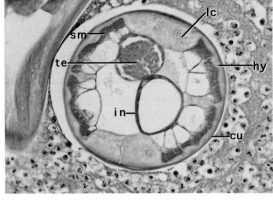

Figure 17.15
Cross section through anterior region of male *O. volvulus* in nodule. Note cuticle (cu), hypodermis (hy), lateral cords (lc), somatic muscle (sm), testis (te), and intestine (in). x215

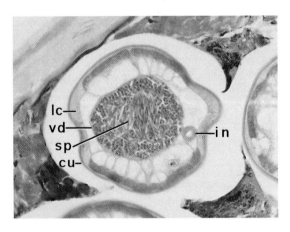

Figure 17.16
Cross section through posterior region of male *O. volvulus* in nodule. Note cuticle (cu), lateral cords (lc), vas deferens (vd), spermatozoa (sp), and intestine (in). x265

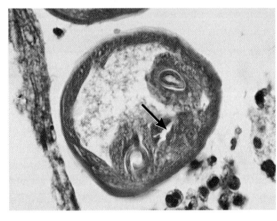

Figure 17.17
Cross section through posterior tip of male *O. volvulus* in nodule, displaying 2 yellow copulatory spicules and the cloaca (arrow). x825

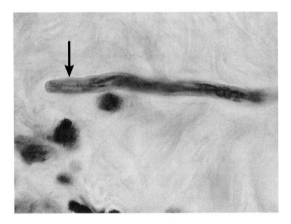

Figure 17.18
Anterior end of *O. volvulus* microfilaria in dermal collagen, demonstrating long cephalic space (arrow) followed by paired nuclei. Giemsa x1060

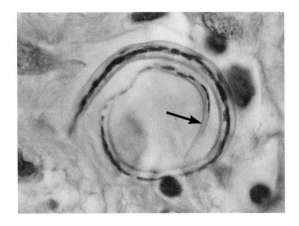

Figure 17.19
Coiled posterior end of microfilaria of *O. volvulus* in onchocercal nodule, demonstrating long caudal space (arrow). Tail tapers to sharp point. x1050

Microfilariae of *O. volvulus* lack a sheath and measure 220 to 360 µm by 5 to 9 µm. At the anterior end there is a 7-to-13-µm cephalic clear space followed by 2 or 3 paired nuclei (Fig 17.18). There are several elongated terminal nuclei followed by a 9-to-15-µm caudal clear space at the posterior end (Fig 17.19). The tail tapers to a sharp point.

For comparison with other filariids, see tables in chapter 13.

Life Cycle and Transmission

The black fly is the intermediate host and humans the definitive host of *O. volvulus*. Infective larvae (L3) enter human skin from the black fly's proboscis. After 2 molts, the larvae develop over 9 to 12 months into adult worms that usually inhabit the dermis and deep fascial planes, and may live for 10 to 15 years.[52]

Male and female worms mate within human tissue. Fertilization usually takes place in the female's seminal receptacles. Each fertile female produces millions of microfilariae during her lifetime. The resulting zygotes develop as they traverse the length of the uteri, progressing through 5 stages: small (8 to 64 cells) and large (more than 64 cells) morula, gastrula, brezel, coiled microfilaria, and stretched microfilaria.[25] Stretched microfilariae exit through the vulva into the host's tissues and begin to migrate. Microfilariae can migrate throughout the body, even entering the eye; however, most live in the upper dermis, where they are accessible to the biting black fly. Only female black flies suck blood and are responsible for transmission.

Black flies have short, coarse mouth parts that rasp and saw into the dermis, producing a pool of cells, blood, and tissue fluid from which the black fly may ingest microfilariae. The saliva of the black fly functions as an anticoagulant and an attractant for microfilariae. Microfilariae that are not damaged in the buccopharyngeal area, or escape entrapment by the peritrophic membrane, penetrate the black fly's gut and migrate to the thoracic (flight) muscles. There they undergo 2 molts, becoming infective larvae (L3) after a period of 6 to 12 days, depending on the ambient temperature. Infective larvae migrate to the fly's proboscis (Fig 17.20), where the cycle may repeat itself when the fly bites again. The black fly bite may leave a bleeding point on the skin, and some-

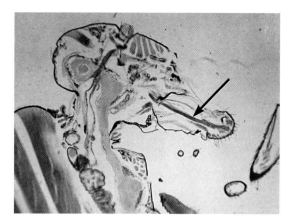

Figure 17.20
Section through anterior portion of female black fly demonstrating infective (L3) larva (arrow) in proboscis.* x38

Figure 17.21
Onchocercal nodule on back of Congolese woman. Nodule is firm, movable, and not tender.*

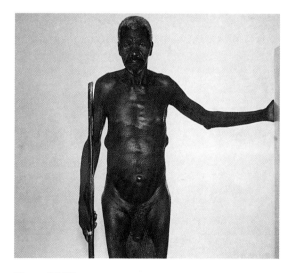

Figure 17.22
African man with conspicuous nodules over bony prominences of elbows, ribs, and trochanters.*

Figure 17.23
Severe onchocercal dermatitis with secondary infection from intense itching and scratching.*

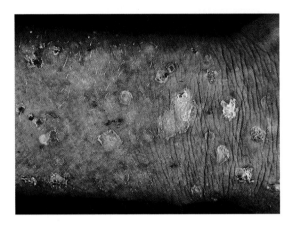

Figure 17.24
Forearm of patient from Guatemala with papular eruption resembling scabies (*gâle filarienne*). Some pustules are scaling.*

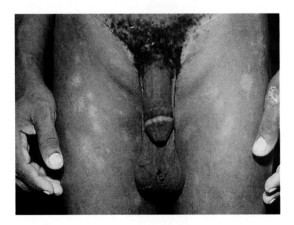

Figure 17.25
African with hypopigmented macules on thighs, caused by microfilariae of *O. volvulus*.*

times surrounding erythema (Fig 17.1).

Transplacental transmission of microfilariae, and possibly of infective larvae, of *O. volvulus* has been reported.[2,43] In heavily infected areas, up to 5% of newborn babies may harbor detectable microfilariae in their skin.

Clinical Features and Pathogenesis

Clinical manifestations of onchocerciasis vary greatly with factors such as the level of infection and geographic region. Very light infection may produce no symptoms at all. Light to moderate infection may only produce an itchy rash. Heavily infected persons, who may harbor up to 200 million microfilariae, may suffer severe morbidity. The evolution of such microfilarial burdens requires years of continuous exposure to infected black flies.

Onchocercomata. Onchocercomata are firm, movable, nontender, subcutaneous nodules that contain adult *O. volvulus* (Fig 17.21). Most onchocercomata are composed of a mass of small, closely aggregated nodules, giving a lobulated pattern on palpation. These aggregated masses range up to several centimeters in diameter and predominate over bony prominences such as the scalp, ribs, elbows, trochanters, iliac crests, coccyx, knees, and ankles (Fig 17.22). Distribution of palpable nodules on the body varies depending upon the endemic area. In Africans, nodules are most common around the pelvic girdle, though many develop in deeper sites and may not be palpable. In Mexican and Guatemalan patients, most nodules are on the upper part of the body, especially the head. With experience it is possible to distinguish onchocercomata from, for example, lipomata, sebaceous cysts, and ganglia, but swollen lymph nodes in children, especially juxta-articular nodes in late yaws, can be difficult to differentiate clinically.

Onchocercal dermatitis. Migrating microfilariae cause nearly all the clinical changes of onchocerciasis. The most common manifestation is dermatitis that usually begins with itching, typically most severe over the lower trunk, pelvis, buttocks, and thighs. Itching is often unilateral and confined to 1 anatomical quarter of the body. Itching is often exquisite; scratching may produce ulcers, bleeding, and secondary infection (Fig 17.23). Occasionally, a papular rash (*gâle filarienne*) develops (Fig 17.24). There are often alterations in skin pigmentation, such as poorly defined areas of hyper- and hypopigmentation or distinct macules (Figs 17.25 and 17.26).

Chronic skin changes are common in African patients, with scaling, edema, depigmentation, and papule formation. The edema can produce a *peau d'orange,* with pitting around hair follicles and sebaceous glands. In later stages there is loss of elasticity, with atrophy of the epidermis and scarring of the dermis, resulting in an aged appearance (presbydermia) (Fig 17.27). The wrinkled skin is typically thinned, with diminished subcutaneous tissue; however, in some individuals epidermal hypertrophy and lichenification persist, resulting in elephantoid skin that can progress to "lizard" skin (Fig 17.28).[12] In the rain forests of West and Central Africa, spotty depigmentation is called "leopard" skin (Fig 17.29).[1] Clinically, this can be mistaken for leprosy.

Latin American patients experience a variety of acute skin lesions, while chronic skin lesions are less common. Acute lesions include *erisípela de la costa,* a macular rash with edema of the face, and *mal morado,* a red-brown discoloration, usually on the trunk and upper limbs.[10]

In Arabia, and less commonly in Ecuador,[36] Guatemala,[57] Sudan, and West Africa, patients can develop a marked darkening of the skin (Fig 17.30). The term *sowda,* an Arabic word meaning black, refers to this change. *Sowda* is usually limited to a single limb. Involved skin is itchy, swollen, darkened, and covered with scaling papules (Fig 17.31). There is usually associated regional lymphadenopathy.

Lymphadenitis. Obstruction of lymphatic vessels can produce adenolymphocele (a pouch of lymphedematous tissue that develops over a cluster of lymph nodes) and possibly elephantiasis.[16] Adenolymphoceles are usually in the inguinal or femoral areas, where they are known as "hanging groin" in males (Fig 17.32) and as a "Hottentot apron" in females. Adenolymphocele and elephantiasis are limited to African patients and are not seen in patients from Central America or Arabia. Many investigators believe that infection with

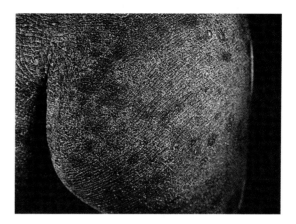

Figure 17.26
Hyperpigmented macules on buttocks of African with onchocercal dermatitis.*

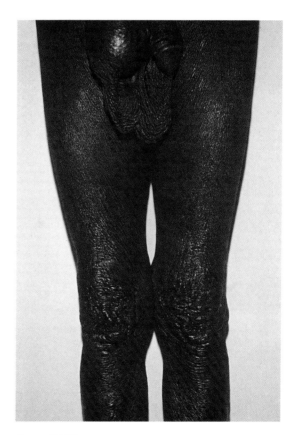

Figure 17.27
African with presbydermia caused by chronic onchocercal dermatitis. Thickened, wrinkled skin hangs in folds over thighs and knees. Note sagging scrotum, probably also related to onchocerciasis.*

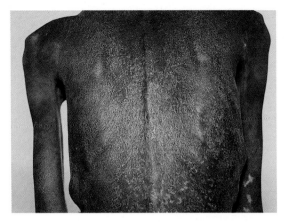

Figure 17.28
African patient with lizard skin, characterized by scaling, depigmentation, and epidermal atrophy. Onchocercal nodules lie over ribs.*

Figure 17.29
Focal depigmentation known as leopard skin. Pigment persists around pores and hair follicles.*

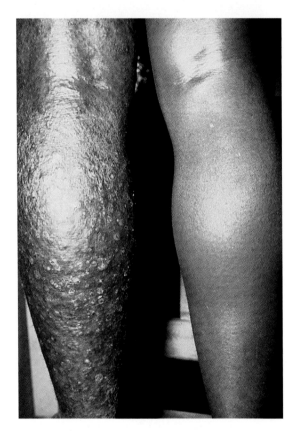

Figure 17.30
Patient from Cameroon with onchocercal dermatitis. Note hyperpigmentation (*sowda*) and papular eruption of left leg.*

filariae other than *O. volvulus* causes the elephantiasis sometimes seen in patients with onchocerciasis. The existence of mixed infections could partially explain this geographic difference. Whether or not *O. volvulus* causes elephantiasis remains unproven.

Ocular lesions. Clinically, ocular invasion by microfilariae can cause photophobia, excessive lacrimation, or pain. Visual impairment ranges from mild to severe. Patients can develop "snowflake" or "fluffy" corneal opacities, sclerosing keratitis, and iridocyclitis leading to secondary glaucoma and cataracts, or chorioretinitis and optic atrophy. Experimental studies suggest that some of the ocular damage is immunologically mediated.[35] Molecular studies indicate that ocular involvement is related to the particular strain of *O. volvulus*.[69] Sera from patients with onchocerciasis contain autoantibodies that react strongly with antigens of surface components, and nucleoli of retinal pigmented epithelial cells and neural retinal cells that could be partially responsible for retinal damage.[68] (See also chapter 18, Ocular Onchocerciasis.)

Pathologic Features

Adult worms produce unsightly nodules, but are usually harmless. Microfilariae migrate through many tissues, causing damaging and progressive lesions. After several months of active life, microfilariae degenerate and die, provoking intense inflammatory reaction.

Nodules. For reasons that are unclear, adult worms eventually become encapsulated and form discrete nodules in the deep dermis and subcutaneous tissues. Palpable onchocercomata are most common over bony prominences (eg, elbows, iliac crests, knees, ribs, sacrum, scapulae, skull, and trochanters); many others are impalpable, lying in deeper sites near joints, muscles, and bones. Some authorities speculate that worms migrating over bony prominences may be traumatized and provoke inflammation that entraps them in fibrous nodules.

Grossly, nodules may be discrete or bound together as conglomerates. The cut surface is white, firm, and reveals a fibrous capsule of variable thickness surrounding scar tissue containing multiple chambers where adult worms reside (Fig 17.33). Older nodules sometimes have areas of calcification and old hemorrhages (Fig 17.34).

Microscopically, nodules contain sections of worms cut at various angles (Fig 17.35). Frequently, nodules contain several female and male worms. Reaction around the worm varies with the immune status of the host and the age of the nodule. Inflammatory reactions to adult worms commonly include variable amounts of suppuration composed of neutrophils and fibrin immediately around the worm (Fig 17.36). The suppuration is in turn surrounded by a granulomatous reaction consisting of epithelioid cells, foamy macrophages, and occasional foreign body giant cells (Fig 17.37). There are lymphocytes at the periphery of the granulomas, and young collagen surrounds the lymphocytes. This collagen eventually hyalinizes, forming a capsule. Amyloid and lipids are sometimes deposited in older lesions, and may be revealed by Congo red and oil red O stains respectively. Adult female worms within nodules are tightly coiled and tangled. Some female worms are fertilized more than once.[25] A few male worms appear to be fixed in nodules, but some observers believe that males may migrate between nodules.

Skin. Early skin changes are minimal and may be limited to slight edema, a few proliferating fibroblasts, lymphocytes, macrophages, plasma cells, and eosinophils around some vessels and appendages. Mast cells may increase, but there are no neutrophils. Microfilariae migrate through the dermal collagen. They may be rare or plentiful, but are most numerous in the upper dermis (Fig 17.38). Sometimes microfilariae invade dermal lymphatics (Fig 17.39). More advanced changes include hyperkeratosis, acanthosis, focal parakeratosis, melanophores in the upper dermis, dilated lymphatics, tortuous blood vessels, and mucin (acid mucopolysaccharide) between dermal collagen fibers.[17] There is destruction of elastic fibers. Chronic fibrosis leads to scarring of papillae, and finally to replacement of dermal collagen by hyalinized scar tissue. Scarring tends to have a concentric arrangement around dermal vessels (Fig 17.40). Of all these changes, fibrosis is the most important because it begins early, persists, and increases until many specialized structures of the skin are ultimately replaced.

The characteristic histologic feature of the papular eruption known as *gâle filarienne* (Fig 17.24) is the intraepidermal abscess containing microfilariae. Leopard skin typically shows loss of melanin pigment from the basal layer and a slightly edematous dermis. Elephantoid skin has a thin epidermis with a few small rete ridges, an edematous dermis infiltrated with numerous macrophages, and a few lymphocytes and plasma cells. Lizard skin has an even thinner, undulating epidermis that lacks rete ridges, and an edematous dermis with increased numbers of elastic fibers. Microfilaricides such as DEC seem to unmask antigens of microfilariae in skin. Thus, dermal changes in onchocercal dermatitis are accentuated soon after treatment with DEC (Figs 17.41 to 17.44).

In patients with *sowda* (Figs 17.30 and 17.31), the epidermis shows hyperkeratosis, parakeratosis, acanthosis, and follicular plugging (Fig 17.45). The dermis manifests edema, melanophages in the upper dermis, fibrosis, chronic perivascular and periappendiceal inflammation with eosinophils and mast cells, dilated lymphatic channels, and tortuous, congested capillaries (Fig 17.46). Neutrophils are scarce and microfilariae are rare or absent (Fig 17.47). *Sowda* probably represents a hyperactive immunologic response.[15]

Lymph nodes. Lymph nodes from African patients are markedly fibrotic, and germinal centers are atrophic or absent (Fig 17.48).[31] The fibrous

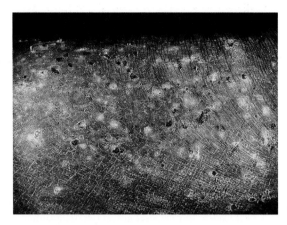

Figure 17.31
Thigh of 11-year-old boy from Yemen, showing maculopapular eruption and hyperpigmentation typical of *sowda*.*

Figure 17.32
African with onchocercal lymphadenitis resulting in hanging groin. The pendulous scrotum is probably a manifestation of onchocerciasis (see also Figure 17.27).*

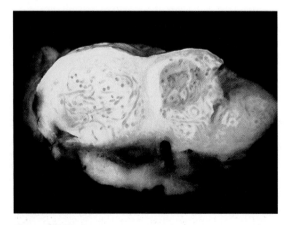

Figure 17.33
Cut surface of conglomerate of nodules dissected from patient with onchocerciasis. Note white, fibrous capsule surrounding scar tissue that encloses several adult worms.* x2.3

tissue has a concentric arrangement around vessels similar to that seen in skin (Fig 17.49). Subcapsular sinusoids and lymphatics in the medulla are dilated. Macrophages collect within the sinusoids. At times, there are large numbers of plasma cells, eosinophils, and mast cells. Lymph nodes contain varying numbers of microfilariae, sometimes in collagen and sometimes in large, tangled clumps (Figs 17.49 and 17.50). Treatment with DEC can cause painful lymphadenitis; in such instances there are often large numbers of degenerating microfilariae in the node (Fig 17.51).

Lymph nodes from patients with *sowda* are vastly different from those of typical African patients. These nodes are large and soft (Fig 17.52), and have follicular hyperplasia (Fig 17.53) rather than follicular atrophy. There is little scarring and the focal inflammatory reaction is minimal. Microfilariae are rare.

Scarcity of autopsy and biopsy specimens has significantly limited our knowledge of deep organ involvement. Microfilariae can invade any organ of the body, including kidney, spleen, lung, peripheral nerves, pancreas, and liver.[46,54] In a Congolese patient, microfilariae caused acute diffuse glomerulitis (Fig 17.54), microabscesses in the lung, and inflammation in the liver sinusoids. In another patient, an incidental finding at autopsy revealed an adult, nongravid female worm in the wall of the aorta (Fig 17.55).[46] This patient, who died of hyperinfection strongyloidiasis, had lepromatous leprosy and was under long-term corticosteroid therapy.

Immunology. Immune response to *O. volvulus* differs widely from patient to patient, leading to a spectrum of clinical disease. Cell-mediated immunity (CMI) to *O. volvulus* antigens varies: patients with strong CMI (for example, those with *sowda*) develop more pronounced cutaneous manifestations. These cutaneous lesions probably reflect active killing of microfilariae lodged in the skin. Microscopically, it is difficult to find viable microfilariae in such lesions (Figs 17.46 and 17.47). Patients with strong CMI may suffer more severe corneal complications. Patients with poor CMI to *O. volvulus* develop more quiescent disease, even though there may be high numbers of microfilariae in the skin (Fig 17.38). Studies in West Africa indicate that patients with onchocerciasis may have a generalized impairment of CMI that can reduce the efficacy of vaccines against other infectious agents.[38,39]

Eosinophils apparently play a central role in killing microfilariae (Figs 17.44, 17.46, and 17.47) and in destroying infective larvae deposited by *Simulium* sp. Eosinophils require antigen-specific antibody to stick to the microfilariae; complement factors greatly enhance this adhesion. The eosinophils' need for a specific antibody emphasizes the importance of a coordinated response by the cellular and humoral immune systems. The most effective sera for destroying microfilariae in vitro are those from patients who are actively killing microfilariae (eg, patients with punctate keratitis), reflecting the importance of antigen-specific antibodies derived from an adaptive immune response. The adaptive immune response does not, however, seem to protect against reinfection; for example, heavily infected patients cleared of microfilariae of *O. volvulus* by chemotherapy sometimes appear to develop reinfections.

We do not yet understand the immune response to adult worms. In vivo, specific antibodies to *O. volvulus* are bound to the cuticle of viable adult worms within nodules and to many structures within degenerating worms. Eosinophils and macrophages commonly cover the surface of adult worms. Sera of patients contain specific immune complexes and circulating antigens of *O. volvulus*; patients with chorioretinal disease have antiretinal antibodies in their sera. The level of immune complex in patients with onchocerciasis varies greatly. To date there is no clear correlation between these findings and clinical features. (See chapter 18 for pathologic changes in ocular disease.)

Diagnosis

The clinical differential diagnosis of the itchy rash typical of early stages of onchocerciasis includes scabies, contact dermatitis, insect bites, food allergies, and prickly heat. Chronic skin lesions resemble those of tertiary yaws, eczema, malnutrition, and old age. Leopard skin mimics other forms of vitiligo, leprosy, and streptocerciasis, all of which can coexist in the same patient.

Microfilariae are detected in skin snips taken

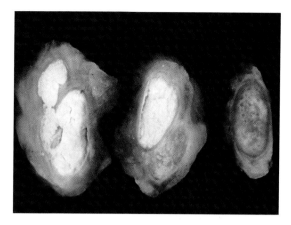

Figure 17.34
Cut surfaces of older nodules with areas of calcification and old hemorrhages. x2.1

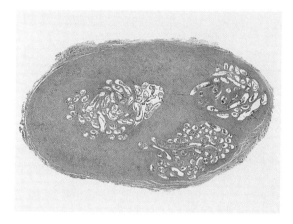

Figure 17.35
Onchocercal nodule showing multiple sections of coiled worms cut at various angles and surrounded by scar tissue. x5

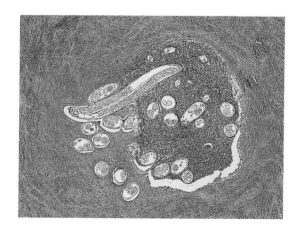

Figure 17.36
Inflammatory reaction to adult worms in onchocercal nodules commonly includes varying amounts of suppuration made up of neutrophils and fibrin immediately around the worm. x11

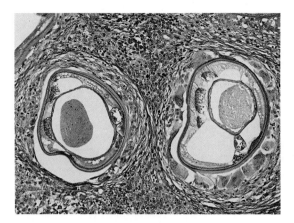

Figure 17.37
Onchocercal nodule showing granulomatous reaction with foreign body giant cells surrounding sections of female worm. x95

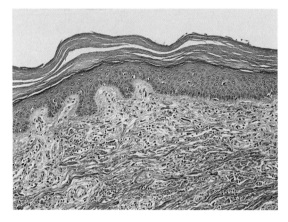

Figure 17.38
Skin of patient heavily infected with *O. volvulus*, demonstrating many microfilariae in dermal collagen and lymphatics. Host reaction includes numerous inflammatory cells, edema, and fibrosis. x95

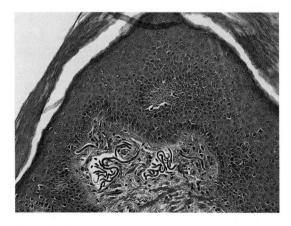

Figure 17.39
Skin of patient with advanced onchocercal dermatitis, demonstrating numerous microfilariae in dilated lymphatics in upper dermis. x130

with a dermal hook and scalpel or sclerodermal punch.[60] For quantitative evaluation, snips are weighed and placed in normal saline on a slide for 3 to 24 hours. The microfilariae that emerge are then counted, and those of *O. volvulus* must be differentiated from all other endemic microfilariae. Diagnosis is also made by identifying microfilariae in skin biopsies.

Immunodiagnosis is less invasive and more sensitive, but not more specific, than skin snips. Microassays and skin tests using *O. volvulus*-specific antigens, as well as antigens of other filarial nematodes, are under study and represent refinements in sensitivity and specificity. However, there are currently no commonly available serologic test kits, and results of the available tests can be misleading. Infections with other helminths such as *Strongyloides* sp and heterologous reactions from exposure to filarial infections of animal origin can cause false-positive reactions.[48]

The oldest diagnostic immunologic tool is the Mazzotti test, which can detect infection even when skin snips do not reveal microfilariae. Infected individuals develop intense itching within hours after a single oral dose of 50 mg of DEC. The itching, with or without accompanying erythema, edema, and papules, is maximal where microfilariae are in highest concentration. These responses seem to be based on immunologic factors. DEC provokes a rise in interleukin-6 and elevations of both specific and nonspecific immune complex levels.[62] This leads to rapid infiltration into tissue of eosinophils and mast cells that degranulate around microfilariae. On receiving DEC, patients with heavy microfilarial loads can develop serious ocular complications, fever, periorbital edema, and arthralgia. On rare occasions, a dose of DEC can be fatal. For this reason, except in lightly infected patients whose eyes are free of microfilariae, the Mazzotti test is a dangerous and inappropriate diagnostic tool.

Both the Nucleopore® filter and Knott's concentration methods can reveal microfilariae in the urine or peripheral blood of heavily infected persons. After treatment with DEC, microfilariae can sometimes be seen in large numbers in blood, urine, sputum, tears, and cerebrospinal fluid.

Slit-lamp examination can demonstrate microfilariae in the cornea and/or anterior chamber and confirm the diagnosis. Prior to examination, the

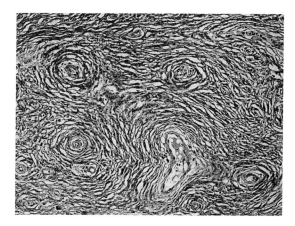

Figure 17.40
Skin of patient with advanced onchocercal dermatitis showing concentric fibrosis oriented around small vessels. x115

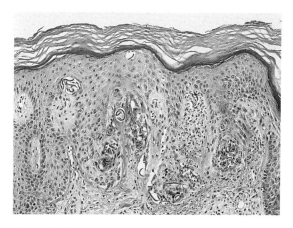

Figure 17.42
Skin of patient with advanced onchocercal dermatitis showing hyperkeratosis and acanthosis. Dermis is edematous and contains dilated lymphatics and numerous microfilariae. Biopsy taken 2 days after treatment with DEC. x110

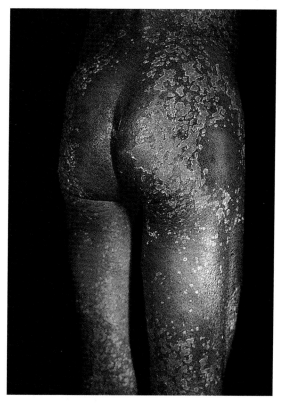

Figure 17.41
African with onchocerciasis showing desquamation 3 days after treatment with DEC.*

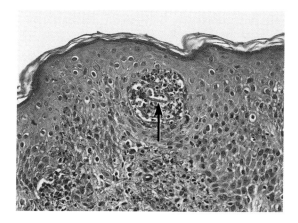

Figure 17.43
Pustular skin from Guatemalan patient demonstrating intraepidermal abscess containing a microfilaria of *O. volvulus* (arrow), 24 hours after treatment with DEC. x190

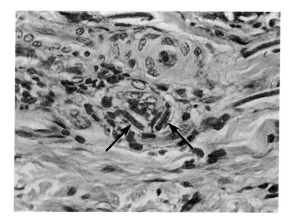

Figure 17.44
Specimen of skin an hour after patient ingested 50 mg of DEC. Note coiled, degenerating microfilaria (arrows) surrounded by eosinophils, some of which are degranulating. x435

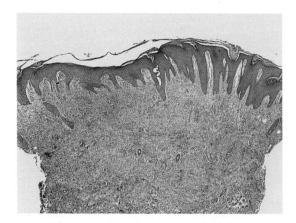

Figure 17.45
Skin of patient with *sowda* demonstrating hyperkeratosis, parakeratosis, acanthosis, and follicular plugging. x19

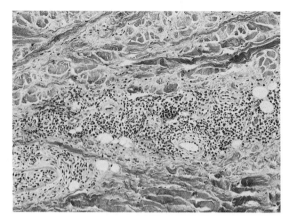

Figure 17.46
Dermis of patient with *sowda*. Note edema, perivascular and periappendiceal inflammation with numerous eosinophils, and tortuous, congested capillaries. x100

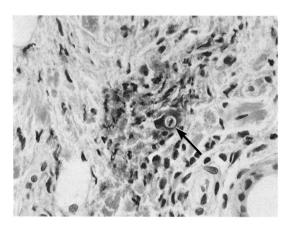

Figure 17.47
Dermis of patient with *sowda* demonstrating edema and rare *O. volvulus* microfilaria (arrow) in the midst of degranulating eosinophils. x475

Figure 17.48
Bisected inguinal lymph node from African with onchocerciasis. Extensive scarring has left only small foci of lymphoid tissue.* x1

patient should sit with his head between his knees for a few minutes, allowing the microfilariae to concentrate behind the central cornea, where they are more readily detectable.

Identifying adult worms in a surgically excised nodule, or fragments of adult worms in aspirates of nodules, confirms the diagnosis.

DNA sequences specific for distinctive members of the *O. volvulus* group have been identified.[27,45,58] These probes are significantly more sensitive than serologic and parasitologic methods for diagnosis.[70]

Gallin et al, in a study in southern Benin, discovered a significant serologic association between HIV-1, HIV-2, and onchocerciasis. Their results suggest *O. volvulus* infection causes false seropositivity for HIV, or may have a direct influence on HIV infection in humans. Polyclonal B-cell activation in onchocerciasis patients may produce antibodies that cross-react with HIV antigens, giving a false-positive result. Alternatively, if the positive antibodies Gallin et al found reflect true HIV infection, this would suggest a synergism between onchocerciasis and HIV infection.[29]

Treatment and Prevention

The oldest known traditional treatment for onchocerciasis was performed by the Bedouin, who bit through the swollen lymph nodes of patients

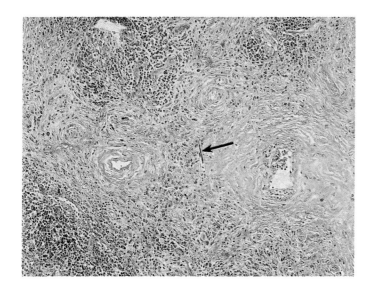
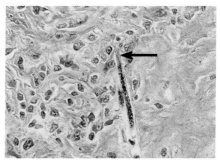

Figure 17.49
Left: Lymph node of African patient with onchocercal lymphadenitis demonstrating concentric fibrosis around vessels. A single microfilaria (arrow) is in collagen. Giemsa x75
Above: Higher magnification of microfilaria showing diagnostic feature of long cephalic space (arrow) in anterior end. x275

with *sowda*.[15] In modern times, therapy has concentrated on nodulectomy and chemotherapy.

Surgical removal of palpable nodules is popular in Guatemala, Ecuador, and Mexico. In Guatemala and Ecuador, nodulectomy campaigns may have reduced the incidence of ocular onchocerciasis.[34] Some believe that nodulectomy is a useful adjunct to chemotherapy.

Suramin was the first chemotherapeutic agent effective against both adults and microfilariae of *O. volvulus*.[63,64] But because of serious side effects, suramin is no longer recommended for treatment of onchocerciasis.

DEC is a microfilaricide with no effect on the adult worm. This drug often induces severe, even life-threatening reactions. Because of these reactions and the advent of ivermectin therapy, DEC is also no longer recommended for onchocerciasis.

Ivermectin, a widely used veterinary gastrointestinal anthelmintic and ectoparasiticide, has dramatically altered the treatment of onchocerciasis in recent years. Registered in France in 1987 as Mectizan®, it is supplied by the manufacturers free of charge for the treatment of onchocerciasis through the Mectizan® Donation Program.[9] In 1996 it was registered for use in the United States under the brand name Stromectol®. Ivermectin produces only minimal Mazzotti reactions.[33] It blocks the exodus of microfilariae from the uteri of gravid worms and kills the microfilariae in human tissues. Blocked microfilariae in the uteri of the worm degenerate and inhibit the production of microfilariae for 3 to 12 months. Two to 3 months after ivermectin therapy, microfilariae usually disappear from the eye, halting the advance of ocular lesions.

Ivermectin is safe for large-scale use when administered according to the manufacturer's instructions, and is proving to be an effective method of preventing river blindness by reducing the human microfilarial reservoir sufficiently to control transmission.[50,61,65] The optimal dose is 150 µg/kg body weight given once orally (usually 2 tablets of 6 mg each for an adult).[22] Multiple dosing at intervals of 3 months appears to produce a steady attrition of both female and male adult worms.[24] Some data suggest that ivermectin may inhibit the maturation of infective larvae.[40]

Amocarzine (CGP 6140), a new oral macrofilaricidal compound, has promising onchocercacidal effects.[67] The majority of adult worms die or are moribund within 4 months post-therapy.

Control of onchocerciasis is based on insecticides that kill black fly larvae. Population coverage with ivermectin somewhat controls transmission, but the major breakthrough will be the development of an inexpensive, nontoxic macrofilaricide deployable on a large scale.[21] A program designed to find immunogens from which a vaccine might be developed is under way.[32]

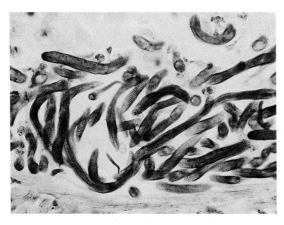

Figure 17.50
Numerous viable microfilariae of *O. volvulus* in lymph node of African patient. Warthin-Starry stain accentuates cuticular features of microfilariae, revealing prominent striations. x485

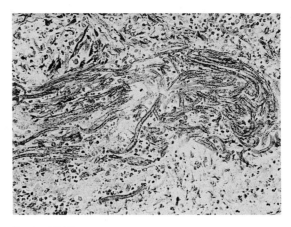

Figure 17.51
Numerous degenerating microfilariae in lymph node following treatment with DEC. Warthin-Starry x190

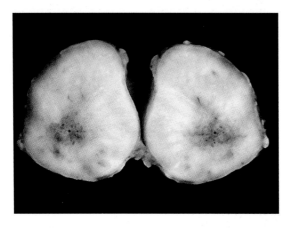

Figure 17.52
Enlarged, soft, homogenous lymph node from Yemenite with *sowda*. x1.5

Figure 17.53
Section of lymph node from patient with *sowda* shown in Figure 17.52. Note follicular hyperplasia. x50

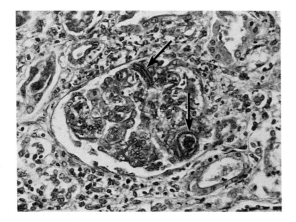

Figure 17.54
Microfilariae (arrows) of *O. volvulus* provoking acute inflammation in renal glomerulus of an African. x230

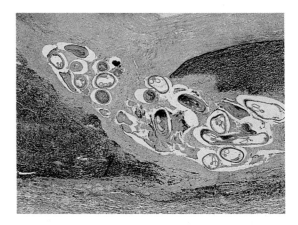

Figure 17.55
Adult, nongravid, degenerating female *O. volvulus* in wall of aorta of patient with lepromatous leprosy. Morphologic features of cuticle of this worm are illustrated in Figure 17.6. Movat x20

References

1. Abanobi OC, Edungbola LD, Nwoke BE, Mencias BS, Nkwogu FU, Njoku AJ. Validity of leopard skin manifestation in community diagnosis of human onchocerciasis infection. *Appl Parasitol* 1994;35:8–11.

2. Anosike JC, Onwuliri CO. A probable case of vertical transmission of Onchocerca volvulus microfilariae. *J Helminthol* 1993;67:83–84.

3. Aziz MA, Diallo S, Diop IM, Lariviere M, Porta M. Efficacy and tolerance of ivermectin in human onchocerciasis. *Lancet* 1982;2:171–173.

4. Baker RH, Guillet P, Seketeli A, et al. Progress in controlling the invasion of windborne vectors into the western area of the Onchocerciasis Control Programme in West Africa. *Philos Trans R Soc Lond B Biol Sci* 1990;328:731–747.

5. Baldwin WR, Duke BO. River blindness [letter]. *Lancet* 1992;339:1178.

6. Beaver PC, Horner GS, Bilos JZ. Zoonotic onchocercosis in a resident of Illinois and observations on the identification of Onchocerca species. *Am J Trop Med Hyg* 1974;23:595–607.

7. Blacklock DB. The development of Onchocerca volvulus in Simulium damnosum. *Ann Trop Med Parasitol* 1926;20:1–48.

8. Brabin L. Factors affecting the differential susceptibility of males and females to onchocerciasis. *Acta Leiden* 1990;59:413–426.

9. Bradshaw H. Onchocerciasis and the Mectizan Donation Programme. *Parasitol Today* 1989;5:63–64.

10. Browne SG. Rare skin lesions in African onchocerciasis. *Dermatol Int* 1968;7:191–195.

11. Brumpt E. *Précis de Parasitologie*. 2nd ed. Paris, France: Masson; 1913.

12. Buck AA, ed. *Onchocerciasis: Symptomatology, Pathology, Diagnosis*. Geneva, Switzerland: World Health Organization; 1974:1–80.

13. Burnham GM. Onchocerciasis in Malawi. 1. Prevalence, intensity and geographical distribution of Onchocerca volvulus infection in the Thyolo highlands. *Trans R Soc Trop Med Hyg* 1991;85:493–496.

14. Connor DH, George GH, Gibson DW. Pathologic changes of human onchocerciasis: implications for future research. *Rev Infect Dis* 1985;7:809–819.

15. Connor DH, Gibson DW, Neafie RC, Merighi B, Buck AA. Sowda-onchocerciasis in north Yemen: a clinicopathologic study of 18 patients. *Am J Trop Med Hyg* 1983;32:123–137.

16. Connor DH, Morrison NE, Kerdel-Vegas F, et al. Onchocerciasis: onchocercal dermatitis, lymphadenitis, and elephantiasis in the Ubangi territory. *Hum Pathol* 1970;1:553–579.

17. Connor DH, Williams PH, Helwig EB, Winslow DJ. Dermal changes in onchocerciasis. *Arch Pathol* 1969;87:193–200.

18. Crosskey RW. Geographical distribution of Simuliidae. In: Laird M, ed. *Black Flies, the Future for Biological Methods in Integrated Control*. New York, NY: Academic Press; 1981:57–68.

19. Davies JB. Sixty years of onchocerciasis vector control: a chronological summary with comments on eradication, reinvasion, and insecticide resistance. *Annu Rev Entomol* 1994;39:23–45.

20. Deas JE, Aguilar FJ, Miller JH. Fine structure of the cuticle of female Onchocerca volvulus. *J Parasitol* 1974;60:1006–1012.

21. Duke BO. Onchocerciasis (river blindness)—can it be eradicated? *Parasitol Today* 1990;6:82–84.

22. Duke BO. Onchocerciasis. *Curr Opin Infect Dis* 1988;1:695–699.

23. Duke BO, Lewis DJ, Moore PJ. Onchocerca-Simulium complexes. I. Transmission of forest and Sudan-savanna strains of Onchocerca volvulus, from Cameroon, by Simulium damnosum from various West African bioclimatic zones. *Ann Trop Med Parasitol* 1966;60:318–326.

24. Duke BO, Zea-Flores G, Castro J, Cupp EW, Munoz B. Effects of three-month doses of ivermectin on adult Onchocerca volvulus. *Am J Trop Med Hyg* 1992;46:189–194.

25. Duke BO, Zea-Flores G, Gannon RT. On the reproductive activity of the female Onchocerca volvulus. *Trop Med Parasitol* 1990;41:387–402.

26. Elson LH, Guderian RH, Araujo E, Bradley JE, Days A, Nutman TB. Immunity to onchocerciasis: identification of a putatively immune population in a hyperendemic area of Ecuador. *J Infect Dis* 1994;169:588–594.

27. Erttmann KD, Unnasch TR, Greene BM, et al. A DNA sequence specific for forest form Onchocerca volvulus. *Nature* 1987;327:415–417.

28. Franz M. Electron microscope study of the cuticle of male and female Onchocerca volvulus from various geographic areas. *Tropenmed Parasitol* 1980;31:149–164.

29. Gallin MY, Adams AZ, Gbaguidi EA, Massougbodji A, Schmitz H, Erttmann KD. The prevalence of antibodies to HIV-1 and HIV-2 in onchocerciasis-endemic rural areas in southern Benin. *AIDS* 1993;7:1534–1536.

30. Garnham PC, McMahon JP. The eradication of Simulium neavei Roubaud, from an onchocerciasis area in Kenya Colony. *Bull Entomol Res* 1947;37:619–628.

31. Gibson DW, Connor DH. Onchocercal lymphadenitis: clinicopathologic study of 34 patients. *Trans R Soc Trop Med Hyg* 1978;72:137–154.

32. Greene BM. Modern medicine versus an ancient scourge: progress toward control of onchocerciasis. *J Infect Dis* 1992;166:15–21.

33. Guderian RH, Anselmi M, Sempertegui R, Cooper PJ. Adverse reactions to ivermectin in reactive onchodermatitis [letter]. *Lancet* 1991;337:188.

34. Guderian RH, Proano R, Beck B, Mackenzie CD. The reduction in microfilariae loads in the skin and eye after nodulectomy in Ecuadorian onchocerciasis. *Trop Med Parasitol* 1987;38:275–278.

35. Haldar JP, Khatami M, Lok JB, Rockey JH, Donnelly JJ. Experimental ocular onchocerciasis: local and systemic antibody and cell-mediated immune responses. *Trop Med Parasitol* 1990;41:234–240.

36. Hay RJ, Mackenzie CD, Guderian R, Noble WC, Proano JR, Williams JF. Onchodermatitis—correlation between skin disease and parasitic load in an endemic focus in Ecuador. *Br J Dermatol* 1989;121:187–198.

37. Hissette J. Sur l'existence d'affections oculaires importantes d'origine filarienne dans certains territoires du Congo. *Ann Soc Belg Med Trop* 1931;11:45–46.

38. Kilian HD, Nielsen G. Cell-mediated and humoral immune responses to BCG and rubella vaccinations and to recall antigens in onchocerciasis patients. *Trop Med Parasitol* 1989;40:445–453.

39. Kilian HD, Nielsen G. Cell-mediated and humoral immune response to tetanus vaccinations in onchocerciasis patients. *Trop Med Parasitol* 1989;40:285–291.

40. Kläger S, Whitworth JA, Post RJ, Chavasse DC, Downham MD. How long do the effects of ivermectin on adult Onchocerca volvulus persist? *Trop Med Parasitol* 1993;44:305–310.

41. Luna RP. Disturbances of vision in patients harboring certain filarial tumors. *Am J Ophthalmol* 1918;1:122–125.

42. Manson P. *The geographical distribution, pathological relations, and life history of Filaria sanguinis-hominis diurna, and of Filaria sanguinis-hominis perstans, in connection with preventive medicine*. London, England: Eyre & Spottiswoode; 1893.

43. Manson-Bahr PE, Apted FI. *Manson's Tropical Diseases*. 18th ed. London, England: Bailliere Tindall; 1982:166.

44. Mazzotti L. Posibilidad de utilizar como medio diagnostico auxiliar en la oncocercosis las reacciones alérgicas consecutivas a la administración del "Hetrazan." *Rev Inst Salubr Enferm Trop* 1948;9:235–237.
45. Meredith SE, Unnasch TR, Karam M, Piessens WF, Wirth DF. Cloning and characterization of an Onchocerca volvulus-specific DNA sequence. *Mol Biochem Parasitol* 1989;36:1–10.
46. Meyers WM, Neafie RC, Connor DH. Onchocerciasis: invasion of deep organs by Onchocerca volvulus. *Am J Trop Med Hyg* 1977;26:650–657.
47. Neafie RC. Morphology of Onchocerca volvulus. *Am J Clin Pathol* 1972;57:574–586.
48. Ogunrinade AF, Kale OO, Chandrashekar R, Weil GJ. Field evaluation of IgG4 serology for the diagnosis of onchocerciasis in children. *Trop Med Parasitol* 1992;43:59–61.
49. O'Neill J. On the presence of a filaria in "craw craw." *Lancet* 1875;1:265–266.
50. Pacque M, Munoz B, Greene BM, Taylor HR. Community-based treatment of onchocerciasis with ivermectin: safety, efficacy, and acceptability of yearly treatment. *J Infect Dis* 1991;163:381–385.
51. Railliet A, Henry A. Les onchocerques, nématodes parasites du tissu conjonctif. *Comp Rend Soc Biol* 1910;68:248–251.
52. Roberts JM, Neumann E, Gockel CW, Highton RB. Onchocerciasis in Kenya 9, 11, and 18 years after elimination of the vector. *Bull World Health Organ* 1967;37:195–212.
53. Robles R. Enfermedad nueva en Guatemala. *La Juventud Med* 1917;17:97–115.
54. Rodhain J, Gavrilov W. Un cas de localisation profonde de "microfilaria volvulus." *Ann Soc Belg Med Trop* 1935;15:551–560.
55. Romeo De Leon J, Duke BO. Experimental studies on the transmission of Guatemalan and West African strains of Onchocerca volvulus by Simulium ochraceum, S. metallicum and S. callidum. *Trans R Soc Trop Med Hyg* 1966;60:735–752.
56. Schulz-Key H, Albiez EJ, Büttner DW. Isolation of living adult Onchocerca volvulus from nodules. *Tropenmed Parasitol* 1977;28:428–430.
57. Schwartz DA, Brandling-Bennett AD, Figueroa H, Connor DH, Gibson DW. Sowda-type onchocerciasis in Guatemala. *Acta Trop* 1983;40:383–389.
58. Shah JS, Karam M, Piessens WF, Wirth DF. Characterization of an Onchocerca-specific DNA clone from Onchocerca volvulus. *Am J Trop Med Hyg* 1987;37:376–384.
59. Status of the eradication/elimination of certain diseases from the Americas. *Bull Pan Am Health Organ* 1992;26:80–86.
60. Taylor HR, Keyvan-Larijani E, Newland HS, White AT, Greene BM. Sensitivity of skin snips in the diagnosis of onchocerciasis. *Trop Med Parasitol* 1987;38:145–147.
61. Taylor HR, Pacque M, Munoz B, Greene BM. Impact of mass treatment of onchocerciasis with ivermectin on the transmission of infection. *Science* 1990;250:116–118.
62. Turner PF, Rockett KA, Ottesen EA, Francis H, Awadzi K, Clark IA. Interleukin-6 and tumor necrosis factor in the pathogenesis of adverse reactions after treatment of lymphatic filariasis and onchocerciasis. *J Infect Dis* 1994;169:1071–1075.
63. van Hoof L, Henrard C, Peel E, Wanson M. Sur la chimiothérapie de l'onchocercose (note préliminaire). *Ann Soc Belg Med Trop* 1947;27:173–177.
64. Voogd TE, Vansterkenburg EL, Wilting J, Janssen LH. Recent research on the biological activity of suramin. *Pharmacol Rev* 1993;45:177–203.
65. Whitworth JA, Morgan D, Maude GH, Downham MD, Taylor DW. A community trial of ivermectin for onchocerciasis in Sierra Leone: clinical and parasitological responses to the initial dose. *Trans R Soc Trop Med Hyg* 1991;85:92–96.
66. World Health Organization. Onchocerciasis and its control: report of a WHO Expert Committee on Onchocerciasis Control. *World Health Organ Tech Rep Ser* 1995;852:1–104.
67. Zak F, Guderian R, Zea-Flores G, Guevara A, Moran M, Poltera AA. Microfilaricidal effect of amocarzine in skin punch biopsies of patients with onchocerciasis from Latin America. *Trop Med Parasitol* 1991;42:294–302.
68. Zhou Y, Dziak E, Unnasch TR, Opas M. Major retinal cell components recognized by onchocerciasis sera are associated with the cell surface and nucleoli. *Invest Ophthalmol Vis Sci* 1994;35:1089–1099.
69. Zimmerman PA, Dadzie KY, De Sole G, Remme J, Alley ES, Unnasch TR. Onchocerca volvulus DNA probe classification correlates with epidemiologic patterns of blindness. *J Infect Dis* 1992;165:964–968.
70. Zimmerman PA, Guderian RH, Araujo E, et al. Polymerase chain reaction-based diagnosis of Onchocerca volvulus infection: improved detection of patients with onchocerciasis. *J Infect Dis* 1994;169:686–689.

18

Ocular Onchocerciasis

Ramon L. Font, Yezid Gutierrez,
Richard D. Semba, and
Aileen M. Marty

Introduction

Definition

Ocular onchocerciasis is infection of the eye by microfilariae of *Onchocerca volvulus*. Microfilariae affect every ocular structure except the lens, producing characteristic clinical features and histologic findings. Damage to the eye is caused by living and dead microfilariae in the cornea, anterior and posterior chambers, iris, retrolental space, vitreous, choroid, retina, sclera, and optic nerve. Ocular damage develops slowly, with blindness usually appearing in adults.

Synonyms

The term river blindness refers to the prevalence of ocular onchocerciasis along rivers. For additional synonyms, see chapter 17.

General Considerations

In 1917, Robles, in Guatemala, was the first to propose that *O. volvulus* produces ocular lesions. This concept was further advanced by Luna, who demonstrated in 1918 that punctate keratitis improved rapidly after removal of nodules. For additional historical information, see chapter 17.

Epidemiology

In some African villages endemic for onchocerciasis, it is common to see children leading blind adults (Fig 18.1). In hyperendemic areas of onchocerciasis, the blindness rate in men over 40 years of age may exceed 40%, making *O. volvulus* a significant cause of blindness. The World Health Organization estimates that over half a million individuals are either visually impaired or blind because of onchocerciasis.[22] The intensity and prevalence of infection depend on ongoing vector control campaigns, chemotherapeutic programs, and ecologic factors. In Africa, the prevalences of ocular lesions in populations living in rain forests and those living in the savanna are different, yet unrelated to transmission rates of onchocerciasis. In the rain forest of Burkina Faso, the prevalence of blindness is only 1% to 3%, but transmission of *O. volvulus* is high. In contrast, in the savanna, where there is a lower transmission rate, the prevalence of ocular lesions is higher and the blindness rate is 5% to 10%.[1,2,3,18] These 2 ecologic areas differ not only in prevalence of blindness, but also in severity of lesions involving the anterior and posterior segments of the eye.[9] Regional variations in the degree of ocular involvement by onchocerciasis are difficult to compare because of a lack of standardized methods, which may explain why a recent study did not confirm such differences (see Table 18.1).[12]

Infectious Agent

Morphologic Description
See chapter 17.

Life Cycle and Transmission
See chapter 17.

Figure 18.1
Adults blinded by onchocerciasis being guided by children.*

Clinical Features and Pathogenesis

Ocular lesions in onchocerciasis include subepithelial punctate keratitis, sclerosing keratitis, iridocyclitis, chorioretinitis, and optic atrophy. If the insult is severe, these lesions may cause blindness.

Subepithelial punctate keratitis

Subepithelial punctate keratitis (stromal opacities) is an early clinical sign common in patients with onchocerciasis (Fig 18.2). Some patients complain of photophobia and lacrimation producing mild to moderate visual impairment; others experience minimal visual symptoms despite having microfilariae in the cornea. In areas where *O. volvulus* infections are severe, lesions of punctate keratitis are common in children under 10 years of age. The stromal opacities, which usually involve the anterior two thirds of the stroma, represent marked inflammatory reactions centered around dead, degenerating microfilariae (Fig 18.25). The opacities are most frequently in the periphery of the cornea. Intrastromal lesions vary in number from 1 to 50 or more, are usually circular, and measure approximately 0.5 mm in diameter. The margins of opacities are ill-defined and not visible in the reflected beam of a slit lamp. They are appropriately called "snowflake" or "fluffy" opacities, and usually disappear without a trace within a few weeks.

Sclerosing keratitis

Sclerosing keratitis, a common precursor of blindness in Central America and the African savanna, results when microfilariae invade the cornea. The lesion begins at the periphery, usually in the nasal or temporal juxtalimbal area (Figs 18.3 and 18.4), appearing first as a white haziness of the anterior one third of the stroma. Melanin-containing cells first infiltrate the limbus and migrate centrally, often accompanied by a prominent fibrovascular pannus (Fig 18.28). The process extends around the peripheral cornea, mainly inferiorly (Fig 18.4), and continues to spread centrally (Fig 18.5). Prominent vessels invade the corneal stroma. If the infiltration involves the pupillary region, vision may be impaired (Fig 18.6). The process may stop before the cornea becomes totally opaque, leaving the superior half of the cornea relatively clear. Careful slit-lamp examination, however, usually reveals numerous microfilariae in the clear area of the cornea.

Onchocercal iridocyclitis

Onchocercal iridocyclitis resembles iridocyclitis of other etiologies. Microfilariae are visible under slit-lamp examination as small, wiggling, silver threads in the anterior chamber (Fig 18.7). They move with the aqueous flow and, when the eye is immobile, sink out of view into the inferior portion of the anterior chamber. A large number of microfilariae may become tangled into a mass resembling Medusa's head. The iridocyclitis produced by *O. volvulus* varies from mild and torpid

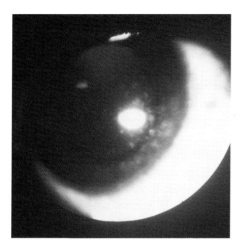

Figure 18.2
Punctate keratitis (stromal opacities of cornea).

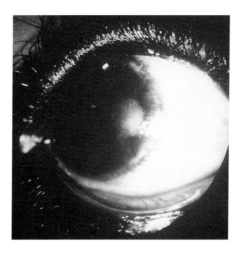

Figure 18.3
Early sclerosing keratitis involving peripheral cornea nasally.

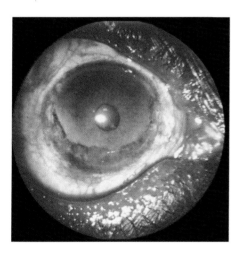

Figure 18.4
Moderately advanced inferior semilunar sclerosing keratitis.

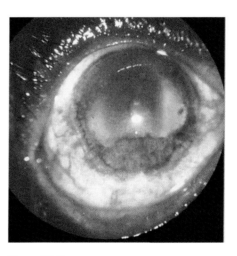

Figure 18.5
Sclerosing keratitis progressing toward central cornea.

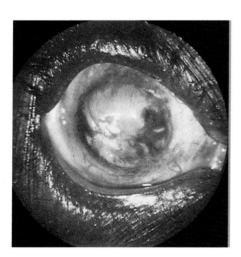

Figure 18.6
Advanced sclerosing keratitis.

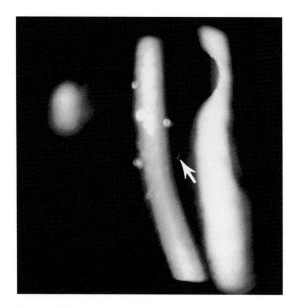

Figure 18.7
Slit-lamp view of microfilaria (arrow) in anterior chamber.*

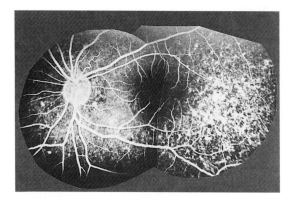

Figure 18.8
Fluorescein angiography revealing mottled areas (right) of hyperfluorescence temporal to the fovea, indicating multiple areas of depigmentation of retinal pigment epithelium. Compare with Figure 18.9.*

to severe and chronic plastic inflammation. Severe iridocyclitis often coexists with sclerosing inferior keratitis, and similarly affects those patients having a heavy microfilarial burden, such as those of the West African savanna. The pupil is often slightly distorted, its margins covered by an exudate rich in fibrin. The lower peripheral iris often displays anterior synechiae; sometimes an organized exudate makes the iris adhere to the cornea. Frequently, after 1 or more episodes of plastic iritis, posterior synechiae develop and the pupil becomes occluded. Peripheral anterior synechiae can cause secondary glaucoma. Massive invasion of microfilariae into the anterior segment can result in secondary open-angle glaucoma.

Chorioretinitis

Chorioretinitis (lesions of the choroid and retina) is frequent in advanced onchocerciasis.[3,4,14] Clinical presentations vary. Lesions are usually bilateral, but the degree of damage in each eye may vary. Early changes in the choroid and retina include migration of pigment-laden macrophages into the outer retina, variable degrees of pigmentary mottling with focal areas of depigmentation, and patchy, mild to moderate atrophy of the retinal pigment epithelium. These changes more commonly develop temporal to the fovea, nasal to the optic disc, and in the peripapillary region. Fluorescein angiography may show areas of mottled hyperfluorescence (Fig 18.8). These lesions develop into a more visible atrophy of the retina, with characteristic clumps of mauve-brown pigment (Fig 18.9) sometimes associated with chorioretinal scarring (Fig 18.10). The defect may present as a circumscribed area of atrophy in the retinal pigment epithelium with foci of proliferation (Figs 18.11 and 18.12).

Small, round or ovoid, pleomorphic white deposits are occasionally seen at all levels of the retina. These deposits appear and disappear over several months, leaving no apparent aftereffects (Fig 18.13).[21] Most of the white intraretinal deposits are 40 to 80 μm in diameter and are not visible with fluorescein angiography. Deposits develop in greater numbers in patients with more intraretinal pigmentation and areas of chorioretinal atrophy. The number and rate of appearance or disappearance of the white intraretinal deposits are the same in patients who receive drug therapy (ivermectin

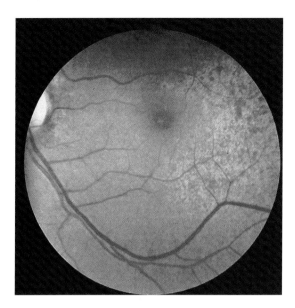

Figure 18.9
Atrophy of retinal pigment epithelium, temporal to macula.

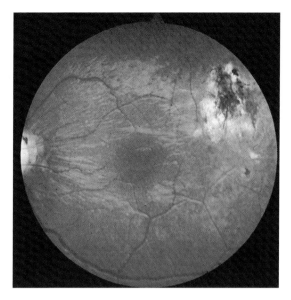

Figure 18.10
Patch of chorioretinal atrophy with scarring supratemporal to macula.

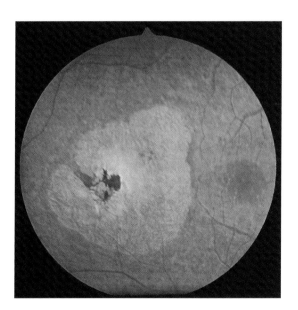

Figure 18.11
Discrete patch of atrophy of retinal pigment epithelium, temporal to macula.

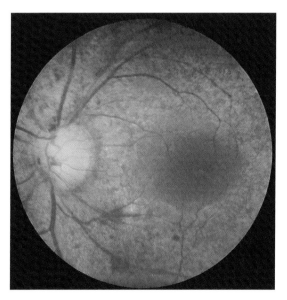

Figure 18.12
Widespread atrophy of retinal pigment epithelium, with hyperpigmented patches involving most of the posterior pole.

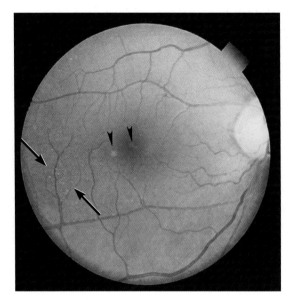

Figure 18.13
Fundus of eye showing white intraretinal deposits (long arrows). Retinal pigment epithelial window defects are also visible (short arrows). The nature of these deposits has not been established.

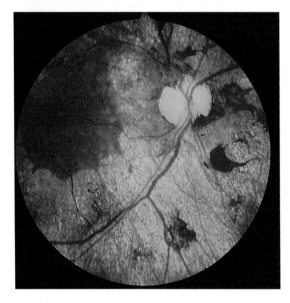

Figure 18.14
Extensive chorioretinal degeneration, known as Ridley fundus, with optic atrophy. The macula has been spared.

or mebendazole) as in those receiving a placebo.[21] Contact lens and slit-lamp biomicroscopy demonstrate actively motile intraretinal microfilariae, appearing as slowly moving, thin, reflective filaments.

In severe lesions there is atrophy of the choriocapillaris. Total atrophy of the retinal pigment epithelium and the choriocapillaris over a wide area of the posterior fundus gives rise to a diffuse chorioretinal scarring with focal proliferation of the retinal pigment epithelium. These changes are known as the Ridley fundus (Fig 18.14). The fully developed lesions surround the optic disc and extend further temporally than nasally. The remainder of the fundus frequently appears normal, and there is often a clear demarcation of the involved structures. Atrophy of the pigmented epithelium allows a clear view of the larger choroidal vessels, which may be red, pink, or white; fluorescein angiography reveals that these vessels are patent (Fig 18.15). Within the involved areas there may be neovascular subretinal membranes and areas of localized or diffuse hyperpigmentation. The macula often has a more normal appearance, and central vision may be maintained. Depigmentation at the edges of chorioretinal scarring may progress at a rate of 200 μm per year, indicating that severe chorioretinal scarring develops over several years.[21]

Optic atrophy

Optic atrophy in onchocerciasis probably results from inflammation of the optic nerve and may be an isolated finding, but more often accompanies chorioretinal lesions. The severity of chorioretinal lesions does not necessarily correlate with optic atrophy. Papillitis is common and fluorescein angiography shows dilatation of the capillaries of the optic disc, with leakage into the juxtapupillary region. In patients with established optic atrophy, excessive pigmentation at the margins of the disc is common (Fig 18.16). Sometimes there is perivascular sheathing, particularly of the arteries, extending variable distances from the optic disc (Fig 18.16). In parts of West Africa, over 80% of blindness originating as lesions of the posterior segment results from advanced optic atrophy rather than from chorioretinal lesions.

	Rain forest	Savanna
Total patients	803	611
Nodules		
All sites	66.8%	53.8%
Head and upper body	15.3%	11.3%
Skin change		
All degrees of involvement	39.4%	38.1%
Moderate to severe	17.7%	13.0%
Eye lesions in persons over 30 years old		
Fluffy opacities	24.7%	20.5%
Sclerosing keratitis	4.3%	3.7%
Iritis	16.1%	8.7%
Chorioretinitis	14.8%	11.3%
Optic atrophy	13.9%	14.2%
Onchocercal blindness	4.3%	4.2%
Severe lesions/blindness	32.2%	24.2%

Table 18.1
Comparative prevalence of some clinical signs and ocular lesions among patients in rain forest and savanna regions.[12]

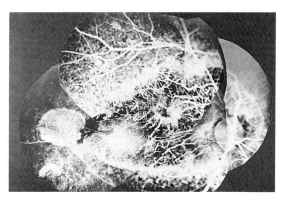

Figure 18.15
Fluorescein angiography showing patchy atrophy of retinal pigment epithelium and choriocapillaris, and large patent choroidal vessels.*

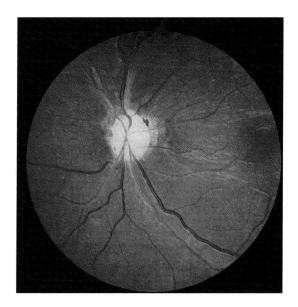

Figure 18.16
Optic atrophy with sheathing of retinal arteries and peripapillary atrophy of pigment epithelium.

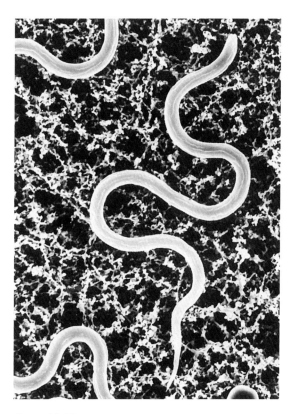

Figure 18.17
Scanning electron micrograph of microfilaria of *O. volvulus* alive at time of fixation, depicting blunt anterior end and pointed tail. x570

Figure 18.18
Top: Microfilaria (arrow) in posterior sclera. x25
Bottom: Same microfilaria at higher magnification.* x345

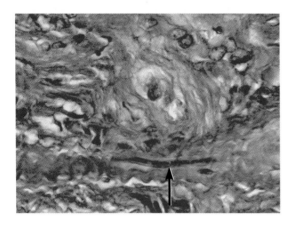

Figure 18.19
Microfilaria (arrow) in lamina cribrosa of optic nerve. x570

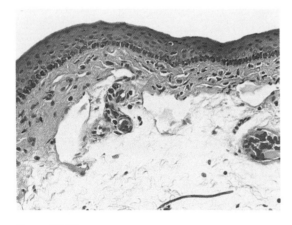

Figure 18.20
Microfilaria within loose conjunctival stroma. x140

Pathologic Features

We do not understand precisely how the parasite damages optic tissue; therefore we only partially understand the mechanisms producing ocular disease. Possible factors include cell-mediated killing of the parasite, primarily by eosinophils in a response dependent on CD4+T cells[7,8]; delayed hypersensitivity to antigens present in living or dead microfilariae[5,7]; and cross-reactivity of onchocercal antigens to autologous proteins such as keratin.[6] As in leishmaniasis,[15] whether the resulting disease is localized or generalized may depend on host factors such as histocompatibility antigens HLA-D and HLA-Q.[13] Some ocular antigens in patients with onchocerciasis cross-react with antigens of the parasite. Patients develop autoantibodies to these antigens, suggesting autoimmunity as a basic mechanism in the development of ocular lesions.[10,11]

The number of microfilariae (Fig 18.17) and inflammatory cells involving the conjunctiva, cornea, anterior and posterior chambers, vitreous, uvea, retina, posterior intrascleral canals (Fig 18.18), and optic nerve head (Fig 18.19) determine the severity of ocular onchocerciasis.

In the bulbar conjunctiva, the microfilariae in the edematous stroma (Figs 18.20 and 18.21) and dilated lymphatics (Fig 18.22) often produce polymorphic inflammatory infiltrates of lymphocytes, plasma cells, eosinophils, and occasional mast cells. Hyperemia and dilatation of vessels are common. Some vessels have thickened walls with perivascular fibrosis. In a study of 42 biopsy specimens of the conjunctiva, we commonly observed lymphangiectasis and live microfilariae, accompanied by a mild to moderate inflammatory reaction. In contrast, dead or degenerating fragments of microfilariae are difficult to recognize, but provoke collections of eosinophils and plasma cells (Figs 18.23 and 18.24) that form small inflammatory nodules. Careful examination of multiple histologic sections, and the use of silver stains, may be helpful in finding microfilariae. Microfilariae appear to migrate through the edematous conjunctival stroma rather than through areas of actinic elastosis of the substantia propria.

In the cornea and limbus, microfilariae are usually in the subepithelial region and anterior half of the stroma (Fig 18.25). Occasionally, there is a coiled microfilaria in the corneal epithelium (Fig 18.26) and in the deep corneal lamellae along the plane of Descemet's membrane (Fig 18.27). Dead microfilariae within the corneal stroma cause opacities of punctate subepithelial keratitis. Florid stromal keratitis results from large numbers of living microfilariae migrating throughout the corneal lamellae. Early inferior sclerosing keratitis may progress to severe scarring with intense chronic inflammation and vascularization (Fig 18.28). Eyes enucleated for advanced secondary glaucoma may reveal microfilariae in the superficial corneal stroma, under large bullae separating the corneal epithelium from the intact Bowman's membrane (Fig 18.29).

Possible routes by which microfilariae enter the anterior chamber (Fig 18.30) include migration from the cornea through Descemet's membrane, direct invasion through the trabecular meshwork, extension along intrascleral canals passing through the adventitia of perforating anterior ciliary vessels, and through the stroma of the iris (Fig 18.31).

Onchocercal iridocyclitis produces a low-grade, chronic, nongranulomatous inflammation containing scattered plasma cells and Russell's bodies. Microfilariae may occupy the stroma of the iris (Fig 18.31) and anterior vitreous along the pars plana ciliaris (Fig 18.32).

There are no studies of the early stages of changes in the posterior segment. There are only a limited number of observations of eyes from patients with long-standing onchocercal blindness. Neumann and Gunders[17] studied an eye with a lesion in the posterior segment. They described fragments of microfilariae and chronic inflammatory cells around intrascleral canals of posterior ciliary arteries, and found live microfilariae in the choroid and retina. They concluded that the chorioretinal scarring resulted from either direct local damage by living microfilariae or from diffusion of toxic products from degenerating microfilariae as they traversed the posterior scleral canals.

Paul and Zimmerman[19] described degenerative changes of the retinal pigment epithelium and obliteration of the choriocapillaris in 1 of 3 eyes (Fig 18.33). We studied a few enucleated eyes from patients treated with diethylcarbamazine (DEC) and were unable to find microfilariae within ocular structures.

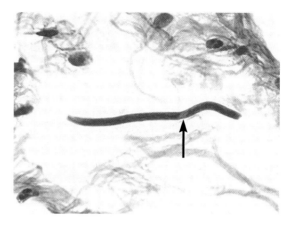

Figure 18.21
Microfilaria in edematous stroma of conjunctiva, showing nerve ring (arrow). x435

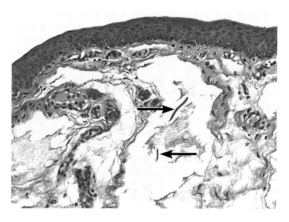

Figure 18.22
Microfilaria (arrows) in dilated lymphatic conjunctiva. x120

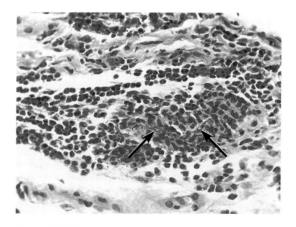

Figure 18.23
Degenerated microfilaria (arrows) in conjunctiva, surrounded by eosinophils and mononuclear cells. x235

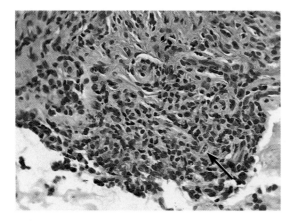

Figure 18.24
Inflammatory nodule in conjunctiva, containing fragment of microfilaria (arrow). x400

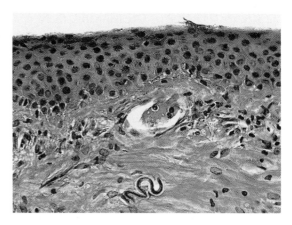

Figure 18.25
Microfilaria coiled in superficial corneal stroma. x225

Figure 18.26
Coiled microfilaria in corneal epithelium. x380

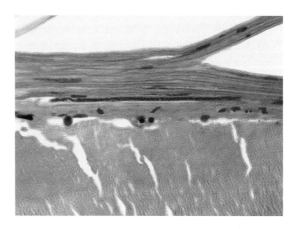

Figure 18.27
Microfilaria in deep corneal stroma along Descemet's membrane. x350

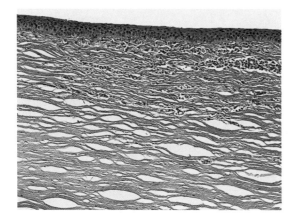

Figure 18.28
Scarring with areas of vascularization and chronic inflammation, involving anterior half of corneal stroma. x70

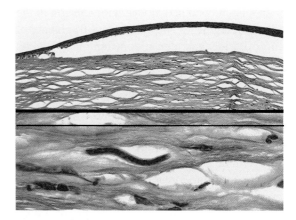

Figure 18.29
Top: Bullous detachment of corneal epithelium. x30
Bottom: Higher magnification of upper image, showing microfilaria in cornea. x440

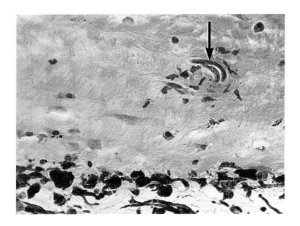

Figure 18.30
Coiled microfilaria (arrow) within organized exudate in anterior chamber. Iris is at lower border of photo. x270

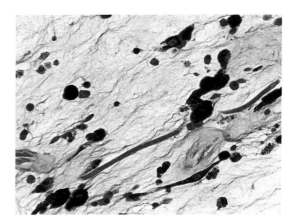

Figure 18.31
Microfilaria in stroma of iris. x295

Diagnosis

Slit-lamp examination can reveal microfilariae in the cornea or anterior chamber and confirm a diagnosis of ocular onchocerciasis. Fluorescein angiography may demonstrate areas of mottled hyperfluorescence and help determine if choroidal vessels are affected. The red-dot card test is an inexpensive, rapid clinical test with 98% specificity for optic nerve disease, and is the recommended screening test for optical pathology.[16] Diagnosis can also be made by identifying *O. volvulus* microfilariae in histologic sections of eye specimens. For additional diagnostic techniques, see chapter 17.

Figure 18.32
Microfilaria in anterior vitreous. Pars plana ciliaris is below. x125

Treatment

The treatment of choice for onchocerciasis is ivermectin (see chapter 17), which reduces the prevalence of ocular lesions. In an endemic area in southern Mexico, a large group of patients with onchocerciasis was treated every 6 months with 150 to 220 µg/kg body weight, after which the number of microfilariae in skin and eye and changes in ocular lesions were assessed.[20] This study showed a significant reduction of microfilariae in the anterior chamber (from 19% to 3%) and of corneal opacities (from 35% to 11%). These reductions paralleled reduced microfilariae skin

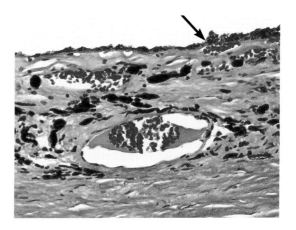

Figure 18.33
Thinning and atrophy of retinal pigment epithelium (arrow) with obliteration of choriocapillaris. Note diffuse hyalinization of choroidal stroma. x160

counts.[20] These observations confirmed other studies showing that severe ocular lesions tend to disappear as the number of microfilariae in skin declines.

Acknowledgment:

Research supported in part by grants from the Retina Research Foundation, Houston, Tex., and Research to Prevent Blindness, Inc., New York, NY.

References

1. Albiez EJ, Ganley JP, Buttner DW. Ocular onchocerciasis in a hyperendemic village in the rain forest of Liberia. *Tropenmed Parasitol* 1981;32:25–28.
2. Anderson J, Fuglsang H, Hamilton PJ, de Marshall TF. Studies on onchocerciasis in the United Cameroon Republic. I. Comparison of populations with and without Onchocerca volvulus. *Trans R Soc Trop Med Hyg* 1974;68:190–208.
3. Anderson J, Fuglsang H, Marshall TF. Studies on onchocerciasis in the United Cameroon Republic. III. A four-year follow-up of 6 rain-forest and 6 Sudan-savanna villages. *Trans R Soc Trop Med Hyg* 1977;70:362–373.
4. Budden FH. Comparative study of ocular onchocerciasis in savannah and rain forest. *Trans R Soc Trop Med Hyg* 1963;57:64–70.
5. Chakravarti B, Herring TA, Lass JH, et al. Infiltration of $CD4^+$ T cells into cornea during development of Onchocerca volvulus-induced experimental sclerosing keratitis in mice. *Cell Immunol* 1994;159:306–314.
6. Chandrashekar R, Curtis KC, Weil GJ. Molecular characterization of a parasite antigen in sera from onchocerciasis patients that is immunologically cross-reactive with human keratin. *J Infect Dis* 1995;171:1586–1592.
7. Elson LH, Calvopina M, Paredes W, et al. Immunity to onchocerciasis: putative immune persons produce a Th1-like response to Onchocerca volvulus. *J Infect Dis* 1995;171:652–658.
8. Lange AM, Yutanawiboonchai W, Scott P, Abraham D. IL-4- and IL-5-dependent protective immunity to Onchocerca volvulus infective larvae in BALB/cBYJ mice. *J Immunol* 1994;153:205–211.
9. Marshall TF, Anderson J, Fuglsang H. The incidence of eye lesions and visual impairment in onchocerciasis in relationship to the intensity of infection. *Trans R Soc Trop Med Hyg* 1986;80:426–434.
10. McKechnie NM, Braun G, Connor V, et al. Immunologic cross-reactivity in the pathogenesis of ocular onchocerciasis. *Invest Ophthalmol Vis Sci* 1993;34:2888–2902.
11. McKechnie NM, Braun G, Klager S, et al. Cross-reactive antigens in the pathogenesis of onchocerciasis. *Ann Trop Med Parasitol* 1993;87:649–652.
12. McMahon JE, Sowa SI, Maude GH, Kirkwood BR. Onchocerciasis in Sierra Leone. 2. A comparison of forest and savanna villages. *Trans R Soc Trop Med Hyg* 1988;82:595–600.
13. Meyer CG, Gallin M, Erttmann KD, et al. HLA-D alleles associated with generalized disease, localized disease, and putative immunity in Onchocerca volvulus infection. *Proc Natl Acad Sci USA* 1994;91:7515–7519.
14. Monjusiau AG, Lagraulet J, d'Haussy R, Goeckel CW. Aspects ophtalmologiques de l'onchocercose au Guatemala et en Afrique occidentale. *Bull World Health Organ* 1965;32:339–355.
15. Muller I, Garcia-Sanz JA, Titus R, Behin R, Louis J. Analysis of the cellular parameters of the immune responses contributing to resistance and susceptibility of mice to infection with the intracellular parasite, Leishmania major. *Immunol Rev* 1989;112:95–113.
16. Murdoch I, Jones BR, Babalola OE, Cousens SN, Bolarin I, Abiose A. Red-dot card test of the paracentral field as a screening test for optic nerve disease in onchocerciasis. *Bull World Health Organ* 1996;74:573–576.
17. Neumann E, Gunders AE. Pathogenesis of the posterior segment lesion of ocular onchocerciasis. *Am J Ophthalmol* 1973;75:82–89.
18. Newland HS, White AT, Greene BM, Murphy RP, Taylor HR. Ocular manifestations of onchocerciasis in a rain forest area of West Africa. *Br J Ophthalmol* 1991;75:163–169.
19. Paul EV, Zimmerman LE. Some observations on the ocular pathology of onchocerciasis. *Hum Pathol* 1970;1:581–594.
20. Rodriguez-Perez MA, Rodriguez MH, Margeli-Perez HM, Rivas-Alcala AR. Effect of semiannual treatments of ivermectin on the prevalence and intensity of Onchocerca volvulus skin infection, ocular lesions, and infectivity of Simulium ochraceum populations in southern Mexico. *Am J Trop Med Hyg* 1995;52:429–434.
21. Semba RD, Murphy RP, Newland HS, Awadzi K, Greene BM, Taylor HR. Longitudinal study of lesions in the posterior segment in onchocerciasis. *Ophthalmology* 1990;97:1334–1341.
22. World Health Organization. WHO Expert Committee on Onchocerciasis. Third report. *World Health Organ Tech Rep Ser* 1987;752:1–167.

American Brugian Filariasis

J. Kevin Baird, Mary K. Klassen-Fischer, Ronald C. Neafie, and Wayne M. Meyers

Introduction

Definition

In the Americas, 3 species of the genus *Brugia* (family Filariidae), a nematode that causes lymphatic filariasis in mammals, have been described. In North America, *Brugia beaveri* is a parasite of the raccoon (*Procyon lotor*) and perhaps of the bobcat and mink.[1,11] *Brugia leporis* is a parasite of swamp rabbits (*Sylvilagus aquaticus*) and eastern cottontails (*Sylvilagus floridanus alacer*).[5] *Brugia guyanensis* is the only known *Brugia* sp in South America. It parasitizes the coatimundi (*Nasua nasua*), a raccoon-like mammal, and the grison (*Grison vittatus*).[14]

Unidentified *Brugia* sp sometimes infect humans. Although the etiologic agents of these infections can be identified as *Brugia* based on their morphologic features in tissue, species identification is not possible. *Brugia beaveri* and *B. leporis* are the most likely causes of North American brugian filariasis.[7,11] In South America, *B. guyanensis* is endemic in areas where brugian infections in humans have been reported.[3] *Brugia* sp adults and microfilariae found in these patients were morphologically consistent with *B. guyanensis,* but the possibility of human infection by undescribed species cannot be excluded.

In Southeast Asia, *Brugia malayi* and *Brugia timori* cause filariasis in humans (see chapter 13 for a description of these worms). A single human infection by a *Brugia* sp, identified by microfilariae in peripheral blood, has been reported from Africa.[13]

Synonyms

Human infection with *Brugia* sp is variously referred to as brugian filariasis, zoonotic brugian filariasis, and brugian zoonosis.

General Considerations

First described in 1962,[16] zoonotic brugian filariasis is a medical curiosity. Even though many infections have probably been undiagnosed or misdiagnosed, reports of only 32 documented cases in 30 years suggest that this infection is truly rare.

Figure 19.1
Longitudinal section of nongravid female *Brugia* sp in lymph node, demonstrating thin, finely striated cuticle. Movat x770

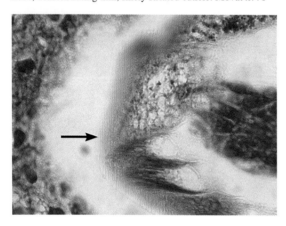

Figure 19.2
Fine striations (arrow) on surface of cuticle of worm shown in Figure 19.1. Movat x800

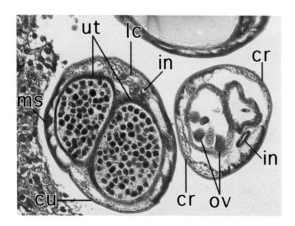

Figure 19.3
Two transverse sections at different levels of mature, nongravid female *Brugia* sp in lymph node, displaying thin cuticle (cu), prominent lateral cords (lc), muscle cells (ms), paired uteri containing globular material (ut), and intestine (in). Section at right shows internal longitudinal cuticular ridge (cr) and several unfertilized ova (ov) in uterus. Movat x315

Epidemiology

In North America, at least 27 brugian infections in humans have been reported since 1962, mostly from rural or suburban areas in the northeastern United States (Pennsylvania, New York, Connecticut, Rhode Island, and Massachusetts).[2,4,15,16] Reports have also come from Michigan, New Jersey, Ohio, North Carolina, Florida, Mississippi, Louisiana, Oklahoma, and California.[6,8,10,15] We report 2 cases from Canada (Figs 19.27 and 19.28). Three other cases originally reported as *Dirofilaria*-, *Dipetalonema*-, or *Brugia*-like have since been reclassified as probable *Brugia* infections.[15]

Since 1965, only 5 human infections by South American *Brugia* sp have been reported from Colombia, Peru, Ecuador, and Brazil.[3,12,15] Most of these patients had recently been in a jungle or rain forest.

Infectious Agent

Morphologic Description

Whole intact worms have rarely been found in humans. Most of our knowledge concerning the anatomy of these zoonotic *Brugia* sp has been obtained through experimental infections or by studying worms in human tissue sections.

In human tissue, zoonotic brugian worms are small filarial nematodes (see Table 19.1 for comparative measurements of zoonotic *Brugia* sp). Female worms are 50 to 130 μm in diameter; male worms are 35 to 55 μm in diameter. Both sexes have a finely striated cuticle that is 1 to 2 μm thick (Figs 19.1 and 19.2). The cuticle is usually slightly thicker in the lateral cord regions, and sometimes appears as a distinct internal ridge (Fig 19.3). Lateral cords vary considerably from inconspicuous to prominent (Figs 19.3 and 19.4). Musculature is coelomyarian with 3 to 6 muscle cells per quadrant. Both contractile and sarcoplasmic portions of muscle are usually readily identified (Figs 19.3 and 19.4). The contractile portion of muscle is especially well-developed in the posterior end of the male (Fig 19.5). The intestine is small and thin-walled, with several nuclei observable in transverse sections. Female worms have paired reproductive tubes and are usually nongravid (Figs 19.3

and 19.4). Gravid worms have only been observed in South American *Brugia* sp. Male worms, which may be immature or mature in humans, have a single reproductive tube (Fig 19.6) and 2 copulatory spicules. Mature males contain sperm.

Zoonotic *Brugia* sp microfilariae have rarely been observed in human peripheral blood. None of the humans infected with adult *Brugia* in North America were found to have circulating microfilariae. There is a case report of a patient from Alabama on steroids who had a single unidentified circulating microfilaria that may have been a *Brugia* sp. The formalin-fixed microfilaria was 250 μm long and 4.7 μm wide. The cephalic space was 9.2 μm long.[9] In another case, an 18-month-old immunodeficient boy from Oklahoma had circulating *Brugia* microfilariae.[17]

Brugia beaveri are tapered anteriorly and have a distinct head bulb. The esophagus is divided into anterior muscular and posterior glandular portions. The cuticle is thin and finely striated. Female worms are 22 to 74 mm long and 60 to 130 μm in diameter. The vulva is slightly anterior to the esophageal-intestinal junction. Reproductive tubes extend posteriorly to the end of the worm. Male worms are 12 to 28 mm long and 55 to 95 μm in diameter. Spicules are unequal and dissimilar. The left spicule is 190 to 270 μm long; the right spicule is 90 to 130 μm long (ratio 2:1 to 2.5:1). *Brugia beaveri* microfilariae are sheathed and approximately 300 by 6 μm. The cephalic space is approximately 16 μm long. The tail has a subterminal nucleus and a terminal nucleus separated by a constriction.[1,11]

Brugia leporis and *B. beaveri* share many morphologic features. *Brugia leporis* females are 39 to 46 mm long and 125 to 134 μm in diameter. The vulva is posterior to the middle of the esophagus. Males are 12 to 19 mm long and 64 to 80 μm in diameter. The left spicule is 210 to 225 μm long; the right spicule is 80 to 90 μm long (ratio 2.3:1 to 2.8:1). *Brugia leporis* microfilariae are sheathed and approximately 300 by 6 μm. The cephalic space is approximately 9 μm long. The tail has a subterminal nucleus and a terminal nucleus separated by a constriction. There is a distinct swelling around both nuclei.[5]

Brugia guyanensis is morphologically very similar to both *B. beaveri* and *B. leporis*. Females are 20 to 30 mm long and 70 to 91 μm in diameter. The

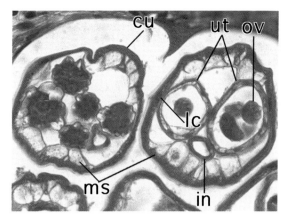

Figure 19.4
Two transverse sections through coiled, mature, nongravid female *Brugia* sp in cervical lymph node. Section at left is from posterior end of worm and contains intestine and 5 sections of ovary. Section at right is closer to midbody and contains intestine (in) and paired uteri (ut) with unfertilized ova (ov). Both sections demonstrate thin cuticle (cu), inconspicuous lateral cords (lc), and typical coelomyarian musculature (ms: left, sarcoplasmic; right, contractile). x560

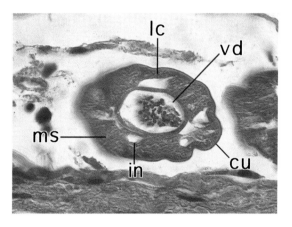

Figure 19.5
Transverse section through posterior region of mature male *Brugia* sp in lymph node. Note thin cuticle (cu), lateral cords (lc), intestine (in), spermatozoa in vas deferens (vd), and prominent contractile portion of muscle cells (ms). x645

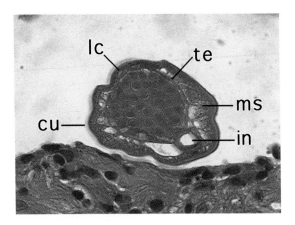

Figure 19.6
Transverse section through anterior portion of mature male *Brugia* sp illustrated in Figure 19.5 at level of testis (te) and intestine (in). Note thin cuticle (cu), lateral cords (lc), and few muscle cells (ms). x680

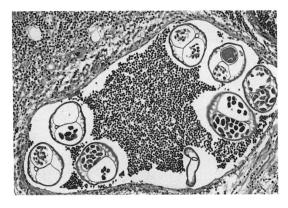

Figure 19.7
Transverse sections at different levels of viable, coiled, gravid South American brugian worm in dilated lymphatic of cervical lymph node. The patient, a 27-year-old woman from New York City, had been camping in the Peruvian rain forest. x115

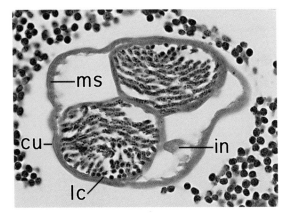

Figure 19.8
Transverse section of gravid South American brugian worm showing 2 uteri filled with microfilariae. Note thin cuticle (cu), inconspicuous lateral cords (lc), low coelomyarian musculature (ms), and fragment of intestine (in). x400

vulva lies slightly anterior to the middle of the esophagus. Males are 13 to 16 mm long and 52 to 60 µm in diameter. The left spicule is 353 to 390 µm long and terminates in a hook shape. The right spicule is 156 to 195 µm long with a tip that tapers to a sharp point. The ratio is 2.1:1. *Brugia guyanensis* microfilariae are sheathed and measure 213 to 232 µm long and 4 to 5 µm in diameter. The cephalic space is approximately 6 µm long. There is a considerable gap with a slight constriction between the last 2 nuclei that extend to the tip of the tail.[14] Baird and Neafie reported a *Brugia* sp, acquired by a woman in Peru, that could not be distinguished morphologically from *B. guyanensis*. They described a gravid female worm 100 µm in diameter (Figs 19.7 to 19.9), a mature male worm 50 µm in diameter (Figs 19.10 to 19.12), and microfilariae with tails characteristic of the genus (Fig 19.13).

Life Cycle and Transmission

Humans almost certainly acquire American brugian worms from mosquitoes which have been infected by feeding on a microfilaremic natural host. Because the species of *Brugia* infecting humans is not known, the natural intermediate and definitive hosts for these worms are also not known. Humans are apparently dead-end hosts for American *Brugia* sp; parasites migrate to lymphatic vessels, usually in lymph nodes, where they mature to adults but do not mate. Microfilariae have been observed in peripheral blood only twice.[9,17]

Clinical Features and Pathogenesis

American brugian filariasis generally produces no systemic signs or symptoms. Most patients present with a single nontender nodule associated with an inflamed lymph node or vessel, usually in the neck or groin. Nodules have also been reported on the chest, lung, limbs, abdominal wall, penis, and conjunctiva.[15,18] Antigens released by dead and dying worms provoke clinical symptoms. In most cases, inflammation is restricted to the involved lymph node. *Brugia* sp in aberrant hosts occasionally stray into blood vessels. In one patient, a female *Brugia* sp initially diagnosed as *Dirofilaria immitis* occluded a pulmonary blood vessel, causing an infarct.[15]

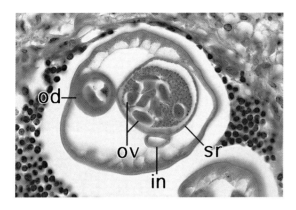

Figure 19.9
Transverse section of gravid South American brugian worm at level of intestine (in), thick-walled oviduct (od), and seminal receptacle (sr) filled with spermatozoa and several ova (ov). x380

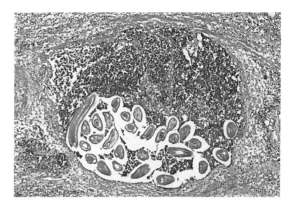

Figure 19.10
Transverse sections at different levels of viable, coiled, mature male South American brugian worm in dilated lymphatic of cervical lymph node in patient described in Figure 19.7. x60

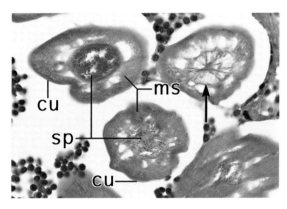

Figure 19.11
Four sections of mature male South American brugian worm. One section is through the esophagus (arrow); the others are in the posterior half of worm and show spermatozoa (sp) in the reproductive tubes. Note slightly thickened cuticle (cu) in lateral cord regions and well-developed contractile fibers of muscle cells (ms). x425

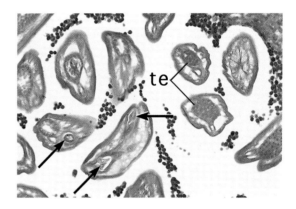

Figure 19.12
Several transverse sections at different levels of mature male South American brugian worm. Two sections are at level of testis (te). Two sections through posterior end of worm demonstrate spicules (arrows). x215

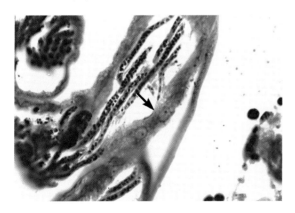

Figure 19.13
Section through gravid South American brugian worm depicting microfilariae-filled uterus. Note slight constriction between terminal nuclei (arrow) in tail of microfilaria. Giemsa x645

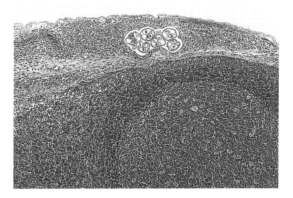

Figure 19.14
North American brugian female worm (same as described in Figure 19.4) in capsule of cervical lymph node from 35-year-old college professor from California. Patient had swollen lymph node for 2 months and 10% eosinophilia. x55

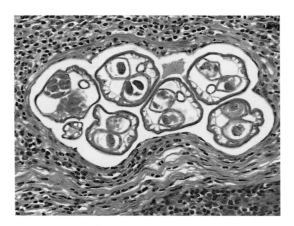

Figure 19.15
Higher magnification of worm in Figure 19.14 showing viable, coiled, mature nongravid female *Brugia* sp causing minimal reaction with a few eosinophils. x230

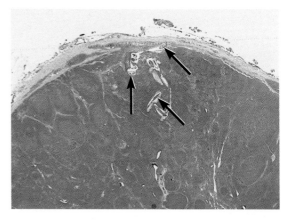

Figure 19.16
North American brugian female worm (arrows) (same as described in Figures 19.1 to 19.3) in submaxillary lymph node of 31-year-old aquatic ecologist from Florida. Movat x7.5

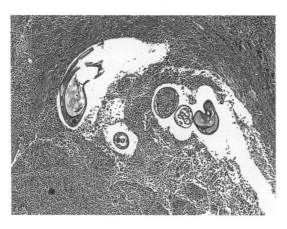

Figure 19.17
Higher magnification of worm in Figure 19.16 showing viable, coiled, mature nongravid female *Brugia* sp in parenchyma of lymph node, causing minimal reaction. x60

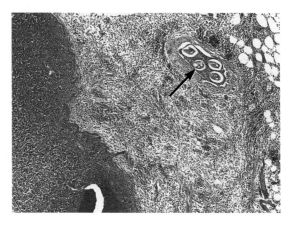

Figure 19.18
North American brugian female worm (arrow) in perinodal tissue of scapular lymph node in 19-year-old man from Florida. x40

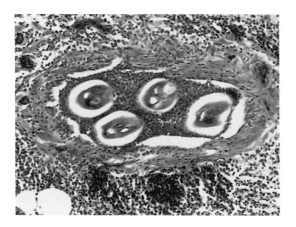

Figure 19.19
Higher magnification of worm in Figure 19.18 showing coiled, viable, nongravid female *Brugia* sp surrounded by inflammatory cells in lumen of blood vessel. Note massive eosinophilia. x115

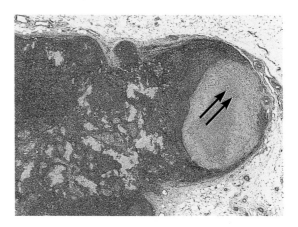

Figure 19.20
North American brugian male worm (arrows) in caseating granuloma of inguinal lymph node of 56-year-old woman from Connecticut. x25

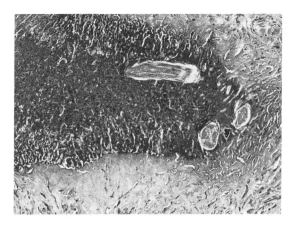

Figure 19.21
Movat-stained section of caseating granuloma pictured in Figure 19.20, showing coiled, mature male *Brugia* sp. Worm is 50 μm in diameter. x120

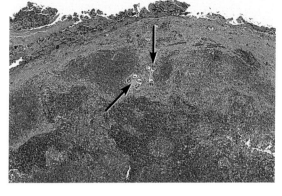

Figure 19.22
North American brugian worm (arrows) in cervical lymph node of 25-year-old welder from Mississippi. The coiled, nongravid female *Brugia* sp is viable and causing minimal reaction. x25

Pathologic Features

Histopathologic changes caused by American *Brugia* infections can involve a minimal reaction, a fibrinopurulent reaction, or a necrotizing granulomatous reaction. Worms are usually found within dilated lymphatic vessel lumens, or within the capsule (Figs 19.14 and 19.15) or parenchyma of a lymph node (Figs 19.16 and 19.17). The lymphatic vessel wall or capsule may be thickened. In some cases the lymph node is reactive, with follicular hyperplasia or sinus histiocytosis, but little or no inflammatory response.

Worms may be surrounded by a dense fibrinous exudate of mixed acute inflammatory cells, including neutrophils and some eosinophils (Figs 19.18, 19.19, and 19.28), or by a caseous granuloma with palisading epithelioid histiocytes and foreign body giant cells (Figs 19.20 and 19.21). Infected lymph nodes sometimes have other necrotizing granulomas with no visible worms (Figs 19.22 and 19.23). Some patients have a spectrum of reactions within a single lymph node (Figs 19.24 to 19.26).

In most zoonotic brugian infections, worms are found in lymph nodes, but other sites have been reported. Orihel and Beaver reported a *Brugia* sp in a nodule from the penis of a 39-year-old Brazilian.[15] Streeten et al described a *Brugia* sp in the conjunctiva of a 64-year-old female from Ecuador.[18] We report cases involving the uterus (Fig 19.27), synovium (Fig 19.28), and conjunctiva (Fig 19.29).

Since adult female *Brugia* are usually nongravid, lesions caused by microfilariae are rare. In one case (Figs 19.7 to 19.13), viable and degenerating microfilariae were observed in a lymph node next to a gravid South American brugian worm. Viable microfilariae (Fig 19.30) were found in the lumen and wall of the dilated lymphatic vessel harboring adult worms. Degenerating microfilariae were seen in the necrotic centers of follicles, in dilated and occluded lymphatic vessels in surrounding adipose tissue, and in granulomas in the parenchyma of the lymph node (Fig 19.31). Microfilariae were not found in the peripheral blood of this patient.

Diagnosis

Serologic or DNA probe diagnostic procedures are not widely available. Diagnosis usually depends on identifying the worm in tissue sections. In many instances the worm's internal features have degenerated. Diagnosis must then be based on cuticular features, location of the worm in lymph nodes, lymphatics, or arteries, and the patient's travel history.

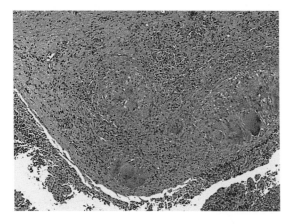

Figure 19.23
Noncaseating granulomatous inflammation and eosinophilia in same lymph node described in Figure 19.22. No worm was found in this area of the lymph node. x60

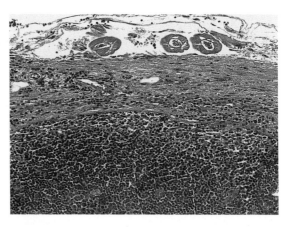

Figure 19.24
Mature, viable, male North American brugian worm (same as described in Figures 19.5 and 19.6) in capsule of inguinal lymph node of 36-year-old woman from Rhode Island, causing minimal reaction. x110

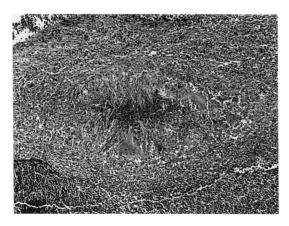

Figure 19.25
Necrotizing granuloma in another area of same lymph node described in Figure 19.24. x75

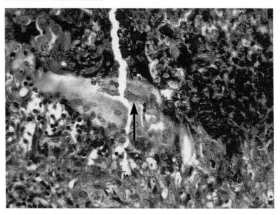

Figure 19.26
Higher magnification of necrotizing granuloma depicted in Figure 19.25 showing fragment of cuticle (arrow) from a degenerating worm. This lymph node probably contains 2 worms. x230

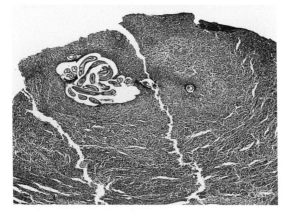

Figure 19.27
North American brugian worm in necrotizing granuloma on surface of uterus in patient from Canada. The coiled, nongravid female worm is 40 μm in diameter. x40

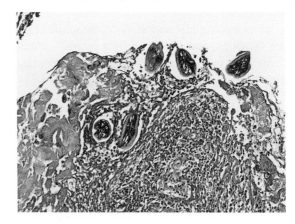

Figure 19.28
North American brugian worm in synovium of right arm of 43-year-old patient from Canada. The coiled worm is in a fibrinopurulent exudate composed of eosinophils, neutrophils, and fibrin. PAS x120

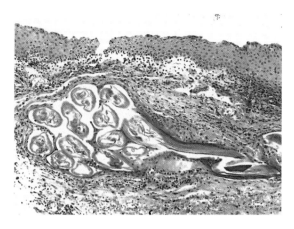

Figure 19.29
South American brugian worm in conjunctiva of patient from Ecuador. The coiled, viable, nongravid female worm is 50 μm in diameter. x95

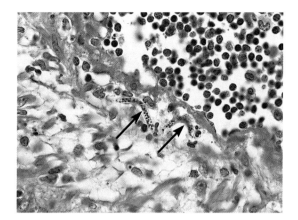

Figure 19.30
Viable microfilaria (arrows) in cervical lymph node. Same patient with South American brugian worm described in Figures 19.7 to 19.13. x425

Treatment

There is a report of diethylcarbamazine (DEC) being used to treat a brugian infection in a 2-year-old patient in Colombia.[12] However, surgical removal of the involved lymph node is curative and the treatment of choice. Clinical intervention beyond simple follow-up is not indicated.

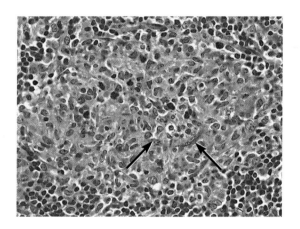

Figure 19.31
Degenerated microfilaria (arrows) in granuloma in parenchyma of cervical lymph node. Same patient with South American brugian worm described in Figures 19.7 to 19.13. x330

	Female			Male				
	Length (mm)	Maximum width	Anterior end to vulva	Length (mm)	Maximum width	Left spicule	Right spicule	Spicule ratio
B. beaveri	22-74 (34)	60-130 (94)	410-890 (577)	12-28 (20)	55-95 (66)	190-270 (237)	90-130 (103)	2-2.5:1
B. leporis	39-46	125-134	532-580	12-19	64-80	210-225	80-90	2.3-2.8:1
B. guyanensis	20-30	70-91	363-374	13-16	52-60	353-390	156-195	(2.1:1)

	Sheathed Microfilaria			
	Length	Width	Length of cephalic space	Length of Innenkörper
B. beaveri	285-325 (303)	4.5-6.5	14-17 (16)	40-45
B. leporis	275-330 (300)	5-7	7-11 (9)	50
B. guyanensis	213-232	4-5	5-6	5-8

Table 19.1
Dimensions of zoonotic *Brugia* sp.[1,5,11,14] All measurements in micrometers except where indicated. Mean in parentheses.

References

1. Ash LR, Little MD. Brugia beaveri sp. n. (Nematoda: Filarioidea) from the raccoon (Procyon lotor) in Louisiana. *J Parasitol* 1964;50:119-123.
2. Baird JK, Alpert LI, Friedman R, Schraft WC, Connor DH. North American brugian filariasis: report of nine infections of humans. *Am J Trop Med Hyg* 1986;35:1205-1209.
3. Baird JK, Neafie RC. South American brugian filariasis: report of a human infection acquired in Peru. *Am J Trop Med Hyg* 1988;39:185-188.
4. Coolidge C, Weller PF, Ramsey PG, Ottesen EA, Beaver PC, von Lichtenberg FC. Zoonotic Brugia filariasis in New England. *Ann Intern Med* 1979;90:341-343.
5. Eberhard ML. Brugia lepori sp. n. (Filarioidea: Onchocercidae) from rabbits (Sylvilagus aquaticus, S. floridanus) in Louisiana. *J Parasitol* 1984;70:576-579.
6. Eberhard ML, DeMeester LJ, Martin BW, Lammie PJ. Zoonotic Brugia infection in western Michigan. *Am J Surg Pathol* 1993;17:1058-1061.
7. Eberhard ML, Telford SR 3d, Spielman A. A Brugia species infecting rabbits in the northeastern United States. *J Parasitol* 1991;77:796-798.
8. Elenitoba-Johnson KS, Eberhard ML, Dauphinais RM, Lammie PJ, Khorsand J. Zoonotic Brugian lymphadenitis. An unusual case with florid monocytoid B-cell proliferation. *Am J Clin Pathol* 1996;105:384-387.
9. Greene BM, Otto GF, Greenough WB. Circulating non-human microfilaria in a patient with systemic lupus erythematosus. *Am J Trop Med Hyg* 1978;27:905-909.
10. Gutierrez Y, Petras RE. Brugia infection in Northern Ohio. *Am J Trop Med Hyg* 1982;31:1128-1130.
11. Harbut CL, Orihel TC. Brugia beaveri: microscopic morphology in host tissues and observations on its life history. *J Parasitol* 1995;81:239-243.
12. Kozek WJ, Reyes MA, Ehrman J, Garrido F, Nieto M. Enzootic Brugia infection in a two-year-old Colombian girl. *Am J Trop Med Hyg* 1984;33:65-69.
13. Menendez MC, Bouza M. Brugia species in a man from Western Ethiopia. *Am J Trop Med Hyg* 1988;39:189-190.
14. Orihel TC. Brugia guyanensis sp. n. (Nematoda: Filarioidea) from the coatimundi (Nasua nasua vittata) in British Guiana. *J Parasitol* 1964;50:115-118.
15. Orihel TC, Beaver PC. Zoonotic Brugia infections in North and South America. *Am J Trop Med Hyg* 1989;40:638-647.
16. Rosenblatt P, Beaver PC, Orihel TC. A filarial infection apparently acquired in New York City. *Am J Trop Med Hyg* 1962;11:641-645.
17. Simmons CF Jr, Winter HS, Berde C, et al. Zoonotic filariasis with lymphedema in an immunodeficient infant. *N Engl J Med* 1984;310:1243-1245.
18. Streeten BW, Beaver PC, Mueller JF. Recurrent episcleritis due to Brugia filariasis. Presented at the Verhoeff Society Meeting; April 13-15, 1989; Boston, Mass.

20

Dracunculiasis

Ronald C. Neafie *and*
Aileen M. Marty

Introduction

Definition

Dracunculiasis is infection by the nematode *Dracunculus medinensis* (Linnaeus, 1758). The main clinical feature is the sudden appearance of a painful blister that ulcerates and slowly releases a gravid worm.

Synonyms

Galen (circa 129 to 199 AD) first named the disease dracontiasis. The infection is also known as dracunculosis. Informal names for the parasite include Guinea worm, Medina worm, dragon worm, and serpent worm. The name Guinea worm was first used by Bruno, who in 1611 made several voyages to West Africa and noted the parasite in inhabitants of the Guinea coast.[10] The species name, *medinensis*, is associated with the ancient focus of infection around the Arabian city of Medina.[26]

General Considerations

Although early observers referred to the etiologic agent as *Filaria medina* or *Filaria medinensis*, *D. medinensis* is a member of the family Dracunculidae, and is not a filarial worm.

Dracunculiasis is among the oldest known parasitic diseases. The well-known Egyptian medical document, the *Papyrus Ebers* (circa 1550 BC),[5] describes the traditional technique of removing the gravid worm by winding it on a stick. Velschius, in 1674, was the first to suggest that the medical emblem of the caduceus may portray this therapeutic custom.[28] Rhazes (circa 900 AD), a Persian doctor, realized that what emerged from the ulcer was a worm, and not, as was previously thought, a protruding nerve. Lind, in 1768, suspected that contaminated drinking water was the source of infection. In 1869 the Russian scientist Fedchenko discovered larvae of *D. medinensis* in the body cavities of copepods (water fleas) found in drinking water. In 1936 Moorthy and Sweet first described the early developmental stages and sexual maturation of *D. medinensis*. A global campaign to eradicate dracunculiasis by the year 1995 began in 1981 and has led to a dramatic decrease in the incidence of infection. In 1980 the World Health Organization estimated the annual incidence of dracunculiasis to be 10 million. In 1993 incidence had fallen to less than 250 000; however, large populations remain at risk of infection.[12] In 1998, dracunculiasis was still prevalent in many countries, especially in West Africa.

Epidemiology

Dracunculiasis is currently found almost exclusively in rural areas of Africa and Asia. The disease is endemic in 16 African countries: Benin, Burkina Faso, Cameroon, Côte d'Ivoire, Chad, Ethiopia, Ghana, Kenya, Mali, Mauritania, Niger, Nigeria, Senegal, Sudan, Togo, and Uganda.[12] In Asia, dracunculiasis was formerly prevalent from India to Turkey, including the Middle East, the Arabian peninsula, and southwestern Russia. Now, however, only India and Pakistan remain endemic, and Pakistan has nearly eradicated the worm. Dracunculiasis is no longer reported in the West Indies or South America. The recent report of a patient with dracunculiasis acquired in Japan remains unexplained.[16] Worms emerge most commonly in the dry season, and infection is more frequent in men than in women. Individuals in their most productive years (10 to 60 years of age) account for over 90% of infections.[20]

Infectious Agent

Morphologic Description

Female *D. medinensis* are long, thin worms (70 to 120 cm by 0.09 to 0.17 cm). Gravid worms are usually white-tan, but are sometimes bright red.[6] The smooth cuticle is 30 to 50 µm thick (Figs 20.1 and 20.2). The hypodermis is barely perceptible, and lateral cords are not usually discernible. There are 2 clearly divided, thick bands of somatic muscle cells on the ventral and dorsal sides of the worm (Fig 20.1). Each band of muscle contains many muscle cells (Fig 20.2). Usually, only the contractile portion of the somatic muscle cell is prominent (Fig 20.2). The intestine atrophies in the gravid worm and is difficult to identify. Eventually, the gravid worm's reproductive organs also atrophy and the entire body cavity fills with rhabditoid larvae (Fig 20.1). These larvae measure 250 to 750 µm by 10 to 25 µm, have a digestive tube, and

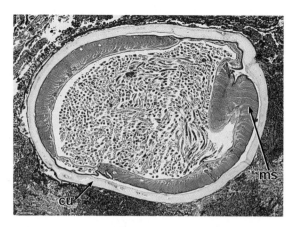

Figure 20.1
Transverse section of gravid *D. medinensis* in lesion from eyelid. Note thick cuticle (cu), 2 prominent bands of somatic muscle cells (ms), and body cavity filled with rhabditoid larvae. Movat x60

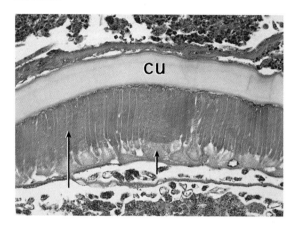

Figure 20.2
Section through body wall of gravid *D. medinensis* showing thick cuticle (cu). Contractile portion of muscle cell (long arrow) is more prominent than sarcoplasmic portion (short arrow). x170

prominent annulations in the cuticle (Figs 20.3 to 20.6). Their bodies taper posteriorly to form a long filiform tail (Figs 20.7 and 20.8). The digestive tube, cuticular annulations, and large size of these larvae distinguish them from microfilariae.

Dracunculus medinensis males are much smaller than females and are rarely seen in humans. One exceptional report describes a single male worm 4 cm long in a man from India.[18] Male worms recovered from experimentally infected dogs measure 1.2 to 2.9 cm by 0.4 mm.[21]

Life Cycle

The life cycle of *D. medinensis* requires 2 hosts (Fig 20.9). Freshwater copepods (Fig 20.10) of the

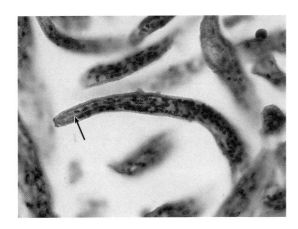

Figure 20.3
Anterior end of rhabditoid larva of *D. medinensis*. Note digestive tube (arrow). x675

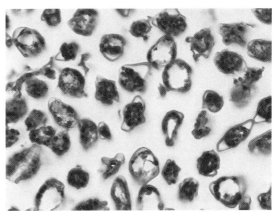

Figure 20.4
Transverse sections of larvae of *D. medinensis* in body cavity of gravid worm. Larvae have maximum diameter of 25 μm. x625

Figure 20.5
Scanning electron micrograph showing numerous rhabditoid larvae within body cavity of gravid *D. medinensis*. Note tapered posterior end and long tail (arrow). x280

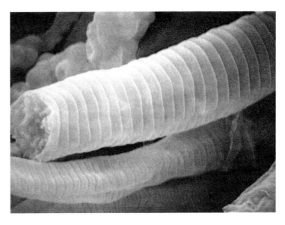

Figure 20.6
Scanning electron micrograph of 2 larvae of *D. medinensis*. Note prominent annulations of cuticle. x1590

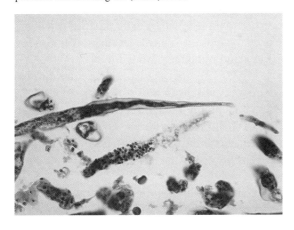

Figure 20.7
Tapered posterior end of rhabditoid larva of *D. medinensis*. x415

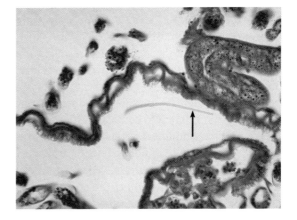

Figure 20.8
Section showing long filiform tail of rhabditoid larva of *D. medinensis*. There is no cytoplasm in tip of tail (arrow). x420

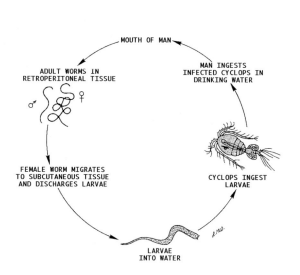

Figure 20.9
Life cycle of *D. medinensis*.

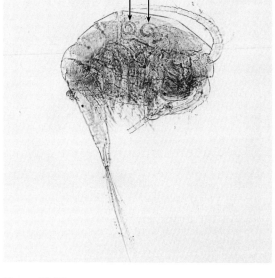

Figure 20.10
Several larvae of *D. medinensis* (arrows) within body cavity of copepod, the intermediate host.* x55

family Cyclopidae (eg, *Acanthocyclops* sp, *Thermocyclops* sp, and *Mesocyclops aequatorialis similis*) serve as intermediate hosts.[27] Many other species of *Dracunculus* infect wild animals worldwide, but do not infect humans. Humans are the only natural definitive hosts of *D. medinensis*.

Humans are infected by drinking water that contains copepods contaminated with larvae of *D. medinensis*. In the gut, larvae leave the copepods, penetrate the intestinal wall, and migrate to the retroperitoneal tissues, where they develop into adult worms in 8 to 12 months. Male worms die soon after mating. Fertilized female worms may migrate throughout the body, some reaching the subcutaneous tissue and skin, where they can discharge their larvae.

A gravid *D. medinensis* usually produces a painful blister in the skin. On contact with water, the blister ruptures and the anterior end of the worm protrudes and discharges hundreds of thousands of larvae into the water (Fig 20.11). When swallowed by a copepod, larvae penetrate the host's tissues and enter the hemocoelom, where they develop into infective third-stage larvae in 2 to 4 weeks. The life cycle is perpetuated when humans ingest these infected copepods.

Clinical Features and Pathogenesis

There are no recognizable symptoms related to either the penetration of the human intestine by the larvae, or to the subsequent development of larvae into adult worms in retroperitoneal tissues. Symptoms of inflammation usually begin when the female worm migrates to the skin 2 to 5 days before a characteristic blister arises from the action of a toxin secreted from the anterior end of the worm. Lesions are typically on the lower leg (Fig 20.12), and frequently around the lateral malleolus (Fig 20.13),[13] but the worm can migrate anywhere and present in any part of the body (Fig 20.14), including breast, scalp, tongue, scrotum (Fig 20.15), penis, buttocks, knee, ankle, wrist, forearm, uterus, broad ligament, Bartholin's gland, pericardium, inguinal lymph nodes, extradural space, eyelid, orbit, or around the sciatic nerve. Frequently, blisters appear on the backs of Indian water carriers, stimulated perhaps by the moisture. The tortuous track of the worm, typically a red, raised, serpentine path, is often visible, and may resemble cutaneous larva migrans.[24] Female worms that do not emerge will die within the human body

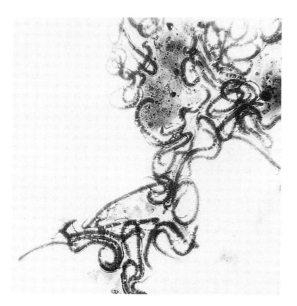

Figure 20.11
This cluster of larvae of *D. medinensis* emerged from a gravid worm into surrounding water, where the larvae can be ingested by copepods.

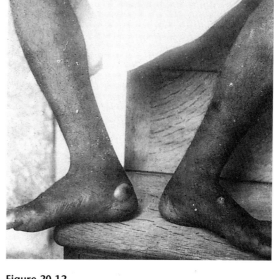

Figure 20.12
Blisters develop in skin, usually on lower extremities, when gravid *D. medinensis* approaches surface of skin.

after about a year. Dead worms become calcified in body tissues, or are absorbed. Clinically, dead or dying worms that form masses in tissues can mimic dermoid cysts, Burkitt's lymphoma,[3] leiomyoma,[1] or a variety of other tumors or infectious processes. While dead worms usually produce no morbidity, symptoms can arise when a worm dies in a critical location (eg, a dead worm in the joint space of the knee can produce aseptic effusion).[19]

In the skin or subcutis, the gravid worm produces induration, edema (Fig 20.16), and eventually a blister. On contact with water this blister ruptures, alleviating pain as the worm releases its larvae. The worm remains viable and continues to discharge larvae each time the blister comes in contact with water. After discharging all its larvae, the worm dies and often calcifies. Radiologically, the calcified worm appears as a beaded, linear, or coiled structure (Fig 20.17).

Urticaria, dyspnea, fainting, giddiness, vomiting, diarrhea, fever, and asthma have all been associated with dracunculiasis. These symptoms develop most often following the rupture of a gravid worm, when eosinophilia may be as high as 40%. Contracture and deformity around joints are frequent complications. Though the sinus tract of the worm usually facilitates drainage, bacterial invasion may cause cellulitis, focal suppuration, gangrene, or tetanus.

The age of the individual, the site and number of ulcers, and the mode of treatment all affect the degree of disability.

Pathologic Features

Lesions develop wherever female worms infiltrate tissues or spaces. Worms that rupture before emerging from the body produce acute cystic swellings (Fig 20.18). Microscopic examination of the cystic area reveals multiple sections of the ruptured, coiled, gravid worm centered in a large necrotizing granuloma (Figs 20.19 and 20.20). The peripheral wall of the granuloma is composed mostly of epithelioid cells, macrophages, and giant cells (Fig 20.21). Larvae released from the ruptured worm into surrounding tissue cause an acute inflammatory response (Fig 20.22). Beyond the granuloma there is perivascular cuffing by lymphocytes and plasma cells (Fig 20.23), and scattered collections of eosinophils (Fig 20.24).

Chronic inflammation and fibrosis surround

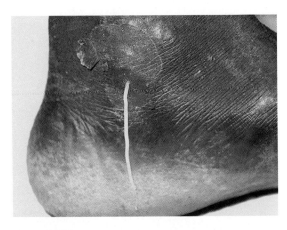

Figure 20.13
Female *D. medinensis* protruding from ruptured blister near malleolus.

Figure 20.14
Female *D. medinensis* protruding from ruptured blister on finger of African.

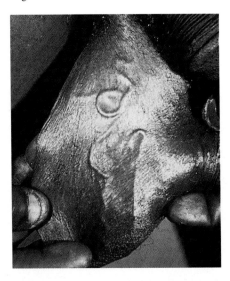

Figure 20.15
Tortuous track of female *D. medinensis* in subcutaneous tissue of scrotum.*

degenerated or calcified worms (Figs 20.25 to 20.31). These worms are often more difficult to identify and may require special staining, such as the modified Russell-Movat pentachrome method (Figs 20.28 and 20.31).

Diagnosis

Diagnosis depends on identifying the adult worm or larvae. Adult worms are identified grossly, radiographically, or in histopathologic sections. Usually, radiographs are helpful only when worms are calcified (Fig 20.17). Using radiopaque material will highlight live worms, but is not usually necessary. When the track of the worm is evident and there are characteristic clinical symptoms, a presumptive diagnosis of dracunculiasis is warranted. ELISA and SDS–PAGE/Western blot techniques are available for seroepidemiologic studies.[2] New attempts at diagnosis in the prepatent period using the Falcon® assay screening test–ELISA, and by enzyme-linked immunoelectrotransfer blot techniques, have proved promising.[9]

Treatment and Prevention

Appropriate topical or systemic pain relievers may be required.[23] Antihistamines can relieve allergic reactions. It is important to prevent or treat secondary bacterial infections that frequently complicate dracunculiasis. Patients at risk for tetanus should receive tetanus toxoid. The ancient remedy of winding the worm on a stick after a sufficient segment of the adult female has emerged through the skin can be effective, but carries significant risk (Fig 20.33). Care must be taken to avoid squeezing larvae back into that portion of the worm still in the skin. The resulting engorgement may rupture the worm, causing a violent tissue reaction and systemic allergic response that may lead to long-lasting disability. Removing the worm by careful surgical dissection is recommended.[25]

No drug is completely effective against dracunculiasis.[7,11] Niridazole, thiabendazole, and metronidazole may kill adult worms and are useful in inducing early expulsion of the worm or to facili-

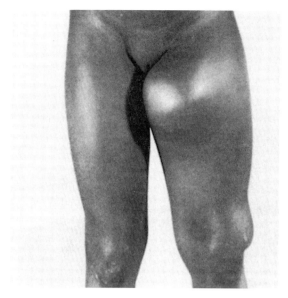

Figure 20.16
Swellings of left thigh and knee of young girl. Each swelling contains an adult female *D. medinensis*. The worms have provoked cysts and cellulitis.

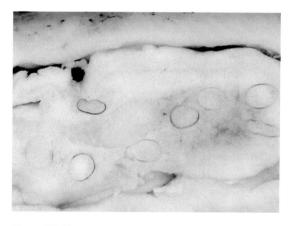

Figure 20.18
Gross specimen of 1-week-old cystic lesion in eyelid containing coiled, adult female *D. medinensis* in necrotic tissue. x5

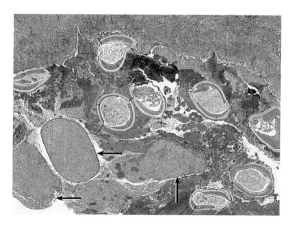

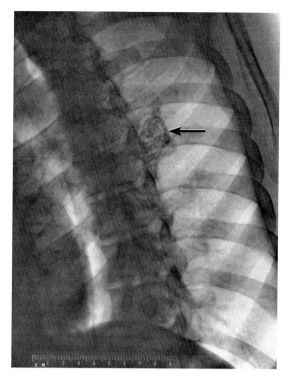

Figure 20.17
Radiograph of chest of patient from Yemen showing coiled, calcified *D. medinensis* (arrow) at fourth anterior intercostal space on left chest wall.*

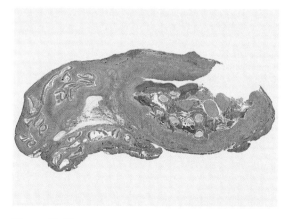

Figure 20.19
Multiple sections of ruptured, coiled, gravid worm shown in Figure 20.18 in large necrotizing granuloma from cystic mass of eyelid. x2.9

Figure 20.20
Left: Higher magnification of Figure 20.19 showing sections of viable worm and well-circumscribed masses of larvae (arrows) from ruptured worm. x11

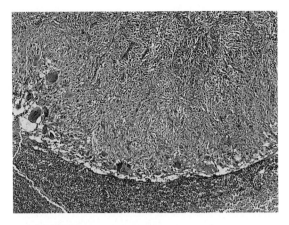

Figure 20.21
Necrotizing granuloma caused by ruptured worm. Numerous neutrophils are adjacent to epithelioid cells, macrophages, and giant cells. x60

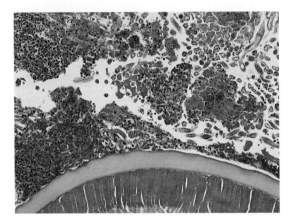

Figure 20.22
Numerous viable and degenerating larvae in acute inflammatory infiltrate outside adult worm. x115

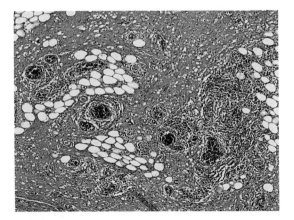

Figure 20.23
Perivascular cuffing by chronic inflammatory cells in wall of cystic mass adjacent to granuloma not visible in this photo. x60

tate surgical or manual removal.[22] Mebendazole and other systemic anthelmintics are contraindicated in mass treatment.[4] Topical agents such as chlortetracycline ointment may be useful in reducing secondary infections.[8]

Patients do not develop a specific immunity, and there is no vaccine. Eradication efforts are directed at preventing infection by interrupting the life cycle of the parasite.[17] Measures include: 1) providing protected sources of drinking water; 2) educating the population to boil drinking water or filter it through a fine-mesh cloth[14] and to avoid contaminating its source; 3) treating infected individuals; and 4) treating contaminated pond water with chemicals that kill copepods and infective larvae but are not harmful to humans.[15]

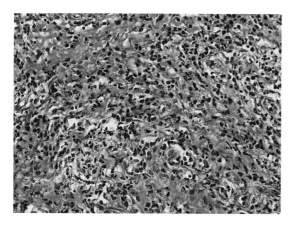

Figure 20.24
Focal collection of eosinophils in wall of cystic mass associated with degenerating *D. medinensis*. x150

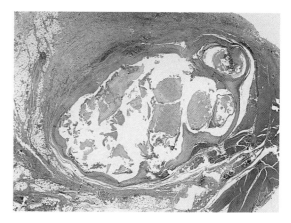

Figure 20.25
Fibrotic nodule in subcutaneous tissue of lumbar area containing degenerated gravid *D. medinensis*. x7

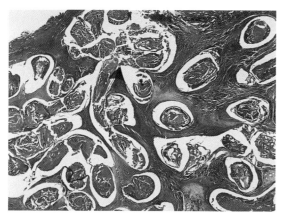

Figure 20.26
Multiple sections of gravid *D. medinensis* embedded in fibrotic tissue in broad ligament of 29-year-old Nigerian. x5.7

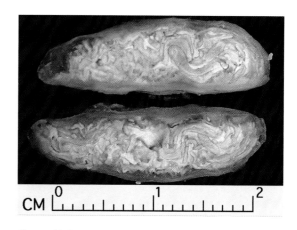

Figure 20.27
Cut section of calcified suprapubic nodule. Numerous sections of coiled *D. medinensis* are evident.

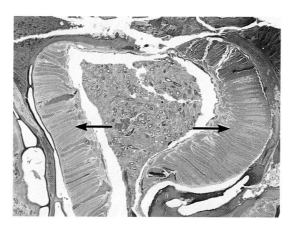

Figure 20.28
Stained transverse section through coiled worm shown in Figure 20.27. Two prominent bands of somatic muscle cells (arrows) and numerous rhabditoid larvae filling body cavity identify worm as a gravid *D. medinensis*. Movat x60

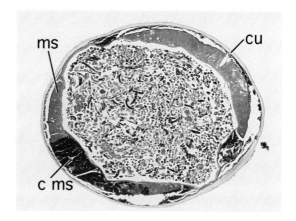

Figure 20.29
Transverse section of partially calcified gravid *D. medinensis* from subcutaneous tissue. Body cavity is filled with larvae, some calcified. Note viable muscle (ms), calcified muscle (c ms), and cuticle (cu). x45

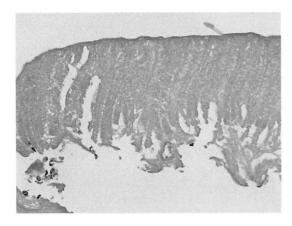

Figure 20.30
Poorly preserved band of somatic muscle of *D. medinensis* from lesion in Figure 20.25 with routine H&E stain. Compare with Movat stain in Figure 20.31. x240

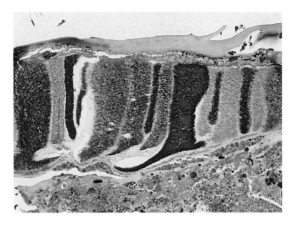

Figure 20.31
Poorly preserved band of somatic muscle of *D. medinensis* similar to that seen in Figure 20.30. Note better detail of individual muscle cells and fibers. Movat x235

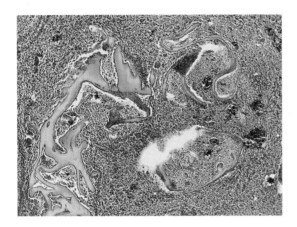

Figure 20.32
Fragments of cuticle from partially calcified female *D. medinensis* in fibrinopurulent exudate. x70

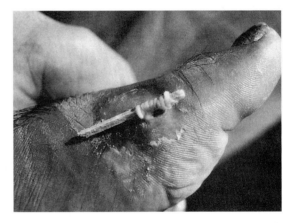

Figure 20.33
Female *D. medinensis* being removed by winding on a stick.*

References

1. Allaire AD, Majmudar B. Dracunculosis of the broad ligament. A case of a "parasitic leiomyoma." *Am J Surg Pathol* 1993;17:937–940.
2. Bloch P, Simonsen PE, Vennervald BJ. The antibody response to Dracunculus medinensis in an endemic human population of northern Ghana. *J Helminthol* 1993;67:37–48.
3. Burnier M Jr, Hidayat AA, Neafie R. Dracunculiasis of the orbit and eyelid. Light and electron microscopic observations of two cases. *Ophthalmology* 1991;98:919–924.
4. Chippaux JP. Mebendazole treatment of dracunculiasis. *Trans R Soc Trop Med Hyg* 1991;85:280.
5. Ebbell B. *The Papyrus Ebers, the Greatest Egyptian Medical Document*. Copenhagen, Denmark: Levin & Munksgaard; 1937:135.
6. Eberhard ML, Rab MA, Dilshad MN. Red Dracunculus medinensis. *Am J Trop Med Hyg* 1989;41:479–481.
7. Eberhard ML, Brandt FH, Ruiz-Tiben E, Hightower A. Chemoprophylactic drug trials for treatment of dracunculiasis using the Dracunculus insignis-ferret model. *J Helminthol* 1990;64:79–86.
8. Eberhard ML, Brandt FH, Kaiser RL. Chlortetracycline for dracunculiasis [letter]. *Lancet* 1991;337:500.
9. Fagbemi BO, Hillyer GV. Immunodiagnosis of dracunculiasis by Falcon assay screening test-enzyme-linked immunosorbent assay (FAST-ELISA) and by enzyme-linked immunoelectrotransfer blot (EITB) technique. *Am J Trop Med Hyg* 1990;43:665–668.
10. Grove DI. *A History of Human Helminthology*. Wallingford, England: CAB Int; 1990:698.
11. Hopkins DR. Eradication of dracunculiasis. In: Bourne P, ed. *Water and Sanitation: Economic and Sociological Perspectives*. Orlando, Fla: Academic Press; 1984:93–112.
12. Hopkins DR, Ruiz-Tiben E, Ruebush T 2d, Agle AN, Withers PC Jr. Dracunculiasis eradication: March 1994 update. *Am J Trop Med Hyg* 1995;52:14–20.
13. Ilegbodu VA, Ilegbodu AE, Wise RA, Christensen BL, Kale OO. Clinical manifestations, disability and use of folk medicine in Dracunculus infection in Nigeria. *J Trop Med Hyg* 1991;94:35–41.
14. Imtiaz R, Anderson JD, Long EG, Sullivan JJ, Cline BL. Monofilament nylon filters for preventing dracunculiasis: durability and copepod retention after long-term field use in Pakistan. *Trop Med Parasitol* 1990;41:251–253.
15. Kaul SM, Saxena VK, Sharma RS, Raina VK, Mohanty B, Kumar A. Monitoring of temephos (abate) application as a cyclopicide under the guineaworm eradication programme in India. *J Commun Dis* 1990;22:72–76.
16. Kobayashi A, Katakura K, Hamada A, et al. Human case of dracunculiasis in Japan. *Am J Trop Med Hyg* 1986;35:159–161.
17. Litvinov SK. How the USSR rid itself of dracunculiasis. *World Health Forum* 1991;12:217–219.
18. Moorthy VN. A re-description of Dracunculus medinensis. *J Parasitol* 1937;23:220–224.
19. Noji EK. Aseptic knee effusion associated with calcified guinea worms. *Ann Emerg Med* 1985;14:1119–1121.
20. Nwoke BE. Behavioural aspects and their possible uses in the control of dracontiasis (guinea-worm) in Igwun river basin area of Imo State, Nigeria. *Angew Parasitol* 1992;33:205–210.
21. Onabamiro SD. The early stages of the development of Dracunculus medinensis (Linnaeus) in the mammalian host. *Ann Trop Med Parasitol* 1956;50:157–166.
22. Pardanani DS, Trivedi VD, Joshi LG, Daulatram J, Nandi JS. Metronidazole (Flagyl) in dracunculiasis: a double blind study. *Ann Trop Med Parasitol* 1977;71:45–52.
23. Rab MA, Khan RN, Atiq A, Ahmed SA. Dracunculiasis: an approach to hasten worm expulsion. *J Trop Med Hyg* 1991;94:325–326.
24. Reddy NB, Srinivasan T. Dracunculus medinensis presenting as larva migrans [letter]. *Trop Doct* 1985;15:148–149.
25. Rohde JE, Sharma BL, Patton H, Deegan C, Sherry JM. Surgical extraction of guinea worm: disability reduction and contribution to disease control. *Am J Trop Med Hyg* 1993;48:71–76.
26. Singer S. Notes on some early references to tropical diseases. *Ann Trop Med Parasitol* 1912;6:386–392.
27. Sullivan JJ, Bishop HS, Hightower AW. Susceptibility of four species of copepods, from areas of endemic Dracunculus medinensis, to the North American D. insignis. *Ann Trop Med Parasitol* 1991;85:637–643.
28. Velschius (Welsch) GH. *Exercitatio de Vena Medinensi, ad mentem Ebnisinae sive de dracunculi veterum. Specimens exhibens novae versiones ex Arabico, cum commentario uberiori, cui accedit altera de vermiculus capillaribus infantium*. Augsburg, Holy Roman Empire: Augustae Vindelicorum, Impensis Theophili Goebelli; 1674:456.

21

Strongyloidiasis

Wayne M. Meyers,
Ronald C. Neafie, *and*
Aileen M. Marty

Introduction

Definition

Strongyloides stercoralis causes classic strongyloidiasis, a nematode infection with cutaneous, pulmonary, and intestinal manifestations. *Strongyloides fuelleborni*, a parasite of nonhuman primates, produces transitory infections in humans. *Strongyloides myopotami, Strongyloides procyonis,* and several other *Strongyloides* sp do not mature in humans but can cause creeping eruption.

Synonyms

Synonyms for strongyloidiasis include threadworm infection and Cochin China diarrhea. None of the older scientific names for *S. stercoralis* (eg, *Anguillula stercoralis, Anguillula intestinalis, Rhabditis stercoralis, Rhabdonema strongyloides,* or *Strongyloides intestinalis*) is in current use.

General Considerations

In 1876, many French troops returned home from Cochin China (Vietnam) with severe diarrhea. Normand, a physician at the Naval Hospital of St. Mandrier in Toulon, France, examined the feces of these troops and found a minute worm that he believed had never been described. Normand gave these worms to Bavay, who determined they were rhabditiform larval nematodes and named them *A. stercoralis*, reflecting their eel-like shape and fecal source. Initially there was skepticism over Normand's discovery, and the complex life cycle of *S. stercoralis* confounded the issue. In 1877, Bavay described the direct development of rhabditiform larvae into filariform larvae in feces. He perceived the invasive filariform larvae in the gut wall as a separate species and called them *A. intestinalis*. Grassi et al pursued this question extensively from 1878 to 1882. From their findings, they proposed that direct transformation from rhabditiform to filariform larvae, and autoinfection, take place in the gut, and concluded that *A. stercoralis* and *A. intestinalis* were 2 forms of the same worm, *S. stercoralis*. From 1899 to 1914, Looss, Durme, Ransom, and Fülleborn established the various stages of the classic life cycle of *S. stercoralis*. In 1919, Thira produced patent infections by rectal administration of filariform larvae, further supporting the concept of internal

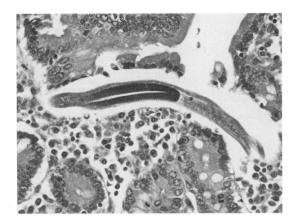

Figure 21.1
Section in anterior fourth of adult female *S. stercoralis* in chronically inflamed mucosa of colon. Note 1 of the reflected ovaries. x240

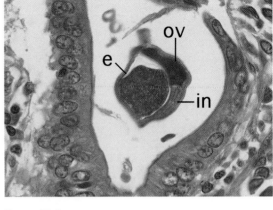

Figure 21.2
Cross section of adult female *S. stercoralis* in crypt of colon. Section contains collapsed intestine (in), ovary (ov), and egg (e) in uterus. x600

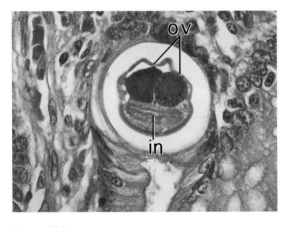

Figure 21.3
Cross section of adult female *S. stercoralis* in crypt of colon. At this level the worm contains 2 sections of ovary (ov) and intestine (in). x790

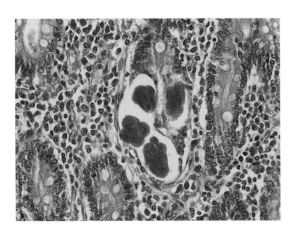

Figure 21.4
Four thin-shelled eggs of *S. stercoralis* in crypt of colon. Chronic inflammatory cells infiltrate surrounding mucosa. x235

autoinfection. In the late 1940s there was a growing awareness of the development of disseminated infections in immunosuppressed persons.

Epidemiology

Strongyloidiasis is distributed worldwide, but is most common in warm, wet regions. In focal rainy areas of some tropical countries, strongyloidiasis infects 85% or more of the population. In temperate climates, prevalence is low except where conditions promote fecal contamination (eg, in institutions for the mentally ill).[27] Infections tend to aggregate in households, suggesting that there are shared risk factors, such as genetic predisposition to infection or lifestyles promoting person-to-person transmission.[9] Infection rates increase with age and are usually higher in males.[3]

Dermatitis from *S. myopotami* and *S. procyonis* is common in swampy areas of Louisiana, where nutria and raccoon (the usual hosts, respectively) are common. There are also rare reports of laboratory workers developing papular and linear skin lesions from filariform larvae of *Strongyloides ransomi*, *Strongyloides papillosus*, and *Strongyloides westeri*.[25] *Strongyloides fuelleborni* infects humans in parts of Asia and Africa.[7]

Infectious Agent

Morphologic Description

Free-living adult females measure 1 mm by 50 to 75 µm; free-living males are 750 by 40 to 50 µm. Males have 2 copulatory spicules, but no caudal alae. Parasitic females measure 2 to 3 mm by 30 to 60 µm and have a 1-to-2-µm-thick cuticle with fine transverse striations. A long cylindrical esophagus occupies the anterior fourth of the worm, and the posterior three fourths contain the intestine and paired reproductive organs. The 2 ovaries reflex on themselves (Fig 21.1). Transverse sections through the posterior region of the worm typically show an intestine and 2 sections of the reproductive tube (Figs 21.2 and 21.3). Each uterus contains eggs in a single row.

Eggs are oval, 50 to 60 µm by 30 to 35 µm, thin-shelled, and embryonated on leaving the female (Fig 21.4). Morphologically, eggs of *S. stercoralis* are virtually identical to those of hookworms.

Rhabditiform larvae measure 200 to 400 µm by 10 to 20 µm and have a short buccal capsule followed by an esophagus with a posterior bulb (Figs 21.5 and 21.6). Filariform larvae measure 300 to 600 µm by 10 to 20 µm, have minute double lateral alae, a long nonbulbous esophagus, and a notched tail (Figs 21.7 to 21.10, and 21.14). The double lateral alae are observed only in cross sections and frequently are not discernible in tissue sections (Fig 21.9).

Life Cycle and Transmission

There are 3 life cycles for *S. stercoralis*: direct, indirect, and autoinfection (Fig 21.11). In the direct development cycle, rhabditiform larvae pass in the feces (Fig 21.12), become filariform larvae while in the soil, and infect humans by penetrating the skin. In the indirect development cycle, rhabditiform larvae molt several times in the soil and mature into free-living male and female adults. This free-living stage of the indirect cycle perpetuates itself as long as the environment is favorable. Not all the favorable factors are known, but temperature and humidity are important. When exposed to unfavorable conditions, rhabditiform larvae change into infective filariform larvae. In the autoinfection cycle, rhabditiform larvae become infective filariform larvae while in the intestine of the host or on the perianal skin, and invade tissues directly.

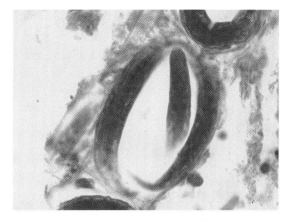

Figure 21.5
Five tangential segments of rhabditiform larvae of *S. stercoralis* in crypt of colon. Anterior end of larva in center demonstrates short buccal capsule. x590

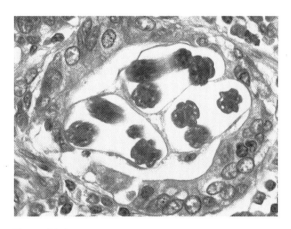

Figure 21.6
Multiple cross sections of rhabditiform larvae of *S. stercoralis* in crypt of colon. Rhabditiform larvae have no alae. x690

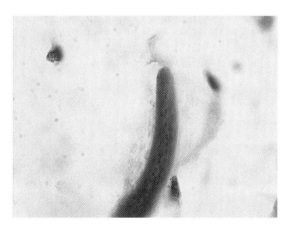

Figure 21.7
Anterior portion of filariform larva of *S. stercoralis* in wall of colon. Note nonbulbous esophagus. x590

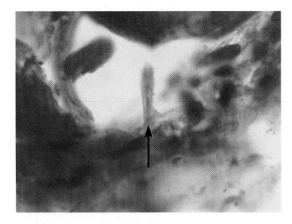

Figure 21.8
Tangential section of posterior end of filariform larva of *S. stercoralis* in lymph node. Note characteristic minute notched tail (arrow) of larva. x945

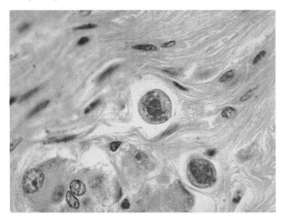

Figure 21.9
Cross sections of filariform larvae of *S. stercoralis* in wall of colon. Section in center is through anterior end at level of esophagus where there are no alae. Note granulomatous reaction adjacent to larvae. x560

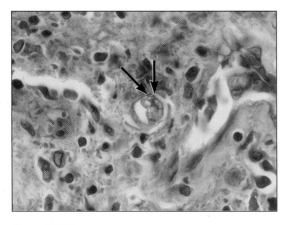

Figure 21.10
Cross section of filariform larva in wall of intestine, with minute double lateral alae (arrows). x690

Once filariform larvae penetrate human skin, they enter dermal vessels and travel to the lungs, where they break out of the pulmonary capillaries into alveolar spaces (Fig 21.13). Here, the larvae migrate up the bronchi and trachea, and are swallowed (Fig 21.14). In rare cases, filariform larvae penetrate the bronchial epithelium and develop into adult female worms (Fig 21.15). In the proximal gastrointestinal tract, the larvae mature into adult females (Fig 21.16) and produce viable eggs by parthenogenesis (Fig 21.4). Some authorities report that parasitic males fertilize females while in the respiratory or intestinal tracts, then pass in feces without penetrating the intestinal mucosa, but most observers have not detected parasitic males.[13] In autoinfection, the filariform larvae penetrate the bowel wall or perianal skin, enter the lymphatic vessels and thoracic duct, spread lymphohematogenously to the lungs, penetrate into the alveoli, and pass into the bronchi. The larvae then migrate upward and enter the gastrointestinal tract by way of the esophagus.[17]

Mother's milk can transmit infective larvae of *S. fuelleborni* in humans[7]; however, transmission of *S. stercoralis* by this means is controversial.

Clinical Features and Pathogenesis

Light infections of *S. stercoralis* in otherwise healthy individuals often are asymptomatic and may be disregarded by overworked clinicians in endemic areas. This attitude may prove disastrous if the patient becomes immunosuppressed.

There are 3 clinical phases of strongyloidiasis in the immunocompetent patient: cutaneous, pulmonary, and intestinal. The latter 2 phases, particularly in autoinfection, may overlap.

A few minutes after contact, filariform larvae penetrate the skin and provoke itching. Within 24 hours, focal edema, urticaria, and petechiae develop at the sites of penetration. The severity of these lesions depends on the number of larvae and the patient's sensitivity. During the cutaneous phase, which lasts only a few days, there may be low-grade fever, mild malaise, and eosinophilia.[4] Approximately a week later, larvae migrating through the lungs and tracheobronchial pathways

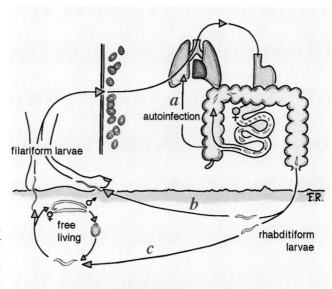

Figure 21.11
Life cycles of *S. stercoralis* and pathways of development in humans: a) autoinfection, b) direct development, and c) indirect development.

may irritate the throat and cause coughing. These symptoms often subside, but hemoptysis, dyspnea, and bronchopneumonia may follow. Eosinophilia is often prominent. Pulmonary involvement may mimic steroid-resistant asthma.[22] Adult female worms in bronchial tissues may cause chronic bronchitis and asthma. In such rare instances, rhabditiform larvae may appear in sputum (Fig 21.14).

Three weeks after infection, larvae enter the intestinal crypts and the infection usually becomes chronic and indolent. At this stage, strongyloidiasis is asymptomatic in up to two thirds of patients. Symptomatic patients have varying degrees of hunger pains, cramping, intermittent diarrhea and constipation, and may have mild anemia, weight loss, and leukocytosis with eosinophilia. Chronic relapsing colitis is a rare complication.[5]

Disseminated hyperinfection usually develops only in immunosuppressed patients, such as those with Hodgkin's disease, lymphocytic leukemia, cachexia and kwashiorkor due to malnutrition, leprosy, lupus erythematosus, burns, radiation sickness, syphilis, pancytopenia, and organ transplants. Many patients with disseminated infection have received steroid therapy.[8,15,16,24]

Although HIV-infected patients do not appear to be unusually susceptible to disseminated strongyloidiasis,[19] there are reports of AIDS patients with systemic strongyloidiasis.[8] Suppression of

Figure 21.12
Rhabditiform larva of *S. stercoralis* in feces, preserved in merthiolate-iodine-formaldehyde (MIF) solution. x345

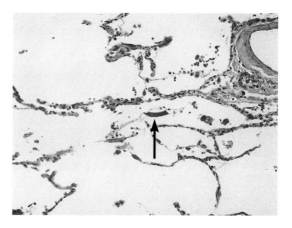

Figure 21.13
Fragment of filariform larva (arrow) in alveoli of lung of patient with strongyloidiasis. x120

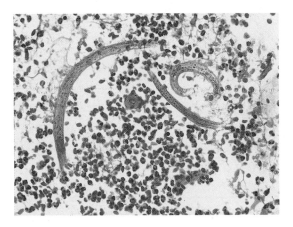

Figure 21.14
Filariform larvae of *S. stercoralis* in sputum. x260

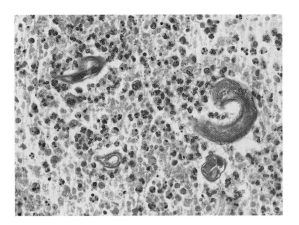

Figure 21.15
Adult female *S. stercoralis* in pulmonary exudate. Note both tangential and cross sections. x230

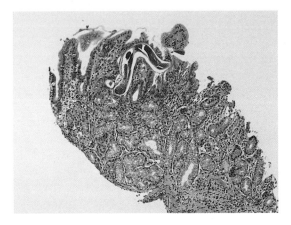

Figure 21.16
Adult female *S. stercoralis* in superficial gastric mucosa. Inflammation is slight. x80

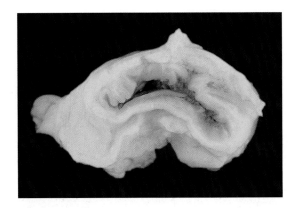

Figure 21.17
Transverse section of stenosed splenic flexure of colon. Note extensive fibrosis of bowel wall, presumably a result of chronic strongyloidiasis. Autopsy specimen is from 35-year-old Congolese female with long history of prednisone therapy for lepromatous leprosy complicated by erythema nodosum leprosum. Cause of death was disseminated strongyloidiasis and bacterial sepsis. x1.6

cell-mediated immunity is common to all of these conditions. Intestinal obstruction by paralytic ileus or stenosis is a common terminal event (Fig 21.17). Destruction of myenteric plexuses by inflammation and fibrosis may enhance hyperinfection by inhibiting peristalsis, giving more time for filariform larvae to develop within the intestine. Larvae may invade the skin internally, causing larva currens, a petechial or linear serpiginous purpuric rash, usually appearing first over the abdomen (Fig 21.18).[29] Patients with hyperinfection frequently have symptoms of malabsorption. Septicemia by enteric organisms is a common cause of death. Urinary, cardiac, and central nervous system symptoms are rare but well-known.[12] Filariform larvae in semen have been associated with infertility.[2]

Pathologic Features

In asymptomatic or mild infections, there are rhabditiform larvae in the stool (Fig 21.12) and eosinophilia is common. Patients without eosinophilia have a poor prognosis and probably have hyperinfection. In more severe infections, total serum proteins are reduced.

Adult female worms inhabit the crypts of the small intestine, where they deposit eggs that hatch and release larvae into the lumen (Figs 21.19 to 21.21). In noninvasive infections, gross lesions are usually minimal or absent, but in severe infections the bowel is edematous and congested.

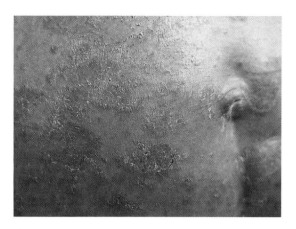

Figure 21.18
Larva currens over abdomen of 45-year-old Congolese woman. Like the woman described in Figure 21.17, this patient had a long history of lepromatous leprosy complicated by erythema nodosum leprosum and disseminated strongyloidiasis, and was under long-term prednisone therapy. Note linear and serpiginous lesions. Patient died of strongyloidiasis and bacterial sepsis several weeks after this photograph was taken.

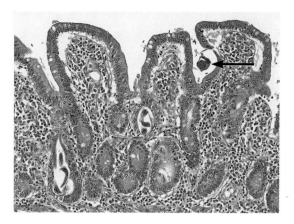

Figure 21.19
Section of colon showing chronic inflammation of mucosa, with adult female (arrow) and rhabditiform larvae of *S. stercoralis* in crypts. x120

In hyperinfection, the mucosal surface has abundant mucus and there may be petechiae and ulceration (Figs 21.22 and 21.23). Long-standing hyperinfection may lead to fibrosis of the bowel wall, usually of the duodenum, jejunum, and proximal ileum, but sometimes of the colon (Fig 21.17). In hyperinfection strongyloidiasis, filariform larvae invade the walls of the small and large intestine (Figs 21.24 to 21.26). Adult worms may invade blood vessels in the bowel wall (Figs 21.27 and 21.28). The pathogenesis of hyperinfection is uncertain, but corticosteroids or other unidentified molting hormones, or ecydsteroids, may play an important role by promoting development of *S. stercoralis* larvae.[13] Filariform larvae often inhabit the liver, spleen, lung, heart, lymph nodes, and many other organs (Figs 21.29 to 21.34). Larvae in tissues may cause no reaction, or may provoke a mixed inflammatory cell infiltrate including lymphocytes, macrophages, giant cells, plasma cells, neutrophils, and eosinophils (Fig 21.35). Degenerating larvae are usually centered in granulomas. Penetration of the bowel wall by filariform larvae may lead to gram-negative septicemia. Untreated patients with disseminated infections have a high mortality rate.

Diagnosis

Microscopic study of feces is the most effective diagnostic method. Direct fecal smear, formalin-ether concentration and filter paper, or agar plate culture may help to identify larvae (usually rhabditiform) or, on rare occasions, adult female worms. Eggs of *S. stercoralis* rarely appear in stools, but are not diagnostically useful since they are microscopically indistinguishable from hookworm eggs.

Rhabditiform larvae of *S. stercoralis* are easily confused with those of hookworms, but can be distinguished by their shorter buccal capsule and larger genital primordium. Filariform larvae of *S. stercoralis* can be distinguished from those of hookworms by their much longer esophagus and notched, rather than sharply pointed, tail. In fresh stools, larvae of *S. stercoralis* move in a whip-like manner, while those of hookworms glide like a snake.[11] In stored stools, or in stool cultures, rhabditiform larvae may develop into filariform larvae and free-living adults. Larvae are occasionally present in bronchial washings (Fig 21.36), sputum (Fig 21.37), pleural effusions, or urine.

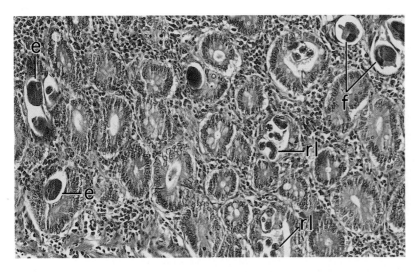

Figure 21.20
Section of colon with chronic inflammation and coiled adult female (f), rhabditoid larvae (rl), and eggs (e) of *S. stercoralis* in crypts. x160

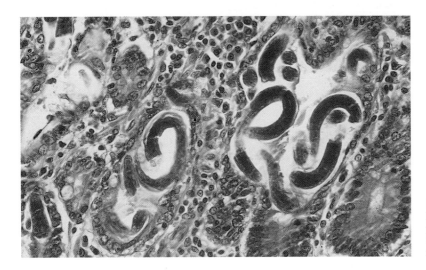

Figure 21.21
Higher magnification of section parallel to that shown in Figure 21.20, demonstrating multiple coiled, rhabditiform larvae in crypts. x315

Figure 21.22
Multiple ulcers in mucosa of colon of 65-year-old Congolese man with lepromatous leprosy complicated by severe protracted erythema nodosum leprosum and neuritis. Patient had received long-term prednisone therapy, and died of disseminated strongyloidiasis with gram-negative bacterial sepsis. x2.4

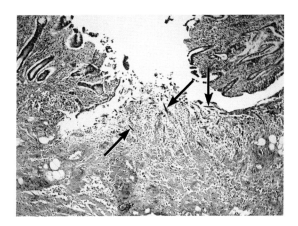

Figure 21.23
Ulceration and inflammation of colon in disseminated strongyloidiasis. Note multiple filariform larvae (arrows) in base of ulcer. x125

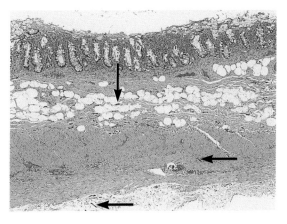

Figure 21.24
Colon of patient with disseminated strongyloidiasis, showing filariform larvae (arrows) at all levels of gut wall. x25

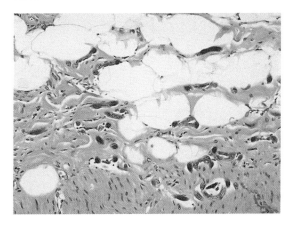

Figure 21.25
Higher magnification of Figure 21.24, revealing large numbers of filariform larvae of *S. stercoralis* in muscle. x120

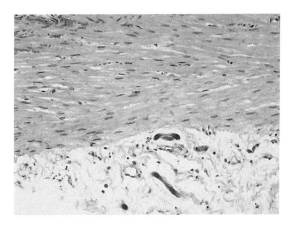

Figure 21.26
Higher magnification of Figure 21.24. Note multiple tangential fragments of filariform larvae of *S. stercoralis* that have invaded serosa of colon. x100

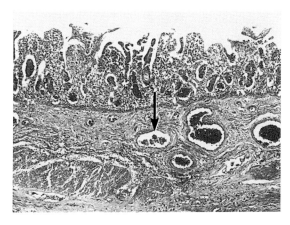

Figure 21.27
Section of ileum showing adult female *S. stercoralis* in blood vessel (arrow) in submucosa. x60

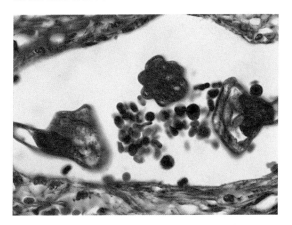

Figure 21.28
Higher magnification of worm shown in Figure 21.27. Adult *S. stercoralis* are rarely seen in the vascular system. x585

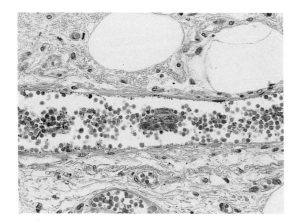

Figure 21.29
Three fragments of *S. stercoralis* filariform larva in lumen of pancreatic vessel. x280

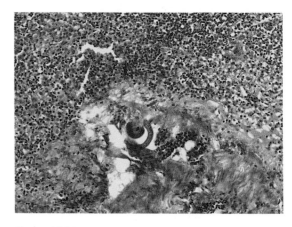

Figure 21.30
Coiled filariform larva in capsule of lymph node from patient with disseminated strongyloidiasis. x115

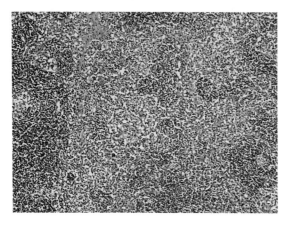

Figure 21.31
Extensive sinus histiocytosis of lymph node shown in Figure 21.30, from patient with disseminated strongyloidiasis. x60

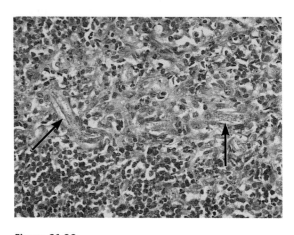

Figure 21.32
Degenerated filariform larva (arrows) in sinus of lymph node, from patient shown in Figure 21.30. x255

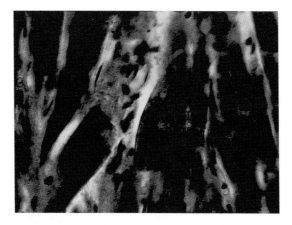

Figure 21.33
Filariform larva between myocardial fibers of patient with congenital heart malformation and disseminated strongyloidiasis. x500

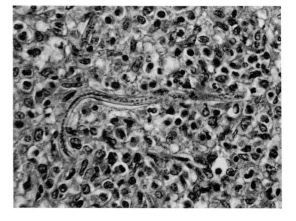

Figure 21.34
Invasion of spleen by filariform larvae in patient described in Figure 21.33. x435

Repeated stool examination is advisable if strongyloidiasis is suspected.

If repeated stool examination reveals no larvae and strongyloidiasis remains a possibility, duodenal aspirates are often productive; endoscopic biopsy is not highly productive.

Enzyme-linked immunosorbent assays (ELISA) for IgG antibodies to antigens of filariform larvae of *S. stercoralis* are highly reliable, but a low percentage of sera from patients with loiasis and ascariasis may cross-react.[1,14] Where available, ELISA is advised for all suspected cases of strongyloidiasis, especially for immunocompromised patients.

Treatment

Thiabendazole given orally for 2 days in noninvasive disease, and for 1 week in disseminated infection, is effective. In patients with bowel obstruction who cannot take oral medications, thiabendazole per rectum may be curative.[6] Albendazole may be as effective as thiabendazole.[21,23]

Recent studies suggest that single-dose oral ivermectin is more effective than albendazole in treating noninvasive strongyloidiasis.[10,18,20] Multiple doses of ivermectin may effectively treat hyperinfection in immunosuppressed hosts, but this requires further study.[28]

Treatment of immunocompromised patients often fails, so weekly stool examinations for up to 3 months are recommended.[8] Prophylactic broad-spectrum antibiotics should be considered as an adjunct to hyperinfection therapy, to prevent gram-negative septicemias.

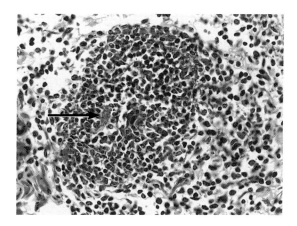

Figure 21.35
Degenerating fragment of filariform larva (arrow) of *S. stercoralis* in inflammatory exudate, composed largely of eosinophils, in wall of colon. x235

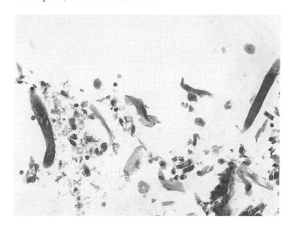

Figure 21.36
Fragments of filariform larvae in bronchial washings of patient with disseminated strongyloidiasis. x235

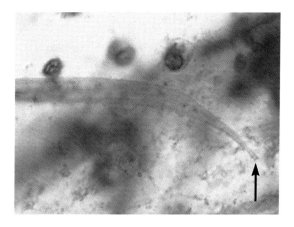

Figure 21.37
Posterior segment of filariform larva in sputum. Note minute notched end of tail (arrow). Pap stain x1075

References

1. Abdul-Fattah MM, Nasr ME, Yousef SM, Ibraheem MI, Abdul-Wahhab SE, Soliman HM. Efficacy of ELISA in diagnosis of strongyloidiasis among the immunocompromised patients. *J Egypt Soc Parasitol* 1995;25:491–498.
2. Agbo K, Deniau M. Anguillospermia resistant to treatment. Apropos of a case diagnosed in Togo [in French]. *Bull Soc Pathol Exot* 1987;80:271–273.
3. Arakaki T, Iwanaga M, Asato R, Ikeshiro T. Age-related prevalence of Strongyloides stercoralis infection in Okinawa, Japan. *Trop Geogr Med* 1992;44:299–303.
4. Barlow N. Clinical notes on infection with Strongyloides intestinalis, based upon a series of twenty-three cases. *Interstate Med J* 1915;22:1201–1208.
5. Berry AJ, Long EG, Smith JH, Gourley WK, Fine DP. Chronic relapsing colitis due to Strongyloides stercoralis. *Am J Trop Med Hyg* 1983;32:1289–1293.
6. Boken DJ, Leoni PA, Preheim LC. Treatment of Strongyloides stercoralis hyperinfection syndrome with thiabendazole administered per rectum. *Clin Infect Dis* 1993;16:123–126.
7. Brown RC, Girardeau MH. Transmammary passage of Strongyloides sp. larvae in the human host. *Am J Trop Med Hyg* 1977;26:215–219.
8. Celedon JC, Mathur-Wagh U, Fox J, Garcia R, Wiest PM. Systemic strongyloidiasis in patients infected with the human immunodeficiency virus. A report of 3 cases and review of the literature. *Medicine (Baltimore)* 1994;73:256–263.
9. Conway DJ, Hall A, Anwar KS, Rahman ML, Bundy DA. Household aggregation of Strongyloides stercoralis infection in Bangladesh. *Trans R Soc Trop Med Hyg* 1995;89:258–261.
10. Datry A, Hilmarsdottir I, Mayorga-Sagastume R, et al. Treatment of Strongyloides stercoralis infection with ivermectin compared with albendazole: results of an open study of 60 cases. *Trans R Soc Trop Med Hyg* 1994;88:344–345.
11. dos Santos Neto JG. Movement of the rhabditiform larva of Strongyloides stercoralis. *Lancet* 1993;342:1310.
12. Dutcher JP, Marcus SL, Tanowitz HB, Wittner M, Fuks JZ, Wiernik PH. Disseminated strongyloidiasis with central nervous system involvement diagnosed antemortem in a patient with acquired immunodeficiency syndrome and Burkitt's lymphoma. *Cancer* 1990;66:2417–2420.
13. Genta RM. Dysregulation of strongyloidiasis: a new hypothesis. *Clin Microbiol Rev* 1992;5:345–355.
14. Genta RM. Predictive value of an enzyme-linked immunosorbent assay (ELISA) for the serodiagnosis of strongyloidiasis. *Am J Clin Pathol* 1988;89:391–394.
15. Genta RM, Miles P, Fields K. Opportunistic Strongyloides stercoralis infection in lymphoma patients. Report of a case and review of the literature. *Cancer* 1989;63:1407–1411.
16. Gentry LO, Zeluff B, Kielhofner MA. Dermatologic manifestations of infectious diseases in cardiac transplant patients. *Infect Dis Clin North Am* 1994;8:637–654.
17. Haque AK, Schnadig V, Rubin SA, Smith JH. Pathogenesis of human strongyloidiasis: autopsy and quantitative parasitological analysis. *Mod Pathol* 1994;7:276–288.
18. Lindo JF, Atkins NS, Lee MG, Robinson RD, Bundy DA. Short report: long-term serum antibody isotype responses to Strongyloides stercoralis filariform antigens in eight patients treated with ivermectin. *Am J Trop Med Hyg* 1996;55:474–476.
19. Lucas SB. Missing infections in AIDS. *Trans R Soc Trop Med Hyg* 1990;84:34–38.
20. Marti H, Haji HJ, Savioli L, et al. A comparative trial of a single-dose ivermectin versus three days of albendazole for treatment of Strongyloides stercoralis and other soil-transmitted helminth infections in children. *Am J Trop Med Hyg* 1996;55:477–481.
21. Mojon M, Nielsen PB. Treatment of Strongyloides stercoralis with albendazole. A cure rate of 86 per cent. *Zentralbl Bakteriol Mikrobiol Hyg [A]* 1987;263:619–624.
22. Pansegrouw D. Strongyloides stercoralis infestation masquerading as steroid-resistant asthma. *Monaldi Arch Chest Dis* 1994;49:399–402.
23. Pitisuttithum P, Supanaranond W, Chindanond D. A randomized comparative study of albendazole and thiabendazole in chronic strongyloidiasis. *Southeast Asian J Trop Med Public Health* 1995;26:735–738.
24. Purtilo DT, Meyers WM, Connor DH. Fatal strongyloidiasis in immunosuppressed patients. *Am J Med* 1974;56:488–493.
25. Roeckel IE, Lyons ET. Cutaneous larva migrans, an occupational disease. *Ann Clin Lab Sci* 1977;7:405–410.
26. Sato Y, Kobayashi J, Toma H, Shiroma Y. Efficacy of stool examination for detection of Strongyloides infection. *Am J Trop Med Hyg* 1995;53:248–250.
27. Schupf N, Ortiz M, Kapell D, Kiely M, Rudelli RD. Prevalence of intestinal parasite infections among individuals with mental retardation in New York State. *Ment Retard* 1995;33:84–89.
28. Torres JR, Isturiz R, Murillo J, Guzman M, Contreras R. Efficacy of ivermectin in the treatment of strongyloidiasis complicating AIDS. *Clin Infect Dis* 1993;17:900–902.
29. von Kuster LC, Genta RM. Cutaneous manifestations of strongyloidiasis. *Arch Dermatol* 1988;124:1826–1830.

22

Ancylostomiasis

Wayne M. Meyers,
Aileen M. Marty, *and*
Ronald C. Neafie

Introduction

Definition

Ancylostomiasis is infection by hookworms. *Ancylostoma duodenale* (Dubini, 1843) Creplin, 1845 and *Necator americanus* (Stiles, 1902) Stiles, 1906, are the most common cause of hookworm infection in humans. Hookworms of cats, hamsters, and dogs, such as *Ancylostoma ceylanicum* Looss, 1911 and *Ancylostoma caninum* (Ercolani, 1858), cause ancylostomiasis in humans less frequently.

Synonyms

Informal terms for ancylostomiasis include hookworm disease, Egyptian chlorosis, tropical chlorosis, tunnel disease, tunnel anemia, brickmaker's anemia, uncinariasis, and miner's anemia.

Some authorities restrict the term ancylostomiasis to infections caused by *A. duodenale*, and call those caused by *N. americanus* necatoriasis or uncinariasis. Ancylostomiasis seems appropriate in either case, however, since both worms are in the family Ancylostomatidae.

The terms "hookworm infection" and "hookworm disease" should be distinguished from each other: "infection" indicates no symptoms; "disease" indicates symptoms, particularly iron deficiency anemia.

General Considerations

Adult hookworms are voracious bloodsuckers that have long plagued humans and animals. The Egyptian *Papyrus Ebers* (circa 1550 BC) contains a description of what was most likely hookworm infection. In 1838, Dubini observed hookworms in the intestine of an Italian, and later wrote the first accurate description of *A. duodenale*. In 1897, Looss noted the infectivity of filariform larvae of *A. duodenale* after accidental self-exposure. He eventually outlined the life cycle of hookworms using both *A. duodenale* in humans and *A. caninum* in dogs as models.

Necator americanus, the indigenous American hookworm, was probably introduced from sub-Saharan Africa. In the early 1900s, Smith found a worm in a stool specimen from the southern United States and sent it to Stiles.[12] Stiles identified this and similar worms from Puerto Rico and Washington, DC, as a previously unknown hookworm;

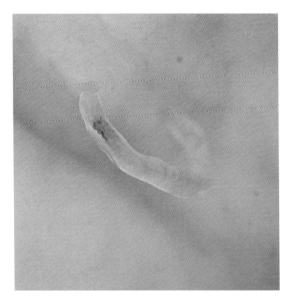

Figure 22.1
Adult *A. duodenale* embedded in mucosa of small intestine. Body of worm is stout and white. Unstained x11

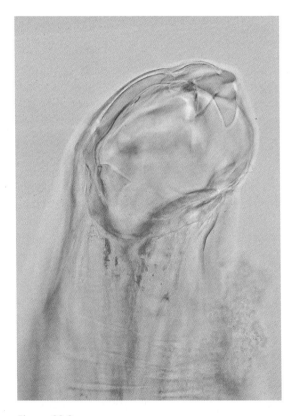

Figure 22.2
Adult female *A. duodenale* demonstrating large buccal capsule. Teeth are large and of nearly equal size. One of 2 pairs is visible. Unstained x240

this species was later named *N. americanus* or, more commonly, the American or New World hookworm. Hookworm infection has produced significant morbidity in troops deployed to endemic areas such as the Pacific, Vietnam, and Grenada.[17]

Epidemiology

Ancylostomiasis is a major helminthic disease. Worldwide, an estimated 1 billion people harbor hookworms; 1.6 million suffer from anemia resulting in 55 000 deaths annually.[4] Ancylostomiasis causes significant morbidity and debility, sapping the vitality and stunting the growth and development of its victims. The International Task Force for Disease Eradication believes that eradicating hookworm infection will be difficult, if not impossible.[21]

Temperature plays a major role in the geographic distribution of the 2 most common species of hookworms infecting humans. Eggs and larvae of *A. duodenale* are more tolerant of low temperatures, surviving in mines and tunnels in Europe, while those of *N. americanus* survive better at higher temperatures and predominate in warmer regions.[30] Hookworm infection is not prevalent in arid regions, but in moist, rural areas where there is inadequate sanitation and people go barefoot. The distribution of hookworms may overlap considerably. Epidemiologic data derived by egg identification generally do not designate the species.

Ancylostoma duodenale is called the Old World hookworm. In Europe, North Africa, and the Middle East, *A. duodenale* predominates, and is present in a few foci in South America, much of sub-Saharan Africa, and Asia.

Necator americanus is indigenous throughout a wide zone of the Americas, stretching from the southern United States to central South America, including the Caribbean islands. In the United States, *N. americanus* is endemic in the southeast and causes the vast majority of hookworm infections. *Necator americanus* also predominates in sub-Saharan Africa, southern Asia, Melanesia, and Polynesia.

Hookworms are the most common helminthic intestinal parasite of humans in the continental

United States and Puerto Rico.[13,16] In 1987, California, Wisconsin, Rhode Island, Colorado, and Washington reported the majority of hookworm infections, mostly in immigrants.[30]

Ancylostoma ceylanicum, an intestinal hookworm of cats, rarely causes intestinal infection in humans.[11] The common habitats of *A. ceylanicum* are the tropical and semitropical zones of the Far East. *Ancylostoma ceylanicum* and *Ancylostoma braziliense*, a common hookworm of cats and dogs, were at one time identified as the same species. However, Biocca outlined their morphologic differences,[3] and Rep et al[28] could not achieve cross-breeding. In the Philippines, Indonesia, New Guinea, Sri Lanka, India, Thailand, southeast Africa, and Brazil, human ancylostomiasis is probably caused by *A. ceylanicum* rather than *A. braziliense,* as has been reported.

Ancylostoma caninum, a hookworm of dogs, foxes, and coyotes, sometimes causes ancylostomiasis in humans.[22,26] A growing body of evidence, mainly serological but also surgical, suggests that *A. caninum* causes eosinophilic enteritis in humans.

Infectious Agents

Morphologic Description

Hookworms are in the order Strongylida.[19] Adults are stout worms about 1 cm long (Fig 22.1). Their bodies are white, red-brown from ingestion of blood, or gray from hemosiderin granules deposited in their intestines. The hookworms' buccal capsules contain either teeth (*Ancylostoma*) or cutting plates (*Necator*) (Figs 22.2 to 22.7). Male worms have an easily identifiable copulatory bursa at the posterior end (Fig 22.8).

Adult *A. duodenale* and *N. americanus* are differentiated by the curvature of the head, structure of the buccal capsule, length of the esophagus, position of the vulva, and configuration of the copulatory bursa.

Adult female *A. duodenale* are 1 to 1.3 cm long with a maximum diameter of 0.7 mm. Males are 0.8 to 1.1 cm long with a maximum diameter of 0.5 mm. In both sexes, the head curves in the same direction as the body. The buccal capsule is large in both sexes, and the anterior edge of the capsule bears 2 pairs of large ventral teeth of nearly equal size

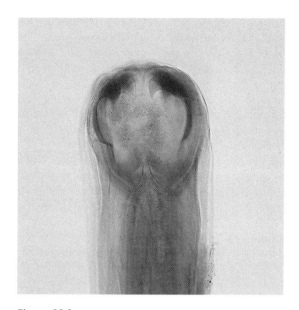

Figure 22.3
Adult *A. caninum* from a dog, showing buccal capsule. There are 3 pairs of teeth; the inner tooth on either side is slightly smaller than the others. Carmine x200

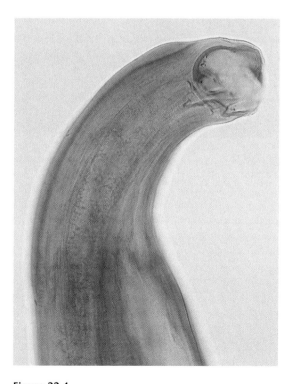

Figure 22.4
Adult male *N. americanus* demonstrating small buccal capsule with no teeth. Head is bent dorsally relative to curvature of body, a characteristic of *N. americanus*. Carmine x185

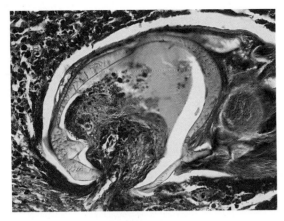

Figure 22.5
Large buccal capsule of adult female *A. duodenale* embedded in submucosa of anus. Two large teeth of nearly equal size are visible. The other pair of teeth is not seen in this section. x230

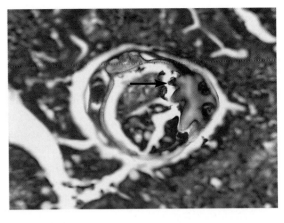

Figure 22.6
Buccal capsule of adult *A. caninum* embedded in intestinal mucosa of a dog. Buccal capsule has 3 pairs of teeth; the inner tooth (arrow) on either side is slightly smaller than the others. Three teeth from 1 side of the buccal capsule are visible. x220

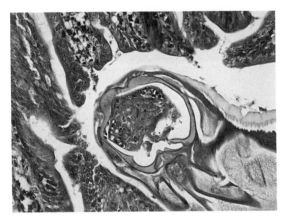

Figure 22.7
Small buccal capsule of adult *N. americanus* attached to mucosa of jejunum. Buccal capsule has no teeth. x225

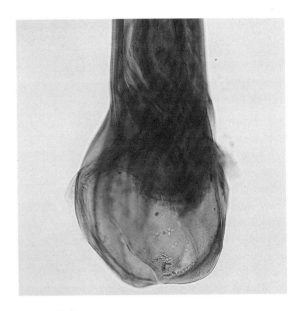

Figure 22.8
Copulatory bursa of adult male *N. americanus* is nearly as broad as long. Carmine x155

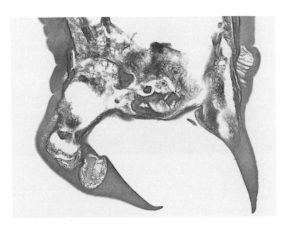

Figure 22.9
Section through copulatory bursa of adult male *A. duodenale* in lumen of small intestine. Copulatory bursa is broader than long. x120

(Figs 22.2 and 22.5). The dorsal buccal capsule wall has a dorsal gutter. In both sexes, the esophagus is long, measuring over 1 mm. In the female, the vulva lies posterior to the middle of the body, and there is usually a 20-μm-long caudal spine at the tip of the tail. The bursa of the male *A. duodenale* is broader than long (Fig 22.9). There are 2 long (approximately 2 mm), slender, sclerotized copulatory spicules at the posterior end of the male (Fig 22.10).

Necator americanus is smaller than *A. duodenale*. Females are 0.9 to 1.1 cm long with a maximum diameter of 0.45 mm; males are 0.5 to 0.9 cm long with a maximum diameter of 0.3 mm. The head in both sexes is curved sharply dorsally (Fig 22.4), so the curvature of the head is counter to the curvature of the body. In both sexes the buccal capsule is small, and the anterior edge of the oral opening has a pair of ventral, semilunar cutting plates (Figs 22.4 and 22.7). The dorsal gutter is replaced by a cone-shaped tooth at the base of the buccal capsule. In both sexes the maximum length of the esophagus is less than 0.8 mm. In the female *N. americanus*, the vulva is slightly anterior to the middle of the body and there is no caudal spine at the tip of the tail. The copulatory bursa of the male *N. americanus* is nearly as broad as long (Fig 22.8). There are 2 long (approximately 1 mm), slender, sclerotized copulatory spicules at the posterior end of the male worm.

Ancylostoma ceylanicum females average 1.05 cm long with a maximum diameter of 0.47 mm, while males average 0.81 cm long with a maximum diameter of 0.4 mm.[31] The buccal capsule of both sexes has a single large tooth and another small, barely perceptible tooth on each side. The copulatory bursa of the male *A. ceylanicum* is about as broad as long.

Ancylostoma caninum selectively seek arterial blood when within the host, so their bodies are often bright red. Females are 1.4 to 1.6 cm long with a maximum diameter of 0.6 mm, and males are 1 to 1.2 cm long with a maximum diameter of 0.4 mm. The buccal capsule of each sex contains 3 pairs of teeth, the innermost pair being the smallest (Figs 22.3 and 22.6). In female *A. caninum* the vulva is near the junction of the second and last thirds of the body, with a small spine at the tip of the tail. The copulatory bursa of male *A. caninum* is nearly as broad as long, and there are 2 sclerotized

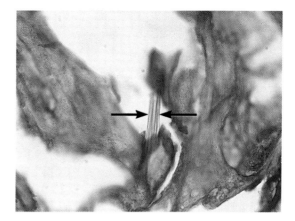

Figure 22.10
Section of copulatory bursa of adult male *A. duodenale* in lumen of small intestine. Note golden color of 2 slender, sclerotized copulatory spicules (arrows). x600

Figure 22.11
Adult female *A. duodenale*, displaying transverse striations of cuticle. Unstained x670

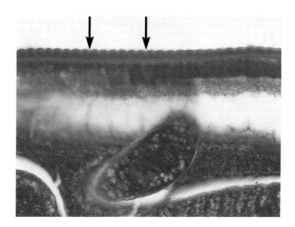

Figure 22.12
Section through stained adult female *A. duodenale* within lumen of small intestine, showing transverse striations of cuticle (arrows). x625

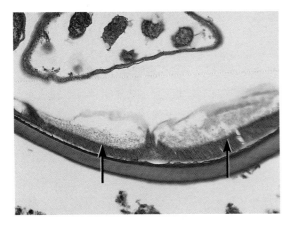

Figure 22.13
Section through adult female *A. duodenale*, demonstrating body wall and thick cuticle. Note 2 large somatic muscle cells. Contractile fibers (arrows) border base of each muscle cell at point of contact with underlying hypodermis. Reticulum stain x235

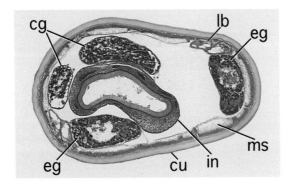

Figure 22.14
Anterior region of adult female *A. duodenale*. There are only a few somatic muscle cells (ms) per quadrant. Intestinal cells (in) have microvilli on their surfaces and multiple nuclei per cell. There is a lateral body (lb) in each lateral cord region. Cephalic (cg) and excretory (eg) glands are seen at this level of worm. Cuticle (cu) is thick. x195

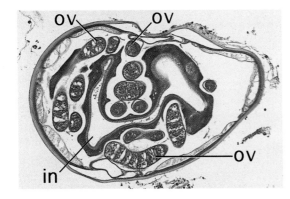

Figure 22.15
Transverse section of adult female *N. americanus* within lumen of jejunum. Many sections of ovaries (ov) coil around centrally located intestine (in). x110

copulatory spicules about 1 mm long.

Morphologic similarities between *A. duodenale* and *N. americanus* are apparent in histologic sections. The cuticle in both worms is usually 10 to 20 μm thick and transversely striated (Figs 22.11 and 22.12). There are great variations in cuticular thickness, but the cuticle of *A. duodenale* is usually thicker than that of *N. americanus*. There are only 3 or 4 somatic muscle cells in each quadrant of the worm. Contractile fibers border only the side of the muscle cell in contact with the hypodermis (Figs 22.13 and 22.14). There is a lateral body at the base of each lateral cord (Fig 22.14). The intestine contains only a few cells, each with microvillous surfaces and multiple nuclei (Figs 22.14 and 22.15). The intestinal tract terminates at the anus at the tip of the tail. The anterior region of the worm contains a pair of cephalic (amphidial) glands and a pair of excretory (cervical) glands (Fig 22.14). Cephalic glands open near the buccal capsule and extend posteriorly to near the midportion of the worm. These glands secrete the anticoagulant that promotes blood flow into the worm. Excretory glands are in the ventral side of the worm and emanate from the excretory pore near the nerve ring. They connect to the tubular excretory system and extend posteriorly to the middle of the worm.

Female reproductive organs are in the posterior two thirds of the worm and consist of a vulva, a vagina, paired ovejectors, uteri, seminal receptacles, oviducts, and ovaries (Figs 22.15 and 22.16). Ovaries are long and encircle the other reproductive organs throughout the entire length of the reproductive tract.

Male reproductive organs are in the posterior two thirds of the worm and consist of a single testis, vas deferens, seminal vesicle, and ejaculatory duct (Fig 22.17). Prominent prostate glands (cement glands) line the ejaculatory duct (Fig 22.18). The ejaculatory duct unites with the intestine in the posterior tip of the worm to form a short cloaca. Male worms have 2 long, slender, sclerotized copulatory spicules.

Eggs of all hookworms infecting humans are morphologically similar; those of *A. duodenale* and *N. americanus* are virtually indistinguishable. Hookworm eggs are oval, have thin, smooth shells, and measure 55 to 75 μm by 35 to 47 μm. In stools, eggs usually contain segmented embryos (Figs 22.19 and 22.20). In constipated patients, first-

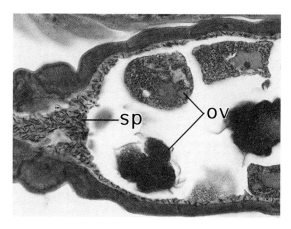

Figure 22.16
Section through adult female *A. duodenale*, demonstrating junction of seminal receptacle filled with sperm (sp) and uterus containing ova (ov). x590

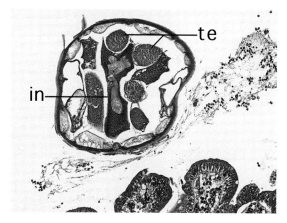

Figure 22.17
Section through middle of adult male *N. americanus* within lumen of jejunum, showing multiple sections of single testis (te) coiled around centrally located intestine (in). Note prominent intestinal microvilli. x115

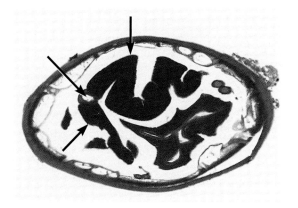

Figure 22.18
Section through posterior region of adult male *A. duodenale*, demonstrating prostate glands (cement glands) (short arrows) lining ejaculatory duct (long arrow). x110

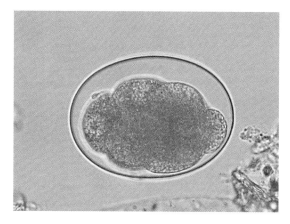

Figure 22.19
Oval hookworm egg from feces. Note smooth shell and segmented embryo. Unstained x1310

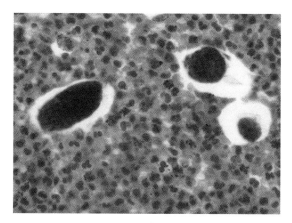

Figure 22.20
Several hookworm eggs embedded in submucosa of small intestine. Embryos in oval eggs are segmented. Eggshells are thin, often making them hard to see in tissue sections, as illustrated here. Acute inflammatory reaction surrounds eggs. x440

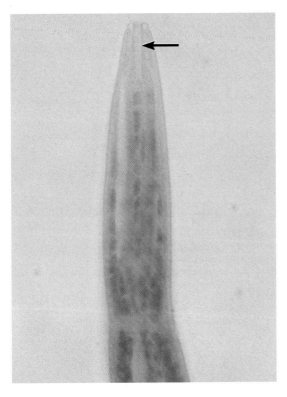

Figure 22.21
Anterior end of hookworm rhabditoid larva, demonstrating long buccal capsule (arrow). x1330

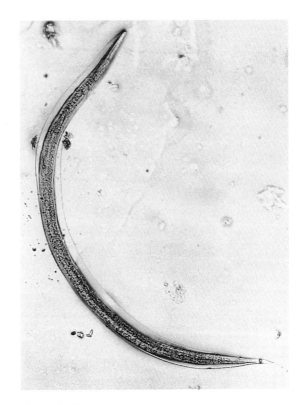

Figure 22.22
Filariform larva of hookworm in feces. Unstained x180

stage larvae can develop within eggs in the intestine. Hookworm rhabditoid larvae measure 250 to 300 µm by 15 to 20 µm and resemble those of *Strongyloides stercoralis*, but have a longer buccal capsule (Fig 22.21) and a much smaller genital primordium. Rhabditoid larvae of *S. stercoralis* move actively, in a whip-like manner, whereas rhabditoid larvae of hookworms glide like snakes.[6] Infective filariform larvae of hookworms are much larger than those of *S. stercoralis*, measuring 850 µm by 35 µm (Fig 22.22).

Life Cycle and Transmission

The major source of hookworm larvae that infect humans is soil contaminated by egg-infested human feces. Most commonly, human infection develops when filariform larvae in soil penetrate the skin. Occasionally, infection may result from eating filariform larvae on fresh vegetables, or from eating dirt (Fig 22.23). Filariform larvae of *A. duodenale* and *A. caninum* can remain in a state of arrested maturation within the host,[22] becoming latent, parenteral (hypobiotic) larvae that can mature many months later into adults.[29] In animals, hypobiotic larvae may migrate into striated muscles, placenta, or milk and serve as a source of infection. In humans, hypobiotic larvae may transmit ancylostomiasis through breast milk, or possibly through the placenta.[28]

Mammals are the definitive hosts of hookworms. No intermediate host is required. Adult hookworms live and mate in the small intestine (Fig 22.24). A single *A. duodenale* can produce 25 000 to 35 000 eggs/day; an *N. americanus* can produce 6000 to 20 000 eggs/day. In warm, moist, shaded soil, eggs from feces hatch in 24 hours, releasing first-stage larvae that develop into second-stage rhabditoid larvae in 3 days. About a week after hatching, infective filariform larvae develop and migrate to the surface of the soil, where they can remain viable for several weeks.

Filariform larvae enter human skin by secreting

hyaluronidase and metalloproteases, then penetrating between epidermal keratinocytes and through the ground substance of the dermis.[14,15] *Necator americanus* penetrates skin more efficiently than *A. duodenale*. Larvae enter venules in the skin and eventually flow into the capillaries of the lung, breaking out into the alveolar sacs. Some authorities believe that larvae may also reach the lungs by lymphatic channels. From the alveoli, larvae migrate up the bronchi and trachea, over the epiglottis, down the esophagus, through the stomach, and into the small intestine. If filariform larvae of *A. duodenale* are swallowed, they mature to adults in the intestine, omitting the pulmonary stage.

In the small intestine, larvae molt and develop a buccal capsule, permitting attachment to the intestinal mucosa. Sometimes this third molt, producing fourth-stage larvae, takes place in the lungs. In the small intestine the fourth-stage larvae molt, become sexually mature, and copulate. Females begin to lay eggs within 2 months after larvae penetrate the skin. About three fourths of the worms that reach the small intestine die within a year, but some remain viable for years.

Clinical Features and Pathogenesis

Hookworm filariform larvae that penetrate the skin, usually of the hands or feet, cause minimal local damage. Within several hours to a few days, an allergic reaction to antigens of the worm produces a local dermatitis known as ground itch. Locally, the skin is red and pruritic (Fig 22.25). With repeated infections, the skin may blister. Scratching often provokes secondary pyogenic infections at these sites. When they occur, vesiculation and pustulation of the skin are usually mild in indigenous patients in the tropics; they are sometimes seen in expatriates.

Occasionally, larvae of *N. americanus* or *A. duodenale* migrate within the epidermis, producing lesions resembling creeping eruption before entering dermal vessels. These lesions are of shorter duration and produce smaller worm paths than those of true creeping eruption caused by *A. braziliense* and other hookworms that do not mature in humans. Filariform larvae may also affect

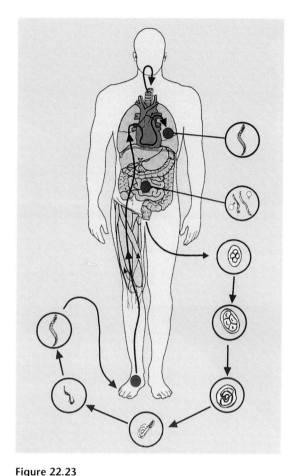

Figure 22.23
Life cycle of hookworm. Adult hookworms live and mate in the small intestine, where females release eggs. Eggs hatch, releasing first-stage rhabditoid larvae. Second-stage rhabditoid larvae develop 3 days later, then mature into infective filariform larvae. Filariform larvae penetrate human skin, enter venules, and travel through pulmonary capillaries into alveolar sacs. Larvae migrate up the trachea, over the epiglottis, down the esophagus, through the stomach, and into the small intestine, where they molt and mature to adult worms. Some larvae may penetrate peripheral muscle and lie dormant as parenteral (hypobiotic) larvae. Later, these larvae may migrate through the lung and into the intestine, where they develop into adults and mate, and where females lay eggs.

Figure 22.24
Numerous adult *A. duodenale* attached to mucosa of duodenum in heavily infected patient.*

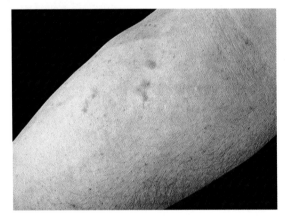

Figure 22.25
Papular erythematous eruption on inner surface of forearm of volunteer exposed experimentally to *N. americanus* filariform larvae.

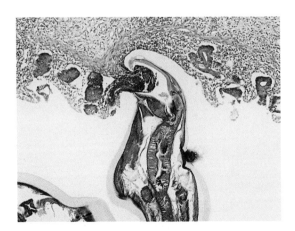

Figure 22.26
Adult *A. duodenale* attached to mucosa of jejunum of heavily infected 12-week-old infant who died of severe anemia. x60

tissues other than the skin. Larvae of *Ancylostoma* sp in the cornea impair vision, and those in muscles produce eosinophilic myositis.[20]

Within 2 weeks, larvae migrate to the lungs. Pulmonary symptoms are generally mild and transitory, but cough is common, and some patients have asthma-like symptoms and bronchitis (Löffler's syndrome). Peripheral eosinophilia frequently begins during these early stages and is more pronounced and persistent in patients infected by *A. duodenale* than in those infected by *N. americanus*.

Severity of symptoms depends on intensity and duration of infection, and natural resistance and nutrition of the host. Most individuals with light infections experience few or no symptoms after larvae reach the small intestine. Patients with heavy infections are more likely to have abdominal symptoms that begin with dyspepsia, nausea, and epigastric discomfort, followed by constipation or diarrhea. Frequently there is an intense desire to eat soil (pica). Specific anthelmintic treatment relieves these symptoms. Stools may contain occult or overt blood. Patients with eosinophilic enteritis caused by *A. caninum* have high eosinophilia and elevated serum IgE.[26]

The most serious consequence of heavy infection (Fig 22.24) is anemia, usually from iron deficiency, but accompanying folate deficiency may be masked by signs of iron deficiency. Iron reserves in the body are central to determining the progress of hookworm disease. In untreated patients, anemia can become intense, producing marked lethargy.[1]

In advanced disease, blood hemoglobin levels can drop to 1 g/dl. Overwhelming infections can cause lassitude, coma, and even death, especially in infants (Fig 22.26).

Heavy, chronic infections cause malnutrition, hypoproteinemia, profound anemia, and growth retardation. Protein loss can produce edema unresponsive to diuretics. Severe anemia leads to cardiac symptoms that include hemic murmurs, tachycardia, cardiomegaly, and cardiac decompensation. Some authorities believe ancylostomiasis also causes nephrosis.

Pathologic Features

In heavy infections, skin at the site of invasion may show vesicles and inflammation with hyperemia edema and infiltrations of neutrophils and eosinophils.

In the lungs, migrating larvae cause small hemorrhages into alveoli. In patients with Löffler's syndrome, larvae induce eosinophilic and mononuclear infiltrates. Lesions of ancylostomiasis in skin and lungs are not often studied since they are seldom biopsied.

Adult hookworms attach to the mucosa of the small intestine by biting and sucking, using their strong buccal capsule and powerful esophageal muscles (Fig 22.26). Worms graze over the intestinal mucosa, attaching and detaching themselves, producing multiple lacerations as they forage. Gross examination of the small bowel reveals focal hemorrhages throughout the mucosa, with numerous attached parasites (Fig 22.24). Previous bite marks appear as multiple pigmented spots.

The worm's buccal capsule attaches to the mucosa (Figs 22.26 and 22.27), and its body hangs free in the lumen (Fig 22.27). In the lamina propria surrounding the buccal capsule, there are neutrophils, eosinophils, and plasma cells.[18] Sometimes the mucosa is atrophic.

The ileum of a patient with *A. caninum* infection shows intense infiltrations of eosinophils in the lamina propria. These infiltrations are less pronounced in tissue at the worm's point of attachment. Some villi are distended by hemorrhage and have collections of red blood cells and eosinophils in the mucinous secretions on the surface. There can be massive eosinophilic enteritis even when a worm is not found attached to or penetrating the mucosa (Figs 22.28 and 22.29). *Ancylostoma duodenale* and *N. americanus* usually produce less intense eosinophilic infiltration of the small intestine.[26]

Occasionally, *N. americanus*, *A. duodenale*, and *A. ceylanicum* penetrate the mucosa, provoking nodular abscesses up to 1 cm in diameter in the mucosa and submucosa (Figs 22.30 and 22.31).[2,7] Nodules are edematous, hemorrhagic, and contain neutrophils and eosinophils. Careful searching in these lesions sometimes reveals adult hookworms, eggs, or both.

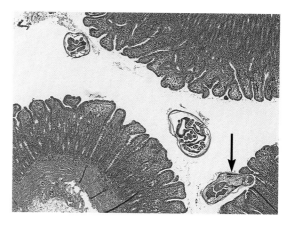

Figure 22.27
Multiple sections of adult male and female *N. americanus* within lumen of jejunum. One worm is attached to mucosa by its buccal capsule (arrow). x22

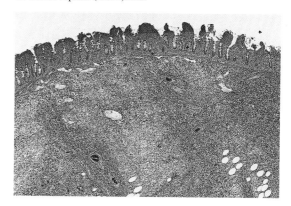

Figure 22.28
Biopsy specimen demonstrating eosinophilic enteritis from hookworm infection of small intestine. Distal ileum is edematous, congested, and has a massive transmural eosinophilic infiltrate. Sections of adult female hookworm are within intestinal lumen, but are not visible in this photo. Species of hookworm was not determined, but massive eosinophilia suggests *A. caninum*. x22

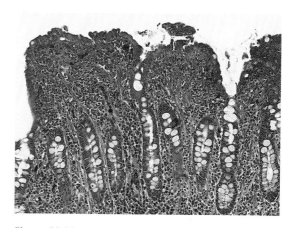

Figure 22.29
Higher magnification of distal ileum of same patient as in Figure 22.28, showing intense eosinophilic enteritis. x115

Figure 22.30
Nodular mucosal abscess of small intestine around adult *N. americanus* (arrow). Abscess is edematous and hemorrhagic, and contains neutrophils and eosinophils. Eggs, although present, are not visible at this magnification. x15

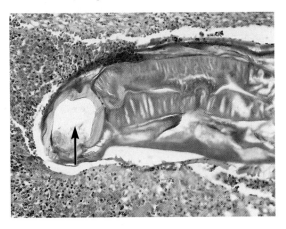

Figure 22.31
Higher magnification of adult *N. americanus* in mucosal abscess illustrated in Figure 22.30. Buccal capsule (arrow) is small and lacks teeth. x115

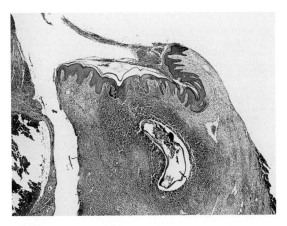

Figure 22.32
Submucosal abscess in anus around adult female *A. duodenale*. Note extensive edema, congestion, and abundant neutrophilic infiltrate. Section of adult worm is centered in abscess. x18

Occasionally, hookworm disease involves the large intestine. *Ancylostoma duodenale*, for example, may penetrate the gut wall and produce submucosal abscesses as far distal as the anus (Fig 22.32).

Death from severe ancylostomiasis is most frequently attributed to complications of profound anemia, such as congestive heart failure with pleural or peritoneal effusions. In such cases, major organs are pale, the heart is dilated and flabby, and the liver shows fatty changes.

Diagnosis

Diagnosis is usually made by identifying hookworm eggs in direct smears or concentrated preparations of stools. Stools of constipated patients, or stools examined after long delays, may contain hookworm larvae, which must be distinguished from those of *S. stercoralis*. Endoscopic observation of changes, or histopathologic identification of hookworms or eggs in surgical or autopsy specimens, may establish the diagnosis.[10] Species-specific antigens have been identified, but are not yet available for diagnosis.[25]

Treatment and Prevention

Mebendazole and albendazole are reliable therapeutic agents.[5,23,24,27] Neither, however, is uniformly curative. Ivermectin reduces worm burden and the number of hookworm eggs (up to 52%).[9] Blood transfusions are sometimes required, but are recommended only for critically ill patients. Iron supplements and improved general nutrition are often indicated. Adequate sanitation and preventive measures are essential to controlling ancylostomiasis.[8]

References

1. Ali AA, Mahmoud LH, el-Zoheiry AA. A study on intestinal helminths causing human anemia in Cairo. *J Egypt Soc Parasitol* 1989;19:251–256.
2. Biagi FF, Villa TS, Alvarez G. Nodulos en la submucosa intestinal producidos por Ancylostoma duodenale (Dubini, 1843). *Rev Biol Trop* 1957;5:35–43.
3. Biocca E. On Ancylostoma braziliense (de Faria, 1910) and its morphological differentiation from A. ceylanicum (Looss, 1911). *J Helminthol* 1951;25:1–10.
4. Bundy DA. Immunoepidemiology of intestinal helminthic infections. 1. The global burden of intestinal nematode disease. *Trans R Soc Trop Med Hyg* 1994;88:259–261.
5. de Silva DG, Hettiarachchi SP, Fonseka PH. Albendazole in the treatment of geohelminth infections in children. *Ceylon Med J* 1989;34:185–189.
6. dos Santos Neto JG. Movement of the rhabditiform larva of Strongyloides stercoralis [letter]. *Lancet* 1993;342:1310.
7. Elmes BG, McAdam IW. Helminthic abscess, a surgical complication of oesophagostomes and hookworms. *Ann Trop Med Parasitol* 1954;48:1–7.
8. Esrey SA, Potash JB, Roberts L, Shiff C. Effects of improved water supply and sanitation on ascariasis, diarrhoea, dracunculiasis, hookworm infection, schistosomiasis, and trachoma. *Bull World Health Organ* 1991;69:609–621.
9. Freedman DO, Zierdt WS, Lujan A, Nutman TB. The efficacy of ivermectin in the chemotherapy of gastrointestinal helminthiasis in humans. *J Infect Dis* 1989;159:1151–1153.
10. Genta RM, Woods KL. Endoscopic diagnosis of hookworm infection. *Gastrointest Endosc* 1991;37:476–478.
11. Goyal N, Gupta S, Katiyar JC, Srivastava VM. NADH oxidase and fumarate reductase of Ancylostoma ceylanicum. *Int J Parasitol* 1991;21:673–676.
12. Harris HF. Ancylostomiasis, the most common of the serious diseases of the southern part of the United States. *Am Med* 1902;4:776.
13. Hillyer GV, Soler de Galanes M, Lawrence S. Prevalence of intestinal parasites in a rural community in north-central Puerto Rico. *Bol Asoc Med P R* 1990;82:111–114.
14. Hotez PJ, Narasimhan S, Haggerty J, et al. Hyaluronidase from infective Ancylostoma hookworm larvae and its possible function as a virulence factor in tissue invasion and in cutaneous larva migrans. *Infect Immun* 1992;60:1018–1023.
15. Hotez P, Haggerty J, Hawdon J, et al. Metalloproteases of infective Ancylostoma hookworm larvae and their possible functions in tissue invasion and ecdysis. *Infect Immun* 1990;58:3883–3892.
16. Kappus KK, Juranek DD, Roberts JM. Results of testing for intestinal parasites by state diagnostic laboratories, United States, 1987. *MMWR CDC Surveill Summ* 1991;40:25–45.
17. Kelley PW, Takafuji ET, Wiener H, et al. An outbreak of hookworm infection associated with military operations in Grenada. *Mil Med* 1989;154:55–59.
18. Layrisse M, Blumenfeld N, Carbonell L, Desenne J, Roche M. Intestinal absorption tests and biopsy of the jejunum in subjects with heavy hookworm infection. *Am J Trop Med Hyg* 1964;13:297–305.
19. Lichtenfels JR. *CIH Keys to the Nematode Parasites of Vertebrates. Keys to Genera of the Superfamily Ancylostomatoidea.* Bucks, England: Commonwealth Agricultural Bureaux; 1980:1-19.
20. Little MD, Halsey NA, Cline BL, Katz SP. Ancylostoma larva in a muscle fiber of man following cutaneous larva migrans. *Am J Trop Med Hyg* 1983;32:1285–1288.
21. MMWR. International task force for disease eradication, 1992. *MMWR* 1992;41:691, 697–698.
22. Noble ER, Noble GA. *Parasitology: The Biology of Animal Parasites.* 5th ed. Philadelphia, Pa: Lea & Febiger; 1982:279–284.
23. Nontasut P, Singhasivanon V, Prarinyanuparp V, et al. Effect of single-dose albendazole and single-dose mebendazole on Necator americanus. *Southeast Asian J Trop Med Public Health* 1989;20:237–242.
24. Pamba HO, Bwibo NO, Chunge CN, Estambale BB. A study of the efficacy and safety of albendazole (Zentel) in the treatment of intestinal helminthiasis in Kenyan children less than 2 years of age. *East Afr Med J* 1989;66:197–202.
25. Pritchard DI, McKean PG, Rogan MT, Schad GA. The identification of a species-specific antigen from Necator americanus. *Parasite Immunol* 1990;12:259–267.
26. Prociv P, Croese J. Human eosinophilic enteritis caused by dog hookworm Ancylostoma caninum. *Lancet* 1990;335:1299–1302.
27. Raccurt CP, Lambert MT, Bouloumie J, Ripert C. Evaluation of the treatment of intestinal helminthiases with albendazole in Djohong (North Cameroon). *Trop Med Parasitol* 1990;41:46–48.
28. Rep BH, Vetter JC, Eijsker M. Cross breeding experiments in Ancylostoma braziliense de Faria, 1910 and A. ceylanicum Looss, 1911. *Trop Geogr Med* 1968;20:367–378.
29. Schad GA. Presidential address. Hooked on hookworm: 25 years of attachment. *J Parasitol* 1991;77:176–186.
30. Smith G, Schad GA. Ancylostoma duodenale and Necator americanus: effect of temperature on egg development and mortality. *Parasitology* 1989;99:127–132.
31. Yoshida Y. Comparative studies on Ancylostoma braziliense and Ancylostoma ceylanicum. I. The adult stage. *J Parasitol* 1971;57:983–989.

23

Creeping Eruption

Douglas J. Wear,
Wayne M. Meyers, *and*
Ronald C. Neafie

Introduction

Definition

Creeping eruption is migratory dermatitis caused by adult or larval nematodes moving through the skin. Worldwide, the most common cause of this condition is infection by filariform larvae of the dog and cat hookworms, *Ancylostoma braziliense* de Faria, 1910 and *Ancylostoma caninum* (Ercolani, 1858) Hall, 1913.[3,5,14] In Japan, type X larvae of the superfamily *Spiruroidea* Railliet and Henry, 1915[7,11] and adult *Anatrichosoma cutaneum* (Swift, Boots, and Miller, 1922) Chitwood and Smith, 1956[2,9,10] also produce serpiginous dermal lesions. The dermatitis provoked by larvae of *Strongyloides* sp, *Ancylostoma duodenale*, *Necator americanus*, *Gnathostoma spinigerum*,[12] and *Pelodera strongyloides*[6] is sometimes identified as creeping eruption (Table 23.1) (see chapters 21, 22, 30, and 36 for a detailed discussion of the dermal lesions caused by these organisms).

Synonyms

Common terms for the dermatitis caused by *A. braziliense* include cutaneous larva migrans, sandworm, plumber's itch, duckhunter's itch, and epidermitis linearis migrans. Synonyms for the hookworm *A. cutaneum* include *Trichosoma cutaneum* and *Capillaria cutanea*.

Epidemiology

Creeping eruption from *A. braziliense* and *A. caninum* results from contact with soil contaminated by the feces of dogs or cats. It is most commmon in the tropics and subtropics, but is also prevalent in the southeastern United States. Children are particularly prone to creeping eruption because of their frequent contact with contaminated soil. The terms sandworm and plumber's itch reflect the heightened risk for people who frequent beaches or crawl beneath beach houses, where animal feces may accumulate. In Japan, where type X larvae of spiruroid nematodes invade marine fish and squid which are often eaten raw, infections peak between late winter and early summer when these commodities are most abundant.[13] The source of human infection by *A. cutaneum* is unknown.

367

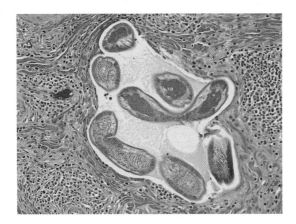

Figure 23.1
Adult male *A. cutaneum* in trachea of monkey. x110

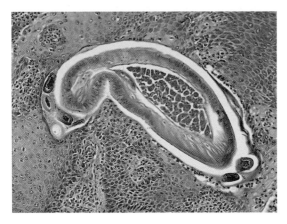

Figure 23.2
Adult female *A. cutaneum* in palm of monkey. x110

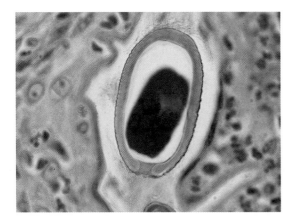

Figure 23.3
Smooth brown egg of *A. cutaneum*, showing 1 of 2 polar opercular plugs. x595

Infectious Agent

Morphologic Description

Infective filariform larvae of *A. braziliense* are 850 µm long and 35 µm in diameter. In processed tissue sections, larvae are only about 20 µm in diameter. Larvae have minute double lateral alae that extend most of their length. All hookworm larvae that cause cutaneous larva migrans in humans, and those of *Strongyloides stercoralis*, are similar morphologically. In contrast, the larvae of *G. spinigerum* are several hundred micrometers in diameter and can be readily identified by their many cuticular spines.[12] Type X larvae of the superfamily *Spiruroidea* are 6.5 to 8.5 mm long and 80 to 105 µm in diameter. Adult female *A. cutaneum*, the only form of this nematode found in humans, are 25 mm long; in diameter, they are 200 to 240 µm at the posterior end and 100 to 110 µm at the anterior end. Eggs are smooth and brown. They have bipolar opercular plugs, a shell 5 to 8 µm thick, and are 56 to 70 µm long and 37 to 49 µm in diameter. Adult male *A. cutaneum* recovered from monkeys were 55 µm in diameter; larvae were 35 to 45 µm in diameter (Figs 23.1 to 23.3).

Life Cycle and Transmission

The life cycles of *S. stercoralis*, hookworms, and *G. spinigerum* are described in chapters 21, 22, and 30. Adult *A. braziliense* and *A. caninum* live in the intestines of dogs and cats, where females lay embryonated eggs that pass in feces. In warm, moist, shaded soil, rhabditiform larvae appear within 24 hours. After feeding, growing, and molting for a week, rhabditiform larvae become infective filariform larvae that can invade human skin. Ordinarily, they penetrate no deeper than the epidermis. In dogs, larvae penetrate the skin within 30 minutes and move into the horny layer, where edges of keratinized cells provide uneven spots. Larvae migrate parallel to the surface into living epidermis, enter the external root sheath of a hair follicle, and exit through the sebaceous hypodermis. In dogs, hyaluronidase appears to sever hyaluronic acid bridges connecting epidermal cells expressing hyaluronic acid receptors.[4] *Ancylostoma braziliense* express higher levels of hyaluronidase than *A. caninum* and can thus directly penetrate the epidermis. Neither species completes its life cycle in humans. Adult

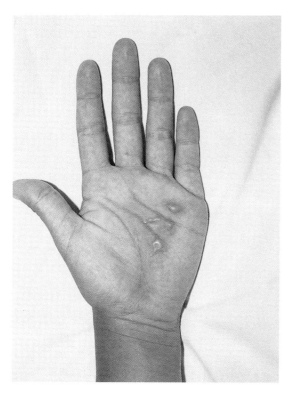

Figure 23.4
Creeping eruption, present for 2 days, on palm of Brazilian child. Tracks are short at this early stage.

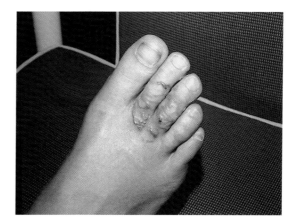

Figure 23.5
Creeping eruption displaying serpiginous track over second and third toes.

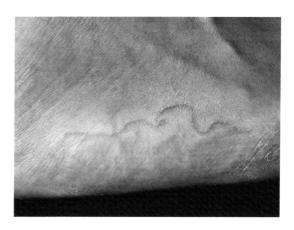

Figure 23.6
Single serpiginous track of creeping eruption on side of foot.

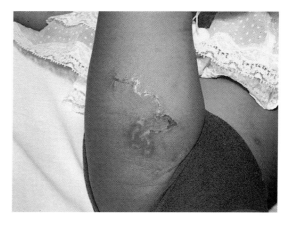

Figure 23.7
Creeping eruption on thigh of 11-month-old Brazilian girl. Note that inferior portion of track is composed of a scar, followed by a crust in the middle portion, then by a more recent lesion on the upper left.

Spiruroidea are thought to live in marine birds or mammals. Type X larvae infect cod (*Theragra chalcogramma*) and firefly squid (*Watasenia scintillans*). When infected cod or squid are ingested raw, larvae may appear in human skin 2 weeks later. The life cycle of *A. cutaneum* is unknown.

Clinical Features and Pathogenesis

A few hours after an *Ancylostoma* larva penetrates the skin, an itchy red papule develops. A serpiginous track arises from the border of the papule as the worm tunnels within the epidermis, sometimes migrating as much as several centimeters a day (Figs 23.4 to 23.7). The surrounding tissue is edematous, and acute inflammatory and healing stages are usually evident, indicating the worm's progress through the skin. The larva itself

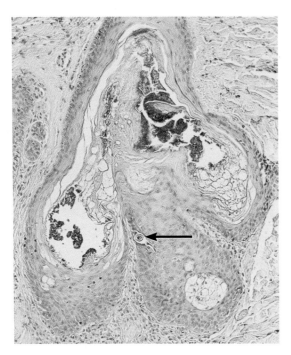

Figure 23.8
Creeping eruption showing *Ancylostoma* sp larva (arrow) in lower layers of epidermis, around hair follicle in skin of back. x95

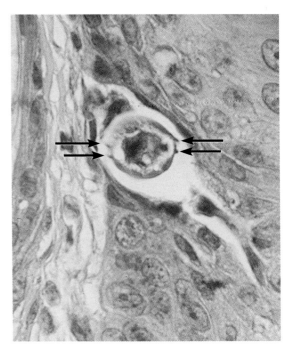

Figure 23.9
Higher magnification of Figure 23.8, showing small double lateral alae (arrows) of *Ancylostoma* sp larva. x820

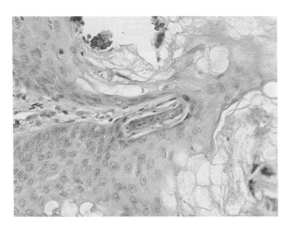

Figure 23.10
Different view of lesion pictured in Figure 23.8, depicting longitudinal section of *Ancylostoma* sp larva. x230

is difficult to pinpoint because it has usually moved beyond the lesion. Itching can be severe, and scratching frequently leads to secondary infection. In heavy infections, itching may be intense enough to prevent sleep or provoke psychosis. Untreated lesions may persist for several months. Type X larvae produce an itchy, sometimes serpiginous dermal eruption with erythema. Infection with *A. cutaneum* causes severe pruritus accompanied by a zigzag inflammation of the skin.

Pathologic Features

Biopsy specimens are usually not taken in *A. braziliense* infection, and even when they are, most specimens contain no larvae and reveal only an infiltration of lymphocytes and eosinophils in the upper dermis. Larvae, which are located in burrows in the deeper layers of the epidermis (Figs 23.8 to 23.10), are most often found in specimens taken from just beyond the leading edge of the

Genus and species	Prevalence	Clinical characteristics
Anatrichosoma cutaneum	Rare	Linear redness, zigzag advance, moves 5-10 mm/day, severe pruritus
Ancylostoma braziliense	Most common	Thread-like, linear burrows, highly pruritic, moves 1-2 cm/day; may persist for months
Ancylostoma caninum	Common	Papular, rarely linear burrows, clears spontaneously in 1-3 weeks
Ancylostoma duodenale	Common	Papulovesicular, pruritic, minimal migration in the skin, clears within 2 weeks
Necator americanus	Common	Same as for *A. duodenale*
Uncinaria stenocephala	Rare	Thread-like, linear burrows, pruritic, moves 1-2 cm/day, may persist for months
Gnathostoma spinigerum	Rare	Deep, wide burrows, furunculoid, may persist for years
Bunostomum phlebotomum	Rare	Papular lesion, minimal migration, clears within 2 weeks
Strongyloides stercoralis (larva currens)	Rare	Band of urticaria and pruritic induration extends up to several cm/day; may persist for weeks; recurrent owing to autoinfection
Strongyloides myopotami	Rare	Papular or linear, thread-like, serpiginous, pruritic lesion
Pelodera strongyloides	Rare	Widespread eruption of papules, pustules, and burrows

Table 23.1
Classification, prevalence, and clinical characteristics of nematodes causing creeping eruption.

track. Identifying the species of parasite from tissue sections is usually not possible.

Host defenses against larvae that cause creeping eruption are not well-understood. Presumably, there is a physiologic incompatibility between humans and these larvae which prohibits their further penetration and development. Some investigators believe larvae lack the specific collagenase necessary to invade human dermis. IgE levels are elevated in patients with creeping eruption, which may play a role in tissue response. Type X larvae leave intraepidermal pustules which are either empty or contain larvae curled in the dermis at the front of the eruption. Adult female *A. cutaneum* lie in the forward terminus of their tracks.

Diagnosis

Diagnosis of *A. braziliense* infection is usually made on the basis of clinical history and the appearance of a characteristic serpiginous track. Biopsy is not recommended, because the larva is usually beyond the obvious lesion and is rarely seen in the specimen. Diagnosis of *Spiruroidea* infection is usually made by identifying type X larvae, though there is a report of 2 patients whose sera reacted against sections of an adult worm.[7] A diagnosis of *A. cutaneum* infection is established by identifying the adult worm in biopsy.

Treatment

For *A. braziliense* and *A. caninum* infections, thiabendazole is effective when applied topically in a 15% liquid with 3% salicylic acid in an anhydrous lanolin ointment,[5] or in a lotion (500-mg tablet dissolved in a water-soluble base) 2 to 3 times a day for 5 days. However, as of 1998 this treatment is not FDA-approved.[3] Albendazole (10 to 15 mg/kg body weight) administered daily for 3 days[8] is therapeutically similar to thiabendazole, but a single dose of ivermectin (12 mg or 150 to 200 µg/kg body weight) seems to be most effective.[1] Penicillin or other appropriate antibiotics should be given for secondary bacterial infections. Surgical excision is the treatment of choice for infections by type X larvae and *A. cutaneum*.

References

1. Caumes E, Carriere J, Datry A, Gaxotte P, Danis M, Gentilini M. A randomized trial of ivermectin versus albendazole for the treatment of cutaneous larva migrans. *Am J Trop Med Hyg* 1993;49:641–644.
2. Chitwood MB, Smith WN. A redescription of Anatrichosoma cynamolgi. *Proc Helminthol Soc* 1958;25:112–117.
3. Davies HD, Sakuls P, Keystone JS. Creeping eruption: a review of clinical presentation and management of 60 cases presenting to a tropical disease unit. *Arch Dermatol* 1993;129:588–591.
4. Hotez PJ, Narasimhan S, Haggerty J, et al. Hyaluronidase from infective Ancylostoma hookworm larvae and its possible function as a virulence factor in tissue invasion and in cutaneous larva migrans. *Infect Immun* 1992;60:1018–1023.
5. Jelinek T, Maiwald H, Nothdurft HD, Loscher T. Cutaneous larva migrans in travelers: synopsis of histories, symptoms, and treatment of 98 patients. *Clin Infect Dis* 1994;19:1062–1066.
6. Jones CC, Rosen T, Greenberg C. Cutaneous larva migrans due to Pelodera strongyloides. *Cutis* 1991;48:123–126.
7. Kagei N. Morphological identification of parasites in biopsied specimens from creeping disease lesions. *Jpn J Parasitol* 1991;40:437–445.
8. Kollaritsch H, Jeschko E, Wiedermann G. Albendazole is highly effective against cutaneous larva migrans but not against Giardia infection: results of an open pilot trial in travellers returning from the tropics. *Trans R Soc Trop Med Hyg* 1993;87:689.
9. Le-Van-Hoa, Duong-Hong-Mo, Nguyen-Luu-Vien. Premier cas de capillariose cutanée humaine. *Bull Soc Pathol Exot* 1963;56:121–126.
10. Morishita K, Tani T. A case of Capillaria infection causing cutaneous creeping eruption in man. *J Parasitol* 1960;46:79–83.
11. Okazaki A, Ida T, Muramatsu T, Shirai T, Nishiyama T, Araki T. Creeping disease due to larva of spiruroid nematoda. *Int J Dermatol* 1993;32:813–814.
12. Pinkus H, Fan J, DeGiusti D. Creeping eruption due to Gnathostoma spinigerum in a Taiwanese patient. *Int J Dermatol* 1981;20:46–49.
13. Taniguchi Y, Ando K, Shimizu M, Nakamura Y, Yamazaki S. Creeping eruption due to larvae of the suborder Spirurina—a newly recognized causative parasite. *Int J Dermatol* 1994;33:279–281.
14. Wong-Waldamez A, Silva-Lizama E. Bullous larva migrans accompanied by Loeffler's syndrome. *Int J Dermatol* 1995;34:570–571.

24

Angiostrongyliasis Cantonensis

Aileen M. Marty *and*
Ronald C. Neafie

Introduction

Definition

Angiostrongyliasis cantonensis is infection by the nematode *Angiostrongylus cantonensis* (Chen, 1935) Dougherty, 1946. The usual manifestations in humans are related to the brain and eyes, but in rare instances the lungs are involved.

Synonyms

Other names for *A. cantonensis*, the rodent lungworm, include *Pulmonema cantonensis*, *Hæmostrongylus ratti*, *Rodentocaulus cantonensis*, and *Parastrongylus cantonensis*.[55] Other names for angiostrongyliasis cantonensis reflect the various forms of the disease, including eosinophilic meningitis, epidemic eosinophilic meningitis, parasitic eosinophilic meningoencephalitis, angiostrongylus eosinophilic meningitis, parastrongylus eosinophilic meningitis, and human ocular angiostrongyliasis.[25]

General Considerations

In 1905, Kamensky established the genus *Angiostrongylus* for the dog lungworm, then named *Strongylus vasorum*.[7,30] In 1945, Nomura and Lin recovered 6 adult worms from the cerebrospinal fluid of a 15-year-old boy from Taiwan. They sent the worms to Yokogawa, who identified them as *Hæmostrongylus ratti*.[9] In 1961, Rosen et al speculated that an unspecified seafood parasite was responsible for the epidemics of eosinophilic meningitis that began in Micronesia in 1944, and spread to New Caledonia and Tahiti.[44] At about the same time, 2 Hawaiian mental patients were reported as dying from meningoencephalitis infections. Chappell, studying these patients' autopsied brains, found fragments of worm in one of them, but was unable to identify the worm from these sections. Some time later, Chitwood, examining the same brain material, found whole worms and identified them as immature adults of *A. cantonensis*. Alicata knew that Ash had discovered *A. cantonensis* in rats in Hawaii,[4] and that terrestrial mollusks serve as intermediate hosts.[32] Based on this information, he postulated that *A. cantonensis* was the likely cause of the epidemics in the Pacific islands. Rosen et al then reported that *A. cantonensis* was the etiologic agent of the eosinophilic meningoencephalitis in the Hawaiian

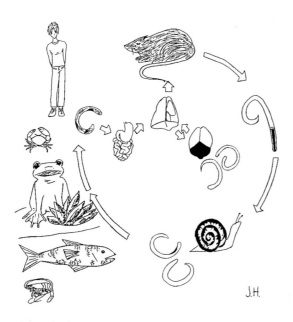

Figure 24.1
Life cycle of *A. cantonensis*. Mollusks serve as intermediate hosts. Crabs, frogs, fish, and prawns are some of the paratenic hosts.*

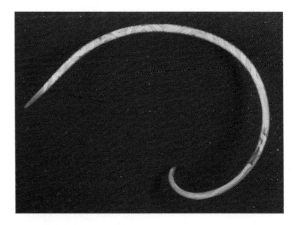

Figure 24.2
Female *A. cantonensis*. Note barber-pole pattern caused by spiral winding of milky-white, paired uterine branches around intestine. Unstained x7

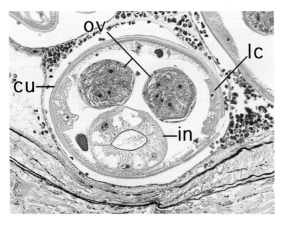

Figure 24.3
Transverse section of female *A. cantonensis* in human lung, displaying cuticle (cu), lateral cords (lc), paired ovaries (ov), and intestine (in). Movat x290

patients studied by Chappell.[43] The crucial work establishing this etiology was done by Wallace and Rosen,[57,58,59,60,61] and Punyagupta et al,[41] who throughout the 1960s conducted epidemiologic studies documenting the infection, the worm's intermediate and paratenic hosts (Fig 24.1), and how transmission varies from region to region. This investigation of variations in the disease in different localities led to the discovery of a new species, *Angiostrongylus malaysiensis*, a closely related worm that is less pathogenic for nonhuman primates and probably for humans as well.[14] Various authors have regrouped worms of the genus *Angiostrongylus* because of minute differences in male worm anatomy. We interpret these minor differences of the bursa as subgenera and retain the genus name *Angiostrongylus*.

Epidemiology

Angiostrongylus cantonensis infects animals in tropical zones throughout the world.[3,22,56] Thousands of human infections are reported annually[53] and are considered a serious public health problem in Southeast Asia and many Pacific islands. Most human infections originate in Asia (Taiwan,[17] China,[28,64] India, Japan,[45] Thailand, Vietnam, Laos), the Indian Ocean region (Sri Lanka,[20] Réunion Island,[6] Comoro Islands[23]), and Oceania (Australia, Fiji,[36] the Philippines, Indonesia, Papua New

Guinea,[47,48] American Samoa[10,27]). Autochthonous human infections have, however, developed in Egypt,[16] Côte d'Ivoire, Cuba,[1] the United States (Louisiana,[34] Hawaii), and possibly Jamaica[8] and Italy.[37]

The less pathogenic *A. malaysiensis* has apparently produced human infections in Malaysia,[31] although initial reports assumed that the agent was *A. cantonensis*.[15,62] People of any age are subject to infection. In Thailand it is more prevalent in young adults (20 to 34 years old), but in Taiwan children are most affected (average 7.6 years old). Susceptibility also varies by gender in different geographic regions. Cultural factors obviously play a role in exposure to infection.

Infectious Agent

Morphologic Description

Angiostrongylus cantonensis are filariform and taper slightly at both ends. Adult female worms from rats are 18.5 to 34 mm by 280 to 560 µm; male worms are 15.5 to 25 mm by 250 to 420 µm. Adult worms recovered from humans are smaller (100 to 260 µm in diameter). In females, the spiral winding of the milky-white, paired uterine and ovarian branches around the dark, blood-filled intestine produces the barber-pole pattern characteristic of the genus (Fig 24.2). Transverse sections through the female typically contain paired reproductive tubes and an intestine (Figs 24.3 and 24.4). A single, thin-walled vagina extends to the vulva, which is 2 to 3 mm from the tip of the tail, near the anal opening. Female *A. cantonensis* have longer vaginas than *A. costaricensis* and *A. malaysiensis*, and no projection at the tip of the tail.[14] The head has 1 dorsal and 2 ventral lips and papillae, but no buccal capsule; the mouth opens directly into the esophagus. The thin cuticle (3 to 6 µm) is smooth, multilayered, and displays fine transverse striations in oblique and longitudinal sections. Lateral cords are conspicuous at most levels (Fig 24.3). The somatic musculature is low coelomyarian, with many somatic muscle cells in each quadrant. There is a subventral gland in the area of the esophageal-intestinal junction. The intestine is large in diameter and contains only a few cuboid, multinucleated cells lined with microvilli (Fig 24.3).

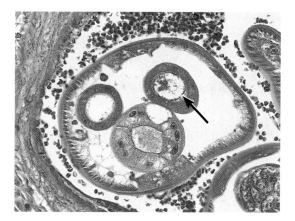

Figure 24.4
Transverse section of female *A. cantonensis* in human lung, displaying intestine and paired uteri. One uterine branch contains an ovum (arrow). x240

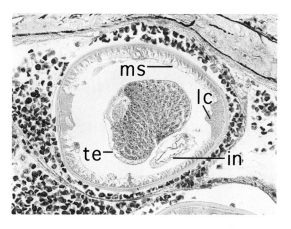

Figure 24.5
Transverse section of male *A. cantonensis* within human pulmonary blood vessel at level of intestine (in) and testis (te). Note coelomyarian muscle (ms) and lateral cords (lc). Movat x320

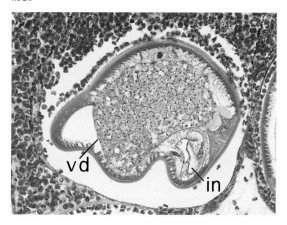

Figure 24.6
Transverse section of male *A. cantonensis* within human pulmonary blood vessel at level of intestine (in) and vas deferens (vd) with sperm. x255

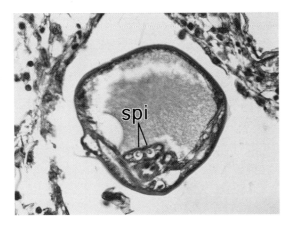

Figure 24.7
Transverse section of male *A. cantonensis* within meninges through posterior end of worm, revealing paired copulatory spicules (spi). x315

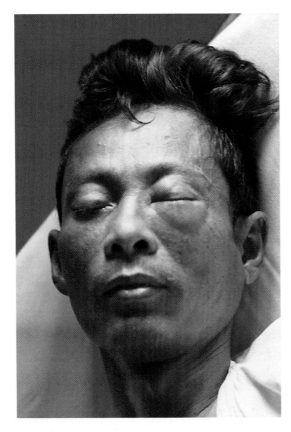

Figure 24.8
Periorbital edema caused by *A. cantonensis* in patient from Thailand.*

Transverse sections of male *A. cantonensis* reveal an intestine and a single reproductive tube (Figs 24.5 and 24.6). The paired golden spicules (Fig 24.7) are unequal and slender, with conspicuous striations throughout, except at their extreme anterior ends.[18] In the bursa, the ventroventral ray separates from the lateroventral ray at one third of the common trunk.[14] The male has a gubernaculum.[12] Cuticle, cords, and musculature are similar to the female (Fig 24.5).

First-stage (L1) larvae are 270 to 300 μm long by 15 to 16 μm wide. They have a short, narrow vestibule leading to a rhabditoid esophagus that forms about half the total gut. The nerve ring lies about 75 μm from the anterior end; the genital rudiment is near the middle of the intestine. The anus is approximately 30 μm from the tip of the tail, which is sharply pointed and has a distinct notch on the dorsal surface. Infective L3 larvae (420 to 490 μm long by up to 25 μm wide) have well-developed chitinous rods with expanded knob-like tips in the vestibule, and a rhabditoid esophagus. The anus is about 40 μm from the tip of the tail. Larger larvae are 30 to 90 μm in diameter.[40]

Angiostrongylus cantonensis eggs are ovoid, thin-shelled, unembryonated when laid, and measure 68 to 74 by 46 to 48 μm. They have not been found in human tissue.

Life Cycle and Transmission

At least 24 species of rodents serve as definitive hosts for *A. cantonensis*, including *Bandicota indica*[17,52] and various *Rattus* sp, especially *Rattus norvegicus* and *Rattus rattus*.[13] Gastropods function as intermediate hosts, particularly slugs (*Laevicaulus alte*, *Vaginulus plebeius*, and *Deroceras (Agriolimax) laeve*) and snails (*Achatina fulica* and *Euglandina rosea*).[60] Animals such as freshwater shrimp, land crabs, land planarians, freshwater cichlid fish, marine carangid fish, monitors,[42] toads, and frogs[5] may serve as paratenic hosts. Humans, nonhuman primates,[22] mice, and other mammals are accidental, dead-end hosts.

In rodents, adult *A. cantonensis* live in pulmonary arteries and in the right side of the heart (Fig 24.1).[11] Eggs lodge in pulmonary capillaries, embryonate, hatch, and release L1 larvae that enter alveoli and migrate up the trachea. Some L1 larvae exit in sputum, but most enter the esophagus and pass in feces. The L1 larvae, which may survive for

2 or more weeks in a moist environment, enter molluscan intermediate hosts either by direct penetration or by ingestion. They molt twice to become infective L3 larvae. A paratenic host may consume L3 larvae, but for the *A. cantonensis* life cycle to continue, a definitive host must ingest L3 larvae from the tissues of a paratenic host or a molluscan intermediate host.

When a rodent swallows tissues containing L3 larvae, its stomach acids release the larvae, which then migrate to the lower ileum, penetrate through the small intestine into mesenteric vessels, and enter the systemic circulation. The infection is blood-borne for at least 24 hours until the L3 larvae find their way to the central nervous system (CNS) and cerebral blood vessels.[61] They tend to congregate in the anterior portion of each cerebral hemisphere and move about without provoking obvious hemorrhage or tissue reaction. L3 larvae molt twice, then migrate to the surface of the cerebrum. The meninges react with dilation of vessels, leukocytic infiltration, and proliferation of the leptomeninges. Immature adult worms penetrate the meningeal veins and reenter the systemic circulation. Soon they reach their definitive site in pulmonary arteries, where they mature sexually and mate, and the female worms begin laying eggs. Excretion of L1 larvae usually begins 42 to 45 days after the rodent consumes L3 larvae. In humans, immature adult worms develop in the brain and migrate to the meninges, but only rarely travel to the lungs, where they must go to reach sexual maturity.[40] There is a report of an *Angiostrongylus* sp producing eggs and L1 larvae in a human lung, but it is not clear whether the worm was *A. cantonensis* or another zoonotic *Angiostrongylus* sp.[37]

Humans are sometimes infected with *A. cantonensis* by eating snails or other mollusks raw. However, a more common source of human infection is the consumption of contaminated paratenic hosts, such as improperly cooked freshwater shrimp, land crabs, frogs, and certain fish from endemic areas.[33,54] In Thailand, eating the raw liver of the yellow tree monitor has led to infection. Humans may inadvertently ingest infected molluscan tissues from unwashed vegetables[21,24] or hands that have handled contaminated gastropods. Another source of infection is drinking water contaminated with *A. cantonensis* L3 larvae.[18,19,20]

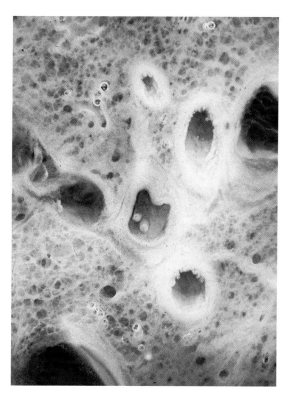

Figure 24.9
Cut surface of lung from autopsy of 6-year-old girl from Taiwan with overwhelming *A. cantonensis* infection. Note 2 adult *A. cantonensis* in lumen of pulmonary vessel.* x12

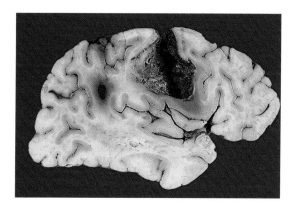

Figure 24.10
Sagittal section of brain depicting several areas of hemorrhage and necrosis caused by *A. cantonensis*.*

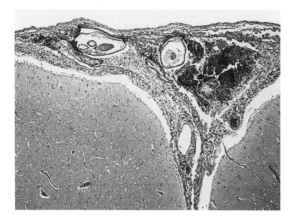

Figure 24.11
Sections of female *A. cantonensis* in meninges eliciting marked eosinophilic response. x40

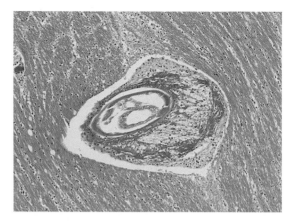

Figure 24.12
Transverse section of female *A. cantonensis* in cerebral blood vessel. Note perivascular neutrophils and fibrin. x70

Clinical Features and Pathogenesis

For humans infected with *A. cantonensis*, the incubation period is usually 12 to 28 days.[2] Prodromal symptoms are uncommon but may include vomiting and subsequent dehydration. The severity of clinical symptoms depends on worm burden; in light infections, the disease may be mild or asymptomatic. Very young children with immature immune systems are most susceptible.[39] There is no evidence that immunity follows infection, since repeat infections are common and well-documented.[41]

In endemic areas, symptomatic patients usually present with 1 of the following clinical combinations: 1) Acute, severe, throbbing headache with or without low fever; 2) symptoms and signs of meningitis or meingoencephalitis with low fever; 3) cranial nerve involvement, especially facial palsy or lateral rectus paralysis, associated with severe headache; 4) ocular manifestations, mainly diplopia of 1 or both sides, with or without headache; or 5) psychosis or sensory impairment associated with severe headache.

Headache is the chief complaint of virtually all patients. Onset may be abrupt or insidious with changing intensity; pain may last 6 weeks or longer. Stiff, painful neck and back are frequent complaints, but Kernig's sign is rare. Percussion of the spine may produce lumbar pain. Intermittent or remittent fever may persist longer than any other symptom. In severe cases, particularly in children, fever may be as high as 40°C. About 80% of patients experience vomiting and/or nausea. Patients frequently present with areas of moderate to severe paresthesia, hyperesthesia, or hypoesthesia such as spontaneous, asymmetric tingling, numbness, or burning sensations. Sometimes paresthesia appears as night paroxysms. The distribution of dysesthetic areas suggests radicular and/or peripheral nerve involvement. Bizarre patterns of sensory symptoms result from combinations of nerve root plexus and peripheral nerve damage.

Patients may have multiple symptoms of cranial nerve involvement. Some develop facial paralysis, often preceded by facial paresthesia, which may be transient or last for several months. Damage to the glossopharyngeal and/or hypoglossal nerves may cause speech difficulties. Infrequently, paresis of the sixth cranial (abducens) nerve causes strabismus with diplopia.

Pain or tenderness in groups of muscles is apparent on deep pressure. Abnormal reflexes, especially of the patella and Achilles tendon, are transient. Interruption of nerve impulses may produce bilateral atrophy of the quadriceps and edema of the eyelid, hand, and ankle.

Severe infection may result in a variety of mental aberrations, disturbed pupillary reflexes, trismus, incontinence, lethargy, or coma and death. In children, angiostrongyliasis cantonensis is more likely to produce fever and lethargy. Focal neurologic signs are less frequent than in adults, but slight retinal edema and pupillary congestion are common. Children with heavy infection may suffer serious brain damage.[39]

White blood cell counts rarely exceed $10 \times 10^3/\mu l$. There is no eosinophilia in 25% of patients, but eosinophilia of 10% to 35% is common, and may exceed 60%. Patients with CNS involvement always have abnormal cerebrospinal fluid (CSF). CSF frequently is cloudy and protein is elevated to more than 50 mg%. Pleocytosis (0.2 to $5 \times 10^3/\mu l$) with eosinophilia (26% to 75%) is a constant finding. Xanthochromia is uncommon, but red blood cells are occasionally present. The chance of recovering live, immature adult *A. cantonensis* in the CSF is highest in small children.[40]

Besides eosinophilic meningitis, *A. cantonensis* may produce eosinophilic myelomeningoencephalitis. Patients have symptoms of lesions in the spinal cord and peripheral nerves or nerve roots, including urinary retention, weakness of the legs, sensory impairment, depressed deep tendon reflexes in the lower limbs, and hyperesthesia of the extremities.[63]

Ocular involvement may accompany CNS disease or may be an isolated manifestation without abnormal CSF findings. Worms may be present in the anterior or posterior chamber of the eye. Patients with CNS disease frequently have photophobia, blurring of vision, and retro-orbital pain. Patients with ocular involvement with or without CNS disease may have periorbital edema (Fig 24.8), exophthalmos, extraocular muscle palsies, optic nerve injury, and/or intraocular damage. Some patients develop permanent blindness.

In the early stage of illness, many patients experience intermittent dry cough with sneezing and rhinorrhea, but serious pulmonary involvement is uncommon. Patients with pneumonitis manifest productive cough and rales. Chest x-rays of infected children show dense opacities with ill-defined margins and segmental distribution in the lower lung fields. These findings differ from those of an apical or axillary lesion of chronic eosinophilic pneumonia.[49]

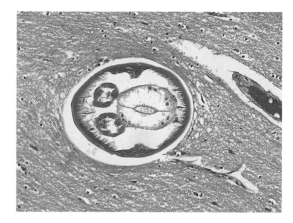

Figure 24.13
Transverse section of viable female *A. cantonensis* in track in brain with no evidence of cellular reaction. x140

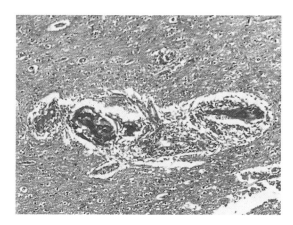

Figure 24.14
Longitudinal section of degenerated and fragmented *A. cantonensis* in track in brain provoking marked acute and chronic cellular reaction. x55

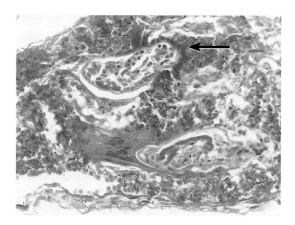

Figure 24.15
Fragments of degenerated *A. cantonensis* in meninges. Note giant cell engulfing fragment of worm, and Splendore-Hoeppli reaction (arrow) forming on edge of another fragment. x235

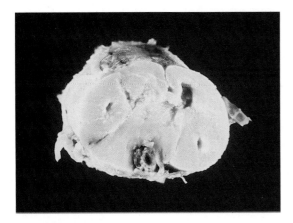

Figure 24.16
Gross transverse section of spinal cord of patient with *A. cantonensis* depicting several hemorrhagic areas of necrosis caused by migrating worms.*

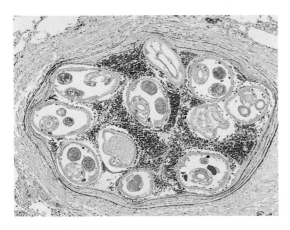

Figure 24.17
Multiple sections of adult male and female *A. cantonensis* in pulmonary vessel. Movat x60

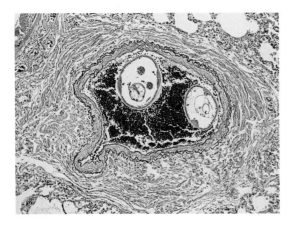

Figure 24.18
Two transverse sections of *A. cantonensis* surrounded by erythrocytes within pulmonary artery. Note extensive collagen formation around vessel. Movat x70

Pathologic Features

It takes about a month for *A. cantonensis* larvae to mature in the CNS of a definitive host. In humans, worms usually remain in the brain, where they cause the most damage, and many die there. But some fully developed larvae migrate to the lungs (Fig 24.9), where they may survive and mature into adults.[51,65]

Gross lesions of the brain include clouding of basal and cerebellar meninges, and minor subdural or subarachnoid hemorrhage with extension to the adjacent cortex (Fig 24.10).[35] Worms may invade the subdural and subarachnoid spaces. Cut surfaces of the brain reveal focal necrosis and hemorrhage corresponding to worm tracks.

Microscopically, living or dead worms are seen in meninges and sometimes in blood vessels or perivascular spaces (Figs 24.11 and 24.12). Cellular reaction to living worms is usually minimal (Fig 24.13) but pronounced around degenerating worms (Fig 24.14). Granulomatous inflammation may be present. Giant cells may engulf degenerating worms, or a Splendore-Hoeppli reaction may form around a degenerating worm (Fig 24.15). Cellular reaction may extend along the meninges and intracerebral vessels well beyond the parasite. Multiple microcavities or tracks of migrating worms are characteristic (Fig 24.14). The tracks disrupt brain tissue and contain debris, gitter cells, cellular infiltrations, and/or hemorrhage. Nonhemorrhagic tracks of *A. cantonensis* are usually less than 150 µm wide. There is pronounced arterial and venous dilation in the subarachnoid space. Nerve cells in areas adjacent to the parasite or tracks may have central chromatolysis and cytoplasmic axonal swelling. Eosinophils predominate in some lesions, and there may be Charcot-Leyden crystals. Eosinophils in the CSF are associated with killing of *A. cantonensis* in brains of nonpermissive hosts and are also responsible for some neurologic disorders.[66] Pathologic findings in the spinal cord (Fig 24.16) are similar to those in the brain.

In the eye, larvae may be found in the anterior chamber and vitreous or below the retina.[50] They tend to localize in the subretinal space and produce retinal detachment. As the parasite moves within the space, it causes further mechanical damage and inflammation. Larvae produce fibrous tracts along the path of migration from the optic nerve into the vitreous.

In patients with significant pulmonary damage, biopsy specimens reveal mature and immature adult *A. cantonensis* in medium to large-sized branches of the pulmonary arteries (Figs 24.17 and 24.18). Viable worms usually provoke little cellular reaction, but degenerating worms cause extensive necrosis (Fig 24.19), hemorrhage, and granuloma formation. Degenerating adult worms may cause endarteritis with mural thrombi that can lead to pulmonary infarcts (Fig 24.20). The endarteritis may extend beyond the worm. Worms may localize in alveolar spaces and incite inflammatory reactions (Fig 24.21).

One patient with mature adult male and gravid female *Angiostrongylus* sp in the lung[37] also had unembryonated and fully embryonated eggs in the capillary bed of the lung, along with free L1 larvae. Well-formed granulomas, consisting of lymphocytes and eosinophils mixed with macrophages and giant cells, were scattered within the interstitium. Giant cells engulfed eggs or larvae. Several granulomas had apparently coalesced to form larger masses with early fibrosis. The number of granulomas increased in direct relation to the proximity of adult worms. The patient also had L1 larvae in a glomerular tuft of the kidney. Because of morphologic differences in the adult worms and the smaller size of eggs, the authors believe this was not *A. cantonensis*.

Diagnosis

Definitive diagnosis is based on identifying the worm in tissue sections or extracting whole worms from the eye, brain, or CSF. Postmortem examination of the brain of any patient suspected of having *A. cantonensis* infection should be extensive. *Angiostrongylus cantonensis* infection presents a very different clinical and histopathologic picture from that of *A. costaricensis* (see chapter 25). Distinguishing *A. cantonensis* from *A. malaysiensis* is difficult in tissue sections, but analyses of isozymes and DNA can differentiate these parasites.[46]

Angiostrongyliasis cantonensis is usually

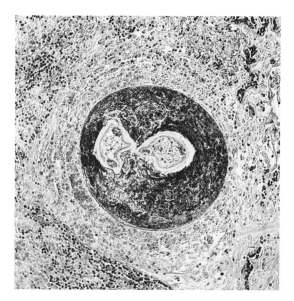

Figure 24.19
Pulmonary vessel containing degenerating *A. cantonensis* provoking extensive fibropurulent reaction and inflammation in surrounding lung tissue. Movat x120

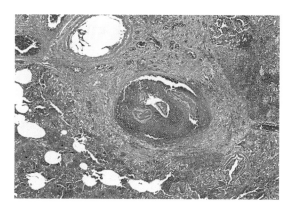

Figure 24.20
Pulmonary infarct with endarteritis and mural thrombi caused by degenerated *A. cantonensis* in pulmonary vessel. x28

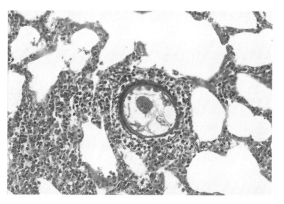

Figure 24.21
Angiostrongylus cantonensis in alveolar space provoking inflammatory reaction. x110

suggested by clinical and epidemiologic features and confirmed with positive serologic tests. Based on clinical features alone, the disease may mimic migraine, brain tumor, or psychoneurosis. CSF and CT findings distinguish purulent meningitis, or tuberculous or viral meningoencephalitis, from *A. cantonensis* infection.

The differential diagnosis of CSF eosinophilia includes other parasitic infections (*Gnathostoma spinigerum*, *Toxocara canis*, *Trichinella spiralis*, *Ascaris lumbricoides*, *Echinococcus granulosus*, *Strongyloides stercoralis*), Hodgkin's disease, lymphocytic choriomeningitis, fungal infections (*Coccidioides immitis*), bacterial infections (neurosyphilis, tuberculous meningitis), foreign bodies (neurosurgical shunts), and allergic reactions.[21,29] The parasitic infection most closely resembling that of *A. cantonensis* is the one caused by *G. spinigerum*. Patients with *G. spinigerum* infection, endemic in parts of Asia, usually present with eosinophilic myeloencephalitis. However, transverse or ascending myelitis is more common in *G. spinigerum* than in *A. cantonensis*, and the brain tracks made by *G. spinigerum* are larger than those made by *A. cantonensis*. Microscopic examination of larvae of both species in tissue or gross specimens shows *G. spinigerum* to be much longer and thicker. Furthermore, most of the body of *G. spinigerum* is covered with spines, which *A. cantonensis* do not have.

Worms in the ocular chambers are best observed by slit lamp. Larvae of *A. cantonensis* in the anterior chamber are more sluggish than those of *G. spinigerum*.[53]

Clinical findings should be verified by serologic testing; however, most available serologic tests have inadequate specificity. Intradermal, indirect hemagglutination and complement fixation tests on sera or spinal fluid are often inconclusive. ELISA has shown promising results, although there may be cross-reactivity with *T. canis*, *Ascaris suum*, and *Metastrongylus apri* infections. Recent immunodiagnostic procedures for detecting circulating antigens have greater specificity.[19]

Treatment and Prevention

Several established anthelmintics have been used to treat angiostrongyliasis cantonensis. None has proven reliably therapeutic,[13] though recent studies with ivermectin[29] and PF1022A have shown promising results.[26] Fortunately, the disease is usually self-limiting and prognosis is generally good. Severe infection may lead to blindness, neurologic damage (especially in children), or chronic pain, but is rarely fatal.

Spinal taps can limit the duration of headache, and nonsteroidal anti-inflammatory agents or narcotics can ease the pain. Systemic corticosteroids may be palliative in severe infections. Larvae in the anterior chamber of the eye can be removed by corneoscleral incision, subretinal parasites can be extracted by scleral incision after precise location and freezing with a cryoprobe, and intravitreous worms can be removed anteriorly or directly through the pars plana.[50]

Adequate preventive measures include not eating raw snails, frogs, monitors, and other intermediate or paratenic hosts, and thorough hand washing after handling raw meat.

References

1. Aguiar PH, Morera P, Pascual J. First record of Angiostrongylus cantonensis in Cuba. *Am J Trop Med Hyg* 1981;30:963–965.
2. Alicta JE, Brown RW. Observations on the method of human infection with Angiostrongylus cantonensis in Tahiti. *Can J Zool* 1962;40:755–760.
3. Andersen E, Gubler DJ, Sorensen K, Beddard J, Ash LR. First report of Angiostrongylus cantonensis in Puerto Rico. *Am J Trop Med Hyg* 1986;35:319–322.
4. Ash LR. The helminth parasites of rats in Hawaii and the description of Capillaria traverae sp. n. *J Parasitol* 1962;48:66–68.
5. Ash LR. The occurrence of Angiostrongylus cantonensis in frogs of New Caledonia with observations on paratenic hosts of metastrongyles. *J Parasitol* 1968;54:432–436.
6. Badiaga S, Levy PY, Brouqui P, Delmont J, Bourgeade A. Eosinophilic meningitis. Review of the literature and a new case originating from Réunion Island [in French]. *Bull Soc Pathol Exot* 1993;86:277–281.
7. Baillet CC. Strongyle des vaisseaux et du coeur du chien Strongylus vasorum (Nobis). *Nouveau Dictionnaire Pratique de Médecine, de Chirurgie et d'Hygiène Vétérinaires, Paris* 1866;8:587–588.
8. Barrow KO, St Rose A, Lindo JF. Eosinophilic meningitis. Is Angiostrongylus cantonensis endemic in Jamaica? *West Indian Med J* 1996;45:70–71.
9. Beaver PC, Rosen L. Memorandum on the first report of Angiostrongylus in man, by Nomura and Lin, 1945. *Am J Trop Med Hyg* 1964;13:589–590.
10. Beck MJ, Cardina TM, Alicata JE. Eosinophilic meningitis due to angiostrongyliasis cantonensis in American Samoa. *Hawaii Med J* 1980;39:254–257.
11. Bhaibulaya M. Comparative studies of the life history of Angiostrongylus mackerrasae Bhaibulaya, 1968 and Angiostrongylus cantonensis (Chen, 1935). *Int J Parasitol* 1975;5:7–20.
12. Bhaibulaya M. Morphology and taxonomy of major Angiostrongylus species of Eastern Asia and Australia. In: Cross JH, ed. *Studies on Angiostrongyliasis in Eastern Asia and Australia.* Taipei, Taiwan; NAMRU2 Special Publication (NAMRU-2-SP-44) 1979:4–13.
13. Bhaibulaya M. Snail borne parasitic zoonoses: angiostrongyliasis. *Southeast Asian J Trop Med Public Health* 1991;22:189–193.
14. Bhaibulaya M, Cross JH. Angiostrongylus malaysiensis (Nematoda: Metastrongylidae), a new species of rat lung-worm from Malaysia. *Southeast Asian J Trop Med Public Health* 1971;2:527–533.
15. Bisseru B, Gill SS, Lucas JK. Human infection with rat lungworm Angiostrongylus cantonensis (Chen, 1935) in West Malaysia. *Med J Malaya* 1972;26:164–167.
16. Brown FM, Mohareb EW, Yousif F, Sultan Y, Girgis NI. Angiostrongylus eosinophilic meningitis in Egypt. *Lancet* 1996;348:964–965.
17. Chen ER. Angiostrongyliasis and eosinophilic meningitis on Taiwan: a review. In: Cross JH, ed. *Studies on Angiostrongyliasis in Eastern Asia and Australia.* Taipei, Taiwan; NAMRU-2 Special Publication (NAMRU-2-SP-44) 1979:57–73.
18. Chen HT. Un nouveau nematode pulmonaire, Pulmonema cantonensis, N.G., n. sp., des rats de Canton. *Ann Parasitol* 1935;13:312–317.
19. Chye SM, Yen CM, Chen ER. Detection of circulating antigen by monoclonal antibodies for immunodiagnosis of angiostrongyliasis. *Am J Trop Med Hyg* 1997;56:408–412.
20. Durette-Desset MC, Chabaud AG, Cassim MH, et al. On an infection of a human eye with Parastrongylus (= Angiostrongylus) sp. in Sri Lanka. *J Helminthol* 1993;67:69–72.
21. Fuller AJ, Munckhof W, Kiers L, Ebeling P, Richards MJ. Eosinophilic meningitis due to Angiostrongylus cantonensis. *West J Med* 1993;159:78–80.
22. Gardiner CH, Wells S, Gutter AE, et al. Eosinophilic meningoencephalitis due to Angiostrongylus cantonensis as the cause of death in captive non-human primates. *Am J Trop Med Hyg* 1990;42:70–74.
23. Graber D, Jaffar-Bandjee MC, Attali T, et al. Angiostrongylosis in infants in Réunion and Mayotte. Apropos of 3 cases of eosinophilic meningitis including 1 fatal radiculo-myeloencephalitis with hydrocephalus [in French]. *Arch Pediatr* 1997;4:424–429.
24. Heyneman D, Lim BL. Angiostrongylus cantonensis: proof of direct transmission with its epidemiological implications. *Science* 1967;158:1057–1058.
25. Jindrak K. Angiostrongyliasis cantonensis (eosinophilic meningitis, Alicata's disease). *Contemp Neurol Ser* 1975;12:133–164.
26. Kachi S, Ishih A, Terada M. Effects of PF1022A on adult Angiostrongylus cantonensis in the pulmonary arteries and larvae migrating into the central nervous system of rats. *Parasitol Res* 1995;81:631–637.
27. Kliks MM, Kroenke K, Hardman JM. Eosinophilic radiculomyeloencephalitis: an angiostrongyliasis outbreak in American Samoa related to ingestion of Achatina fulica snails. *Am J Trop Med Hyg* 1982;31:1114–1122.
28. Ko RC, Chan SW, Chan KW, et al. Four documented cases of eosinophilic meningoencephalitis due to Angiostrongylus cantonensis in Hong Kong. *Trans R Soc Trop Med Hyg* 1987;81:807–810.
29. Koo J, Pien F, Kliks MM. Angiostrongylus (Parastrongylus) eosinophilic meningitis. *Rev Infect Dis* 1988;10:1155–1162.
30. Leiper RT. On the round worm genera Protostrongylus and Angiostrongylus of Kamensky, 1905. *J Helminthol* 1926;4–5:203–207.
31. Lim BL, Ramachandran CP. Ecological studies on Angiostrongylus malaysiensis. In: Cross JH, ed. *Studies on Angiostrongyliasis in Eastern Asia and Australia.* Taipei, Taiwan; NAMRU-2 Special Publication (NAMRU2-SP-44) 1979:26–48.
32. Mackerras MJ, Sandars DF. The life history of the rat lung-worm, Angiostongylus cantonensis (Chen) (Nematoda: Metastongylidae). *Aust J Zool* 1955;3:1–21.
33. Montseny JJ, Chauveau P, Pallot JL, Kleinknecht D. Eosinophilic meningitis (angiostrongyloidosis). Dangers of Asian cooking [letter] [in French]. *Presse Med* 1995;24:238.
34. New D, Little MD, Cross J. Angiostrongylus cantonensis infection from eating raw snails [letter]. *N Engl J Med* 1995;332:1105–1106.
35. Nye SW, Tangchai P, Sundarakiti S, Punyagupta S. Lesions of the brain in eosinophilic meningitis. *Arch Pathol* 1970;89:9–19.
36. Paine M, Davis S, Brown G. Severe forms of infection with Angiostrongylus cantonensis acquired in Australia and Fiji. *Aust N Z J Med* 1994;24:415–416.
37. Pirisi M, Gutierrez Y, Minini C, et al. Fatal human pulmonary infection caused by an Angiostrongylus-like nematode. *Clin Infect Dis* 1995;20:59–65.
38. Prakash S. Angiostrongylus cantonensis abscess in the brain; what do we learn? [letter]. *J Neurol Neurosurg Psychiatry* 1992;55:982.
39. Prociv P, Tiernan JR. Eosinophilic meningoencephalitis with permanent sequelae. *Med J Aust* 1987;147:294–295.
40. Punyagupta S. Angiostrongyliasis: clinical features and human pathology. In: Cross JH, ed. *Studies on Angiostrongyliasis in Eastern Asia and Australia.* Taipei, Taiwan; NAMRU-2 Special Publication (NAMRU2SP-44) 1979:138–150.
41. Punyagupta S, Bunnag T, Juttijudata P, Rosen L. Eosinophilic meningitis in Thailand. Epidemiologic studies of 484 typical cases and the etiologic role of Angiostrongylus cantonensis. *Am J Trop Med Hyg* 1970;19:950–958.

42. Radomyos P, Tungtrongchitr A, Praewanich R, et al. Occurrence of the infective stage of Angiostrongylus cantonensis in the yellow tree monitor (Varanus bengalensis) in five provinces of Thailand. *Southeast Asian J Trop Med Public Health* 1994;25:498–500.
43. Rosen L, Chappell R, Laqueur GL, Wallace GD, Weinstein PP. Eosinophilic meningoencephalitis caused by a metastrongylid lungworm of rats. *JAMA* 1962;179:620–624.
44. Rosen L, Laigret J, Bories S. Observations on an outbreak of eosinophilic meningitis on Tahiti, French Polynesia. *Am J Hyg* 1961;74:26–42.
45. Sato, Y, Otsuru M. Studies on eosinophilic meningitis and meningoencephalitis caused by Angiostrongylus cantonensis in Japan. *Southeast Asian J Trop Med Public Health* 1983;14:515–524.
46. Sawabe K, Makiya K. Genetic variability in isozymes of Angiostrongylus malaysiensis. *Southeast Asian J Trop Med Public Health* 1994;25:728–736.
47. Scrimgeour EM. Distribution of Angiostrongylus cantonensis in Papua New Guinea. *Trans R Soc Trop Med Hyg* 1984;78:776–779.
48. Scrimgeour EM, Welch JS. Angiostrongylus cantonensis in East New Britain, Papua New Guinea. *Trans R Soc Trop Med Hyg* 1984;78:774–775.
49. Shih SL, Hsu CH, Huang FY, Shen EY, Lin JC. Angiostrongylus cantonensis infection in infants and young children. *Pediatr Infect Dis J* 1992;11:1064–1066.
50. Singalavanija A, Wangspa S, Teschareon S. Intravitreal angiostrongyliasis. *Aust N Z J Ophthalmol* 1986;14:381–384.
51. Sonakul D. Pathological findings in four cases of human angiostrongyliasis. *Southeast Asian J Trop Med Public Health* 1978;9:220–227.
52. Stafford EE, Sukeri S, Sutanti T. The bandicoot rat, a new host record for Angiostrongylus cantonensis in Indonesia. *Southeast Asian J Trop Med Public Health* 1976;1:41–44.
53. Teekhasaenee C, Ritch R, Kanchanaranya C. Ocular parasitic infection in Thailand. *Rev Infect Dis* 1986;8:350–356.
54. Thobois S, Broussolle E, Aimard G, Chazot G. L'ingestion du poisson cru: une cause de méningite à eosinophile due à Angiostrongylus cantonensis après un voyage à Tahiti. *Presse Med* 1996;25:508.
55. Ubelaker JE. Systematics of species referred to the genus Angiostrongylus. *J Parasitol* 1986;72:237–244.
56. Vargas M, Gomez Perez JD, Malek EA. First record of Angiostrongylus cantonensis (Chen, 1935) (Nematoda: Metastrongylidae) in the Dominican Republic. *Trop Med Parasitol* 1992;43:253–255.
57. Wallace GD, Rosen L. Experimental infection of Pacific Island mollusks with Angiostrongylus cantonensis. *Am J Trop Med Hyg* 1969;18:13–19.
58. Wallace GD, Rosen L. Studies on eosinophilic meningitis. 2. Experimental infection of shrimp and crabs with Angiostrongylus cantonensis. *Am J Epidemiol* 1966;84:120–131.
59. Wallace GD, Rosen L. Studies on eosinophilic meningitis. 4. Experimental infection of fresh-water and marine fish with Angiostrongylus cantonensis. *Am J Epidemiol* 1967;85:395–402.
60. Wallace GD, Rosen L. Studies on eosinophilic meningitis. 5. Molluscan hosts of Angiostrongylus cantonensis on Pacific Islands. *Am J Trop Med Hyg* 1969;18:206–216.
61. Wallace GD, Rosen L. Studies on eosinophilic meningitis. 6. Experimental infection of rats and other homoiothermic vertebrates with Angiostrongylus cantonensis. *Am J Epidemiol* 1969;89:331–344.
62. Watts MB. Five cases of eosinophilic meningitis in Sarawak. *Med J Malaya* 1969;24:89–93.
63. Witoonpanich R, Chuahirun S, Soranastaporn S, Rojanasunan P. Eosinophilic myelomeningoencephalitis caused by Angiostrongylus cantonensis: a report of three cases. *Southeast Asian J Trop Med Public Health* 1991;22:262–267.
64. Yang SQ, Yu BW, Chen YS, et al. The first human case of angiostrongyliasis cantonensis in the mainland of China. *Chin Med J (Engl)* 1988;101:783–786.
65. Yii CY, Chen CY, Fresh JW, Chen T, Cross JH. Human angiostrongyliasis involving the lungs. *Chin J Microbiol* 1968;1:148–150.
66. Yoshimura K, Sugaya H, Kawamura K, Kumagai M. Ultrastructural and morphometric analyses of eosinophils from the cerebrospinal fluid of the mouse and guinea-pig infected with Angiostrongylus cantonensis. *Parasite Immunol* 1988;10:411–423.

Angiostrongyliasis Costaricensis

Pedro Morera,
Ronald C. Neafie, and
Aileen M. Marty

Introduction

Definition

Angiostrongyliasis costaricensis is infection by the nematode *Angiostrongylus costaricensis*, a common parasite of rats and other rodents. Humans are not considered to be definitive hosts. The infection in humans is characterized by a granulomatous inflammatory reaction with heavy infiltration of eosinophils in the intestinal wall, especially in the ileocecal region. Inflammatory reaction and vascular lesions may cause subocclusion, occlusion, or perforation of the intestine. Along with the abdominal changes, *A. costaricensis* may also produce lesions in lymph nodes, liver, and testicle.

Synonyms

The most common synonym for this disease is abdominal angiostrongyliasis. In Latin America, it is sometimes called Morera's disease. *Angiostrongylus costaricensis* is also known as *Morerastrongylus costaricensis*,[4] *Pulmonela costaricensis*, and *Parastrongylus costaricensis*.[37]

General Considerations

In the 1950s and '60s, physicians in Costa Rica treated a growing number of patients who presented with pain in the right lower quadrant, often associated with a palpable mass, and eosinophilia. Lesions discovered in surgical specimens excised from these patients led to an initial diagnosis of malignancy. However, histopathologic study of these lesions showed an inflammatory reaction caused by nematode eggs and, in some instances, fragments of adult nematode within vessels of the intestinal wall. In 1967, Céspedes et al described the clinical and pathologic features of 31 of these patients.[3] That same year, Morera published a description of the nematode.[18] In 1971, after extensive studies, Morera and Céspedes reported their identification of a new species of nematode, which they named *Angiostrongylus costaricensis*.[23] Morera later identified the definitive hosts of *A. costaricensis*[19] and outlined its life cycle.[20] Morera and Ash identified the slug *Vaginulus plebeius* as

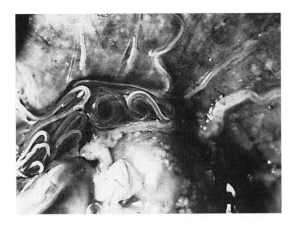

Figure 25.1
Arteries in upper part of flexure of rat cecum containing coiled adult *A. costaricensis*. White spots in serosal surface are clusters of eggs.*

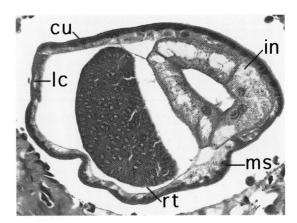

Figure 25.2
Transverse section of adult male *A. costaricensis* in lumen of artery in wall of cecum. Note thin cuticle (cu), small lateral cords (lc), low coelomyarian musculature (ms), thick-walled intestine with large diameter (in), and single reproductive tube (rt). x380

an intermediate host.[21] In 1995, Mota et al further refined the life cycle of *A. costaricensis*.

Epidemiology

In 1972, an autochthonous infection of abdominal angiostrongyliasis was reported in an 11-year-old boy from Honduras, extending the known distribution of the infection.[34] Since then, reports of autochthonous human infections have come from Argentina,[5] Brazil,[43] Colombia, Dominica, Ecuador,[13] El Salvador,[33] Guadeloupe,[12] Guatemala,[30] Martinique,[11] Mexico,[42] Nicaragua,[6] Panama,[32] Puerto Rico,[27] the United States,[8] and Venezuela.[40] There is a report of a 25-year-old African man with an apparent case of abdominal angiostrongyliasis.[2] The parasite recovered from this patient is morphologically indistinguishable from *A. costaricensis*, although the worm has not been reported in rodents or other definitive hosts in Africa. There are reports of natural infection in rodents throughout most of the Americas,[24,38] including the United States and several Caribbean islands. A closely related worm, *Angiostrongylus siamensis*, parasitic in mesenteric arteries of rodents in Asia, has caused angiostrongyliasis in a monkey[28] and is a potential parasite of humans.

As the worm's name suggests, the great majority of cases of human angiostrongyliasis costaricensis are patients from Costa Rica, where the disease is distributed from sea level to an altitude of 2000 meters. In 1991, authorities in Costa Rica reported over 500 new human infections in a population of 3 million, a rate of 16.6/100 000. This high rate of human infection is accompanied by a high prevalence in cotton rats in Costa Rica (43.2%) and Panama (35%).[36]

Small children, especially boys, are most frequently afflicted, perhaps because of their propensity to put things in their mouths. In a recent survey of angiostrongyliasis costaricensis in a pediatric hospital in Costa Rica, 53% of patients were school age, 37% were preschool age, and 10% were infants. Of these patients, 64% were boys.[14]

In endemic areas where clinical knowledge and diagnostic methods have improved, angiostrongyliasis costaricensis is emerging as a common parasitic disease.

Infectious Agent

Morphologic Description
Adults

Angiostrongylus costaricensis is a nematode in the superfamily Metastrongyloidea and the family

Angiostrongylidae. Adult worms live in the mesenteric arteries of the definitive host. In Costa Rica, the cotton rat (*Sigmodon hispidus*) is the most important definitive host (Fig 25.1). The size of the worm varies with the host,[10] but female *A. costaricensis* are generally 28 to 42 mm by 320 to 350 μm in maximum diameter. Maximum width at the base of the esophagus is 150 μm. Male worms are 17 to 22 mm by 280 to 310 μm in maximum diameter. Maximum width at the base of the esophagus is 110 to 140 μm.

Adult male and female *A. costaricensis* share many common features: 3 small lips around the oral opening, but no buccal capsule; a nerve ring anterior to the middle of the esophagus; and an excretory pore on the anterior end, slightly posterior to the esophageal-intestinal junction. Both worms taper toward the ends; the caudal end of both male and female worms curves ventrally.[19] The cuticle is transparent and smooth, about 2 μm thick, and finely striated toward both ends (Fig 25.2). Striations are difficult to discern in tissue sections. Lateral cords are small and dome-like (Fig 25.3). Somatic musculature is polymyarian and low coelomyarian (Fig 25.4). The thick-walled intestine, composed of only a few cells, has a large diameter (Fig 25.4). Intestinal cells have thin microvilli. Nuclei are generally located in the middle of the cytoplasm, giving the appearance of being lined up in a row (Fig 25.4).[34] Female *A. costaricensis* have a barber-pole appearance typical of the genus, caused by the spiral winding of the milky-white, paired uterine branches around the darker intestine. Vagina, vulva, and anus are near the caudal end, which is somewhat conical and slightly curved. There is an inconspicuous projection at the tip of the tail. Gravid females produce eggs that are in various stages of development in the reproductive tract. Males have a well-developed, symmetrical copulatory bursa and 2 slender, striated spicules 318 to 330 μm long. Cuticular structures fuse to form a 45-μm-long gubernaculum.

Larvae

First stage (L1) larvae of *A. costaricensis* observed in the feces of cotton rats are 260 to 290 μm by 14 to 15 μm. In human tissue, most larvae are 8 to 11 μm wide (Fig 25.5). The tail is pointed, with a notched tip on the dorsal surface (Fig 25.6). Scanning electron microscopy reveals single, long,

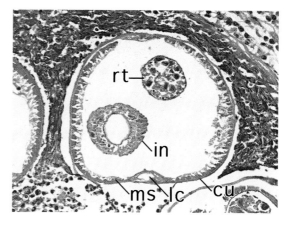

Figure 25.3
Transverse section of adult female *A. costaricensis* in wall of small intestine, demonstrating thin cuticle (cu), dome-like lateral cords (lc), low coelomyarian musculature (ms), thick-walled intestine with large diameter (in), and single reproductive tube (rt). x220

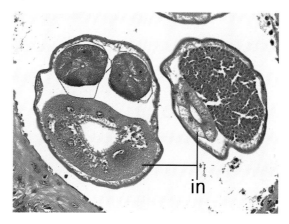

Figure 25.4
Transverse sections of adult female (*left*) and male (*right*) worms in vessel lumen in wall of cecum. Note large, thick-walled intestines (in) and nuclei, and low coelomyarian musculature. x120

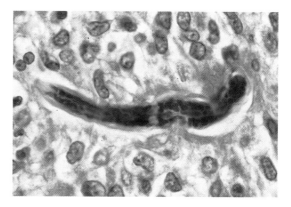

Figure 25.5
L1 larva of *A. costaricensis* in lymphatic vessel in wall of cecum. Patient, a 42-year-old man from Puerto Rico, presented with severe right lower quadrant abdominal pain. x880

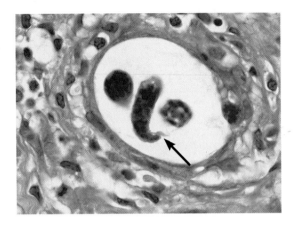

Figure 25.6
Egg containing coiled L1 larva in wall of cecum of 25-year-old African man. Note notched tail of larva (arrow). x1065

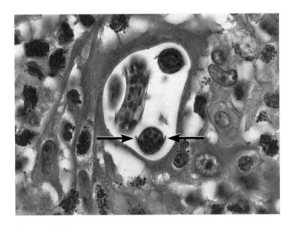

Figure 25.7
Coiled L1 larva within egg in wall of cecum. Note minute single lateral alae (arrows). x810

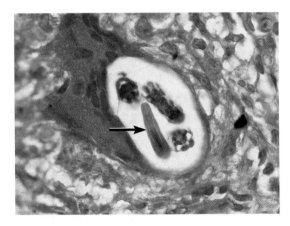

Figure 25.8
Coiled L1 larva within egg engulfed by giant cell in wall of colon. Note esophagus (arrow) in anterior end of larva. x700

lateral alae that begin at the anterior end and extend almost to the tip of the tail.[9] These minute lateral alae are not easily identified in tissue sections (Fig 25.7). The nerve ring is 40 μm from the cephalic end. A club-shaped, 128-μm-long esophagus (Fig 25.8) leads to the finely granular intestine. The anus is on the ventral surface, 26 μm from the tip of the tail; the genital primordium is 85 μm from the tip of the tail.

Eggs

Eggs of *A. costaricensis* are 85 by 50 μm in maximum size, but are usually 60 to 65 by 40 to 45 μm. The thin shells are usually not discernible in tissue sections (Fig 25.9). The ovoid eggs, which resemble hookworm eggs, are unembryonated to fully embryonated when deposited in tissue (Figs 25.6 and 25.10).

Life Cycle and Transmission

The life cycle of *A. costaricensis* requires 2 hosts. The cotton rat (*Sigmodon hispidus*) is the most common definitive host for *A. costaricensis* (Fig 25.1), but 11 other rodent species (including the ubiquitous black rat, *Rattus rattus*), dogs (in Costa Rica),[1] monkeys (*Saguinus mystax*),[35] and sometimes the coati (*Nasua narcia*)[16] can be natural definitive hosts. Slugs (various veronicellid species from Brazil, Costa Rica,[17] and Ecuador) and certain aquatic and terrestrial snails are natural intermediate hosts.

In rodents, adult worms live within ileocecal branches of the anterior mesenteric artery (Fig 25.1). When the worms mate, females release eggs into the bloodstream. These eggs are carried to the rodent's intestine, where they lodge in the intestinal wall, embryonate, and hatch 4 days later. L1 larvae migrate through the intestinal wall to the lumen and pass in feces. In human hosts, adult worms reach the egg-laying stage, but intense inflammatory reaction traps eggs and/or L1 larvae in the intestinal wall, preventing their appearance in feces (Figs 25.11 and 25.12).[15]

A molluscan intermediate host (slug or snail) consumes L1 larvae in contaminated soil. In the mollusk's fibromuscular tissue, larvae molt twice, maturing to infective L3 larvae in 16 to 19 days. L3 larvae may remain within the mollusk for several months, or may exit in mucous secretions.

When ingested by a definitive host, infective L3

larvae migrate rapidly to the ileocecal region, where they adhere to and depress epithelial cells, and penetrate the intestinal wall. At this point, most larvae enter lymphatic vessels, but some enter intestinal veins. Larvae entering lymphatic vessels travel to mesenteric lymph nodes, molt to L4 stage, and exit the lymph nodes. L4 larvae pass through the thoracic duct to venous vessels, enter pulmonary veins, and travel through the heart into the systemic arterial circulation. They may migrate through several organs (brain, kidney, ovary, spleen, stomach), but most settle in mesenteric, pancreatic, and ileocolic arteries, where they mature to adult worms. Egg laying starts by the 18th day; L1 larvae appear in feces within 24 days.[19] The few L3 larvae that enter intestinal veins travel to intrahepatic portal veins, where they molt twice to become adult worms. Eggs laid in hepatic veins disseminate to portal venous branches and may lead to pulmonary embolism.[26]

Rodents usually acquire *A. costaricensis* by eating infected mollusks. Humans are usually infected by ingesting infective L3 larvae in molluscan secretions left on fruits and vegetables, or by eating fragments of infected slugs or snails on unwashed produce. Children may accidentally or deliberately consume infected mollusks while playing in the dirt.

Clinical Features and Pathogenesis

Symptoms vary with the number and location of worms, tissue damage caused by the worms, and the host's susceptibility and immunologic response. When worms lodge in the ileocecal region, most patients complain of pain in the right iliac fossa and right flank. Palpation in this area often causes discomfort; rectal examination is painful in about 50% of patients. Most patients present with fever ranging from 38° to 38.5°C, but rarely with chills. Intestinal bleeding can mimic the symptoms of Meckel's diverticulum. In chronic infection, a mild fever may persist for several weeks. Approximately 50% of patients experience anorexia, vomiting, and constipation. A tumor-like mass, usually palpable in the right lower quadrant, may lead to a preliminary diagnosis of malignancy. Leukocytosis and

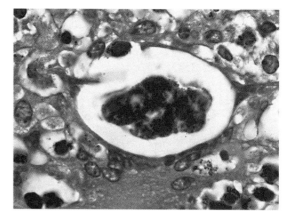

Figure 25.9
Early segmented egg engulfed by giant cell in wall of cecum. Thin shell is not discernible. x760

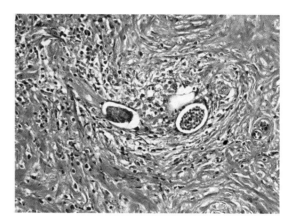

Figure 26.10
Unembryonated and embryonated eggs of *A. costaricensis* in wall of colon. x170

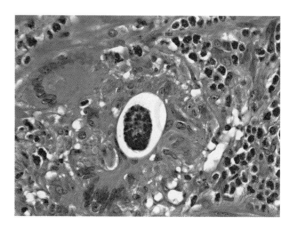

Figure 25.11
Segmented egg engulfed by giant cells in wall of cecum. Note prominent eosinophilia. x310

eosinophilia are usually present. Roentgenographic examination may reveal changes in the terminal ileum, cecum, appendix, and/or ascending colon (Fig 25.13). A contrast medium may reveal filling defects and irritability in involved areas, and reduction of lumen size caused by thickening of the intestinal wall.

Some patients complain of pain in the right upper quadrant; they usually have enlarged, tender livers. Laparoscopy reveals small yellow spots on the hepatic surface. Most patients with hepatic involvement also have intestinal angiostrongyliasis.

With testicular involvement, the most significant findings are acute pain, a red or purple discoloration, and leukocytosis with eosinophilia. According to reports, all angiostrongyliasis costaricensis patients with testicular necrosis were originally misdiagnosed as having testicular torsion, and were correctly diagnosed only after surgical intervention.

Left untreated, symptoms of *A. costaricensis* infection may persist for weeks or months. In some cases, infection can be fatal. Seroepidemiologic studies are uncovering many subclinical infections in endemic areas.

Pathologic Features

There are 2 major pathogenic mechanisms in angiostrongyliasis costaricensis. First, adult *A. costaricensis* living within mesenteric arteries damage the endothelium, inducing thrombosis and, consequently, necrosis of tissues formerly supplied by the vessel. Second, eggs, embryos, larvae, and excretory-secretory products incite inflammatory reactions.

In humans and in natural hosts (Fig 25.1), adult *A. costaricensis* live primarily within mesenteric arteries. The majority of lesions are found in the ileocecal region, but may also appear in the small intestine, hepatic flexure, descending colon, regional lymph nodes, omentum, liver, or testicle.

Gross examination of surgical specimens reveals 2 nonexclusive patterns, both showing segmental distribution: a hypertrophic-pseudoneoplastic pattern (Fig 25.14) and an ischemic-congestive pattern (Fig 25.15). The pseudoneoplastic pattern presents as a hardened and

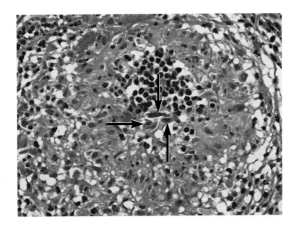

Figure 25.12
Coiled L1 larva (arrows) trapped in granuloma in wall of cecum. Note numerous eosinophils. x240

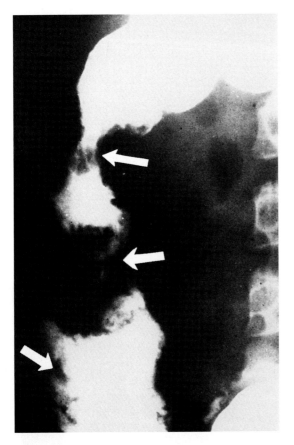

Figure 25.13
X-ray of ascending colon during barium enema. Filling defects (arrows) represent areas where intestinal wall has thickened.

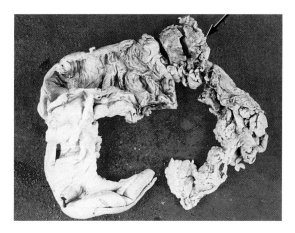

Figure 25.14
Hypertrophic-pseudoneoplastic pattern (arrow) in ascending colon of 32-year-old Guatemalan woman, a result of *A. costaricensis* infection.

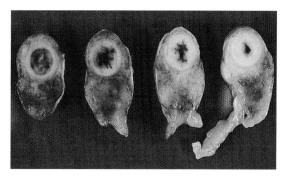

Figure 25.15
Ischemic-congestive pattern, with perforation and peritonitis, in 10-year-old child with acute appendicitis caused by *A. costaricensis*.

thickened intestinal wall (resulting from intense fibrin deposition), with yellow foci on the serosal surface and mesentery. The intestinal lumen is reduced, sometimes causing partial or complete obstruction. The ischemic-congestive pattern presents segments with necrotic and/or congested areas. When necrotic areas are present, there may also be perforation and peritonitis (Fig 25.15).

Lesions may contain unsegmented eggs, segmented eggs, eggs with mature larvae, larvae free in the tissue, and adult worms. Histopathologic examination reveals 3 fundamental findings. First, a massive infiltration of eosinophils in all layers of the intestinal wall, but especially heavy in the mucosa (Fig 25.16) and submucosa (Fig 25.17); involvement of the serosa and muscular layers is often less intense. In some instances, there is extensive infiltration of the muscular layer with dissociation of myofibers, denoting severe myositis. Second, a granulomatous reaction composed predominantly of epithelioid cells and giant cells (Figs 25.18 and 25.19) with eosinophils, frequently distributed around blood vessels. Eggs engulfed by giant cells may appear in the walls of blood vessels (Fig 25.20). Third, eosinophilic vasculitis affecting arteries (Fig 25.21), veins, lymphatics, and capillaries. Changes vary from isolated PAS-positive deposits on endothelial surfaces to severe vasculitis that may include focal necrosis and fibrinoid deposits. Viable adult worms may be present within the lumens of arteries (Figs 25.22 to 25.24). The arterial wall closest to a viable worm

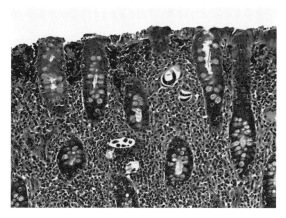

Figure 25.16
Intestinal mucosa of 39-year-old American woman showing intense inflammatory reaction, especially with eosinophils, entrapping L1 larvae of *A. costaricensis* and preventing their passage into feces. Patient, a medical student in Dominica, developed severe headache, stiff neck, and rectal bleeding a year after arriving on the island. x115

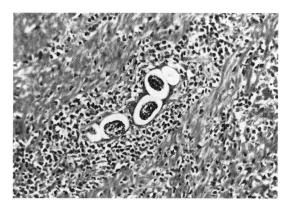

Figure 25.17
Several embryonated *A. costaricensis* eggs in vessel within submucosa of cecum from patient described in Figure 25.5. Shells are thin and difficult to discern. Note marked diffuse eosinophilia and perivascular inflammation. x550

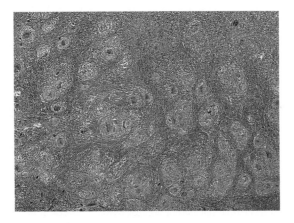

Figure 25.18
Miliary granulomatous pattern in intestinal wall of patient described in Figure 25.16. x25

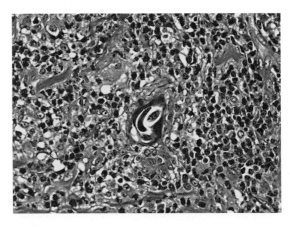

Figure 25.19
Higher magnification of lesion shown in Figure 25.18, depicting coiled L1 larva in giant cell provoking an intense eosinophilia. x250

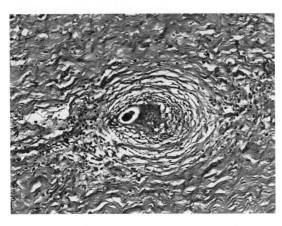

Figure 25.20
Egg of *A. costaricensis* engulfed by giant cell in blood vessel wall in cecum. Note vasculitis and perivascular inflammation. x115

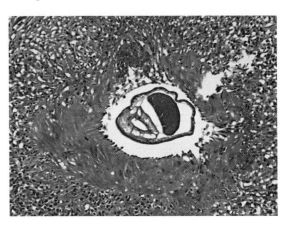

Figure 25.21
Viable adult male *A. costaricensis* in artery, causing eosinophilic vasculitis. x115

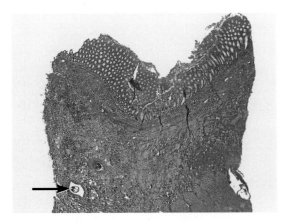

Figure 25.22
Section of colon from 32-year-old Guatemalan woman who presented with fever and right lower quadrant pain. Note inflammation and adult female worm in vessel lumen (arrow). x11

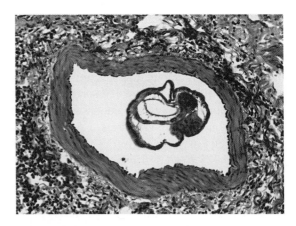

Figure 25.23
Higher magnification of lesion described in Figure 25.22 showing viable adult female worm in lumen of artery. x120

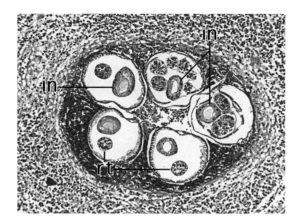

Figure 25.24
Multiple cross sections of viable adult female *A. costaricensis* in lumen of artery in intestine, stimulating intense eosinophilic vasculitis. Note reproductive tubes (rt) and large diameter of intestine (in). x80

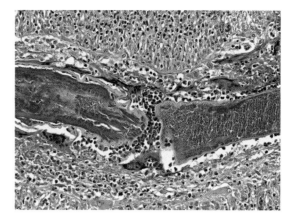

Figure 25.25
Longitudinal section of degenerating adult male *A. costaricensis* within vessel from patient described in Figure 25.5. Vessel wall has been destroyed and giant cells are devouring the worm. Note massive eosinophilia. x385

is usually less affected than segments downstream, a result of the downstream drainage of antigens. Degenerated parasites may incite extensive vasculitis (Figs 25.25 and 25.26) and sometimes thrombosis, especially in arteries. Large necrotic areas result from arterial thrombosis.

Epithelial ulceration is common (Figs 25.27 to 25.29). Moderate plasmacytosis in the upper mucosa and an increase in mast cells are common. Eggs and larvae appear in small cavities lined by endothelium (Figs 25.5 and 25.16). Unfertilized eggs usually degenerate and are difficult to discern. Larvae free in the tissue may incite an intense eosinophilia (Figs 25.30 and 25.31). Immunohistochemical techniques can easily identify these structures, as well as excretory-secretory antigens. Similar histopathologic changes may be seen in the appendix (Fig 25.32) which can clinically mimic appendicitis.

Mesenteric lymph nodes may contain eggs and embryos accompanied by follicular hyperplasia, sinusoidal histiocytosis, eosinophilic infiltrates, and isolated giant cells. Adult worms release eggs and larvae that enter marginal and lymphatic sinuses of mesenteric lymph nodes, stimulating the eosinophilic infiltrate.

The liver is dark and hard. Hepatic lesions resemble those caused by *Toxocara canis*[25] (see chapter 27), but unlike toxocariasis, *A. costaricensis* lesions include not only larvae, but also eggs (Fig 25.33) and even adult worms. There are focal granulomas with giant cells containing ghosts

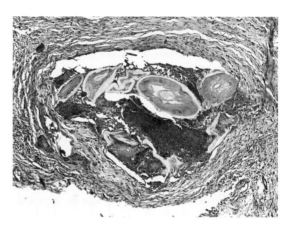

Figure 25.26
Intense vasculitis caused by degenerated adult male *A. costaricensis* in wall of colon from patient described in Figure 25.6. x80

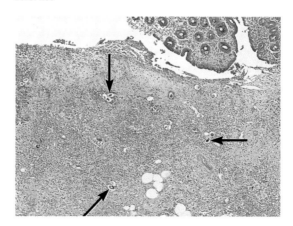

Figure 25.27
Ulcerated cecum from patient described in Figure 25.5 showing inflammation and eggs (arrows). x35

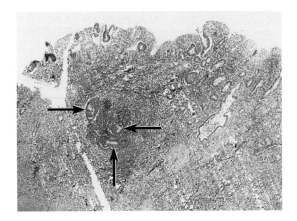

Figure 25.28
Ulcerated colon from Costa Rican patient depicting coiled, degenerated adult male *A. costaricensis* (arrows) in inflamed submucosa. x25

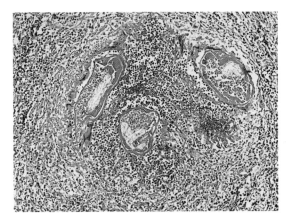

Figure 25.29
Higher magnification of degenerated male worm shown in Figure 25.28. Note intense eosinophilia. x70

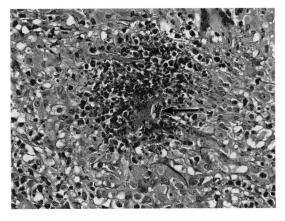

Figure 25.30
Fragment of L1 larva (arrow) free in wall of cecum centered among degranulating eosinophils. x260

of liver cells which surround necrotic centers filled with eosinophils, lymphocytes, and cellular debris. A moderate eosinophilic and lymphocytic infiltrate enlarges portal tracts. Fibrillar connective tissue containing eosinophils may obliterate hepatic arteries, with granulomas surrounding damaged arteries.[39]

Testicular involvement produces extensive hemorrhaging; necrosis results from adult worms obstructing the arteries of the spermatic cord and from obstructive organized thrombi.[31] In some patients, there are areas where eosinophils surround testicular tubules, and other areas where unfertilized eggs appear as pink hyaline masses scattered throughout the testicular parenchyma. Lesions may extend to the spermatic cord and epididymis.

Diagnosis

Differential clinical diagnosis includes appendicitis, Meckel's diverticulum, Crohn's disease,[13] and, in ectopic lesions, visceral larva migrans and testicular torsion. Barium enema may show filling defects of the colon that resemble malignancy (Fig 25.13).

Neither eggs nor L1 larvae have been observed in human feces, as they have been in rodents. Demonstrating a specific and acute antibody response is instructive, but definitive diagnosis depends on identifying the parasite in tissue sections. Various immunologic tests for *A. costaricensis* are available, including immunoelectrophoresis, Ouchterlony immunodiffusion,[33] and latex bead agglutination. In a study conducted in Cost Rica, over 97% of potential patients tested with latex bead agglutination had no cross reactions or false positive/false negative results.

Treatment and Prevention

Surgery is the treatment of choice for acute angiostrongyliasis costaricensis. Less severe infections are frequently self-limiting and may be treated with careful observation and well-controlled palliatives. Some reports suggest that treatment with diethylcarbamazine, thiabendazole, and albendazole resulted in remission of symptoms. None of these reports, however, offered objective evidence that the "cure" was attributable to the drugs. In other studies, cotton rats infected with *A. costaricensis* were placed on various regimens of thiabendazole, levamisole, and diethylcarbamazine. Adult worms within the rats were not killed by the drugs; in fact, they continued to produce L1 larvae after treatment. Thiabendazole in particular seemed to stimulate the parasites, causing erratic migrations and worsening of lesions.[22] Chemotherapy is therefore not recommended until more effective drugs are available. Nonsurgical patients should be monitored for either remission of symptoms or development of an acute syndrome requiring surgery.

In the absence of satisfactory medical treatment, the logical strategy is to control transmission of the disease. Irradiation of commercial foodstuffs is an option in some areas.[29] Where that is not available, washing fruits and vegetables in a vinegar or salt solution is highly effective. Zanini and Graeff-Teixeira found that cleaning foods in solutions of sodium hypochlorite, sodium chloride, or vinegar dramatically reduced the risk of infection.[41]

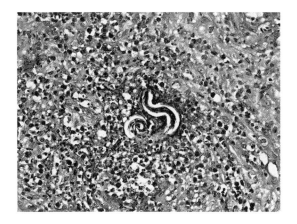

Figure 25.31
L1 larva free in wall of colon provoking intense eosinophilia. x175

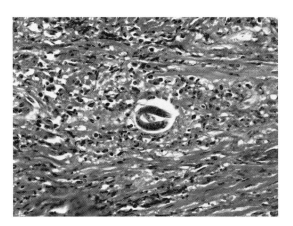

Figure 25.32
Coiled L1 larva in egg causing inflammation of appendix wall. x235

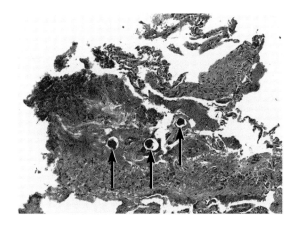

Figure 25.33
Embryonated *A. costaricensis* eggs (arrows) in necrotic liver. x60

References

1. Arroyo R, Rodriguez F, Berrocal A. Abdominal angiostrongylosis in Canis familiaris [in Spanish]. *Parasitol al Dia* 1988;12:181–185.
2. Baird JK, Neafie RC, Lanoie L, Connor DH. Abdominal angiostrongylosis in an African man: case study. *Am J Trop Med Hyg* 1987;37:353–356.
3. Céspedes R, Salas J, Mekbel S, Troper L, Müllner F, Morera P. Granulomas entéricos y linfáticos con intensa eosinofilia tisular producidos por un estrongilídeo (Strongylata). 1. Patología. *Acta Med Cost* 1967;10:235–255.
4. Chabaud AG. Description of Stefanskostrongylus dubosti n. sp., parasite of Potamogale, and attempt at classification of Angiostrongylinae nematodes [in Spanish]. *Ann Parasitol Hum Comp* 1972;47:735–744.
5. Demo OJ, Pessat OA. Angiostrongilosis abdominal. Primer caso humano encontrado en Argentina. *Prensa Med Argent* 1986;73:732–738.
6. Duarte Z, Morera P, Vuong PN. Abdominal angiostrongyliasis in Nicaragua: a clinico-pathological study on a series of 12 case reports. *Ann Parasitol Hum Comp* 1991;66:259–262.
7. Graeff-Teixeira C, Camillo-Coura L, Lenz HL. Histopathological criteria for the diagnosis of abdominal angiostrongyliasis. *Parasitol Res* 1991;77:606–611.
8. Hulbert TV, Larsen RA, Chandrasoma PT. Abdominal angiostrongyliasis mimicking acute appendicitis and Meckel's diverticulum: report of a case in the United States and review. *Clin Infect Dis* 1992;14:836–840.
9. Ishih A, Rodriguez BO, Sano M. Scanning electron microscopic observations of first- and third-stage larvae and adults of Angiostrongylus costaricensis. *Southeast Asian J Trop Med Public Health* 1990;21:568–573.
10. Ishii AI, Sano M. Strain-dependent differences in susceptibility of mice to experimental Angiostrongylus costaricensis infection. *J Helminthol* 1989;63:302–306.
11. Jeandel R, Fortier G, Pitre-Delaunay C, Jouannelle A. Intestinal angiostrongyliasis caused by Angiostrongylus costaricensis. Apropos of a case in Martinique [in French]. *Gastroenterol Clin Biol* 1988;12:390–393.
12. Juminer B, Roudier M, Raccurt CP, Pujol HP, Gerry F, Bonnet R. Presence of abdominal angiostrongylosis in Guadeloupe. Apropos of 2 cases [in French]. *Bull Soc Pathol Exot* 1992;85:39–43.
13. Liacouras CA, Bell LM, Aljabi MC, Piccoli DA. Angiostrongylus costaricensis enterocolitis mimics Crohn's disease. *J Pediatr Gastroenterol Nutr* 1993;16:203–207.
14. Loría-Cortés R, Lobo-Sanahuja JF. Clinical abdominal angiostrongylosis. A study of 116 children with intestinal eosinophilic granuloma caused by Angiostrongylus costaricensis. *Am J Trop Med Hyg* 1980;29:538–544.
15. Mojon M. Human angiostrongyliasis caused by Angiostrongylus costaricensis [in French]. *Bull Acad Natl Med* 1994;178:625–631.
16. Monge E, Arroyo R, Solano E. A new definitive natural host of Angiostrongylus costaricensis (Morera and Céspedes, 1971). *J Parasitol* 1978;64:34.
17. Morera P. Angiostrongiliasis abdominal: transmision y observaciones sobre su posible control. *Serie de Copublicaciones de la OPS* 1985;1:230–235.
18. Morera P. Granulomas entéricos y linfáticos con intensa eosinofilia tisular producidos por un estrongilídeo (Strongylata: Railliet y Henry, 1913). 2. Aspecto parasitológico (nota previa). *Acta Med Cost* 1967;10:257–265.
19. Morera P. Studies on the definitive host of Angiostrongylus costaricensis [in Spanish]. *Bol Chil Parasitol* 1970;25:133–134.
20. Morera P. Life history and redescription of Angiostrongylus costaricensis Morera and Céspedes, 1971. *Am J Trop Med Hyg* 1973;22:613–621.
21. Morera P, Ash LR. Studies on the intermediate host of Angiostrongylus costaricensis [in Spanish]. *Bol Chil Parasitol* 1970;25:135.
22. Morera P, Bontempo I. Accion de algunos antihelminticos sobre Angiostrongylus costaricensis. *Rev Med Hosp Nal Ninos Costa Rica* 1985;20:165–174.
23. Morera P, Céspedes R. Angiostrongylus costaricensis n. sp. (Nematoda: Metastrongyloidea), a new lungworm occurring in man in Costa Rica. *Rev Biol Trop* 1970;18:173–185.
24. Morera P, Lazo R, Urquizo J, Llaguno M. First record of Angiostrongylus costaricensis Morera and Céspedes, 1971 in Ecuador. *Am J Trop Med Hyg* 1983;32:1460–1461.
25. Morera P, Perez F, Mora F, Castro L. Visceral larva migrans-like syndrome caused by Angiostrongylus costaricensis. *Am J Trop Med Hyg* 1982;31:67–70.
26. Mota EM, Lenzi HL. Angiostrongylus costaricensis life cycle: a new proposal. *Mem Inst Oswaldo Cruz* 1995;90:707–709.
27. Neafie RC, Marty AM. Unusual infections in humans. *Clin Microbiol Rev* 1993;6:34–56.
28. Oku Y, Kudo N, Ohbayashi M, Narama I, Umemura T. A case of abdominal angiostrongyliasis in a monkey. *Jpn J Vet Res* 1983;31:71–75.
29. Ooi HK, Ishii K, Inohara J, Kamiya M. Effect of irradiation on the viability of Angiostrongylus cantonensis and A. costaricensis infective larvae. *J Helminthol* 1993;67:238–242.
30. Pena GP, Andrade Filho J, de Assis SC. Angiostrongylus costaricensis: first record of its occurrence in the State of Espirito Santo, Brazil, and a review of its geographic distribution. *Rev Inst Med Trop Sao Paulo* 1995;37:369–374.
31. Ruiz PJ, Morera P. Spermatic artery obstruction caused by Angiostrongylus costaricensis Morera and Céspedes, 1971. *Am J Trop Med Hyg* 1983;32:1458–1459.
32. Sánchez GA. Intestinal perforation by Angiostrongylus costaricensis. A report of 2 cases [in Spanish]. *Rev Med Panama* 1992;17:74–81.
33. Sauerbrey M. A precipitin test for the diagnosis of human abdominal angiostrongyliasis. *Am J Trop Med Hyg* 1977;26:1156–1158.
34. Sierra E, Morera P. Angiostrongilosis abdominal: primer caso humano encontrado en Honduras (Hospital Evangélico de Siguatepeque). *Acta Med Cost* 1972;15:95–99.
35. Sly DL, Toft JD 2d, Gardiner CH, London WT. Spontaneous occurrence of Angiostrongylus costaricensis in marmosets (Saguinus mystax). *Lab Anim Sci* 1982;32:286–288.
36. Tesh RB, Ackerman LJ, Dietz WH, Williams JA. Angiostrongylus costaricensis in Panama. Prevalence and pathological findings in wild rodents infected with the parasite. *Am J Trop Med Hyg* 1973;22:348–356.
37. Ubelaker JE. Systematics of species referred to the genus Angiostrongylus. *J Parasitol* 1986;72:237–244.
38. Ubelaker JE, Hall NM. First report of Angiostrongylus costaricensis Morera and Céspedes, 1971 in the United States. *J Parasitol* 1979;65:307.
39. Vázquez JJ, Sola JJ, Boils PL. Hepatic lesions induced by Angiostrongylus costaricensis. *Histopathology* 1994;25:489–491.
40. Zambrano Paredes Z. Pseudotumoral eosinophilic ileocolitis of parasitic origin [in Spanish]. *Rev Latinoam Patol* 1973;12:43–50.
41. Zanini GM, Graeff-Teixeira C. Abdominal angiostrongyliasis: its prevention by the destruction of infecting larvae in food treated with salt, vinegar or sodium hypochlorite [in Portuguese]. *Rev Soc Bras Med Trop* 1995;28:389–392.
42. Zavala Velázquez J, Ramírez Baquedano W, Reyes Pérez A, Bates Flores M. Angiostrongylus costaricensis. First cases in Mexico [in Spanish]. *Rev Invest Clin* 1974;26:389–394.
43. Ziliotto A Jr, Künzle JE, Fernandes LA, Prates-Campos JC, Britto-Costa R. Angiostrongyliasis: report of a probable case [in Spanish]. *Rev Inst Med Trop Sao Paulo* 1975;17:312–318.

26

Ascariasis

Ronald C. Neafie *and*
Aileen M. Marty

Introduction

Definition
Ascariasis is infection by the nematode *Ascaris lumbricoides* (Linnaeus, 1758). The adult worm lives in the small intestine of infected humans. Worm burden is usually low and patients are asymptomatic; however, occasionally a bolus of worms obstructs the intestine, or a single worm migrates into a critical site, causing sudden and severe symptoms. Infrequently, immature *Ascaris suum* (Goeze, 1782), the ascarid of swine, develop in human intestines and cause disease.[21]

Synonym
Ascaris lumbricoides is often called the large roundworm of the small intestine.

General Considerations
Ancient texts describe ascariasis; however, the Romans and other authors of antiquity sometimes confused this large roundworm with the common earthworm. Prehistoric human remains from Europe[4] and the Americas, and Egyptian mummies,[10] harbor eggs of *A. lumbricoides*.

Epidemiology

Ascariasis is cosmopolitan, but is most common in moist, tropical environments. *Ascaris lumbricoides* infects over a billion people, joining *Enterobius vermicularis*, *Trichuris trichiura*, and hookworms as one of the most common human helminths. Prevalence varies depending largely upon climate, population density, and personal hygiene. The majority of patients live in Asia (73%), followed by Africa (12%) and Central and South America (8%). In the United States, there are about 4 million infected individuals, mostly in the southern states. In the Commonwealth of Independent States, prevalence is low where the climate is hot and dry, but infection may reach 80% to 90% of the population in agricultural areas with wet soil.

Ascariasis prevails in overcrowded rural communities and shantytowns, where inadequate sewerage allows human feces to contaminate the environment. Infections are heavier and more frequent in children than in adults. Children acquire ascariasis from soil, fingers, toys, and other objects

Figure 26.1
Large bolus of adult *A. lumbricoides* recovered from necrotic small intestine, stomach, esophagus, gallbladder, and intrahepatic and extrahepatic bile ducts of 2-year-old South African girl.*

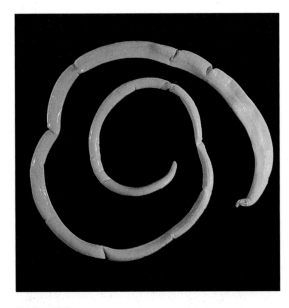

Figure 26.2
Adult female *A. lumbricoides* passed per anum by 10-year-old Congolese girl. Worm measured 57 cm long by 1 cm in maximum diameter.*

contaminated with feces, but adults usually become infected by ingesting contaminated raw vegetables or drinking water.

Infectious Agent

Morphologic Description

Ascaris lumbricoides is the largest parasitic nematode infecting the human intestine. The worm is cylindrical, white-pink to tan-gray, and has tapered ends (Figs 26.1 to 26.3). The anterior end has a terminal mouth with 3 conspicuous, finely serrated lips. A white lateral line often extends along the entire length of the worm. The cuticle is thick, multilayered, and has both annulations and striations (Fig 26.4). A hypodermis lies just beneath the cuticle and projects into the body cavity to form lateral cords. Numerous somatic muscle cells in each quarter of the worm project prominently into the body cavity (Fig 26.5), and contractile fibers extend along the sides of the muscle cells. The esophagus is prominent, but has no posterior bulb (Fig 26.6). Tall columnar cells with prominent microvilli line the intestine.

Most female worms are 20 to 35 cm long by 3 to 6 mm in diameter, but worms may reach up to 60 cm in length (Fig 26.2). The ovaries, oviducts, and uteri are in the posterior two thirds of the body (Figs 26.7 to 26.9). At some levels, the only reproductive organ is the egg-filled uterus (Fig 26.9). The vulva is located about one third the length of the body from the anterior end.

Male worms measure 15 to 31 cm long and 2 to 4 mm in diameter and differ grossly from female worms by the ventrad curvature of the posterior end. The single reproductive tube is composed of the testis, vas deferens, and ejaculatory duct, and is confined to the posterior half of the body (Figs 26.10 and 26.11). The ejaculatory duct and intestine unite in the posterior tip of the worm to form a short cloaca (Fig 26.12). There are 2 copulatory spicules.

Third-stage larvae of *A. lumbricoides* in the human lung may reach 2 mm in length and 60 μm in diameter; however, some early fourth-stage larvae inhabiting the lung may measure up to 75 μm in diameter.[7] These larvae have prominent lateral alae, large paired excretory columns at least

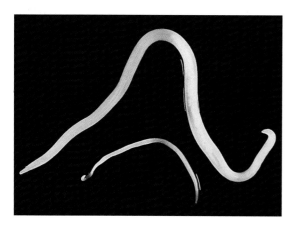

Figure 26.3
Gross specimens of juvenile male and female *A. lumbricoides*. Note tightly coiled posterior end of smaller male worm. x0.8

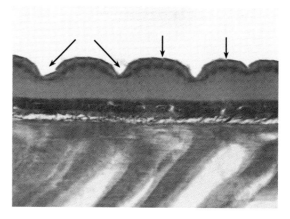

Figure 26.4
Longitudinal section of cuticle of adult male *A. lumbricoides*. Note both annulations (long arrows) and striations (short arrows) on cuticle. x505

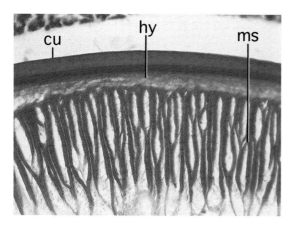

Figure 26.5
Transverse section of body wall of adult male *A. lumbricoides* demonstrating cuticle (cu), hypodermis (hy), and contractile portion of muscle (ms). x280

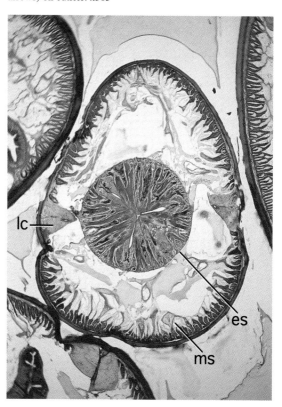

Figure 26.6
Transverse section through anterior region of adult female *A. lumbricoides* at level of esophagus (es). Esophagus has triradiate lumen. Note prominent lateral cords (lc) and clearly visible sarcoplasmic portion of muscle (ms). x40

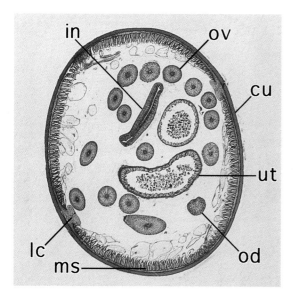

Figure 26.7 ◄
Transverse section through midbody of gravid *A. lumbricoides*. Section displays cuticle (cu), contractile portion of muscle (ms), lateral cord (lc), ovary (ov), oviduct (od), intestine (in), and uterus (ut). x12

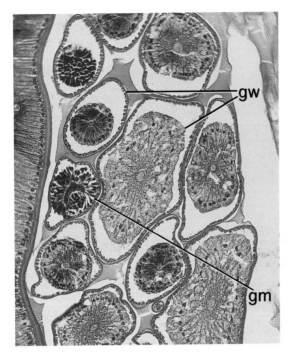

Figure 26.8
Transverse sections of growth regions (gw) and germinal region (gm) of ovary in adult female *A. lumbricoides*. x150

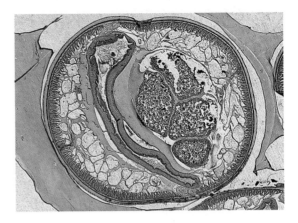

Figure 26.9
Transverse section of gravid *A. lumbricoides* demonstrating intestine and several sections of egg-filled uteri. Movat x16

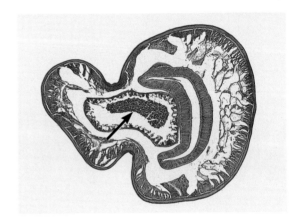

Figure 26.10
Transverse section through posterior region of adult male *A. lumbricoides* demonstrating vas deferens containing sperm (arrow), and intestine. x40

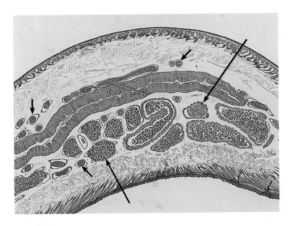

Figure 26.11
Longitudinal section of adult male *A. lumbricoides* showing intestine, testis (short arrows), and vas deferens (long arrows). x23

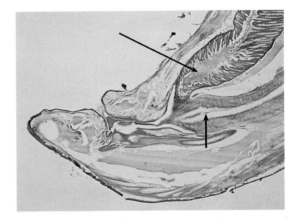

Figure 26.12
Posterior tip of adult male *A. lumbricoides* demonstrating junction of intestine (short arrow) and ejaculatory duct (long arrow). x40

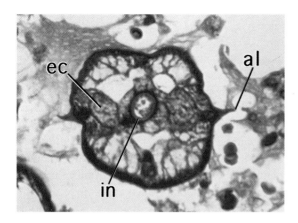

Figure 26.13
Third-stage larva of *A. lumbricoides* in lung. Larva has prominent lateral alae (al). A large excretory column (ec) occupies each lateral cord area. Body cavity contains intestine (in) but no reproductive organs. x890

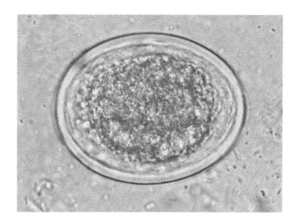

Figure 26.14
Fertilized decorticated egg of *A. lumbricoides* from feces. Unstained x935

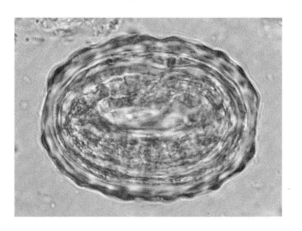

Figure 26.15
Fertilized mamillated egg of *A. lumbricoides* from feces. Unstained x860

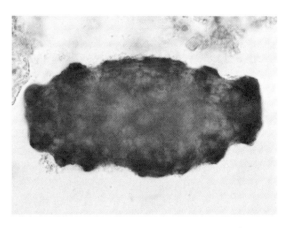

Figure 26.16
Elongate, unfertilized, mamillated egg of *A. lumbricoides* from feces, preserved in merthiolate-iodine-formaldehyde (MIF) solution. x965

equal to the diameter of the intestine, and a well-developed intestine (Fig 26.13). The cuticle is thin and transversely striated. In the posterior third of the worm there is a small genital primordium, but no well-developed reproductive organs. There are 3 or 4 muscle cells in each quadrant of a cross section of the larvae.[20]

Fertilized eggs measure 45 to 75 μm by 35 to 50 μm, are round to oval, thick-shelled, and golden brown from bile staining (Fig 26.14). Most often, fertilized eggs are mamillated, but they can be smooth (decorticated) (Fig 26.15). Unfertilized eggs, 88 to 94 μm by 44 μm, are usually elongate (Fig 26.16), with a thin middle layer of shell, and often little or no outer mamillated layer.

Adult worms, larvae, and eggs of *A. suum* are morphologically indistinguishable from those of *A. lumbricoides*.[13]

Life Cycle and Transmission

To complete its life cycle, *A. lumbricoides* needs a single host—humans (Fig 26.17). Adult male and female worms live in the small intestine, usually the jejunum, where each gravid worm lays 200 000 to 250 000 eggs daily. Fertilized, nonembryonated eggs pass in feces and require variable periods of incubation to become infective. The shortest incubation periods (3 to 4 weeks)

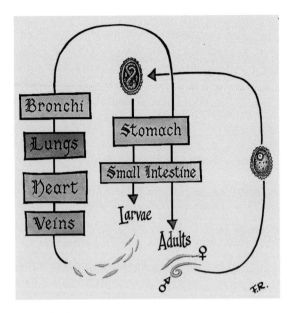

Figure 26.17
Life cycle of *A. lumbricoides*. Note that worm passes through intestine twice, once during each of 2 stages in life cycle.*

Figure 26.18
Numerous adult *A. lumbricoides* obstructing jejunum in this surgical specimen from 13-year-old Congolese boy.* x2.5

take place in moist, shady locations with temperatures of 22° to 33°C. *Ascaris* eggs resist freezing, drying, and many chemicals, and can remain viable for years. Eggs survive better in argillaceous soil than in sandy soil.

Eggs containing second-stage rhabditoid larvae are infective. Some authorities, however, believe there can be 2 molts within the egg, resulting in a third-stage filariform larva. Ingested infective eggs hatch in the stomach and upper small intestine. The emerging rhabditoid larvae are about 260 μm in length and 14 μm in diameter. These larvae penetrate the intestinal wall, enter the portal vein or intestinal lymphatic vessels, migrate through the liver to the heart, and are pumped through the pulmonary arteries to the lungs, where they break out of pulmonary capillaries into the air spaces. During their course through the lungs, rhabditoid larvae molt into third-stage filariform larvae. Third-stage larvae are up to 2 mm in length and 60 μm in diameter (Fig 26.13). These larvae may migrate up the bronchi and trachea and enter the esophagus, or be coughed up and swallowed. In the intestine the larvae molt twice and develop into sexually mature adults. Adult worms usually live in the lumen of the small intestine for about a year before being ejected from the anus.

Clinical Features and Pathogenesis

Adult *A. lumbricoides* are well-adapted to humans. Nonallergic older children, and adults with a small worm burden, seldom have symptoms. The nature and severity of clinical symptoms are related to the stage of infection, number of infecting worms, host response, and location of worms. Adult worms, larvae, or eggs can cause severe complications, and even death.[22]

Pulmonary symptoms are most common in previously exposed patients who are hypersensitive to *Ascaris* antigens. These patients develop *Ascaris* pneumonitis, an asthma-like reaction to migrating larvae. Patients with *Ascaris* pneumonitis present with Löffler's syndrome,[17] an acute eosinophilic pneumonia with tracheitis. The manifestations of Löffler's syndrome range from a mild, transient cough to severe pneumonitis lasting 2 to

3 weeks. Symptoms can include coughing, wheezing, dyspnea, low-grade fever, cyanosis, tachycardia, pressure or pain over the chest, and mucoid or bloody sputum. There is often transient eosinophilia and the sputum may contain numerous eosinophils, Charcot-Leyden crystals, or third-stage larvae. Löffler's syndrome is marked by crepitant rales; radiologically, there are prominent peribronchial markings and diffuse mottled infiltrates resembling miliary tuberculosis or viral pneumonia. Severe symptoms last 7 to 10 days and usually resolve spontaneously when larvae migrate out of the lungs, but *Ascaris* pneumonia can be fatal.[22] Löffler's syndrome is rare in hyperendemic areas of ascariasis, but common in patients living in temperate areas where there is only seasonal transmission.[25]

Patients harboring only a few adult worms in the intestine have only vague abdominal pain. In heavy infections, however, adult worms may cause nausea, vomiting, abdominal discomfort, and anorexia. In small children, numerous adult ascarides in the small intestine cause loss of appetite and failure to thrive. Entangled masses of adult worms can produce partial or complete intestinal obstruction (Fig 26.18), intussusception, or volvulus.[5,9] Intestinal obstruction is most common at the terminal ileum and ileocecal valve. Ascariasis may mimic an acute abdomen.[24] Adult worms may also invade the appendix and cause acute appendicitis (Fig 26.19).

Fever and chills, medications, and general anesthesia may stimulate migration of adult or juvenile *A. lumbricoides,* producing serious complications such as obstruction of the ampulla of Vater, pancreatic duct, and bile ducts (Fig 26.20).[22] Adult worms in bile ducts cause acute, agonizing, throbbing epigastric and right upper quadrant pain that may radiate to the shoulder, back, or hypogastrium.[12] Bacteria transported into ducts may cause severe or even fatal suppurative cholangitis and abscesses in the liver.[2,16] Dead worms and eggs in the common bile duct and gallbladder are potential sites for stones.[15] Abscesses in the liver surrounding adult ascarides may spread through the diaphragm, causing empyema and other pulmonary complications.[8] Obstruction of the pancreatic ducts produces pancreatitis.[6] Worms migrating into the stomach may induce vomiting. An adult worm in the trachea can cause choking and even asphyxiation.[19] Adult *A.*

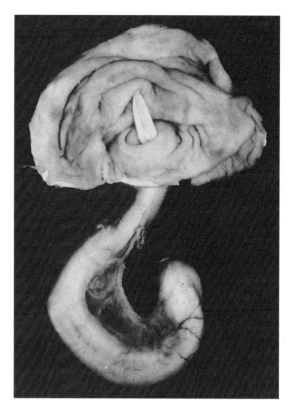

Figure 26.19
Adult *A. lumbricoides* protruding from orifice of appendix.*

Figure 26.20
Coronal section of liver with numerous adult *A. lumbricoides* in dilated bile ducts. x0.5

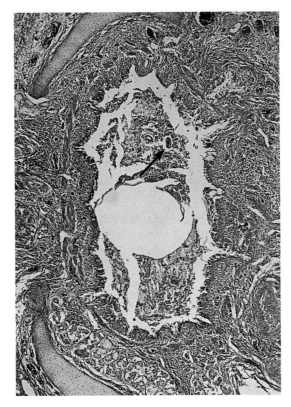

Figure 26.21
Coiled larva (arrow) of *A. lumbricoides* within lumen of bronchus of patient who died of *Ascaris* pneumonia. Bronchus is inflamed; lumen is filled with eosinophils and fibrin. x45

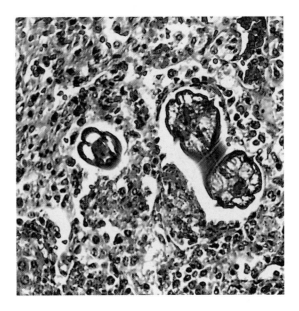

Figure 26.22
Three transverse sections of coiled *A. lumbricoides* larva in lung, causing bronchopneumonia. x340

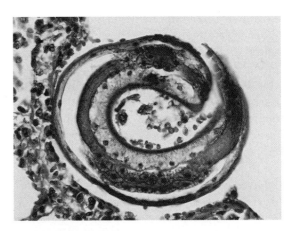

Figure 26.23
Coiled *A. lumbricoides* larva disrupting alveolar wall, which is thickened by interstitial infiltrate of eosinophils and macrophages. x325

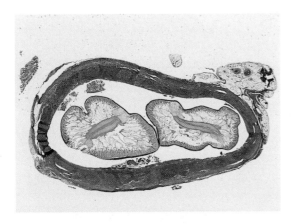

Figure 26.24
Two transverse sections of coiled adult *A. lumbricoides* within lumen of appendix. x5.5

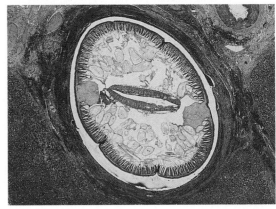

▶ **Figure 26.25**
Transverse section of adult *A. lumbricoides* within dilated hepatic duct. x10.5

lumbricoides can migrate into the eustachian tube, provoking suppurative otitis media, and ultimately exit via the auditory canal.[14] Migrating adult worms may emerge directly from the mouth, nose, and lacrimal glands, or perforate the inguinal canal or umbilicus.

A migrating gravid worm can deposit eggs in any human tissue and stimulate granulomas or abscesses. The liver, biliary tree, peritoneum, and lungs are the most frequent sites for these tissue reactions. Adult worms can perforate the digestive tract and deposit eggs in the peritoneum, leading to the rare complication of *Ascaris*-induced granulomatous or purulent peritonitis.[18]

Other conditions associated with ascariasis include fever, malaise, eosinophilia, central nervous system disorders, pylephlebitis, and fistulas. Severe ascariasis may lead to malnutrition, causing weight loss, edema, and emaciation.[11,26]

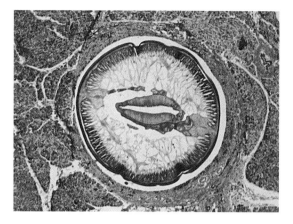

Figure 26.26
Transverse section of adult male *A. lumbricoides* within pancreatic duct. x28

Pathologic Features

Ascaris pneumonia produces small, patchy consolidations, most numerous in the lower lobes; the bronchioles contain macrophages and large numbers of eosinophils. There may be interstitial pneumonitis with thickening of alveolar walls. The larvae of *A. lumbricoides* may be numerous within alveolar walls, air sacs, bronchioles, and bronchi (Fig 26.21), and can also cause bronchopneumonia (Figs 26.22 and 26.23).

Adult *A. lumbricoides* in the small intestine rarely cause significant lesions; however, the crypts may be longer and the villi shorter than normal, and there can be increased mononuclear cells in the lamina propria. The intestine reverts to its normal structure after deworming. Adult worms in the appendix (Fig 26.19 and 26.24) or in a Meckel's diverticulum may obstruct the lumen and initiate acute inflammation with congestion. The mucosa of the appendix becomes necrotic; acute inflammation can be transmural and produce perforation and peritonitis.

Adult worms in the biliary tree dilate the ducts (Figs 26.25 and 26.26). They also transport bacteria from the intestine that incite acute inflammation in the mucosa of the bile duct, producing suppurative cholangitis (Fig 26.27) and liver

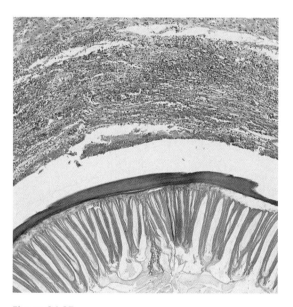

Figure 26.27
Adult *A. lumbricoides* within hepatic duct, surrounded by acute inflammation. Exudate contains many bacteria not visible at this magnification. x65

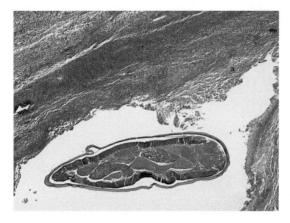

Figure 26.28
Adult *A. lumbricoides* in necrotic liver abscess, with extensive acute inflammation containing clumps of bacteria. In other areas of specimen, eggs of *A. lumbricoides* incited abscesses (see Figure 26.29). x13

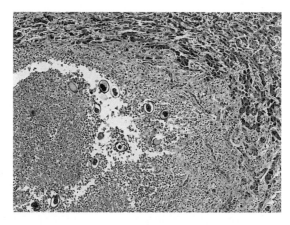

Figure 26.29
Eggs of *A. lumbricoides* in liver abscess of same patient depicted in Figure 26.28. x60

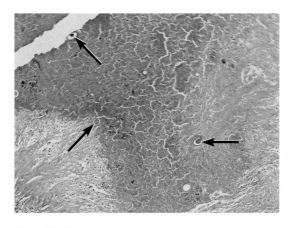

Figure 26.30
Eggs of *A. lumbricoides* (arrows) stimulated this fibrocaseous nodule in liver. x55

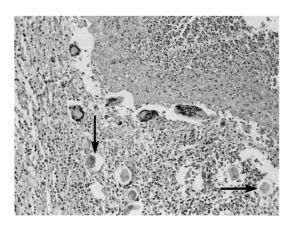

Figure 26.31
Several eggs of *A. lumbricoides* (arrows) causing suppurative and granulomatous cholangitis. x120

abscesses (Fig 26.28). Sections of biliary stones from patients with ascariasis may reveal remnants of mineralized worm.

Dead worms or eggs from migrating worms may cause abscesses or fibrocaseous granulomas (Figs 26.29 and 26.30). Centers of these granulomas show coagulative necrosis, neutrophils, and eggs of *A. lumbricoides* (Figs 26.31 and 26.32). Giant cells in the granuloma may contain eggs (Fig 26.33). Lesions caused by eggs, which can resemble inflammatory pseudotumors,[1] are most common in the liver and biliary tree, but a variety of other sites may be involved (Figs 26.34 to 26.37).

Diagnosis

Diagnosis is usually made by identifying eggs of *A. lumbricoides* in feces. Less frequently, feces contain adult worms, or worms emerge from body orifices. Occasionally, sputum or gastric aspirates will contain larvae of *A. lumbricoides*. Sometimes diagnosis is established by identifying eggs of *A. lumbricoides* (Fig 26.38), larvae, or adult worms in biopsy or autopsy specimens. Radiographs, CT scans (Fig 26.39), and ultrasound may be useful in diagnosis. For example, intestinal gas may outline worms in radiographs, and sonography can reveal tubular, echogenic filling defects that sometimes exhibit slow movements and contain a central sonolucent line.[3]

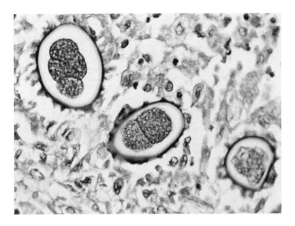

Figure 26.32
Oval, mamillated, and segmented eggs of *A. lumbricoides* in liver abscess. x375

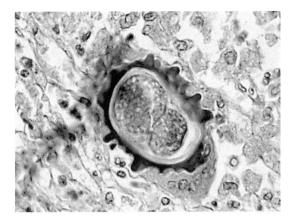

Figure 26.33
Oval, mamillated, and fertilized egg of *A. lumbricoides* at 2-cell stage, within giant cell in liver. x515

A clinical history and comparative chest radiograms taken 2 to 3 days apart will suggest a diagnosis of Löffler's syndrome. Recovering larvae of *A. lumbricoides* in the sputum will confirm the diagnosis.

Treatment and Prevention

Mebendazole is effective in treating intestinal ascariasis and does not produce significant side effects. Piperazine citrate administered as a syrup is also effective. Other useful drugs include pyrantel pamoate, cyclobendazole, and low-dose albendazole (400 mg). Complications such as obstruction of the intestine and biliary passages must be treated surgically; however, *A. lumbricoides* have been removed from the bile duct by strong suction during endoscopy. Antispasmodic drugs to relax Oddi's sphincter can help release a retained worm. Increasing hydrostatic pressure by infusion of normal saline solution through a T-tube in the bile duct may be helpful.[23]

Hand washing and use of appropriate sanitary facilities are the most effective preventive measures. Other precautions include washing fruits and vegetables and not using human excrement as fertilizer.

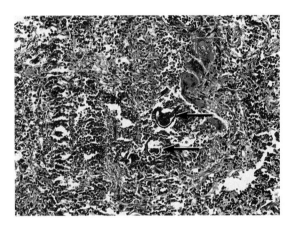

Figure 26.34
Two collapsed eggs of *A. lumbricoides* (arrows) in lung, causing acute and chronic pneumonia. x120

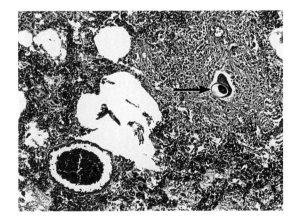

Figure 26.35
Egg of *A. lumbricoides* (arrow) within granuloma in lung. x82

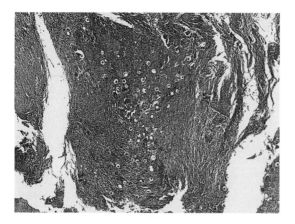

Figure 26.36
Eggs of *A. lumbricoides* provoking suppurative granuloma in connective tissue of abdominal wall in 13-year-old Brazilian girl. x25

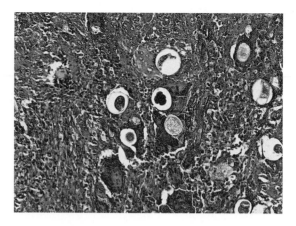

Figure 26.37
Higher magnification of eggs of *A. lumbricoides* in suppurative granuloma of abdominal wall illustrated in Figure 26.36. Decorticated eggs in various stages of development are within giant cells and scattered among inflammatory cells. x125

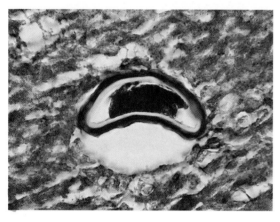

Figure 26.38
Collapsed, mamillated egg of *A. lumbricoides* within lumen of appendix. Movat x585

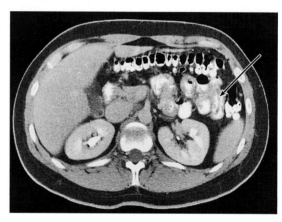

Figure 26.39
Computed tomogram of abdomen depicting filling defect produced by *A. lumbricoides* in small intestine (arrow) of 20-year-old man.*

References

1. Adebamowo CA, Akang EE, Ladipo JK, Ajao OG. Ascarid granuloma presenting as pseudotumor. *Trop Geogr Med* 1993;45:86–88.
2. Adedeji SO, Ogunba EO. Ascaris lumbricoides as a vehicle of bacterial infections. *Afr J Med Med Sci* 1986;15:85–92.
3. Aslam M, Dore SP, Verbanck JJ, De Soete CJ, Ghillebert GG. Ultrasonographic diagnosis of hepatobiliary ascariasis. *J Ultrasound Med* 1993;12:573–576.
4. Aspock H, Flamm H, Picher O. Intestinal parasites in human excrements from prehistoric salt-mines of the Hallstatt period (800–350 B.C.)[in German]. *Zentralbl Bakteriol* 1973;223:549–558.
5. Baird JK, Mistrey M, Pimsler M, Connor DH. Fatal human ascariasis following secondary massive infection. *Am J Trop Med Hyg* 1986;35:314–318.
6. Baldwin M, Eisenman RE, Prelipp AM, Breuer RI. Ascaris lumbricoides resulting in acute cholecystitis and pancreatitis in the Midwest. *Am J Gastroenterol* 1993;88:2119–2121.
7. Beaver PC, Danaraj TJ. Pulmonary ascariasis resembling eosinophilic lung. *Am J Trop Med Hyg* 1958;7:100–111.
8. Beaver PC, Jung RC, Cupp EW. *Clinical Parasitology*. 9th ed. Philadelphia, Pa: Lea & Febiger; 1984:316.
9. Blumenthal DS, Schultz MG. Incidence of intestinal obstruction in children infected with Ascaris lumbricoides. *Am J Trop Med Hyg* 1975;24:801–805.
10. Cockburn A, Barraco RA, Reyman TA, Peck WH. Autopsy of an Egyptian mummy. *Science* 1975;187:1155–1160.
11. Crompton DW. Ascariasis and childhood malnutrition. *Trans R Soc Trop Med Hyg* 1992;86:577–579.
12. de Andrade Junior DR, Karam JA, Warth M do P, et al. Massive infestation by Ascaris lumbricoides of the biliary tract: report of a successfully treated case. *Rev Inst Med Trop Sao Paulo* 1992;34:71–75.
13. Douvres FW, Tromba FG, Malakatis GM. Morphogenesis and migration of Ascaris suum larvae developing to fourth stage in swine. *J Parasitol* 1969;55:689–712.
14. Fagan JJ, Prescott CA. Ascariasis and acute otitis media. *Int J Pediatr Otorhinolaryngol* 1993;26:67–69.
15. Khuroo MS, Zargar SA. Biliary ascariasis. A common cause of biliary and pancreatic disease in an endemic area. *Gastroenterology* 1985;88:418–423.
16. Khuroo MS, Zargar SA, Mahajan R. Hepatobiliary and pancreatic ascariasis in India. *Lancet* 1990;335:1503–1506.
17. Löffler W. Zur Differentialdiagnose der Lungen-infiltrierungen; über Flüchtige Succedan-infiltrate (mit Eosinophilie). *Beit Klin Tuberk* 1932;79:368–82.
18. Mello CM, Briggs M do C, Venancio ES, Brandao AB, Queiroz Filho CC. Granulomatous peritonitis by ascaris. *J Pediatr Surg* 1992;27:1229–1230.
19. Mittal VK, Dhaliwal R, Yadav R, Sahariah S. Fatal respiratory obstruction due to a roundworm. *Med J Aust* 1976;2:210–212.
20. Nichols RL. The etiology of visceral larva migrans. II. Comparative larval morphology of Ascaris lumbricoides, Necator americanus, Strongyloides stercoralis, and Ancylostoma caninum. *J Parasitol* 1956;42:363–399.
21. Phills JA, Harrold AJ, Whiteman GV, Perelmutter L. Pulmonary infiltrates, asthma, and eosinophilia due to Ascaris suum infestation in man. *N Engl J Med* 1972;286:965–970.
22. Piggott J, Hansbarger EA Jr, Neafie RC. Human ascariasis. *Am J Clin Pathol* 1970;53:223–234.
23. Rezaul Karim M. Biliary ascariasis. *Int Surg* 1991;76:27–29.
24. Sasaki J, Seidel JS. Ascariasis mimicking an acute abdomen. *Ann Emerg Med* 1992;21:217–219.
25. Spillmann RK. Pulmonary ascariasis in tropical communities. *Am J Trop Med Hyg* 1975;24:791–800.
26. Thein-Hlaing, Thane-Toe, Than-Saw, Myat-Lay-Kyin, Myint-Lwin. A controlled chemotherapeutic intervention trial on the relationship between Ascaris lumbricoides infection and malnutrition in children. *Trans R Soc Trop Med Hyg* 1991;85:523–528.
27. Tripathy K, Duque E, Bolanos O, Lotero H, Mayoral LG. Malabsorption syndrome in ascariasis. *Am J Clin Nutr* 1972;25:1276–1281.

27

Toxocariasis

Aileen M. Marty

Introduction

Definition

Larvae of the canine ascarid *Toxocara canis* (Werner, 1782) Johnston, 1916, and the feline ascarid *Toxocara cati* (Schrank, 1788) Brumpt, 1927, cause toxocariasis. Infection by either organism may generate a systemic syndrome of sustained eosinophilia, pneumonitis, and hepatomegaly. These larvae are a major cause of blindness,[36] and may provoke rheumatic, neurologic, or asthmatic symptoms.

Synonyms

Ascaris canis is an archaic synonym for *T. canis*. Synonyms for *T. cati* include *Ascaris mystax*, *Belascaris mystax*, *Toxocara mystax*, and *Belascaris cati*. Toxocariasis is often called visceral larva migrans. Depending on geographic location, degree of eosinophilia, and pulmonary signs, the terms Weingarten's disease, Frimodt-Möller's syndrome, and eosinophilic pseudoleukemia have been applied to systemic toxocariasis. Other terms for ocular toxocariasis include nematode ophthalmitis and ocular larva migrans.

General Considerations

In 1782, Werner described a parasitic nematode of dogs and named it *A. canis*. Johnston determined that it was a member of the genus *Toxocara*, established by Stiles in 1905. Fülleborn, in 1921, speculated that *T. canis* larvae might cause granulomatous nodules in humans.[2,11] In 1947, Perlingiero and György described the earliest case of what was probably toxocariasis. Their patient, a 2-year-old boy from Florida, had a fever of 40°C, Löffler-like syndrome, vomiting, diarrhea, hypergammaglobulinemia, anemia, and hepatomegaly with eosinophilic necrotizing granulomas in the liver.[33] In 1950, Campbell-Wilder studied the eyes of patients with endophthalmitis, Coats' disease, or pseudoglioma.[46] In 24 of 46 specimens, she found nematode larvae in eosinophilic granulomas, and called this condition nematode ophthalmitis.[43]

In 1952, Beaver et al described 3 young children with hepatomegaly, anemia, and eosinophilia, and introduced the term visceral larva migrans (VLM). In the liver of one of these children they identified a larva of *Toxocara* sp, probably *T. canis*.[3] Nichols' exhaustive study of *Toxocara* larvae in tissues, and a comparative study of larvae of *Ascaris lumbricoides, Necator americanus, Strongyloides stercoralis*, and *Ancylostoma caninum*,[29,30] established *T. canis* and *T. cati* as the predominant cause of VLM.

Initially, VLM was synonymous with toxocariasis, but it was later discovered that many species of helminth larvae cause this clinical syndrome.[2] A similar syndrome can develop from other nematodes (eg, *Gnathostoma spinigerum, Baylisascaris procyonis, Ancylostoma braziliense*, and *Ancylostoma caninum*); cestodes (sparganum); and a trematode (*Alaria*).

In 1975, de Savigny contributed significantly to the immunobiology of *Toxocara* and diagnosis of toxocariasis by developing a method of maintaining L2 larvae in tissue culture for the study of host parasite interaction and serologic tests.[7] The creation in 1979 of *T. canis* excretory-secretory (TEX)-based ELISA[6] led to seroepidemiologic surveys, establishing toxocariasis as a cosmopolitan disease.

Only *T. canis* and *T. cati* are known to cause toxocariasis, but the other 9 *Toxocara* sp, and *Toxascaris leonina*, are potential human parasites.

Epidemiology

Toxocara canis and *T. cati* are perhaps the most ubiquitous gastrointestinal helminths of domestic dogs and cats.[32]

Toxocariasis is one of the most common zoonotic infections of children in the United States. Infection rates worldwide are highest in poor and rural populations, where serologic prevalence is as high as 84% (see Table 27.1).[24,42] Dog trainers, public health inspectors, and outdoor or soil-related workers are at increased risk.

A few milligrams of contaminated soil can contain hundreds of infective eggs, which may explain why systemic toxocariasis is mainly a disease of toddlers. Ocular toxocariasis affects a minimum of 750 patients in the United States each year.[36] This disease tends to appear around age 12, but can develop later.[12] Older individuals acquire light infections by ingesting small amounts of contaminated soil, eggs or larvae in fresh, unwashed vegetables,[5] or infective larvae in improperly cooked tissue of paratenic hosts (Fig 27.1).[28]

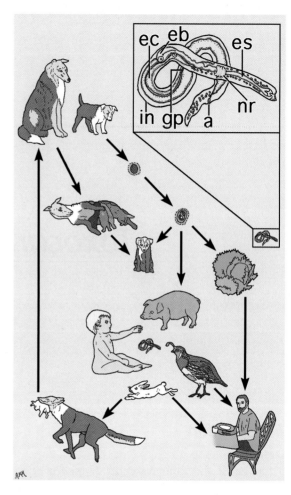

Figure 27.1
Life cycle of *T. canis*. Humans acquire toxocariasis through contact with contaminated soil and vegetables, or poorly cooked meat containing infective larvae. *Inset*: Note esophagus (es), nerve ring (nr), esophageal bulb (eb), excretory column (ec), intestine (in), genital primordium (gp), and anus (a).

Infectious Agent

Morphologic Description
Adult

Female *T. canis* (Fig 27.2) can be up to 20 cm long by 3 mm wide; male *T. canis* can be up to 11 cm long by 2.5 mm wide. *Toxocara cati* are smaller; female worms are 4 to 12 cm long and male worms 3 to 7 cm long. An adult *Toxocara* has characteristic cervical alae (Fig 27.3) and an arrow-shaped head. Cross sections of the anterior end reveal a triangular acellular spicule within lateral cervical

Spain 1.7%–2.2%	Italy 3.9%–13.3%
Japan 3.5%	Jordan 15.8%
Ethiopia 4%	Canada 17%
New Zealand 4.4%	France 22%
Russia 5.3%	USA 9%–54%
Netherlands 7.1%	Venezuela 1.8%–65%
Iraq 7.3%	St. Lucia 84%

Table 27.1
Seroprevalence of toxocariasis in humans.[10,16]

alae (Fig 27.4). The cuticle is multilayered and the musculature polymyarian and coelomyarian, with a prominent contractile portion and minimal sarcoplasmic portion (Figs 27.5 and 27.6). The esophagus is triradiate (Fig 27.4), but, unlike other Ascaridoidea, it has a distinct, muscular, posterior ventriculus. The intestine has a single layer of ciliated columnar epithelium. Females are oviparous, didelphic, and have a muscular vagina. Male worms have paired spicules and characteristic finger-shaped perianal papillae.

Larvae

Toxocara canis and *T. cati* second-stage (L2) larvae range from 357 to 445 μm in length, but tissue processing causes shrinkage. Larvae of these 2 species are distinguishable only by their diameter: *T. canis* larvae vary from 18 to 21 μm, while *T. cati* larvae vary from 15 to 17 μm (see Table 27.2).[29] The anterior end has a 3-lipped, dorsally inclined mouth (Fig 27.1 inset); the posterior end tapers more abruptly. Lateral alae extend along the sides, terminating about 20 μm from each extremity. Most cross sections, except of the most anterior and posterior ends, reveal minute, thin, single lateral alae (Fig 27.7). Larvae have 3 general body regions: anterior, middle, and posterior (Figs 27.1 and 27.8). In the anterior region (Figs 27.7 and 27.9), the esophagus extends for approximately one third the total length of the larva, has a cuticular-lined lumen, and is divided into 4 zones: procorpus, metacorpus, isthmus, and terminal bulb. In longitudinal sections, basophilic ganglionic nuclei may obliterate the esophagus, unless the cut is median sagittal (Fig 27.9). The middle region contains the intestine and H-shaped excretory cell (Fig 27.10). The excretory cell forms 2 lateral columns extending anteriorly to just beyond the esophageal

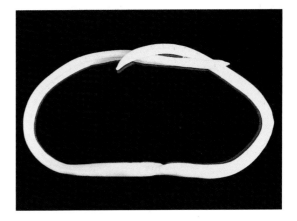

Figure 27.2
Adult female *T. canis* recovered from dog feces. Unstained

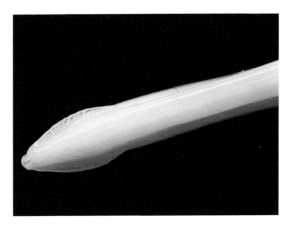

Figure 27.3
Anterior end of adult female *T. canis* recovered from dog feces. Note long, narrow cervical alae. Unstained

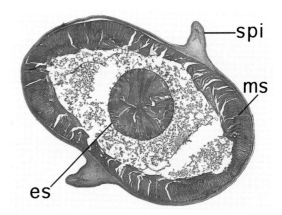

Figure 27.4
Cross section of anterior end of adult *T. canis* showing lateral cervical alae with triangular acellular spicule (spi). Musculature (ms) is polymyarian and coelomyarian. Note triradiate esophagus (es). Movat x60

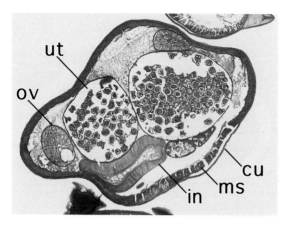

Figure 27.5
Transverse section through gravid *T. canis* showing cuticle (cu), muscle (ms), intestine (in), uteri with eggs (ut), and ovaries (ov). x30

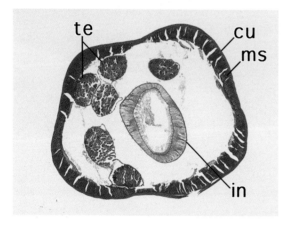

Figure 27.6
Transverse section through mature male *T. canis* demonstrating cuticle (cu), muscle (ms), intestine (in), and multiple sections of testis (te). x45

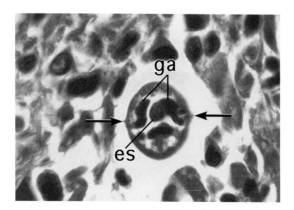

Figure 27.7
Anterior region of *T. canis* larva in eosinophilic granuloma, demonstrating esophagus (es), ganglionic cells (ga), and minute lateral alae (arrows). x1180

bulb and posteriorly on either side of the intestine. These columns strongly compress the intestine laterally, pushing it dorsad (Fig 27.8). The intestine is without an apparent lumen, and consists of 7 elongate cells containing opaque fat globules arranged linearly. Thus, cross sections can never show more than a single intestinal cell. The posterior region contains the genital primordium, the posterior portion of the intestine, and the anal pore. Here the intestine expands and fills the body cavity (Fig 27.8). The seventh intestinal cell terminates 20 to 25 µm above the anal pore, to which it connects by a thin cuticular tube. Posterior to the intestine, only scattered ganglion cells fill the pseudocoelom (Fig 27.11).

Eggs

Eggs of *T. canis* are 85 by 75 µm, dark brown or gray-brown, subspherical, with a thick, pitted shell (Figs 27.12 and 27.13). Eggs of *T. cati* are 75 by 65 µm with surface pitting finer than in *T. canis*.[1] Eggs of both species are unembryonated, both in the intestine of a definitive host and when passed in dog and cat feces.

Life Cycle and Transmission

Definitive hosts for *T. canis* include dogs, coyotes, wolves, jackals, dingoes, hyenas, and foxes. Definitive hosts for *T. cati* include domestic and wild cats. Many animals, including rodents, rabbits, sheep, cattle, pigs, goats, monkeys, birds (chickens, quail, pigeons), earthworms, and humans, may serve as paratenic hosts.[9,27,32] Infection of canids and felids may take any of several routes: direct transmission, transplacental (canids only) and transmammary (amphiparatenic host) transmission, and paratenic host transmission (Fig 27.1).

Adult female *T. canis* in the canine intestines produce 25 000 to 85 000 unembryonated eggs a day. Infective L2 larvae develop in eggs in about 10 to 20 days, depending on temperature and humidity. Eggs with L2 larvae contaminate vegetables or remain in the soil until consumed by a definitive or paratenic host.

After a canid consumes infective eggs of *T. canis*, or a felid ingests infective eggs of *T. cati*, the eggs hatch in the gastrointestinal tract, releasing larvae that invade the small intestine, enter the liver via the portal system, tunnel into the liver, and then pass through the heart to pulmonary capillaries. From there, larvae either migrate to the trachea and mature, or circulate to somatic tissue and remain infective larvae.

Tracheal migration leads to development of adult worms in dog or cat intestine. Worms begin to pass eggs 30 to 34 days after infection. For *T. canis*, the probability of tracheal migration is highest in a newborn puppy. For *T. cati*, the probability of tracheal migration remains high throughout the life of the cat.

Larvae that do not enter the alveoli undergo somatic migration. They continue through the pulmonary vein and heart to the systemic circulation and lodge in somatic tissue, where they encyst as arrested infective larvae. Such larvae accumulate in an infected puppy's tissues as it matures. Arrested infective larvae may also invade an adult female's unborn or newborn puppies through the placenta or mammary glands. In the last gestational trimester, arrested larvae reactivate and migrate through the placenta to fetal liver.[11] Puppies infected prenatally may pass eggs in feces by 21 days of age. Transmammary transmission in dogs is a relatively minor route of infection. In cats, there is no transplacental migration, but transmammary infection is common in kittens.[32]

Somatic migration results in accumulation of infective L2 larvae in humans and other paratenic hosts. These larvae mature when ingested by the definitive host. In this way, *Toxocara* exploits the predator-prey relationship. When humans or other paratenic hosts consume an infected paratenic host, arrested larvae reactivate and somatic migration continues in the new host (Fig 27.1). Reported infections in humans by other *Toxocara* sp are questionable.

Clinical Features and Pathogenesis

Most infections are mild or subclinical, but provoke positive serologic tests and, sometimes, persistent eosinophilia (occult toxocariasis).[40] Clinical illness manifests in any combination of systemic toxocariasis, ocular toxocariasis, and a variety of less severe forms.[17] Reactions to excretory and secretory larval materials differ in various

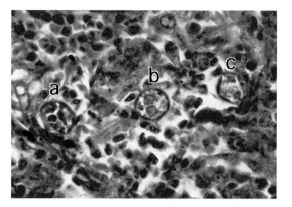

Figure 27.8
Pulmonary granuloma containing 3 transverse sections of *T. canis* larvae: (a) esophageal procorpus where loosely organized ganglionic nuclei surround tiny esophagus; (b) middle region demonstrating 2 lateral excretory columns compressing intestine laterally and dorsad; and (c) interface between middle and posterior region. Note intestine expanding to fill body cavity. x530

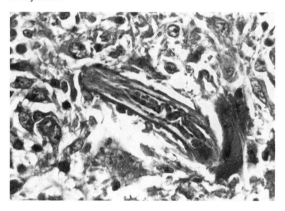

Figure 27.9
Anterior end of *T. canis* larva within pulmonary granuloma. x530

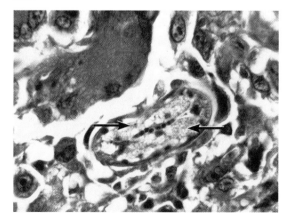

Figure 27.10
Tangential section of middle region of *T. canis* larva in pulmonary granuloma demonstrating 2 prominent excretory columns (arrows). The intestine is not visible in this section. x725

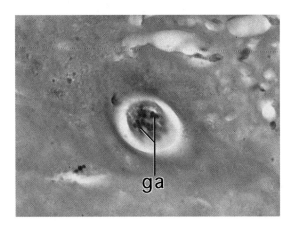

Figure 27.11
Ocular granuloma containing cross section of *T. canis* larva posterior to intestine, showing scattered ganglion cells (ga) in pseudocoelom. x610

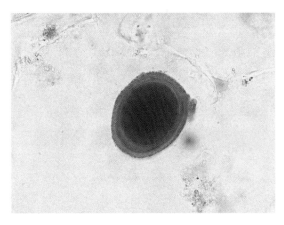

Figure 27.12
Toxocara canis egg from feces. Egg is dark brown, unembryonated, and has thick pitted shell. Unstained x315

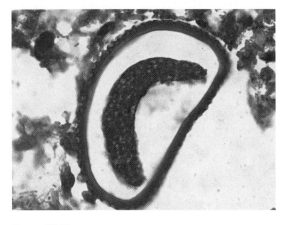

Figure 27.13
Section of unembryonated *T. canis* egg in lumen of dog intestine. x610

organs: endophthalmitis in the eye (Fig 27.14); hepatomegaly in the liver; pneumonia, bronchitis, asthma, or eosinophilic pleurisy in the lung; and urticaria in the skin.[18] Symptoms in other areas vary: in the intestine—abdominal pain and anorexia; in the brain and central nervous system—epileptic seizures or neuropsychologic changes; in muscle—myalgia or chronic weakness; and in the joints—rheumatic changes.[22,44] Many individuals have allergic manifestations. Pregnant women may be at increased risk of miscarriage.[41]

The most severe form of infection is systemic toxocariasis (VLM),[3] which commonly develops in young children who ingest large numbers of eggs. Beaver et al originally emphasized hepatic disease; however, respiratory and pulmonary manifestations often predominate. Patients present with Löffler's syndrome, including cough, dyspnea, wheezing, anorexia, and weight loss or failure to thrive. Chest x-rays reveal scattered, migrating opacities. Even patients without respiratory symptoms often have radiologic evidence of pulmonary infiltrates. The majority of children also have fever (as high as 40°C), hepatosplenomegaly, lymphadenopathy, pallor, and lassitude.[38] Laboratory findings include anemia, leukocytosis, and eosinophilia (up to 90%).[4] Hyperglobulinemia (4-7g/dl), mainly IgE, is common in severe infections. Eosinophilia, splenomegaly, invasion of the liver by antigen-stimulated lymphocyte proliferation, and antitoxocaral antibody production (IgG and IgM) are proportional to the larval burden.

Ocular toxocariasis results from invasion by L2 larvae (Fig 27.14). Most patients with ocular disease have no other history of VLM. Patients who had occult toxocariasis in childhood may have dormant larvae that later reactivate. Uncommonly, patients have a history of or simultaneous development of VLM. Menarche may stimulate larval reactivation, perhaps explaining why typical patients with ocular toxocariasis are older (over 12 years) than those with systemic toxocariasis (under 3 years).

Ocular toxocariasis presents a spectrum of disease.[37,47] Symptoms are usually unilateral. Patients may have overt signs of ocular inflammation, or present with strabismus or poor vision after inflammation has subsided. Occasionally ocular involvement is asymptomatic or produces nonspecific symptoms. Lesions may include diffuse

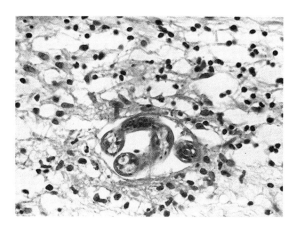

Figure 27.14
Coiled *T. canis* larva in vitreous membrane. x400

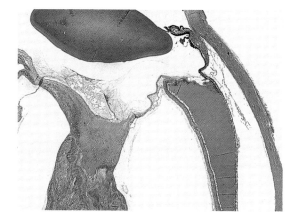

Figure 27.15
Eye with toxocaral ophthalmitis. Note vitreous membrane, subretinal serous exudate, and retinal detachment. x25

endophthalmitis, posterior retinochoroiditis, peripheral retinochoroiditis, optic papillitis, optic neuritis,[21] keratitis, or phakitis, and possibly conjunctivitis, diffuse unilateral subacute neuroretinitis (DUSN), and motile chorioretinal nematode syndrome.

Diffuse toxocaral endophthalmitis, the most common form of ocular toxocariasis, produces little pain or photophobia. External examination reveals only minimal inflammation, usually with no ciliary flush. In acute stages, slit-lamp examination reveals corneal precipitates, aqueous flare and cells, posterior synechiae, and hypopyon. The vitreous is hazy, the retina often contains a yellow-white mass that resembles an endophytic retinoblastoma, and there may be exudative retinal detachment (Fig 27.15). Sometimes inflammation subsides, leaving a white retrolental mass and a dense, gray-white membrane (cyclitic membrane) extending from the ciliary body to the retina. The cyclitic membrane may grow across the posterior surface of the lens, causing posterior synechiae, with forward bulging of the iris (iris bombé), and a posterior subcapsular cataract. The cataract may degenerate into an opaque lens, producing a white pupillary reflex (leukokoria). Phthisis bulbi or glaucoma may ensue.

Clinically, posterior retinochoroiditis is a hazy, ill-defined, white lesion, often involving the center of the macula, with overlying inflammatory cells in the vitreous. Peripheral retinochoroiditis appears as a hazy, white reaction in the peripheral

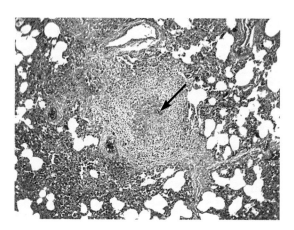

Figure 27.16
Granuloma containing *T. canis* larva (arrow) in lung. Note diffuse thickening of alveolar septae. x40

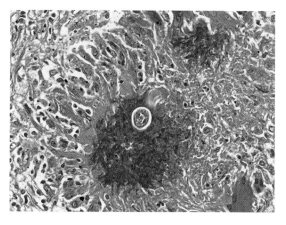

Figure 27.17
Hepatic granuloma containing *T. canis* larva surrounded by eosinophils, neutrophils, and giant cells. x230

Larva	Diameter at midgut	Lateral alae	Intestine	Excretory columns
Toxocara canis (L2)	18–21 μm	Single minute lateral alae	Single gut cell, no lumen	Posterior excretory columns large, well-defined; minute central excretory canaliculi often present
Toxocara cati (L2)	15–17 μm	Single minute lateral alae	Single gut cell, no lumen	Posterior excretory columns large, well-defined; minute central excretory canaliculi often present
Baylisascaris procyonis	50–65 μm	Single prominent lateral alae	7–10 intestinal cells form functional lumen	Paired excretory columns smaller than intestine; prominent, conical
Ascaris lumbricoides (advanced L2)	14–23 μm	Single minute lateral alae	Intestinal cells form 2 rows bordering narrow lumen	Excretory columns small, well-defined
Ascaris lumbricoides (L2–L3)	26–55 μm	Single prominent lateral alae	Intestine functional; particulate matter in lumen	Cross-sectional area of excretory columns greater than that of intestine
Strongyloides stercoralis (early L3)	14–16 μm	Double minute lateral alae	Intestinal cells form 2 rows bordering unseen, flattened lumen	At this level, excretory columns reduced or missing
Necator americanus (early L3)	22–26 μm	Double minute lateral alae	2 intestinal cells filling body cavity	Excretory cells form sacculate columns
Necator americanus (late L3)	27–50 μm	No lateral alae	Intestinal cells accumulate large amounts of pigment and compress excretory columns	Excretory columns compressed into flat lateral bands
Ancylostoma caninum (early L3)	22–24 μm	Double minute lateral alae	Single intestinal cell with narrow lumen	Excretory columns large, well-defined. Left column terminates near end of intestine. Right column terminates 60–80 μm posterior to genital primordium. No central excretory canaliculi

Table 27.2
Comparison of *Toxocara* and other larvae at level of midgut.

fundus. In late stages, the inflammation may lead to macular heterotopia and severe loss of vision. Optic papillitis causes an elevation of the optic disc and telangiectasia of blood vessels, sometimes with subretinal exudation. Larvae may lodge in end-arteries of the limbus and produce keratitis. Slit-lamp examination readily reveals motile larvae in the peripheral cornea. Phakitis, often with cataract formation, is an occasional complication.[19,43] There is a recorded instance of a *T. canis* larva being extracted from the eye of a 9-year-old Brazilian boy with DUSN,[1] but in most cases of DUSN, and of motile chorioretinal nematode syndrome, it is uncertain whether *Toxocara, Baylisascaris,* or another larva is the cause.[2]

Neurologic toxocariasis may cause seizures; reactors to toxocaral skin tests were 3 times as common among epileptics as among controls.[45] Infected children may have impaired mental and behavioral development.[13]

Uncommonly, the initial presentation resembles juvenile rheumatoid arthritis.[4] Cutaneous involvement may appear as urticaria or hemorrhagic necrotic skin lesions. Osseous lesions are uncommon.[38]

In humans and other paratenic hosts, migrating *Toxocara* larvae neither molt nor grow, but are metabolically active, producing excretory and secretory materials and shedding cuticular substances. These substances are a continuing source of larval antigens that stimulate both Th1 and Th2 subsets of T helper cells. The major response involves Th2 cells, which secrete Il-4, Il-5, and Il-10, and regulate humoral immune responses which cause eosinophilia and hyperglobulinemia (mainly IgE). Neither the eosinophilia nor the hyperglobulinemia eliminates L2 larvae. Th1 cells secrete Il-2 and interferon-γ, mediating delayed hypersensitivity and granuloma formation (Fig 27.16).[20]

Pathologic Features

After a certain time, larvae stop migrating. *Toxocara* larvae invading the central nervous system

Figure 27.18
Subretinal mass of inflammatory tissue and organized hemorrhage with cholesterol slits in ocular toxocariasis, clinically suspected to be Coats' disease. x85

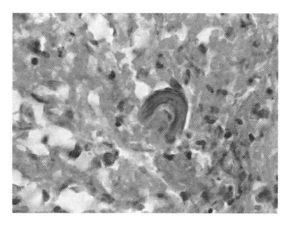

Figure 27.19
Fragment of *T. canis* larva in older lesion of ocular toxocariasis. x410

do not elicit an inflammatory reaction. In all other tissues, however, Th1 cells provoke granuloma formation around larvae, often with surrounding collagenous capsules. Larvae initiate parenchymal disruption, stimulating repair of tissues, followed by a predominantly eosinophilic infiltrate. Eosinophils attach to larval surfaces but are shed and do not kill the larvae. Within 7 to 14 days, lymphocytes and macrophages replace the eosinophils. In about 6 months, the granulomas contract and contain primarily epithelioid cells. Eventually the lesions become well-formed eosinophilic granulomas with multinucleated foreign body giant cells, epithelioid cells, and lymphocytes, with eosinophils and neutrophils immediately around the larvae (Fig 27.17). Larvae can burrow out of these granulomas and migrate elsewhere, provoking similar lesions. Thus, the cellular composition of the inflammatory reaction does not indicate the duration of infection. When a larva migrates repeatedly, the granulomatous reaction surrounds a tortuous, irregular track filled with necrotic material. Granulomas without a parasite slowly resorb (Fig 27.18) and heal completely. Larvae may also remain dormant within a granuloma throughout the host's life.

Grossly, the liver is slightly enlarged and soft, with white, randomly distributed, 0.5-mm-to-1-cm umbilicated and nonumbilicated nodules on the surface and throughout the parenchyma. Patients with systemic toxocariasis tend to accumulate *Toxocara* larvae in the liver, causing extensive hepatocellular unrest, focal necrosis, empty tracks, and numerous eosinophilic granulomas with and without larvae (Fig 27.17). Larvae may lodge in the walls of veins or arteries and provoke focal granulomatous vasculitis. Fibroblasts eventually organize the granuloma, and mineralization may follow. Plasma cells, lymphocytes, and eosinophils infiltrate portal tracts.

Toxocara larvae in the heart and lung tend to cause hypersensitivity. Continued antigen stimulation of the immune system, including eosinophilia, may permanently alter normal cardiopulmonary tissues.[20] Persistent pneumonitis with diffuse infiltrations of macrophages and eosinophils causes thickening of alveolar septae and petechial hemorrhages (Fig 27.16). Patients with Löffler's syndrome have scattered eosinophilic granulomas (Figs 27.16 and 27.8). The myocardium may be softened; myocarditis is usually focal and edematous. Initially there are predominantly eosinophilic infiltrates, which progress to granulomas containing necrotic debris, and sometimes infective larvae.[31] Later, there is collagen deposition, followed by fibrosis and hemosiderin-laden macrophages.

Sections of brain show only empty tracks consisting of small necrotic foci with a few inflammatory cells, mainly eosinophils and lymphocytes. Larvae are rare.

Optic damage from *Toxocara* varies with the

location and severity of the infection. Ocular lesions frequently contain viable larvae (Figs 27.11 and 27.14) that stimulate a florid inflammatory reaction. Gross examination reveals intraocular disorganization, often with total retinal detachment (Fig 27.15). There may be focal inflammatory thickening of the periphery of the detached retina. Subretinal fluid usually becomes gelatinous. A dense cyclitic membrane sometimes incorporates the detached retina. Microscopically, there are focal eosinophilic granulomas associated with retinal detachment. Serial sections may reveal well-developed larvae (Fig 27.14), remnants of degenerated larvae, or residual hyaline capsules of larvae (Fig 27.19) Eyes with retinal granulomas without severe endophthalmitis have a retinochoroidal mass (Fig 27.18), most often around the macula. Vitreoretinal traction and retinal detachment are common, but generally less severe than in patients with pronounced endophthalmitis.

Diagnosis

Enzyme-linked immunosorbent assays (ELISA) absorbed with *A. lumbricoides* antigen are 85% to 90% specific.[14] A serum titer of 1:32 is the cutoff for diagnosis of systemic toxocariasis. Most ophthalmologists consider a 1:8 titer positive. Some accept a titer as low as 1:2 if the patient has symptoms compatible with ocular toxocariasis.[34] If clinical suspicion is high, but serum titers are negative, analyzing anterior chamber fluid by paracentesis may facilitate diagnosis. ELISA testing of vitreous fluid removed at vitrectomy is useful.[37] Western blot is more specific than ELISA and diminishes the problems of cross-reactivity with other members of Ascaridoidea.[26]

Definitive diagnosis is based on histologic identification of *Toxocara* larvae in biopsy or autopsy tissues. Random tissue sections seldom show larvae, but a study of serial sections is often successful. Usually, attention to maximum width, lateral alae, excretory cell, and intestine is adequate to specifically identify larval nematodes in tissue (see Table 27.2).[29,30] Crushing tissues between glass slides, or digesting tissues with pepsin, may reveal greater numbers of motile intact larvae in unfixed biopsy specimens.[38]

Treatment and Prevention

Albendazole is the preferred therapy for toxocariasis worldwide.[39] Besides being effective, it has a lower incidence of adverse reactions than other recommended regimens, including mebendazole,[23] thiabendazole, or diethylcarbamazine.[25] Concomitant use of corticosteroids may be necessary in severe infections. Ocular toxocariasis usually calls for a combination of anthelmintics and other interventions (cycloplegic agents, corticosteroids, ocular surgery, and laser photocoagulation)[37] to help restore vision. Treatment of asymptomatic, ELISA-positive individuals is not indicated unless there is hypereosinophilia.

Eggs in feces of domestic dogs and cats is the primary source of infection. Effective sanitation is difficult to achieve because *Toxocara* eggs survive for years, even in adverse environments. Eggs are highly resistant to cold temperatures and a wide range of chemicals, including formalin, acids, alkalis, and commercial disinfectants. Ultraviolet light, aqueous iodine, and high temperatures, on the other hand, effectively destroy eggs. Puppies are a major source of *T. canis* eggs, excreting up to 2 million eggs per day. Dogs and cats should be dewormed at 3 weeks of age, repeated 3 times at 2-week intervals, and every 6 months thereafter. If an infected, lactating bitch is treated daily with fenbendazole, her puppies should be worm-free. Deworming pets and removing animal feces from the environment are essential safeguards of human health.[35]

References

1. Ash LR, Orihel TC. *Atlas of Human Parasitology*. 4th ed. Chicago, Ill: ASCP Press; 1997:231–233.
2. Beaver PC. The nature of visceral larva migrans. *J Parasitol* 1969;55:3–12.
3. Beaver PC, Snyder CH, Carrera GM, Dent JH, Lafferty JW. Chronic eosinophilia due to visceral larva migrans. Report of three cases. *Pediatrics* 1952;9:7–19.
4. de Corral VR, Lozano-Garcia J, Ramos-Corona LE. Una presentación poco usual de toxocariasis sistémica. *Bol Med Hosp Infant Mex* 1990;47:841–844.
5. de Oliveira CA, Germano PM. Estudo da ocorrencia de enteroparasitas em hortalicas comercializadas na regiao metropolitana de Sao Paulo, SP, Brazil. I—Pesquisa de helmintos. *Rev Saude Publica* 1992;26:283–289.
6. de Savigny DH, Voller A, Woodruff AW. Toxocariasis: serological diagnosis by enzyme immunoassay. *J Clin Pathol* 1979;32:284–288.
7. de Savigny DH. In vitro maintenance of Toxocara canis larvae and a simple method for the production of Toxocara ES antigen for use in serodiagnostic tests for visceral larva migrans. *J Parasitol* 1975;61:781–782.
8. de Souza EC, Nakashima Y. Diffuse unilateral subacute neuroretinitis: report of transvitreal surgical removal of a subretinal nematode. *Ophthalmology* 1995;102:1183–1186.
9. Dubinsky P, Havasiová-Reiterová K, Petko B, Hovorka I, Tomasovicová O. Role of small mammals in the epidemiology of toxocariasis. *Parasitology* 1995;110:187–193.
10. Embil JA, Tanner CE, Pereira LH, Staudt M, Morrison EG, Gualazzi DA. Seroepidemiologic survey of Toxocara canis infection in urban and rural children. *Public Health* 1988;102:129–133.
11. Fülleborn F. Askarisinfektion durch Verzehren eingekapselter Larven und uber gelungene intrauterine Askarisinfektion. *Archiv Schiffs Tropenhygiene* 1921;25:367–375.
12. Gillespie SH, Dinning WJ, Voller A, Crowcroft NS. The spectrum of ocular toxocariasis. *Eye* 1993;7:415–418.
13. Glickman LT, Magnaval JF. Zoonotic roundworm infections. *Infect Dis Clin North Am* 1993;7:717–732.
14. Glickman LT, Schantz P, Dombroske R, Cypress R. Evaluation of serodiagnostic tests for visceral larva migrans. *Am J Trop Med Hyg* 1978;27:492–498.
15. Goldberg MA, Kazacos KR, Boyce WM, Ai E, Katz B. Diffuse unilateral subacute neuroretinitis. Morphometric, serologic, and epidemiologic support for Baylisascaris as a causative agent. *Ophthalmology* 1993;100:1695–1701.
16. Gueglio B, de Gentile L, Nguyen JM, Achard J, Chabasse D, Marjolet M. Epidemiologic approach to human toxocariasis in western France. *Parasitol Res* 1994;80:531–536.
17. Humbert P, Buchet S, Barde T. Toxocariasis. A cosmopolitan parasitic zoonosis [in French]. *Allerg Immunol (Paris)* 1995;27:284–291.
18. Jeanfaivre T, Cimon B, Tolstuchow N, de Gentile L, Chabasse D, Tuchais E. Pleural effusion and toxocariasis. *Thorax* 1996;51:106–107.
19. Karel I, Peleska M, Uhlíková M, Hübner J. Larva migrans lentis. *Ophthalmologica* 1977;174:14–19.
20. Kayes SG. Human toxocariasis and the visceral larva migrans syndrome: correlative immunopathology. *Chem Immunol* 1997;66:99–124.
21. Komiyama A, Hasegawa O, Nakamura S, Ohno S, Kondo K. Optic neuritis in cerebral toxocariasis. *J Neurol Neurosurg Psychiatry* 1995;59:197–198.
22. Kraus A, Valencia X, Cabral AR, de la Vega G. Visceral larva migrans mimicking rheumatic diseases. *J Rheumatol* 1995;22:497–500.
23. Krcméry V Jr, Gould I, Sobota K, Spánik S. Two cases of disseminated toxocariasis in compromised hosts successfully treated with mebendazole. *Chemotherapy* 1992;38:367–368.
24. Lynch NR, Hagel I, Vargas V, et al. Comparable seropositivity for ascariasis and toxocariasis in tropical slum children. *Parasitol Res* 1993;79:547–550.
25. Magnaval JF. Comparative efficacy of diethylcarbamazine and mebendazole for the treatment of human toxocariasis. *Parasitology* 1995;110:529–533.
26. Magnaval JF, Fabre R, Maurieres P, Charlet JP, de Larrard B. Application of the western blotting procedure for the immunodiagnosis of human toxocariasis. *Parasitol Res* 1991;77:697–702.
27. Maruyama S, Yamamoto K, Katsube Y. Infectivity of Toxocara canis larvae from Japanese quails in mice. *J Vet Med Sci* 1994;56:399–401.
28. Nagakura K, Tachibana H, Kaneda Y, Kato Y. Toxocariasis possibly caused by ingesting raw chicken. *J Infect Dis* 1989;160:735-736.
29. Nichols RL. The etiology of visceral larva migrans. I. Diagnostic morphology of infective second-stage Toxocara larvae. *J Parasitol* 1956;42:349–362.
30. Nichols RL. The etiology of visceral larva migrans. II. Comparative larval morphology of Ascaris lumbricoides, Necator americanus, Strongyloides stercoralis, and Ancylostoma caninum. *J Parasitol* 1956;42:363–399.
31. Orihel TC, Ash LR. *Parasites in Human Tissues*. Chicago, Ill: ASCP Press; 1995:94-99.
32. Parsons JC. Ascarid infections of cats and dogs. *Vet Clin North Am Small Anim Pract* 1987;17:1307–1339.
33. Perlingiero JG, György P. Chronic eosinophilia: report of case with necrosis of the liver, pulmonary infiltrations, anemia and ascaris infestation. *Am J Dis Child* 1947;73:34–43.
34. Pollard ZF. Long-term follow-up in patients with ocular toxocariasis as measured by ELISA titres. *Ann Ophthalmol* 1987;19:167–169.
35. Polley L. Visceral larva migrans and alveolar hydatid disease. Dangers real or imagined. *Vet Clin North Am* 1978;8:353–378.
36. Shantz PM. Of worms, dogs, and human hosts: continuing challenges for veterinarians in prevention of human disease. *J Am Vet Med Assoc* 1994;204:1023–1028.
37. Shields JA. Ocular toxocariasis. A review. *Surv Ophthalmol* 1984;28:361–381.
38. Snyder CH. Visceral larva migrans: ten years' experience. *Pediatrics* 1961;28:85–91.
39. Stürchler D, Schubarth P, Gualzata M, Gottstein B, Oettli A. Thiabendazole vs. albendazole in treatment of toxocariasis: a clinical trial. *Ann Trop Med Parasitol* 1989;83:473–478.
40. Taylor MR, Keane CT, O'Connor P, Girdwood RW, Smith H. Clinical features of covert toxocariasis. *Scand J Infect Dis* 1987;19:693–696.
41. Taylor MR, O'Connor P, Hinson AR, Smith HV. Toxocara titres in maternal and cord blood. *J Infect* 1996;32:231–233.
42. Thompson DE, Bundy DA, Cooper ES, Schantz PM. Epidemiological characteristics of Toxocara canis zoonotic infection of children in a Caribbean community. *Bull World Health Organ* 1986;64:283–290.
43. Wilder HC. Nematode endophthalmitis. *Trans Am Acad Ophthalmol Otolaryngol* 1950;55:99–109.
44. Wolfrom E, Chêne G, Boisseau H, Beylot C, Géniaux M, Taïeb A. Chronic urticaria and Toxocara canis [letter]. *Lancet* 1995;345:196.
45. Woodruff AW, Bisseru B, Bowe JC. Infection with animal helminths as a factor in causing poliomyelitis and epilepsy. *Br Med J* 1966;1:1576–1579.
46. Zimmerman LE. The registry of ophthalmic pathology: past, present and future. *Trans Am Acad Ophthalmol Otolaryngol* 1961;65:51–113.
47. Zygulska-Mach H, Krukar-Baster K, Ziobrowski S. Ocular toxocariasis in children and youth. *Doc Ophthalmol* 1993;84:145–154.

28

Anisakiasis

Ellen M. Andersen *and*
J. Ralph Lichtenfels

Introduction

Definition

Anisakiasis is infection with larval nematodes belonging to several genera within the family Anisakidae. Humans are infected by ingesting raw, pickled, or salted fish contaminated with third- or fourth-stage larvae. Marine mammals are the normal definitive hosts.

General Considerations

Human anisakiasis was first described in the Netherlands in 1960 by Rodenburg and Wielinga[13] and by Kuipers et al.[8] The anisakid larvae causing these infections came from salted herring and were classified as *Eustoma rotundatum*. From a study of additional specimens, also in 1960, van Thiel et al reclassified these larvae as *Anisakis marina*.[17] The first patients from Japan and the first description of the pathology of anisakiasis were reported in 1965.[1] Davey renamed the species *Anisakis simplex* in 1971.[3]

Epidemiology

Human infection is found wherever raw, poorly cooked, salted, or pickled fish are consumed, particularly in Japan (sushi and sashimi), the Netherlands (green herring), and Latin America (ceviche). Other reports of human anisakiasis have come from Belgium, Canada, Denmark, France, Germany, New Zealand, Norway, Switzerland, Thailand, the United Kingdom, and the United States.

Adult anisakids, which produce larvae that are serious human pathogens, infect marine mammals (Fig 28.1) of the Atlantic and Pacific oceans and the North Sea. Several kinds of fish (cod, herring, mackerel, salmon, tuna, and yellowtail) and squid may transmit anisakid larvae to humans.[2,6,16] In places where eating sushi, sashimi, and ceviche has recently become popular, especially Canada, Europe, and the United States, the incidence of anisakiasis has increased.

423

Infectious Agents

At least 4 species are implicated in human anisakiasis, including *A. simplex* (Rudolphi, 1809), *Anisakis* type II, *Pseudoterranova* (*Phocanema*) *decipiens*, and *Contracaecum* sp. *Anisakis simplex* and *P. decipiens* cause most human infections. Based on genetic and morphometric analysis, *A. simplex* and *P. decipiens* may each represent a complex of several sibling species.[4,11,12]

Morphologic Description

Anisakid larvae from humans are stout, white or cream-colored worms (Fig 28.2) with a finely striated cuticle, 1 dorsal and 2 subventral reduced lips surrounding a mouth, and a triangular boring tooth. Larvae measure 10 to 50 mm long by 0.3 to 1.2 mm wide. Distinctive morphologic features of the digestive tract help to identify a specific worm (Fig 28.3). *Anisakis* sp have an esophagus and ventriculus, and an intestine with no projections; *Pseudoterranova* sp have an anterior projecting intestinal cecum (Fig 28.4); and *Contracaecum* sp have an isodiametric ventriculus with a posterior projecting ventricular appendix and an anterior projecting intestinal cecum (Figs 28.5 and 28.6). In *Contracaecum* sp, the ventricular appendix is in the anterior intestinal region. The excretory pore of all 3 types is at the base of the lips.

Microscopically, several features distinguish anisakid larvae from other invasive nematodes. The cuticle is multilayered and varies considerably in thickness from 5 to 50 μm (Figs 28.7 and 28.8). In cross section, larvae measure 250 to 800 μm in diameter. Muscle cells are large, numerous, U-shaped, and separated into quadrants of 60 to 90 cells each. A large excretory gland runs subventrally to ventrally, with its widest point at the level of the esophagus (Figs 28.8 and 28.9). Transverse intestinal sections show tall columnar cells: 60 to 80 cells in *A. simplex* (Fig 28.10) and more than 100 cells in *P. decipiens* (Fig 28.11) and *Contracaecum* sp. Lateral cords are prominent and either Y-shaped with a narrow base (*A. simplex*, Fig 28.12) or butterfly-shaped with a wide base (*Contracaecum* sp and *P. decipiens*, Fig 28.13). Diagnostic features of all 3 anisakid larvae are summarized in Table 28.1.

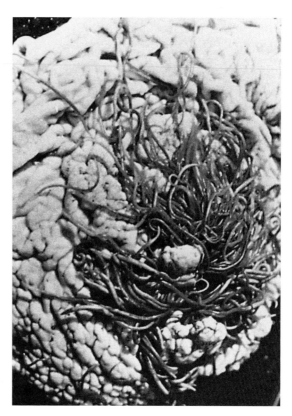

Figure 28.1
Adult *A. simplex* worms attached to first stomach of a dolphin.

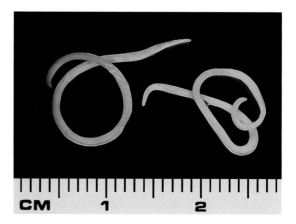

Figure 28.2
Two whole *Pseudoterranova* sp larvae collected from fresh market cod.

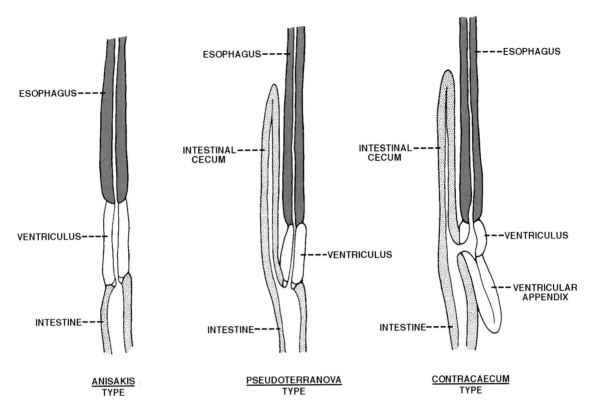

Figure 28.3
Comparative morphologic features of digestive tracts of anisakid larvae of *Anisakis, Pseudoterranova (Phocanema)*, and *Contracaecum* types.

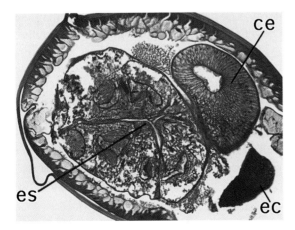

Figure 28.4
Cross section of *Pseudoterranova* sp larva. Note large esophagus (es), smaller intestinal cecum (ce), and small portion of excretory gland cell (ec). x75

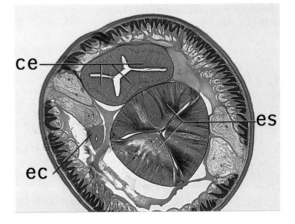

Figure 28.5
Cross section of *Contracaecum spiculigerum* larva showing centrally located muscular esophagus (es), intestinal cecum (ce), and portion of excretory gland cell (ec). ZN x25

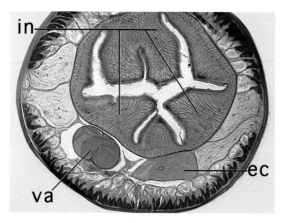

Figure 28.6
Cross section of *C. spiculigerum* larva demonstrating large central intestine (in) with numerous intestinal cells, ventricular appendix (va), and excretory gland cell (ec). ZN x30

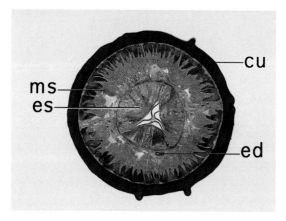

Figure 28.8
Cross section of anterior portion of *C. spiculigerum* larva showing thick cuticle (cu), central esophagus (es), prominent musculature (ms), and small posterior excretory duct (ed). The 3 cuticular projections are presumed to be artifacts. x35

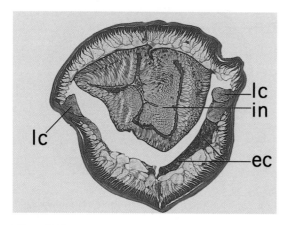

Figure 28.9
Cross section of *Pseudoterranova* sp larva. Note large central intestine (in), lateral cords (lc), and large excretory gland cell (ec). x175

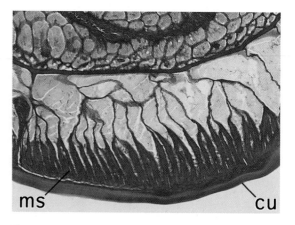

Figure 28.7
Body wall of *Pseudoterranova* sp larva showing multilayered, 7-to-9-µm-thick cuticle (cu), and prominent muscle cells (ms). x230

Life Cycle and Transmission

The life cycle of anisakids (Fig 28.14) involves marine crustaceans, fish, and mammals. Adult worms attached to the stomach or intestinal lining of marine mammals (dolphins, sea lions, seals, walrus, and whales) deposit eggs into the lumen of the host's gut (Fig 28.1), which are then passed in the feces. Larvae develop into the second or possibly third stage within the eggs.[7] When the eggs hatch, larvae are released and ingested by krill, such as *Euphausia* sp, or other marine crustaceans.[9] Fish or squid ingest the infected crustaceans and serve as paratenic hosts, wherein the larvae invade the body cavity or musculature, but develop no further. Larvae then pass up the food chain from fish to fish, potentially concentrating in large numbers within a single host. If a marine mammal eats an infected fish or squid, the larvae develop into adults in the mammal's gastrointestinal tract, mate, and produce eggs to complete their life cycle.

Although adult anisakids may develop in the wall of a human stomach, larvae usually do not mature in humans, so humans are dead-end hosts.[9]

Clinical Features and Pathogenesis

Clinical features depend on whether anisakid larvae only attach to the mucosa of the gastrointestinal tract, or proceed to invade tissues. After

Figure 28.10
Cross section of *A. simplex* larva from human omentum showing intestine with fewer than 80 columnar cells. x115

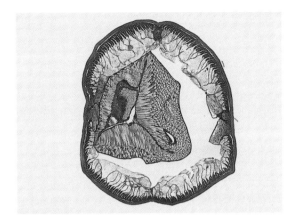

Figure 28.11
Cross section of *P. decipiens* larva at level of intestine. Note more than 100 intestinal cells. x160

invading the stomach, larvae sometimes migrate up the esophagus and attach to the throat, causing coughing or a tickling sensation, and may be expectorated or passed in the stool. When larvae penetrate the stomach wall, they provoke gastritis, causing severe epigastric pain, nausea, vomiting, hematemesis, diarrhea, urticaria, and chest pain. Symptoms usually develop 1 to 12 hours after eating infected fish. Gastric aspirate or vomitus may contain occult blood. Left undiagnosed, symptoms can last for months and mimic acute or chronic gastritis, gastric ulcer, or a gastric tumor.

The usual clinical presentation of intestinal anisakiasis, most frequently of the small bowel or cecum, is sudden abdominal pain with nausea, vomiting, and peritoneal irritation. Diarrhea may alternate with constipation and the stool may contain occult blood. Onset is usually within 48 hours of ingesting larvae. Leukocytosis is moderate. Eosinophilia is variable, ranging up to 40%. Intestinal invasion can mimic appendicitis or regional ileitis. The colon is seldom affected beyond the cecum.

Some of the earliest symptoms of anisakiasis may represent hypersensitivity (anaphylactic) reactions. Fernandez de Corres et al[5] reported 28 patients not infected with *A. simplex* who developed angioedema and urticaria soon after eating fish. One patient experienced respiratory arrest. All patients had positive skin tests, specific IgE reaction, and histamine release using *A. simplex* extracts. The allergen resisted cooking and deep freezing.

Figure 28.12
Cross section of *A. simplex* larva showing prominent Y-shaped lateral cord with narrow base (arrow). x583

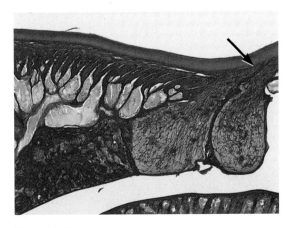

Figure 28.13
Cross section of *P. decipiens* larva showing prominent, butterfly-shaped lateral cords with wide base (arrow). x225

Species	Intestine	Number of intestinal cells in cross section	Lateral cords
A. simplex	Simple; no cecum or appendix	60–80	Y-shaped
P. decipiens	Cecum	> 100	Butterfly-shaped
Contracaecum sp	Cecum and ventricular appendix	> 100	Butterfly-shaped

Table 28.1
Diagnostic features of *A. simplex*, *P. decipiens*, and *Contracaecum* sp.

Pathologic Features

Pathologic changes differ depending on the stage of infection. In early infection, suppuration may surround a well-preserved larva. These neutrophilic infiltrates may contain a few eosinophils and giant cells, accompanied by minimal edema, hemorrhage, and erosion. This stage is followed by a more intensive infiltration of eosinophils, intermixed with lymphocytes, monocytes, neutrophils, and plasma cells (Figs 28.15 and 28.16). Viable larvae can cause a massive tissue eosinophilia and invade blood vessels (Figs 28.17 and 28.18). There may be ulceration, edematous thickening of the mucosa, and hemorrhage (Fig 28.19).

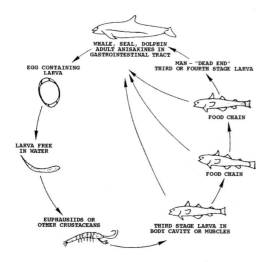

Figure 28.14
Life cycle of anisakids. 1) Eggs laid by adult worms in stomach lumen of a marine mammal are passed in feces and embryonate in seawater. 2) Larvae develop to second or possibly third stage. 3) Eggs hatch. 4) Larvae are ingested by small crustaceans such as krill. 5) Parasitized krill are ingested by marine fish or squid and larvae encyst at third stage in musculature or body cavity of fish. 6) and 7) Larvae are passed up food chain from fish to fish without further development, until ingested by a marine mammal.

When the larva dies, an abscess with abundant eosinophils and peripheral granulation tissue forms around it (Figs 28.20 and 28.21). Eventually, a granuloma develops, containing eosinophils and perhaps remnants of the degenerating larva (Fig 28.22).

Diagnosis

Although histologic sections are helpful in diagnosis, they are not as definitive as an entire larva retrieved through endoscopy. Because larvae rather than adults are infective, identification can only be certain at the generic level. Most infections are caused by *A. simplex*, *P. decipiens*, or *Contracaecum* sp.

Several clinical diagnostic techniques are available. Endoscopy, whereby larvae can be directly visualized and removed (Fig 28.23), is extremely useful in diagnosing and treating anisakiasis.[10] Ultrasonography may show thickened loops of bowel, luminal narrowing, and decreased peristalsis.[14] Radiologic findings in the bowel include thickening of the wall, narrowing of the lumen, and even obstruction of affected areas (Fig 28.24).

Many serologic assays have been evaluated, including immunofluorescent antibody tests, enzyme-linked immunosorbent assays, radioallergosorbent tests, and complement fixation tests. Cross-reactivity with other nematodes such as *Ascaris* sp or *Toxocara* sp limits the specificity of immunologic assays. A microenzyme-linked immunosorbent assay using monoclonal antibodies which specifically recognize *A. simplex* larvae has been reported.[18] Several polypeptides from excretory-secretory and somatic antigens have been reported to react with IgG in sera from patients with anisakiasis, but not with sera from patients

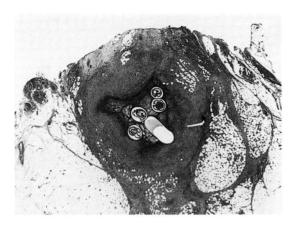

Figure 28.15
Anisakis sp in nodule in omentum. Note multiple sections of coiled viable larva walled off by chronic inflammation. Movat x5.9

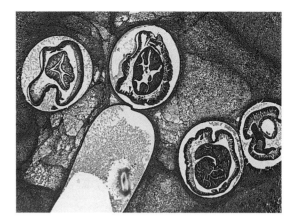

Figure 28.16
Higher magnification of specimen shown in Figure 28.15. Note central suppuration and morphologic features of anisakid larva. x35

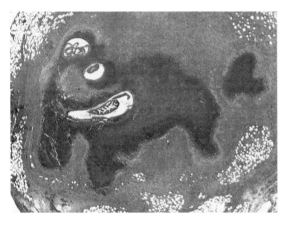

Figure 28.17
Anisakiasis in 27-year-old Russian fisherman operated on for possible appendicitis. Surgery revealed viable larva provoking eosinophilic abscess in omentum. x11

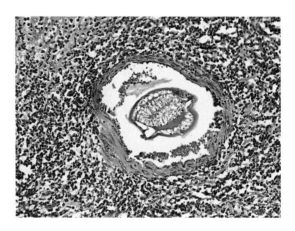

Figure 28.18
Anisakiasis in 31-year-old patient from Thailand. Note massive tissue eosinophilia in wall of ileum. Viable larva has penetrated vessel lumen. x115

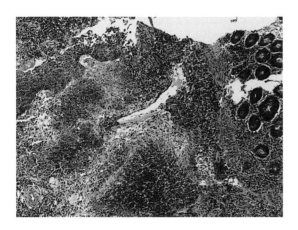

Figure 28.19
Cecal mass from patient with anisakiasis. Note obliteration of mucosal folds, and necrosis, edema, and prominent eosinophilia. x60

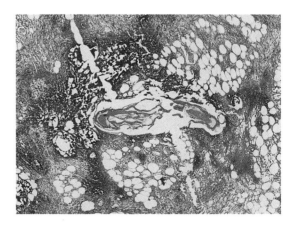

Figure 28.20
Cecal mass from patient with anisakiasis. Note degenerated *Anisakis* sp larva in pericolic fat. x22

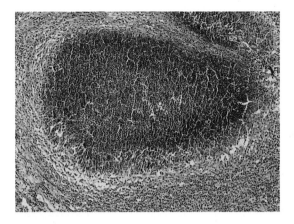

Figure 28.21
Cecal mass from patient with anisakiasis depicting eosinophilic granuloma in submucosa. Parasite is not visible at this level. x60

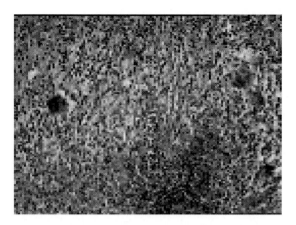

Figure 28.22
Higher magnification of wall of eosinophilic granuloma in Figure 28.21. Note necrosis, epithelioid cells, giant cells, and eosinophils. x115

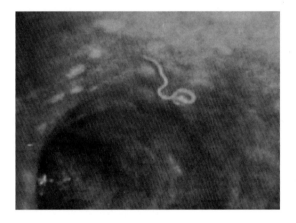

Figure 28.23
Anisakid larva in prepyloric mucosa of stomach as seen at gastroscopy.

with other nematode infections.[15] The usefulness of these tests for specific diagnosis has not been determined.

Treatment

Early removal of invading larvae is curative and prevents the development of chronic inflammation. In gastric or intestinal anisakiasis, larvae may be removed endoscopically or by surgical resection. Corticosteroids may decrease the inflammatory response to larvae, but no effective anthelmintic drugs are available.

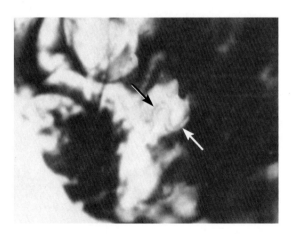

Figure 28.24
Anisakid larva (arrows) on gastric mucosa, revealed radiographically by air contrast.

References

1. Asami K, Watanuki T, Sakai H, Imano H, Okamoto R. Two cases of stomach granuloma caused by Anisakis-like larval nematodes in Japan. *Am J Trop Med Hyg* 1965;14:119–123.
2. Chitwood M. Nematodes of medical significance found in market fish. *Am J Trop Med Hyg* 1970;19:599–602.
3. Davey JT. A revision of the genus Anisakis Dujardin, 1845 (Nematoda: Ascaridata). *J Helminthol* 1971;45:51–72.
4. di Deco MA, Orecchia P, Paggi L, Petrarca V. Morphometric stepwise discriminant analysis of three genetically identified species within Pseudoterranova decipiens (Krabbe, 1878) (Nematoda: Ascaridida). *Sys Parasitol* 1994;29:81–88.
5. Fernandez de Corres L, Audicana M, Del Pozo MD, et al. Anisakis simplex induces not only anisakiasis: report on 28 cases of allergy caused by this nematode. *J Investig Allergol Clin Immunol* 1996;6:315–319.
6. Jackson GJ, Bier JW, Payne WL, Gerding TA, Knollenberg WG. Nematodes in fresh market fish of the Washington, D.C. area. *J Food Protection* 1978;41:613–620.
7. Koie M, Fagerholm HP. Third-stage larvae emerge from eggs of Contracaecum osculatum (Nematoda, Anisakidae). *J Parasitol* 1993;79:777–780.
8. Kuipers FC, van Thiel PH, Roskam ET. Eosinophilic phlegmon of the small intestine caused by a worm not adapted to the human body [in German]. *Ned Tijdschr Geneeskd* 1960;104:422–427.
9. Maejima J, Fukumoto S, Yazaki S, Hirai K, Hasegawa H, Takagi H. Morphological features of an adult of Pseudoterranova decipiens (Krabbe, 1878) found in human stomach wall. *Jpn J Parasitol* 1992;41:420–424.
10. Namiki M, Yazaki Y. Treatment of gastric anisakiasis with acute symptoms. In: Ishikura H, Namiki M, eds. *Gastric Anisakiasis in Japan: Epidemiology, Diagnosis, Treatment*. Tokyo, Japan: Springer-Verlag; 1989:129–132.
11. Nascetti G, Paggi L, Orecchia P, Mattiucci S, Bullini L. Two sibling species within Anisakis simplex (Ascaridida: Anisakidae). *Parassitologia* 1983;25:306–307.
12. Paggi L, Nascetti G, Cianchi R, et al. Genetic evidence for three species within Pseudoterranova decipiens (Nematoda, Ascaridida, Ascaridoidea) in the North Atlantic and Norwegian and Barents Sea. *Int J Parasitol* 1991;21:195–212.
13. Rodenburg W, Wielinga WJ. Eosinophilic phlegmon of the small intestine caused by a worm [in Dutch]. *Ned Tijdschr Geneeskd* 1960;104:417–421.
14. Shirahama M, Koga T, Ishibashi H, Uchida S, Ohta Y, Shimoda Y. Intestinal anisakiasis: US in diagnosis. *Radiology* 1992;185:789–793.
15. Sugane K, Sun SH, Matsuura T. Radiolabelling of the excretory-secretory and somatic antigens of Anisakis simplex larvae. *J Helminthol* 1992;66:305–309.
16. Sun SZ, Koyama T, Kagei N. Anisakidae larvae found in marine fishes and squids from the Gulf of Tongking, the East China Sea and the Yellow Sea. *Jpn J Med Sci Biol* 1991;44:99–108.
17. van Thiel PH, Kuipers FC, Roskam RT. A nematode parasitic to herring causing acute abdominal syndromes in man. *Trop Geogr Med* 1960;12:97–113.
18. Yagihashi A, Sato N, Takahashi S, Ishikura H, Kikuchi K. A serodiagnostic assay by microenzyme-linked immunosorbent assay for human anisakiasis using a monoclonal antibody specific for Anisakis larvae antigen. *J Infect Dis* 1990;161:995–998.

29

Enterobiasis

Juvady Leopairut, Ronald C. Neafie,
Wayne M. Meyers, and
Aileen M. Marty

Introduction

Definition
Enterobiasis is infection by the nematode *Enterobius vermicularis* (Linnaeus, 1758) Leach, 1853.

Synonyms
Older names for *E. vermicularis*, the pinworm or seatworm, are *Ascaris vermicularis* and *Oxyuris vermicularis*. *Enterobius gregorii* Hugot, 1983 probably represents a balanced polymorphism of *E. vermicularis*. Enterobiasis is synonymous with human oxyuriasis, pinworm disease, and threadworm infection.

General Considerations
Throughout history, enterobiasis has been the subject of countless clinical, parasitological, and epidemiologic studies. Archaeological studies confirm that enterobiasis has pestered humanity for over 10 000 years.[21] The Egyptian *Papyrus Ebers* (circa 1550 BC) describes "*Herxetef*," a worm that may represent *E. vermicularis*. Many ancient Greek, Roman, and Arabic writers mention this worm,[23] including Hippocrates, who described the cardinal clinical aspects of enterobiasis (perianal itching and involvement of the female genital tract) over 2500 years ago.[25] In 1758 Linnaeus classified this parasite as *A. vermicularis*; Bremser recognized it as an oxyurid in 1819. In 1865 Leuckart and 3 of his students swallowed mature *Enterobius* eggs and subsequently developed severe infections, proving that an intermediate host was unnecessary and settling a protracted dispute on the life cycle of *Enterobius*. In 1937 Hall introduced the so-called NIH swab for the diagnosis of enterobiasis. In 1956 investigators identified the first efficacious medication, pyrvinium pamoate. Brugmans et al in 1971 demonstrated the therapeutic value of mebendazole.[11]

From his observations on Grégoire Hugot's collection of human pinworm infections in France, J.P. Hugot, in 1983, concluded that *E. vermicularis* has a relative species, which he described and named *E. gregorii*.[26] Studies from around the world

Figure 29.1
Gravid *E. vermicularis* (6 mm by 400 μm). Note egg-filled uteri taking up most of the body cavity, vulva (arrow) in anterior portion, and long pointed tail. x35

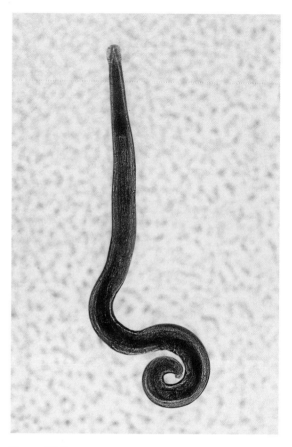

Figure 29.2
Adult male *E. vermicularis* (3.5 mm by 170 μm) with posterior end curved ventrally. Tail is blunt. x45

report that most enterobiasis patients have dual infections with *E. vermicularis* and *E. gregorii*, leading some investigators to speculate that differences between males of the putative 2 species are spurious: the smaller worms are merely younger, and variations in spicules represent developmental stages.[24] These differences also suggest the possibility of a balanced polymorphism with 2 male morphologies.[5] Other observers conclude that the 2 species live sympatrically in humans.[16] There is a single report of another oxyurid infecting a human. In 1920 Riley described an infection with the mouse pinworm *Syphacia obvelata* (Rudolphi, 1802) in a young American girl living in the Philippines.[39] In 1957 Hussey clearly differentiated *S. obvelata* from *Syphacia muris* (Yamaguti, 1935), the rat pinworm.[27] *Syphacia muris* and another oxyurid, *Aspiculuris tetraptera* (Nitzsch, 1821) Schulz, 1924, are potential human parasites.[2,3]

Epidemiology

Enterobius vermicularis is cosmopolitan and the most common parasitic helminth of humans in temperate, developed countries.[28] In colder climates, factors such as less exposure to sunlight, heavy clothing, and fewer baths lead to a higher prevalence of enterobiasis, especially in children. In the United States, *E. vermicularis* affects approximately 30% of children and 16% of adults.[43] Infection tends to spread in social groups such as families, summer campers, and institutionalized persons. Group infections usually reflect poor personal hygiene, overcrowding, and substandard sanitation, but at any socioeconomic level, families with 2 or more children can expect at least 1 bout of enterobiasis.

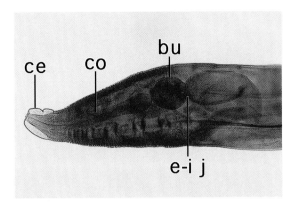

Figure 29.3
Anterior end of adult female *E. vermicularis* displaying 2 vesicular cephalic expansions (ce), corpus (co), bulb (bu), and esophageal-intestinal junction (e-i j). Carmine x60

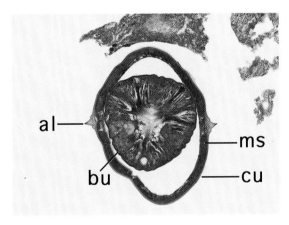

Figure 29.5
Adult male *E. vermicularis* in lumen of appendix illustrating alae (al), cuticle (cu), muscle (contractile portion) (ms), and esophageal bulb (bu). x245

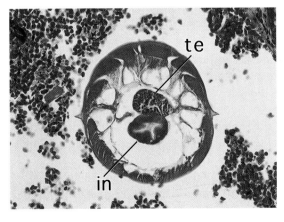

Figure 29.7
Anterior end of adult male *E. vermicularis* in lumen of appendix at level of intestine (in) and testis (te). x235

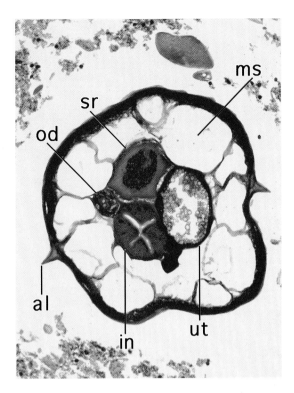

Figure 29.4
Adult nongravid female *E. vermicularis* in colon demonstrating lateral alae (al), meromyarian muscle (sarcoplasmic portion) (ms), intestine (in), uterus (ut), seminal receptacle (sr), and oviduct (od). x195

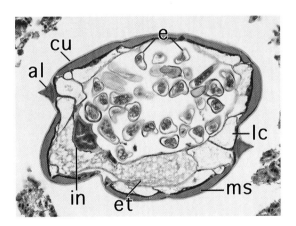

Figure 29.6
Gravid *E. vermicularis* in lumen of appendix. Section is through anterior half of worm and demonstrates alae (al), cuticle (cu), lateral cords (lc), muscle (contractile portion) (ms), excretory tissue (et), intestine (in), and uterus with eggs (e). Movat x190

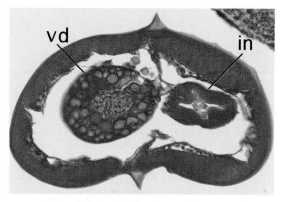

Figure 29.8
Adult male *E. vermicularis* in lumen of appendix through posterior half of worm at level of intestine (in) and vas deferens (vd) containing sperm. x530

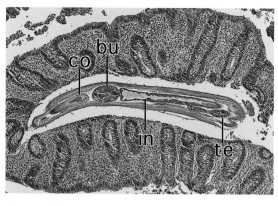

Figure 29.9
Longitudinal section of anterior region of adult male *E. vermicularis* in lumen of appendix depicting corpus (co), esophageal bulb (bu), intestine (in), and testis (te). x55

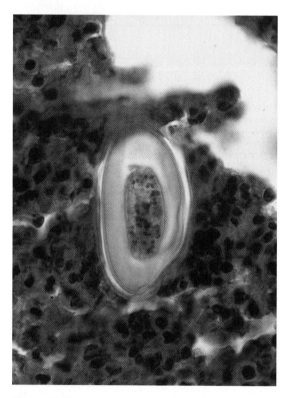

Figure 29.10
Egg of *E. vermicularis* in area of necrosis in fallopian tube. Egg is thick-shelled, flattened on 1 side, and contains a larva. x830

Infectious Agent

Morphologic Description

Adult *E. vermicularis* are small, spindle-shaped, white to yellow roundworms. Adult females (Fig 29.1) are 8 to 13 mm by 0.3 to 0.5 mm; adult males (Fig 29.2) are 2 to 5 mm by 0.1 to 0.2 mm. The anterior end lacks a true buccal capsule, but has 3 lips and 2 vesicular cephalic expansions (Fig 29.3). The striated cuticle is 1 to 5 µm thick and contains narrow lateral alae that extend most of the length of the worm (Figs 29.4 and 29.5). The esophagus is composed of a corpus, isthmus, and prominent bulb (Fig 29.3). The intestine varies from cuboid to columnar cells, each with a single large nucleus and conspicuous microvilli. Lateral cords are readily identified, but are usually vacuolated and inconspicuous. Somatic muscles are platymyarian and meromyarian, with 2 or 3 muscle cells in each quadrant of the worm.

Female worms have paired reproductive organs at the midsection of the body that are conspicuous in the nongravid worm (Fig 29.4). In gravid worms (Fig 29.6), eggs distend the uteri, which fill most of the body cavity and distort other reproductive organs. The vulva is located in the anterior half of the worm, posterior to the esophageal-intestinal junction (Fig 29.1).

Male worms have a single reproductive tube (Figs 29.7 and 29.8) that begins slightly posterior to the esophageal-intestinal junction (Fig 29.9) and continues posteriorly as a straight tube to the end of the worm, where it unites with the intestine to form the cloaca. The posterior end of the male has a marked ventral curve (Fig 29.2). The male has a single copulatory spicule about 70 µm long.

Eggs of *E. vermicularis* are 50 to 60 µm by 20 to 30 µm, elongate, and flattened on 1 side (Figs 29.10 and 29.11). The shell is colorless, smooth, thick, and composed of 2 layers, an outer albuminous covering and an inner, embryonic, lipoid membrane. Eggs are embryonated when laid and reach the infective stage within a few hours.

Life Cycle and Transmission

Enterobius vermicularis takes about 6 weeks to complete its life cycle (Fig 29.12). Humans are the natural host for *E. vermicularis* and there is no intermediate host. *Enterobius vermicularis* has no tissue phase; adult worms rarely penetrate human tissue.

When ingested, infective eggs hatch in the stomach, releasing larvae that pass into the upper small intestine where they grow rapidly, molt, and develop into adult worms. Worms in all stages of development inhabit the distal small intestine, cecum, and appendix. Male worms rarely migrate from the jejunum and ileum, and die soon after copulation. Gravid worms migrate to the rectum to deposit embryonated eggs on the perianal and perineal skin. External migration begins soon after the patient is asleep and lasts about 3 hours. Worms that wander into the female genitalia apparently do not return to the bowel. Eggs remain viable for 2 weeks under normal conditions and are resistant to most disinfectants.

Infective *E. vermicularis* eggs are transmitted in 3 ways. First, anal pruritus and subsequent scratching promote direct anal-oral transmission by contaminated fingers. Second, the outer albuminous layer of the shell allows eggs to adhere to any surface, promoting contamination of food, dust, pet fur, bedding, clothing, and other fomites that come in contact with human hands. Third, inhaling or swallowing airborne eggs dislodged from fomites may lead to infection. There is no convincing evidence of internal autoinfection.

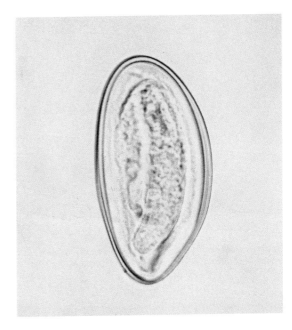

Figure 29.11
Egg of *E. vermicularis* in feces. Egg is thick-shelled, flattened on 1 side, and contains a larva. Unstained x1140

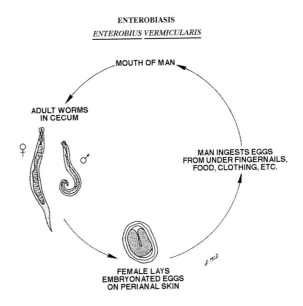

Figure 29.12
Simplified life cycle of *E. vermicularis*. No intermediate host is necessary.

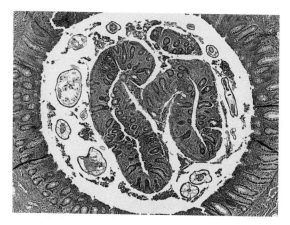

Figure 29.13
Thirteen sections of adult male and female *E. vermicularis* in lumen of appendix. Section of mucosa in lumen is an artifact. x25

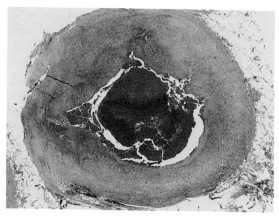

Figure 29.14
Eosinophilic coagulum in lumen of appendix with degenerated gravid *E. vermicularis*. Note mucosal perforation and transmural eosinophilic inflammation. x8

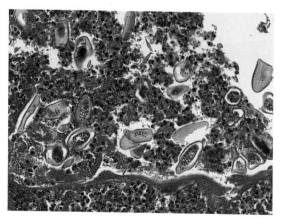

Figure 29.15
Eggs of *E. vermicularis* in lumen of appendix from patient described in Figure 29.14. Note eosinophilia and Charcot-Leyden crystals. x245

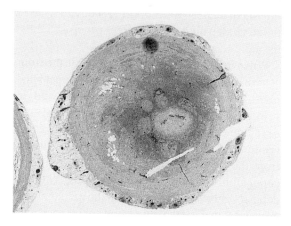

Figure 29.16
Degenerated gravid *E. vermicularis* provoking granuloma that totally obstructs lumen of appendix. x10

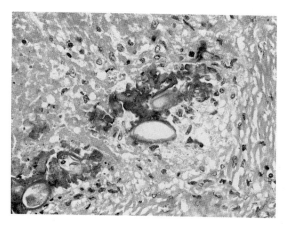

Figure 29.17
Eggs from degenerated gravid *E. vermicularis* in center of granuloma from patient in Figure 29.16. Note Splendore-Hoeppli reaction around eggs. x245

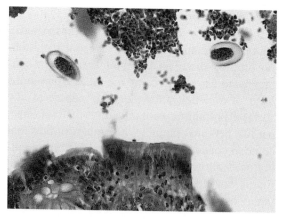

Figure 29.18
Two eggs of *E. vermicularis* lying free in lumen of appendix. x240

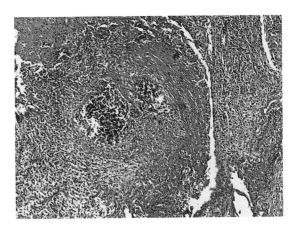

Figure 29.19
Fallopian tube obstructed by acute and chronic inflammation caused by several eggs of a degenerating, gravid *E. vermicularis*. x60

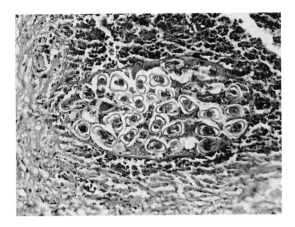

Figure 29.20
Fallopian tube obstructed by degenerating gravid *E. vermicularis* from patient in Figure 29.19. x175

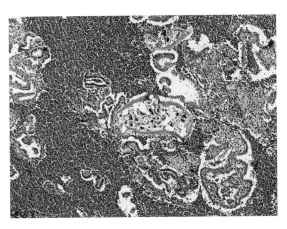

Figure 29.21
Endometrial curettage with marked inflammation caused by gravid *E. vermicularis*. x60

Figure 29.22
Cervical os obstructed by necrotizing granuloma caused by eggs from degenerated gravid *E. vermicularis*. x25

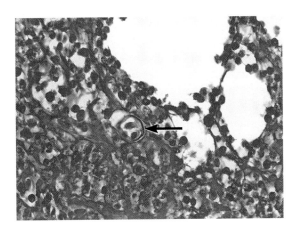

Figure 29.23
Higher magnification of cervical tissue in Figure 29.22 revealing egg (arrow) of *E. vermicularis* centered in eosinophilic coagulum. x310

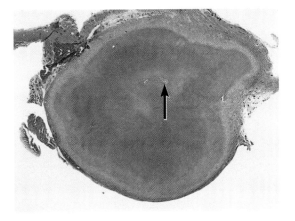

Figure 29.24
Subserosal uterine nodule caused by degenerating gravid *E. vermicularis* (arrow). x5.4

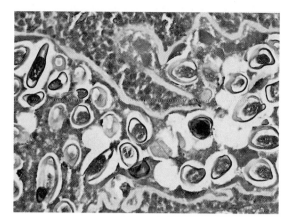

Figure 29.25
Higher magnification of degenerating gravid *E. vermicularis* in Figure 29.24. x230

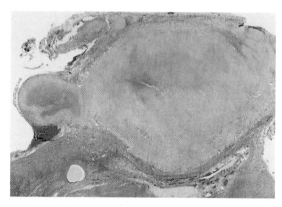

Figure 29.26
Large necrotizing, eosinophilic granuloma in ovary caused by eggs from degenerated gravid *E. vermicularis*. x5.3

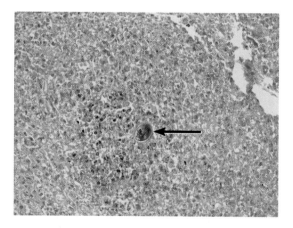

Figure 29.27
Higher magnification of Figure 29.26 demonstrating egg (arrow) of *E. vermicularis* in necrosis. x150

Clinical Features and Pathogenesis

Patients commonly experience anal pruritus and eczematous dermatitis caused by the migration of gravid worms onto the perianal and perineal skin. Some patients suffer rectal colic, restless sleep, irritability, emotional instability, allergic reactions, inattentiveness, and enuresis. Adult worms typically inhabit the lumen of the lower small intestine, cecum, and upper colon. Worms in the lumen of the intestine provoke no clinical symptoms. Worms near the appendix may obstruct or penetrate the appendiceal lumen.[50] Pinworms in the lumen can cause appendiceal pain, especially if there is a high worm load, but there is no clear relationship to acute appendicitis; worms are found more frequently in noninflamed than in inflamed appendices.[5,49] However, studies do reveal a substantially higher incidence of *E. vermicularis* in patients with chronic appendicitis.[12] Worms sometimes invade the intestinal mucosa, causing an inflamed appendix with petechial hemorrhage. Worms may also infiltrate the appendiceal mucosa.[22,47]

Adult *E. vermicularis* or their eggs can sometimes localize to sites other than the lumen of the intestine. Ectopic enterobiasis may be asymptomatic or symptomatic, depending on the tissue or organ involved, and is most serious in girls and women. Female worms sometimes migrate from the gastrointestinal tract into the female genital tract and deposit eggs. Eggs and/or worms in these sites can cause vulvitis,[45] vaginitis with vaginal discharge,[36] cervicitis,[14] endometritis,[37,40,44] salpingitis,[4] oophoritis,[8,33] peritonitis,[10,13,32] and infertility.[38] In pregnant women, *E. vermicularis* can pass through the genital tract and enter the embryo.[34] Invasion of the peritoneal area through the female genital tract may produce a significant peripheral eosinophilia. In females, the urethra is more susceptible to infection than in males. Worms in the urethra can cause dysuria from urethritis. Migrating worms carry enteric bacteria into the urinary bladder, causing urinary tract infection.[30,41]

Pulmonary involvement may incite a noncalcified coin lesion visible on chest radiograph.[7] Patients with infection of the intestinal wall,[15,48] esophagus, and/or stomach present with intestinal symptoms. Direct penetration by *E. vermicularis* of the colonic wall damaged, for example, by an adenocarcinoma may cause peritoneal granuloma. Invasion of perianal tissues may produce folliculitis[20] and subcutaneous abscess.[42] Infection may also develop in the liver,[17,31,35] spleen,[47] or lymph node.[18] One teenager repeatedly discharged adult worms from the conjunctival sac.[19] While ectopic lesions in men and boys are rare, patients with involvement of the urethra,[1] urinary bladder,[30] kidney, or ureter may present with urinary tract infection.[41] Worms invading the prostate gland may produce prostatitis.[46]

Clinical disease varies with worm load, which may be in the tens of thousands.[9] Most patients present with a single nocturnally migrating worm.

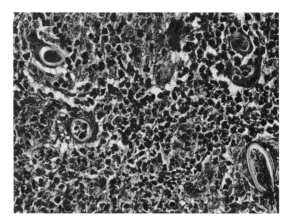

Figure 29.28
Granuloma in omentum demonstrating 3 eggs of *E. vermicularis* in giant cells. Note massive eosinophilia. x235

Pathologic Features

Enterobius vermicularis rarely causes serious lesions; an incidental finding in the lumen of the appendix is most common (Fig 29.13). The worm often produces minute ulcers with catarrhal inflammation of the intestinal mucosa. Though the role of *E. vermicularis* in appendicitis is controversial, there are cases on file at AFIP, and reports in the literature, of rare instances when *E. vermicularis* has invaded the intestinal wall and purportedly caused appendicitis. In these instances, there is significant eosinophilia with mucosal perforation and transmural inflammation of the appendix (Figs 29.14 and 29.15). Degenerated adult worms and their eggs may provoke granulomas that obstruct the lumen of the appendix and lead to appendicitis (Figs 29.16 and 29.17). Necrotizing granulomas containing degenerated worms and their eggs in necrotic centers are composed of palisading epithelioid cells surrounding a necrotic infiltrate of neutrophils, lymphocytes, macrophages, plasma cells, eosinophils, and worm elements (Fig 29.17). Intact and degranulating eosinophils are abundant and may produce numerous Charcot-Leyden crystals. Splendore-Hoeppli material may form around eggs (Fig 29.17).[33] Rarely, *E. vermicularis* eggs are free in the lumen of the appendix (Fig 29.18). Worms within the intestinal wall, with accompanying pyogenic bacteria, produce suppurative inflammation with abscesses. A macroscopic, circumscribed nodule may form in

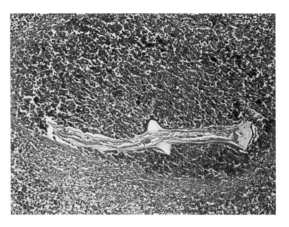

Figure 29.29
Degenerated female *E. vermicularis* in necrotic lesion of omentum. x120

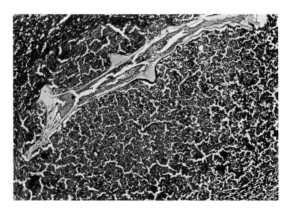

Figure 29.30
Section of female *E. vermicularis* depicted in Figure 29.29 demonstrating green staining of lateral alae. Movat x120

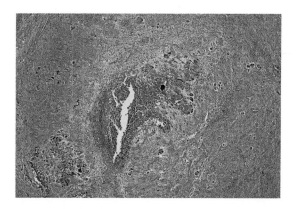

Figure 29.31
Perianal abscess with granulomatous component caused by eggs of *E. vermicularis*. x15

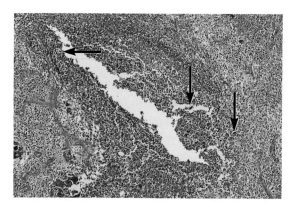

Figure 29.32
Higher magnification of perianal abscess in Figure 29.31 showing eggs of *E. vermicularis* in eosinophilic infiltrate (arrows). x35

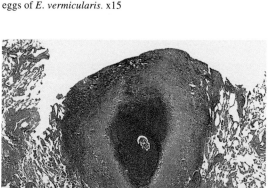

Figure 29.33
Degenerating gravid *E. vermicularis* causing necrotizing granuloma in lung. x15

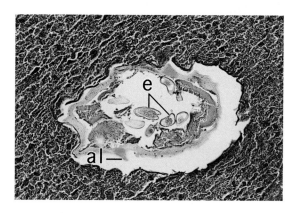

Figure 29.34
Higher magnification of degenerating gravid *E. vermicularis* in lung (Figure 29.33) revealing green staining of lateral alae (al) and eggs in uterus (e). Movat x160

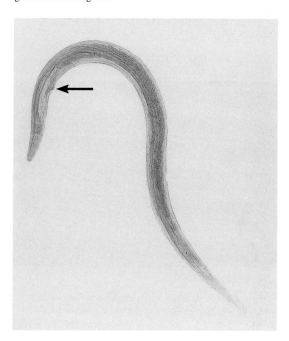

Figure 29.35 ◀
Gravid *Syphacia obvelata* (4.3 mm by 200 μm) with protruding vagina (arrow). x35

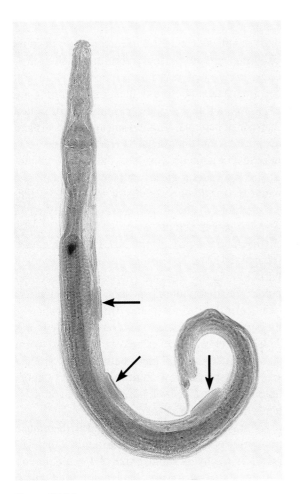

Figure 29.36
Adult male *S. obvelata* (1.1 mm by 68 μm). Note position of 3 cuticular mamelons (arrows) and curved tail. x180

Figure 29.37
Gravid *Syphacia muris* (3 mm by 115 μm) with protruding vagina (arrows). Carmine x40

chronic infections. Sclerotic nodules may develop in long-standing infections.

Serious complications may ensue when adult female *E. vermicularis* migrate into the human female genital tract, where they can incite significant tissue reactions that lead to infertility (Figs 29.19 to 29.23). Degenerating adults and eggs provoke lesions similar to foreign body granulomas (Figs 29.24 and 29.25). The eggs are in the center of granulomas consisting of epithelioid cells, multinucleated giant cells, macrophages, and eosinophils (Figs 29.26 and 29.27). In some lesions there is a heavy infiltration of eosinophils with no granuloma.[37]

The characteristic appearance of eggs (Fig 29.28) and the lateral alae in degenerating adult worms help identify *E. vermicularis* in ectopic locations (Figs 29.29 and 29.30). Lateral alae contain acid mucopolysaccharides that are bright green with the Movat stain (Fig 29.6), which is especially helpful in identifying alae in degenerated worms (Fig 29.30). In most ectopic lesions only eggs remain, since they are more resistant to degeneration. Eggs in ectopic sites indicate a gravid worm either close by or recently degenerated. On the other hand, some observers suggest that most granulomas form around a mass of eggs laid while the worm is in transit to more distant sites. Gravid worms or their eggs may be present in the peritoneum, omentum (Figs 29.28 to 29.30), urinary bladder, lymph node, spleen, liver, ovary, kidney, or perianal skin (Figs 29.31 and 29.32). In very rare instances, gravid worms can cause pulmonary lesions (Figs 29.33 and 29.34).

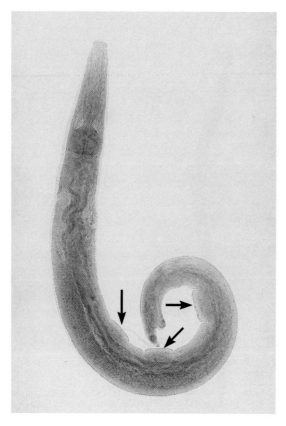

Figure 29.38
Adult male *S. muris* (0.75 mm by 60 µm). Note position of 3 cuticular mamelons (arrows) and curved tail. x215

Diagnosis

Clinically, a diagnosis of enterobiasis is usually suggested by perianal itching. Graham's Scotch® tape method is the standard diagnostic technique for identifying eggs, adult worms, or fragments of adult worms in the perianal region. Pressing a 3-inch length of tape attached sticky-side-out to the rounded end of a tongue depressor or microscope slide against the perianal skin captures a specimen. The tape is then placed sticky-side-down on a clean slide on which there is a drop of toluene, which clears the specimen.[6] Low-power microscopic examination reveals typical eggs (flattened on 1 side and containing a larva) or adult worms. Alternative methods such as the pestle or petroleum jelly-paraffin swab technique are preferable for men with abundant perianal hair.[51] A single sample detects infection only half the time, but 5 daily samples detect 99% of infections.[29] In some children, eggs are present intermittently for 1 to 4 days, so repeated sampling is essential. Morning specimens taken before defecation or bathing are most productive. Eggs are present in feces in less than 5% of patients. Vaginal smears, endocervical aspirations, and endometrial curettage (Fig 29.21) may reveal eggs or worms in the female genital tract. Histopathologic examination is diagnostic when *E. vermicularis* adults or eggs are present. Tissue biopsy is the only means of identifying *E. vermicularis* in ectopic sites.

Differentiating *E. vermicularis* from its recently recognized sister species, *E. gregorii*, is based on variations in the size and shape of the male spicule.[26] The distal tubular portion of the spicule is identical in these forms (about 70 µm long), whereas the basal portion in *E. vermicularis* is much longer.[24] This difference, however, may be related to worm size and maturation. Hugot believes that spicule size is independent of adult worm size.[26] Others note that adult *E. vermicularis* males are consistently larger than *E. gregorii* males.[5]

Adult female *S. obvelata* (Fig 29.35), the mouse pinworm, is distinguished from adult female *E. vermicularis* by its smaller size (3.4 to 5.8 mm by 240 to 400 µm), protruding vagina, and much larger eggs (118 to 153 µm by 33 to 55 µm). Adult male *S. obvelata* (Fig 29.36) differs from adult male *E. vermicularis* by its smaller size (1.1 to 1.5 mm by 120 to 140 µm) and the presence of 3 cuticular mamelons on the ventral surface.

Adult female *S. muris* (Fig 29.37), the rat pinworm, differs from adult female *E. vermicularis* by its smaller size (2.8 to 3.4 mm by 180 to 250 µm), protruding vagina, and larger eggs (72 to 82 µm by 25 to 36 µm). Adult male *S. muris* (Fig 29.38) differs from adult male *E. vermicularis* by its smaller size (1.2 to 1.3 mm by 100 µm) and the presence of 3 cuticular mamelons on the ventral surface. The anterior mamelon in the male *S. muris* is near the middle of the body, whereas in the male *S. obvelata* the center mamelon is near the middle of the body.

Treatment and Prevention

Treatment for pinworm infection is indicated if there is evidence that the infection is continuing. The drug of choice is mebendazole (Vermox®) given as a single dose of 100 mg. Pyrantel pamoate, pyrvinium pamoate, and piperazine citrate are effective but have more adverse side effects. Mebendazole and pyrvinium pamoate remove worms at all stages of development, whereas pyrantel pamoate and piperazine are ineffective against larvae.[23] Thiabendazole is indicated when enterobiasis is accompanied by strongyloidiasis, cutaneous larva migrans, or visceral larva migrans. Occasionally, enterobiasis patients treated with thiabendazole require additional therapy. Lesions around the anus caused by scratching are treated with palliative ointments. Treating all family members is recommended.

Neither chlorinated water nor fumigation destroys *E. vermicularis* eggs. Sunlight and ultraviolet radiation kill eggs in the environment, and dry heat sterilizes nonwashable items such as toys. Anything that improves personal and group hygiene lessens or eliminates the risk of infection: hand washing (especially after stool), cleaning of clothing and living quarters, wearing nightclothes that offer full coverage, and keeping fingernails short and clean.

References

1. Al-Allaf GA, Hayatee ZG. Recto-urethral migration of Enterobius vermicularis. *Trans R Soc Trop Med Hyg* 1977;71:351.
2. Anya AO. Studies on the biology of some oxyurid nematodes. 1. Factors in the development of eggs of Aspiculuris tetraptera Schulz. *J Helminthol* 1966;40:253–260.
3. Anya AO. Studies on the biology of some oxyurid nematodes. 2. The hatching of eggs and development of Aspiculuris tetraptera Schulz, within the host. *J Helminthol* 1966;40:261–268.
4. Arthur HR, Tomlinson BE. Oxyuris granulomata of the fallopian tube and peritoneal surface of an ovarian cyst. *J Obstet Gynecol Br Emp* 1958;65:996–997.
5. Ashford RW, Hart CA, Williams RG. Enterobius vermicularis infection in a children's ward. *J Hosp Infect* 1988;12:221–224.
6. Beaver PC. Methods of pinworm diagnosis. *Am J Trop Med* 1949;29:577–587.
7. Beaver PC, Kriz JJ, Lau TJ. Pulmonary nodule caused by Enterobius vermicularis. *Am J Trop Med Hyg* 1973;22:711–713.
8. Beckman EN, Holland JB. Ovarian enterobiasis—a proposed pathogenesis. *Am J Trop Med Hyg* 1981;30:74–76.
9. Bijlmer J. An exceptional case of oxyuriasis of the intestinal wall. *J Parasitol* 1946;32:359–366.
10. Brooks TJ Jr, Goetz CC, Plauche WC. Pelvic granuloma due to Enterobius vermicularis. *JAMA* 1962;179:492–494.
11. Brugmans JP, Thienpont DC, van Wijngaarden I, Vanparijs OF, Schuermans VL, Lauwers HL. Mebendazole in enterobiasis. Radiochemical and pilot clinical study in 1,278 subjects. *JAMA* 1971;217:313–316.
12. Budd JS, Armstrong C. Role of Enterobius vermicularis in the aetiology of appendicitis. *Br J Surg* 1987;74:748–749.
13. Campbell CG, Bowman J. Enterobius vermicularis granuloma of pelvis. *Am J Obstet Gynecol* 1961;81:256–258.
14. Campbell JS, Mandavia S, Threlfall W, McCarthy P, Loveys CN. Pinworm granuloma of cervix uteri—incidental observation following IUD use and cone biopsy. *J Trop Med Hyg* 1981;84:215–217.
15. Chandrasoma PT, Mendis KN. Enterobius vermicularis in ectopic sites. *Am J Trop Med Hyg* 1977;26:644–649.
16. Chittenden AM, Ashford RW. Enterobius gregorii Hugot 1983; first report in the U.K. *Ann Trop Med Parasitol* 1987;81:195–198.
17. Daly JJ, Baker GF. Pinworm granuloma of the liver. *Am J Trop Med Hyg* 1984;33:62–64.
18. Deeds DD. Migration of Oxyuris vermicularis to lymph node of round ligament. *Am J Obstet Gynecol* 1947;54:890–892.
19. Dutta LP, Kalita SN. Enterobius vermicularis in the human conjunctival sac. *Indian J Ophthalmol* 1976;24:34–35.
20. Fiumara NJ, Tang S. Folliculitis of the buttocks and pinworms. A case report. *Sex Transm Dis* 1986;13:45–46.
21. Fry GF, Moore JG. Enterobius vermicularis: 10,000-year-old human infection. *Science* 1969;166:1620.
22. Gordon H. Appendical oxyuriasis and appendicitis, based on study of 26,051 appendixes. *Arch Pathol* 1933;16:177–194.
23. Grove, DI. *A History of Human Helminthology*. Wallingford, England: CAB Int; 1990:439–454.
24. Hasegawa H, Takao Y, Nakao M, Fukuma T, Tsuruta O, Ide K. Is Enterobius gregorii Hugot, 1983 (Nematoda: Oxyuridae) a distinct species? *J Parasitol* 1998;84:131–134.

25. Hippocrates. *The Works of Hippocrates*. Jones WH, Whithington ET, trans. London, England: Heinemann; 1948–1953. Loeb Classical Library; Book 2, sec. 1.
26. Hugot JP. Enterobius gregorii (Oxyuridae, Nematoda), a new human parasite [in French]. *Ann Parasitol Hum Comp* 1983;58:403–404.
27. Hussey KL. Syphacia muris vs. S. obvelata in laboratory rats and mice. *J Parasitol* 1957;43:555–559.
28. Jones JE. Pinworms. *Am Fam Physician* 1988;38:159–164.
29. Knuth KR, Fraiz J, Fisch JA, Draper TW. Pinworm infestation of the genital tract. *Am Fam Physician* 1988;38:127–130.
30. Kropp KA, Cichocki GA, Bansal NK. Enterobius vermicularis (pinworms), introital bacteriology and recurrent urinary tract infection in children. *J Urol* 1978;120:480–482.
31. Little MD, Cuello CJ, D'Alessandro A. Granuloma of the liver due to Enterobius vermicularis. Report of a case. *Am J Trop Med Hyg* 1973;22:567–569.
32. Mayayo E, Mestres M, Sarmiento J, Camblor G. Pelvic oxyuriasis. *Acta Obstet Gynecol Scand* 1986;65:805–806.
33. McMahon JN, Connolly CE, Long SV, Meehan FP. Enterobius granulomas of the uterus, ovary and pelvic peritoneum. Two case reports. *Br J Obstet Gynaecol* 1984;91:289–290.
34. Mendoza E, Jordà M, Rafel E, Simón A, Andrada E. Invasion of human embryo by Enterobius vermicularis. *Arch Pathol Lab Med* 1987;111:761–762.
35. Mondou EN, Gnepp DR. Hepatic granuloma resulting from Enterobius vermicularis. *Am J Clin Pathol* 1989;91:97–100.
36. Mossop RT. Threadworm vaginitis. *Cent Afr J Med* 1978;24:10–11.
37. Nairn RC, Duguid HLD. Oxyuris granuloma of the endometrium. *J Clin Pathol* 1954;7:228–230.
38. Neri A, Tadir Y, Grausbard G, Pardo J, Ovadia J, Braslavsky D. Enterobius (Oxyuris) vermicularis of the pelvic peritoneum—a cause of infertility. *Eur J Obstet Gynecol Reprod Biol* 1986;23:239–241.
39. Riley WA. A mouse oxyurid, Syphacia obvelata, as a parasite of man. *J Parasitol* 1920;6:89–92.
40. Schenken JR, Tamisiea J. Enterobius vermicularis (pinworm) infection of the endometrium. A case report. *Am J Obstet Gynecol* 1956;72:913–914.
41. Simon RD. Pinworm infestation and urinary tract infection in young girls. *Am J Dis Child* 1974;128:21–22.
42. Sinniah B, Leopairut J, Neafie RC, Connor DH, Voge M. Enterobiasis: a histopathological study of 259 patients. *Ann Trop Med Parasitol* 1991;85:625–635.
43. Smith JW, Gutierrez Y. Medical parasitology. In: Henry JB, ed. *Clinical Diagnosis and Management by Laboratory Methods*. 17th ed. Philadelphia, Pa: WB Saunders Co; 1984:1245–1247.
44. Sogbanmu MO. Pelvic inflammatory disease associated with Enterobius vermicularis in the endometrium. *East Afr Med J* 1976;53:702–706.
45. Sun T, Schwartz NS, Sewell C, Lieberman P, Gross S. Enterobius egg granuloma of the vulva and peritoneum: review of the literature. *Am J Trop Med Hyg* 1991;45:249–253.
46. Symmers WC. Two cases of eosinophilic prostatitis due to metazoan infestation (with Oxyuris vermicularis and with a larva of Linguatula serrata). *J Path Bact* 1957;73:549–555.
47. Symmers WC. Pathology of oxyuriasis, with special reference to granulomas due to presence of Oxyuris vermicularis (Enterobius vermicularis) and its ova in tissues. *Arch Pathol* 1950;50:475–516.
48. Vinuela A, Fernandez-Rojo F, Martinez-Merino A. Oxyuris granulomas of pelvic peritoneum and appendicular wall. *Histopathology* 1979;3:69–77.
49. Wiebe BM. Appendicitis and Enterobius vermicularis. *Scand J Gastroenterol* 1991;26:336–338.
50. Williams DJ, Dixon MF. Sex, Enterobius vermicularis and the appendix. *Br J Surg* 1988;75:1225–1226.
51. Wolfe MS. Oxyuris, trichostrongylus and trichuris. *Clin Gastroenterol* 1978;7:201–217.

30

Gnathostomiasis

Peter A.S. Johnstone, Parsotam R. Hira,
Ronald C. Neafie, and
Mary K. Klassen-Fischer

Introduction

Definition

Gnathostomiasis is infection by nematodes of the genus *Gnathostoma*. *Gnathostoma spinigerum* Owen, 1836 causes most infections in humans. *Gnathostoma hispidum* Fedchenko, 1872, *Gnathostoma nipponicum* Yamaguti, 1941, and *Gnathostoma doloresi* Tubangui, 1925 occasionally infect humans. Humans are aberrant hosts for *Gnathostoma* sp. The worms rarely reach sexual maturity in a human host, but larval stages may migrate to any part of the body, causing symptoms of cutaneous or visceral larva migrans.

Synonyms

In Asia, where infection is most prevalent, gnathostomiasis is also known as Yangtze River edema (China), *Choko-fushu* (Japan), *Tua chid* (Thailand), consular disease (Nanjing), and Shanghai's rheumatism. In other parts of the world, synonyms include nodular eosinophilic panniculitis (Ecuador) and Woodbury bug (Australia).

Epidemiology

The various species of *Gnathostoma* infect a variety of animals in many countries of Asia and other parts of the world. *Gnathostoma spinigerum*, with the widest distribution, is found in Asia, Africa, South America, North America, Australia, and the CIS. The worm inhabits the stomach wall of its animal hosts, most commonly cats and dogs. Human gnathostomiasis caused by *G. spinigerum* is endemic to Southeast Asia, especially Thailand[17] and Japan, but cases have also been reported from Tanzania, Ecuador, and Mexico. Humans usually become infected by eating raw or undercooked fowl (Thailand) or fish (Japan).[7]

Gnathostoma hispidum generally inhabits the stomach wall of wild and domesticated pigs in Asia, Europe, and Australia. Two *G. hispidum* infections in humans have been reported.[5,13] A 43-year-old Japanese male presented with clinical features of creeping eruption of the left thenar eminence. A young female worm, 5 cm long, was recovered from the lesion. The other patient was a 45-year-old Chinese male with an ocular lesion

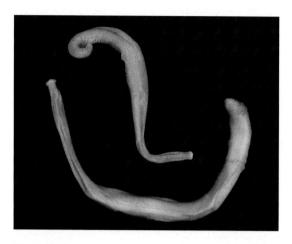

Figure 30.1
Adult male and female *G. spinigerum* demonstrating anterior head bulb and thick posterior end. Male worm is smaller and curved ventrad.* x4

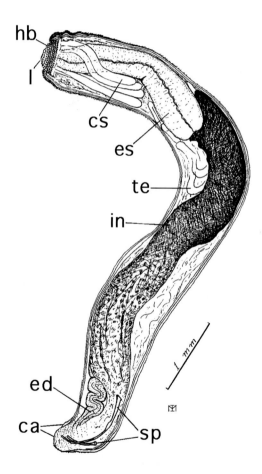

Figure 30.2
Lateral depiction of adult male *G. spinigerum* that emerged spontaneously from skin of patient. Note retracted head bulb (hb), lip (l), cervical sac (cs), esophagus (es), testis (te), intestine (in), ejaculatory duct (ed), spicules (sp), and caudal alae (ca).*

from which a 2.5-mm-long juvenile *G. hispidum* was extracted. Humans are probably infected by eating raw or undercooked freshwater fish, or other vertebrates serving as *G. hispidum*'s second intermediate or paratenic hosts.

Adult *G. nipponicum* lives only in Japan, where it inhabits the esophageal wall of weasels. Humans acquire infection by eating raw freshwater fish. There have been at least 7 confirmed cases of human infection by *G. nipponicum*; all were Japanese patients who presented with creeping eruption.[3,25,27]

Gnathostoma doloresi lives in the gastric wall of pigs and wild boars in several countries of Southeast Asia, including Thailand, Japan, and India. Fewer than 10 cases of human infection by *G. doloresi* have been reported.[15,19] The first proven case involved a 61-year-old Japanese man who presented with creeping eruption over his abdomen.[15] A length of worm 2.7 mm long was recovered and identified as *G. doloresi*. The patient probably was infected by eating raw freshwater fish.

Infectious Agents

Morphologic Description

Gnathostoma sp are spirurid nematodes. All gnathostomes, regardless of species, sex, or maturity, share certain basic morphologic features that are readily identified in histologic sections. (In this chapter, many of the characteristics shared by all species are illustrated only in *G. spinigerum*.) The most significant of these shared characteristics are an anterior head bulb armed with cephalic hooklets and a body armed with cuticular spines. Though variations occur with the maturity of the worm, the number of rows of hooklets on the head bulb, the number of hooklets per row, and the distribution and configuration of cuticular spines are taxonomically significant. In humans, most *Gnathostoma* sp infections have been caused by third-stage (L3) larvae. A monograph on the genus *Gnathostoma* and gnathostomiasis in Thailand, published by Daengsvang, describes in great detail the morphologic features of the worms and eggs of *G. spinigerum*, *G. hispidum*, and *G. doloresi*.[6] Much of the morphologic data presented below on these 3 species is from this monograph.

Gnathostoma spinigerum

Adult *G. spinigerum* males measure 16 to 40 by 1 to 3 mm; adult females are 13 to 55 by 1 to 3 mm (Figs 30.1 and 30.2).[6] Head bulbs of both sexes have 7 to 9 transverse rows of cephalic hooklets, with 20 to 131 hooklets per row (Figs 30.3 to 30.6). The anterior half to two thirds of the body is covered with numerous rows of spines that gradually diminish in density and size toward the posterior end until they are barely perceptible or absent (Figs 30.7 to 30.9). Cuticular spines vary in size and shape depending on their location on the body (Figs 30.7 and 30.10 to 30.13). They measure up to 62 µm long and have 1 to 5 points. The cuticle is 5 to 10 µm thick and transversely striated. The hypodermis is thin except for readily discernible lateral cords. Somatic muscle cells are coelomyarian, with many muscle cells per quadrant. The esophagus has a cuticular-lined, triradiate lumen and is composed of an anterior muscular portion and a longer posterior glandular portion. Intestinal cells are lined with prominent microvilli. They contain pigmented granules and have 2 to 8 nuclei (Fig 30.14).[1] Adults of both sexes have 2 pairs of cervical sacs in the anterior end. Adult male worms have a single testis, 2 unequal copulatory spicules (Fig 30.15), and paired papillae (Fig 30.16). Adult female worms have paired ovaries, and a vulva in the posterior half of the body.

Eggs of *G. spinigerum* are oval and unembryonated when laid. They have a mucoid plug at 1 end and measure 56 to 79 by 34 to 43 µm (Fig 30.17).

Third-stage larvae (Fig 30.18) are morphologically similar to adults, but are smaller (3 to 4 mm by 630 µm) and usually have only 4 transverse rows of cephalic hooklets[11] with an average of 42 to 49 hooklets per row (Figs 30.19 and 30.20).[12] The body of the larva is covered with more than 200 transverse rows of small, single-pointed spines that diminish in size and density toward the posterior end.[12] Transverse sections through the anterior region of *G. spinigerum* L3 larvae reveal a cuticular-lined glandular esophagus and 2 pairs of thick-walled cervical sacs (Fig 30.21). Further posteriorly, the intestine follows the esophagus and the cervical sacs disappear (Fig 30.22). The cuticle is 5 to 10 µm thick and usually bears spines (Fig 30.23). The hypodermis is thin except for readily discernible lateral cords (Fig 30.22). Somatic muscle cells are coelomyarian with many muscle cells per quad-

Figure 30.3
Anterior end of adult female *G. spinigerum* showing head bulb with 9 transverse rows of cephalic hooklets and cuticular spines on body.

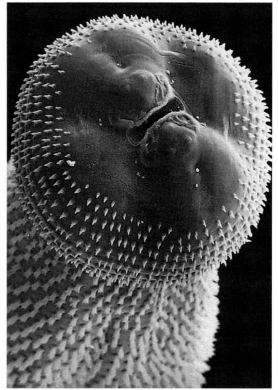

Figure 30.4
Scanning electron micrograph of head bulb of adult *G. spinigerum* showing lateral lips and 8 transverse rows of cephalic hooklets.* x415

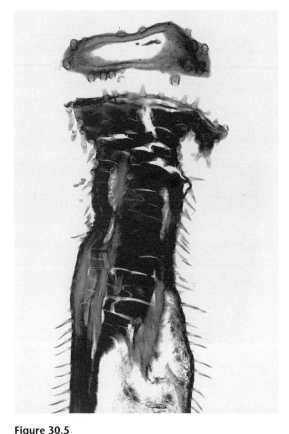

Figure 30.5
This young adult, nongravid female *G. spinigerum* (over 1 cm long) was viable when spontaneously expelled through skin of penis of 18-year-old Laotian patient. (Figures 30.5, 30.6, and 30.10 to 30.14 describe the same worm.) Head bulb of worm was 600 µm wide and armed with 8 transverse rows of cephalic hooklets. Patient presented with fever, weight loss, migratory pleuritic chest pain, arthralgia, headache, and peripheral eosinophil count of 20%. Patient also had history of *Clonorchis sinensis* infection. x170

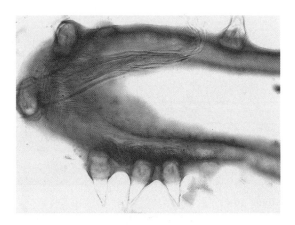

Figure 30.6
Higher magnification of head bulb of worm illustrated in Figure 30.5. Cephalic hooklets are broadbased, single-pointed, and measure 25 by 10 µm. x570

rant of worm (Fig 30.23). Reproductive organs are usually rudimentary, but advanced male L3 larvae may have 2 unequal copulatory spicules (Fig 30.24).

Gnathostoma hispidum

Adult *G. hispidum* males average 19.7 mm long by 1.7 mm in diameter and females average 26.2 mm long by 2.3 mm in diameter.[6] The head bulb of both sexes has 9 to 12 transverse rows of cephalic hooklets. The body circumference increases significantly immediately behind the head bulb, forming a conical anterior end. The body is entirely covered with transverse rows of cuticular spines that have 1 to 10 points and vary in size and density. The largest spines average 108 µm in length.

Eggs average 66 by 38 µm and are morphologically indistinguishable from those of *G. spinigerum*. Advanced L3 larvae measure 1.2 to 3.5 by 0.2 to 0.3 mm.[6] The head bulb has 4 transverse rows of cephalic hooklets, with an average of 33 to 39 hooklets per row. The entire body is covered with many rows of small, single-pointed spines. Transverse sections through *G. hispidum* L3 larvae reveal an intestine composed of columnar epithelial cells that usually have a single large nucleus near the center.[1]

Gnathostoma nipponicum

Adult *G. nipponicum* males are 30 mm in maximum length; females are 42 mm in maximum length.[28] Both sexes are 1 to 1.6 mm in diameter. Head bulbs of both sexes have 7 to 11 transverse rows of cephalic hooklets. Cuticular spines cover the anterior half to two thirds of the body and usually have 3 to 5 points. The largest spines are 110 µm long. The male worm has 2 unequal and dissimilar copulatory spicules. The subglobular spermatozoa are 18 to 21 by 10 to 16 µm and have a rough surface.[28] In the female, the vulva is situated at the posterior two thirds of the worm.

Eggs of *G. nipponicum* are elongate to oval and measure 63 to 78 by 38 to 48 µm. They are unembryonated and have a mucoid plug at 1 end.[28]

Advanced L3 larvae are 1 to 1.5 mm by 0.12 to 0.18 µm.[25] The head bulb has 3 transverse rows of cephalic hooklets with 30 to 45 hooklets per row.[2,25] The entire body is covered with numerous rows of minute, single-pointed spines.[28] Transverse sections through *G. nipponicum* L3 larvae

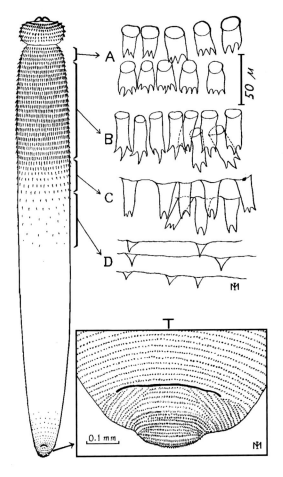

Figure 30.7
Adult female *G. spinigerum* showing distribution and density of cuticular spines. Gnathostomes extirpated or spontaneously escaping through human skin are sexually immature, stunted adults no more than 1 cm long. Minute, poorly developed cuticular spines cover entire body surface. *Inset*: Magnified ventral view of terminal end covered by minute spines.*

Figure 30.8
Surface of adult male *G. spinigerum* in area of body where most spines have 3 points (arrows). x

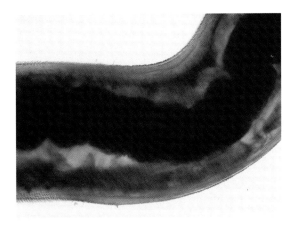

Figure 30.9
Adult male *G. spinigerum* in transitional area where cuticular spines diminish in number. x45

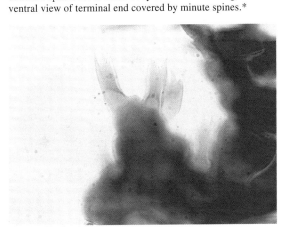

Figure 30.10
Cuticular spines on body near head bulb of adult female *G. spinigerum* described in Figure 30.5. Most spines on this part of the body are 3-pointed. Spines shown here measure 27 by 10 μm. x615

Figure 30.11
Loose cuticular spines from body of adult female *G. spinigerum* described in Figure 30.5. Some spines have 4 points and measure 35 by 10 μm. Movat x650

Figure 30.12
Single-pointed cuticular spines on anterior portion of adult female *G. spinigerum* described in Figure 30.5. Cuticle is 5 μm thick; spines are 45 μm long. x255

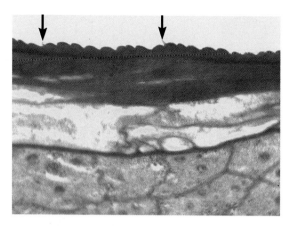

Figure 30.13
Posterior portion of adult female *G. spinigerum* described in Figure 30.5. On this part of the body, cuticular spines (arrows) greatly diminish in number and size, and cuticle is distinctly annulated. x325

show an intestine composed of columnar epithelial cells that usually have 1 or 2 nuclei per cell.[25]

Gnathostoma doloresi

Adult *G. doloresi* males are 9 to 37 by 1 to 3 mm and females are 15 to 63 by 1 to 4.5 mm.[6] In both sexes, the anterior half is much thinner than the posterior half. The head bulb has 7 to 12 transverse rows of cephalic hooklets with 5 to 149 hooklets per row. The body is covered with cuticular spines that have 1 to 7 points. The largest spines are up to 69 μm long. The male has a tail that curves ventrally, a pair of narrow caudal alae, and 2 stout, curved, unequal spicules. In the female, the tail curves ventrally and the vulva is posterior to the middle of the body.

Eggs are oval with a pitted shell and a cap at each end. They average 64 by 32 μm. Advanced L3 larvae are 1.83 to 3.99 by 0.26 to 0.49 μm.[6] The head bulb usually has 4 transverse rows of cephalic hooklets, but a fifth row is sometimes seen. Each transverse row averages 36 to 38 hooklets. The body of the larva is covered with transverse rows of minute cuticular spines that diminish in size and density toward the posterior end.[15] Transverse sections through *G. doloresi* L3 larvae reveal an intestine composed of columnar epithelial cells that are usually binucleate.[1,15,19]

Life Cycle and Transmission

Typical definitive hosts for adult gnathostomes

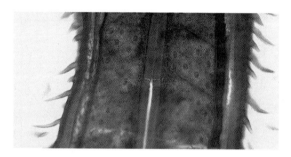

Figure 30.14
Section through body of adult female *G. spinigerum* described in Figure 30.5 showing intestine. Intestinal cells are lined with microvilli, contain numerous pigmented granules, and have 3 or more nuclei per cell. x215

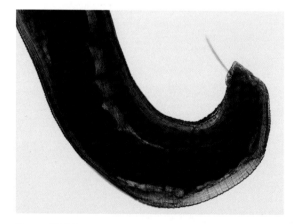

Figure 30.15
Posterior end of adult male *G. spinigerum* showing 1 of 2 copulatory spicules. Cuticle at this level is striated but aspinous. x30

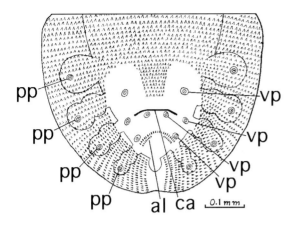

Figure 30.16
Terminal end of adult male *G. spinigerum*, distinguished from that of female by paired papillae and absence of spines around the cloaca. Anterior end is identical to that of female. Note large pedunculate papillae (pp), small ventral papillae (vp), cloacal aperture (ca), and spinous arched lines (al).*

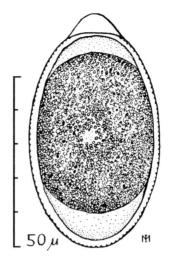

Figure 30.17
Fertilized, unsegmented egg of *G. spinigerum*. Note mucoid plug at upper end.*

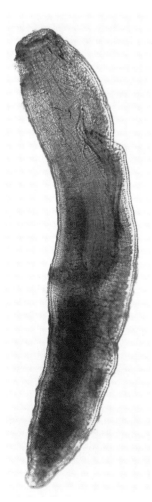

Figure 30.18
L3 larva of *G. spinigerum* (3.2 mm by 550 μm) removed from patient described in Figure 30.39. Transversely striated cuticle of body has numerous single-pointed spines. x40

Figure 30.19
Higher magnification of head bulb of larva shown in Figure 30.18. Head bulbs of L3 *G. spinigerum* larvae usually have 4 transverse rows of cephalic hooklets. x120

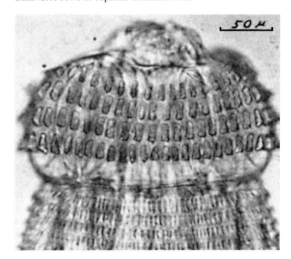

Figure 30.20
Head bulb of *G. spinigerum* L3 larva encircled by 4 transverse rows of cephalic hooklets. Size, form, and number of hooklets are useful in species identification.* x290

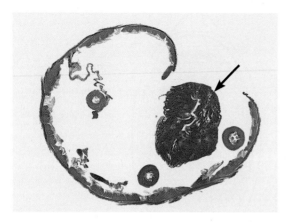

Figure 30.21
Transverse section through anterior portion of *G. spinigerum* L3 larva surgically removed from anterior chamber of eye of 8-year-old Vietnamese boy. Worm was 2 mm by 500 μm. Glandular esophagus (arrow) and 3 of 4 thick-walled cervical sacs are discernible at this level. (Figures 30.21 to 30.23 describe the same worm.) x125

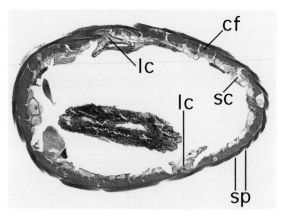

Figure 30.22
Transverse section through *G. spinigerum* L3 larva described in Figure 30.21, at level of intestine. Intestinal cells are lined with microvilli and contain numerous pigmented granules. Note cuticle with single-pointed spines (sp), lateral cords (lc), and both contractile fibers (cf) and sarcoplasm (sc) of muscle cells. x155

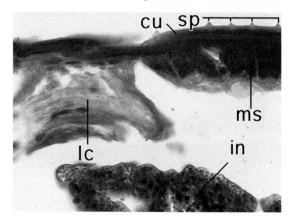

Figure 30.23
Higher magnification of body wall of *G. spinigerum* L3 larva described in Figure 30.21. Note cuticle (cu) with 2-μm-long single-pointed spines (sp), lateral cord (lc), somatic muscle cells (ms), and intestine with pigmented granules (in). x590

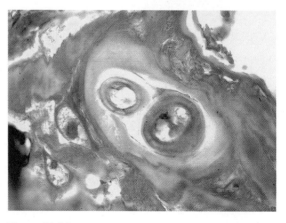

Figure 30.24
Section through immature male *G. spinigerum* demonstrating transverse sections of both copulatory spicules. x610

are domestic dogs and cats, wild and domestic pigs, and many wild carnivores such as leopards, tigers, lions, mink, opossum, raccoons, and otter. Adult worms live and mate in nodules in the gastric wall of a definitive host (Fig 30.25). Unembryonated eggs are discharged into the lumen of the stomach and passed in the feces. In the presence of water, first-stage larvae develop within eggs that hatch in approximately 7 days. Minute *Cyclops* copepod crustaceans ingest the L1 larvae and serve as first intermediate hosts. Larvae penetrate the gastric wall of the copepod and mature into second- and early third-stage larval forms in the body cavity. When second intermediate hosts (fish, eel, frogs, snakes, chickens, pigs, or other small mammals) ingest infected copepods, larvae penetrate the gastric wall of the new host, mature into advanced L3 larvae, and encyst in the musculature (Fig 30.26). The larval life cycle takes about a year to complete. When a definitive host ingests an infected second intermediate host, the larvae excyst in the stomach, penetrate the gastric wall, and migrate to the liver, connective tissue, and muscles. After 4 weeks, the larvae migrate back to the gastric wall, where they mature into adults in 6 to 8 months. Eggs appear in feces several months later.

Humans are infected by eating raw or inadequately cooked flesh of second intermediate hosts. Ingesting water contaminated with infected *Cyclops* sp may also be a source of infection in humans. Some humans, such as food preparers who

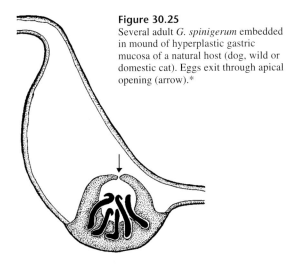

Figure 30.25
Several adult *G. spinigerum* embedded in mound of hyperplastic gastric mucosa of a natural host (dog, wild or domestic cat). Eggs exit through apical opening (arrow).*

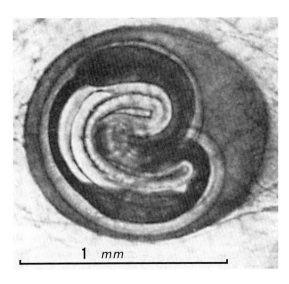

Figure 30.26
L3 larva of *G. spinigerum* encysted in flesh of freshwater fish (*Ophicephalus argus*).* x55

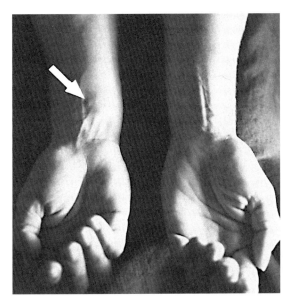

Figure 30.27
Patient with chronic gnathostomiasis initially experienced acute onset of right upper quadrant pain, urticaria, and 50% eosinophilic leukocytosis, lasting approximately 7 days. Four weeks later, a circumscribed swelling gradually migrated down the right arm (arrow). Recurrent migrating edema of forearm, wrist, and hand lasted 12 years, with intervals between attacks gradually lengthening to 6 months. After 12 years, sudden onset of carpal tunnel syndrome required surgical release of median nerve compression and a course of diethylcarbamazine citrate (Hetrazan®). Two years after therapy, there had been no recurrence of symptoms.*

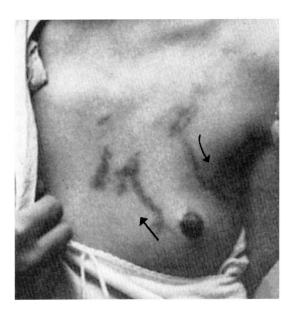

Figure 30.28
Creeping eruption caused by parasites migrating through skin, an unusual form of gnathostomiasis. Migration has been recorded at 1 cm per hour. Ability of parasite to migrate out of operative field makes surgical removal difficult.*

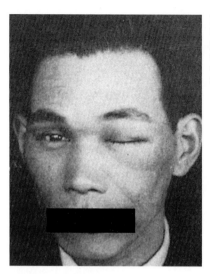

Figure 30.29
Marked palpebral edema due to ocular gnathostomiasis.*

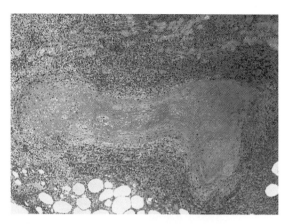

Figure 30.30
Thrombosed necrotic vessel and adjacent eosinophilic infiltrate caused by migrating *G. spinigerum* L3 larva removed from patient described in Figure 30.39. Lesion was a painless recurrent lump on left anterior chest. The parasite was not found in this biopsy specimen, but was surgically removed 5 months later, following treatment with albendazole. x60

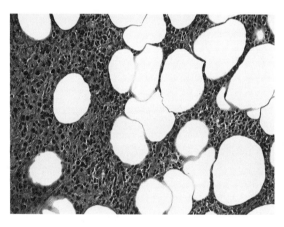

Figure 30.31
Prominent eosinophilic infiltrate in subcutis adjacent to thrombosed vessel described in Figure 30.30. x200

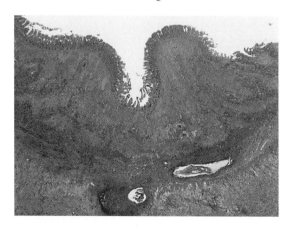

Figure 30.32
Section through ileum of 43-year-old Thai patient who presented with acute pain of 2 days duration in right iliac fossa. Note immature male *G. spinigerum* centered in eosinophilic abscess in muscularis externa. (Figures 30.32 to 30.35 describe the same worm.) x6

handle raw flesh, may possibly be infected by dermal penetration. Three cases of presumed perinatal transmission have been reported.[21] It is not certain that larvae infecting humans achieve sexual maturity and produce eggs.

Clinical Features and Pathogenesis

In humans, larvae do not migrate back into the gastric wall and mature into adults, as they do in natural definitive hosts, but continue to migrate throughout subcutaneous tissues. Symptoms result from the combined effects of host reaction, mechanical damage, and toxins excreted by gnathostomes, including an acetylcholine-like substance, hyaluronidase, a proteolytic enzyme, and a hemolytic agent. Twenty-four to 48 hours after ingesting larvae, patients may develop generalized malaise, fever, urticaria, anorexia, nausea, vomiting, diarrhea, and epigastric pain. Eosinophilia develops when larvae penetrate the gastric or intestinal wall.

Symptoms caused by migrating worms often develop 3 to 4 weeks after ingestion, but may take months or years to appear.[24] Symptoms caused by visceral or cutaneous migration of L3 larvae are collectively referred to as gnathostomiasis externa. Migrating worms produce intermittent subcutaneous swellings that are well-circumscribed, nonpitting, hard, red, and painful or pruritic (Fig 30.27). Swellings generally last 1 to 2 weeks, occurring at intervals of 2 to 4 weeks. They may appear in any part of the body, but are most common on the upper extremities. They may subside in one site only to reappear in another. As infection persists, these episodes become milder and less frequent. Episodes of migratory swelling may recur for 10 to 12 years, although some such instances may actually be reinfection. Cutaneous gnathostomiasis less commonly presents as a creeping eruption (Fig 30.28). Spontaneous extrusion of the worm from subcutaneous tissue has been described.

Rarely, gnathostomes migrate into the central nervous system (CNS), eye, or ear, or into respiratory, gastrointestinal, or genitourinary tracts. Worms may be expelled spontaneously in sputum, urine, or vaginal discharge.[12,16,21] Migration into nerve roots causes burning pain that lasts 1 to 5 days and cannot be relieved by analgesics. This pain may be followed by weakness, urinary retention, sensory disturbances, or paralysis of the extremities. Cranial nerve involvement commonly follows paralysis of the extremities, a pattern thought to be caused by the parasite migrating upward after invading the spinal cord. Invasion of the CNS results in headache, vomiting, or impaired consciousness due to cerebral hemorrhage or transitory obstructive hydrocephalus.[26] Subarachnoid hemorrhage presents as sudden, severe headache with meningeal signs, sometimes progressing to coma. Mortality ranges from 8% to 25%.[20]

Ocular invasion may cause periorbital edema (Fig 30.29), subconjunctival edema, intraocular hemorrhage, uveitis (usually anterior), iritis, increased intraocular pressure, retinal scarring and detachment, and blindness. All cases involving the ear and throat have reported migratory facial swelling; some patients experience hearing loss, tinnitus, and ultimate extrusion of the worm from the external auditory canal. Gnathostomes have been found in the gingiva, tongue, and salivary glands,[24] and may exit from the tip of the tongue, soft palate, cheek, and tympanic membrane.

Migration of larvae through the liver may produce right upper quadrant pain; penetration of the diaphragm may cause pleuropulmonary symp-

Figure 30.33
Higher magnification of eosinophilic abscess described in Figure 30.32. x12

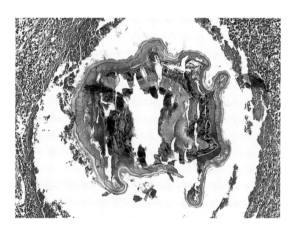

Figure 30.34
Transverse section through immature male *G. spinigerum* described in Figure 30.32. Worm is surrounded by numerous eosinophils. At this level, pigmented granules in intestinal cells are readily observable. x100

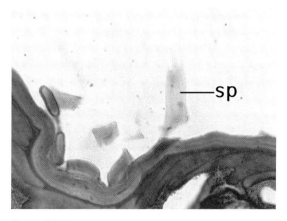

Figure 30.35
Higher magnification of body wall of immature male *G. spinigerum* described in Figure 30.32. At this level, most cuticular spines (sp) have 3 points. x675

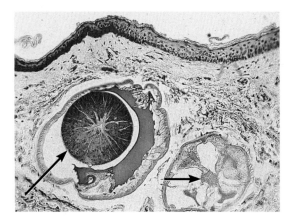

Figure 30.36
Photomicrograph of larval gnathostome in upper dermis. Transverse section of worm in lower right corner is far anterior, at level of muscular esophagus with triradiate lumen (short arrow) and cuticle with single-pointed spines. Large transverse section at left is more posterior, at level of glandular esophagus with triradiate lumen (long arrow). Cuticular spines are not discernible in this section.* x100

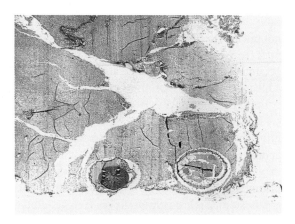

Figure 30.37
Section of brain showing 2 transverse sections of adult *G. spinigerum*. Section of worm at left is through anterior end, illustrating glandular esophagus. Section at right is more posterior, at level of intestine. Worm has provoked minimal chronic inflammation. x13

toms.[24] Pulmonary gnathostomiasis presents as cough, pleuritic chest pain, dyspnea, hemoptysis, lobar consolidation or collapse, pleural effusions, or pneumothorax.[14] In several cases, symptoms resolved after expectoration of the worm. Gastrointestinal gnathostomiasis presents as a mass and/or acute pain in the right lower quadrant accompanied by fever, mimicking appendicitis, or as intestinal obstruction.[24] It has also been an asymptomatic, incidental finding at surgery. Gnathostomiasis of the genitourinary tract may present as perineal swelling, burning suprapubic pain, costovertebral pain, fever, adnexal mass, vaginal bleeding, cervicitis, or balanitis. Four infections manifested only when a gnathostome was passed in the urine.[24]

Pathologic Features

Migrating worms cause edema, track-like necrosis, hemorrhage, and an inflammatory infiltrate consisting of eosinophils, neutrophils, lymphocytes, and plasma cells (Figs 30.30 and 30.31). Worms may provoke an eosinophilic abscess (Figs 30.32 to 30.35).[9] Degenerating larvae may be circumscribed by granulomatous inflammation with histiocytes, foreign body giant cells, and fibrosis. Some worms cause minimal inflammation (Fig 30.36). Gnathostomes that invade the CNS (Figs 30.37 and 30.38) may cause hemorrhage and perivascular infiltrates of eosinophils, plasma cells, and lymphocytes, without granulocytes or parasitic fragments. Immunohistochemical staining of L3 larvae in formalin-fixed, paraffin-embedded tissue has been reported.[22]

Diagnosis

Gnathostomiasis is a likely diagnosis in a patient with migratory subcutaneous swellings, peripheral eosinophilia, and a history of residence or travel in an endemic area. Differential diagnosis includes visceral toxocariasis, ectopic fascioliasis, paragonimiasis, sparganosis, Calabar swellings of loiasis, *Necator americanus* infection, *Ancyclostoma duodenale* infection, or myiasis. Definitive diagnosis depends on identifying adult worms or larvae that have been surgically excised (Fig 30.39) or expelled spontaneously.[16,21] Though gnathostome eggs have not been unequivocally identified in human feces,[4] for identification purposes it is useful to note that eggs of *G. spinigerum* have a single mucoid plug, whereas some *Gnathostoma* sp have a mucoid plug at each end (Fig 30.40).

Because most gnathostomes that infect humans

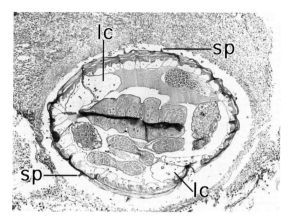

Figure 30.38
Higher magnification of a section of adult *G. spinigerum* described in Figure 30.37. Intestinal cells have pigmented granules and 3 or more nuclei. Lateral cords (lc) are large. Though not clearly seen in this photo, cuticular spines (sp) have multiple points. Body cavity contains coiled reproductive tubes. x55

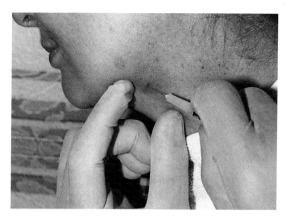

Figure 30.39
Vesicle on anterior neck of 57-year-old woman. Patient developed recurrent nodules on chest, chin, and arm after consuming raw catfish in Tanzania. Gnathostome serology was positive. The worm surgically removed from this vesicle after 3 weeks of treatment with albendazole is also pictured in Figures 30.18 and 30.19. An earlier biopsy specimen from this patient's left anterior chest is shown in Figures 30.30 and 30.31.*

are less than 1 cm long, diagnostic imaging techniques are generally not useful. Skin tests and serologic tests (including radioimmunoassay, Ouchterlony gel diffusion, indirect immunofluorescence, indirect hemagglutination, microprecipitation reactions, immunoblots, and ELISA) can detect elevated serum IgE and IgG in infected patients, but are not sensitive and specific enough to be generally useful.[8,26]

Antibodies, parasite antigen, and immune complexes may be detected in cerebrospinal fluid.[18] In gnathostomiasis involving the CNS, cerebrospinal fluid is often bloody or xanthochromic with eosinophilia.[20,26] Computed tomography can reveal lesions in the CNS, but other supportive data are necessary for a specific diagnosis. In Southeast Asia, the differential diagnosis includes infection with *Angiostrongylus cantonensis* or *Cysticercus cellulosae*. The atypical location of subarachnoid hemorrhage may be helpful in distinguishing gnathostomiasis from a ruptured aneurysm (aneurysms are usually in the basal cisterns).

Ocular gnathostomiasis may resemble the bilateral periorbital edema of trichinosis. Other ocular parasitic infections endemic to Southeast Asia may mimic gnathostomiasis. However, an examiner familiar with parasite morphology will note that *A. cantonensis* is longer and thinner than *Gnathostoma* sp and often comma-shaped, and that cysticerci appear as translucent white cysts.

Figure 30.40
Egg of *Gnathostoma* sp obtained from rat feces in Indonesia. Egg is unembryonated, elongate, and measures 67 by 35 μm. Note thick pitted shell and mucoid plugs at both ends. Trichrome x650

Treatment and Prevention

Surgical excision of the worm is the treatment of choice. Blind biopsies of subcutaneous areas of edema are generally not successful. In ocular gnathostomiasis, the worm is usually in the anterior chamber. A worm in the vitreous should be induced to migrate into the anterior chamber by placing the patient in a prone position and dilating the pupil before resection. Surgical removal with vitrectomy is recommended for a worm that remains in the vitreous.

Although some patients have been successfully treated with albendazole[10,23] and diethylcarbamazine,[14] chemotherapy is generally ineffective. Nor is there any demonstrated benefit from administration of oral prednisolone or intravenous dexamethasone to patients with CNS gnathostomiasis. Anti-inflammatory agents such as antihistamines and corticosteroids may provide temporary relief of cutaneous symptoms.

Preventive measures should be aimed at educating endemic populations to thoroughly cook the flesh of potentially infected animals.

References

1. Akahane H, Sano M, Mako T. Morphological difference in cross sections of the advanced third-stage larvae of Gnathostoma spinigerum, G. hispidum and G. doloresi. *Jpn J Parasitol* 1986;35:465-467.

2. Ando K, Sato Y, Miura K, Matsuoka H, Chinzei Y. Migration and development of the larvae of Gnathostoma nipponicum in the rat, second intermediate or paratenic host, and the weasel, definitive host. *J Helminthol* 1994;68:13-17.

3. Ando K, Tanaka H, Taniguchi Y, Shimizu M, Kondo K. Two human cases of gnathostomiasis and discovery of a second intermediate host of Gnathostoma nipponicum in Japan. *J Parasitol* 1988;74:623-627.

4. Chandler AC. *Introduction to Parasitology with Special Reference to the Parasites of Man*. 9th ed. New York, NY: John Wiley & Sons; 1955:489.

5. Chen HT. A human ocular infection by Gnathostoma in China. *J Parasitol* 1949;35:431-433.

6. Daengsvang S. *A Monograph on the Genus Gnathostoma and Gnathostomiasis in Thailand*. Tokyo, Japan: Southeast Asian Medical Information Center, International Medical Foundation of Japan; 1980:21.

7. Daengsvang S. An experimental study on the life cycle of Gnathostoma hispidum Fedchenko 1872 in Thailand with special reference to the incidence and some significant morphological characteristics of the adult and larval stages. *Southeast Asian J Trop Med Public Health* 1972;3:376-389.

8. Dharmkrong-at A, Migasena S, Suntharasamai P, Bunnag D, Priwan R, Sirisinha S. Enzyme-linked immunosorbent assay for detection of antibody to Gnathostoma antigen in patients with intermittent cutaneous migratory swelling. *J Clin Microbiol* 1986;23:847-851.

9. Hira PR, Neafie R, Prakash B, Tammim L, Behbehani K. Human gnathostomiasis: infection with an immature male Gnathostoma spinigerum. *Am J Trop Med Hyg* 1989;41:91-94.

10. Kraivichian P, Kulkumthorn M, Yingyourd P, Akarabovorn P, Paireepai CC. Albendazole for the treatment of human gnathostomiasis. *Trans R Soc Trop Med Hyg* 1992;86:418-421.

11. Maleewong W, Sithithaworn P, Tesana S, Morakote N. Scanning electron microscopy of the early third-stage larvae of Gnathostoma spinigerum. *Southeast Asian J Trop Med Public Health* 1988;19:643-647.

12. Miyazaki I. On the genus Gnathostoma and human gnathostomiasis, with special reference to Japan. *Exp Parasitol* 1960;9:338-370.

13. Morishita KO. A pig nematode, Gnathostoma hispidum, Fedchenko, as a human parasite. *Ann Trop Med Parasitol* 1924;18:23-26.

14. Nagler A, Pollack S, Hassoun G, Kerner H, Barzilai D, Lengy J. Human pleuropulmonary gnathostomiasis: a case report from Israel. *Isr J Med Sci* 1983;19:834-837.

15. Nawa Y, Imai J, Ogata K, Otsuka K. The first record of a confirmed human case of Gnathostoma doloresi infection. *J Parasitol* 1989;75:166-169.

16. Nitidandhaprabhas P, Hanchansin S, Vongsloesvidhya Y. A case of expectoration of gnathostoma spinigerum in Thailand. *Am J Trop Med Hyg* 1975;24:547-548.

17. Nitidandhaprabhas P, Sirimachan S, Charnvises K. A case of penile gnathostomiasis in Thailand. *Am J Trop Med Hyg* 1978;27:1282-1283.

18. Nopparatana C, Setasuban P, Chaicumpa W, Tapchaisri P. Purification of Gnathostoma spinigerum specific antigen and immunodiagnosis of human gnathostomiasis. *Int J Parasitol* 1991;21:677-687.

19. Ogata K, Imai J, Nawa Y. Three confirmed and five suspected human cases of Gnathostoma doloresi infection found in Miyazaki Prefecture, Kyushu. *Jpn J Parasitol* 1988;37:358-364.

20. Punyagupta S, Bunnag T, Juttijudata P. Eosinophilic meningitis in Thailand. Clinical and epidemiological characteristics of 162 patients with myeloencephalitis probably caused by Gnathostoma spinigerum. *J Neurol Sci* 1990;96:241-256.

21. Radomyos P, Daengsvang S. A brief report on Gnathostoma spinigerum specimens obtained from human cases. *Southeast Asian J Trop Med Public Health* 1987;18:215-217.

22. Rojekittikhun W, Saito S, Yamashita T, Watanabe T, Sendo F. Immunohistochemical localization of Gnathostoma spinigerum larval antigens by monoclonal antibodies: 1. Light microscopy. *Southeast Asian J Trop Med Public Health* 1993;24:494-500.

23. Ruiz-Maldonado R, Mosqueda-Cabrera MA. Human gnathostomiasis (nodular migratory eosinophilic panniculitis). *Int J Dermatol* 1999;38:56-57.

24. Rusnak JM, Lucey DR. Clinical gnathostomiasis: case report and review of the English-language literature. *Clin Infect Dis* 1993;16:33-50.

25. Sato H, Kamiya H, Hanada K. Five confirmed human cases of gnathostomiasis nipponica recently found in northern Japan. *J Parasitol* 1992;78:1006-1010.

26. Schmutzhard E, Boongird P, Vejjajiva A. Eosinophilic meningitis and radiculomyelitis in Thailand, caused by CNS invasion of Gnathostoma spinigerum and Angiostrongylus cantonensis. *J Neurol Neurosurg Psychiatry* 1988;51:80-87.

27. Taniguchi Y, Hashimoto K, Ichikawa S, Shimizu M, Ando K, Kotani Y. Human gnathostomiasis. *J Cutan Pathol* 1991;18:112-115.

28. Yamaguti S. Studies on the helminth fauna of Japan. XXXV. Mammalian nematodes II. *Jpn J Zool* 1941;9:409-440.

31

Trichuriasis

Aileen M. Marty *and*
Ronald C. Neafie

Introduction

Definition

Trichuriasis is infection by roundworms of the genus *Trichuris*. *Trichuris trichiura* (Linnaeus, 1771) causes nearly all human trichuriasis, although there are reports of infection with *Trichuris vulpis* (Froel, 1789).[21] Adult worms infecting the cecum and colon produce colitis. Heavily infected patients may experience tenesmus and rectal prolapse.[4] Severe or neglected infections can be fatal.

Synonyms

Trichocephaliasis is an infrequently used synonym for trichuriasis. Whipworm is the popular term for members of the family Trichuridae, including *T. trichiura*. Older synonyms for *T. trichiura* include *Trichocephalus hominis*, *Trichocephalus dispar*, and *Mastigodes hominis*.

Epidemiology

Trichuriasis is among the most common helminthic infections of humans, along with ascariasis, enterobiasis, and hookworms. The World Health Organization estimates global prevalence at 800 million infections.[5] *Trichuris trichiura* is cosmopolitan, but infections are most common and most severe in the tropics and subtropics. School-age children have the highest prevalence. Poor sanitation and personal hygiene, shortage of clean drinking water, use of night soil as fertilizer, overcrowding, and inadequate health education promote the spread of whipworm, a fecal-orally transmitted helminth. Similar modes of transmission for other geohelminths result in a high rate of multiple helminthic infections in patients with trichuriasis.[6] Mentally retarded patients are at an increased risk of heavy infections from ingesting soil or feces.[1] Houseflies can transport eggs of *T. trichiura*, making them a possible mechanical vector of trichuriasis.[15]

Infectious Agent

Morphologic Description

Adult female *T. trichiura* are 3.5 to 5 cm long and pink-gray; males are slightly shorter (3 to 4.5 cm). Adult whipworms have 2 distinct body regions (Fig 31.1): a slender, thread-like anterior segment (three fifths to two thirds of the worm) forms the lash of the "whip"; a thick posterior segment is the handle. The anterior portion is 100 to 150 μm in diameter and contains the esophagus and stichosome, and a single bacillary band. The esophagus has 2 parts: a short, anterior, muscular region with a triradiate lumen (Fig 31.2), and a long tubular region, 5 to 10 μm in diameter, that passes through the stichosome. The stichosome consists of a long chain of large glandular cells, known individually as stichocytes (Figs 31.3 and 31.4), that partially or completely enclose the esophagus. The bacillary band is a subcuticular structure composed of tall columnar cells with pore-like openings[9]; this band probably has a respiratory function (Figs 31.3 and 31.4). The posterior portion of the worm is 400 to 700 μm in diameter and contains the intestine and reproductive organs. Both male and female adult *T. trichiura* have a single set of reproductive organs.

The cuticle is 5 to 10 μm thick and has fine transverse striations (Fig 31.3) in the slender anterior region. The thick posterior region of the worm has annulations as well as striations in some areas of the cuticle (Figs 31.5 and 31.6). A hypodermis lies immediately beneath the cuticle (Fig 31.7). This thin hypodermal layer is 1 to 3 μm thick throughout the circumference of the worm. Lateral cords are inconspicuous, and hypodermal nuclei are randomly distributed throughout the entire circumference of the hypodermis (Fig 31.7). Readily identifiable somatic muscle, consisting of numerous small, tightly packed cells, lies beneath the hypodermis at all levels of the worm. The proportion of contractile fibers for each area of sarcoplasm in the muscle cells varies considerably. The intestine is usually large in diameter and contains columnar cells with microvilli.

In the female *T. trichiura,* the vulva lies near the esophageal-intestinal junction. The single reproductive tube is located in the thick posterior portion of the worm and consists of a vulva, vagina, uterus, seminal receptacle, oviduct, and ovary (Figs 31.8 to 31.10). The posterior end of the female is bluntly rounded and uncoiled.

The male *T. trichiura* has a coiled posterior extremity (Fig 31.1) readily distinguishable from the uncoiled posterior extremity of the female. A single reproductive tube lies in the thick posterior portion of the worm and consists of a testis, vas

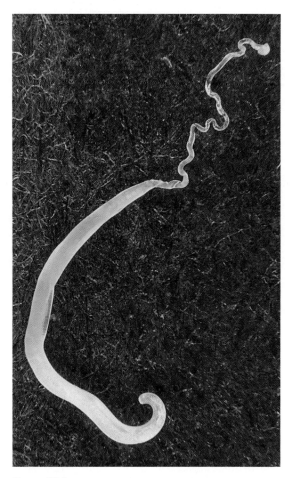

Figure 31.1
Adult male *T. trichiura* with 2 distinct body parts: thin (whip) anterior end and thick (handle) posterior end. Coiled posterior extremity identifies worm as a male. x10.5

Figure 31.2
Transverse section through anterior muscular region of *T. trichiura* esophagus, showing triradiate lumen. x750

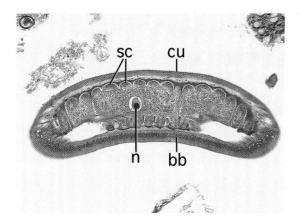

Figure 31.3
Tangential section through slender anterior portion of *T. trichiura*. Note stichocytes (sc), stichosome nucleus (n), finely striated cuticle (cu), and bacillary band (bb). x145

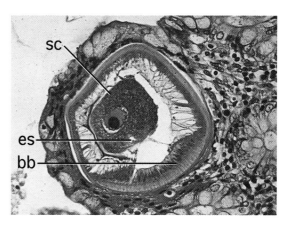

Figure 31.4
Transverse section through slender anterior portion of *T. trichiura* in colonic mucosa. Note bacillary band (bb), stichocyte (sc), and esophagus (es). x375

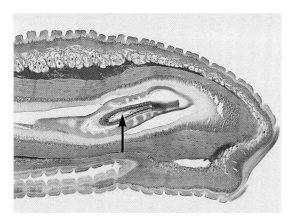

Figure 31.5
Posterior end of male *T. trichiura* through region of spicule (arrow). Note annulations of cuticle. x120

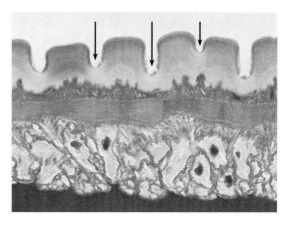

Figure 31.6
Higher magnification of Figure 31.5, showing cuticle of male *T. trichiura*. Note prominent annulations (arrows). x660

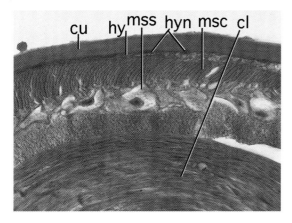

Figure 31.7
Transverse section of body wall in posterior end of male *T. trichiura*. Note cuticle (cu), hypodermis (hy), hypodermal nuclei (hyn), contractile portion of muscle (msc), sarcoplasmic portion of muscle with nuclei (mss), and muscular portion of cloaca (cl). x815

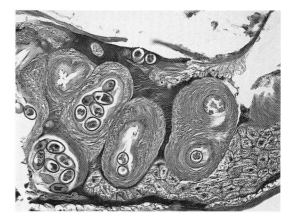

Figure 31.8
Tangential section of female *T. trichiura* at level of coiled, muscular, egg-filled vagina. x135

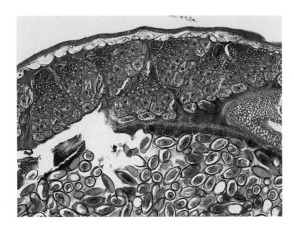

Figure 31.9
Tangential section of female *T. trichiura* at level of ovary and egg-filled uterus. x140

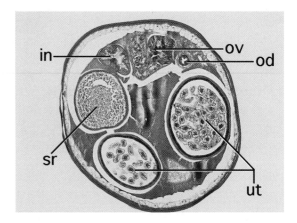

Figure 31.10
Transverse section of thick posterior portion of adult female *T. trichiura*. Note seminal receptacle (sr), intestine (in), ovary (ov), oviduct (od), and uteri (ut). x85

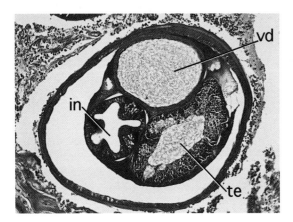

Figure 31.11
Transverse section of posterior portion of adult male *T. trichiura* in lumen of appendix. Note testis (te), intestine (in), and vas deferens (vd). x95

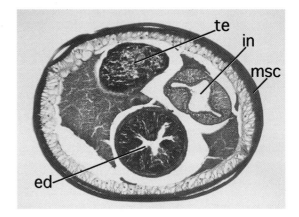

Figure 31.12
Transverse section of posterior portion of adult male *T. trichiura*. Note ejaculatory duct (ed), testis (te), intestine (in), and contractile portion of muscle (msc). x105

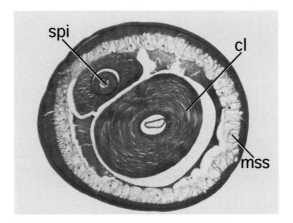

Figure 31.13
Transverse section of posterior extremity of adult male *T. trichiura*. Note spicule (spi), cloaca (cl), and sarcoplasmic portion of muscle (mss). x120

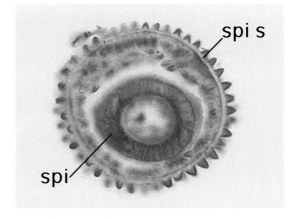

Figure 31.14
Transverse section of spicule (spi) and spiny spicular sheath (spi s) of adult male *T. trichiura*. x865

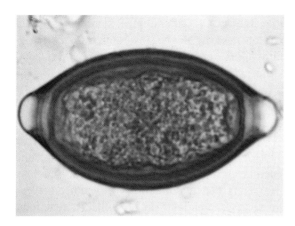

Figure 31.15
Egg of *T. trichiura* in feces, showing brown-tinged shell and bipolar plugs. Unstained x1285

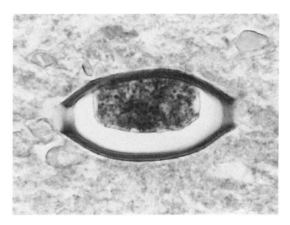

Figure 31.16
Section of appendix showing unsegmented egg of *T. trichiura* in lumen. Movat x1000

deferens, and ejaculatory duct (Figs 31.11 and 31.12). The ejaculatory duct and intestine join in the posterior extremity to form the cloaca (Fig 31.13). The reproductive tube makes a single hairpin loop, so that most sections of the posterior portion of the worm contain 2 sections of reproductive tube. A single spicule about 2.5 mm long and 50 μm in diameter protrudes through a retractile spiny sheath (Fig 31.14).[22]

In feces, *T. trichiura* eggs are barrel-shaped and unsegmented. They measure 50 to 56 μm by 22 to 23 μm and have a thick, brown-tinged shell with prominent bipolar plugs (Figs 31.15 and 31.16). Sometimes, usually following treatment, *T. trichiura* eggs up to 80 μm long are found in human feces.[13]

Life Cycle

Trichuris trichiura has a direct life cycle (Fig 31.17). Adult *T. trichiura* live in the human cecum and upper colon, where they can survive up to 3 years. Female worms lay unembryonated eggs that pass in the feces. In moist soil, infective first-stage larvae develop within the eggs in 2 to 3 weeks. Ingestion of embryonated eggs in contaminated food or soil leads to infection. Larvae break out of the eggs in the lumen of the small intestine and penetrate the mucosa of the ileum, where they develop and then reenter the intestinal lumen. Eventually, the worms migrate to the colon and become sexually mature.

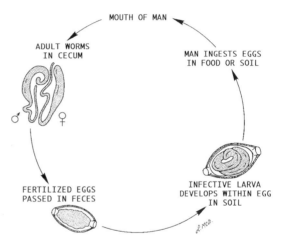

Figure 31.17
Life cycle of *T. trichiura*.

Clinical Features and Pathogenesis

Most infections are light and tend to be asymptomatic. Interference with intestinal functions (eg, blockage of crypts), production of toxic irritants, or a combination of these factors may provoke symptoms. Diarrhea, abdominal pain, and tenesmus characterize heavy infections. Severe tenesmus may cause rectal prolapse. Increased peristalsis, seen in heavily infected patients, may lead to intussusception[10] and contribute to rectal prolapse. Stools are mucoid and may be blood-streaked.

Heavy infections (Figs 31.18 to 31.20) can lead

Figure 31.18
Colonic mucosa heavily infected with *T. trichiura*. Note extensive mucosal hemorrhage.

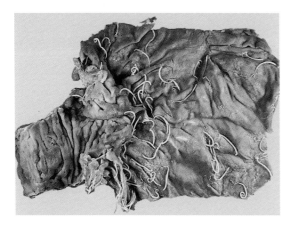

Figure 31.19
Section of colon from patient with fatal trichuriasis, showing colonic mucosa heavily infected with *T. trichiura*.

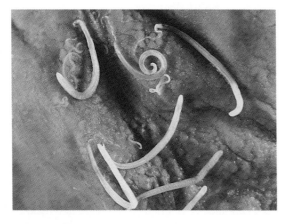

Figure 31.20
Higher magnification of specimen in Figure 31.19. Note thin anterior ends of *T. trichiura* embedded in colonic mucosa. x2.7

to malnutrition.[7] Other clinical symptoms include nervousness, headache, insomnia, impaired cognition, loss of appetite, lower right quadrant pain, vomiting, and abdominal distention.[16,17] If diarrhea persists, weakness, dehydration, and emaciation ensue. Patients sometimes exhibit mild eosinophilia, hypoproteinemia, and blood loss in stools that can lead to anemia and even death (Fig 31.18).[11,20] Pica, often related to malnutrition and anemia, may aggravate the patient's disease by provoking heavier infections with other geohelminths. Some patients develop allergic reactions, with edema of the face and hands, dyspnea, cardiac dilatation, and convulsions. Chronic infections in children may retard growth.

Adult *T. trichiura* in the appendix can obstruct the lumen and cause appendicitis (Fig 31.21). Worms undergo no obligatory tissue cycle before localizing in the intestine; however, they have been found attached to the peritoneal surface (Figs 31.22 and 31.23), presumably from a prior intestinal perforation.

Pathologic Features

Worms tend to concentrate in the cecum (Fig 31.19), but small numbers may localize in any portion of the colon, including the appendix. An adult worm embeds its narrow anterior end into the intestinal mucosa (Figs 31.20 and 31.24 to 31.26). Flattened epithelial cells surround the embedded portion of the worm (Fig 31.25). Host tissue response is usually mild, with local chronic inflammation, eosinophilia, and increased mitotic activity of cells lining the crypts (Fig 31.27). In patients with dysentery related to trichuriasis, there are increased numbers of IgM-positive plasma cells in the lamina propria and decreased T cells in the epithelium.[14] Crypts may become dilated with mucus, and sometimes with fibrin and neutrophils (Figs 31.27 and 31.28). In heavy infections, adult worms may completely coat the luminal surface of the colon, and the bowel wall may be edematous and friable. Away from the worms, there is no significant inflammatory response in the colonic mucosa, even in heavily infected patients.[14] Removing the worms, such as during colonoscopy, produces small, subepithelial petechiae. At the site of attachment there are chronic inflammatory infiltrates consisting of lymphocytes, plasma cells, and eosinophils, which do not extend beyond the muscularis mucosa. Lymphoid tissue of the involved intestinal region may become hyperplastic.

Diagnosis

Identifying adult *T. trichiura* or their eggs in the stool establishes a diagnosis of trichuriasis (Fig 31.29). Concentration procedures, especially the Kato-Katz technique, are helpful in detecting light infections. The eggs of *T. vulpis* are larger and broader (72 to 90 μm by 32 to 40 μm) than those of *T. trichiura*, and have more prominent bipolar plugs (Figs 31.30 and 31.31). *Trichuris trichiura* eggs are sometimes identified in routine cytologic specimens.[12] It is relatively easy to recognize adult worms attached to the cecal or colonic mucosa during colonoscopy,[18] or in biopsy or autopsy specimens. Charcot-Leyden crystals are often found in the stool, and a mild to moderate peripheral eosinophilia is common.

Treatment

The benzimidazoles (mebendazole and albendazole)[19] are effective against whipworm infections, mebendazole being the drug of choice.[3,8] Exercise caution when giving mebendazole to patients with multiple geohelminthic infections (Fig 31.32); mebendazole can stimulate erratic migration of *Ascaris*, with serious consequences. Thiabendazole is effective, but frequently produces serious side effects. Benzimidazoles do not affect immature worms. Hexylresorcinol enemas were once a common therapy,[2] but they are troublesome and no longer recommended.

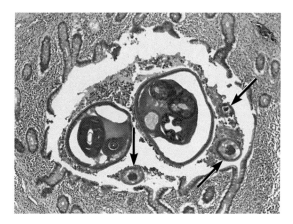

Figure 31.21
Cross section of adult male *T. trichiura* in lumen of appendix. There are a total of 5 sections: 3 through thin anterior portion (arrows) and 2 through thick posterior portion of worm. x45

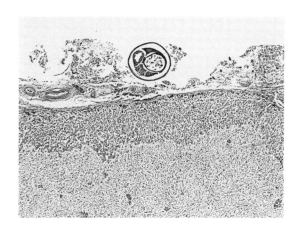

Figure 31.22
Posterior section of gravid *T. trichiura* in exudate on peritoneal surface. x25

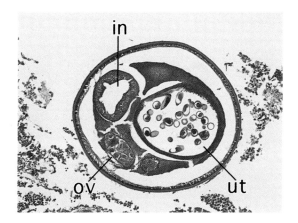

Figure 31.23
Higher magnification of Figure 31.22. Note intestine (in), ovary (ov), and egg-filled uterus (ut) in this section of *T. trichiura*. x85

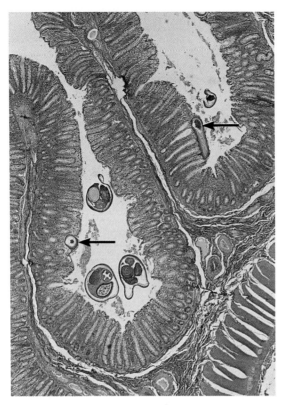

Figure 31.24
Several sections of adult male and female *T. trichiura* in colon, showing narrow anterior portion of worm within mucosa (arrows) and thicker posterior portion free in lumen. x18

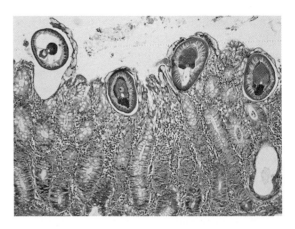

Figure 31.25
Four transverse sections of anterior region of *T. trichiura* embedded in superficial portion of colonic mucosa. Epithelial cells surrounding worm are flattened. x65

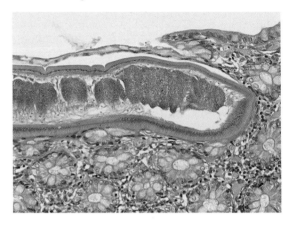

Figure 31.26
Tangential section of anterior end of *T. trichiura* embedded in superficial colonic mucosa. Epithelium covering worm is atrophic. x150

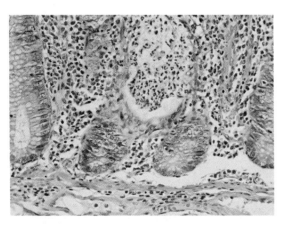

Figure 31.27
Inflamed colonic mucosa in patient with trichuriasis. Exudate contains acute and chronic inflammatory cells, including eosinophils. Fibrin and neutrophils have collected in dilated crypt. x155

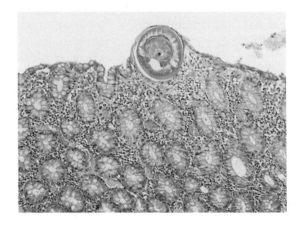

Figure 31.28
Transverse section of *T. trichiura* embedded in superficial colonic mucosa. x85

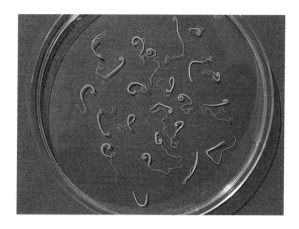

Figure 31.29
Numerous adult male and female *T. trichiura* recovered from stool. Note coiled posterior extremity of male worms.

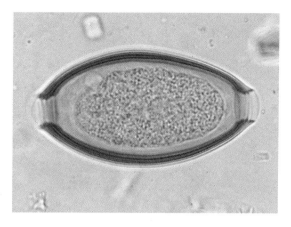

Figure 31.30
Egg of *T. vulpis* in dog feces. Unstained x440

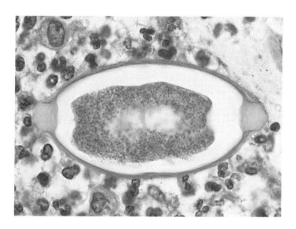

Figure 31.31
Unsegmented *T. vulpis* egg in colonic lumen of dog. Note prominent bipolar plugs. x815

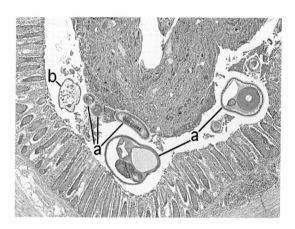

Figure 31.32
Appendix of patient infected with 3 geohelminths. Note multiple sections of adult male *T. trichiura* (a), single section of adult female *Enterobius vermicularis* (b), and eggs of *T. trichiura* and *Ascaris lumbricoides* in the lumen. Eggs are less obvious at this magnification. Movat x25

References

1. Allen KD, Green HT. An outbreak of Trichuris trichiura in a mental handicap hospital. *J Hosp Infect* 1989;13:161–166.
2. Alvarez Chacon R, Rodriguez Rodriguez M, Cob Sosa CE. Evaluation of the therapeutic effectiveness of hexylresorcinol and of thiabendazole in massive trichocephalosis in children [in Spanish]. *Bol Med Hosp Infant Mex* 1974;31:1125–1147.
3. Bartoloni A, Guglielmetti P, Cancrini G, et al. Comparative efficacy of a single 400 mg dose of albendazole or mebendazole in the treatment of nematode infections in children. *Trop Geogr Med* 1993;45:114–116.
4. Bundy DA, Cooper ES. Trichuris and trichuriasis in humans. *Adv Parasitol* 1989;28:107–173.
5. Crompton DW, Savioli L. Intestinal parasitic infections and urbanization. *Bull World Health Organ* 1993;71:1–7.
6. Forrester JE, Scott ME, Bundy DA, Golden MH. Clustering of Ascaris lumbricoides and Trichuris trichiura infections within households. *Trans R Soc Trop Med Hyg* 1988;82:282–288.
7. Gilman RH, Chong YH, Davis C, Greenberg B, Virik HK, Dixon HB. The adverse consequences of heavy Trichuris infection. *Trans R Soc Trop Med Hyg* 1983;77:432–438.
8. Kan SP. Efficacy of single doses of mebendazole in the treatment of Trichuris trichiura infection. *Am J Trop Med Hyg* 1983;32:118–122.
9. Kaur M, Sood ML. In vitro effect of albendazole and fenbendazole on the histochemical localization of some enzymes of Trichuris globulosa (Nematoda: Trichuroidea). *Angew Parasitol* 1992;33:33–45.
10. Kerrigan KR. Ileocolic intussusception complicating heavy Trichuris trichiura infection. *Trop Doct* 1991;21:134–135.
11. Layrisse M, Aparcedo L, Martinez-Torres C, Roche M. Blood loss due to infection with Trichuris trichiura. *Am J Trop Med Hyg* 1967;16:613–619.
12. Learmonth GM, Murray MM. Helminths and protozoa as an incidental finding in cytology specimens. *Cytopathology* 1990;1:163–170.
13. Little MD. A strain of Trichuris trichiura having large eggs. In: *Program and Abstracts, 43rd Annual Meeting, the American Society of Parasitologists*. Madison: Univeristy of Wisconsin Press; 1968:153.
14. MacDonald TT, Choy MY, Spencer J, et al. Histopathology and immunohistochemistry of the caecum in children with the Trichuris dysentery syndrome. *J Clin Pathol* 1991;44:194–199.
15. Monzon RB, Sanchez AR, Tadiaman BM, et al. A comparison of the role of Musca domestica (Linnaeus) and Chrysomya megacephala (Fabricius) as mechanical vectors of helminthic parasites in a typical slum area of metropolitan Manila. *Southeast Asian J Trop Med Public Health* 1991;22:222–228.
16. Nokes C, Bundy DA. Trichuris trichiura infection and mental development in children [letter]. *Lancet* 1992;339:500.
17. Nokes C, Grantham-McGregor SM, Sawyer AW, Cooper ES, Robinson BA, Bundy DA. Moderate to heavy infections of Trichuris trichiura affect cognitive function in Jamaican school children. *Parasitology* 1992;104:539–547.
18. Okamura S, Washida Y, Iesaki K, Hayashi S. Colonoscopic diagnosis of whipworm infection [letter]. *Gastrointest Endosc* 1993;39:215–216.
19. Ramalingam S, Sinniah B, Krishnan U. Albendazole, an effective single dose, broad spectrum anthelmintic drug. *Am J Trop Med Hyg* 1983;32:984–989.
20. Robertson LJ, Crompton DW, Sanjur D, Nesheim MC. Haemoglobin concentrations and concomitant infections of hookworm and Trichuris trichiura in Panamanian primary schoolchildren. *Trans R Soc Trop Med Hyg* 1992;86:654–656.
21. Singh S, Samantaray JC, Singh N, Das GB, Verma IC. Trichuris vulpis infection in an Indian tribal population. *J Parasitol* 1993;79:457–458.
22. Zaman V. *Scanning Electron Microscopy of Medically Important Parasites*. Boston, Mass: ADIS Health Science Press; 1983:80–83.

32

Trichinosis

Ronald C. Neafie,
Aileen M. Marty, *and*
Ellen M. Andersen

Introduction

Definition

Trichinosis is infection by any of 3 species of roundworm: *Trichinella spiralis*, *Trichinella pseudospiralis*, or *Trichinella britovi*. The most common roundworm, *T. spiralis*, has at least 3 subspecies which may infect humans, but which are primarily parasites of other mammals: 1) *T. spiralis spiralis* in domestic pigs in temperate regions; 2) *T. spiralis nativa* in polar bears and walrus in arctic regions; and 3) *T. spiralis nelsoni* in wild carnivores, particularly pigs, in Africa, southern Europe, and middle Asia. Enzymatic polymorphism and genetic differences separate the subspecies, though they are nearly identical morphologically.[8,9] Some authorities believe these subspecies merit designation as separate species.[19,21]

Trichinella pseudospiralis, a parasite of raccoons, cats, mice, and birds, and *T. britovi*, a parasite of foxes, may infect humans.[2,22] Both adult and larval worms produce disease in all infected hosts, including humans. Acute infection from migrating and encysting larvae produces the 4 cardinal features of trichinosis, as described by Beeson: fever, myalgia, palpebral and facial edema, and eosinophilia.[4] Trichinosis can be a mild, localized, self-limiting infection or a severe systemic disease, depending on the worm burden and subspecies involved.

Synonyms

Trichinellosis and trichiniasis are occasionally used as synonyms for trichinosis.

General Considerations

Trichinella spiralis was considered a single species until the early 1960s,[18] when Nelson et al observed differences in the virulence of isolates from Africa and Europe. Subsequent studies confirmed and extended these findings, revealing a highly complex pattern of speciation within the genus.[5]

Epidemiology

Consuming raw or improperly cooked meat containing encysted first-stage larvae of *T. spiralis* causes trichinosis, a zoonosis involving numer-

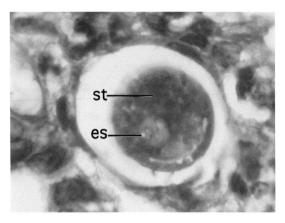

Figure 32.1
Transverse section of anterior portion of adult *T. spiralis* in crypt of rat small intestine. Note stichocyte (st) filling body cavity and enclosing narrow esophagus (es). x1430

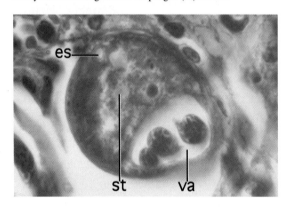

Figure 32.2
Transverse section through anterior portion of gravid adult *T. spiralis* in crypt of rat small intestine. Stichocyte (st) is displaced by vagina (va). Narrow esophagus (es) is visible within stichocyte. Note 3 transverse sections of immature first-stage larva within vagina, just posterior to vulva. x1080

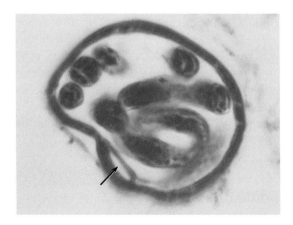

Figure 32.3
Transverse section through midbody of gravid adult *T. spiralis* in lumen of rat small intestine. Larvae-filled uterus occupies majority of body cavity and displaces intestine (arrow). x1250

ous animal species. Swine, wild boar, bear, horse, and walrus meat are the major sources of human infection.[12,14] Trichinosis is cosmopolitan, but most common in temperate climates where pork is a popular food.[20] Endemic areas are: 1) Africa; 2) the Americas (Brazil, Chile, Guatemala, Mexico, United States, Uruguay, and Venezuela); and 3) Eurasia (China, Commonwealth of Independent States, France, Italy, Korea, Lebanon, Lithuania, Poland, Spain, Thailand, and Yugoslavia). While trichinosis remains one of the more common food-borne helminthic zoonoses worldwide, incidence in the United States is steadily declining.[24]

Infectious Agent

Morphologic Description

Adult *T. spiralis* are small, thread-like worms whose posterior portion is slightly thicker than the anterior portion. The anterior portion makes up about one third of the body length of the female and about one half the length of the male, and contains the esophagus and stichosome (Fig 32.1). The esophagus has 2 divisions, a short, anterior, muscular region with a triradiate lumen, and a long, thin, tubular region that passes through the stichosome. The stichosome consists of a long chain of large glandular cells, known individually as stichocytes, that partially or completely enclose the esophagus (Fig 32.1). In both the male and female, the cuticle has transverse striations.[15]

Female worms are 2.2 to 4 mm long and 60 µm in diameter, and have a single reproductive tube consisting of a vulva, vagina, uterus, seminal receptacle, oviduct, and ovary. The ovary is located in the posterior end of the worm. The vulva opens in the anterior fifth of the body alongside the stichosome (Fig 32.2). Females are viviparous, discharging immature first-stage larvae (Fig 32.3).

Males are 1.2 to 1.6 mm long and 40 µm in diameter. The single reproductive tube consists of a testis, vas deferens, seminal vesicle, and ejaculatory duct (Fig 32.4). The ejaculatory duct joins the intestine in the posterior end of the worm, forming a cloaca. There are 2 copulatory appendages at the posterior tip (Fig 32.5), but no spicule.

When ejected from the vulva, first-stage larvae are immature and measure 80 to 160 µm in length and 5 to 7 µm in diameter. They have a short

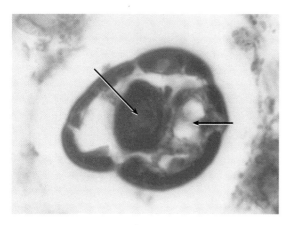

Figure 32.4
Transverse section through posterior end of adult male *T. spiralis* in lumen of rat small intestine. Note ejaculatory duct (long arrow) and intestine (short arrow). x1585

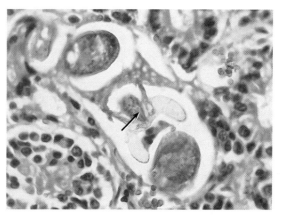

Figure 32.5
Three sections through adult male *T. spiralis* in rat small intestine. Section in center displays 2 copulatory appendages and cloaca (arrow). x525

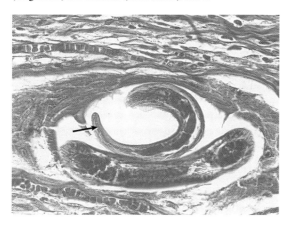

Figure 32.6
Anterior end of coiled, mature first-stage *T. spiralis* larva in skeletal muscle. Longitudinal section of anterior tip contains muscular esophagus (arrow), followed by stichosome. Cuticle is finely striated. x250

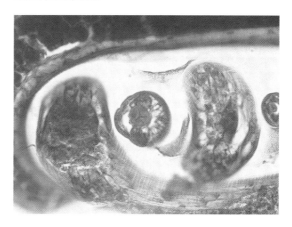

Figure 32.7
Multiple sections of coiled, mature first-stage *T. spiralis* larva in skeletal muscle. Section in center of photograph displays anterior tip of worm containing muscular esophagus with its triradiate lumen. x585

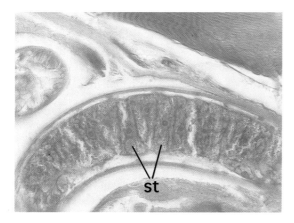

Figure 32.8
Longitudinal section of anterior end of coiled, mature first-stage *T. spiralis* larva in skeletal muscle. Note stichocytes (st). x600

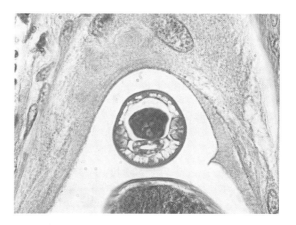

Figure 32.9
Transverse section of posterior end of mature first-stage *T. spiralis* larva in nurse cell. Body cavity contains reproductive tube and intestine. Lateral cords are prominent in this region. x570

Figure 32.10
Life cycle of *T. spiralis*.*

esophagus, an oblique nerve-ring space, and numerous nuclei of undifferentiated cells (Fig 32.2). First-stage larvae that penetrate skeletal muscle fiber mature without molting, growing to 0.8 to 1.3 mm in length and 30 to 40 μm in diameter, with a finely striated cuticle (Fig 32.6). The anterior tip of an encysted larva contains a muscular esophagus (Figs 32.6 and 32.7). The remainder of the larva is equally divided into anterior and posterior portions. The anterior portion contains a series of large stichocytes (Fig 32.8) that partially or completely enclose the esophagus. The posterior portion contains the intestine and immature reproductive tube (Fig 32.9). Larvae of *T. pseudospiralis* do not encapsulate, but otherwise are morphologically similar to *T. spiralis*.[1]

Life Cycle

Adult worms and encysted larvae of *T. spiralis* develop within a single warm-blooded host. Each infected mammal serves first as a definitive host, and then as a potential intermediate host. A second warm-blooded host is needed to perpetuate the life cycle of the parasite (Fig 32.10).

Adult *T. spiralis* mate in the small intestine; embryogenesis starts immediately after the oocytes are fertilized. Adult male worms die soon after copulation, but adult female worms live within the crypts of the small intestine for about 30 days (Fig 32.11). Each female gives birth to thousands of immature first-stage larvae that penetrate the small intestine, enter the portal system, and disseminate throughout the body through vascular channels. When first-stage larvae reach skeletal muscle they penetrate individual fibers, where they grow and mature, but do not molt. These encysted larvae remain infective for many years.

The *T. spiralis* life cycle is perpetuated when another mammal consumes infected flesh. First-stage larvae excyst in the stomach by the action of digestive juices on the capsule. Larvae then migrate to the duodenum and jejunum, where they attach to the mucosa, rapidly go through 4 molts, and develop into adults within 36 hours.

Clinical Features and Pathogenesis

Symptoms of trichinosis change as the worm develops in the host. Three stages in the life cycle of *T. spiralis* can produce 3 phases of disease in humans: the enteric phase, while adult worms are mating in the small bowel; the invasive phase, when larvae are migrating; and the encystment phase.

The enteric phase lasts about a week. Symptoms may be mild or severe, depending on the worm burden and the subspecies of *Trichinella* involved. Enteric symptoms may include nausea, vomiting, diarrhea alternating with constipation, and abdominal pain. Severe infections may cause anorexia, oliguria, disabling fatigue, and weakness.

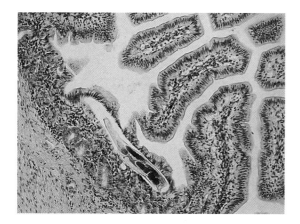

Figure 32.11
Adult *T. spiralis* in jejunal mucosa. Note edema and chronic inflammation. x85

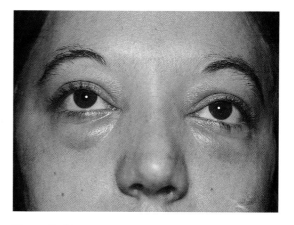

Figure 32.12
Patient with trichinosis showing swelling of eyelids and facial edema.

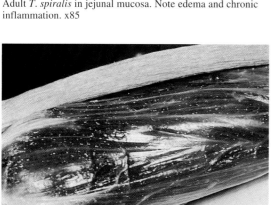

Figure 32.13
Numerous encysted larvae in gross specimen of skeletal muscle; each white spot represents at least 1 encysted larva.

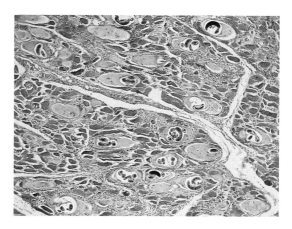

Figure 32.14
Numerous nurse cells containing encysted *T. spiralis* larvae in skeletal muscle of patient who died of trichinosis. Note extensive chronic inflammation. x27

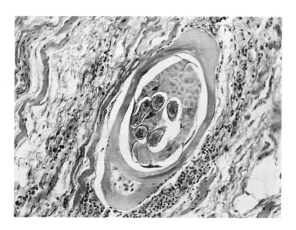

Figure 32.15
Nurse cell containing encysted, coiled, mature first-stage *T. spiralis* larva in skeletal muscle. Chronic inflammatory cells surround nurse cell. x75

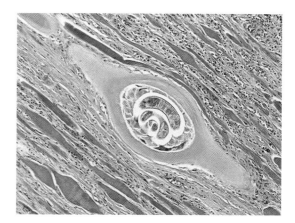

Figure 32.16
Mature, coiled, first-stage *T. spiralis* larva in nurse cell. Note hyaline appearance of infected muscle cell immediately around larva. Fibrous tissue and numerous chronic inflammatory cells infiltrate skeletal muscle. x90

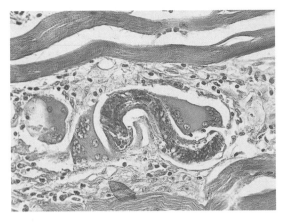

Figure 32.17
Degenerating mature first-stage *T. spiralis* larva within giant cell in skeletal muscle. Note extensive chronic inflammation in nurse cell. x235

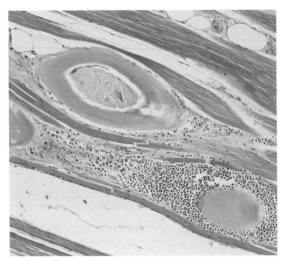

Figure 32.18
Amputation specimen in which trichinosis was discovered incidentally in skeletal muscle. Note 2 degenerated nurse cells. Inflammatory cells have penetrated larger nurse cell (upper left) and replaced its contents. x110

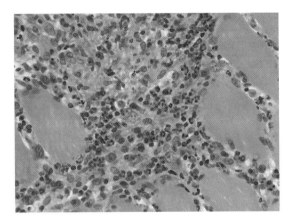

Figure 32.19
Skeletal muscle from patient with trichinosis showing numerous eosinophils, an infrequent finding in trichinosis. x235

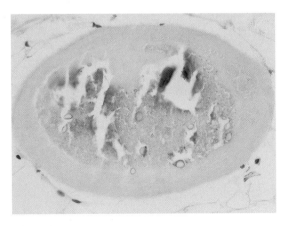

Figure 32.20
Calcified nurse cell with degenerated larva. x230

At its height, the invasive phase provokes an allergic reaction causing high fever, myalgia, palpebral and facial edema (Fig 32.12), and marked eosinophilia, known together as Beeson's 4 cardinal features of trichinosis. The high fever, which may reach 40°C, is not seen in most other helminthic diseases. In very light infections, patients may have only eosinophilia. Heavily infected patients may have splinter hemorrhages in the fingernail beds, beneath the conjunctiva, or in the retina. Systemic symptoms are largely the result of ischemia.[10]

Larvae passing through the central nervous system can cause meningitis and intracerebral hemorrhages that lead to ataxia, dizziness, and psychoses. When massive numbers of larvae invade the brain, extensive hemorrhage may lead to seizures, monoparesis, and even coma. Computed tomography of the brain may show small rarefactions in the cerebral hemispheric white matter or cortex.

Larvae migrating through the eye frequently cause conjunctivitis. Other ocular complications include painful restriction of eye movement, diplopia, dilation of pupils, nystagmus, and, less commonly, anterior and posterior uveitis.

As they migrate through the myocardium, larvae can produce cardiac arrhythmia, myocarditis, or sudden death. Pericardial effusion is common. Myocardial injury is reflected in abnormal electrocardiograms and plasma creatine phosphokinase isoenzyme changes. Bronchopneumonia or nephritis may develop in severe infections and cause death, often between the fourth and sixth weeks after infection.

Larvae encysting in skeletal muscle (Fig 32.13) incite a strong inflammatory response. Patients complain of muscular pain, swelling, weakness, and tenderness, and may experience hoarseness, dysphagia, dyspnea, and peripheral and facial edema. The myalgia and severe proximal muscle weakness may resemble polymyositis.[23]

Pathologic Features

Adult worms in the intestine cause edema and chronic inflammation (Fig 32.11). The small bowel is hyperemic and dilated, with petechiae on the serosa, many eosinophils in the lamina propria, swollen intestinal villi, and lymphoid nodules. During this phase of infection, symptoms are generally mild and nonspecific; therefore, biopsy specimens of the intestine that might reveal adult *T. spiralis* are rarely obtained.

First-stage larvae encyst only in skeletal muscle (Figs 32.14 and 32.15), most frequently that of the limbs and diaphragm, followed by the tongue, masseter, intercostal, extrinsic ocular, laryngeal, and paravertebral muscles. First-stage larvae encysting in skeletal muscle provoke a strong host response (Fig 32.16). The region immediately around the larvae becomes amorphous as the sarcomeres disappear. Muscle fibers become edematous, and there is a mixed inflammatory infiltrate of neutrophils, lymphocytes, and macrophages. Eventually, a broad, dense, outer zone develops around the worm (presumably composed of metabolic products), which appears as a capsule within the muscle fiber. The infected muscle cell, known as the nurse cell, contains 1 or more encysted larvae, an enlarged nucleus, and increased numbers of mitochondria. The nurse cell is ovoid, refractile, and has pointed ends (Fig 32.16). Each nurse cell usually contains a single larva, but large nurse cells may contain as many as 7 larvae. Macrophages and giant cells may surround the

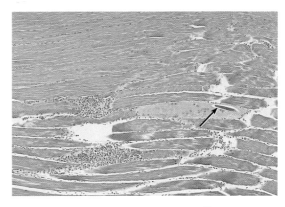

Figure 32.21
Anterior end of immature first-stage larva of *T. spiralis* (arrow) in nurse cell in skeletal muscle. Note extensive chronic inflammation and edema around nurse cell and surrounding skeletal muscle. Larvae in this specimen have a maximum diameter of 20 µm. x60

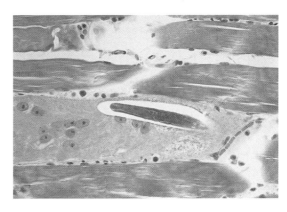

Figure 32.22
Higher magnification of Figure 32.21 showing anterior end of immature first-stage larva of *T. spiralis*. Internal structures are not as well-defined as in mature first-stage larvae. x250

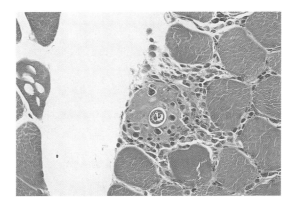

Figure 32.23
Transverse section of posterior end of immature first-stage larva of *T. spiralis* in nurse cell, containing primitive reproductive organ and intestine. Note chronic inflammation in and around nurse cell. x225

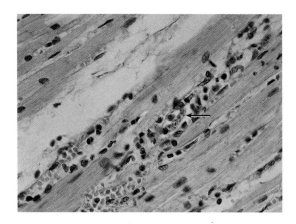

Figure 32.24
Myocardium of patient who died of trichinosis. Note migrating, immature first-stage larva of *T. spiralis* (arrow), edema, and chronic inflammation within myocardium. Larvae do not encyst within cardiac muscle. x290

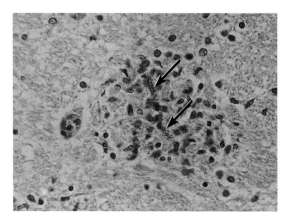

Figure 32.25
Brain of patient described in Figure 32.24. Note degenerating, immature first-stage larva of *T. spiralis* (arrows) and extensive chronic inflammation in brain. Larva is not encapsulated. x270

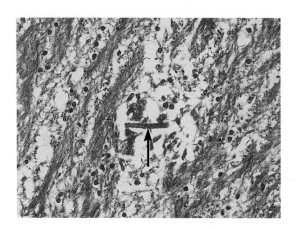

Figure 32.26
Immature, unencapsulated first-stage larva of *T. spiralis* (arrow) in brain, producing cerebral edema and chronic inflammation. x260

nurse cell or encysted larvae (Fig 32.17). Phagocytes occasionally penetrate the nurse cell and replace its contents (Fig 32.18); uncommonly, there are many eosinophils within the muscle (Fig 32.19). After approximately 10 months, leukocytic infiltrates and calcification thicken the outer surface of infected cells (Fig 32.20).

Biopsy samples of skeletal muscle generally show mature first-stage larvae, but occasionally reveal immature first-stage larvae in nurse cells. These immature first-stage larvae are smaller (10 to 30 µm in diameter), and their internal structures less clearly defined, than those of mature first-stage larvae (Figs 32.21 to 32.23).

First-stage larvae can migrate throughout the body and invade any host tissue, but they remain immature and unencapsulated outside of skeletal muscle. Larvae may invade cardiac muscle, but do not encyst within it; they do, however, provoke myocarditis with focal infiltration of neutrophils, eosinophils, and lymphocytes (Fig 32.24). These changes regress without scarring.

In the brain, immature first-stage larvae cause hyperemia, edema, gliosis, and chronic inflammation, including granuloma formation (Figs 32.25 and 32.26). Infected bone marrow may reveal marked eosinophilia and hyperplasia of red marrow.

Trichinella pseudospiralis develops in skeletal muscle without inducing a nurse cell.

Diagnosis

During the enteric phase, adult worms or larvae rarely appear in feces, but they can be recovered from the intestinal tract by biopsy or duodenal aspiration. Diagnosis is usually made by identifying encysted first-stage larvae in striated muscle from biopsy or autopsy specimens of superficial skeletal muscle. In a specimen of approximately 1 cm^3 of muscle, usually from the calf, larvae can be detected by 1) digestion in 1% pepsin and 1% hydrochloric acid and direct examination, 2) compression of part of the fresh muscle biopsy between glass slides and examination (Fig 32.27), or 3) histologic study of paraffin-embedded tissue.

Some serologic assays using purified and highly antigenic components of *T. spiralis* are sensitive and specific, and may be used for immunodiagnosis of trichinosis in humans and other mammals. Enzyme-linked immunosorbent assays (ELISA), using either the excretory-secretory (ES) antigen of stichocytes or crude somatic antigens from adult or first-stage larvae, are useful and popular. The major constituent of the ES antigen is a 45 kd protein.[11] An IgG-ELISA using larval ES antigen-based assays appears to be highly sensitive and specific when tested against sera from patients 57 days and 120 days post-infection; however, sera collected at 23 days and 700 days post-infection are less reactive and cross-react with other helminth antigens.[16,17]

For practical purposes, histopathologic changes in skeletal muscle of nonhuman hosts infected with subspecies of *T. spiralis* are virtually indistinguishable from those in humans (Figs 32.28 and 32.29).

Treatment and Prevention

There is no specific treatment for encysted *Trichinella* larvae. In experimental infections, thiabendazole, pyrantel pamoate, and mebendazole were all effective anthelmintics for adult worms in the intestine, and prevented the production of larvae.

Treating humans with anthelmintics for the intestinal phase of trichinosis must be done cautiously because of possible adverse reactions provoked by killing unencysted larvae in tissue, especially in the brain. When they die, these larvae release antigens that intensify tissue reaction. This increases the severity of symptoms and may be life-threatening in patients with heavy infections. Anti-inflammatory drugs, particularly prednisone, decrease tissue damage, but their use is controversial.[3] Mebendazole, along with corticosteroids, is generally recommended in heavily infected patients with encysted larvae of *T. spiralis*, but steroids are not recommended in patients with mild or moderate infections. In such cases, steroids may diminish the host's response to adult worms in the intestine and prolong larval production.

In view of the difficulties of early diagnosis and effective therapy, prevention is particularly

Figure 32.27
Specimen from skeletal muscle compressed between glass slides showing coiled larva of *T. spiralis*. Unstained x150

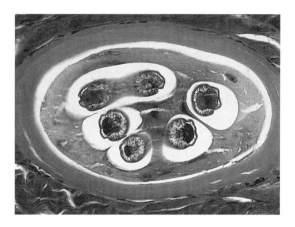

Figure 32.28
Mature, encapsulated, coiled first-stage larva of *T. spiralis* in nurse cell in skeletal muscle of a raccoon. x230

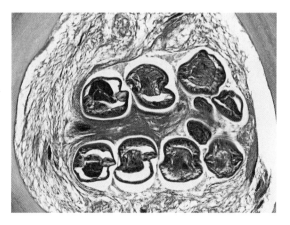

Figure 32.29
Transverse sections of mature, encapsulated, coiled first-stage larva of *T. spiralis* in nurse cell in skeletal muscle of a black bear. A man who ate meat from this bear later developed trichinosis. x365

important. Avoiding raw or poorly cooked meat, especially pork, pork products, boar, walrus, horse, or dog meat, is essential. Raw meat should not be sampled during food preparation, and microwave cooking of pork is not recommended. Larvae of *T. spiralis* in infected pork can be destroyed by freezing the meat to below -25°C for a minimum of 8 days[13]; however, *T. spiralis nativa* may survive freezing for over a year.[7] The best way to prevent human infection is to curtail infection in pigs, including cooking swill and controlling rodent and wild game populations in pigpens.[6]

References

1. Al Karmi TO, Faubert GM. Comparative analysis of mobility and ultrastructure of intramuscular larvae of Trichinella spiralis and Trichinella pseudospiralis. *J Parasitol* 1981;67:685–691.

2. Andrews JR, Ainsworth R, Abernethy D. Trichinella pseudospiralis in humans: description of case and its treatment. *Trans R Soc Trop Med Hyg* 1994;88:200–203.

3. Azab ME, Sanad MM, Kamel AM, Nasr ME. Immunopathological studies on the encystation phase of experimental trichinosis after cortisone and cyclophosphamide treatment. *J Egypt Soc Parasitol* 1992;22:177–188.

4. Beeson PB. Factors influencing the prevalence of trichinosis in man. *Proc Roy Soc Med* 1941;34:585–594.

5. Bryant C. Molecular variation in Trichinella. *Acta Trop* 1993;53:319–330.

6. Compton SJ, Celum CL, Lee C, et al. Trichinosis with ventilatory failure and persistent myocarditis. *Clin Infect Dis* 1993;16:500–504.

7. Dick TA. Infectivity of isolates of Trichinella and the ability of an arctic isolate to survive freezing temperatures in the raccoon, Procyon lotor, under experimental conditions. *J Wildl Dis* 1983;19:333–336.

8. Dick TA, Chadee K. Interbreeding and gene flow in the genus Trichinella. *J Parasitol* 1983;69:176–180.

9. Flockhart HA, Harrison SE, Dobinson AR, James ER. Enzyme polymorphism in Trichinella. *Trans R Soc Trop Med Hyg* 1982;76:541–545.

10. Fourestie V, Douceron H, Brugieres P, Ancelle T, Lejonc JL, Gherardi RK. Neurotrichinosis. A cerebrovascular disease associated with myocardial injury and hypereosinophilia. *Brain* 1993;116:603–616.

11. Homan WL, Derksen AC, van Knapen F. Identification of diagnostic antigens from Trichinella spiralis. *Parasitol Res* 1992;78:112–119.

12. Hou HW. Survey of an outbreak of trichinosis caused by eating roast dog meat [in Chinese]. *Chung Hua Yu Fang I Hsueh Tsa Chih* 1983;17:109–110.

13. Hulinska D, Figallova V, Shaikenov B. Effects of low temperatures on larvae of the genus Trichinella. *Folia Parasitol (Praha)* 1985;32:211–216.

14. Kim CW. The significance of changing trends in trichinellosis. *Southeast Asian J Trop Med Public Health* 1991;22:316–320.

15. Kozek WJ. Trichinella spiralis: morphological characteristics of male and female intestine-infecting larvae. *Exp Parasitol* 1975;37:380–387.

16. Mahannop P, Chaicumpa W, Setasuban P, Morakote N, Tapchaisri P. Immunodiagnosis of human trichinellosis using excretory-secretory (ES) antigen. *J Helminthol* 1992;66:297–304.

17. Morakote N, Khamboonruang C, Siriprasert V, Suphawitayanukul S, Marcanantachoti S, Thamasonthi W. The value of enzyme-linked immunosorbent assay (ELISA) for diagnosis of human trichinosis. *Trop Med Parasitol* 1991;42:172–174.

18. Nelson GS, Rickman R, Pester FR. Feral trichinosis in Africa. *Trans R Soc Trop Med Hyg* 1961;55:514–517.

19. Pozio E. Present knowledge of the taxonomy, distribution and biology of genera of Trichinella (Nematoda, Trichinellidae) [in Italian]. *Ann Ist Super Sanita* 1989;25:615–623.

20. Pozio E, La Rosa G. General introduction and epidemiology of trichinellosis. *Southeast Asian J Trop Med Public Health* 1991;22:291–294.

21. Pozio E, Rossi P, Amati M, Mancini Barbieri F. Genetic differentiation of the Trichinella genus with isoenzymatic analysis [in Italian]. *Parassitologia* 1987;29:49–62.

22. Pozio E, Varese P, Morales MA, Croppo GP, Pelliccia D, Bruschi F. Comparison of human trichinellosis caused by Trichinella spiralis and by Trichinella britovi. *Am J Trop Med Hyg* 1993;48:568–575.

23. Santos Duran-Ortiz J, Garcia-de la Torre I, Orozco-Barocio G, Martinez-Bonilla G, Rodriguez-Toledo A, Herrera-Zarate L. Trichinosis with severe myopathic involvement mimicking polymyositis. Report of a family outbreak. *J Rheumatol* 1992;19:310–312.

24. Schantz PM, McAuley J. Current status of food-borne parasitic zoonoses in the United States. *Southeast Asian J Trop Med Public Health* 1991;22:65–71.

33

Capillariasis
Intestinal and Hepatic

John H. Cross *and*
Ronald C. Neafie

■ Intestinal Capillariasis

Introduction

Definition

Intestinal capillariasis is infection by the nematode *Capillaria philippinensis*. This parasite is capable of producing several generations of adult worms within a single host. As the infection worsens, diarrhea, abdominal pain, and borborygmi develop. Protein-losing enteropathy, weight loss, weakness, anorexia, edema, and cachexia evolve as the disease progresses. Left untreated, the infection can be fatal.

Synonyms

Intestinal capillariasis was first called Pudoc's disease or mystery disease of Pudoc, and later, capillariasis philippinensis.

General Considerations

The first case of intestinal capillariasis was reported in 1964, in a man from northern Luzon in the Philippines who died in a hospital in Manila.[7] At autopsy there were thousands of worms in the small intestine. Chitwood et al[7] described this worm as a new species. The disease was next seen in 1965 in Pudoc West, a barrio along the western coast of central Luzon. In the epidemic of 1967 to 1970, there were over 1400 infections, with 95 deaths.[17] A new endemic focus was identified in the central Philippines in 1980. Recently, only a few cases per year have been seen in northern Luzon. Intestinal capillariasis was identified in Thailand in 1970, and in the 1980s infections were reported in Iran, Egypt, Japan, and Taiwan, and in Indonesia in 1992.[6,9,10]

Epidemiology

Epidemiologic data are available only from northern Luzon, where there have been over 1900 cases with 115 deaths. Men acquire the infection at twice the rate of women; most patients are middle-aged. The source of infection appears to be uncooked freshwater fish eaten by certain population groups in endemic areas.[15] In countries reporting the disease, most infected persons have consumed raw fish. Fish caught in Philippine lagoons or purchased in the marketplace have produced infections when fed to Mongolian gerbils.[11] Monkeys and fish-eating birds have also been experimentally infected.[4,14] *Capillaria philippinensis* is now considered a parasite of migratory fish-eating birds, with a natural life cycle involving birds as definitive hosts and fish as intermediate hosts. Birds disperse the parasitosis.[14]

Infectious Agent

Morphologic Description

Capillaria philippinensis is a tiny worm with the distinguishing characteristic of the trichurid nematodes–a stichosome composed of stichocytes (Fig 33.1). *Capillaria philippinensis* is further characterized by 3 bacillary bands extending most of the length of the worm (Figs 33.2 and 33.3).

Adult female *C. philippinensis* are 2.5 to 4.3 mm long and 20 to 50 μm in diameter; adult males are 2.3 to 3.17 mm long and 20 to 30 μm in diameter. Adult worms are small, slender, and of uniform diameter. Adult worms have 2 body regions: an anterior portion containing a single row of large glandular cells (stichocytes) which partially or completely enclose the esophagus (Figs 33.1 and 33.3), and a posterior portion containing the intestine and reproductive organs. The cuticle is smooth and 1 to 2 μm thick.

The female *C. philippinensis* is divided almost equally into anterior and posterior regions. The stichosome is 1.06 to 1.85 mm in length. The vulva is immediately posterior to the esophageal-intestinal junction (Fig 33.4). The single reproductive tube is confined to the posterior region of the worm and consists of a vulva, vagina, seminal receptacle, oviduct, and ovary (Figs 33.5 and 33.6). The oviduct reflexes twice near the middle of the posterior region of the body.

In male *C. philippinensis,* the anterior segment is almost one and a half times the length of the posterior segment. The single reproductive tube is confined to the posterior region of the body and consists of a testis, vas deferens, and ejaculatory duct (Figs 33.7 and 33.8). The ejaculatory duct joins the intestine in the posterior end of the worm to form the cloaca. The single spicule is 230 to 300

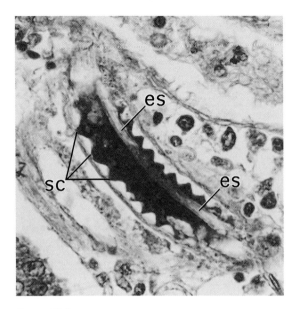

Figure 33.1
Longitudinal section through anterior region of adult *C. philippinensis* in small intestine, depicting single row of stichocytes (sc) and enclosed tubular esophagus (es). x890

μm in length and has a long nonspinous sheath (Fig 33.9).

Capillaria philippinensis eggs are 36 to 45 μm long, 21 μm wide, and have bipolar plugs. Eggs may be unembryonated with a thick, striated shell (Figs 33.10 and 33.11) or embryonated and thin-shelled (Figs 33.12 and 33.13). Embryos without shells, 5 to 10 μm in diameter, are frequently seen in the uteri of female worms, sometimes along with embryonated eggs (Fig 33.14). Various stages of *C. philippinensis* larvae may be found in the host intestine. They can be distinguished from adult worms by their smaller size, stichocytes frequently in a double row, and their immature reproductive tubes (Fig 33.15).

Life Cycle and Transmission

The life cycle of *C. philippinensis* was established experimentally in monkeys and gerbils; it is presumably similar in humans (Fig 33.16).

Some female worms in the small intestine of a host produce eggs and others produce larvae; some produce both. Larvae remain in the intestine, but eggs are passed in feces. Those that reach water embryonate in 5 to 10 days. Eggs that are ingested by fish hatch in the intestine and release embryos that develop into infective larvae in about 3 weeks. When infected fish are fed to monkeys or

Mongolian gerbils, the larvae develop into adult worms in 12 to 14 days. When these worms mate, the first-generation females produce larvae that remain in the intestine and develop into second-generation females, which then produce eggs that are passed in feces. A few female worms remain in the intestine, producing larvae that sustain the infection. In humans, this autoinfection leads to hyperinfection with massive numbers of worms. As the worm population increases, the disease becomes more severe and, if untreated, may be fatal. At autopsy, as many as 200 000 worms have been recovered from a liter of intestinal contents.

Clinical Features and Pathogenesis

At the onset of infection with *C. philippinensis*, patients usually have dull epigastric or generalized abdominal pain, chronic diarrhea, and loud borborygmi.[22] As the disease progresses, malaise, anorexia, nausea, and vomiting develop. Intractable diarrhea leads to ascites, weight loss, cachexia, and death. As a rule, the natural course of infection is from 3 to 6 months.

Stools, passed as often as 5 to 10 times a day, are watery and voluminous and contain increased amounts of fat. Cachexia is marked in late stages of the disease. Muscular wasting and loss of subcutaneous fat may be so severe that hyperperistalsis is visible. Other clinical signs include hypotension, distant heart sounds, gallop rhythm, pulsus alternans, abdominal distention, epigastric tenderness, edema, hyporeflexia, and hypoproteinemia. Metabolic studies reveal malabsorption of fats and sugar; loss of protein; low levels of carotene, potassium, and calcium; decreased levels of plasma protein; and sometimes decreased excretion of xylose. Protein loss in stools is up to 15 times higher than normal. IgE levels increase while IgM and IgG levels decrease. Patients with low levels of immunoglobulins may develop bacterial enteritis or bacterial septicemia. Death is generally a result of severe malnutrition complicated by infections, especially of the lung. Mortality rates may reach 20%, but treatment leads to dramatic improvement. In a study covering a 3-year period, the mortality rate was 6.7%.

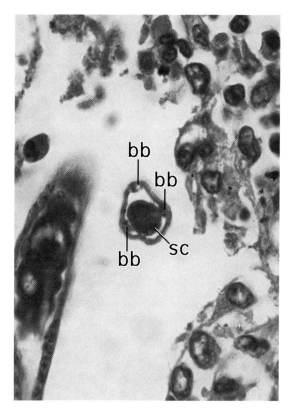

Figure 33.2
Transverse section through anterior region of adult *C. philippinensis* in small intestine, showing stichocyte (sc) and 3 bacillary bands (bb). x1100

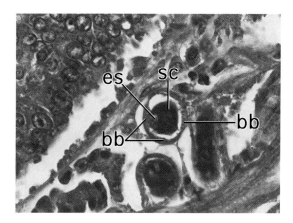

Figure 33.3
Transverse section through anterior region of adult *C. philippinensis* in small intestine, demonstrating esophagus (es) within a stichocyte (sc), and 3 bacillary bands (bb). x615

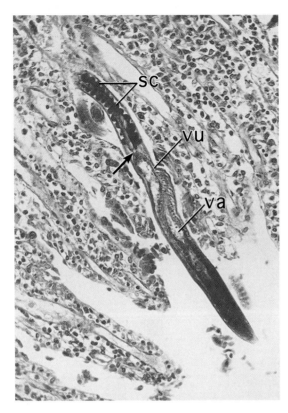

Figure 33.4
Longitudinal section through adult female *C. philippinensis* in small intestine at level of esophageal-intestinal junction (arrow). Stichocytes (sc), vulva (vu), and vagina (va) are also visible. x200

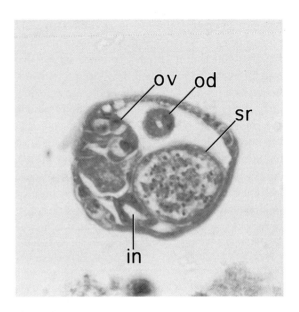

Figure 33.5
Transverse section through posterior region of adult female *C. philippinensis* in small intestine, demonstrating oviduct (od), ovary (ov), seminal receptacle (sr), and intestine (in). x1525

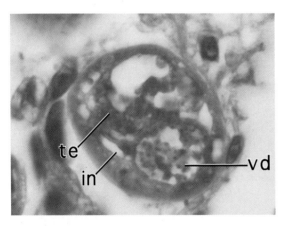

Figure 33.7
Transverse section through posterior region of adult male *C. philippinensis* in small intestine, demonstrating testis (te), vas deferens (vd), and intestine (in). x1625

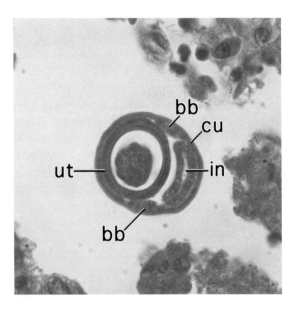

Figure 33.6
Transverse section through posterior region of gravid *C. philippinensis* in small intestine. Note intestine (in), uterus with single egg (ut), thin cuticle (cu), and 2 lateral bacillary bands (bb). x825

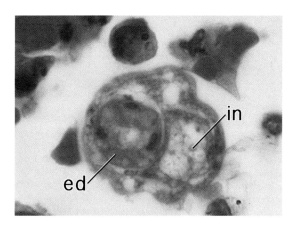

Figure 33.8
Transverse section through posterior region of adult male *C. philippinensis* in small intestine. Note thick-walled ejaculatory duct (ed) and intestine (in). x1475

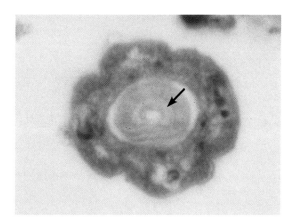

Figure 33.9
Transverse section through posterior end of adult male *C. philippinensis* in small intestine, displaying single spicule (arrow). x2185

Pathologic Features

The pathologic features of intestinal capillariasis were described at autopsy by Canlas et al[5] and Fresh et al.[18] The small intestine is thickened, indurated, congested, and distended with fluid (Fig 33.17). Microscopically, intestinal villi tend to be flattened and atrophic. The small intestine, especially the jejunum, usually contains many adults and larvae of *C. philippinensis* (Figs 33.18 and 33.19). Worms are most commonly found in the lumens of, or the crypts of the mucosa of, the jejunum, upper ileum, and duodenum. The anterior ends of the worms sometimes penetrate into the lamina propria. Worms frequently go unnoticed, but usually there is chronic inflammation. The inflammatory cellular infiltrate is composed predominantly of plasma cells and lymphocytes. Eosinophils are usually sparse, but can be abundant. Degenerating worms may produce microabscesses containing many neutrophils and eosinophils (Fig 33.20).

How such clinical and histopathologic changes are produced is unknown; however, damage to the mucosa, as seen by electron microscopy, probably contributes to protein, fluid, and electrolyte loss.[21] Larvae are occasionally identified in the liver (Fig 33.21). An infected liver may show secondary changes, including congestion, atrophy, deposition of hemosiderin, and fatty metamorphosis.

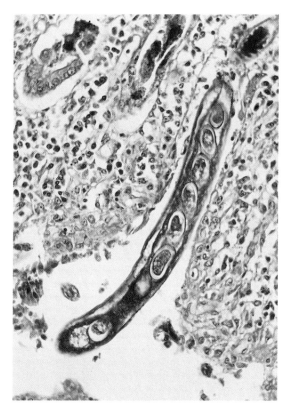

Figure 33.10
Longitudinal section through gravid *C. philippinensis* in small intestine, showing a row of thick-shelled, unembryonated eggs in the vagina. x270

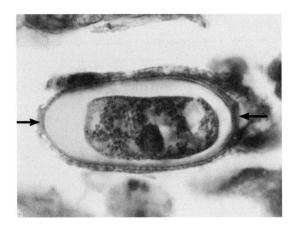

Figure 33.11
Section of small intestine showing unembryonated *C. philippinensis* egg with thick, striated shell and bipolar plugs (arrows). x1370

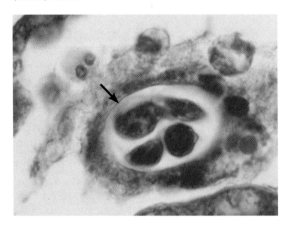

Figure 33.13
Section of small intestine demonstrating thin-shelled (arrow), embryonated egg of *C. philippinensis*. x1525

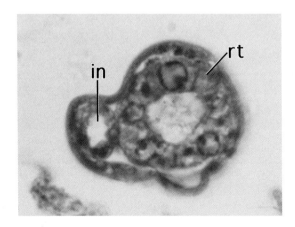

Figure 33.15
Section of small intestine showing larva of *C. philippinensis* in lumen. Note intestine (in) and immature reproductive tube (rt). x1800

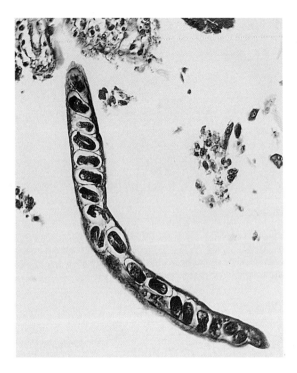

Figure 33.12
Longitudinal section through gravid *C. philippinensis* in small intestine, depicting a row of thin-shelled, embryonated eggs in the uterus. x238

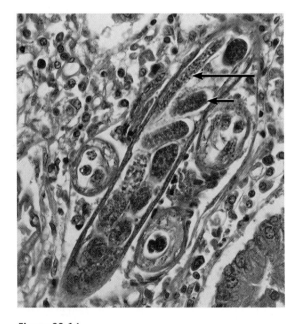

Figure 33.14
Longitudinal section through gravid *C. philippinensis* in small intestine. Uterus contains both thin-shelled, embryonated eggs (short arrow) and embryos without shells (long arrow). x395

Congestion and hyperplasia of the red pulp of the spleen have also been described.[18]

Diagnosis

Diagnosis is made by identifying eggs, larvae, or adults of *C. philippinensis* in the stool or in tissue specimens of the small intestine. In stool specimens, eggs of *C. philippinensis* (Fig 33.22) resemble those of *Trichuris trichiura*. Eggs of *C. philippinensis* are distinguished from those of *T. trichiura* by their nonprotruding plugs, more quadrate shape and slightly smaller size, and shells that are usually pitted. In biopsy specimens, adult female *Strongyloides stercoralis* are about the same diameter as adult female *C. philippinensis*; however, *S. stercoralis* lack a stichosome and bacillary bands. Available serologic tests for intestinal capillariasis are not dependable.[2,16]

Treatment

Treatment consists of electrolyte replacement (especially potassium), antidiarrheics, a high-protein diet, and prolonged administration of an anthelmintic. The drugs of choice are mebendazole, 400 mg a day in 2 divided doses for 20 days; albendazole at the same doses for 10 days; or flubendazole, 200 mg twice daily for 30 days.[13] Strict compliance with this regimen is essential. Relapses will occur if the drugs are discontinued too soon. Early treatment can prevent death.[12]

■ Hepatic Capillariasis

Introduction

Definition
Hepatic capillariasis is infection by the nematode *Capillaria hepatica* (Bancroft, 1893).

Synonyms
Trichocephalus hepaticus, *Trichosoma tenuissimum*, *Trichosoma tenue*, *Capillaria leidyi*, *Hepaticola hepatica*, *Hepaticola soricicola*, *Hepaticola anthropopitheci*, and *Calodium hepaticum* are all synonyms for *C. hepatica*.

Epidemiology

Hepatic capillariasis is a common liver infection of rats worldwide. The *C. hepatica* nematode has also been reported in mice, many wild rodents, domestic rabbits, hares, cottontails, cats, dogs, shrews, coyotes, pronghorn deer, pigs, beavers, muskrats, gerbils, meerkats, opossums, spotted skunks, peccaries, ground squirrels, prairie dogs, monkeys, and chimpanzees. Humans acquire the infection by ingesting infective eggs in contaminated soil, water, or food, making unsanitary conditions the most important factor in human infections. There have been 28 confirmed infections in humans from many geographic locations. Thirteen of these infections were in patients under 10 years of age, some of whom habitually ate dirt.[3,19]

Infectious Agent
Morphologic Description
In adult worms, the posterior part of the body is not distinctly wider than the anterior part, a characteristic that places *C. hepatica* in the subfamily Capillariinae.

Capillaria hepatica females are 52 to 104 mm long and 78 to 184 µm in diameter, and males are 22 mm long and 26 to 78 µm in diameter. Adult worms have 2 body regions of about equal length. The anterior region contains the stichosome enclosing the esophagus, and 3 bacillary bands (Fig 33.23).[23] The posterior region contains the intestine and a single reproductive tube (Fig 33.24). Usually only 2 of the 3 bacillary bands are visible in the posterior region of the worm. The cuticle is thin and finely striated. Muscle cells are poorly developed in both sexes. Females are oviparous and lay unembryonated eggs. The vulva is near the esophageal-intestinal junction. Males have a single spicule with an unarmed spicular sheath.

Capillaria hepatica eggs are 48 to 66 µm by 28 to 36 µm and have bipolar plugs (Fig 33.25). They are barrel-shaped and have a thick, 2-layered shell, the outer layer of which is distinctly striated.

Life Cycle and Transmission

By studying infections in rats and mice, Luttermoser was able to delineate the life cycle of *C. hepatica* in 1953. The life cycle is direct and requires only a single host. Adult worms invade the liver of the host, usually a rodent, where they lay unembryonated eggs in the surrounding parenchyma. The eggs remain in the liver of the host until it dies and decomposes, or until it is eaten by another animal. Eggs are not passed in the host's feces and are noninfective at this stage. Eggs require air and damp soil to embryonate and become infective; under favorable conditions this usually occurs within 30 days. If the infected liver of the animal host is eaten by a predator or scavenger, the nonembryonated eggs pass unchanged through the animal's intestinal tract and are disseminated within a few days. Some insects may also disseminate eggs, which will then embryonate if they reach suitable soil. The cycle repeats itself when infective eggs are ingested by suitable mammalian hosts, including humans. In humans, the infective eggs hatch in the intestine, releasing larvae that penetrate the intestinal wall and migrate to the liver via the portal vein. Sometimes, larvae migrate to lungs, kidneys, or other organs. Larvae develop into adult worms that mate in the liver within 4 weeks. Mating pairs then die, but not before the female has laid thousands of eggs in the liver. Cannibalism among small rodents is a prime factor in transmission of *C. hepatica*. Humans are infected by ingesting infective eggs in contaminated water, food, or soil.

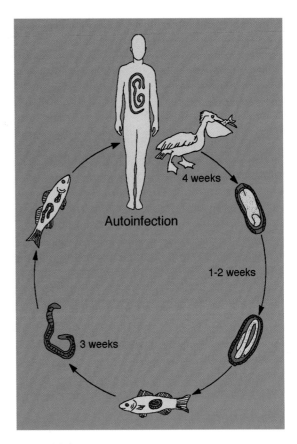

Figure 33.16
Proposed life cycle of *C. philippinensis* based on experimental infections in monkeys and gerbils.

Clinical Features and Pathogenesis

Clinical features vary considerably depending upon the worm burden. Single worms rarely cause clinical symptoms; lesions are usually an incidental finding presenting as solitary granulomas in the liver. A heavy worm burden may cause severe symptoms and even death, especially when female worms are laying eggs. In these patients, hepatic capillariasis presents as a complex of variable symptoms: persistent fever, hepatomegaly, leukocytosis with hypereosinophilia, anemia, and pulmonary involvement.[8] Similar clinical symptoms may occur in visceral larva migrans.

Pathologic Features

Capillaria hepatica is the only adult capillarid reported in the liver of humans. Hepatic capillariasis is characterized by the formation of granulomas and abscesses (Fig 33.26), usually in the right lobe. Granulomas are 0.5 to 3.5 cm in diameter, necrotizing, and frequently stellate (Fig 33.27). The central area of necrosis is surrounded by connective tissue, macrophages, eosinophils, lymphocytes, plasma cells, and, occasionally, foreign body giant cells (Fig 33.28). Eosinophils and Charcot-Leyden crystals may be abundant (Fig 33.29). In solitary granulomas of the liver, the adult worms are usually highly degenerated and can be easily overlooked (Figs 33.30 and 33.31). Worms range from 26 to 184 µm in diameter. Even degenerated

Figure 33.17
Mucosal surface of small intestine, showing edema and diffuse hyperemia. x1

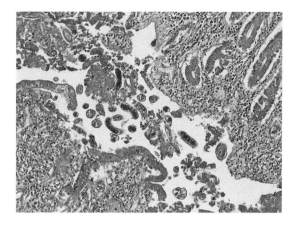

Figure 33.18
Jejunum from patient infected with *C. philippinensis*. Note diffuse inflammation, atrophy of villi, and adult worms and larvae in lumen and crypts. x100

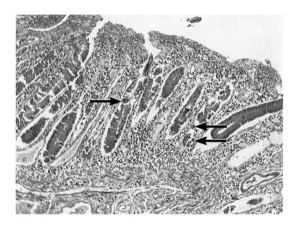

Figure 33.19
Three transverse sections of *C. philippinensis* larvae (arrows) embedded in intestinal glands. x70

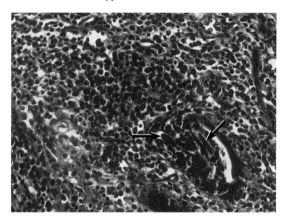

Figure 33.20
Degenerating larva (arrows) of *C. philippinensis* provoking a microabscess composed of eosinophils in small intestine. x225

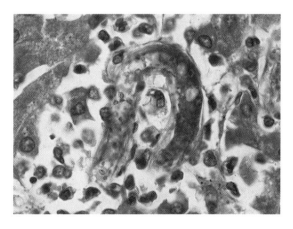

Figure 33.21
Coiled, degenerated larva of *C. philippinensis* provoking chronic inflammation in liver. x610

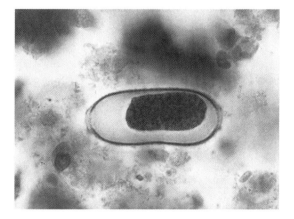

Figure 33.22
Typical egg of *C. philippinensis* in feces. Egg is unembryonated and has bipolar plugs. Trichrome x770

C. hepatica can often be identified by the persisting stichosome and/or bacillary bands (Fig 33.31). *Capillaria hepatica* eggs may stimulate formation of fibrotic nodules (Figs 33.32 and 33.33). There may be extensive chronic inflammation in liver parenchyma adjacent to the granulomas.

Diagnosis

Diagnosis is made by identifying adult worms or eggs in biopsy or autopsy specimens of liver (Figs 33.30 to 33.33). A needle biopsy may be diagnostic.[1] When liver destruction is widespread and *C. hepatica* eggs are visible, diagnosis is fairly simple. In cases involving a solitary granuloma in the liver, a single worm, and no eggs, it may be necessary to study many sections to identify the etiologic agent. Finding eggs of *C. hepatica* in human feces is not evidence of infection. Eggs of *Capillaria* (Eucoleus) *aerophila*, discussed in chapter 36, are rarely found in human sputum or feces.

Treatment and Prevention

Various anthelmintic drugs have been used to treat *C. hepatica* infections, including antimony sodium gluconate, dithiazanine iodide, diethylcarbamazine, disophenol, pyrantel tartrate, levamisole, ivermectin, and thiabendazole, the drug of choice.[20] Steroids may inhibit further granulomatous inflammation in the liver.[20] Habitual or accidental ingestion of contaminated soil, the primary method of acquiring *C. hepatica* infections, must be avoided.

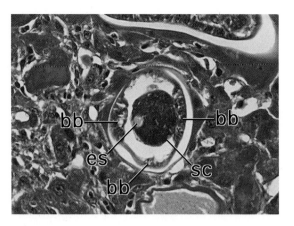

Figure 33.23
Transverse section through anterior region of adult *C. hepatica* in rat liver. Three bacillary bands (bb) and a stichocyte (sc) enclosing the esophagus (es) are illustrated. x350

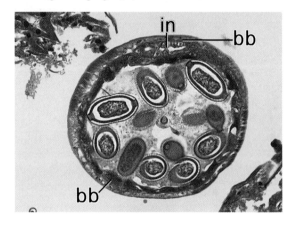

Figure 33.24
Transverse section through posterior region of gravid *C. hepatica* in rat liver. Two of the 3 bacillary bands (bb), the intestine (in), and uterus with eggs are depicted. x230

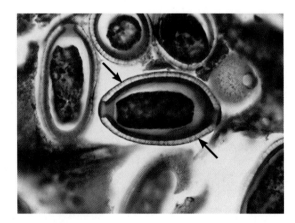

Figure 33.25
Characteristic eggs of *C. hepatica* in rat liver. Egg is unembryonated, and has bipolar plugs and a thick, striated shell (arrows). Movat x590

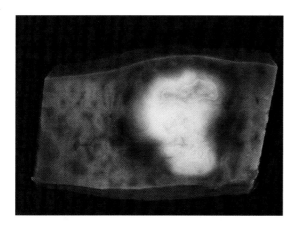

Figure 33.26
Necrotizing stellate granuloma in liver of 43-year-old man from Oregon. The lesion was an incidental finding at autopsy and contained an adult male *C. hepatica*.* x4.5

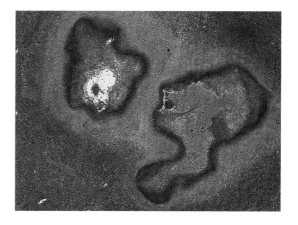

Figure 33.27
Section through lesion in liver described in Figure 33.26 showing stellate areas of necrosis. x8

Figure 33.28
Lesion in liver of patient infected by adult male *C. hepatica*, demonstrating area of necrosis, foreign body giant cell (arrow), and chronic inflammation. x105

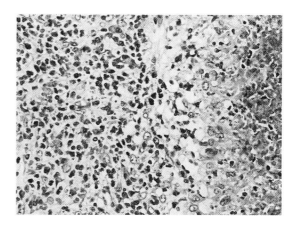

Figure 33.29
Prominent eosinophilia in liver from patient infected with *C. hepatica*. Biebrich scarlet x310

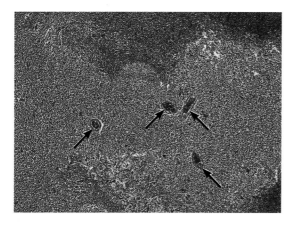

Figure 33.30
Necrotic granuloma in liver showing 4 fragments (arrows) of degenerated male *C. hepatica*. x40

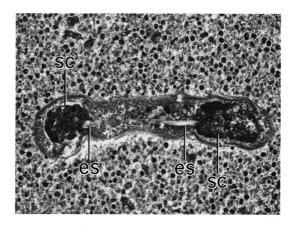

Figure 33.31
Anterior region of adult male *C. hepatica* in necrotizing granuloma of liver. Tubular esophagus (es) is enclosed by stichocytes (sc). x180

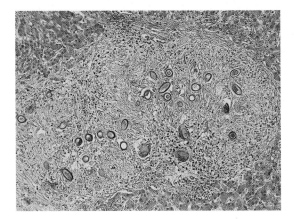

Figure 33.32
Granulomatous and fibrotic nodule containing eggs of *C. hepatica* in liver of infected patient. x100

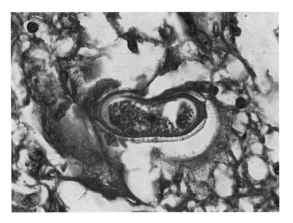

Figure 33.33
Higher magnification of lesion in Figure 33.32, depicting typical *C. hepatica* egg. Egg is unembryonated, and has bipolar plugs and a thick, striated shell. x650

References

1. Attah EB, Nagarajan S, Obineche EN, Gera SC. Hepatic capillariasis. *Am J Clin Pathol* 1983;79:127–130.
2. Banzon TC, Lewert RM, Yogore MG. Serology of Capillaria philippinensis infection: reactivity of human sera to antigens prepared from Capillaria obsignata and other helminths. *Am J Trop Med Hyg* 1975;24:256–263.
3. Berger T, Degremont A, Gebbers JO, Tonz O. Hepatic capillariasis in a 1-year-old child. *Eur J Pediatr* 1990;149:333–336.
4. Bhaibulaya M, Indra-Ngarm S. Amaurornis phoenicurus and Ardeola bacchus as experimental definitive hosts for Capillaria philippinensis in Thailand. *Int J Parasitol* 1979;9:321–322.
5. Canlas BC, Cabrera BD, Davis U. Human intestinal capillariasis. 2. Pathological features. *Acta Med Philipp* 1967;4:84–91.
6. Chichino G, Bernuzzi AM, Bruno A, et al. Intestinal capillariasis (Capillaria philippinensis) acquired in Indonesia: a case report. *Am J Trop Med Hyg* 1992;47:10–12.
7. Chitwood MB, Valesquez C, Salazar NG. Capillaria philippinensis sp. n. (Nematoda: Trichinellida) from the intestine of man in the Philippines. *J Parasitol* 1968;54:368–371.
8. Choe G, Lee HS, Seo JK, et al. Hepatic capillariasis: first case report in the Republic of Korea. *Am J Trop Med Hyg* 1993;48:610–625.
9. Cross JH. Intestinal capillariasis. *Clin Microbiol Rev* 1992;5:120–129.
10. Cross JH. Intestinal capillariasis. *Parasitol Today* 1990;6:26–28.
11. Cross JH, Banzon T, Singson C. Further studies on Capillaria philippinensis: development of the parasite in the Mongolian gerbil. *J Parasitol* 1978;64:208–213.
12. Cross JH, Basaca-Sevilla V. Intestinal capillariasis. *Prog Clin Parasitol* 1989;1:105–119.
13. Cross JH, Basaca-Sevilla V. Albendazole in the treatment of intestinal capillariasis. *Southeast Asian J Trop Med Public Health* 1987;18:507–510.
14. Cross JH, Basaca-Sevilla V. Experimental transmission of Capillaria philippinensis to birds. *Trans R Soc Trop Med Hyg* 1983;77:511–514.
15. Cross JH, Bhaibulaya, M. Intestinal capillariasis in the Philippines and Thailand. In: Croll NA, Cross JH, eds. *Human Ecology and Infectious Diseases*. New York, NY: Academic Press; 1983:103–136.
16. Cross JH, Chi JC. The ELISA test in the detection of antibodies to some parasitic diseases in Asia. In: *Proceedings, 18th SEAMEO TROPMED Seminar: Current Concepts in the Diagnosis and Treatment of Parasitic and Other Tropical Diseases*. Kuala Lumpur, Malaysia; 1978:178–182.
17. Detels R, Gutman L, Jaramillo J, et al. An epidemic of intestinal capillariasis in man. A study in a barrio in Northern Luzon. *Am J Trop Med Hyg* 1969;18:676–682.
18. Fresh JW, Cross JH, Reyes V, Whalen GE, Uylangco CV, Dizon JJ. Necropsy findings in intestinal capillariasis. *Am J Trop Med Hyg* 1972;21:169–173.
19. Kokai GK, Misic S, Perisic VN, Grujovska S. Capillaria hepatica infestation in a 2-year-old girl. *Histopathology* 1990;17:275–277.
20. Pannenbecker J, Miller TC, Muller J, Jeschke R. Severe liver involvement by Capillaria hepatica [in German]. *Monatsschr Kinderheilkd* 1990;138:767–771.
21. Sun SC, Cross JH, Berg HS, et al. Ultrastructural studies of intestinal capillariasis—Capillaria philippinensis in human and gerbil hosts. *Southeast Asian J Trop Med Public Health* 1974;5:524–533.
22. Whalen GE, Rosenberg EB, Strickland GT, Gutman RA, Cross JH, Watten RH. Intestinal capillariasis. A new disease in man. *Lancet* 1969;1:13–16.
23. Wright KA. Cytology of the bacillary bands of the nematode Capillaria hepatica (Bancroft, 1893). *J Morphol* 1963;112:233–259.

34

Halicephalobiasis

Chris H. Gardiner,
Wayne M. Meyers, *and*
Ronald C. Neafie

Introduction

Definition
Halicephalobiasis is infection by nematodes of the genus *Halicephalobus*. *Halicephalobus deletrix* is generally accepted as the etiologic agent of halicephalobiasis, but many authorities suggest that until more definitive taxonomic studies are completed, these nematodes should remain generic.

Synonyms
Halicephalobus was called *Micronema* until 1973, when it was discovered that the name *Micronema* already applied to a genus of fish, prompting a redesignation of the nematode as *Halicephalobus*. *Micronema deletrix* was the original name for the *Halicephalobus* sp that infects humans. Halicephalobiasis was formerly known as micronemiasis.

Epidemiology

Halicephalobiasis is primarily a disease of horses. It is distributed worldwide, with infections reported in Colombia, Egypt, Germany, Japan, the Netherlands, Switzerland, the United Kingdom, and the United States.[5] There have been 3 reported infections in humans, 1 each from Canada,[5] Texas,[9] and Washington, DC,[3] all of which were fatal.

Infectious Agent

Morphologic Description
Nematodes of the genus *Halicephalobus* (order Rhabditida) are small, rhabditoid, free-living saprophytes that thrive in soil, manure, and humus. Adult female worms are 250 to 460 µm by 15 to 25 µm (Figs 34.1 and 34.2). Larval stages are smaller. The cuticle is thin, with fine transverse striations. Eggs measure 30 to 45 µm by 10 to 15 µm. Worms in tissue can be identified by their characteristic rhabditiform esophagus consisting of a corpus, isthmus, and bulb (Fig 34.2). The

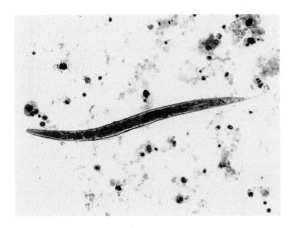

Figure 34.1
Female *Halicephalobus* sp recovered from 5-year-old boy who died of halicephalobiasis. Worm measured 290 by 16 μm. Note sharply pointed tail. Unstained x215

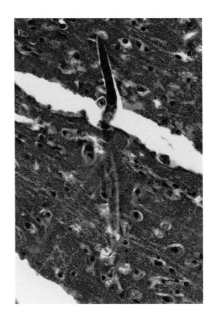

Figure 34.2
Longitudinal section through entire length of female *Halicephalobus* sp in brain of patient described in Figure 34.1. Worm measured 255 by 15 μm. x270

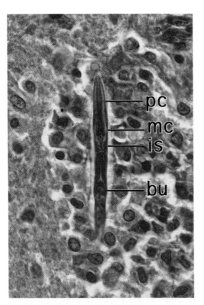

Figure 34.3
Longitudinal section of anterior end of mature female *Halicephalobus* sp in cerebral exudate. Rhabditiform esophagus is composed of a procorpus (pc), metacorpus (mc), isthmus (is), and bulb (bu). x585

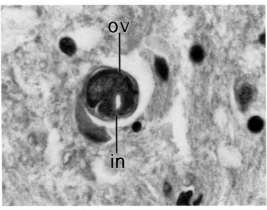

Figure 34.4
Transverse section of female *Halicephalobus* sp in brain showing intestine (in) and portion of single ovary (ov). x1030

single ovary of the adult female (Fig 34.3) reflexes dorsad (Fig 34.4); the remaining reproductive structures are in the posterior half of the worm. In adults and larvae, the tail tapers to a point (Fig 34.1). Gravid females usually have a single egg (Fig 34.5), which may contain a first-stage larva. The wide range in length of adult females and the diverse arrangements of esophageal components suggest that more than 1 species of *Halicephalobus* may infect both humans and horses.

Life Cycle and Transmission

Halicephalobus sp are parthenogenetic when parasitic, so infected tissues usually contain many parasites, eggs, developing larvae, and adult females. The parasite is thought to enter a host through open lesions in the skin or oral cavity. One infection followed extensive trauma to the body of a 5-year-old boy who passed through the rotating spiked cylinders and fan assembly of a manure spreader.[5] Another human infection may have been acquired through a decubitus ulcer[3]; the origin of the third human infection is unknown.[9]

Clinical Features and Pathogenesis

Inflammation of the central nervous system (CNS) is a constant feature of halicephalobiasis in

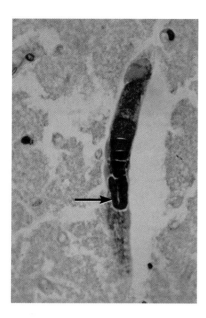

Figure 34.5
Longitudinal section of female *Halicephalobus* sp in brain showing reflected ovary (arrow). x440

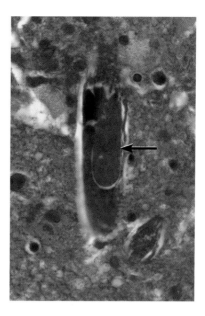

Figure 34.6
Longitudinal section of gravid *Halicephalobus* sp in brain. Single egg (arrow) measures 40 by 14 µm. x600

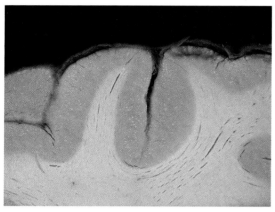

Figure 34.7
Gross specimen of cerebral cortex of patient with halicephalobiasis. Note thickened leptomeninges and foci of inflammation (white spots) in gray matter. x2.8

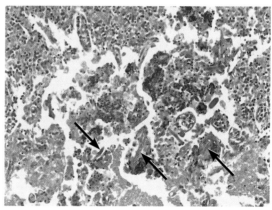

Figure 34.8
Inflammatory exudate in cerebellum of patient with halicephalobiasis. Exudate is composed of lymphocytes and neutrophils and contains fragments of worms (arrows). x125

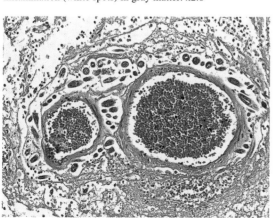

Figure 34.9
Meningeal vessels surrounded by numerous fragments of *Halicephalobus* sp. x125

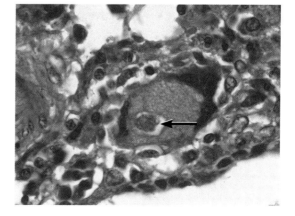

Figure 34.10
Multinucleated giant cell in brain containing fragment of *Halicephalobus* sp (arrow). x575

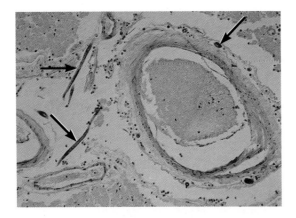

Figure 34.11
Fragments of *Halicephalobus* sp (arrows) near blood vessels in brain. x90

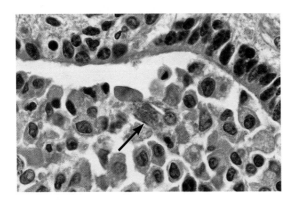

Figure 34.12
Central canal of spinal cord containing inflammatory exudate and a fragment of *Halicephalobus* sp (arrow). x625

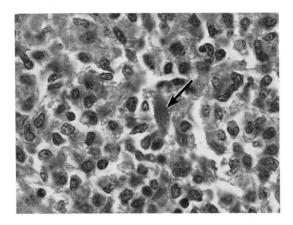

Figure 34.13
Pituitary gland of patient with halicephalobiasis. Note fragment of *Halicephalobus* sp (arrow) in chronic inflammatory exudate. x575

both horses and humans.[1,4,5] After entering the body through the skin or oral tissues, parasites migrate to the CNS, where they multiply and provoke inflammatory infiltrates that produce progressive mental confusion, lethargy, coma, and death. Infection sometimes localizes outside the CNS. In horses, affected tissues may include kidney, eye, oral and nasal cavities, lymph node, lung, adrenal gland, bone, joint, and skin.[2,6,7,8,10,11,12] In 1 human, the parasite was found in the liver and heart[3] as well as in the CNS.

In the 3 human infections, symptoms included fever and a variety of neurological signs, including headache, muscle rigidity, mental confusion, lethargy, and coma. Examination of cerebrospinal fluid revealed pleocytosis up to $.3 \times 10^3/\mu l$,[3] but no parasite was found antemortem. One patient developed decerebrate rigidity and died 29 days after onset of symptoms; another died of meningoencephalitis 24 days after accidental infection, and 6 days after onset of relevant symptoms. The site of infection and onset of specific symptoms could not be established in the third patient.

Pathologic Features

At autopsy, significant gross changes are confined to the brain. The meninges are thickened and congested and the brain swollen with foci of inflammation (Fig 34.6). Cut surfaces show areas of softening and hemorrhage in the cerebrum, cerebellum, midbrain, brain stem, and upper spinal cord.

Microscopically there is widespread meningoencephalomyelitis (Figs 34.7 and 34.8). Inflammatory exudates contain lymphocytes, macrophages, and occasionally giant cells, neutrophils and eosinophils. Portions of worm may be engulfed by giant cells (Fig 34.9). Worms are generally found in exudates (Fig 34.2) and around vessels in the meninges (Fig 34.8) and brain (Fig 34.10). However, worms also invade the choroid plexus and may be free in the subarachnoid space, lumina of ventricles, and central canal of the spinal cord (Fig 34.11). Other organs that may be involved include the pituitary (Fig 34.12), liver (Fig 34.13), heart (Fig 34.14), and lung.

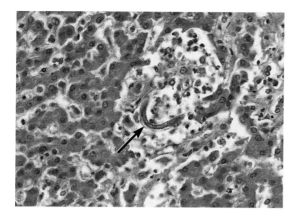

Figure 34.14
Microabscess in liver of patient with halicephalobiasis containing single fragment (arrow) of *Halicephalobus* sp. x230

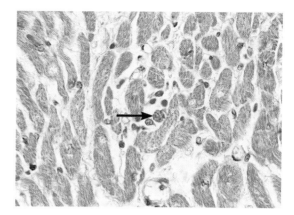

Figure 34.15
Fragment of *Halicephalobus* sp (arrow) between myocardial fibers, with no significant inflammatory response. x400

Diagnosis

None of the 3 known human infections with *Halicephalobus* sp was diagnosed antemortem. Because parasites are present in the subarachnoid space and spinal cord, lumbar puncture may reveal them in spinal fluid. There are no reagents to confirm the identification of suspected worms, and no serologic tests for diagnosis.

Treatment

Because no patient has been diagnosed antemortem, no anthelmintic regimen has been instituted or studied.

References

1. Darien BJ, Belknap J, Nietfeld J. Cerebrospinal fluid changes in two horses with central nervous system nematodiasis (Micronema deletrix). *J Vet Intern Med* 1988;2:201–205.
2. Dunn DG, Gardiner CH, Dralle KR, Thilsted JP. Nodular granulomatous posthitis caused by Halicephalobus (syn. Micronema) sp in a horse. *Vet Pathol* 1993;30:207–208.
3. Gardiner CH, Koh DS, Cardella TA. Micronema in man: third fatal infection. *Am J Trop Med Hyg* 1981;30:586–589.
4. Hoogstraten J, Connor DH, Neafie RC. Micronemiasis. In: Binford CH, Connor DH, eds. *Pathology of Tropical and Extraordinary Diseases.* Vol II. Washington, DC: AFIP; 1976:468–470.
5. Hoogstraten J, Young WG. Meningo-encephalomyelitis due to the saprophagous nematode, Micronema deletrix. *Can J Neurol Sci* 1975;2:121–126.
6. Kreuder C, Kirker-Head CA, Rose P, Gliatto J. What is your diagnosis? Severe granulomatous osteomyelitis associated with Micronema deletrix infection in a horse. *J Am Vet Med Assoc* 1996;209:1070–1071.
7. Rames DS, Miller DK, Barthel R, et al. Ocular Halicephalobus (syn. Micronema) deletrix in a horse. *Vet Pathol* 1995;32:540–542.
8. Ruggles AJ, Beech J, Gillette DM, Midla LT, Reef VB, Freeman DE. Disseminated Halicephalobus deletrix infection in a horse. *J Am Vet Assoc* 1993;203:550–552.
9. Shadduck JA, Ubelaker J, Telford VQ. Micronema deletrix meningoencephalitis in an adult man. *Am J Clin Pathol* 1979;72:640–643.
10. Simpson RM, Hodgin EC, Cho DY. Micronema deletrix-induced granulomatous osteoarthritis in a lame horse. *J Comp Pathol* 1988;99:347–351.
11. Spalding MG, Greiner EC, Green SL. Halicephalobus (Micronema) deletrix infection in two half-sibling foals. *J Am Vet Med Assoc* 1990;196:1127–1129.
12. Teifke JP, Schmidt E, Traenckner CM, Bauer C. Halicephalobus (Micronema) deletrix as a cause of granulomatous gingivitis and osteomyelitis in a horse [in German]. *Tierarztl Prax Ausg G Grosstiere Nutztiere* 1998;26:157–161.

35

Oesophagostomiasis and Ternidenamiasis

Ronald C. Neafie *and*
Aileen M. Marty

Introduction

Definition

Oesophagostomiasis is infection of the intestine or abdominal cavity by strongylid nematodes of the genus *Oesophagostomum* Molin, 1861. Ternidenamiais is infection of the same tissues by another member of the superfamily Strongyloidea, *Ternidens deminutus* Railliet and Henry, 1909. Adult worms of these species normally inhabit the large intestine of ruminants, primates, and swine. The main clinical presentations are painful abdominal masses, dysentery, or intussusception.

Synonyms

Synonyms for oesophagostomiasis include helminthoma, helminthic abscess, nodular worm infection, pimply gut, *tumeur de Dapaong, koulkoul, tougnale,* and, specifically for *T. deminutus*, bowel worm infection.

General Considerations

Railliet and Henry,[17] in 1905, first described oesophagostomiasis in humans, based on the discovery of immature worms in tumors of the cecum and colon of an inhabitant of southern Ethiopia. In 1911, Leiper[11] reported the first finding of adult worms in humans after he discovered 6 of them in the stool of a Nigerian. Until recently, oesophagostomiasis was considered an uncommon human disease, presenting mainly as immature worms in nodules. But in 1987, Gigase et al[4] reported large numbers of patients in northern Togo. By 1991, Polderman et al[15] had clearly established that oesophagostomiasis is common in humans in West Africa, especially Togo and Ghana.

Several species of *Oesophagostomum* that typically infect simians also infect humans. Identifying these parasites in specimens is often difficult because the entire worm is seldom available and the taxonomy of these species is uncertain.

Members of the subfamily Oesophagostominae belong to the family Chabertiidae. Another member of the family Chabertiidae, *T. deminutus*, an intestinal parasite of nonhuman primates, can also cause helminthomas in humans.[6]

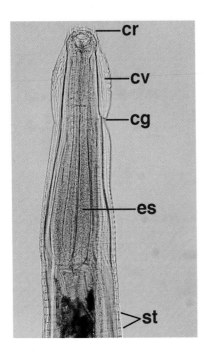

Figure 35.1 Anterior end of male *Oesophagostomum* sp recovered from a rhesus monkey. Note corona radiata (cr), cephalic vesicle (cv), cervical groove (cg), esophagus (es), and transverse striations (st).* x150

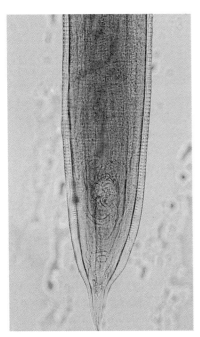

Figure 35.2 Posterior end of female *Oesophagostomum* sp from a rhesus monkey. Note sharply pointed tip. x100

Epidemiology

Oesophagostomiasis of domestic animals is broadly distributed but more common in the tropics and subtropics.

There have been isolated reports of humans with oesophagostomiasis from Brazil,[20] Malaysia,[7] and Indonesia,[12] but all other patients have been from tropical Africa. Rural areas have the highest prevalence. Oesophagostomiasis afflicts both sexes and all ages, but prevalence is higher in females and lower in children under 5 years of age.[15] The species seen most frequently in humans is *Oesophagostomum bifurcum*, previously described as *Oesophagostomum apiostomum*, a parasite of monkeys. *Oesophagostomum bifurcum* produces disease in up to 50% of the population of villages of northern Togo and northern Ghana.[8,15] *Oesophagostomum stephanostomum*, a parasite of monkeys in Brazil, French Guiana, and West Africa, and *Oesophagostomum aculeatum*, a parasite of animals in Asia, rarely infect humans.

The only *Ternidens* sp known to infect humans, *T. deminutus*, is a frequent parasite of nonhuman primates in Africa, Indonesia, and India. In Zimbabwe, the prevalence in humans can be as high as 80%.[5]

Infectious Agents

Morphologic Description

Oesophagostomum sp are strongylid nematodes that resemble hookworms. Adult *Oesophagostomum* sp differ from hookworms in that the buccal capsule opens straight forward, lacks teeth or cutting plates, and contains a corona radiata (Fig 35.1). Also unlike hookworms, *Oesophagostomum* sp have a ventral cervical groove (Fig 35.1). Anterior to the ventral cervical groove, the cuticle of some *Oesophagostomum* sp is inflated. This portion of the cuticle is called the cephalic vesicle. Mature adult female *O. bifurcum* are 8.5 to 10.5 mm long and 200 to 300 µm in diameter. Males are 8 to 10 mm long and 300 to 350 µm in diameter. Sharply pointed leaf crowns surround the buccal opening at the cephalic end. Reproductive organs are confined to the posterior two thirds of the worm. The female's tail is sharply pointed (Fig

35.2), with the vulva slightly anterior to the anus. Male worms have paired spicules and a copulatory bursa (Fig 35.3).

Though similar to adults in size and morphologic features, worms in nodules are immature. The anterior end of the worm in tissue sections sometimes shows the cephalic vesicle (Fig 35.4). The 5-to-10-µm-thick cuticle has transverse striations 10 µm apart (Fig 35.5). Lateral cords have dilated spaces and an inconspicuous lateral body at the base. There are 2 to 3 muscle cells per quadrant. The intestine contains only a few multinucleated cells lined with microvilli. Gonads consist of small tubular structures (Fig 35.6).

Infective third-stage (L3) filariform larvae of *Oesophagostomum* sp are 20 to 29 µm in diameter and approximately 800 µm long. The sheath has prominent transverse striations that are relatively thick compared to the cuticle, and ends in long, drawn-out filaments. Within the sheath, the posterior end of the larva is rounded. The distance from the tip of the tail to the end of the sheath is greater than the distance from the anus to the tip of the tail. In fresh specimens, 16 to 30 triangular intestinal cells, arranged in a distinctive zigzag pattern, are often visible.[15]

Adult *T. deminutus* closely resemble *Oesophagostomum* sp and hookworms. They differ from hookworms in that their buccal capsule includes a corona radiata and opens straight forward, and they have a ventral cervical groove. *Ternidens deminutus* differ from *Oesophagostomum* sp by having a buccal capsule with 3 serrated teeth, and a minimal cephalic vesicle. Female *T. deminutus* are 12 to 16 mm long, 650 to 730 µm in diameter, and have a vulva just anterior to the anus. Males are smaller (about 9.5 mm long and 0.56 mm in diameter) and have a branching copulatory bursa.

Except for their greater diameter, immature *T. deminutus* in nodules are morphologically indistinguishable from immature *Oesophagostomum* sp (Fig 35.7).

Eggs of both genera resemble those of hookworms.[2] Eggs of *O. bifurcum* are 60 to 63 µm by 24 to 40 µm and have a colorless, smooth shell. There are usually 8 to 32 blastomeres in eggs from fresh stools.[14] Eggs of *T. deminutus* are 84 by 51 µm, transparent, and segmented.

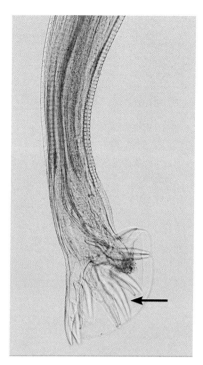

Figure 35.3
Posterior end of male *Oesophagostomum* sp recovered from a rhesus monkey, demonstrating copulatory bursa (arrow). x115

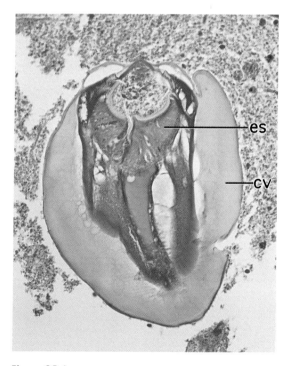

Figure 35.4
Anterior end of immature female *Oesophagostomum* sp in nodule. Note cephalic vesicle (cv) and esophagus (es). x315

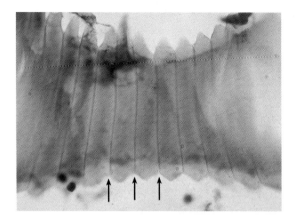

Figure 35.5
Cuticle of immature female *Oesophagostomum* sp in nodule, showing transverse striations 10 μm apart (arrows). x570

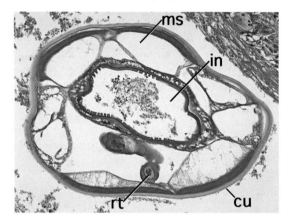

Figure 35.6
Transverse section of immature female *Oesophagostomum* sp in nodule, demonstrating cuticle (cu), reproductive tube (rt), intestine (in), and sarcoplasmic portion of muscle (ms). x130

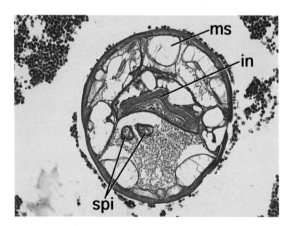

Figure 35.7
Transverse section of immature male *T. deminutus* in omental mass, demonstrating sarcoplasmic portion of muscle (ms), intestine (in), and paired spicules (spi). x135

Life Cycle and Transmission

Adult worms of both genera live in the intestinal lumen in their definitive hosts (monkeys, cattle, sheep, swine, and goats; humans also are definitive hosts for *O. bifurcum* and *T. deminutus*) (Fig 35.8).[8]

Adult worms in the intestinal lumen deposit eggs in the host's feces. Eggs hatch in contaminated soil, releasing rhabditiform larvae that molt twice to become infective L3 filariform larvae. Animals, and probably humans, become infected by consuming filariform larvae in contaminated food or water. When ingested by a definitive host, L3 larvae burrow into the submucosa of the small or large intestine and induce submucosal cysts. Within the cysts, larvae molt once to become fourth-stage (L4) larvae. Later, L4 larvae return to the lumen of the large intestine, where they molt into adult worms and remain attached by their mouths to the mucosa. Eggs appear in feces 30 to 40 days after consumption of infective larvae. In a study in Lotogou, northern Togo, Krepel and Polderman found that infected humans may harbor 12 to 300 adult *O. bifurcum*, which can release up to 5055 eggs per day.[10]

Sometimes, *Oesophagostomum* sp cannot complete their life cycle in humans, limiting infection to L4 larvae in nodules.

Clinical Features and Pathogenesis

Symptoms vary according to severity of infection. Light infections produce minimal symptoms and may escape clinical detection. In 1966, Welchman[21] described 4 clinical presentations:

1. *Intestinal obstruction or intussusception.* Nodules infrequently cause intestinal obstruction, but can initiate intussusception or incarceration of a hernia.

2. *Abscess of the abdominal wall in children with a short history of pain and fever.*

3. *Acute abdomen in adults.* This common presentation can mimic appendicitis: low-grade fever and localized tenderness, commonly in the right lower quadrant. Onset of pain is gradual (1 to 5 days). Vomiting, anorexia, and diarrhea are uncommon. Physical examination usually reveals a

tender mass in the abdominal cavity.

4. *Painless abdominal or cutaneous masses in adults.* Abdominal masses can be impressive in size. Gigase[4] described enormous masses in 16 patients, 9 of whom presented merely because the masses were disfiguring.

Radiologic examination usually reveals the mass, or masses, most often with a smooth, intact mucosa overlying the lesion (Fig 35.9). Rarely, a barium spike delineates the track through which the worm penetrated the wall of the gut (Fig 35.10). Sometimes, flat films of the abdomen reveal residual calcification years after aspiration of the nodule.

Laparotomy may reveal a single mass in, or attached to, the wall of the ileum or colon, commonly in the ileocecal region.[1] Adherent mesentery may hide the mass. The worm may be in the abdominal wall or omentum, or stuck to the surface of other abdominal viscera. Patients often have multiple lesions; one had 187 distinct nodules.[20] Nodules vary in size from 3 mm to 6 cm, and at surgery may mimic carcinoma, tuberculosis, or Crohn's disease.

In rare instances, *O. bifurcum* perforate the wall of the bowel and cause purulent peritonitis, or migrate into the skin and produce a nodule.[3,7] Ross et al reported oesophagostomiasis in an 18-year-old patient with a subcutaneous nodule in his back. The patient had no abdominal symptoms, but the nodule enlarged over 8 to 9 months. When incised, the nodule exuded greenish-brown pus and a 1-cm-long living, immature female *O. bifurcum*.[18,19]

Peripheral eosinophilia varies and is nonspecific, especially where other helminthic infections are common.

Pathologic Features

Oesophagostomiasis lesions may be soft, spongy, or hard, and usually have smooth, opaque, gray-black surfaces that can be rough or covered with inflammatory exudate or adhesions (Fig 35.11). There may be a single abdominal mass, or the surface of the colon may be studded with well-defined nodules firmly seated between the external muscular layer and the peritoneum. Nodules also arise in the colonic submucosa and generally

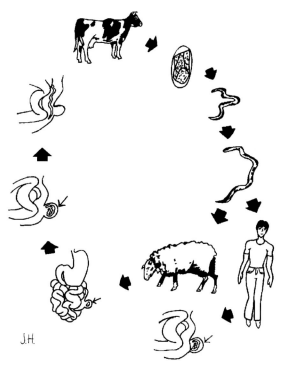

Figure 35.8
Life cycle of *Oesophagostomum* sp. Common livestock such as cattle and sheep are typical definitive hosts. Humans become infected by ingesting infective L3 larvae, and can also serve as definitive hosts.*

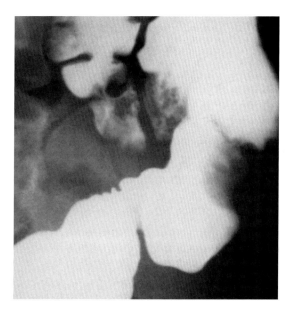

Figure 35.9
Radiograph showing large filling defect in distal descending colon of African woman, corresponding to palpable mass.*

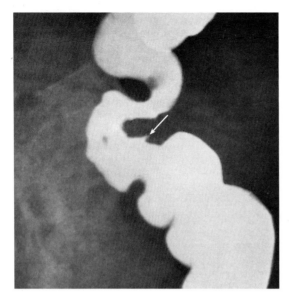

Figure 35.10
Spicule of barium (arrow) corresponding to worm track in colon of expatriate European woman.*

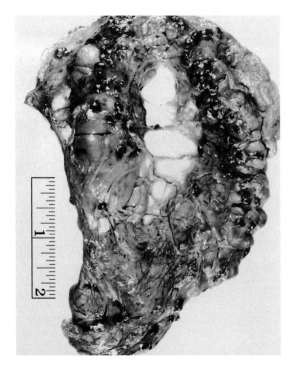

Figure 35.11
Specimen from a rhesus monkey with oesophagostomiasis. Bowel segment is studded with opaque, gray-black nodules.*

lie transversely or diagonally.[7] The overlying mucosa is generally not affected. The immature worm inside the nodule is either coiled (Fig 35.12) or extended. Dissection of the gross specimen, or multiple sectioning, may reveal a worm track 0.5 to 2 mm in diameter leading from the nodule to the submucosa. Lesions may adhere to neighboring structures, with foci of fat necrosis in the mesentery and enlarged mesenteric and paracolic lymph nodes. Enlarged lymph nodes show reactive hyperplasia with prominent germinal centers. At times, sinuses are dilated by mononuclear B cells, some of which may have xanthomatous cytoplasm. Occasionally, lesions contain a few eosinophils.[13]

Anthony and McAdam[1] have outlined 3 classes of lesions:

Class 1: *Acute lesion.* The center of the nodule is an abscess containing yellow-green or brown odorless fluid of varying viscosity. The ill-defined abscess or track is filled with a fibrinous exudate containing fat globules and very few inflammatory cells, mainly macrophages and eosinophils. These may be early lesions because, although sharply demarcated, they have only tenuous fibrous walls, or none at all. Sectioning the entire mass usually reveals immature worms (Fig 35.12).

Class 2: *Abscess walled off by macrophages, epithelioid cells, foreign body giant cells, and fibroblasts.* Contents of the abscess become caseous, and may contain granular calcification from dead worms undergoing resorption. There are no recognizable worms.

Class 3: *Chronic fibrosis with epithelioid and giant cell granulomas, and varying numbers of eosinophils.* There can be numerous Charcot-Leyden crystals, calcified bodies, and granular pigment.

Histopathologic features of lesions of *T. deminutus* are identical to those of *Oesophagostomum* sp (Figs 35.13 and 35.14). Older, empty Class 2 or Class 3 nodules eventually shrink or calcify.

Diagnosis

Simultaneous oesophagostomiasis and hookworm infection is common in endemic areas,

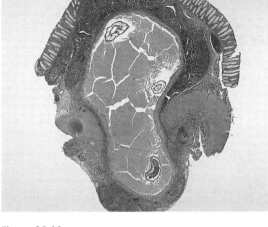

Figure 35.12
Coiled immature female *Oesophagostomum* sp within nodule. Necrotic nodule presented as paraumbilical mass in 5-year-old Nigerian girl. Lesion arose in ascending colon and was adherent to muscle and fascia of anterior abdominal wall. x7

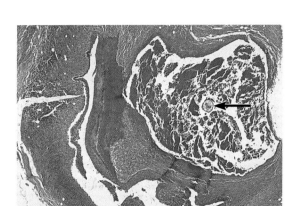

Figure 35.13
Coiled immature male *T. deminutus* (arrow) within omental abscess of 33-year-old Thai woman. Patient had abdominal pain and palpable 6-cm mass in right lower quadrant. Nodule was movable, had a smooth surface and firm-to-hard consistency, and was adherent to anterior mesenteric border of terminal ileum. x26

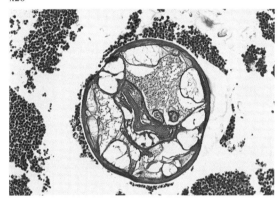

Figure 35.14
Higher magnification of worm in Figure 35.13. Note extensive acute inflammation around worm. x110

complicating diagnosis. Eggs of *Oesophagostomum* sp, *T. deminutus*, and hookworms are morphologically indistinguishable. Stool specimens containing over 170 000 eggs/ml usually indicate *Oesophagostomum* sp or *T. deminutus* infection; specimens with fewer than 150 000 eggs/ml suggest hookworm infection.[14] Eggs of *Oesophagostomum* sp can be distinguished from hookworm eggs only after they have been incubated in stool culture to produce L3 larvae, which are morphologically distinguishable. Radiologic examination may also be diagnostic (Figs 35.9 and 35.10).

Older lesions that retain worm tracks but yield no parasites may suggest, but do not establish, a diagnosis. Finding an intact worm during surgery or in a biopsy specimen provides a definitive diagnosis. If oesophagostomiasis or ternidenamiasis is suspected, the diagnosis may be confirmed by immediately opening or aspirating the nodule and extracting the worm from the thick yellow pus (Fig 35.15). This procedure is ordinarily safe because the pus is usually free of bacteria. Identifying an intact worm is easier than studying serial or step sections of the worm in tissue, but both diagnostic methods are effective.

Clinically, oesophagostomiasis must be differentiated from, for example, carcinomatosis, intestinal tuberculosis, Crohn's disease, ameboma, and schistosomal granuloma of the bowel. Polderman

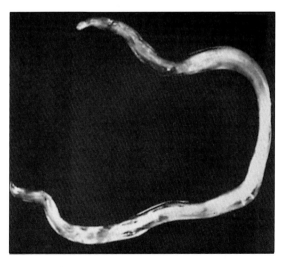

Figure 35.15
Immature *Oesophagostomum* sp surgically extracted from nodule, confirming diagnosis of oesophagostomiasis.

et al[16] developed an ELISA test that, in a trial in northern Togo, appeared sensitive and specific for oesophagostomiasis.

Treatment

Albendazole is the drug of choice for oesophagostomiasis, but pyrantel pamoate is also effective.[9] Thiabendazole is relatively ineffective. Surgical resection of the involved bowel, or removal of worms from nodules, is curative. The traditional practice of applying heat until the abscess bursts is not recommended.

References

1. Anthony PP, McAdam IW. Helminthic pseudotumors of the bowel: thirty-four cases of helminthoma. *Gut* 1972;13:8–16.
2. Blotkamp J, Krepel HP, Kumar V, Baeta S, Van't Noordende JM, Polderman AM. Observations on the morphology of adults and larval stages of Oesophagostomum sp isolated from man in northern Togo and Ghana. *J Helminthol* 1993;67:49–61.
3. Chabaud AG, Lariviere M. Sur les Oesophagostomes parasites de l'homme. *Bull Soc Pathol Exot* 1958;51:384–393.
4. Gigase P, Baeta S, Kumar V, Brandt J. Frequency of symptomatic human oesophagostomiasis (helminthoma) in Northern Togo. In: Geerts S, Kumar V, Brandt J, eds. *Helminth Zoonoses*. Boston, Mass: Martinus Nijhoff Pub; 1987:228–236.
5. Goldsmid JM. Cited by Rogers S, Goldsmid JM. Preliminary studies using the indirect fluorescent antibody test for the serological diagnosis of Ternidens deminutus infection in man. *Ann Trop Med Parasitol* 1977;71:503–504.
6. Goldsmid JM. Ternidens deminutus (Railliet and Henry, 1909) and hookworm in Rhodesia and a review of the treatment of human infections with T. deminutus. *Cent Afr J Med* 1972;18:1–14.
7. Karim N, Yang CO. Oesophagostomiasis in man: report of the first Malaysian case with emphasis on its pathology. *Malays J Pathol* 1992;14:19–24.
8. Krepel HP, Baeta S, Polderman AM. Human Oesophagostomum infection in northern Togo and Ghana: epidemiological aspects. *Ann Trop Med Parasitol* 1992;86:289–300.
9. Krepel HP, Haring T, Baeta S, Polderman AM. Treatment of mixed Oesophagostomum and hookworm infection: effect of albendazole, pyrantel pamoate, levamisole and thiabendazole. *Trans R Soc Trop Med Hyg* 1993;87:87–89.
10. Krepel HP, Polderman AM. Egg production of Oesophagostomum bifurcum, a locally common parasite of humans in Togo. *Am J Trop Med Hyg* 1992;46:469–472.
11. Leiper RT. The occurrence of Oesophagostomum apiostomum as an intestinal parasite of man in Nigeria. *J Trop Med Hyg* 1911;14:116–118.
12. Lie Kian Joe. Helminthiasis of the intestinal wall caused by Oesophagostomum apiostomum (Willach, 1891), Railliet and Henry, 1905. *Doc Neerl Indones Morbis Trop* 1949;1:75–80.
13. Pagés A, Kpodzro K, Baeta S, Akpo-Allavo K. Dapaong "tumor". Helminthiasis caused by Oesophagostomum [in French]. *Ann Pathol* 1988;8:332–335.
14. Piekarski G. *Medical Parasitology*. New York, NY: Springer-Verlag; 1987:232.
15. Polderman AM, Krepel HP, Baeta S, Blotkamp J, Gigase P. Oesophagostomiasis, a common infection of man in northern Togo and Ghana. *Am J Trop Med Hyg* 1991;44:336–344.
16. Polderman AM, Krepel HP, Verweij JJ, Baeta S, Rotmans JP. Serological diagnosis of Oesophagostomum infections. *Trans R Soc Trop Med Hyg* 1993;87:433–435.
17. Railliet A, Henry A. Encore un nouveau sclérostomien (Oesophagostomum brumpti nov. sp.) parasite de l'homme. *Comp Rend Soc Biol* 1905;58:643–645.
18. Ross RA, Gibson DI, Harris EA. Cutaneous oesophagostomiasis in man. *J Helminthol* 1989;63:261–265.
19. Ross RA, Gibson DI, Harris EA. Cutaneous oesophagostomiasis in man. *Trans R Soc Trop Med Hyg* 1989;83:394–395.
20. Thomas HW. The pathological report of a case of oesophagostomiasis in man. *Ann Trop Med Parasitol* 1910;4:57–88.
21. Welchman JM. Helminthic abscess of the bowel. *Br J Radiol* 1966;39:372–376.

36

Miscellaneous Nematodiases

Aileen M. Marty, Ronald C. Neafie,
Mary K. Klassen-Fischer, Lawrence R. Ash, *and*
Douglas J. Wear

Introduction

Definition

This chapter discusses species of 28 genera of nematodes that either cause disease or are spurious in humans. The diseases they induce are diverse and range from mild to fatal. Some of the worms described are very rare, others are quite common. Some are focal, while others are distributed worldwide. Animals are the usual definitive hosts for most of these nematodes. Humans are infected only through unusual circumstances; under certain conditions, some free-living worms in soil or vegetation become pathogenic to humans.

The nematodes considered in this chapter belong to 2 superfamilies of subclass Adenophorea (Aphasmidia) or 12 superfamilies of subclass Secernentea (Phasmidia). Clinical or histologic features of 10 genera are illustrated in figures accompanying this chapter, some of which are cited only in Table 36.4.

Synonyms

See Table 36.1.

General Considerations

Tables 36.1 to 36.4 describe 28 genera in 14 superfamilies of miscellaneous species of nematodes. Historical aspects of a few species in 3 superfamilies are discussed below.

Dioctophymatoidea

First described by de Clamorgam in 1570 in the kidneys of foxes, infections by the giant kidney worm were also reported in humans in 1674 and 1737 in Amsterdam, and by Moublet in France in 1758. In 1802, Collet-Meygret proposed naming this worm *Dioctophyme* to emphasize the 8 pairs of tubercles at the cephalic and caudal ends. By 1941, a committee of the American Society of Parasitologists had accepted the designation *Dioctophyma renale* (Figs 36.3 and 36.4).

The basic life cycle of *D. renale* was worked out in the early 1950s, the role of paratenic hosts in the early 1960s, and the various developmental stages in the early 1970s. A related genus, *Eustrongylides*, had been established in 1909, 16 species of which are parasites of fish-eating birds and do not mature in humans; however, *Eustrongylides* larvae have been reported in humans (Figs 36.21 to 36.24).

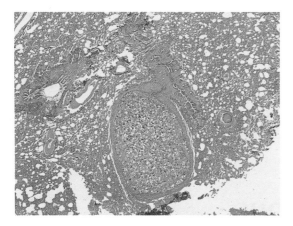

Figure 36.1
Eggs of *C. aerophila* in encapsulated pulmonary nodule of an opossum. x20

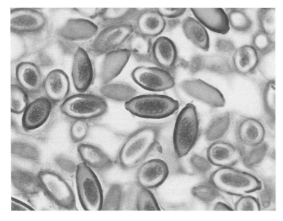

Figure 36.2
Higher magnification of Figure 36.1 showing bipolar-plugged eggs of *C. aerophila*. x230

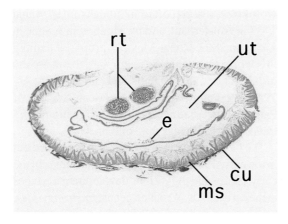

Figure 36.3
Transverse section of adult female *D. renale* from dog kidney. Note cuticle (cu), muscle (ms), uterus (ut) with eggs (e), and reproductive tubes (rt). x7

The first reports came in 1982, when 3 fishermen from the Baltimore area presented with *Eustrongylides* infections. In each of these patients, characteristic red larvae were found burrowing through the abdominal wall into the abdominal cavity. Other reported human infections have been more difficult to classify. In 1984, a larval *Dioctophyma*-like nematode was extracted from the chest wall of a 14-year-old Thai boy. This case is considered a *D. renale* infection, even though the larva was too immature to distinguish definitively from *Eustrongylides* sp.

Ascaridoidea

Of the 5 species of *Lagochilascaris,* only *Lagochilascaris minor* infects humans (Fig 36.7). First described in 1971 in abscesses of 2 patients in Trinidad, *Lagochilascaris* was named for the characteristic longitudinal groove on the internal surface of the lips, which gives the worm a harelip appearance. *Lagochilascaris minor* are natural parasites of cats and dogs, but white mice have been experimentally infected. Although there are only 70 reports of *L. minor* in humans, prevalence may in fact be much higher, since most patients live in undeveloped, inaccessible areas. Brazil, for example, where the disease was unknown until 1968, now reports 74% of all cases of human lagochilascariasis.

Filarioidea

In 1928, Peruzzi reported finding adult filariae in the central nervous system (CNS) of a talapoin monkey from Uganda. Orihel and Esslinger, in 1973, recovered filariae from the CNS of 3 more talapoin monkeys from Equatorial Guinea. They named the filariae *Meningonema peruzzii*, and recognized similarities to filariae described by Dukes et al[15] in a European soldier stationed in Zimbabwe and in a 43-year-old African. In 1995, Boussinesq et al[9] described an asymptomatic Cameroonian harboring a larva of *M. peruzzii* in his cerebrospinal fluid. These few known infections suggest that zoonosis with *M. peruzzii* may be underreported, perhaps because microfilariae of *M. peruzzii* greatly resemble those of *Mansonella perstans*.

Two separate investigators in 1965 discovered that adult *Onchocerca* sp can produce zoonotic onchocerciasis in humans. First, an *Onchocerca*

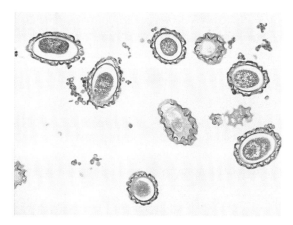

Figure 36.4
Eggs of *D. renale* within gravid worm. Eggs are ellipsoid and brown-yellow. Outer layer of shell is deeply pitted, except at poles. x240

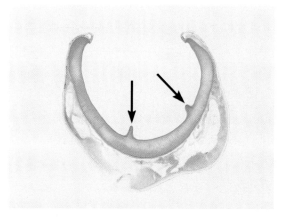

Figure 36.5
Buccal cavity of adult female *M. laryngeus* demonstrating 2 internal ridges (arrows). x115

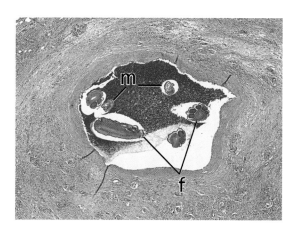

Figure 36.6
Transverse sections of male (m) and female (f) *Metastrongylus* sp in lumen of vessel of a marmoset. x24

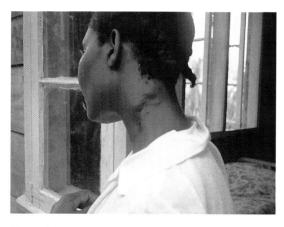

Figure 36.7
Lesion on neck caused by *L. minor*.

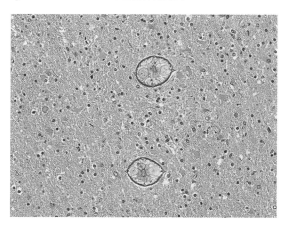

Figure 36.8
Two transverse sections of viable larva in brain of 10-month-old Caucasian with fatal *B. procyonis* infection. Note lack of tissue reaction. x120

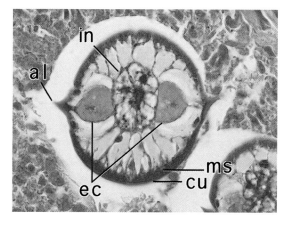

Figure 36.9
Transverse section of third-stage larva of *B. procyonis* in lymph node from infant described in Figure 36.8. Note lateral alae (al), cuticle (cu), muscles (ms), excretory columns (ec), and intestine (in). x600

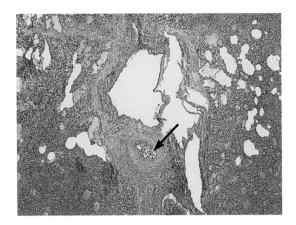

Figure 36.10
Degenerated third-stage larva (arrow) of *B. procyonis* in pulmonary granuloma of infant described in Figure 36.8. Note extensive hemorrhage, edema, and chronic inflammation. x25

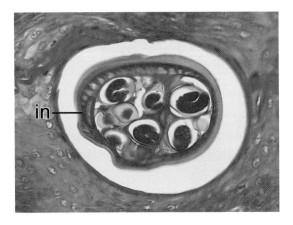

Figure 36.11
Transverse section of gravid *Gongylonema* sp in tongue of tree shrew. Note intestine (in) and uterine tube with eggs. x240

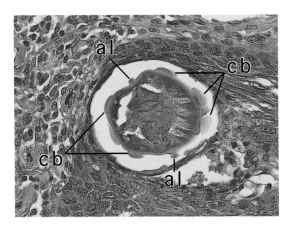

Figure 36.12
Transverse section through male *Gongylonema* sp at level of esophagus. Cuticle has 2 alae (al) and multiple cuticular bosses (cb). x240

sp, most likely *Onchocerca gutturosa,* was discovered in a nodule removed from the knee of a 25-year-old Swiss woman. That same year, a similar worm was removed from the cornea of a 15-year-old Ukrainian girl. In 1973, a third worm was found to have induced a nodule on the wrist of a Canadian woman. The following year, a fourth worm was identified in a woman from Illinois with no history of travel outside the United States. Two recent reports from Japan and 2 additional cases of wrist lesions of zoonotic onchocerciasis on file at the Armed Forces Institute of Pathology (AFIP) make a total of 8 reported human infections (Figs 36.15 to 36.18). The morphologic features of the worms reported by AFIP were most similar to *O. gutturosa* (Table 36.3). Both *O. gutturosa* and *O. cervicalis* are geographically widespread in cervical ligaments of cattle and horses (Table 36.2).

Epidemiology
See Table 36.2.

Infectious Agents
Morphologic Description
See Table 36.3.

Life Cycle and Transmission
See Table 36.2.

Clinical and Pathologic Features
See Table 36.4.

Treatment

Ivermectin is highly effective against *Capillaria aerophila* in the airways of infected animals, and against *L. minor*.[6] Levamisole combined with surgical removal of the parasite cured a 7-year-old Colombian girl of lagochilascariasis.

Surgical extraction is the treatment of choice for *D. renale, Gongylonema pulchrum,* and zoonotic onchocerciasis. *Thelazia* sp are removed from the

Subclass: superfamily	Genus/species	Original description, synonyms
Adenophorea:		
Trichuroidea	*Capillaria aerophila*[2]	(Creplin, 1939) Travassos, 1915; fox lungworm; *Eucoleus aerophilum, E. aerophilus, Trichosoma aerophilus, T. aerophilumn; Thominx aerophilan*
Dioctophymatoidea	*Dioctophyma renale*[5,37]	(Goeze, 1782), Collet-Meygret, 1802, Lamouroux 1824; giant kidney worm; *Ascaris renalis, A. canis, A. martis, A. visceralis, Fusaria visceralis, Lumbricus gulonissibirici, L. sanguineus, Dioctophyme renale, D. skrjabini, D. visceralis, Eustrongylus visceralis, E. gigante, E. renalis, E. gigas, Strongylus gigas, S. renalis*
	Eustrongylides sp[29,39]	Jaegerskiöld, 1909
Secernentea:		
Rhabditoidea	*Pelodera strongyloides*[23]	(Schneider, 1860) Schneider, 1866; *Rhabditis axei strongyloides, R. strongyloides*
	Rhabditis sp[40]	Dujardin, 1845; *Anguilla leptodera* for *Rhabditis nielly*
	Diploscapter coronata[13]	(Cobb, 1893) Cobb, 1913; *Rhabditis coronata, R. bicornis*
	Turbatrix aceti[12]	(Müller, 1783) Peters, 1927; vinegar eel; *Rhabditis pellio, Anguillula aceti*
Tylenchoidea	*Ditylenchus dipsaci*[11]	bulb eelworm, red clover eelworm
	Meloidogyne javanica[26]	*Heterodera radiciola, H. radicicola, H. incognita, H. marioni, Oxyuris incognita*
Strongyloidea	*Mammomonogamus laryngeus*[17,31]	(Railliet, 1899) Ryjikov, 1948; gape worm; *Mammongamus laryngeus, Syngamus laryngeus, S. kingi, Cyathostoma*
Ancylostomatoidea	*Cyclodontostomum purvisi*[7]	Adams, 1933; *Ancistronema coronatum*
Trichostrongyloidea	*Trichostrongylus* sp[10]	10 species: *T. orientalis, T. colubriformis, T. probolurus, T. calcaratus, T. vitrinus, T. axei, T. skrjabini, T. lerouxi, T. capricola, T. brevis*
	Haemonchus contortus[18]	(Rudolphi, 1803) Cobb 1898; sheep wireworm; *Strongylus contortus, S. haemonchus, Haemonchus cervinus, H. placei*
	Ostertagia ostertagi[18]	(Stiles,1892) Ransom,1907; *Stronylus convolutus, S. ostertagi, S. harkeri*
Metastrongyloidea	*Metastrongylus apri*[32]	(Gmelin, 1790) Vostokov 1905; swine lungworm; *Metastrongylus elongatus, M. longevaginatus, Ascaris apri, Strongylus suis, S. paradoxus, Gordius pulmonalis apri*
Ascaridoidea	*Lagochilascaris minor*[35,38]	Leiper, 1909; harelip ascarid
	Baylisascaris sp[21,27]	Large, small intestinal roundworms of raccoon
Oxyuroidea	*Enterobius gregorii*[14]	Hugot, 1983; possibly a balanced male polymorphism of *E. vermicularis*
	Syphacia obvelata[34]	(Rudolphi, 1802); mouse pinworm; *Ascaris obvelata*
Spiruroidea	*Cheilospirura* sp[1]	Diesing, 1861; gizzard worm; *Acuaria*
	Spirocerca lupi[8]	(Rudolphi, 1809) Railliet and Henry 1911; canine esophageal tumor worm; *Spirocerca sanguinolenta*
	Gongylonema pulchrum[22]	Molin 1857; esophageal worm, gullet worm, scutate threadworm; *Filaria labialis, Gongylonema scutatum, G. hominis*
	Rictularia sp[24,28]	Frölich, 1802; some reclassified as *Pterygodermatites* sp
Physalopteroidea	*Physaloptera caucasica*[30]	Von Linstow, 1902; *Physaloptera mordens, Abbreviata caucasica*
Thelazioidea	*Thelazia callipaeda*[20]	Railliet and Henry, 1910; Oriental eye worm
	Thelazia californiensis[25]	Price, 1930; Kofoid and Williams, 1935; California eye worm
Filarioidea	*Microfilaria bolivarensis*[19]	Godoy et al, 1980; proposed: *Onchocerca bolivarensis*
	Microfilaria rodhaini[33]	Peel and Chardome, 1946; proposed: *Mansonella rodhaini*
	Microfilaria semiclarum[16]	Fain, 1974; *Dipetalonema semiclarum*; proposed: *Mansonella semiclarum*
	Meningonema peruzzii[9,15]	Orihel and Esslinger, 1973
	Onchocerca gutturosa[3,36]	*Onchocerca lienalis*
	Onchocerca cervicalis[3,4]	

Parentheses around original discoverer (ie, Creplin, 1939) means organism has been reclassified.

Table 36.1
Classification of miscellaneous nematodes, original descriptions, and synonyms.

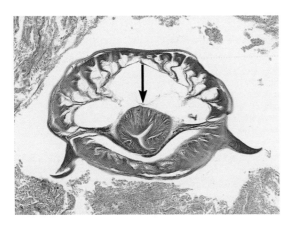

Figure 36.13
Transverse section of *Rictularia* sp in rat at level of esophagus (arrow). Note 2 subventral cuticular combs. x150

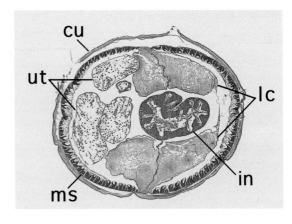

Figure 36.14
Transverse section through *Physaloptera turgida*. Note cuticle (cu), muscle (ms), lateral cords (lc), intestine (in), and uteri with eggs (ut). x24

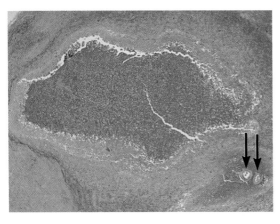

Figure 36.15
Two sections of adult nongravid female *O. gutturosa* (arrows) at edge of large suppurative granuloma from wrist of patient in Alabama. x17

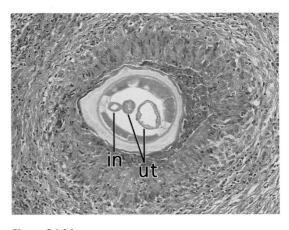

Figure 36.16
Transverse section of *O. gutturosa* from granuloma described in Figure 36.15. Note irregular thickness of cuticle. Internal organs consist of intestine (in) and 2 uteri (ut). x120

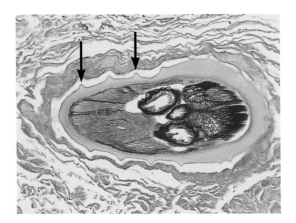

Figure 36.17
Section through adult nongravid female *O. gutturosa* in subcutaneous lesion in wrist of 13-year-old girl from Wyoming. Note prominent transverse cuticular ridges (arrows). PAS x115

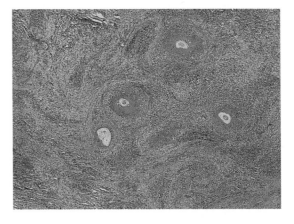

Figure 36.18
Four sections of coiled nongravid female *O. gutturosa* in subcutaneous lesion in wrist of 42-year-old woman from Pennsylvania. Worm is in necrotizing granuloma. x25

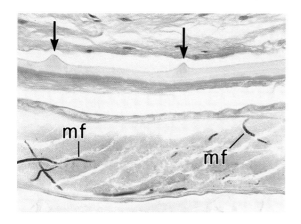

Figure 36.19
Longitudinal section of adult gravid *O. gutturosa* from nuchal ligament of cow. Note prominent transverse cuticular ridges (arrows) and microfilariae (mf). Giemsa x240

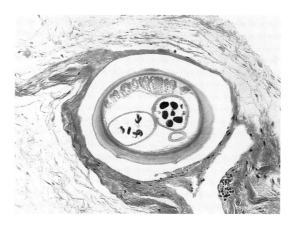

Figure 36.20
Transverse section of gravid *O. gutturosa* from nuchal ligament of cow. Note irregular thickness of cuticle. Giemsa x115

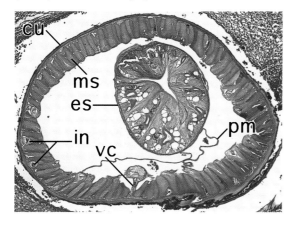

Figure 36.21
Transverse section through *Eustrongylides* sp L4 larva in colonic mesentery of 52-year-old patient from New Jersey. Note esophagus (es), cuticle (cu), muscle (ms), ventral cord (vc), intercordal nuclei (in), and pseudocoelomic membranes (pm). x60

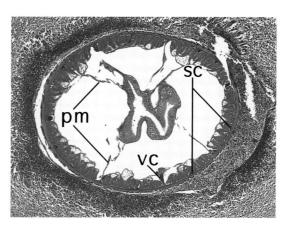

Figure 36.22
Transverse section through intestine of *Eustrongylides* sp larva described in Figure 36.21 showing ventral cord (vc), small cords (sc), and pseudocoelomic membranes (pm). Note intense inflammation surrounding worm. x50

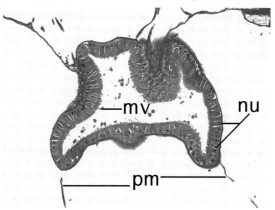

Figure 36.23
Transverse section through intestine of *Eustrongylides* sp larva described in Figure 36.21 showing microvilli (mv)-lined columnar cells with large nuclei (nu) near base of cells and pseudocoelomic membranes (pm). x120

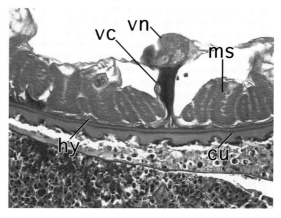

Figure 36.24
Higher magnification of *Eustrongylides* sp larva described in Figure 36.21 depicting cuticle (cu), hypodermis (hy), ventral cord (vc), ventral nerve (vn), and muscle cells (ms). x255

eye with forceps following topical anesthesia. *Mammomonogamus laryngeus* may be removed by bronchofibroscopy.

Santonin or pyrantel combined with mebendazole relieves urinary tract symptoms of *Rhabditis* sp infection and rids the host of the parasite. Thiabendazole and mebendazole are both effective against *M. laryngeus*. Bephenium, thiabendazole, mebendazole, and pyrantel are effective against *Trichostrongylus*. Thiabendazole may benefit patients with *Baylisascaris procyonis* infections confined to the eye; as an alternative, laser treatment can destroy intraretinal larvae and prevent further migratory damage. There is no effective therapy for *Baylisascaris* larva migrans in humans. *Enterobius gregorii* respond to piperazine.

In animals, *Haemonchus contortus*, *Ostertagia* sp, and *Metastrongylus apri* are treated with benzimidazoles (albendazole, fenbendazole, or oxfendazole) or the avermectins. All of these drugs are effective against developing larvae and adult worms. The avermectins are also effective against *Thelazia* sp in animals.

Genus and species	Life cycle in humans	Source of infection	Reported cases/distribution
Capillaria aerophila	E»GI¬L4»lung¬Adult¬E»GI»feces	E»soil, oligochaetes carry	10/worldwide
Dioctophyma renale	L3»GI»SQ.or»kidney¬Adult¬E»urine or Adult»urethra	E»aquatic oligochaetes¬L2–3, amphibians & fish carry L3	27/worldwide
Eustrongylides sp	L3»GI»peritoneal cavity.	E»aquatic oligochaetes¬L3» fish or amphibians; in fish-eating birds GI¬Adult¬E	6/North America, Ukraine
Pelodera sp	L3»skin.	In moist soil L2–4¬Adult¬E¬L2, rodents & farm animals carry L3	3/Americas, Sri Lanka, Poland
Rhabditis sp	L3»skin. ?Adults»GI¬E»feces	In feces L2–3¬Adult¬E¬L2, soil	60/worldwide
Diploscapter coronata	Adult»GI anhydrochloria, females parthenogenic¬E	In decaying plants & organic matter L1–4¬Adult¬E¬L1, soil	13/worldwide, especially warm areas
Turbatrix aceti	Adult or L4 in douche»vagina»urine	In vinegar L4¬Adult, douche	1/Mexico, worldwide
Ditylenchus dipsaci	Adult in onion»GI»vomit	In onion stem/bulb L2–4¬Adult¬E	1/Europe, worldwide
Meloidogyne javanica	E in plant»GI»feces	In plant stem/root L2–4¬Adult¬E	Many spurious/worldwide
Mammomonogamus laryngeus	?Adult»GI»trachea¬E»GI»feces	E»soil¬L3, oligochaetes & snails carry, L3¬Adult in cattle & goats	94/tropical America, Asia
Cyclodontostomum purvisi	L4»GI¬Adult»feces	E¬L1–4»green vegetables» skin(rat)»lung»GI¬Adult¬E	1/Southeast Asia, Australia
Trichostrongylus sp	L3»GI¬Adult¬E»feces	E»soil¬L1–4, on plants & grasses	Sparse-prevalent/ worldwide
Haemonchus contortus	L3»GI¬Adult¬E»feces	E»soil¬L2–3, on vegetation	5/worldwide
Ostertagia ostertagi	L3»GI¬Adult¬E»feces	E»soil¬L2–3, on vegetation, from meat ?spurious	3/worldwide
Metastrongylus apri	L3»GI»lacteals»LN»lung¬Adult¬E»GI»feces	E»soil»oligochaetes¬L2–3	3/worldwide
Lagochilascaris minor	?L3»GI, tonsil, soft tissue neck¬ Adult ¬E¬L2-4 or E»feces. or Adult & E»fistula	E»soil»¬L3 in meat wild animals	70/South and Central America
Baylisascaris sp	E»GI¬L3-4»brain, lung, heart	E»soil»raccoons»GI¬L4¬Adult¬E» feces, wild/domestic animals carry	Rare worldwide
Enterobius gregorii	E»GI¬L3¬Adult»perianal skin¬E	E»bedclothes»inhaled, E on fingers or in food	Very common/Europe, Africa, Asia
Syphacia obvelata	E»GI¬L2-4¬Adult»perianal skin¬E	E»food, mouse-tainted fingers	1/worldwide
Cheilospirura sp	?L3»esophagus»pharynx»tear duct» conjunctiva	E»arthropod¬L2–3»bird»gizzard¬ Adult¬E»GI»feces	1/worldwide
Spirocerca lupi	L3»GI»placenta»fetus»GI¬Adult¬E	E»feces»dung beetle¬L2–3»GI carnivore¬Adult, birds/rodents carry L3	1/tropical and subtropical areas
Gongylonema pulchrum	L3»GI»esophagus»mouth¬Adult¬E»GI»feces	E»dung beetle or cockroach¬L2–3	49/worldwide
Rictularia sp	?L3»GI¬Adult in appendix¬E	E»invertebrate¬L3»lizard or snake» rodent or bat¬Adult	1/worldwide
Physaloptera caucasica	L3»GI»mesentery blood vessels. or »esophagus & stomach or liver¬Adult ¬E»GI»feces	E»cockroach or grasshopper or cricket¬L2–3»GI carnivore¬Adult	Common/worldwide, especially tropics
Thelazia callipaeda	?L3»conjunctiva¬Adult¬E»tears	E tears»flies¬L2–3»dog eye	250/Asia
Thelazia californiensis	?L3»conjunctiva¬Adult¬E»tears	E tears»flies¬L2–3»dog or deer eye	8/North America
Microfilaria bolivarensis	?L3¬Adult¬Mf blood	?Mf»fly (Simulium)¬L3	2/Venezuela
Microfilaria rodhaini	?L3¬Adult¬Mf skin snips	?Mf»biting fly¬L3¬primates	46/Gabon, Africa
Microfilaria semiclarum	?L3¬Adult¬Mf blood	?Mf»biting fly¬L3¬primate or carnivore¬Adult	52/Democratic Republic of Congo
Meningonema peruzzii	?L3¬subarachnoid space¬Adult¬Mf CSF	?Mf»fly»L2–3»CSF talapoin & green monkeys¬Adult¬Mf»blood	3/West Africa
Onchocerca gutturosa Onchocerca cervicalis	?L3»skin¬Adult	?Mf»fly Simulium/Culicoides¬ L3»cow/horse ligaments¬Adult	8/worldwide, California

Terms and symbols: oligochaetes = class of annelids; carry = transported unchanged in paratenic host; E = egg; CSF = cerebrospinal fluid; GI = gastrointestinal tract; LN = lymph node; L1–4 = larval stage; Mf = microfilaria; SQ = subcutaneous; >> = enters, travels to, or exits with; ¬ = stage change; . = development halts; ? = unknown or unconfirmed; underlined = source of infection.
Sample translation: E>>GI¬L4>>lung¬Adult¬E>>GI>>feces (see *Capillaria aerophila* in table) means:
Egg enters gastrointestinal tract. Larva excysts and matures to stage 4. L4 travels to lung and matures to adult. Adult lays eggs in lung. Eggs are coughed up and swallowed, entering gastrointestinal tract. Eggs exit with feces.

Table 36.2
Life cycle and epidemiology of miscellaneous nematodes.

Genus and species	Stages in humans	Adult size (mm) female/male	E/Mf size (mm) Larva size (mm)	Distinguishing features
Capillaria aerophila	L4 A E	40 by 0.2/25 by 0.1	E 79	E: bipolar plugs
Dioctophyma renale	L3 A E	1050 by 12/450 by 6	E 80; L3 12 by 0.4	A: red E: pitted shells
Eustrongylides sp	L3 L4		L3 126 by 0.4 L4 150 by 1	L: pseudocoelomic membranes, prominent ventral cord
Pelodera sp	L3		L3 0.6 by 0.04	Minute double lateral alae
Rhabditis sp	L3		L3 1.7 by 0.07	Similar to *S. stercoralis*
Diploscapter coronata	A E	0.4 by 0.03/0.4 by 0.03		A: parthenogenic female
Turbatrix aceti	L4 A	2.4 long/1.5 long	L 0.2 long	Uteri directed anteriorly in female
Ditylenchus dipsaci	A	1.1 by 0.034/1 by 0.027		One lateral excretory canal
Meloidogyne javanica	E		E 120	Asymmetric
Mammomonogamous laryngeus	A E	23.5 by 0.6/6.3 by 0.38	E 95	A: permanently joined in copula
Cyclodontostomum purvisi	A E	11.5 by 0.4/9 by 0.35	E 80	A:16 curved ventral teeth in buccal capsule
Trichostrongylus sp	A E	8.6 by 0.06/7.7 by 0.08	E 95	E: similar to hookworm
Haemonchus contortus	A E	30 by 0.5/20 by 0.4	E 95	E: similar to hookworm
Ostertagia ostertagi	A E	9.2 by 0.16/7.5 by 0.15	E 80; Mf 928 long	A: external longitudinal cuticular ridges
Metastrongylus apri	L3 A E	60 by 0.45/26 by 0.23	E 57	E: rough, thick shell, embryonated
Lagochilascaris minor	L2-4 A E	24 by 0.8/17 by 0.6	E 65	A: lateral alae E: spherical, pitted shell
Baylisascaris procyonis	L3-4		L3-4 2 by 0.06	L: single lateral alae, excretory column diameter<intestine
Enterobius gregorii	L3 A E	13 by 0.5/2.4 by 0.2	E 60	E: similar to *E. vermicularis*
Syphacia obvelata	L2-4 A E	5.8 by 0.4/1.6 by 0.14	E 153	E: similar to *E. vermicularis* but more elongate
Cheilospirura sp	L3			
Spirocerca lupi	A E	80 by 2.5/54 by 0.8	E 38	A: red, coiled in spiral configuration
Gongylonema pulchrum	A E	145 by 0.5/62 by 0.3	E 70	A: cuticle with alae and bosses
Rictularia sp	A E	30 by 0.7/9 long	E 45	A: subventral cuticular combs
Physaloptera caucasica	L3 A E	100 by 2.8/50 by 1	E 65	A: large cephalic collarette
Thelazia callipaedia	A E	17 by 0.85/13 by 0.75	E 60	E: ovoid and embryonated
Thelazia californiensis	A E	19 long/13 long	E 51	E: ovoid and embryonated
Microfilaria bolivarensis	Mf		Mf 280 by 8	Unsheathed, pointed tail void of nuclei
Microfilaria rodhaini	Mf		Mf 332 by 2.5	Unsheathed, nuclei to tip of tail
Microfilaria semiclarum	Mf		Mf 242 by 5.2	Unsheathed, middle clear zone
Meningonema peruzzii	Mf		Mf 181 by 5	Clear sheath, nuclei to tip of tail
Onchocerca gutturosa	A	500 by 0.2/55 by 0.1		Cuticle asymmetric in thickness with transverse ridges
Onchocerca cervicalis	A	500 by .04/70 by 0.15		Cuticle with transverse ridges

Terms and symbols: E = egg; L2-4 = larval stage; A = adult; Mf = microfilaria.

Table 36.3
Sizes and morphologic features used to identify adults and eggs of miscellaneous nematodes.

Genus and species	Clinical findings	Pathology
Capillaria aerophila (Figs 36.1 and 36.2)	Cough, rales, asthma-like syn, respiratory distress	Interstitial pneumonia, eosinophils, RBCs in bronchi
Dioctophyma renale	Renal colic, hematuria, eosinophilia, fever	Renal abscess, eosinophilic granuloma
Eustrongylides sp	Visceral larva migrans-like syn, mimic appendicitis	Uncoils from fish, penetrates wall GI tract
Pelodera strongyloides	Cutaneous larva migrans-like syn; pruritus, papules; hypopigmented papulonodules; alopecia	Dermal eosinophils & fibroblasts, larva associated with hair follicles
Rhabditis sp	Vaginitis; itchy papules, ulcers, bleeding; chyluria, hematuria; leg edema	Vagina: adult & larvae induce albumin, pus, & RBCs in urine; larvae intraepidermal abscesses
Diploscapter coronata	Achlorhydric stomach, many worms; acute pyelitis	Alkaline urine with nematodes
Turbatrix aceti	Worms passed per vaginal os	None
Ditylenchus dipsaci	GI tract (contaminant): worms vomited	None
Meloidogyne javanica	GI tract (contaminant): worms passed in feces	None
Mammomonogamus laryngeus (Fig 36.5)	Nocturnal & chronic cough, asthma-like syn, hemoptysis	Edema at attachment site
Cyclodontostomum purvisi	Incidental finding in patient with GI capillariasis	Incidentally found in feces
Trichostrongylus sp	Eosinophilia, GI discomfort, diarrhea	Traumatic damage SI mucosa, hemorrhage
Haemonchus contortus	Anemia, weight loss, weakness, GI disturbance	Worms attach to intestinal mucosa
Ostertagia ostertagi	Asymptomatic	Unknown
Metastrongylus apri (Fig 36.6)	Lung and GI tract, incidental findings	Dead worms produce white nodular patches in lungs
Lagochilascaris minor	Painful abscess SQ, fistulas mastoid & neck; nasal obstruction, fever, tonsillitis, pneumonitis; headache, coma; infection eye, ear, dental alveoli	Worms in granulomas & abscesses, bronchopneumonia, sinusitis; necrosis cerebellum, petechia, no inflammation near worms
Baylisascaris sp (Figs 36.8 to 36.10)	Systemic visceral larva migrans-like syn, carditis, neuroretinitis; macular rash, pneumonitis; fever, lethargy, nuchal rigidity, obtundation, CSF eosinophils	Granulomas with eosinophils, plasma cells, & lymphocytes, lung, eye & heart (pseudotumor); cerebral edema & atrophy, cerebellar herniation, gray meninges; marked follicular hyperplasia LN
Enterobius gregorii	Asymptomatic, pruritus ani & abdominal pain likely	None described
Syphacia obvelata	Crawling sensation, pruritus perianal & perineal skin	Scratching, scarring; inflammation colonic submucosa
Cheilospirura sp	Chronic catarrhal conjunctivitis and keratitis	Inflammation
Spirocerca lupi	Intestinal obstruction, peritonitis; aberrant migration	Granulomas around worms in small intestine
Gongylonema pulchrum (Figs 36.11 and 36.12)	Unusual sensations in mouth; vomiting, fever, pharyngitis, stomatitis, esophageal erosions	Eggs in tunnels in mucosa with inflammation
Rictularia sp (Fig 36.13)	Asymptomatic infection of appendix	No significant pathology
Physaloptera caucasica	Vomiting, catarrhal gastritis, bowel infarction, malaise	Erosions at attachment, gastritis, enteritis; ischemic necrosis lamina propria, eosinophils, neutrophils, macrophages, plasma cells, lymphocytes & larvae
Thelazia callipaeda	Asymptomatic; sensation of foreign body, conjunctivitis, photophobia; paralysis eyelid, uveitis, glaucoma, blindness	Conjunctival edema, keratitis; corneal ulceration, scarification, & opacity; iridocyclitis
Thelazia californiensis	Symptoms similar to *T. callipaeda*	Similar to *T. callipaeda*
Microfilaria bolivarensis	Bloodstream: asymptomatic	No known pathology
Microfilaria rodhaini	Bloodstream: asymptomatic	No known pathology
Microfilaria semiclarum	Bloodstream: all known infections mixed	No significant pathology
Meningonema peruzzii	Headache, drowsiness, fatigue; encephalomyelitis	Adult worms coiled in subarachnoid spaces along dorsum of brain stem at level of medulla oblongata
Onchocerca gutturosa *Onchocerca cervicalis* (Figs 36.15 to 36.20)	Fibrous nodules in wrist, conjunctival sac, knee joint, & sole of foot	Suppurative epithelioid granuloma with giant cells & eosinophils, containing nongravid worm

Terms and symbols: CSF = cerebrospinal fluid; GI = gastrointestinal tract; LN = lymph node; RBCs = red blood cells; SI = small intestine; SQ = subcutaneous; syn = syndrome

Table 36.4
Clinical findings and pathology in infections of miscellaneous nematodes.

References

1. Africa CM, Garcia EY. Cited by: Beaver PC, Jung RC, Cupp EW. *Clinical Parasitology*. 9th ed. Philadelphia, Pa: Lea & Febiger; 1984:346–347.
2. Aftandelians R, Raafat F, Taffazoli M, Beaver PC. Pulmonary capillariasis in a child in Iran. *Am J Trop Med Hyg* 1977;26:64–71.
3. Bain O. Redescription de cinq especes d'Onchocerques. *Ann Parasitol Hum Comp* 1975;50:763–788.
4. Beaver PC, Horner GS, Bilos JZ. Zoonotic onchocercosis in a resident of Illinois and observations on the identification of Onchocerca species. *Am J Trop Med Hyg* 1974;23:595–607.
5. Beaver PC, Theis JH. Dioctophymatid larval nematode in a subcutaneous nodule from man in California. *Am J Trop Med Hyg* 1979;28:206–212.
6. Bento RF, Mazza C do C, Motti EF, Chan YT, Guimarães JR, Miniti A. Human lagochilascariasis treated successfully with ivermectin: a case report. *Rev Inst Med Trop Sao Paulo* 1993;35:373–375.
7. Bhaibulaya M, Indrangarm S. Man, an accidental host of Cyclodontostomum purvisi (Adams, 1933), and the occurrence in rats in Thailand. *Southeast Asian J Trop Med Public Health* 1975;6:391–394.
8. Biocca E. Infestazione umana prenatale da Spirocerca lupi (Rud., 1809). *Parassitologia* 1959;1:137–142.
9. Boussinesq M, Bain O, Chabaud AG, Gardon-Wendel N, Kamgno J, Chippaux JP. A new zoonosis of the cerebrospinal fluid of man probably caused by Meningonema peruzzii, a filaria of the central nervous system of Cercopithecidae. *Parasite* 1995;2:173–176.
10. Boreham RE, McCowan MJ, Ryan AE, Allworth AM, Robson JM. Human trichostrongyliasis in Queensland. *Pathology* 1995;27:182–185.
11. Botkin, SP. Cited by: Beaver PC, Jung RC, Cupp EW. *Clinical Parasitology*. 9th ed. Philadelphia, Pa: Lea & Febiger; 1984:265.
12. Caballero E. Cited by: Beaver PC, Jung RC, Cupp EW. *Clinical Parasitology*. 9th ed. Philadelphia, Pa: Lea & Febiger; 1984:264–265.
13. Chandler AC. Diploscapter coronata as a facultative parasite of man, with a general review of vertebrate parasitism by Rhabditoid worms. *Parasitology* 1938;30:44–55.
14. Chittenden AM, Ashford RW. Enterobius gregorii Hugot 1983; first report in the U.K. *Ann Trop Med Parasitol* 1987;81:195–198.
15. Dukes DC, Gelfand M, Gadd KG, Clarke V de V, Goldsmid JM. Cerebral filariasis caused by Acanthocheilonema perstans. *Cent Afr J Med* 1968;14:21–27.
16. Fain A. Dipetalonema semiclarum sp. nov. from the blood of man in the Republic of Zaire (Nematoda: Filarioidea). *Ann Soc Belg Med Trop* 1974;54:195–207.
17. Gardiner CH, Schantz PM. Mammomonogamus infection in a human, report of a case. *Am J Trop Med Hyg* 1983;32:995–997.
18. Ghadirian E, Arfaa F. First report of human infection with Haemonchus contortus, Ostertagia ostertagi, and Marshallagia marshalli (family Trichostrongylidae) in Iran. *J Parasitol* 1973;59:1144–1145.
19. Godoy GA, Orihel TC, Volcan GS. Microfilaria bolivarensis: a new species of filaria from man in Venezuela. *Am J Trop Med Hyg* 1980;29:545–547.
20. Hong ST, Park YK, Lee SK, et al. Two human cases of Thelazia callipaeda infection in Korea. *Korean J Parasitol* 1995;33:139–144.
21. Huff DS, Neafie RC, Binder MJ, De Leon GA, Brown LW, Kazacos KR. Case 4. The first fatal Baylisascaris infection in humans: an infant with eosinophilic meningoencephalitis. *Pediatr Pathol* 1984;2:345–352.
22. Jelinek T, Loscher T. Human infection with Gongylonema pulchrum: a case report. *Trop Med Parasitol* 1994;45:329–330.
23. Jones CC, Rosen T, Greenberg C. Cutaneous larva migrans due to Pelodera strongyloides. *Cutis* 1991;48:123–126.
24. Kenney M, Eveland LK, Yermakov V, Kassouny DY. A case of Rictularia infection of man in New York. *Am J Trop Med Hyg* 1975;24:596–599.
25. Kirschner BI, Dunn JP, Ostler HB. Conjunctivitis caused by Thelazia californiensis [letter]. *Am J Ophthalmol* 1990;110:573–574.
26. Kofoid CA, White AW. A new nematode infection of man. *JAMA* 1919;72:567–569.
27. Küchle M, Knorr HL, Medenblik-Frysch S, Weber A, Bauer C, Naumann GO. Diffuse unilateral subacute neuroretinitis syndrome in a German most likely caused by the raccoon roundworm, Baylisascaris procyonis. *Graefes Arch Clin Exp Ophthalmol* 1993;231:48–51.
28. Montali RJ, Gardiner CH, Evans RE, Bush M. Pterygodermatites nycticebi (Nematoda: Spirurida) in golden lion tamarins. *Lab Anim Sci* 1983;33:194–197.
29. Narr LL, O'Donnell JG, Libster B, Alessi P, Abraham D. Eustrongylidiasis—a parasitic infection acquired by eating live minnows. *J Am Osteopath Assoc* 1996;96:400–402.
30. Nicolaides NJ, Musgrave J, McGuckin D, Moorhouse DE. Nematode larvae (Spirurida: Physalopteridae) causing infarction of the bowel in an infant. *Pathology* 1977;9:129–135.
31. Nosanchuk JS, Wade SE, Landolf M. Case report of and description of parasite in Mammomonogamus laryngeus (human syngamosis) infection. *J Clin Microbiol* 1995;33:998–1000.
32. Rainey G. Entozoon found in the larynx. *Trans Pathol Soc Lond* 1855;6:370–372.
33. Richard-Lenoble D, Kombila M, Bain O, Chandenier J, Mariotte O. Filariasis in Gabon: human infections with Microfilaria rodhaini. *Am J Trop Med Hyg* 1988;39:91–92.
34. Riley WA. A mouse oxyurid, Syphacia obvelata, as a parasite of man. *J Parasitol* 1919;6:89–92.
35. Sprent JF. Speciation and development in the genus Lagochilascaris. *Parasitology* 1971;62:71–112.
36. Takaoka H, Bain O, Tajimi S, et al. Second case of zoonotic Onchocerca infection in a resident of Oita in Japan. *Parasite* 1996;3:179–182.
37. Tuur SM, Nelson AM, Gibson DW, et al. Liesegang rings in tissue. How to distinguish Liesegang rings from the giant kidney worm, Dioctophyma renale. *Am J Surg Pathol* 1987;11:598–605.
38. Volcán GS, Medrano CE, Payares G. Experimental heteroxenous cycle of Lagochilascaris minor Leiper, 1909 (Nematoda: Ascarididae) in white mice and in cats. *Mem Inst Oswaldo Cruz* 1992;87:525–532.
39. Wittner M, Turner JW, Jacquette G, Ash LR, Salgo MP, Tanowitz HB. Eustrongylidiasis—a parasitic infection acquired by eating sushi. *N Engl J Med* 1989;320:1124–1126.
40. Yamaguchi N, Takeuchi T, Kobayashi S, et al. Health status of Indochinese refugees in Japan: statistical analyses on anemia, eosinophilia and serum alkaline phosphatase. *Southeast Asian J Trop Med Public Health* 1984;15:209–216.

37

Acanthocephaliasis

Ronald C. Neafie *and*
Aileen M. Marty

Introduction

Definition

Acanthocephaliasis is infection with a helminth of the Acanthocephala phylum. These worms embed their thorny heads into the intestinal wall, producing mild to fatal disease. Species of the genera *Macracanthorhynchus* and *Moniliformis* are the principle agents of acanthocephaliasis in humans. Most of these infections are induced by just 2 species: *Macracanthorhynchus hirudinaceus* (Pallas, 1781) Travassos, 1917 and *Moniliformis moniliformis* (Bremser, 1811) Travassos, 1915. Acanthocephalans of the *Bolbosoma, Acanthocephalus*, and *Corynosoma* genera rarely infect humans.

Synonyms

Common names for adult acanthocephalans are spiny-headed worms, *Hakenwürmer*, hooked worms, and thorny-headed worms, referring to the thorny armature on the proboscis. Infection by *Macracanthorhynchus* sp is sometimes called human macracanthorhynchosis, and by *Moniliformis* sp, human echinorhynchosis. Outdated names for *M. hirudinaceus*, the giant leech-like acanthocephalan, include *Taenia hirudinaceus, Echinorhynchus hirudinaceus, Gigantorhynchus hirudinaceus*, and *Hormorhynchus gigas*. Some older names for *M. moniliformis*, the moniliform acanthocephalan, are *Echinorhynchus moniliformis, Gigantorhynchus moniliformis, Hormorhynchus moniliformis, Moniliformis cestodiformis*, and *Moniliformis dubius*.

General Considerations

In 1684, Redi provided the earliest account of worms having a proboscis armed with hooks. Leeuwenhoek in 1695 described and illustrated 2 kinds of acanthocephalans found in the intestine of an eel. In 1771, Koelreuther placed the Acanthocephala into a distinct group of helminths, the Acanthocephali. Several years later, Müller, unaware of Koelreuther's work, described a number of species under the name *Echinorhynchus*; for many years this was a general name for all acanthocephalans. Gmelin was the first to recognize the discrete genera and species of Acanthocephala and included them in his 13th edition of *Systema Naturae* (1788 to 1793). Meyer in 1931 classified

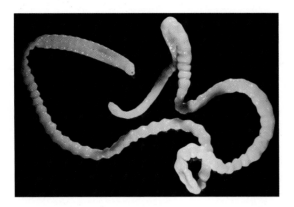

Figure 37.1
Adult female *M. moniliformis* passed in stool of 15-month-old boy living in Florida. Note pseudosegmentation. x10

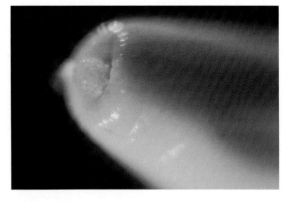

Figure 37.2
Anterior end of adult female *M. moniliformis* depicted in Figure 37.1, demonstrating partially evaginated proboscis.

all known species of Acanthocephala and proposed that they were aschelminths, a group that includes the nematodes and rotiferas. Meyer's classification was refined by Van Cleave in 1936, Yamaguti in 1963, and Bullock in 1969. Garey et al in 1996 conducted phylogenetic studies of the morphologic characteristics and 18S rRNA of Acanthocephala and established that it is a parasitic sister group of the free-living Rotifera of the subclass Bdelloidea.[16] Most acanthocephalans are intestinal parasites of marine vertebrates.[18,34] *Macracanthorhynchus* sp and *Moniliformis* sp are parasitic in land mammals.

There are 3 species of *Macracanthorhynchus*: *M. hirudinaceus*, a cosmopolitan parasite of swine; *Macracanthorhynchus catulinus*, an Old World species that parasitizes dogs, foxes, and other carnivora; and *Macracanthorhynchus ingens*, a parasite of raccoons.[13] Two of these species, *M. hirudinaceus* and *M. ingens*, cause infections in humans. Lambl reported the first infection by an acanthocephalan in humans in 1859.[24] He recovered an acanthocephalan from a boy who died of leukemia in Prague, and identified the worm as *Echinorhynchus hominis*; Meyer, in 1933, identified this organism as *M. hirudinaceus*.[24,28] In 1865, Lindemann reported frequent human infections with *M. hirudinaceus* in the Volga Valley of Russia, where Schneider found that eating raw beetles was common.[26,43] Dingley, in 1984, published the first report of a human infection with *M. ingens*.[12]

In 1888 Calandruccio and Grassi described the clinical manifestations of Calandruccio's self-inflicted infection with larvae of *M. moniliformis*.[20] Beck, in 1959, reported a possible case of *M. moniliformis* in an adult in Florida.[7] The first unequivocal infection of a human by *M. moniliformis* in the United States was in a 15-month-old male from Florida.[11]

Tada et al, in 1983, reported infection with a *Bolbosoma* sp in a human. The genus *Bolbosoma* includes 16 species which normally parasitize marine animals. The species of *Bolbosoma* involved in human infection is not known.

In addition to the infections mentioned above, there are a few records of single infections in humans with 3 other acanthocephalans. Four adult male *Acanthocephalus bufonis*, a common parasite of amphibians in Asia, were recovered at autopsy from the small intestine of an Indonesian man. *Acanthocephalus rauschi*, a species whose natural host is not known, was recovered from the peritoneum of an Alaskan Eskimo. And an immature *Corynosoma strumosum*, a common intestinal parasite of seals that uses many species of fish as paratenic hosts, was recovered from the stool of an Alaskan Eskimo after anthelmintic treatment for other parasites.[42]

Epidemiology

Acanthocephalans are geographically widespread, but human acanthocephaliasis is most common in areas where people eat arthropods as a delicacy or for nutritional and medicinal purposes.

Macracanthorhynchus hirudinaceus is the most frequent agent of human acanthocephaliasis. Most of the several hundred infections documented in humans are from China[25,51]; others come from the Czech Republic,[24] Bulgaria, Thailand,[10,23,36,39,47] Vietnam,[44] Papua New Guinea,[4] Australia,[38] CIS,[26,45] and Madagascar.[50] Screening of stools from 1236 people in Brazil revealed eggs of *M. hirudinaceus* in 2 individuals.[19] These were most likely spurious parasitisms from eating the intestines of parasitized swine.

The only known human infection with *M. ingens* was in a 10-month-old girl in Texas who ingested arthropods.[13] She passed 6 female and 3 male worms.

There are reports of human infection with *M. moniliformis* from Italy, Israel,[8] Bangladesh,[15] CIS,[29] Iran,[30,41] Iraq,[1] Nigeria, Zimbabwe,[17] Sudan, Zambia, Java, Australia,[38] Belize, Colombia, and Florida.[7,11] There is indirect evidence of prehistoric infection with *Moniliformis*. Analyses of dried feces of prehistoric humans from Utah have revealed eggs of *Moniliformis* which could represent either true or spurious infections.[32]

The only reported cases of human *Bolbosoma* sp infection were from Japan.[6,46]

Infectious Agents

Morphologic Description
Adults

Acanthocephalans are pseudocoelomates that are usually white or cream-colored, but can be yellow, orange, or red. They are wrinkled and somewhat flat, but become cylindrical and turgid in water.[13] Grossly, many acanthocephalans have constrictions and are pseudosegmented (Fig 37.1). They have 2 major body regions: the short, slender presoma and the much longer, stouter metasoma. The presoma (Figs 37.2 to 37.5) is composed of the armed proboscis, proboscis receptacle, cerebral ganglion, lemnisci, various muscles, and an unarmed neck. The retractile proboscis is armed with rows of recurved hooks that can invaginate into a proboscis sheath, or a receptacle made up of 1 or 2 muscle layers within a thin sheath. Proboscis hooks are arranged longi-

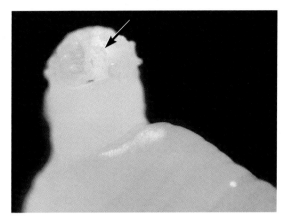

Figure 37.3
Armed proboscis and short neck of infective cystacanth larva of *M. hirudinaceus*. One hook is clearly visible (arrow).

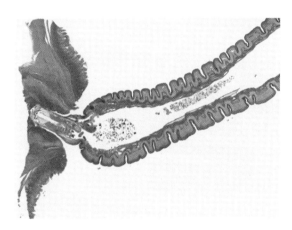

Figure 37.4
Anterior end of adult gravid *M. hirudinaceus* with proboscis embedded in intestine of a pig. x5.8

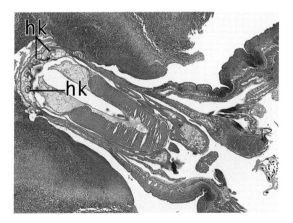

Figure 37.5
Higher magnification of Figure 37.3, depicting attached proboscis of *M. hirudinaceus* and demonstrating hooks (hk). x25

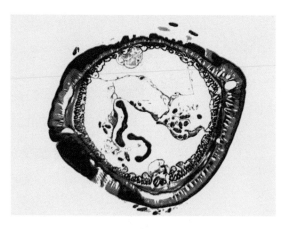

Figure 37.6
Transverse section through adult male *M. hirudinaceus*, showing body wall and internal reproductive structures within pseudocoele. x12

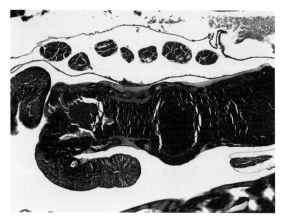

Figure 37.7
Higher magnification of adult male *M. hirudinaceus* in Figure 37.6, depicting reproductive structures. x60

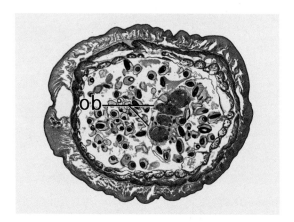

Figure 37.8
Gravid *M. moniliformis* showing body wall and pseudocoele containing numerous eggs and ovarian balls (ob). Movat x40

tudinally or in a spiral pattern.

The metasoma, or body proper, contains ligament sacs, reproductive organs (Figs 37.6 to 37.8), and excretory organs within a pseudocoele. Acanthocephalans have no digestive tract. However, studies of the embryonic development of acanthocephalans have shown that the ligament sacs in adults are vestigial intestines, and that in both sexes the terminal parts of the genital system were once part of a cloaca.[22,27] The sexes are separate: posteriorly, males have a copulatory bursa and females have a gonopore.

The body wall (Figs 37.9 and 37.10) consists of 3 distinct sections: a thin outer tegument, a thick middle epidermis (hypodermis), and an inner group of muscles. The tegument is 1 µm thick and barely perceptible. Various writers describe the tegument of *Macracanthorhynchus* sp or *Moniliformis* sp as either a cuticle with a cuticular zone, or a plasma membrane with a subplasma membrane. Electron microscopy of the body wall demonstrates that the tegument is a syncytium. The thick epidermis has 3 components: a thin outer layer of parallel radial fibers perpendicular to the surface, a thick middle layer of fibers running in several directions, and a thick inner layer of radial fibers containing a lacunar system of channels. Cavities of the lacunar system interconnect and form specific patterns in some species. A very thin basement layer (dermis) separates the epidermis from the underlying muscles. Muscles are arranged in an outer circular layer and an inner longitudinal layer.

The reproductive system of female acanthocephalans includes ovaries that break up into ovarian balls (Fig 37.11) that produce oocytes and eventually eggs (shelled acanthors) (Figs 37.12 to 37.15). Ligament sacs of female worms, when they persist, usually contain developing eggs. The remainder of the female reproductive system includes an efferent duct system composed of a muscular uterine bell (selective apparatus), uterus, and vagina. Eggs exit through the terminal gonopore.

Male acanthocephalans are considerably smaller than females. The male reproductive system (Figs 37.6 and 37.7) includes 2 ovoid testes, each with a vas deferens, vasa efferentia, seminal vesicle, several cement glands, cement reservoir, Saefftigen's pouch, piriform glands, campanulate copulatory bursa, and penis. The genital pore is at the posterior extremity. A suspensory ligament anchors the male organs in the body cavity.[14]

Macracanthorhynchus hirudinaceus

Macracanthorhynchus hirudinaceus varies from milky-white (Fig 37.16) or pinkish-white to red. The body tapers from a broad, rounded anterior portion to a slender posterior segment.[21] Females measure 18.5 to 65 cm by 4 to 10 mm, and males 5 to 10 cm by 3 to 5 mm.[39] Specimens in humans tend to be smaller: female worms are only 12 to 31.5 cm by 3 to 6 mm, and male worms are 6.9 to 7.5 cm by 3.5 to 4 mm.[51] The worm is slightly flattened dorsoventrally and pseudosegmented.[36] The proboscis in the infective cystacanth larva (Fig 37.3) and the adult is 900 µm long by 1 mm wide and has 5 or 6 alternating (spiral) rows of recurved hooks. Hooks range from 120 to 360 µm in length.[49] In tissue, longitudinal sections at the anterior end may reveal hooks on the proboscis (Figs 37.4 and 37.5). The posterior end of the female is bluntly rounded, while the posterior end of the male forms a campanulate bursa.

Microscopically, a transverse section through the metasoma reveals the classic features of an acanthocephalan (Figs 37.6, 37.7, and 37.9). In males, the ligament sac encloses the testes. In females, the ligmanent sac has a very thin wall and visibly surrounds balls of ovarian tissue in younger worms. Eventually, the ligament sac ruptures and these balls lie free in the pseudocoele (Fig 37.17).

Macracanthorhynchus hirudinaceus eggs are symmetrically ovoid (80 to 100 µm by 40 to 50 µm), and contain a fully developed infective acanthor when laid (Fig 37.18). Eggs have 3 envelopes: a thin outer membrane, a thick, 2-layered, mottled shell, and an inner fertilization membrane (Fig 37.12).

Macracanthorhynchus ingens

Macracanthorhynchus ingens are wrinkled and somewhat flat, but become cylindrical and turgid in water. Female worms are 9 to 20 cm long and male worms are 5 to 7 cm long.[13] The proboscis is subglobular, about 500 µm long by 580 µm wide, and bears 6 rows of stout hooks with 6 hooks in each row. Hooks range from 66 to 187 µm in length.[31] In males, the testes are in tandem in the middle third of the body.

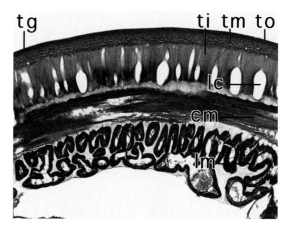

Figure 37.9
Body wall of adult male *M. hirudinaceus,* showing thin outer tegument (tg). Note 3 layers of epidermis (thin outer (to), thick middle (tm), and thick inner (ti) containing lacunar channels (lc)), and outer circular (cm) and inner longitudinal muscles (lm). x60

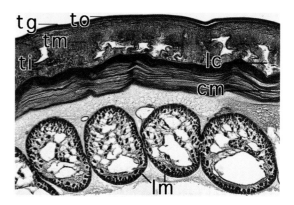

Figure 37.10
Body wall of female *M. moniliformis* showing thin outer tegument (tg). Note 3 layers of epidermis (thin outer (to), thick middle (tm), and thick inner (ti) containing lacunar channels (lc)), and outer circular (cm) and inner longitudinal muscles (lm). Movat x220

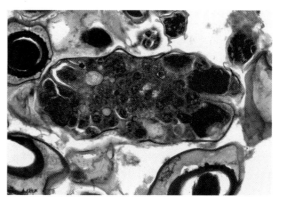

Figure 37.11
Floating ovarian ball in pseudocoele of gravid *M. moniliformis*. Movat x355

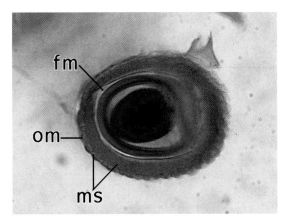

Figure 37.12
Symmetrically ovoid egg within gravid *M. hirudinaceus*, demonstrating 3 embryonic envelopes surrounding an infective acanthor. Note thin outer membrane (om), thick, 2-layered, mottled shell (ms), and inner fertilization membrane (fm). x600

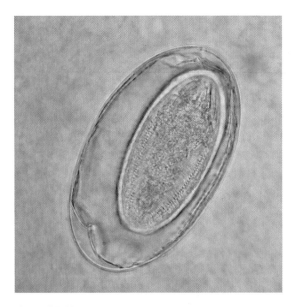

Figure 37.13
Moniliformis moniliformis egg passed in human feces. Unstained x580

Macracanthorhynchus ingens eggs are ellipsoid (90 to 106 μm by 53 to 59 μm), dark brown, and of 3 types: eggs with a mottled brown middle coat, eggs with an incomplete middle coat at one end, and eggs with a lacy middle membrane with strands forming elongate fields oriented parallel to the long axis of the egg.[13]

Moniliformis moniliformis

Moniliformis moniliformis is chalky or creamy-white. Female worms are 10 to 27 cm long by 1.25 to 1.5 mm wide; male worms are 4 to 10 cm long. Worms are cylindrical and conspicuously pseudosegmented (Fig 37.1). The cylindrical proboscis (Fig 37.2) bears 12 to 15 rows of recurved hooks with 7 to 8 hooks in each row. *Moniliformis moniliformis* is distinct from other acanthocephalans in that the muscles outside the proboscis sheath are spiral. Ligament sacs of *M. moniliformis* females fuse to form a single dorsal ligament sac, which terminates in the uterine bell in a specialized area called the ligament attachment syncytium. The vagina consists of an outer layer of muscle that extends from the uterus, and a middle smooth muscle layer organized as a sphincter that contains a narrow lumen. In *M. moniliformis*, the vagina joins the uterus at a 45-degree angle ventrally. The vaginal sphincter opens and closes the gonopore.[2,3]

Microscopic examination of transverse sections reveal the classic features of a female acanthocephalan (Figs 37.8, 37.10, and 37.11). The thin-walled ligmanent sac surrounds balls of ovarian tissue in younger *M. moniliformis*, but is not visible in very mature female worms. (A ligament sac also encloses the testes in male worms.)

Moniliformis moniliformis eggs are ellipsoid (85 to 118 μm by 40 to 52 μm) (Fig 37.13). They have 3 coverings: an outer membrane, a middle smooth or mottled shell, and an inner fertilization membrane (Figs 37.14 and 37.15). There is abundant mucopolysaccharide material between the outer membrane and middle shell.

Bolbosoma

Bolbosoma sp have a distinctive expanded trunk bulb on the anterior extremity of the body just posterior to the presoma. An immature female *Bolbosoma* sp recovered from a human was 13.5 mm long; the proboscis was 0.715 mm long and 0.455 mm in maximum width. There were 16 longitudinal rows of hooks on the proboscis, each with 8 or 9 hooks. The bulb had 2 spiny zones, one around the anterior portion of the anterior cone and the other at the intermediate ring-bulb region. There were no spines on the posterior cone of the bulb. Spines are usually arranged as 2 separate collars or spine fields.[48] The maximum width of

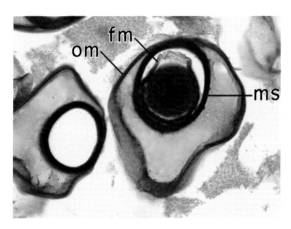

Figure 37.14
Egg in gravid *M. moniliformis*, showing the 3 coverings surrounding an infective acanthor. Note outer membrane (om), middle shell (ms), and inner fertilization membrane (fm). Note also conspicuous mucopolysaccharide material between outer membrane and middle shell. Movat x600

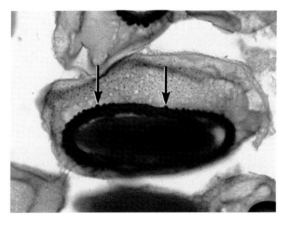

Figure 37.15
Longitudinal section of *M. moniliformis* egg in gravid worm, illustrating conspicuous mottling of middle shell (arrows). Movat x560

the body at the trunk bulb was 0.65 mm. There were spines on the posterior portion of the body, including the caudal end. The only other *Bolbosoma* sp removed from a human was studied microscopically in a biopsy specimen. The largest section of this worm was 1.1 mm in diameter. It had a thin tegument and a thick epidermis consisting of 3 layers. Under the epidermis, resting on a distinct basement membrane, was a thick outer layer of circular muscles and a thin inner layer of longitudinal muscles. Muscle fibers of the inner, noncontractile part stained poorly, giving a hollow appearance to individual fibers in sections. Transverse sections of the terminal end of the parasite contained vagina, uterus, and uterine bell. In other sections, numerous balls of ovarian tissue were between longitudinal and circular layers of somatic muscles, and between extensions of the circular fibers reaching to the longitudinal fibers.

Life Cycle and Transmission

Acanthocephalans are parasitic during all but their egg stage. Vertebrates serve as definitive hosts. Various invertebrates, including crustaceans, annelids, and larval arthropods, serve as intermediate hosts. Humans are typically infected by eating arthropods containing the cystacanth stage (infective acanthor). Some humans are infected by eating fish, amphibians, or reptiles, which serve as paratenic hosts for several species of acanthocephalans.[42]

Macracanthorhynchus

Swine (pig, peccary, wild boar) are the usual definitive hosts for *M. hirudinaceus*.[37] Accidental hosts include dogs, monkeys, and humans.[34] Various species of Scarabaeidae beetles and even an aquatic insect, *Tropisternus collaris*, serve as intermediate hosts. Eggs in feces of a definitive host contain a fully developed acanthor that does not develop further until it is within an appropriate intermediate host. Within the hemocoelom of a beetle, the acanthor molts into the second larval stage, or acanthella.[21] After 6 weeks to 3 months, the developing worm reaches the infective juvenile stage and remains quiescent until ingested by a definitive host. Pigs usually acquire the infection when rooting for beetle grubs, but an infected adult beetle is also a source of cystacanths.[9] In a suitable definitive host, liberated juveniles attach to the intestinal wall, where in 60 to 80 days they develop to maturity and mate. Fertile female worms then generate fully developed, shelled acanthors in the feces of the definitive host.[23] In humans, however, the worms seldom mature, and even when they do, they generally do not produce eggs. Thus, eggs are rarely found in the feces of infected humans.

In certain rural districts of China, children ingest

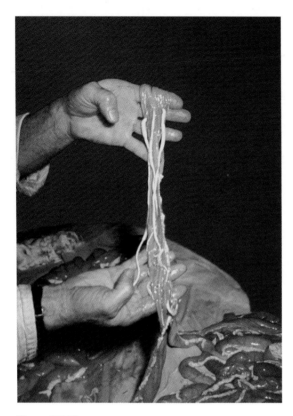

Figure 37.16
Several adult *M. hirudinaceus* attached to intestine of pig.

various kinds of roasted beetles either voluntarily or as an indigenous medication to treat conditions such as nocturia or asthma.[51] The roasting is often insufficient to kill the juvenile worm within the beetle.[25] Similarly, in Thailand, various adult beetles are consumed raw or broiled as food or traditional remedies.[23,39]

Macracanthorhynchus ingens uses millipedes of the genus *Narceus* as intermediate hosts. Because millipedes give off a potent defensive secretion when attacked, raccoons roll the millipede in dust to exhaust its supply of defensive secretion.[9]

Moniliformis
Rodents are the usual definitive hosts for *M. moniliformis*.[41] The Oriental and American cockroach, and beetles, serve as intermediate hosts.[34,40]

Bolbosoma
Cetaceans (whales and dolphins) are the natural definitive hosts for *Bolbosoma* sp.[35] The first intermediate hosts are microcrustacea; fish serve as paratenic hosts. Patients infected with *Bolbosoma* sp had eaten raw marine fish as sashimi.[6,46]

Clinical Features and Pathogenesis

Acanthocephalans can cause severe clinical symptoms in humans. The most frequent manifestations of macracanthorhynchosis, in decreasing frequency, are abdominal pain, abdominal distension, fever, impaired appetite, intestinal perforation, nausea, ascites (serosanguineous fluid), vomiting, weight loss, diarrhea, abdominal mass, constipation, and bloody stools.[51] Nearly all patients with *M. hirudinaceus* infection present with acute abdominal colic.[25] Some patients have chronic, long-standing ulcers of the small intestine, provoked by attachment of the worm's proboscis to the intestinal wall.[36] Leukocytosis with neutrophilia is common.[47] Eosinophilia may be as high as 21%.[39] Without prompt diagnosis and treatment, intestinal obstruction, intussusception, acute peritonitis, intra-abdominal abscess, and even death may result.[51]

In the only reported infection with *M. ingens,* the juvenile patient was asymptomatic and spontaneously passed worms. Laboratory exam revealed a 14% eosinophilia and hematocrit of 34%.[13]

Children with *M. moniliformis* infection present with anorexia and vomiting, sometimes accompanied by foamy diarrhea. The abdomen may protrude and there may be slight anemia and marked eosinophilia (23%).[41]

There are 2 reported infections with *Bolbosoma* sp in humans.[6,46] One patient, a 51-year-old Japanese fisherman, had right lower quadrant pain and a peripheral leukocytosis of $19 \times 10^3/\mu l$.[46] The other patient, a 16-year-old Japanese schoolboy, presented with severe abdominal pain and leukocytosis of $18.5 \times 10^3/\mu l$.[6] Radiographs revealed gas in the small intestine.

Pathologic Features

Sections of ileum or jejunum removed from patients infected with *M. hirudinaceus* reveal

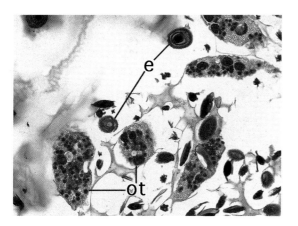

Figure 37.17
Section through gravid *M. hirudinaceus*, showing balls of ovarian tissue (ot) and eggs (e) free in pseudocoele. x120

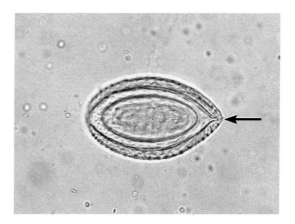

Figure 37.18
Infective acanthor in symmetrically ovoid egg of *M. hirudinaceus* in pig stool. Note raphe (arrow). Unstained x335

multiple perforations and contain mature or immature worms. Sometimes there is necrosis[25] or ulceration of the intestine.[36]

Grossly, the intestine is thick and edematous, the lumen narrowed, and the mucosa ulcerated. Microscopically, there is eosinophilic enteritis.[21] In the mucosa surrounding the edge of the ulcer there is a heavy infiltration of neutrophils and eosinophils, with massive edema of the submucosa surrounding the ulcer. Ulcers contain necrotic tissue and serofibrinopurulent exudates. The anterior end of the worm is sometimes embedded in the intestinal wall.[36] Near the ulcer, intestinal villi may show clubbing, and nearby lymphoid follicles may become hyperactive. There is dilatation of lacteals in the lamina propria, and areas of hyperemia and hemorrhage in the intestinal wall are common. Patchy hemorrhagic areas in the lamina propria may resemble early, preconfluent stages of segmental necrotizing enteritis. The muscularis externa may be disrupted and diffusely infiltrated by the same type of inflammatory cells that involve the serosa just beneath the ulcers. If separation of the muscular wall is extensive, the intestine can perforate, leading to localized acute peritonitis. In some patients there is only chronic, nonspecific inflammation, with eosinophils, lymphocytes, and plasma cells heavily infiltrating all layers of the intestinal wall. Secondary infection at sites of previous worm attachment may develop. Mesenteric lymph nodes have follicular hyperplasia, and many eosinophils enter lymph node sinuses, causing dilatation.[23]

The Japanese patients with *Bolbosoma* sp infections presented with peritonitis. The fisherman had a perforation of the jejunum, and a portion of the jejunum was covered with a purulent mass containing the worm. The schoolboy had an obstruction caused by strangulating fibrous adhesions extending from the intestine to the greater omentum, and a tumor on the serosa of the ileum whose cut surface showed a gray-white mass protruding outward from the serosa. Microscopic sections revealed an eosinophilic granuloma around a knotted worm, and marked edema with chronic inflammatory cells and vascularization of the adjacent submucosa.

Diagnosis

For persons presenting with acute abdomen, especially children from rural areas with a history of eating raw or undercooked insects, acanthocephaliasis should be considered.[25]

Identifying eggs or worms in feces establishes the diagnosis.[33] The metasoma (body proper) of adult acanthocephalans is flattened in situ and pseudosegmented, giving these worms a superficial resemblance to tapeworms. Thus, the first step in preparing specimens for identification is to place them in water, wherein the metasoma becomes cylindrical. Turgor forces the retractile proboscis to evaginate, revealing the shape and number of hooks.[9] The armed, retractile proboscis distinguishes intact adult acanthocephalans from other helminths.

Occasionally the diagnosis is made by identifying worms in biopsy specimens. In stained transverse sections, the epidermis of acanthocephalans is much thicker than the tegument and contains lacunar channels. Transverse sections also reveal 2 layers of somatic muscles (outer circular and inner longitudinal), and no digestive tract.

A skin test using an antigen from acanthocephalan eggs is reportedly positive in 90.9% of infected individuals.[51]

Treatment and Prevention

The best treatment includes early detection and prompt administration of an anthelmintic such as piperazine citrate, tetramisole, or bithionol to expel the worm.[25,51] N-(2' chloronitrophenyl) 5 chlorosalicylamide eliminated *M. moniliformis* in an 18-month-old Iranian boy,[41] and mebendazole was effective in treating a 12-month-old Caucasian girl in Zimbabwe.[17] Surgery is usually necessary for patients with acute abdomen.

Avoiding deliberate or accidental ingestion of arthropods can prevent infection by acanthocephalans. Chemical insecticides such as dioxacarb and chlorpyrifos that kill roaches and other arthropods, or deter them from nesting, are also helpful.[5]

References

1. Al-Rawas AY, Mirza MY, Shafig A, Al-Kindy, L. First finding of Moniliformis moniliformis (Bremser, 1811) Travassos, 1915 (Acanthocephala: Oligacanthorhynchidae) in Iraq from a human child. *J Parasitol* 1977;63:396–397.
2. Asaolu SO. Morphology of the reproductive system of female Moniliformis dubius (Acanthocephala). *Parasitology* 1980;81:433–446.
3. Asaolu SO, Whitfield PJ, Crompton DW, Maxwell L. Observations on the development of the ovarian balls of Moniliformis (Acanthocephala). *Parasitology* 1981;83:23–32.
4. Barnish G, Misch KA. Unusual cases of parasitic infections in Papua New Guinea. *Am J Trop Med Hyg* 1987;37:585–587.
5. Baumholtz MA, Parish LC, Witkowski JA, Nutting WB. The medical importance of cockroaches. *Int J Dermatol* 1997;36:90–96.
6. Beaver PC, Otsuji T, Otsuji A, Yoshimura H, Uchikawa R, Sato A. Acanthocephalan, probably Bolbosoma, from the peritoneal cavity of a man in Japan. *Am J Trop Med Hyg* 1983;32:1016–1018.
7. Beck JW. Research notes: report of a possible human infection with the acanthocephalan Moniliformis moniliformis (syn. M. dubius). *J Parasitol* 1959;45:510.
8. Belding DL. *Textbook of Clinical Parasitology.* 3rd ed. New York, NY: Appleton Century-Crofts; 1965:559–563.
9. Bowman DD. *Georgis' Parasitology for Veterinarians.* 6th ed. Philadelphia, Pa: WB Saunders Co; 1995:229–245.
10. Chaiyaporn V. Discovering of two kinds of parasites in Thailand. *J Med Assoc Thai* 1967;50:834–838.
11. Counselman K, Field C, Lea G, Nickol B, Neafie RC. Moniliformis moniliformis from a child in Florida. *Am J Trop Med Hyg* 1989;41:88–90.
12. Dingley D. Human infection with Macracanthorhynchus sp. in Austin, Texas: report of a case. *Public Health Lab* 1984;42:9–11.
13. Dingley D, Beaver PC. Macracanthorhynchus ingens from a child in Texas. *Am J Trop Med Hyg* 1985;34:918–920.
14. Faust EC, Beaver PC, Jung RC. *Animal Agents and Vectors of Human Disease.* 3rd ed. Philadelphia, Pa: Lea & Febiger; 1968:305–309.
15. Faust EC, Russel PF. *Clinical Parasitology.* 7th ed. Philadelphia, Pa; Lea & Febiger; 1964:701.
16. Garey JR, Near TJ, Nonnemacher MR, Nadler SA. Molecular evidence for Acanthocephala as a subtaxon of Rotifera. *J Mol Evol* 1996;43:287–292.
17. Goldsmid JM, Smith EW, Fleming F. Human infection with Moniliformis sp. in Rhodesia. *Ann Trop Med Parasitol* 1974;68:363–364.
18. Golvan YJ, De Buron I. Hosts of Acanthocephala. II–Definitive hosts. 1. Fishes [in French]. *Ann Parasitol Hum Comp* 1988;63:349–375.
19. Gonzaga A, Gaviao. Contribuiçao ao saneamento de Nordeste. *Brasil Med* 1921;35:134–135.
20. Grassi G, Calandruccio S. Üeber einen Echinorhynchus, welcher auch im Menschen parasitirt und dessen Zwischenwirth ein Blaps ist. *Zentralbl Bakteriol* 1888;3:521–528.
21. Hemsrichart V, Pichyangkura C, Chitchang S, Vutichamnong U. Eosinophilic enteritis due to Macracanthorhynchus hirudinaceus infection: report of 3 cases. *J Med Assoc Thai* 1983;66:303–310.
22. Hyman LH. The pseudocoelomate Bilateria—phylum Acanthocephala. In: *The Invertebrates: Acanthocephala, Aschelminthes, and Entoprocta: The Pseudocoelomate Bilateria.* New York, NY: McGraw-Hill Book Co, Inc; 1951:1–52.
23. Kliks M, Tantachamrun T, Chaiyaporn V. Human infection by an acanthocephalan Macracanthorhynchus hirudinaceus in Thailand: new light on a previous case. *Southeast Asian J Trop Med Public Health* 1974;5:303–309.

24. Lambl W. Mikroskopische Untersuchungen der Darm-Excrete. *Vierteljahrschrift für die Praktische Heilkunde, Herausgegeben* 1859;61:1–59.

25. Leng YJ, Huang WD, Liang PN. Human infection with Macracanthorhynchus hirudinaceus Travassos, 1917 in Guangdong Province, with notes on its prevalence in China. *Ann Trop Med Parasitol* 1983;77:107–109.

26. Lindemann K. Zur Anatomie der Acanthocephalen. *Bull Soc Imp Nat Moscou* 1865;38:484–498.

27. Malakhov VV. Classification of the pseudocoelomates. In: Hope WD, ed. *Nematodes: Structure, Development, Classification, and Phylogeny*. Washington, DC: Smithsonian Institution; 1994:224–253.

28. Meyer A. Acanthocephala. *Braun's Klassen und Ordnung des Tierreichs* 1932-1933;582:383.

29. Mizgireva MF. A rare case of Moniliformis moniliformis parasitization in man [in Russian]. *Med Parazitol (Mosk)* 1962;31:612–613.

30. Moayedi B, Izadi M, Maleki M, Ghadirian E. Human infection with Moniliformis moniliformis (Bremser, 1811) Travassos, 1915 (syn. Moniliformis dubius). Report of a case in Isfahan, Iran. *Am J Trop Med Hyg* 1971;20:445–448.

31. Moore DV. Studies on the life history and development of Macracanthorhynchus ingens Meyer, 1933, with a redescription of the adult worm. *J Parasitol* 1946;32:387–399.

32. Moore JG, Fry GF, Englert E Jr. Thorny-headed worm infection in North American prehistoric man. *Science* 1969;163:1324–1325.

33. Neafie RC, Marty AM. Unusual infections in humans. *Clin Microbiol Rev* 1993;6:34–56.

34. Neveu-Lemaire M. Classe: Acanthocephala, Rudolphi 1808. In: *Traité d'helmintholgie médicale et vétérinaire*. Paris, France: Vigot Frères; 1936:1330–1361.

35. Pendergraph GE. First report of the acanthocephalan, Bolbosoma vasculosum (Rudolphi, 1819), from the pigmy sperm whale, Kogia breviceps. *J Parasitol* 1971;57:1109.

36. Pradastsundarasar A, Pechranond K. Human infection with the acanthocephalan Macracanthorhynchus hirudinaceus in Bangkok: report of a case. *Am J Trop Med Hyg* 1965;14:774–776.

37. Prestwood AK, Kellogg FE, Pursglove SR, Hayes FA. Helminth parasitisms among intermingling insular populations of white-tailed deer, feral cattle, and feral swine. *J Am Vet Med Assoc* 1975;166:787–789.

38. Prociv P, Walker J, Crompton LJ, Tristram SG. First record of human acanthocephalan infections in Australia. *Med J Aust* 1990;152:215–216.

39. Radomyos P, Chobchuanchom A, Tungtrongchitr A. Intestinal perforation due to Macracanthorhynchus hirudinaceus infection in Thailand. *Trop Med Parasitol* 1989;40:476–477.

40. Roth LM, Willis ER. *The Medical and Veterinary Importance of Cockroaches*. Washington, DC: Smithsonian Institution; 1957. *Smithsonian Miscellaneous Collections*; vol 134:1–137.

41. Sahba GH, Arfaa F, Rastegar M. Human infection with Moniliformis dubius (Acanthocephala) (Meyer, 1932) (syn. M. moniliformis (Bremser, 1811) (Travassos, 1915)) in Iran. *Trans R Soc Trop Med Hyg* 1970;64:284–286.

42. Schmidt GD. Acanthocephalan infections of man, with two new records. *J Parasitol* 1971;57:582–584.

43. Schneider AF. On the development of Echinorhynchus gigas. *Ann Mag Nat Hist* 1871;4:441–443.

44. Segal DB, Humphrey JM, Edwards SJ, Kirby MD. Parasites of man and domestic animals in Vietnam, Thailand, Laos, and Cambodia. *Exp Parasitol* 1968;23:412–464.

45. Skrinnik MR, Likhotinskaya MV, Ocheret AM. A case of Macracanthorhynchus infection in man. *Med Parazitol (Mosk)* 1958;27:450–451.

46. Tada I, Otsuji Y, Kamiya H, Mimori T, Sakaguchi Y, Makizumi S. The first case of a human infected with an acanthocephalan parasite, Bolbosoma sp. *J Parasitol* 1983;69:205–208.

47. Tesana S, Mitrchai J, Chunsuttwat S. Acute abdominal pain due to Macracanthorhynchus hirudinaceus infection: a case report. *Southeast Asian J Trop Med Public Health* 1982;13:262–264.

48. Van Cleave HJ. *Acanthocephala of North American Mammals*. Urbana: University of Illinois Press; 1953:97–103.

49. Van Cleave HJ. Parasitological reviews: some host-parasite relationships of the acanthocephala, with special reference to the organs of attachment. *Exp Parasitol* 1952;1:305–330.

50. Voelckel J, Cathalna G. Parasitisme humain par Macracanthorhynchus hirudinaceus (Pallas 1781) á Madagascar. Cited by: Dingley D, Beaver PC. Macracanthorhynchus ingens from a child in Texas. *Am J Trop Med Hyg* 1985;34:918–920.

51. Zhong HL, Feng LB, Wang CX, et al. Human infection with Macracanthorhynchus hirudinaceus causing serious complications in China. *Chin Med J* 1983;96:661–668.

INDEX

Page references for clinical and pathologic features of each disease are listed inclusively under each disease heading. A partial list of representative references to particular signs and symptoms and pathologic changes is presented under the main headings "Clinical features of helminthiases" and "Pathologic features of helminthiases," respectively.

Abdominal angiostrongyliasis (*see* Angiostrongyliasis costaricensis)
Acanthella, 525-526
Acanthocephala, phylum of, 9-10, 519
Acanthocephalans (*see* Acanthocephaliasis)
Acanthocephaliasis, 519-529
    clinical features of, 526
    definition of, 519
    diagnosis of, 527-528
    epidemiology of, 520-521
    etiology of, 519
    history of, 519-520
    life cycle and transmission of, 525-526
    pathologic features of, 526-527
    synonyms for, 519
    treatment and prevention of, 528
Acanthocephalus bufonis, 520
Acanthocephalus rauschi, 520
Acanthocephalus sp, 519
Acanthocheilonema perstans, 245
Acanthocheilonema streptocerca, 245
Acanthocyclops sp (*see* Cyclopidae)
Acanthor (*see* Cystacanth)
Acetylcholine-like substance, in pathogenesis of gnathostomiasis, 456
Achillurbainia nouveli and *A. recondita*, 111-112
    life cycle and transmission of, 112
    morphology of adults of, 111-112
    morphology of eggs of, 112
Achillurbainiidae trematodiasis, 111-112
    clinical features of, 112
    diagnosis of, 112
    epidemiology of, 111
    etiology of, 111
    history of, 111
    pathologic features of, 112
    treatment of, 112
Acoelomates, 3-6
    definition of, 3-4
    Platyhelminthes, phylum of, 4-6
        Cestoda, class of, 5-6
        Trematoda, class of, 4-5
Adenophorea, subclass of, 511
Adiaconidia, as mimic of parasite, 19
Aedes polynesiensis, as vector for dirofilariasis, 280
Aedes samoanus, as vector for dirofilariasis, 280
Aedes sierrensis, as vector for dirofilariasis, 280
Aedes sollicitans, as vector for dirofilariasis, 280
Aedes sp, as vectors for lymphatic filariasis, 222
Aedes taeniorhynchus, as vector for dirofilariasis, 280
Africa, ancylostomiasis in, 355
    angiostrongyliasis costaricensis in, 386
    ascariasis in, 397
    Bertiella sp in, 201
    coenurosis in, 186
    cysticercosis in, 118
    Dicrocoelium hospes in, 88
    dirofilariasis in, 276
    dracunculiasis in, 330
    Enterobius gregorii in, 515
    fascioliasis in, 82
    gnathostomiasis in, 447
    hydatidosis in, 146
    Hymenolepis diminuta in, 201
    Hymenolepis nana in, 199
    intestinal trematodiases in, 96, 97
    lymphatic filariasis in, 219
    mansonelliasis perstans in, 245
    Microfilaria rodhaini in, 515
    oesophagostomiasis in, 499, 500
    Poikilorchis congolensis trematodiasis in, 111
    streptocerciasis in, 251
    strongyloidiasis in, 342
    ternidenamiasis in, 500
    trichinosis in, 472
African eye worm (*see* Loiasis)
Agamofilaria streptocerca, 250
Alae, 343, 355, 386, 398, 412, 436, 516
Alaria alata, 107-110
Alaria americana, 107-110
Alaria arisaemoides, 107-110
Alaria marcianae, 107
Alaria mustelae, 107-110
Alaria sp, 107-110
    life cycle and transmission of, 107-108
    morphology of adults of, 108
    morphology of larvae of, 108
Alaria sp trematodiasis, 107-110
    clinical features of, 108-109
    diagnosis of, 109
    epidemiology of, 107
    etiology of, 107
    history of, 107
    pathologic features of, 109
    treatment of, 109-110
Albendazole, in *Alaria* sp trematodiasis, 110
    in ancylostomiasis, 364
    in ascariasis, 407
    in creeping eruption, 371
    in gnathostomiasis, 460
    in *Haemonchus contortus* infection (in animals), 514
    in hydatidosis, 161
    in intestinal capillariasis, 487
    in loiasis, 273
    in lymphatic filariasis, 238
    in *Metastrongylus apri* infection (in animals), 514
    in neurocysticercosis, 132
    in oesophagostomiasis, 506
    in *Ostertagia* sp infection (in animals), 514
    in strongyloidiasis, 351
    in taeniasis, 131
    in ternidenamiasis, 506
    in toxocariasis, 420

in trichuriasis, 467
Allergic granulomatosis, in differential diagnosis of lymphatic
	filariasis, 236
Alligator, as paratenic host for *Alaria* sp, 108
Alveolar hydatid disease, 155
Ameboma, in differential diagnosis of oesophagostomiasis and
	ternidenamiasis, 505
American brugian filariasis, 319-328
	clinical features of, 322
	definition of, 319
	diagnosis of, 325
	epidemiology of, 320
	etiology of, 319
	history of, 319
	pathologic features of, 325
	synonyms for, 319
	treatment of, 327
American Samoa, angiostrongyliasis cantonensis in, 375
Americas, *Alaria* sp trematodiasis in, 107-110
	ancylostomiasis in, 354
	Ascaris lumbricoides in prehistoric remains in, 397
	fascioliasis in, 82
	intestinal trematodiases in, 96, 97
	Mammomonogamus laryngeus in, 515
	Pelodera sp in, 515
	trichinosis in, 472
Amerindians, mansonelliasis ozzardi in, 258
Amocarzine (CGP 6140), in onchocerciasis, 303
Amphimerus noverca, 89-90
	life cycle and transmission of, 89-90
	morphology of adults of, 89
	morphology of eggs of, 89
Amphimerus pseudofelineus, 89-90
	life cycle and transmission of, 89-90
	morphology of adults of, 89
	morphology of eggs of, 89
Amphiparatenic transmission of toxocariasis in canids and
	felids, 414
Anatrichosoma cutaneum, 367-372
	morphology of adults of, 368
	morphology of eggs of, 368
Ancylostoma braziliense, in ancylostomiasis, 355
	in creeping eruption, 367-372
	morphology of larvae of, 368
Ancylostoma caninum, in ancylostomiasis, 353
	in creeping eruption, 367-372
	life cycle and transmission of, 360-361
	morphology of adults of, 357-358
	morphology of eggs of, 358
	morphology of larvae of, 360
Ancylostoma ceylanicum, cat as natural host for, 353, 355
	life cycle and transmission of, 360-361
	morphology of adults of, 357
	morphology of eggs of, 358
	morphology of larvae of, 360
Ancylostoma duodenale, in creeping eruption, 367, 371
	epidemiology of, 354-355
	life cycle and transmission of, 360-361
	morphology of adults of, 355, 357-358
	morphology of eggs of, 358
	morphology of larvae of, 360
Ancylostomatidae (*see* Ancylostomiasis; Creeping eruption)
Ancylostomatoidea, superfamily of, 511
Ancylostomiasis, 353-365
	clinical features of, 361-362
	definition of, 353
	diagnosis of, 364
	epidemiology of, 354-355
	etiology of, 353
	history of, 353-354
	in differential diagnosis of gnathostomiasis, 458
	pathologic features of, 363-364
	synonyms for, 353
	treatment and prevention of, 364
Angiostrongyliasis cantonensis, 373-384
	clinical features of, 378-379
	definition of, 373
	diagnosis of, 381-382
	epidemiology of, 374-375
	etiology of, 373
	history of, 373-374
	in differential diagnosis of gnathostomiasis, 382
	pathologic features of, 380-381
	synonyms for, 373
	treatment and prevention of, 382
Angiostrongyliasis costaricensis, 385-396
	clinical features of, 389-390
	definition of, 385
	diagnosis of, 394
	epidemiology of, 386
	etiology of, 585
	history of, 385-386
	pathologic features of, 390-391, 393-394
	synonyms for, 385
	treatment and prevention of, 394-395
Angiostrongylus cantonensis, 373-384
	life cycle and transmission of, 376-377
	morphology of adults of, 375-376
	morphology of eggs of, 376
	morphology of larvae of, 376
Angiostrongylus costaricensis, 387-389
	life cycle and transmission of, 388-389
	morphology of adults of, 386-387
	morphology of eggs of, 388
	morphology of larvae of, 387-388
Angiostrongylus malaysiensis, 374-384
Angiostrongylus siamensis, 386
Angola, loiasis in, 262
	onchocerciasis in, 288
	streptocerciasis in, 251
Anisakiasis, 423-431
	clinical features of, 426-427
	definition of, 423
	diagnosis of, 428-430
	epidemiology of, 423
	etiology of, 423, 424
	history of, 423
	pathologic features of, 428
	treatment of, 430
Anisakis simplex, life cycle and transmission of, 426
	morphology of larvae of, 424
Anisakis type II, life cycle and transmission of, 426
	morphology of larvae of, 424
Annelida, phylum of, 10
	as intermediate hosts in acanthocephaliasis, 525
Anopheles bradleyi, as vector for dirofilariasis, 280
Anopheles sp, as vectors for *Wuchereria bancrofti*, 222
Anoplocephala arvicanthidis, 198
Antelope, as host for *Echinococcus* sp, 152
Antibiotics, in creeping eruption, 371

in dracunculiasis, 334, 336
   in hyperinfection strongyloidiasis, 351
Antimony sodium gluconate, in hepatic capillariasis, 490
Ants, in life cycle of *Raillietina celebensis*, 208
Appendicitis, in differential diagnosis of gnathostomiasis, 458
Arabia, onchocerciasis in, 288
Areoles, 3, 4
Argentina, angiostrongyliasis costaricensis in, 386
   Bertiella sp in, 201
   diphyllobothriasis in, 167
   dipylidiasis in, 138
   hydatidosis in, 148
   Hymenolepis diminuta in, 201
   mansonelliasis ozzardi in, 258
   mansonelliasis perstans in, 245
Arthropods, 11, 520-521
   as intermediate hosts for acanthocephalans, 525-526
   as intermediate hosts for *Mesocestoides lineatus*, 208
   in intestinal trematodiases, 97
Ascariasis, 397-409
   clinical features of, 402-403, 405
   definition of, 397
   diagnosis of, 406-407
   epidemiology of, 397-398
   etiology of, 397
   history of, 397
   pathologic features of, 405-406
   synonym for, 397
   treatment and prevention of, 407
Ascaridoidea, superfamily of, 508, 511
Ascaris canis, 411
Ascaris lumbricoides, 397-409
   in differential diagnosis of angiostrongyliasis
      cantonensis, 382
   life cycle of, 401-402
   morphology of adults of, 398
   morphology of eggs of, 401
   morphology of larvae of, 398, 401
Ascaris mystax, 411
Ascaris suum, 397, 401
Ascaris vermicularis (*see* Enterobiasis)
Asia, ancylostomiasis in, 354
   angiostrongyliasis cantonensis in, 374
   ascariasis in, 397
   cysticercosis in, 118
   dirofilariasis in, 276
   dracunculiasis in, 330
   Enterobius gregorii in, 515
   Eurytrema pancreaticum in, 110
   fascioliasis in, 82
   gnathostomiasis in, 447
   hydatidosis in, 146
   Hymenolepis diminuta in, 201
   intestinal trematodiases in, 96
   lymphatic filariasis in, 219
   Mammomonogamus laryngeus in, 515
   Poikilorchis congolensis trematodiasis in, 111
   Strongyloides fuelleborni in, 342
   Thelazia callipaeda in, 515
   trichinosis in, 472
Asian Taenia, 117
Asian *Taenia saginata*-like tapeworm, 117
Aspergillosis, in differential diagnosis of lymphatic filariasis, 236
Aspiculuris tetraptera, in enterobiasis, 434

Australia, acanthocephaliasis in, 521
   angiostrongyliasis cantonensis in, 374
   Cyclodontostomum purvisi in, 515
   dipylidiasis in, 138
   dirofilariasis in, 276
   fascioliasis in (Tasmania), 82
   gnathostomiasis in, 447
   hydatidosis in, 146, 148
   intestinal trematodiases in, 97
   Raillietina celebensis in, 202
Austria, dipylidiasis in, 138
   hydatidosis in, 148
Autoimmunity, in ocular onchocerciasis, 314
Avermectins, in *Metastrongylus apri* infection (in animals), 514
   in *Thelazia* sp infection, 514
Bacillary band, 461, 472, 482, 487
Balantidium coli cyst, as mimic of helminth egg, 17
Balkans, intestinal trematodiases in, 96
Baltic states, diphyllobothriasis in, 167
Bancroftian filariasis (*see* Lymphatic filariasis)
Bancroft's filariasis (*see* Lymphatic filariasis)
Bangladesh, acanthocephaliasis in, 521
   intestinal trematodiases in, 96, 97
Barbados leg (*see* Lymphatic filariasis)
Baylisascaris sp and *B. procyonis*, clinical features of
      infection by, 517
   distinguishing features of, 516
   distribution of, 515
   life cycle of, 515
   morphology of larvae of, 516
   pathologic features of infection by, 517
   source of infection by, 515
   synonym for, 511
   treatment of infection by, 510, 514
Bear, dirofilariasis in, 276
Beef tapeworm (*see Taenia saginata*)
Beeson's cardinal features of trichinosis, 471, 476
Beetle, in *Hymenolepis diminuta* life cycle, 207
Belascaris cati, 411
Belascaris mystax, 411
Belgium, anisakiasis in, 423
   Hymenolepis diminuta in, 201
Belize, acanthocephaliasis in, 521
Benin, dracunculiasis in, 330
   onchocerciasis in, 288
Benzimidazoles (*see* Mebendazole; Albendazole; Thiabendazole)
Bephenium, in *Trichostrongylus* sp infection, 514
Bertia satyri, 197
Bertiella cercopitheci, 197
Bertiella conferta, 197
Bertiella mucronata, 197
   life cycle and transmission of, 208
   morphology of adults of, 204
   morphology of eggs of, 204
   morphology of larvae of, 204
Bertiella polyorchis, 197
Bertiella satyri, 197
Bertiella studeri, 197
   life cycle and transmission of, 208
   morphology of adults of, 204
   morphology of eggs of, 204
   morphology of larvae of, 204
Biliary distomiasis (*see* Clonorchiasis and opisthorchiasis)

Biliary trematodiasis (*see* Clonorchiasis and opisthorchiasis)
Birds, as paratenic hosts for *Alaria* sp, 108
Bison, as host for *Echinococcus* sp, 152
Bithionol, in acanthocephaliasis, 528
    in diphyllobothriasis, 180
    in paragonimiasis, 65
    in taeniasis, 131
Bithynia sp (*see* Snail; Clonorchiasis and opisthorchiasis)
Biting midge, in *Mansonella ozzardi*, 258
    in *Mansonella perstans*, 245
Black fly, in dirofilariasis, 280
    in mansonelliasis ozzardi, 258
    in *Microfilaria bolivarensis* infection, 515
    in *Onchocerca gutturosa; O. cervicalis* infection, 515
    in onchocerciasis, 287-288, 290
Black rat (*see Rattus rattus*)
Bladder worm, 185
Bladder worm infection (*see* Taeniasis and cysticercosis)
Blinding filarial disease (*see* Onchocerciasis; Ocular onchocerciasis)
Bobcat, *Brugia beaveri* in, 319
    dirofilariasis in, 277
Bolbosoma sp, 519, 521-525
    life cycle and transmission of, 521, 526
    morphology of adults of, 521-525
Bolivia, fascioliasis in, 82
    hydatidosis in, 148
    mansonelliasis ozzardi in, 258
Bone and joint, hydatidosis in, 154
    toxocariasis in, 411, 418
Bone marrow, trichinosis in, 478
Bosses, 262, 516
Bothria, 167-168, 171
Bothriocephalus anemia, differentiation from pernicious anemia, 172
    diphyllobothriasis and, 172
Bowel worm infection (*see* Oesophagostomiasis)
Bowman's membrane, in ocular onchocerciasis, 315
Brachylaima sp, life cycle and transmission of, 103
    morphology of adults of, 101
    morphology of eggs of, 101
Brachylaimidae, family of, 93, 97
Brachylaimidiasis (*see* Fasciolopsiasis)
Brain, angiostrongyliasis cantonensis in, 378-380
    coenurosis in, 186, 190
    cyclophyllidean larval infection in, 211
    cysticercosis in, 127-128
    halicephalobiasis in, 496
    hydatidosis in, 154
    loiasis in, 269, 271
    paragonimiasis in, 62-63
    schistosomiasis in, 42
    sparganosis in, 174-175
    trematodiases in, 104
    trichinosis in, 476, 478
Braunia sp and *B. jasseyensis*, 165-166
Brazil, acanthocephaliasis in, 521
    American brugian filariasis in, 320
    ancylostomiasis in, 355
    angiostrongyliasis costaricensis in, 386
    Bertiella sp in, 201
    dipylidiasis in, 138
    hydatidosis in, 148
    Hymenolepis diminuta in, 201
    intestinal trematodiases in, 97

    mansonelliasis ozzardi in, 258
    oesophagostomiasis in, 500
    onchocerciasis in, 288
    trichinosis in, 472
Breast, paragonimiasis in, 61
Bricklayer's anemia (*see* Ancylostomiasis)
Brood capsule, in *Echinococcus* sp, 152
Brugia beaveri, life cycle and transmission of (*see Brugia* sp)
    morphology of adults of, 321
    morphology of microfilariae of, 321
Brugia guyanensis, life cycle and transmission of (*see Brugia* sp)
    morphology of adults of, 321-322
    morphology of microfilariae of, 321-322
Brugia leporis, life cycle and transmission of (*see Brugia* sp)
    morphology of adults of, 321
    morphology of microfilariae of, 321
Brugia malayi, life cycle and transmission of, 222-224
    morphology of adults of, 220-221
    morphology of microfilariae of, 221-222
Brugian filariasis (*see* American brugian filariasis; Lymphatic filariasis)
Brugian zoonosis (*see* American brugian filariasis)
Brugia sp, *Dipetalonema* infection reclassified as, 320
    Dirofilaria infection reclassified as, 320
    life cycle and transmission of, 322
    morphology of adults of, 320-321
    morphology of microfilariae of, 321-322
Brugia timori, life cycle and transmission of, 222-224
    morphology of adults of, 221
    morphology of microfilariae of, 222
Bulb eelworm (*see Ditylenchus dipsaci*)
Bulgaria, acanthocephaliasis in, 521
Bulge-eye (*see* Mansonelliasis perstans)
Bullfrog, as paratenic host for *Alaria* sp, 108
Bung-eye (*see* Mansonelliasis perstans)
Burkina Faso, dracunculiasis in, 330
    onchocerciasis in, 288
    paragonimiasis in, 52
Burkitt's lymphoma, in differential diagnosis of dracunculiasis, 333
Burundi, onchocerciasis in, 288
Bush dog, hydatidosis in, 146, 148
Calabar swelling, in differential diagnosis of gnathostomiasis, 458
California eye worm (*see Thelazia californiensis*)
Calodium hepaticum, 487
Cambodia (Kampuchea), intestinal trematodiases in, 96
    paragonimiasis in, 52
Camel, as host for *Eurytrema pancreaticum*, 110
Cameroon, dracunculiasis in, 330
    loiasis in, 262
    onchocerciasis in, 288
    paragonimiasis in, 52
Canada, *Alaria* sp trematodiasis in, 107
    American brugian filariasis in, 320
    anisakiasis in, 423
    coenurosis in, 186
    dirofilariasis in, 276
    halicephalobiasis in, 493
    hydatidosis in, 146, 148
    Metorchis conjunctus in, 89
    paragonimiasis in, 52
Canine esophageal tumor worm (*see Spirocerca lupi*)
Capillaria aerophila, clinical features of infection by, 517

distinguishing features of, 516
distribution of, 515
history of, 511
life cycle of, 515
morphology of adults of, 516
pathologic features of infection by, 517
source of infection by, 515
synonyms for, 511
treatment of infection by, 510, 514
Capillaria hepatica, life cycle and transmission of, 488
morphology of adults of, 487
morphology of eggs of, 487
Capillaria leidyi, 487
Capillaria philippinensis, life cycle and transmission of, 482
morphology of adults of, 482
morphology of eggs of, 482
stichocytes of, 482
Capillariasis (*see* Capillariasis, hepatic; Capillariasis, intestinal)
Capillariasis, hepatic, 487-492
clinical features of, 488
definition of, 487
diagnosis of, 490
epidemiology of, 487
etiology of, 487
pathologic features of, 488, 490
synonyms for, 487
treatment and prevention of, 490
Capillariasis, intestinal, 481-487
clinical features of, 483
definition of, 481
diagnosis of, 487
etiology of, 481
epidemiology of, 481-482
history of, 481
pathologic features of, 485, 487
synonyms for, 481
treatment of, 487
Carcinoma, *Clonorchis sinensis* infection and, 77
in differential diagnosis of dirofilariasis, 282
of urinary bladder in schistosomiasis, 41
Carcinomatosis, in differential diagnosis of oesophagostomiasis and ternidenamiasis, 505
Cardiovascular system, in onchocerciasis, 298
Caribbean, angiostrongyliasis costaricensis in (Puerto Rico), 386
Bertiella sp in, 201
dipylidiasis in, 138
fascioliasis in, 82
Hymenolepis diminuta in, 201
lymphatic filariasis in, 219
Casoni skin test, 159
Cat (domestic), ancylostomiasis in, 353, 355
Capillaria hepatica in, 487
clonorchiasis and opisthorchiasis in, 71
dirofilariasis in, 276
Gnathostoma sp in, 454
hydatidosis in, 146, 148
lymphatic filariasis in, 222
Mesocestoides lineatus in, 201
Taenia taeniaeformis in, 202
Toxocara cati in, 412
Trichinella pseudospiralis in, 471
Cattle, as hosts for *Eurytrema pancreaticum*, 110
as hosts for *Oesophagostomum bifurcum*, 502

as hosts for *Taenia saginata*, 121, 122
as hosts for *Ternidens deminutus*, 502
Ceguera de Los Ríos (*see* Onchocerciasis)
Central Africa, mansonelliasis perstans in, 245
Poikilorchis congolensis trematodiasis in, 111
streptocerciasis in, 251
Central African Republic, loiasis in, 262
onchocerciasis in, 288
paragonimiasis in, 52
Central America, ascariasis in, 397
cysticercosis in, 118
hydatidosis in, 148
Hymenolepis diminuta in, 201
Hymenolepis nana in, 199
Lagochilascaris minor in, 515
mansonelliasis ozzardi in, 258
Central nervous system (CNS), angiostrongyliasis cantonensis in, 379
coenurosis in, 186, 190
gnathostomiasis in, 457
halicephalobiasis in, 494, 496
hydatidosis in, 154
strongyloidiasis in, 346
trematodiases in, 104
Cephalic gland, 355
Cerebrospinal fluid (CSF), in angiostrongyliasis cantonensis, 379
in coenurosis, 190
in loiasis, 269, 271
in onchocerciasis, 300
Cestoda, class of, 5-6
Ceviche, as food source of anisakiasis, 423
Chad, dracunculiasis in, 330
loiasis in, 262
onchocerciasis in, 288
Charcot-Leyden crystals (*see also* Pathologic features of helminthiases), 86, 403, 441, 488
Cheilospirura sp, clinical features of infection by, 517
distribution of, 515
history of, 511
life cycle of, 515
pathologic features of infection by, 517
source of infection by, 515
synonyms for, 511
Chicken, as intermediate host for *Gnathostoma* sp, 454
Chile, diphyllobothriasis in, 167
dipylidiasis in, 138
hydatidosis in, 148
trichinosis in, 472
Chimpanzee, *Capillaria hepatica* in, 487
mansonelliasis perstans in, 245
streptocerciasis in, 251
China, acanthocephaliasis in, 521
Achillurbainiidae trematodiasis in, 111
angiostrongyliasis cantonensis in, 374
dirofilariasis in, 276
Eurytrema pancreaticum in, 110
gnathostomiasis in, 447
Hymenolepis diminuta in, 201
intestinal trematodiases in, 96-97
lymphatic filariasis in, 219
paragonimiasis in, 52
Raillietina celebensis in, 202
sparganosis in, 167
trichinosis in, 472

Chinese liver fluke infection (*see* Clonorchiasis and opisthorchiasis)
Choko-fushu (*see* Gnathostomiasis)
Cholangiography, 81
Chronic eosinophilic pneumonia, in differential diagnosis of lymphatic filariasis, 236
Chrysops centurionis, in simian loiasis, 262
Chrysops dimidiata (*see* Tabanid fly)
Chrysops distinctipennis (*see* Tabanid fly)
Chrysops langi, in simian loiasis, 262
Chrysops longicornis (*see* Tabanid fly)
Chrysops silacea (*see* Tabanid fly)
Chrysops sp (*see* Tabanid fly)
Chrysops zahrai (*see* Tabanid fly)
CIS (Commonwealth of Independent States),
    acanthocephaliasis in, 521
    ascariasis in, 397
    diphyllobothriasis in, 167
    Eurytrema pancreaticum in, 110
    gnathostomiasis in, 447
    hydatidosis in, 146, 148
    Hymenolepis diminuta in, 201
    Hymenolepis nana in, 199
    intestinal trematodiases in, 96, 97
    opisthorchiasis in, 71
    paragonimiasis in, 52
    trichinosis in, 472
Clinical features of helminthiases (combined for all diseases in this volume)
    abdomen, acute, 502
    abdominal involvement, 56, 172, 362, 503, 526
    abscess, abdominal, 502, 526
        hepatic, 403
    adenolymphoceles, 295
    allergic symptoms, 155, 466, 476
    anaphylaxis, 155
    anemia, 60, 84, 125, 172, 362, 416, 466, 488
    angioedema, transient migratory, 265
    anorexia, 77, 84, 141, 389, 403, 416, 474, 526
    aortic embolism, 154
    appendicitis, 125, 403, 440, 466
    appendicitis-like symptoms, 458
    arthralgia, 247
    arthritis, 269
        juvenile rheumatoid arthritis-like symptoms, 418
    ascites, 77, 84, 483
    asthma, 84, 345
    asthma-like symptoms, 362
    asymptomatic infection, 30, 77, 88, 112, 208, 224, 247, 258, 281, 294, 322, 344, 362, 402, 458, 465, 488, 526
    B_{12}, low levels of, 172
    biliary colic, 85
    biliary involvement, 153
    biliary obstruction, 85, 172, 403
    blindness, 190, 287, 307-318, 382, 411, 416-417, 457
    blister, on leg, 332
    borborygmi, 483
    bronchopneumonia, 345
    bung-eye, 247
    cachexia, 483
    Calabar swellings, 265
    cancer, biliary tract, 77
        urinary bladder, 41
    cardiac involvement, 154, 362
    cardiomyopathy, 269
    cellulitis, 225
    cerebral involvement, 58, 60, 476
    cervicitis, 440
    cholangiocarcinoma, 77
    cholangitis, 77, 125, 153, 403
    cholecystitis, 77, 125, 153
    cholelithiasis, 79
    chorioretinitis, 310
    chyluria, 225
    CNS involvement, 30, 42, 58, 104, 190, 269, 457, 496
    coin lesions, 281
    colic, 103
        biliary, 85
    congestive heart failure, 154
    conjunctival involvement, 265, 281
    conjunctivitis, 113
    constipation, 122, 362, 389, 526
    convulsions, 208
    cor pulmonale, 231
    cough, 29, 57, 84, 153, 225, 416, 427
    cranial nerve involvement, 127, 378
    CSF involvement, 60, 190, 496
    cutaneous manifestations, 29, 56, 60, 109, 127, 189-190, 208, 253, 265, 295, 332-333, 344, 369-370, 416, 503
    cyst, asymptomatic, 153
        breast, 61
        mastoid, 60-61
        postauricular, 112
        subcutaneous, 112
    cystitis, 41, 440
    death, 79, 109, 155, 362, 483, 496
    debility, 333
    dermatitis, 258
        cercarial, 42
        eczematous, 440
        ground itch, 361
        onchocercal, 294-295
        perianal and perineal, 440
    diarrhea, 77, 84, 88, 103, 122, 362, 465, 474, 483, 526
    dysphagia, 477
    dyspnea, 153
    dysuria, 39
    edema, 77, 362, 483
        cerebral, 60
        facial, 103, 295, 476
        palpebral, 476
    elephantiasis, 227, 295-296
    empyema, 154
    encephalitis, 269
    endometritis, 440
    endophthalmitis, 417
    enuresis, 440
    eosinophilia, 29, 56, 77, 85, 88, 103, 125, 141, 172, 211, 224, 227, 258, 269, 281, 333, 344, 362, 390, 403, 416, 427, 456, 466, 476, 488, 503, 526
        tropical pulmonary, 227
    epididymitis, 225
    epilepsy, jacksonian, 42, 60, 127, 154, 269
    erisípela de la costa, 295
    failure to thrive, 140
    fatigue, 474
    fever, 29, 56, 77, 84, 225, 258, 281, 378, 416, 476, 488, 502, 526

(Clinical features, continued)
- folate deficiency, 362
- funiculitis, 225, 269
- *gâle filarienne*, 295
- gallbladder disease, 88, 403
- gallbladder involvement, 247
- gastric intrinsic factor, low levels of, 172
- gastritis, 427
- gastrointestinal involvement, 56, 84, 208, 345, 458
- genitourinary involvement, 30, 33, 457-458
- glaucoma, 310, 417
- glomerulonephritis, 33
- growth retardation, 33, 362
- headache, 258, 378
- hematemesis, 427
- hematuria, 39, 154, 227
- hemiparesis, 127, 154
- hemoptysis, 57, 153, 281
- hemorrhage, cerebral, 60
    - splinter, 476
- hepatic involvement, 247, 390, 457
- hepatoma, 33
- hepatomegaly, 77, 231, 416, 488
- hepatosplenomegaly, 30, 84, 88
- hives, 29
- hunger pain, 122
- hydatiduria, 154
- hydrocele, 225, 269
- hydrocephalus, 175
- hyperglobulinemia, 416
- hypersensitivity reactions, 427
- hypochondrodynia, 111
- hypoproteinemia, 362, 483
- IgE response, 362, 483
- IgG response, 58, 85, 483
- IgM response, 483
- ileus, 103, 346
- indigestion, 77
- infertility, 42, 346, 440
- insomnia, 125, 370, 440
- intestinal bleeding, 389
- intestinal involvement, 344, 440
- intestinal obstruction, 103, 403, 502, 526
- intestinal perforation, 526
- intra-abdominal fistulas, 154
- intracranial pressure, 154
- intussusception, 465, 502, 526
- iridocyclitis, 308, 310
- irritability, 208, 440
- jaundice, 77, 84, 88, 153
- keratitis, sclerosing, 308
    - subepithelial punctate, 308
    - superficial, 113
- larva currens, 346
- laryngitis, 85, 113
- lassitude, 88
- leopard skin, 295
- lethargy, 362
- leukocytosis, 60, 77, 85, 103, 172, 389, 416, 427, 526
- lithiasis of biliary system, 85
- lizard skin, 295
- Löffler's syndrome, 236, 362, 402-403, 416, 419
- lymphadenitis, 295
- lymphadenopathy, 29, 225, 253, 258
- lymphangitis, 225
- lymphatic involvement, 56
- lymphatics, alterations of, 224
- lymphedema, 225
- macules, cutaneous, 253
- malabsorption, 346, 483
- malaise, 77, 281, 483
- *mal morado*, 295
- malnutrition, 362, 465-466
- mass, abdominal, 503, 526
    - cutaneous, 127, 503
    - soft tissue, 127
- Mazzotti reaction, 300
- Mazzotti-like reaction, 253
- mediastinitis, fibrosing, 225
- meningitis, 60, 127
    - eosinophilic, 378-379
- meningoencephalitis, 496
- mossy foot (lymphostatic verrucosa), 227
- mucoid stools, 465
- myalgia, 476
- myocarditis, 104
- nausea, 88, 104, 122, 474
- nephropathy, 269
- nephrosis, 362
- neurologic involvement, 125, 418
- nodules, cutaneous, 247, 253
    - in breast, 281
    - in conjunctiva, 247
    - in groin, 322
    - in neck, 322
    - in penis, 247
    - in scrotum, 281
    - lymphatic, nontender, 322
    - migratory, 60
    - periorbital, 281
    - subcutaneous, 174, 281
- ocular involvement, 127, 154, 174, 190, 247, 296, 362, 378-379, 415-418, 457, 476
- oliguria, 474
- onchocercomata, 294
- oophoritis, 440
- optic atrophy, 154, 312
- optic neuritis, 154
- orchitis, 269, 390
- otitis media, 405
- pain, abdominal, 84, 88, 103, 141, 247, 403, 458, 465, 474, 483, 502, 526
    - back, 154
    - chest, 153, 281
    - epigastric, 427
    - joint, 258
    - muscular, 476
    - on rectal examination, 389
    - periorbital, 127
- pancreatitis, 403
- papilledema, 127
- paralysis, 269
- paraplegia, 154
- paresthesia, 269
- *peau d'orange*, 295
- pericarditis, 247
    - constrictive, 225
- peripheral nerve involvement, 127
- peritonitis, 154, 405, 440, 503, 526
    - granulomatous, 125

(Clinical features, continued)
   phthisis bulbi, 154, 417
   pica, 362
   pleocytosis, 496
   pleural effusion, 58, 225, 247, 269
   pleuritis, 247
   pneumonia, 416
   pneumonitis, 378-379
      Ascaris pneumonitis, 402
   pneumothorax, 57, 154
   polyps, schistosomal, in colon, 30-31
   portal cirrhosis, 77
   portal hypertension, 153
   presbydermia, 295
   proptosis, 154
   prostatitis, 441
   proteinuria, 227
   prurigo ani, 125
   prurigo nodularis (allergic), 125
   pruritus, 85, 247, 253, 258, 294
      anal, 122, 140, 440
   psoriasis, 269
   psychosis, 370, 378-379
   pulmonary embolism, 154
   pulmonary infarction, 281
   pulmonary infiltrates, 269
   pulmonary involvement, 42, 57-58, 84, 153-154, 322, 344-345, 362, 402, 457-458, 488
   rash, macular, 295
      papular, 295
   rectal prolapse, 465
   renal involvement, 154
   respiratory involvement, 125
   salpingitis, 440
   scrotitis, 269
   seizures, epileptiform, 125
   septicemia, bacterial, 346
   shock, 483
   sowda, 295
   splenic involvement, 154
   splenomegaly, 77, 225
   sterility, 42
   subclinical infection, 415
   subcutaneous manifestations, 189-190, 456-457
   superinfection, bacterial and fungal, 227
   tenesmus, 465
   testicular torsion, 390
   tetany, 77
   urethritis, 440
   urinary bladder involvement, 41
   urinary tract infection, 440
   urticaria, 84, 125, 265
   vaginitis, 440
   vertigo, 258
   visceral larva migrans, 109, 416
   vomiting, 88, 104, 122, 141, 389, 403, 474, 483
   vulvitis, 440
   weakness, 474
   weight loss, 77, 122, 141, 208, 416, 526
   worm burden, symptoms related to, 139, 172, 402, 441
   worm migration, symptoms related to, 456-457
   worms in body secretions, 457
   xylose absorption, reduced, 211
Clinostomum complanatum, life cycle and transmission of, 113
   morphology of adults of, 112-113
Clinostomum complanatum trematodiasis, 112-113
   clinical features of, 113
   epidemiology of, 112
   etiology of, 112
   history of, 112
Clonorchiasis and opisthorchiasis, 69-92
   clinical features of, 77, 79
   definition of, 69
   diagnosis of, 81
   epidemiology of, 70-71
   etiology of, 69-70
   history of, 70
   pathologic features of, 79-80
   synonyms for, 70
   treatment and prevention of, 90
Clonorchis sinensis, 69-81
   differentiation of eggs from *Paragonimus*, 64
   life cycle and transmission of, 75-77
   morphology of adults of, 71, 73
   morphology of eggs of, 73
Coati, American brugian filariasis in, 319
   angiostrongyliasis costaricensis in, 388
Coccidioides immitis infection, in differential diagnosis of angiostrongyliasis cantonensis, 382
Cochin China diarrhea (*see* Strongyloidiasis)
Cockroach, as intermediate host for *Hymenolepis diminuta*, 207
   as intermediate host for *Moniliformis moniliformis*, 526
Codfish, in transmission of anisakid larvae, 423
Coelomates (nonhelminths), 10-11
   Annelida, phylum of, 10
   Arthropoda, phylum of, 11
   Pentastomida, phylum of, 10-11
Coenurosis, 185-196
   clinical features of, 189-190
   definition of, 185
   diagnosis of, 192, 194
   epidemiology of, 186
   etiology of, 185
   history of, 186
   pathologic features of, 191-192
   synonyms for, 185
   treatment and prevention of, 194
Coenurus (*see also* Bladder worm; Coenurosis)
   in differential diagnosis of sparganum, 180
Colombia, acanthocephaliasis in, 521
   American brugian filariasis in, 320
   angiostrongyliasis costaricensis in, 386
   halicephalobiasis in, 493
   hydatidosis in, 148
   Hymenolepis diminuta in, 201
   mansonelliasis ozzardi in, 258
   onchocerciasis in, 288
   paragonimiasis in, 52
Commonwealth of Independent States (*see* CIS)
Comoro Islands, angiostrongyliasis cantonensis in, 374
   Inermicapsifer madagascariensis in, 201
Complement fixation test, for angiostrongyliasis cantonensis, 382
   for hydatidosis, 146
   for loiasis, 273
   for paragonimiasis, 65
Computed tomography, 58, 65, 81, 88, 131, 159, 180, 194, 236, 382, 406

Congo, Democratic Republic of (Zaire), *Dicrocoelium hospes* in, 88
    loiasis in, 262
    Microfilaria semiclarum in, 515
    onchocerciasis in, 288
    paragonimiasis in, 52
    streptocerciasis in, 251
Congo, Republic of (Brazzaville-Congo), loiasis in, 262
    onchocerciasis in, 288
    paragonimiasis in, 52
    streptocerciasis in, 251
Consular disease (*see* Gnathostomiasis)
Contracaecum sp, life cycle and transmission of, 426
    morphology of larvae of, 424
Copepods, as intermediate hosts for *Diphyllobothrium* sp, 171
    as intermediate hosts for *Dracunculus medinensis*, 332
    as intermediate hosts for *Gnathostoma* sp, 454
    as intermediate hosts for *Spirometra* sp, 171
Cordiella inflata (*see* Snail; Clonorchiasis and opisthorchiasis)
Cordiella leachi (*see* Snail; Clonorchiasis and opisthorchiasis)
Cordiella troscheli (*see* Snail; Clonorchiasis and opisthorchiasis)
Corsica, fascioliasis in, 82
Corticosteroids, in *Alaria* sp trematodiasis, 110
    in angiostrongyliasis cantonensis, 382
    in anisakiasis, 430
    in cysticercosis, 132
    in development of *Strongyloides stercoralis*, 347
    in loiasis, 273
    in lymphatic filariasis, 238
    in paragonimiasis, 65
    in promotion of hyperinfection strongyloidiasis, 345
    in toxocariasis, 420
    in trichinosis, 479
Corynosoma sp (*see* Acanthocephaliasis)
Corynosoma strumosum (*see* Acanthocephaliasis)
Cosmopolitan infections
    Achillurbainiidae trematodiasis (in mammals), 111
    ancylostomiasis, 354
    ascariasis, 397
    Baylisascaris sp, 515
    Capillaria aerophila, 515
    Cheilospirura sp, 515
    cysticercosis, 118
    Dicrocoelium dendriticum, 88
    Dioctophyma renale, 515
    Diploscapter coronata, 515
    dirofilariasis, 276
    Ditylenchus dipsaci, 515
    enterobiasis, 434
    Gongylonema pulchrum, 515
    Haemonchus contortus, 515
    hepatic capillariasis (in rats), 487
    intestinal trematodiases, 94
    Meloidogyne javanica, 515
    Metastrongylus apri, 515
    Onchocerca gutturosa, 515
    Ostertagia ostertagi, 515
    Physaloptera caucasica, 515
    Rhabditis sp, 515
    Rictularia sp, 515
    strongyloidiasis, 342
    Syphacia obvelata, 515
    taeniasis, 118
    toxocariasis, 412

    Trichostrongylus sp, 515
    trichuriasis, 461
    Turbatrix aceti, 515
Costa Rica, angiostrongyliasis costaricensis in, 386
    hydatidosis in, 148
    paragonimiasis in, 52
Cotton rat (*see Sigmodon hispidus*)
Cottontail, *Capillaria hepatica* in, 487
Coyote, ancylostomiasis in, 355
    Capillaria hepatica in, 487
Craw-craw (*see* Onchocerciasis)
Creeping eruption, 367-372
    clinical features of, 369-370
    definition of, 367
    diagnosis of, 371
    epidemiology of, 367
    etiology of, 367
    in differential diagnosis of gnathostomiasis, 447
    life cycle and transmission of etiologic agents of, 368-369
    pathologic features of, 370-371
    synonyms for, 367
    treatment of, 371
Crohn's disease, in differential diagnosis of angiostrongyliasis costaricensis, 394
    in differential diagnosis of oesophagostomiasis and ternidenamiasis, 505
Crustaceans, as intermediate hosts for acanthocephalans, 525
    in intestinal trematodiases, 97
CT (*see* Computed tomography)
Ctenocepnalides felis, as intermediate host for *Dipylidium caninum*, 139
Cuba, angiostrongyliasis cantonensis in, 375
    Bertiella sp in, 201
    Hymenolepis diminuta in, 201
    Inermicapsifer madagascariensis in, 201
    paragonimiasis in, 52
    Raillietina sp in, 202
Culex pipiens, 216
Culex quinquefasciatus, 216
Culex sp, 222
Culicoides austeri, 246-247
Culicoides grahami, 247, 253
Culicoides milnei (*see also* Biting midge), 253
Culicoides sp (*see also* Biting midge)
Curse of St. Thomas (*see* Lymphatic filariasis)
Cutaneous larva migrans (*see* Creeping eruption)
Cuticular combs, 516
Cyclobendazole, in ascariasis, 407
Cyclodontostomum purvisi, clinical features of infection by, 517
    distinguishing features of, 516
    distribution of, 515
    history of, 511
    life cycle of, 515
    morphology of adults of, 516
    morphology of eggs of, 516
    pathologic features of infection by, 517
    source of infection by, 515
    synonym for, 511
Cyclopidae, as intermediate hosts for *Dracunculus medinensis*, 332
    as intermediate hosts for *Gnathostoma* sp, 454
Cyclops sp, as hosts for *Diphyllobothrium latum*, 171
    as intermediate hosts for *Spirometra* sp, 172
    as intermediate hosts for *Gnathostoma* sp, 454
Cyprus, fascioliasis in, 82

Cystacanth, 525
Cysticercosis (see Taeniasis and cysticercosis)
Cysticercus, in differential diagnosis of gnathostomiasis, 459
    in differential diagnosis of sparganum, 180
    in differential diagnosis of tuberculosis, 131
Cysticercus bovis, 117
    morphology of, 131
Cysticercus cellulosae, 117
    morphology of, 120-121, 131
Cysticercus fasciolaris, 198
Cysticercus racemosus, 118
    cestode species of, 121
    morphology of, 121
Cysticercus viscerotropica, 117
    morphology of, 131
Cystic hydatid disease (see Hydatidosis; *Echinococcus granulosus*)
Cytokines, in development of *Schistosoma mansoni*, 3
Czech Republic, acanthocephaliasis in, 521
Davainea celebensis, 198
Davainea formosana, 198
Davainea madagascariensis (see also *Inermicapsifer madagascariensis*), 198
Dead-end hosts, humans as, for American *Brugia* sp, 322
    for *Angiostrongylus cantonensis*, 376
    for *Anisakis*, 426
    for *Dirofilaria* sp, 275
    for *Gnathostoma* sp, 452, 456
Deer, as host for *Echinococcus* sp, 152
Demodex folliculorum, 9
Denmark (see also Greenland), anisakiasis in, 423
Dermatobia hominus, 7, 8
Deroceras (Agriolimax) laeve (see Slug)
Descemet's membrane, ocular onchocerciasis in, 315
Diaptomus sp (copepods), as hosts for *Diphyllobothrium latum*, 171
Dicrocoeliasis, 88-89
    clinical features of, 88
    definition of, 88
    diagnosis of, 89
    epidemiology of, 88
    etiology of, 88
    pathologic features of, 88
    synonyms for, 88
    treatment and prevention of, 90
Dicrocoelium dendriticum, 88-89
    morphology of adults of, 88, 98, 101
    morphology of eggs of, 88, 98, 101
Dicrocoelium hospes, 88-89
    morphology of adults of, 88
    morphology of eggs of, 88
Diethylcarbamazine, in American brugian filariasis, 327
    in gnathostomiasis, 460
    in hepatic capillariasis, 490
    in loiasis, 273
    in lymphatic filariasis, 237
    in mansonelliasis ozzardi, 259
    in mansonelliasis perstans, 250
    in onchocerciasis, 303
    in streptocerciasis, 256
    in toxocariasis, 420
Digramma sp and *D. brauni*, 165-166
Dingo, hydatidosis in, 146
Dioctophyma renale, clinical features of infection by, 517
    distinguishing features of, 516
    distribution of, 515
    history of, 511
    life cycle of, 515
    morphology of adults of, 516
    morphology of eggs of, 516
    morphology of larvae of, 516
    pathologic features of infection by, 517
    source of infection by, 515
    synonyms for, 511
    treatment of infection by, 510
Dioctophymatoidea, superfamily of, 507, 508, 511
Dipetalonema perstans (see *Mansonella perstans*)
Dipetalonema streptocerca (see *Mansonella streptocerca*)
Dipetalonemiasis perstans (see Mansonelliasis perstans)
Diphyllobothriasis (see also Sparganosis), 165-183
    clinical features of, 172-175
    definition of, 165
    diagnosis of, 177, 180
    epidemiology of, 167
    etiology of, 165
    history of, 166-167
    pathologic features of, 175-176
    synonyms for, 165-166
    treatment and prevention of, 180
Diphyllobothriidae, classification and morphology of, 167
Diphyllobothrium alascense, 165
Diphyllobothrium cordatum, 165
Diphyllobothrium dalliae, 165
Diphyllobothrium dendriticum, 165
Diphyllobothrium klebanovskii, 165
Diphyllobothrium lanceolatum, 165
Diphyllobothrium latum, 165
    differentiation of eggs of from *Paragonimus*, 64
    life cycle of, 171-172
    morphology of adults of, 167-168
    morphology of eggs of, 169
    synonyms for, 165
Diphyllobothrium nihonkaiense, 165, 168-169
    morphology of adults of, 168-169
Diphyllobothrium pacificum, 165
Diphyllobothrium sp, 165
    morphology of larvae of, 169, 171
Diphyllobothrium ursi, 165
Diphyllobothrium yonagoensis, 165
Diplogonoporus balaenopterae, 166
Diplogonoporus sp, 165
Diploscapter coronata, clinical features of infection by, 517
    distinguishing features of, 516
    distribution of, 515
    history of, 511
    life cycle of, 515
    morphology of adults of, 516
    pathologic features of infection by, 517
    source of infection by, 515
    synonyms for, 511
Diplostomidae, family of, 97
Diplostomidiasis (see Fasciolopsiasis)
Dipylidiasis, 137-144
    clinical features of, 139, 141
    definition of, 137
    diagnosis of, 141-142
    epidemiology of, 137-138
    etiology of, 137
    history of, 137
    pathology and pathogenesis of, 141

synonyms for, 137
treatment and prevention of, 142
Dipylidium caninum, fleas as intermediate hosts of, 137, 139
   life cycle and transmission of, 139
   morphology of adults of, 138
   morphology of eggs of, 138-139
   morphology of oncospheres of, 139
Dirofilaria conjunctivae, 275
Dirofilaria Dirofilaria immitis (*D. immitis*), 275
   life cycle and transmission of, 280
   morphology of adults of, 277
Dirofilaria Dirofilaria sp, 275
Dirofilaria Nochtiella sp, 275
   life cycle and transmission of, 280
   morphology of adults of, 277-278
Dirofilaria repens, 275
   life cycle and transmission of, 280
   morphology of adults of, 277-278
Dirofilariasis, 275-285
   clinical features of, 281
   definition of, 275
   diagnosis of, 282-283
   epidemiology of, 276-277
   etiology of, 275
   history of, 275-276
   pathologic features of, 281-282
   synonyms for, 275
   treatment and prevention of, 283
Dirofilaria striata, 275
   life cycle and transmission of, 280
   morphology of adults of, 277, 279-280
Dirofilaria subdermata, 275
   life cycle and transmission of, 280
   morphology of adults of, 277, 279
Dirofilaria tenuis, 275
   life cycle and transmission of, 280
   morphology of adults of, 277-278
Dirofilaria ursi, 275
   life cycle and transmission of, 280
   morphology of adults of, 277-279
Dirofilaria ursi-like worms, 275
   life cycle and transmission of, 280
   morphology of adults of, 277-279
Disophenol, in hepatic capillariasis, 490
Dithiazanine iodide, in hepatic capillariasis, 490
Ditylenchus dipsaci, clinical features of infection by, 517
   distinguishing features of, 516
   distribution of, 515
   life cycle of, 515
   pathologic features of infection by, 517
   source of infection by, 515
   synonyms for, 511
DNA analysis, epidemiological use in hydatidosis, 159
   in angiostrongyliasis cantonensis, 381
   in diagnosis of onchocerciasis, 290, 302
   in loiasis, 273
   in lymphatic filariasis, 235-236
   in *Taenia saginata; T. solium*, 131
Dog, ancylostomiasis in, 355
   Capillaria hepatica in, 487
   clonorchiasis and opisthorchiasis in, 71
   dirofilariasis in, 276
   Gnathostoma sp in, 454
   hydatidosis in, 146, 148
   Mesocestoides lineatus in, 201

   Toxocara canis in, 412
Dog heartworm (*see* Dirofilariasis)
Dolphin, anisakids in, 426
   as definitive host for *Bolbosoma* sp, 526
Dominica (*see* Lesser Antilles)
Dracontiasis (*see* Dracunculiasis)
Dracunculiasis, 329-339
   clinical features of, 332-333
   definition of, 329
   diagnosis of, 334
   epidemiology of, 330
   etiology of, 329
   history of, 329-330
   pathologic features of, 333-334
   synonyms for, 329
   treatment and prevention of, 334, 336
Dracunculidae (*see* Dracunculiasis)
Dracunculus medinensis, 329-339
   life cycle and transmission of, 332, 339
   morphology of adults of, 330
   morphology of larvae of, 330
Dwarf tapeworm (*see* *Hymenolepis nana*)
Ear, gnathostomiasis in, 457
Earthworm, 10
Eastern cottontail, American brugian filariasis in, 319
Eastern mesocestoides worm (*see* *Mesocestoides lineatus*)
Echinococcosis (*see* Hydatidosis)
Echinococcus granulosus, in differential diagnosis of angiostrongyliasis cantonensis, 382
   life cycle and transmission of, 152-153
   morphology of adults of, 148-149, 152
   morphology of metacestode larvae of, 149-150
Echinococcus multilocularis, morphology of adults of, 149, 152
   morphology of metacestode larvae of, 151
Echinococcus oligarthrus, life cycle and transmission of, 152-153
   morphology of adults of, 148-149
   morphology of metacestode larvae of, 151
Echinococcus sp, life cycle and transmission of, 152-153
   morphology of adults of, 148, 149, 152
   primary and daughter cysts of, 152
   variant strains of, 148
Echinococcus vogeli, life cycle and transmission of, 152-153
   morphology of adults of, 149, 152
   morphology of metacestode larvae of, 151
Echinostoma ilocanum, life cycle and transmission of, 103
   morphology of adults of, 98
   morphology of eggs of, 98
Echinostomatidae, family of, 96
Echinostomiasis (*see* Fasciolopsiasis)
Ecuador, American brugian filariasis in, 320
   angiostrongyliasis costaricensis in, 386
   gnathostomiasis in, 447
   hydatidosis in, 148
   Hymenolepis diminuta in, 201
   mansonelliasis ozzardi in, 258
   onchocerciasis in, 288
   paragonimiasis in, 52
   Raillietina sp in, 202
Ecydsteroids, in development of *Strongyloides stercoralis*, 347
Eel, as intermediate host for *Gnathostoma* sp, 454
Egret, *Clinostomum complanatum* in, 113
Egypt (*see also* Egyptian mummies), angiostrongyliasis cantonensis in, 375

halicephalobiasis in, 493
Hymenolepis nana in, 199
intestinal capillariasis in, 481
intestinal trematodiases in, 96, 97
Egyptian chlorosis (*see* Ancylostomiasis)
Egyptian mummies, *Ascaris lumbricoides* eggs in, 397
Ehrlichia sp, transmission of by intestinal trematodes, 95
Elephantiasis (*see* Lymphatic filariasis)
Elephantiasis arabum (*see* Lymphatic filariasis)
ELISA, in angiostrongyliasis cantonensis, 382
    in anisakiasis, 428
    in clonorchiasis and opisthorchiasis, 81
    in cysticercosis, 130-131
    in dracunculiasis, 334
    in fascioliasis, 87-88
    in gnathostomiasis, 459
    in lymphatic filariasis, 235
    in oesophagostomiasis, 505-506
    in paragonimiasis, 65
    in sparganosis, 177, 180
    in strongyloidiasis, 351
    in toxocariasis, 412, 420
    in trichinosis, 479
Elk, as host for *Echinococcus* sp, 152
El Salvador, angiostrongyliasis costaricensis in, 386
    paragonimiasis in, 52
Emmonsonia parva var *crescens*, as mimic of helminth egg, 19
Endoscopy, 364, 428
Enfermedad de Robles (*see* Onchocerciasis)
Enterobiasis, 433-446
    clinical features of, 440-441
    definition of, 433
    diagnosis of, 444
    epidemiology of, 434
    etiology of, 433-434
    history of, 433-434
    pathologic features of, 441, 443
    synonyms for, 433
    treatment and prevention of, 445
Enterobius gregorii (*see also* Enterobiasis), clinical features of infection by, 517
    distinguishing features of, 516
    distribution of, 515
    history of, 511
    life cycle of, 515
    morphology of adults of, 516
    morphology of eggs of, 516
    pathologic features of infection by, 517
    relation to *E. vermicularis*, 433-434
    source of infection by, 515
    treatment of infection by, 514
Enterobius vermicularis, differential identification of, 433-434, 443
    high prevalence of, 397
    life cycle and transmission of, 437
    morphology of adults of, 436-437
    morphology of eggs of, 437
Enzyme immunoassay (EIA), in paragonimiasis, 65
Enzyme-linked immunosorbent assay (*see* ELISA)
Eosinophilic lung (*see* Lymphatic filariasis)
Eosinophilic meningitis (*see* Angiostrongyliasis cantonensis)
Eosinophilic pseudoleukemia (*see* Toxocariasis)
Epidermitis linearis migrans (*see* Creeping eruption)
Equatorial Guinea, loiasis in, 262
    onchocerciasis in, 288

paragonimiasis in, 52
Erisípela de la costa (*see* Onchocerciasis)
Esophageal worm (*see* Gongylonema pulchrum)
Estonia, diphyllobothriasis in, 167
Ethiopia, dracunculiasis in, 330
    oesophagostomiasis in, 499
    onchocerciasis in, 288
    Pseudamphistomum aethiopicum in, 69
Euphausia sp, as intermediate hosts for anisakids, 426
Europe, *Alaria* sp trematodiasis in, 107
    ancylostomiasis in, 354
    Ascaris lumbricoides in prehistoric remains in, 397
    cysticercosis in, 118
    dirofilariasis in, 276
    Ditylenchus dipsaci in, 515
    Enterobius gregorii in, 515
    fascioliasis in, 82
    hydatidosis in, 146
    Hymenolepis diminuta in, 201
    Hymenolepis nana in, 199
    intestinal trematodiases in, 96
    opisthorchiasis in, 71
    Pseudamphistomum truncatum in, 89
    trichinosis in, 472
Eurytrema pancreaticum trematodiasis, clinical features of, 111
    diagnosis of, 111
    epidemiology of, 110
    etiology of, 110
    history of, 110
    life cycle and transmission of etiologic agent of, 111
    morphology of adults of etiologic agent of, 110
    morphology of eggs of etiologic agent of, 110
    pathologic features of, 111
    treatment of (in sheep), 111
Eustrongylides sp, clinical features of infection by, 517
    distinguishing features of, 516
    distribution of, 515
    history of, 511
    life cycle of, 515
    morphology of larvae of, 516
    pathologic features of infection by, 517
    source of infection by, 515
Excretory columns, 355, 398, 412, 516
Excretory gland, 355
Excretory gland cell, 424
Eye, American brugian filariasis in, 322, 325
    ancylostomiasis in, 362
    angiostrongyliasis cantonensis in, 378-381
    coenurosis in, 190-192
    cysticercosis in, 128, 130
    dracunculiasis in, 332
    gnathostomiasis in, 457
    hydatidosis in, 154
    loiasis in, 265, 271
    onchocerciasis in, 287-318
    paragonimiasis in, 60, 63
    sparganosis in, 174
    toxocariasis in, 415-420
    trichinosis in, 476
Far East, ancylostomiasis in, 355
Fasciola gigantica, life cycle and transmission of, 83-84
    morphology of adults of, 83
    morphology of eggs of, 83
Fasciola hepatica, life cycle and transmission of, 83-84
    morphology of adults of, 82-83

morphology of eggs of, 83
Fascioliasis, 81-88
    clinical features of, 84-85
    definition of, 81
    diagnosis of, 86-88
    epidemiology of, 82
    etiology of, 81
    history of, 82
    in differential diagnosis of gnathostomiasis, 458
    pathologic features of, 85-86
    synonyms for, 81
    treatment and prevention of, 90
Fasciolidae, family of, 69, 96
Fasciolopsiasis and other intestinal trematodiases, 93-105
    clinical features of, 103-104
    definition of, 93
    diagnosis of, 104
    epidemiology of, 94-97
    etiology of, 93, 96-97
    history of, 93-94
    pathologic features of, 104
    synonyms for, 93, 96-97
    treatment and prevention of, 104-105
    trematodes as transmitters of *Ehrlichia* sp, 95
Fasciolopsis buski, life cycle and transmission of, 103
    morphology of adults of, 95, 98-101
    morphology of eggs of, 95, 98-101
Fenbendazole, in *Haemonchus contortus* infection (in animals), 514
    in *Metastrongylus apri* infection (in animals), 514
Fiji, angiostrongyliasis cantonensis in, 374
Filaria bancrofti (see also *Wuchereria bancrofti*; *Brugia malayi*), 215
Filaria demarquayi, 258
Filarial hypereosinophilia (see Lymphatic filariasis)
Filaria malayi (see also *Brugia malayi*), 215
Filaria nocturna, 215
Filaria ozzardi, 258
Filaria sanguinis hominis (see also *Wuchereria bancrofti*), 215
Filaria volvulus, 287
Filaria wuchereria, 215
Filariidae, *Brugia* sp, 319-328
Filarioidea, superfamily of, 508, 511
Filariopheresis, in loiasis, 273
Finland, diphyllobothriasis in, 167
Fish, as intermediate hosts for *Capillaria philippinensis*, 482
    as intermediate hosts for *Clinostomum complanatum*, 113
    as intermediate hosts in clonorchiasis and opisthorchiasis, 75
    as intermediate hosts for *Gnathostoma* sp, 454
    as intermediate hosts in intestinal trematodiases, 96, 97
Fish-eating birds, as definitive hosts for *Capillaria philippinensis*, 482
    as hosts for *Clinostomum complanatum*, 113
Flea, *Ctenocephalides felis* (cat flea), 139
    in *Dipylidium caninum* infection, 137
    in *Hymenolepis diminuta* infection, 207
    Pulex irritans (human flea), 139
Flubendazole, in intestinal capillariasis, 487
Fluorescein angiography, in ocular onchocerciasis, 312, 317
Folate deficiency, in ancylostomiasis, 362
Foodborne helminthic zoonoses, anisakiasis, 423
    trichinosis, 472
Fox, ancylostomiasis in, 355

    dirofilariasis in, 276
    hydatidosis in, 146, 148
    Mesocestoides lineatus in, 201
    Trichinella britovi in, 471
Fox lungworm (see *Capillaria aerophila*)
France, anisakiasis in, 423
    coenurosis in, 186
    hydatidosis in, 146, 148
    trichinosis in, 472
French Guiana, oesophagostomiasis in, 500
French Polynesia, *Raillietina celebensis* in, 202
Frog, as intermediate host for *Alaria* sp, 108
    as intermediate host for *Gnathostoma* sp, 454
    as intermediate host in intestinal trematodiases, 96-97
Fungal infection, in differential diagnosis of dirofilariasis, 282
Gabon, loiasis in, 262
    Microfilaria rodhaini in, 515
    onchocerciasis in, 288
    paragonimiasis in, 52
Gâle filarienne (see Onchocerciasis)
Gape worm (see *Mammomonogamus laryngeus*)
Garlic, in *Hymenolepis nana* infection, 212
Gastrodiscidae, family of, 97
Gastrodiscoidiasis (see Fasciolopsiasis)
Gastrodiscus hominis, life cycle and transmission of, 103
    morphology of adults of, 98
    morphology of eggs of, 98
Gastropods (see also Slug; Snail), as intermediate hosts for *Angiostrongylus cantonensis*, 376
Geohelminths, *Ascaris lumbricoides*, 397
    Capillaria hepatica, 487
    Trichuris sp, 461
Gerbil, *Capillaria hepatica* in, 487
Germany, *Alaria* sp trematodiasis in, 107
    anisakiasis in, 423
    halicephalobiasis in, 493
    hydatidosis in, 146, 148
Ghana, *Dicrocoelium hospes* in, 88
    dracunculiasis in, 330
    oesophagostomiasis in, 499-500
    onchocerciasis in, 288
Giant intestinal fluke (see Fasciolopsiasis)
Giant kidney worm (see *Dioctophyma renale*)
Gizzard worm (see *Cheilospirura* sp)
Gnathostoma doloresi, life cycle and transmission of, 452, 454, 456
    morphology of adults of, 452
    morphology of eggs of, 452
    morphology of larvae of, 452
Gnathostoma hispidum, life cycle and transmission of, 452, 454, 456
    morphology of adults of, 450
    morphology of eggs of, 450
    morphology of larvae of, 450
Gnathostoma nipponicum, life cycle and transmission of, 452, 454, 456
    morphology of adults of, 450
    morphology of eggs of, 450
    morphology of larvae of, 450, 452
Gnathostoma spinigerum, in creeping eruption, 367-368, 371, 447
    in differential diagnosis of angiostrongyliasis cantonensis, 382
    life cycle and transmission of, 452, 454, 456
    morphology of adults of, 449

morphology of eggs of, 449
morphology of larvae of, 368, 449
Gnathostomiasis, 447-460
    clinical features of, 456-458
    definition of, 447
    diagnosis of, 458-459
    epidemiology of, 447-448
    etiology of, 447
    pathologic features of, 458
    synonyms for, 447
    treatment and prevention of, 460
Gnathostomiasis externa, 456
Goat, as host for *Echinococcus* sp, 152
    as host for *Eurytrema pancreaticum*, 110
    as host for *Oesophagostomum bifurcum*, 502
    as host for *Ternidens deminutus*, 502
Gongylonema pulchrum, clinical features of infection by, 517
    distinguishing features of, 516
    distribution of, 515
    history of, 511
    life cycle of, 515
    morphology of adults of, 516
    morphology of eggs of, 516
    pathologic features of infection by, 517
    source of infection by, 515
    synonyms for, 511
Gordiids, 15
Gorilla, mansonelliasis perstans in, 245
    streptocerciasis in, 251
Graham's Scotch® tape method, 444
Grasshopper, *Eurytrema pancreaticum* trematodiasis in, 111
Great Britain (*see* United Kingdom)
Greece, hydatidosis in, 146
Green herring, anisakiasis in, 423
Greenland (*see also* Denmark), intestinal trematodiases in, 97
    Mesocestoides lineatus in, 201
    Metorchis conjunctus in, 89
Grenada, *Hymenolepis diminuta* in, 201
Grison, American brugian filariasis in, 319
Grison vittatus (*see* Grison)
Ground itch (*see* Ancylostomiasis)
Ground squirrel, *Capillaria hepatica* in, 487
Guatemala, angiostrongyliasis costaricensis in, 386
    onchocerciasis in, 288
    paragonimiasis in, 52
    trichinosis in, 472
Guinea, onchocerciasis in, 288
    paragonimiasis in, 52
Guinea-Bissau, onchocerciasis in, 288
Guinea worm (*see Dracunculus medinensis*)
Gullet worm (*see Gongylonema pulchrum*)
Gulls, *Clinostomum complanatum* in, 113
Gymnophallidae, family of, 97
Gymnophallidiasis (*see* Fasciolopsiasis)
Gynecophoral canal, 24-25
Haemonchus contortus, clinical features of infection by, 517
    distinguishing features of, 516
    distribution of, 515
    history of, 511
    life cycle of, 515
    morphology of adults of, 516
    morphology of eggs of, 516
    pathologic features of infection by, 517
    source of infection by, 515
    synonyms for, 511

treatment of infection by (in animals), 514
Hair follicle mite (*see Demodex folliculorum*)
Hakenwürmer (*see* Acanthocephaliasis)
Halicephalobiasis, 493-497
    clinical features of, 494, 496
    definition of, 493
    diagnosis of, 497
    epidemiology of, 493
    etiology of, 493
    pathologic features of, 496
    synonyms for, 493
    treatment of, 497
Halicephalobus deletrix (*see* Halicephalobiasis)
Halicephalobus sp, 493-497
    life cycle of, 494
    morphology of adults of, 493-494
    morphology of eggs of, 493
    morphology of larvae of, 493-494
Halzoun, in differential diagnosis of *Clinostomum complanatum* trematodiasis, 113
Hanging groin (*see* Onchocerciasis)
Hare, *Capillaria hepatica* in, 487
Harelip ascarid (*see Lagochilascaris minor*)
Head bulb, 521-524
Heart, ancylostomiasis in, 362
    halicephalobiasis in, 496
    loiasis in, 269, 271
    lymphatic filariasis in, 225
    strongyloidiasis in, 346-347
    trematodiases in, 104
    trichinosis in, 476-478
Helicosporous fungi, 19, 222
Helminthic abscess (*see* Oesophagostomiasis)
Helminthic infections, fundamental features of, 2
    histologic patterns of, 2, 3
Helminths, classification of (*see* Acoelomates; Pseudocoelomates), 3-10
Helminths, phyla of, 4-10
    Acanthocephala, 9-10
    Aschelminthes, 8-10
    Platyhelminthes, 4-6
Hemagglutination test, in gnathostomiasis, 459
Hemolytic agent, in pathogenesis of gnathostomiasis, 456
Hepaticola anthropopitheci, 487
Hepaticola hepatica, 487
Hepaticola soricicola, 487
Hepatic trematodiases (*see also* Clonorchiasis and opisthorchiasis; Fascioliasis; Dicrocoeliasis), 69-92
    Dicrocoeliidae in, 69
    Fasciolidae in, 69
    Opisthorchiidae in, 69
Hepatitis B, Symmers' fibrosis in, 33
Heron, *Clinostomum complanatum* in, 113
Herring, anisakid larvae in, 423
Heterophyes heterophyes, life cycle and transmission of, 103
    morphology of adults of, 98
    morphology of eggs of, 98
    source of metacercariae of, 96
Heterophyiasis (*see* Fasciolopsiasis)
Heterophyidae, family of, 96-97
Hexylresorcinol, in intestinal trematodiases, 104
    in trichuriasis, 467
Histocompatibility factors (HLA-D and HLA-Q), in ocular onchocerciasis, 314
Histopathologic patterns of helminthic infections, 2, 3

HIV (*see* Human immunodeficiency virus)
Hodgkin's disease, in differential diagnosis of
    angiostrongyliasis cantonensis, 382
Holland (*see* Netherlands)
Honduras, *Achillurbainia recondita* trematodiasis in, 111
    angiostrongyliasis costaricensis in, 386
    paragonimiasis in, 52
Hong Kong, *Eurytrema pancreaticum* in, 110
Hooked worm (*see* Acanthocephaliasis)
Hookworm (*see* Ancylostomiasis; Creeping eruption)
Hookworm disease (*see* Ancylostomiasis)
Hookworm infection (*see* Ancylostomiasis)
Horse, halicephalobiasis in, 493-497
    trichinosis in, 472
Horsehair worm, 15
Hottentot apron (*see* Onchocerciasis)
Housefly, as vector for trichuriasis, 461
Human echinorhynchosis (*see* Acanthocephaliasis)
Human immunodeficiency virus (HIV), false positive ELISA
    in taeniasis, 131
    in onchocerciasis, 302
    in strongyloidiasis, 345
Human macracanthorhyncosis (*see* Acanthocephaliasis)
Hungary, intestinal trematodiases in, 96
Hyaluronidase, in pathogenesis of ancylostomiasis, 360-361
    in pathogenesis of creeping eruption, 368
    in pathogenesis of gnathostomiasis, 456
Hydatid disease (*see* Hydatidosis)
Hydatidosis (Echinococcosis), 145-164
    clinical features of, 153-155
    definition of, 145
    diagnosis of, 159
    epidemiology of, 146, 148
    etiology of, 145
    history of, 145-146
    pathologic features of, 155, 157
    synonyms for, 145
    treatment and prevention of, 159, 161
Hydatid Pott's disease, in hydatidosis, 154
Hydatid sand, 146, 150
Hydatid tapeworm (*see Echinococcus granulosus*)
Hydatigera taeniaeformis, 198
Hyena, hydatidosis in, 146
Hymenolepiasis and miscellaneous cyclophyllidiases, 197-214
    clinical features of, 208, 211
    definition of, 197
    diagnosis of, 212
    epidemiology of, 199, 201-202
    etiology of, 197
    history of, 198-199
    pathologic features of, 211-212
    synonyms for, 197-198
    treatment and prevention of, 212
Hymenolepis diminuta, life cycle and transmission of, 207-208
    morphology of adults of, 202
    morphology of eggs of, 202
Hymenolepis fraterna (*see Hymenolepis nana*)
Hymenolepis nana, in differential diagnosis of dipylidiasis, 142
    life cycle and transmission of, 207
    morphology of adults of, 202
    morphology of eggs of, 202
Hypobiotic (parenteral) larvae, 360
Iceland, fascioliasis in, 82
Id reaction, in hymenolepiasis, 208
Immunoassays, in angiostrongyliasis cantonensis, 394
    in anisakiasis, 428
    in dracunculiasis, 334
    in fascioliasis, 87-88
    in taeniasis and cysticercosis, 130-131
Immunofluorescence, indirect test, in fascioliasis, 87
    in gnathostomiasis, 459
Immunologic responses, general considerations in helminthic
    diseases, 3
    in anisakiasis, 427
    in coenurosis, 192
    in creeping eruption, 371
    in fascioliasis, 85, 87-88
    in gnathostomiasis, 459
    in hydatidosis, 155
    in hymenolepiasis, 208
    in intestinal capillariasis, 483
    in lymphatic filariasis, 224
    in ocular onchocerciasis, 314
    in onchocerciasis, 298-299
    in toxocariasis, 416
    Mazzotti reaction, 253, 288, 300
Immunosuppression, in hymenolepiasis, 208, 211
    in hyperinfection strongyloidiasis, 345
India, ancylostomiasis in, 355
    angiostrongyliasis cantonensis in, 374
    Bertiella sp in, 201
    Clinostomum complanatum in, 112
    dipylidiasis in, 138
    dracunculiasis in, 330
    hydatidosis in, 146
    Hymenolepis diminuta in, 201
    Hymenolepis nana in, 199
    intestinal trematodiases in, 96, 97
    lymphatic filariasis in, 219
    paragonimiasis in, 52
    ternidenamiasis in, 500
Indonesia, acanthocephaliasis in, 520
    ancylostomiasis in, 355
    angiostrongyliasis cantonensis in, 374
    Bertiella sp in, 201
    intestinal capillariasis in, 481
    intestinal trematodiases in, 96, 97
    lymphatic filariasis in, 219
    oesophagostomiasis in, 500
    paragonimiasis in, 52
    taeniasis in, 118
    ternidenamiasis in, 500
Inermicapsifer arvicanthidis, 198
Inermicapsifer cubanensis, 198
Inermicapsifer madagascariensis, 197
    in differential diagnosis of dipylidiasis, 142
    life cycle and transmission of, 208
    morphology of adults of, 206
    morphology of eggs of, 206
Innenkörper, in *Brugia malayi* microfilariae, 221-222
    in *Brugia timori* microfilariae, 222
    in *Wuchereria bancrofti* microfilariae, 221
Interleukins, in toxocariasis, 418
International Task Force for Disease Eradication, for
    ancylostomiasis, 354
    for lymphatic filariasis, 238
Intestinal capillariasis (*see* Capillariasis, intestinal)
Intestinal involvement, in acanthocephaliasis, 526-527
    in ancylostomiasis, 362-364
    in angiostrongyliasis costaricensis, 389-391, 393-394

in anisakiasis, 426-428
in ascariasis, 401-403, 405
in hymenolepiasis, 211-212
in intestinal capillariasis, 483, 485-486
in loiasis, 269
in strongyloidiasis, 344-347
in *Taenia* tapeworm infection, 122, 125, 127-128
in trichuriasis, 465-467
Intestinal roundworm, in raccoon (*see Baylisascaris* sp and *B. procyonis*)
Intestinal trematodiases, miscellaneous, other than fasciolopsiasis (*see* Fasciolopsiasis)
Iran, acanthocephaliasis in, 521
    hydatidosis in, 146
    Hymenolepis diminuta in, 201
    intestinal capillariasis in, 481
    intestinal trematodiases in, 96
    Raillietina sp in, 202
Iraq, acanthocephaliasis in, 521
    hydatidosis in, 146
Isozymes, in angiostrongyliasis cantonensis, 381
Israel, acanthocephaliasis in, 521
    Clinostomum complanatum in, 112
    intestinal trematodiases in, 96
    Philophthalmus sp trematodiasis in, 113
Italy, acanthocephaliasis in, 521
    angiostrongyliasis cantonensis in, 375
    coenurosis in, 186
    dipylidiasis in, 138
    dirofilariasis (in animals) in, 276
    hydatidosis in, 146, 148
    Hymenolepis diminuta in, 201
    intestinal trematodiases in, 96
    trichinosis in, 472
Ivermectin, in ancylostomiasis, 364
    in angiostrongyliasis cantonensis, 382
    in *Capillaria aerophila* infection, 510
    in creeping eruption, 371
    in dirofilariasis (in animals), 283
    in hepatic capillariasis, 490
    in lagochilascariasis, 510
    in loiasis, 273
    in lymphatic filariasis, 237-238
    in mansonelliasis perstans, 250
    in ocular onchocerciasis, 317
    in onchocerciasis, 303
    in streptocerciasis, 256
    in strongyloidiasis, 351
Ivory Coast (Côte d'Ivoire), angiostrongyliasis cantonensis in, 375
    dracunculiasis in, 330
    onchocerciasis in, 288
    paragonimiasis in, 52
Jackal, hydatidosis in, 146
Jaguar, hydatidosis in, 146
Jamaica, angiostrongyliasis cantonensis in, 375
    Hymenolepis diminuta in, 201
Japan, acanthocephaliasis in, 521
    angiostrongyliasis cantonensis in, 374
    anisakiasis in, 423
    Clinostomum complanatum in, 112
    creeping eruption in, 367
    diphyllobothriasis in, 167
    dirofilariasis in, 276
    Eurytrema pancreaticum in, 110
    fascioliasis in, 82
    gnathostomiasis in, 447
    halicephalobiasis in, 493
    hydatidosis in, 148
    Hymenolepis diminuta in, 201
    intestinal trematodiases in, 96, 97
    lymphatic filariasis in, 219
    Mesocestoides lineatus in, 201
    paragonimiasis in, 52
    Philophthalmus sp trematodiasis in, 113
    Raillietina celebensis in, 202
    sparganosis in, 167
Juvenile tapeworm, 130
Kampala eye worm (*see* Mansonelliasis perstans; Loiasis)
Kampuchea (*see* Cambodia)
Kangaroo, as host for *Echinococcus* sp, 152
Kato-Katz technique, in trichuriasis, 467
Kato method, in schistosomiasis, 46
Kazakhstan, intestinal trematodiases in, 97
Kenya, dracunculiasis in, 330
    Inermicapsifer madagascariensis in, 201
Kernig's sign, in angiostrongyliasis cantonensis, 378
Kidney, enterobiasis in, 443
    hydatidosis in, 154
    loiasis in, 269, 271
    onchocerciasis in, 298
    paragonimiasis in, 64
    schistosomiasis in, 33, 39, 41
    sparganosis in, 175
Knott's concentration technique, in lymphatic filariasis, 235
Korea, *Clinostomum complanatum* in, 112
    Eurytrema pancreaticum in, 110
    intestinal trematodiases in, 96, 97
    lymphatic filariasis in, 219
    paragonimiasis in, 52
    taeniasis in, 118
    trichinosis in, 472
Koukoul (*see* Oesophagostomiasis)
Krill, as intermediate hosts for anisakids, 426
Laboratory analysis, principles of, 2
Laevicaulus alte (*see* Slug)
Lagochilascaris minor, clinical features of infection by, 517
    distinguishing features of, 516
    distribution of, 515
    history of, 511
    life cycle of, 515
    morphology of adults of, 516
    morphology of eggs of, 516
    pathologic features of infection by, 517
    source of infection by, 515
    synonym for, 511
    treatment of infection by, 510
Laminated membrane, 148-149
Laos, angiostrongyliasis cantonensis in, 374
    intestinal trematodiases in, 96, 97
    paragonimiasis in, 52
Laparoscopy, in angiostrongyliasis costaricensis, 390
Larva currens, in strongyloidiasis, 346
Larval taeniasis (*see* Taeniasis and cysticercosis)
Laser therapy, in *Alaria* sp trematodiasis, 109
    in ocular *Baylisascaris* sp infection, 514
Lateral body, 355, 500
Latin America, anisakiasis in, 423
Lebanon, hydatidosis in, 146
    trichinosis in, 472

Lecithodendriidae, family of, 97
Lecithodendriidiasis (*see* Fasciolopsiasis)
Leech, 7
Leeward Islands, angiostrongyliasis costaricensis in, 386
Lentil pneumonia, as mimic of parasitic infection, 18
Leopard, *Gnathostoma* sp in, 454
Leprosy, in differential diagnosis of streptocerciasis, 250, 253, 256
Lesser Antilles, angiostrongyliasis costaricensis in (Dominica), 386
    Bertiella sp in, 201
    Hymenolepis diminuta in (Martinique), 201
Levamisole, in hepatic capillariasis, 490
    in lagochilascariasis, 510
Liberia, onchocerciasis in, 288
    paragonimiasis in, 52
Liesegang rings, as mimics of parasites, 11
Ligula sp and *L. intestinalis*, 165, 166
Limnatis paluda (*see* Leech)
Lion, *Gnathostoma* sp in, 454
Lipid pseudomembrane, as mimic of parasite, 10
Lithuania, diphyllobothriasis in, 167
    trichinosis in, 472
Little liver fluke (*see Dicrocoelium dendriticum*)
Liver, ancylostomiasis in, 364
    angiostrongyliasis costaricensis in, 390, 393
    ascariasis in, 403, 405-406
    enterobiasis in, 443
    fascioliasis in, 84-86
    halicephalobiasis in, 496-497
    hepatic trematodiases in, 69-92
    hydatidosis in, 153
    intestinal capillariasis in, 485
    hyperinfection strongyloidiasis in, 347
    loiasis in, 271
    lymphatic filariasis in, 235
    onchocerciasis in, 298
    paragonimiasis in, 63
    schistosomiasis in, 31-33
    Taenia taeniaeformis larvae in, 212
Liver rot (in sheep), 81
Loa loa, life cycle and transmission of, 264
    morphology of adults of, 262-263
    morphology of microfilariae of, 263
    periodicity of microfilariae of, 264
Loa worm (*see* Loiasis)
Loiasis, 261-274
    clinical features of, 265, 269
    definition of, 261
    diagnosis of, 271, 273
    epidemiology of, 262
    etiology of, 261
    history of, 261-262
    pathologic features of, 269, 271
    synonyms for, 261
    treatment and prevention of, 273
Louse (*Trichodectes canis*), as intermediate host for *Dipylidium caninum*, 137, 139
Lung, American brugian filariasis in, 322
    ancylostomiasis in, 362
    angiostrongyliasis cantonensis in, 379, 381
    ascariasis in, 402, 405
    cysticercosis in, 125
    dirofilariasis in, 281
    enterobiasis in, 440, 443

    fascioliasis in, 84
    gnathostomiasis in, 457
    halicephalobiasis in, 496
    hydatidosis in, 153-154
    loiasis in, 269, 271
    lymphatic filariasis in, 227, 231, 233
    onchocerciasis in, 298
    paragonimiasis in, 57-58, 62
    schistosomiasis in, 42
    sparganosis in, 175
    strongyloidiasis in, 344-345, 347
    toxocariasis in, 416
Lycopodium clavatum and *L. granuloma*, as mimics of parasites, 16, 17
Lymnaea auricularia (*see also* Snail), 84
Lymnaea natalensis (*see also* Snail), 84
Lymnaeidae, family of (*see also* Snail), 84
Lymphatic filariasis, 215-243
    clinical features of, 224-225, 227, 231
    definition of, 215
    diagnosis of, 235-237
    epidemiology of, 219
    etiology of, 215
    history of, 215-216, 219
    pathologic features of, 231, 233, 235
    synonyms for, 215
    treatment and prevention of, 237-238
Lymph nodes, acanthocephaliasis in, 527
    American brugian filariasis in, 322, 325
    angiostrongyliasis costaricensis in, 393
    enterobiasis in, 443
    hyperinfection strongyloidiasis in, 347
    loiasis in, 269, 271
    lymphatic filariasis in, 235
Lymphoscintigraphy, in lymphatic filariasis, 224
Lysozyme, reduced activity of in hymenolepiasis, 211
Mackerel, in transmission of anisakid larvae, 423
Macracanthorhynchus catulinus, 520
Macracanthorhynchus hirudinaceus, 521-526
    life cycle and transmission of, 525-526
    morphology of adults of, 521-523
    morphology of eggs of, 523
    morphology of larvae of, 523
Macracanthorhynchus ingens, 520
    life cycle and transmission of, 525-526
    morphology of adults of, 521-524
    morphology of eggs of, 524
Madagascar (Malagasy Republic), acanthocephaliasis in, 521
    hydatidosis in, 146
    Inermicapsifer madagascariensis in, 201
    Raillietina celebensis in, 202
Magnetic resonance imaging (MRI), for cerebral paragonimiasis, 65
    for cerebral sparganosis, 180
    for coenurosis, 194
Malawi, *Hymenolepis diminuta* in, 201
    Inermicapsifer madagascariensis in, 201
    onchocerciasis in, 288
Malayan filariasis (*see* Lymphatic filariasis)
Malaysia, *Bertiella* sp in, 201
    Eurytrema pancreaticum in, 110
    intestinal trematodiases in, 96
    lymphatic filariasis in, 219
    oesophagostomiasis in, 500
Mali, dracunculiasis in, 330

onchocerciasis in, 288
Mal morado (*see* Onchocerciasis)
Mammomonogamus laryngeus, clinical features of infection by, 517
    distinguishing features of, 516
    distribution of, 515
    history of, 511
    life cycle of, 515
    morphology of adults of, 516
    morphology of eggs of, 516
    pathologic features of infection by, 517
    source of infection by, 515
    synonyms for, 511
    treatment of infection by, 514
Mango (mangrove) fly (*see* Tabanid fly)
Mansonella ozzardi, 245, 256-259
    life cycle and transmission of, 258
    morphology of adults of, 258
    morphology of microfilariae of, 258
Mansonella perstans, 245-250
    life cycle and transmission of, 246-247
    morphology of adults of, 246
    morphology of microfilariae of, 246
Mansonella streptocerca, 245, 250-256
    life cycle and transmission of, 253
    morphology of adults of, 251
    morphology of microfilariae of, 251, 253
Mansonelliasis, 245-260
    definition of, 245
    etiology of, 245
Mansonelliasis ozzardi, 256-259
    clinical features of, 258
    definition of, 256
    diagnosis of, 259
    epidemiology of, 258
    etiology of, 256
    history of, 258
    pathologic features of, 259
    synonyms for, 256, 258
    treatment of, 259
Mansonelliasis perstans, 245-250
    clinical features of, 247
    definition of, 245
    diagnosis of, 249
    epidemiology of, 245
    etiology of, 245
    history of, 245
    pathologic features of, 247, 249
    synonyms for, 245
    treatment of, 250
Mansonellosis (*see* Mansonelliasis ozzardi)
Mansonia sp (*see* Lymphatic filariasis)
Margay, dirofilariasis in, 277
Martinique (*see* Lesser Antilles; Windward Islands)
Mauritania, dracunculiasis in, 330
Mauritius, *Bertiella* sp in, 201
    Eurytrema pancreaticum in, 110
    Inermicapsifer madagascariensis in, 201
Mazzotti reaction, in onchocerciasis, 288, 300
    in streptocerciasis, 253
Mebendazole, in acanthocephaliasis, 528
    in ancylostomiasis, 364
    in ascariasis, 407
    in dracunculiasis, 336
    in enterobiasis, 445

    in hydatidosis, 161
    in intestinal capillariasis, 487
    in loiasis, 273
    in *Mammomonogamus laryngeus* infection, 514
    in *Rhabditis* sp infection, 514
    in taeniasis, 131
    in toxocariasis, 420
    in trichinosis, 479
    in *Trichostrongylus* sp infection, 514
    in trichuriasis, 467
Meckel's diverticulum, in differential diagnosis of angiostrongyliasis costaricensis, 389
Mediterranean, dirofilariasis in, 276
    intestinal trematodiases in, 96
Meerkat, *Capillaria hepatica* in, 487
Meggittia celebensis (*see Raillietina celebensis*)
Melanesia, ancylostomiasis in, 354
Melanoides tuberculatus (*see also* Snail), 71
Meloidogyne javanica, clinical features of infection by, 517
    distinguishing features of, 516
    distribution of, 515
    life cycle of, 515
    morphology of eggs of, 516
    pathologic features of infection by, 517
    source of infection by, 515
    synonyms for, 511
Meningonema peruzzii (*see also* Mansonelliasis perstans)
    clinical features of infection by, 517
    distinguishing features of, 516
    distribution of, 515
    history of, 511
    life cycle of, 515
    morphology of microfilariae of, 516
    pathologic features of infection by, 517
    source of infection by, 515
Mermithids, 2-3, 8
Mesocestoides leptothylacus (*see Mesocestoides lineatus*; Red fox)
Mesocestoides lineatus, 197
    life cycle and transmission of, 208
    morphology of adults of, 203
    morphology of eggs of, 203
    morphology of larvae of, 204
Mesocestoides variabilis (*see* Hymenolepiasis; *Mesocestoides lineatus*)
Mesocyclops aequatorialis similis (*see* Cyclopidae)
Metacestode larva, *Echinococcus* sp, 152
Metagonimiasis (*see* Fasciolopsiasis)
Metagonimus yokogawai, life cycle and transmission of, 103
    morphology of adults of, 98
    morphology of eggs of, 98
Metalloproteases, in pathogenesis of ancylostomiasis, 361
Metastrongyloidea, superfamily of, 511
    Angiostrongylus cantonensis in, 373
    Angiostrongylus costaricensis in, 386
Metastrongylus apri, clinical features of infection by, 517
    distinguishing features of, 516
    distribution of, 515
    history of, 511
    life cycle of, 515
    morphology of adults of, 516
    morphology of eggs of, 516
    pathologic features of infection by, 517
    source of infection by, 515
    synonyms for, 511

treatment of infection by (in animals), 514
Metorchis albidus, 69
Metorchis conjunctus, 89-90
Metrifonate, in schistosomiasis, 47
Metronidazole, in dracunculiasis, 336
Mexico, *Alaria* sp trematodiasis in, 107
    angiostrongyliasis costaricensis in, 386
    cysticercosis in, 118
    dipylidiasis in, 138
    gnathostomiasis in, 447
    Hymenolepis nana in, 199
    onchocerciasis in, 288
    paragonimiasis in, 52
    trichinosis in, 472
    Turbatrix aceti in, 515
Microfilaria bolivarensis (*see also* Mansonelliasis perstans)
    clinical features of infection by, 517
    distinguishing features of, 516
    distribution of, 515
    history of, 511
    life cycle of, 515
    morphology of microfilariae of, 516
    pathologic features of infection by, 517
    source of infection by, 515
    synonym for, 511
Microfilaria malayi (*see also Brugia malayi*), 215
Microfilaria rodhaini, 249
    clinical features of infection by, 517
    distinguishing features of, 516
    distribution of, 515
    history of, 511
    life cycle of, 515
    morphology of microfilariae of, 516
    pathologic features of infection by, 517
    source of infection by, 515
    synonym for, 511
Microfilaria semiclarum, 249
    clinical features of infection by, 517
    distinguishing features of, 516
    distribution of, 515
    history of, 511
    life cycle of, 515
    morphology of microfilariae of, 516
    pathologic features of infection by, 517
    source of infection by, 515
    synonyms for, 511
Micronema deletrix (*see also* Halicephalobiasis), 493
Micronemiasis (*see* Halicephalobiasis)
Microphallidae, family of, 97
Microphallidiasis (*see* Fasciolopsiasis)
Middle East, ancylostomiasis in, 354
    intestinal trematodiases in, 96
Millipede, as intermediate host for *Macracanthorhynchus ingens*, 526
Mimics of parasites, 10, 11, 15-21
Miner's anemia (*see* Ancylostomiasis)
Mink, *Brugia beaveri* in, 319
    coenurosis in, 186
    Gnathostoma sp in, 454
Minor Opisthorchiidae infections, clinical features of, 90
    definition of, 89
    epidemiology of, 89
    etiology of, 89
    history of, 89
    pathologic features of, 90

    synonyms for, 89
Mite, as intermediate host for *Bertiella studeri*, 208
    as intermediate host for *Inermicapsifer madagascariensis*, 208
Mollusk, intestinal trematodiases in, 96-97, 103
Mongolia, *Eurytrema pancreaticum* in, 110
Mongolian gerbil, intestinal capillariasis in, 482
Moniliformis moniliformis, life cycle and transmission of, 525-526
    morphology of adults of, 524
    morphology of eggs of, 524
Monkey (*see also Saguinus mystax*), as host for *Eurytrema pancreaticum*, 107
    as host for *Oesophagostomum bifurcum*, 502
    as host for *Ternidens deminutus*, 502
    Capillaria hepatica in, 487
    intestinal capillariasis in, 481
    loiasis in, 262
Moose, as host for *Echinococcus* sp, 152
Morera's disease (*see* Angiostrongyliasis costaricensis)
Mosquito, in transmission of American brugian filariasis, 322
    in transmission of dirofilariasis, 280
Mouse, *Capillaria hepatica* in, 487
    Syphacia obvelata (mouse pinworm) in, 434
    Trichinella pseudospiralis in, 471
Mouse pinworm (*see Syphacia obvelata*)
MRI (*see* Magnetic resonance imaging)
Muskrat, *Capillaria hepatica* in, 487
Myanmar (Burma), intestinal trematodiases in, 96, 97
    paragonimiasis in, 52
    Raillietina celebensis in, 202
Myenteric plexuses, destruction of in strongyloidiasis, 346
Myiasis, identification of in tissue sections, 7, 8, 11
    in differential diagnosis of gnathostomiasis, 459
Mystery disease of Pudoc (*see* Capillariasis, intestinal)
N-(2' chloronitrophenyl) 5 chlorosalicylamide, in acanthocephaliasis, 528
Nanophyetiasis (*see* Fasciolopsiasis)
Nanophyetidae, family of, 97
Nanophyetus salmincola, differentiation of eggs from *Diphyllobothrium* sp, 177
    differentiation of eggs from *Paragonimus*, 64
    life cycle and transmission of, 103
    morphology of adults of, 98, 101
    morphology of eggs of, 101
Nasua narcia (*see* Coati)
Nasua nasua (*see* Coati)
Necator americanus, epidemiology of, 354-355
    in creeping eruption, 367, 371
    life cycle and transmission of (*see Ancylostoma duodenale*)
    morphology of adults of, 357
    morphology of eggs of, 358
    morphology of larvae of, 360
Necatoriasis (*see* Ancylostomiasis)
Nematoda, class of, 8, 9
Nematode ophthalmitis (*see* Toxocariasis)
Nematodiases, miscellaneous, 507-518
    clinical features of, 517
    definition of, 507
    epidemiology of, 515
    etiology of, 511
    history of, 507-508, 511
    life cycles of etiologic agents of, 515
    pathologic features of, 517

treatment of, 510, 514
Neodiplostomum seoulense, life cycle and transmission of, 103
 morphology of adults of, 98, 101
 morphology of eggs of, 98, 101
Nepal, paragonimiasis in, 52
Netherlands (Holland), *Alaria* sp trematodiasis in, 107-110
 anisakiasis in, 423
 halicephalobiasis in, 493
New Guinea, acanthocephaliasis in, 521
 ancylostomiasis in, 355
 cysticercosis in, 118
 fascioliasis in, 82
 paragonimiasis in, 52
New World hookworm (*Ancylostoma duodenale*) (*see* Ancylostomiasis)
New Zealand, anisakiasis in, 423
 fascioliasis in, 82
 hydatidosis in, 146, 148
Nicaragua, angiostrongyliasis costaricensis in, 386
 Hymenolepis diminuta in, 201
 paragonimiasis in, 52
Niclosamide, in diphyllobothriasis, 180
 in dipylidiasis, 137, 142
 in *Hymenolepis nana* and cyclophyllidean cestode infection, 212
 in intestinal trematodiases, 105
 in taeniasis, 132
Niger, dracunculiasis in, 330
 onchocerciasis in, 288
Nigeria, acanthocephaliasis in, 521
 Dicrocoelium hospes in, 88
 dracunculiasis in, 330
 loiasis in, 262
 mansonelliasis perstans in, 247
 oesophagostomiasis in, 499
 onchocerciasis in, 288
 paragonimiasis in, 52
NIH swab, in diagnosis of enterobiasis, 433
Niridazole, in dracunculiasis, 336
Nodular eosinophilic panniculitis (*see* Gnathostomiasis)
North Africa, ancylostomiasis in, 354
 intestinal trematodiases in, 96
North America, American brugian filariasis in, 320
 diphyllobothriasis eradication program in, 167
 dirofilariasis in, 276
 Eustrongylides sp in, 515
 gnathostomiasis in, 447
 hydatidosis in, 146
 Hymenolepis nana in, 201
 intestinal trematodiases in, 96
 Pseudamphistomum truncatum in, 89
Norway, anisakiasis in, 423
 hydatidosis in, 146
Nucleopore® filtration technique, in loiasis, 271
 in lymphatic filariasis, 235
 in onchocerciasis, 300
Nurse cell (*see* Trichinosis)
Nutria, *Strongyloides myopotami* in, 342
Occult filariasis (*see* Lymphatic filariasis)
Oceania, fascioliasis in, 82
 hydatidosis in, 146
 Raillietina celebensis in, 202
Ocelot, dirofilariasis in, 277
Ocular larva migrans (*see* Toxocariasis)
Ocular onchocerciasis (*see also* Onchocerciasis), 307-318
 clinical features of, 308-314
 definition of, 307
 diagnosis of, 317
 epidemiology of, 307-308
 etiology of, 307
 history of, 307
 pathologic features of, 314-315
 synonyms for, 307
 treatment of, 317-318
Oesophagostomiasis and ternidenamiasis, 499-506
 clinical features of, 502-503
 definition of, 499
 diagnosis of, 504-506
 epidemiology of, 500
 etiology of, 499
 history of, 499
 pathologic features of, 503-504
 synonyms for, 499
 treatment of, 506
Oesophagostomum aculeatum, 500
Oesophagostomum apiostomum, 500
Oesophagostomum bifurcum, 499-506
 life cycle and transmission of, 502
 morphology of adults of, 500-501
 morphology of eggs of, 501
 morphology of larvae of, 501
Oesophagostomum sp, as parasites of monkeys, 499
Oesophagostomum stephanostomum, as parasite of monkeys and humans, 500
Old World hookworm (*Necator americanus*) (*see* Ancylostomiasis)
Onchocerca cervicalis, clinical features of infection by, 517
 distinguishing features of, 516
 life cycle of, 515
 morphology of adults of, 516
 pathologic features of infection by, 517
 source of infection by, 515
Onchocerca gutturosa, clinical features of infection by, 517
 distinguishing features of, 516
 distribution of, 515
 life cycle of, 515
 morphology of adults of, 516
 pathologic features of infection by, 517
 source of infection by, 515
 synonym for, 511
Onchocerca volvulus, biologic differences in strains, 290
 life cycle and transmission of, 293-294
 morphology of adults of, 290-291
 morphology of microfilariae of, 293
Onchocerciasis (*see also* Ocular onchocerciasis), 287-306
 clinical features of, 294-296
 definition of, 287
 diagnosis of, 299-300, 302
 epidemiology of, 288, 290
 etiology of, 287
 history of, 287-288
 pathologic features of, 296-299
 synonyms for, 287
 treatment and prevention of, 302-303
Onchocerciasis Control Program (WHO) (OCP) (*see* Onchocerciasis)
Onchosphere, in *Echinococcus* sp, 152
Opisthorchis felineus, 69-81
 epidemiology of, 70, 71
 life cycle and transmission of, 75-77

morphology of adults of, 75
morphology of eggs of, 75
Opisthorchis viverrini, 69-81
    epidemiology of, 71
    life cycle and transmission of, 75-77
    morphology of adults of, 75
    morphology of eggs of, 75
Opossum, as paratenic host for *Alaria* sp, 108
    Capillaria hepatica in, 487
    Gnathostoma sp in, 454
Orient, clonorchiasis in, 70-71
    Raillietina celebensis in, 202
Oriental eye worm (*see Thelazia callipaeda*)
Oriental liver fluke infection (*see* Clonorchiasis and opisthorchiasis)
Ostertagia ostertagi, clinical features of infection by, 517
    distinguishing features of, 516
    distribution of, 515
    history of, 511
    life cycle of, 515
    morphology of adults of, 516
    morphology of eggs of, 516
    morphology of larvae of, 516
    pathologic features of infection by, 517
    source of infection by, 515
    synonyms for, 511
Otter, dirofilariasis in, 276
    Gnathostoma sp in, 454
Ouchterlony gel diffusion test, in gnathostomiasis, 459
Ovarian balls, 521-524
Oxamiquine, in schistosomiasis, 47
Oxfendazole, in *Haemonchus contortus* infection (in animals), 514
    in *Metastrongylus apri* infection (in animals), 514
Oxyuris vermicularis (*see* Enterobiasis)
Oxyuroidea, superfamily of, 511
Ozzard's filariasis (*see* Mansonelliasis ozzardi)
Paca, as host for *Echinococcus* sp, 152
Pacific Islands, dirofilariasis in, 276
    lymphatic filariasis in, 219
Pakistan, dracunculiasis in, 330
    intestinal trematodiases in, 96
Panama, angiostrongyliasis costaricensis in, 386
    hydatidosis in, 148
    mansonelliasis perstans in, 245
    paragonimiasis in, 52
Pancreas, onchocerciasis in, 298
Panther (Florida), dirofilariasis in, 277
Papua New Guinea, angiostrongyliasis cantonensis in, 374
Papyrus Ebers, ancylostomiasis in, 353
    dracunculiasis in, 329
    enterobiasis in, 433
Parafossarulus manchouricus (*see also* Snail), 76
Paragonimiasis, 49-67
    clinical features of, 56-58, 60-61
    definition of, 49
    diagnosis of, 64-65
    epidemiology of, 51-52
    etiology of, 49
    history of, 50-51
    in differential diagnosis of gnathostomiasis, 458
    pathologic features of, 61-65
    synonyms for, 49
    treatment and prevention of, 65
Paragonimus africanus, 49, 61

Paragonimus congolensis, 61
Paragonimus ecuardoriensis, 49
Paragonimus heterotremus, 49, 60
Paragonimus hueitungensis, 49, 60-61
Paragonimus kellicotti, 49
Paragonimus mexicanus, 49, 52
Paragonimus miyazakii, 49, 58
Paragonimus pulmonalis, 49
Paragonimus skrjabini, 49, 60
Paragonimus sp, life cycle of, 53, 55-56
    morphology of adults of, 52-53
    morphology of eggs of, 53
    synonyms for, 49-50
Paragonimus uterobilateralis, 49, 52, 61
Paragonimus westermani, 49
Paragonimus westermani filipinus, 52
Paragordius sp, 4, 9
Paraguay, *Alaria* sp trematodiasis in, 107-110
    Bertiella sp in, 201
Paramphistomatidiasis (*see* Fasciolopsiasis)
Parasitic helminths, classification of, 3-11
    Acanthocephala, phylum of, 9-10
    Cestoda, class of, 5, 6
    Nematoda, class of, 8, 9
    Trematoda, class of, 4, 5
Paratenic hosts, for acanthocephalans, 526
    for *Angiostrongylus cantonensis*, 376
    for anisakids, 426
    for *Corynosoma strumosum*, 520
    for *Diphyllobothrium* sp, 171
    for *Mesocestoides lineatus*, 208
    for *Spirometra* sp, 172
    for toxocariasis, 414
Parauterine organ, 203
Paromomycin sulfate, in diphyllobothriasis, 180
Parthenogenesis, in *Strongyloides stercoralis*, 344
Pathologic features of helminthiases (combined for all diseases in this volume)
    abscess, 62-63, 249, 504
        around dead microfilariae, 233
        eosinophilic, 458
        intestinal, 104
        intestinal wall, 441
        hepatic, 80, 405-406, 488
        micro, around degenerating worms, 485
        nodular, in intestines, 363
    acanthocephalans in tissue, 527
    anemia, bothriocephalus, 175
    anthelmintic therapy, changes after, 235
    arachnoiditis, chronic, 62
    atrophy, of hepatocytes, 80
        of intestinal mucosa, 485
        retinal, 128
        villus, 104
    bile duct changes, 79, 85, 86, 88
    bilharziomas, 31
    biliary cirrhosis, 63
    bone changes, 157
    bone marrow changes, 175, 478
    Calabar swellings, 269
    calcareous corpuscles, 128, 157, 176
    calcification, 86, 128, 233
        of cysts, 478
    cancer, cholangiocarcinoma, 80
        urinary bladder, 41

(Pathologic features, continued)
- cardiopulmonary changes, 42
- caseation, 231, 504
- cell-mediated immunity, 298-299, 314
- cerebrospinal changes, 42, 62-63, 104, 128, 176, 191, 271, 380, 419, 478
- cestode larval infection, aberrant, 211
  - cyclophyllidean, 211
- Charcot-Leyden crystals, 86, 441, 488
- cholangiocarcinoma, in *C. sinensis* and *O. viverrini* infection, 80
- cholangitis, secondary, 80
- cholelithiasis, 86
- CNS changes, 458, 496
- colonic changes, 31, 364, 503
- conjunctival changes, 314
- corneal changes, 314-315
- cor pulmonale, schistosomal, 42
- cutaneous changes, 363
  - post-DEC therapy for filariasis, 254
- cyst, debris, identification of, 155
  - development, 155
  - in peritoneal cavity, 112
  - membrane, 155
  - wall, 112
- cystitis cystica, 39
- dermal changes, 269, 297
- ectopic lesions, of enterobiasis, 443
  - of fascioliasis, 86
- eggs, in CNS, 30
  - in intestinal lumen, 441
  - in lesions, 128
- enteritis, catarrhal, 211
  - eosinophilic, 527
  - necrotizing, 527
- enzymatic factors in migration of larvae, 176
- eosinophilia, 29, 249, 254, 334, 346, 391, 428
  - hepatic, 488
  - in skin, 370
  - tissue, 157, 325, 363, 419, 441, 466
  - tropical, changes in, 233
  - variable, 485
- eosinophilic infiltrates, 325
- fibrosis, 128, 334, 504
  - around bile ducts, 86
  - colonic, 31
  - of pancreatic ducts, 111
  - periductal, 80
  - portal, 31
  - Symmers', 31, 42
- flukes, in cysts, 112
  - in intrahepatic ducts, 79
- *gâle filarienne*, 297
- gallbladder changes, 85, 247
- gastrointestinal changes, 269
- genital lesions, 41-42
- glaucoma, 128
- gliosis, 62-63, 128
- granulomas, 62-63, 104, 112, 128, 155, 249, 281, 347, 380-381, 391, 428, 458, 478, 504
  - around eggs, 86
  - around larvae, 419
  - caseous, 325
  - circumoval, 29
  - fibrocaseous, 406
  - hepatic, 488
  - in intestine, 441
  - in lymphatics, 231
  - in pancreatic duct walls, 111
  - necrotizing, 325, 441
  - suppurative subcutaneous, 112
- granulomatous tracks, 419
- hematuria, 39
- hemorrhage, 109, 458
- hepatic changes, 32, 235, 247, 364, 393-394, 419, 485
  - gross findings, 79, 85
- hepatoma, 33
- hepatomegaly, 79
- hepatosplenic changes, 31
- hernia sac, invasion of, 249
- histocompatibility factors, 314
- hooklets in tissue, 155
- humoral antibody response, 299
- hydrocele, 269
- hydrocephalus, 174-175
- hyperinfection, 347
- hyperplasia, adenomatous, of biliary duct epithelium, 80
  - glandular epithelial, 85
  - of lymphoid tissue of intestine, 467
  - of mesenteric lymph nodes, 504, 527
- IgE, elevated, 371
- immunologic response and disease form, 154-157
- immunosuppression, 192
- incontinence of melanin, 254
- infarction of tissue, 281
- inflammation, acute, 128, 333
  - around degenerating worms, 231, 249, 269, 458
  - catarrhal, of intestinal lumen, 441
  - chronic, 128, 176, 466-467, 485, 490
  - of colon, 466-467
  - of female genital tract, 443
  - of intestine, at site of worm attachment, 104, 127
  - of mucosa and lamina propria, 363
  - of skin, 254, 259
  - suppurative, in intestinal wall, 441
- intestinal changes, 30, 211, 249, 405
  - gross, 485
  - mucosal, 128
  - perforation, 391, 526-527
  - strangulation, 127
  - ulceration, 104, 527
- intestinal disease, noninvasive, 346
- intestinal hypertrophic-pseudoneoplastic pattern, 390-391
- intestinal ischemic-congestive pattern, 390-391
- intussusception, 127
- iridocyclitis, 128
- lymphadenitis, 192, 235, 254, 269, 271, 297-298, 325, 393
  - of porta hepatis, 85
- lymphatics, dilatation of, 231
  - obliteration of, 231
- meningeal disease, acute, 62
- meningoencephalomyelitis, 496
- mesenteric artery changes, 390
- Meyers and Kouwenaar bodies, 235
- microemboli of worms, 104
- microfilaremia, 249, 269
- microfilariae, in anterior chamber, 315
  - in dermal vessels, 269

(Pathologic features, continued)
    in skin, 254, 297
    mucosal changes, gastrointestinal tract, 428
    muscle changes, 477-478
    myocarditis, 104, 478
    necrosis, ischemic, 281
        of hepatocytes, 80
        track-like, 109, 458
    nodules, formation related to trauma, 297
    nurse cells, 477
    ocular changes, 128, 191-192, 249, 314-315, 380-381, 419-420
    optic nerve changes, 62
    pancreatic changes, 111, 247
    panniculitis, eosinophilic, 282
    peritonitis, 391, 527
    placental changes, 128
    pleural lesions, 62
    pneumonia, 405
    polycystic hydatid disease, 157
    polyps, colonic, 31
    proteases, in immune evasion, 61
    pseudotumors, inflammatory, 406
    pulmonary changes, 62, 157, 363, 381, 419
        hypertension, 42
    pustules, intraepidermal, with type X larvae, 371
    pyelonephritis, 41
    renal changes, 41, 247
    schistosomal pigment, 30
    septicemia, gram-negative, 347
    small intestine, gross changes in, 363
        inflammation of mucosa and lamina propria, 363
    sowda, 297
    spinal cord changes, 380
    Splendore-Hoeppli phenomenon, 29, 380, 441
    splenic changes, 235, 487
    splenomegaly, 33
    suppuration around larvae, 428
    Symmers' fibrosis, 31, 42
    testicular changes, 394
    tracks of migrating parasite, 86, 109
    ultrasonographic changes, 32
    uremia, 41
    ureteritis cystica, 39
    urinary bladder changes, 39
    urogenital changes, 33
    uropathy, obstructive, 39, 41
    vasculitis, 109, 391
        and microfilariae, 233
        necrotizing, 86
    villus changes, 104, 211
    volvulus, 127
    worms (adult), degenerating, in lesions, 281
        encapsulated, in brain, 104
        in body cavities, 259
        in CNS, 42
        in colon, 466
        in connective tissue, 247
        in dermis, 254
        in intestinal lumen, 441
        in intestinal mucosa, 466
        in intestines, 104, 477
        in lesions, 488, 504
        in lymphatic vessels and lymph nodes, 233, 325
        in nodules, 297
        metastatic, 211

Peccary, *Capillaria hepatica* in, 487
Pelodera sp and *P. strongyloides*, clinical features of infection by, 517
    distinguishing features of, 516
    distribution of, 515
    history of, 511
    in creeping eruption, 367, 371
    life cycle of, 515
    morphology of larvae of, 516
    pathologic features of infection by, 517
    source of infection by, 515
    synonyms for, 511
Pentastomida, phylum of, 7, 10
Peptic ulcer disease, in differential diagnosis of fasciolopsiasis, 103
Peru, American brugian filariasis in, 320
    hydatidosis in, 148
    mansonelliasis ozzardi in, 258
    paragonimiasis in, 52
PF1022A, in angiostrongyliasis cantonensis, 382
Phaneropsolus bonnei, life cycle and transmission of, 103
    morphology of adults of, 98, 101
    morphology of eggs of, 98, 101
Pheromermis californica (see Mermithids)
Philippines, ancylostomiasis in, 355
    angiostrongyliasis cantonensis in, 374
    Bertiella sp in, 201
    dipylidiasis in, 138
    Eurytrema pancreaticum in, 110
    fascioliasis in, 82
    Hymenolepis diminuta in, 201
    intestinal capillariasis in, 481
    intestinal trematodiases in, 96, 97
    lymphatic filariasis in, 219
    paragonimiasis in, 52
    Raillietina sp in, 202
    Syphacia obvelata (pinworm) in, 434
    taeniasis in, 118
Philophthalmus gralli, 113
Philophthalmus lacrymosus, 113
Philophthalmus palpebrarum, 113
Philophthalmus sp trematodiasis, clinical features of, 113
    epidemiology of, 113
    etiology of, 113
    history of, 113
    life cycle and transmission of etiologic agent of, 113
    morphology of adults of etiologic agent of, 113
    morphology of eggs of etiologic agent of, 113
    treatment of infection by, 113
Phylogenetics and taxonomic classification, general, 1
Physaloptera caucasica, clinical features of infection by, 517
    distinguishing features of, 516
    distribution of, 515
    history of, 511
    life cycle of, 515
    morphology of adults of, 516
    morphology of eggs of, 516
    pathologic features of infection by, 517
    source of infection by, 515
    synonyms for, 511
Physalopteroidea, superfamily of, 511
Pica, in ancylostomiasis, 362
    relationship to trichuriasis, 466
Pig (see Swine)

Pimply gut (*see* Oesophagostomiasis)
Pinworm (*see* Enterobiasis)
Piperazine citrate, in acanthocephaliasis, 528
    in ascariasis, 407
    in enterobiasis, 445
    in *Enterobius gregorii*, 514
Plagiorchiidae, family of, 97
Plagiorchiidiasis (*see* Fasciolopsiasis)
Plagiorchis sp, life cycle and transmission of, 103
    morphology of adults of, 101
    morphology of eggs of, 101
Plerocercoid larvae (*see* Sparganosis)
Plumber's itch (*see* Creeping eruption)
Poikilorchis congolensis, 111
Poland, *Pelodera* sp in, 515
    trichinosis in, 472
Polar bear, *Trichinella spiralis nativa* in, 471
Polycystic hydatid disease, 155
Polymyositis, in differential diagnosis of trichinosis, 477
Polynesia, ancylostomiasis in, 354
Porcupine, dirofilariasis in, 276
Pork tapeworm (*see Taenia solium*)
Portugal, cysticercosis in, 118
    dipylidiasis in, 138
Prairie dog, *Capillaria hepatica* in, 487
Praziquantel, in Achillurbainiidae trematodiasis, 112
    in *Alaria* sp trematodiasis, 110
    in clonorchiasis and opisthorchiasis, 90
    in coenurosis, 194
    in dicrocoeliasis, 90
    in diphyllobothriasis, 180
    in dipylidiasis, 142
    in *Eurytrema pancreaticum* (in sheep), 111
    in hydatidosis, 161
    in *Hymenolepis nana* and cyclophyllidean cestode infection, 212
    in intestinal trematodiases, 105
    in neurocysticercosis, 132
    in paragonimiasis, 65
    in taeniasis, 131
    in schistosomiasis, 47
Primary and daughter cysts (*see Echinococcus* sp)
Primates, nonhuman, *Bertiella* sp in, 201, 208
    Strongyloides fuelleborni in, 341
Proboscis, 448
Procyon lotor (*see* Raccoon)
Pronghorn deer, *Capillaria hepatica* in, 487
Prosthodendrium molenkampi, life cycle and transmission of, 103
    morphology of adults of, 98, 101
    morphology of eggs of, 98, 101
Proteolytic enzymes, in pathogenesis of gnathostomiasis, 456
Pseudamphistomum truncatum, life cycle and transmission of, 89-90
    morphology of adults of, 89
    morphology of eggs of, 89
Pseudocoelomates, 6-10
    Acanthocephala, phylum of, 9-10
    Aschelminthes, phylum of, 8-9
        Nematoda, class of, 8-9
        Nematomorpha, class of, 9
Pseudocoelomic membranes, 516
Pseudophyllidean tapeworms, life cycle and transmission of, 171-172
Pseudoterranova (Phocanema) decipiens, life cycle and
    transmission of, 426
    morphology of larvae of, 424
Pudoc's disease (*see* Capillariasis, intestinal)
Puerto Rico (*see* Caribbean)
Pulex irritans, as intermediate host for *Dipylidium caninum*, 139
Puma, hydatidosis in, 146
Pyrantel pamoate, in ascariasis, 407
    in enterobiasis, 445
    in oesophagostomiasis, 506
    in *Rhabditis* sp infection, 514
    in *Trichostrongylus* sp infection, 514
Pyrantel tartrate, in hepatic capillariasis, 490
Pyrvinium pamoate, in enterobiasis, 445
Quinacrine, in diphyllobothriasis, 180
    in dipylidiasis, 142
    in taeniasis, 131
Rabbit, as host for *Eurytrema pancreaticum*, 110
    Capillaria hepatica in, 487
    coenurosis in, 186
Raccoon, acanthocephaliasis in, 520, 526
    American brugian filariasis in, 319
    as paratenic host for *Alaria* sp, 108
    Baylisascaris sp in, 511
    dirofilariasis in, 276
    Gnathostoma sp in, 454
    Mesocestoides lineatus in, 201
    Strongyloides procyonis in, 342
    Trichinella pseudospiralis in, 471
Radioimmunoassay, in gnathostomiasis, 458
Raillietina celebensis, life cycle and transmission of, 208
    morphology of adults of, 206
    morphology of eggs of, 206-207
Raillietina cubensis (*see also* Inermicapsifer madagascariensis), 198
Raillietina davainei (*see also* Inermicapsifer madagascariensis), 198
Raillietina demerariensis (*see also R. celebensis*), 197
Raillietina garrisoni (*see also R. celebensis*), 198
Raillietina kouridovali (*see also* Inermicapsifer madagascariensis), 198
Raillietina loeche-salavezi (*see also* Inermicapsifer madagascariensis), 198
Raillietina madagascariensis (*see also* Inermicapsifer madagascariensis), 198
Raillietina murium (*see also R. celebensis*), 198
Raillietina sinensis (*see also R. celebensis*), 198
Raillietina siriraji (*see also R. celebensis*), 198
Raillietina sp, in differential diagnosis of dipylidiasis, 142
Rattus norvegicus, *Angiostrongylus cantonensis* in, 376
Rattus rattus, *Angiostrongylus cantonensis* in, 376
    Angiostrongylus costaricensis in, 388
Rattus sp, *Angiostrongylus cantonensis* in, 376
    Syphacia muris (rat pinworm) in, 434
Red clover eelworm (*see Ditylenchus dipsaci*)
Red-dot card test, in ocular onchocerciasis, 317
Red fly (*see* Tabanid fly)
Red fox, *Mesocestoides leptothylacus* infection in, 197
Reptiles, as intermediate hosts for *Mesocestoides lineatus*, 208
Réunion Island, angiostrongyliasis cantonensis in, 374
Rhabditis sp, clinical features of infection by, 517
    distinguishing features of, 516
    distribution of, 515
    history of, 511
    life cycle of, 515

morphology of larvae of, 516
  pathologic features of infection by, 517
  source of infection by, 515
  synonyms for, 511
  treatment of infection by, 514
Rhabditoidea, superfamily of, 511
Rictularia sp, clinical features of infection by, 517
  distinguishing features of, 516
  distribution of, 515
  history of, 511
  life cycle of, 515
  morphology of adults of, 516
  morphology of eggs of, 516
  pathologic features of infection by, 517
  source of infection by, 515
  synonyms for, 511
Ridley fundus, in ocular onchocerciasis, 312
River blindness (*see* Onchocerciasis; Ocular onchocerciasis)
Rodents, as definitive hosts for *Inermicapsifer madagascariensis*, 208
  as paratenic hosts for *Alaria* sp, 108
Romania, dipylidiasis in, 138
  intestinal trematodiases in, 96
Russia (*see* CIS)
Rwanda, *Inermicapsifer madagascariensis* in, 201
  loiasis in, 262
  Mesocestoides lineatus in, 201
Saguinus mystax, as definitive host for *Angiostrongylus costaricensis*, 388
Saint Kitts, *Bertiella* sp in, 201
Salmon, in transmission of anisakid larvae, 423
Salmonella infection, and schistosomiasis, 33, 41
Samoa, paragonimiasis in, 52
Santonin, in *Rhabditis* sp infection, 514
Sardinia, fascioliasis in, 82
Sashimi, as food source of anisakiasis, 423
Saudi Arabia, onchocerciasis in, 288
Scandinavia, diphyllobothriasis in, 167
Scarabaeidae beetles, as intermediate hosts for acanthocephalans, 525
Schistosoma haematobium, 23-25
Schistosoma intercalatum, 23-25
Schistosoma japonicum, 23-25
Schistosoma mansoni, 23-25
Schistosoma mattheei, 23-25
Schistosoma mekongi, 23-25
Schistosomal granuloma of bowel, in differential diagnosis of oesophagostomiasis and ternidenamiasis, 505
Schistosomiasis, 23-48
  clinical features of, 29-30
  definition of, 23
  diagnosis of, 46-47
  epidemiology of, 24
  etiology of, 23-25
  history of, 23
  life cycle and transmission of etiologic agent of, 28
  pathologic features of, 30-33, 39, 41-42
  Salmonella infections and, 33, 41
  synonyms for, 23
  treatment of, 47
Sclerotized cuticular openings, 10
Scutate threadworm (*see Gongylonema pulchrum*)
Sea lion, anisakids in, 426
  dirofilariasis in, 276
Seal, acanthocephalans in, 520
  anisakids in, 426
Seatworm (*see* Enterobiasis)
Secernentea, subclass of, 511
Senegal, dracunculiasis in, 330
  mansonelliasis perstans in, 245
  onchocerciasis in, 288
Serologic tests, for angiostrongyliasis cantonensis, 382
  for dirofilariasis, 282
  for gnathostomiasis, 458-459
  for hydatidosis, 159
  for intestinal trematodiases, 104
  for lymphatic filariasis, 235-237
  for trichinosis, 479
Shanghai's rheumatism (*see* Gnathostomiasis)
Sheep, as host for *Echinococcus* sp, 152
  as host for *Eurytrema pancreaticum*, 110
  as host for *Oesophagostomum bifurcum*, 502
  as host for *Ternidens deminutus*, 502
  coenurosis in, 189
  vaccination of for *Taenia multiceps*, 194
Sheep wireworm (*see Haemonchus contortus*)
Shrew, *Capillaria hepatica* in, 487
Sicily, fascioliasis in, 82
Sierra Leone, *Dicrocoelium hospes* in, 88
  onchocerciasis in, 288
  paragonimiasis in, 52
Sigmodon hispidus, as definitive host for *Angiostrongylus costaricensis*, 388
Simulium amazonicum (*see also* Black fly), 258
Simulium damnosum (*see also* Black fly), 287, 290
Simulium exiguum (*see also* Black fly), 290
Simulium guianense (*see also* Black fly), 290
Simulium metallicum (*see also* Black fly), 290
Simulium neavei (*see also* Black fly), 288, 290
Simulium ochraceum (*see also* Black fly), 290
Simulium oyapockense (*see also* Black fly), 290
Simulium quadrivittatum (*see also* Black fly), 290
Simulium sanguineum (*see also* Black fly), 258, 290
Simulium sp (*see also* Black fly), 280, 288, 290
Simulium venustum (*see also* Black fly), 280
Singapore, *Bertiella* sp in, 201
  intestinal trematodiases in, 96
Skin and subcutaneous tissue, Achillurbainiidae trematodiases in, 112
  Alaria sp trematodiasis in, 109
  ancylostomiasis in, 361
  coenurosis in, 189
  creeping eruption in, 367-372
  dracunculiasis in, 332-334
  fascioliasis in, 86
  gnathostomiasis in, 456-458
  loiasis in, 265, 269
  mansonelliasis ozzardi in, 258-259
  onchocerciasis in, 294-296
  paragonimiasis in, 60
  schistosomiasis in, 29
  sparganosis in, 174
  streptocerciasis in, 253-254
  strongyloidiasis in, 344-346
  toxocariasis in, 418
Skin test, Casoni, for hydatidosis, 159
  for acanthocephaliasis, 528
  for angiostrongyliasis cantonensis, 382
  for fascioliasis, 87
  for gnathostomiasis, 458-459

for intestinal trematodiases, 104
for loiasis, 273
for onchocerciasis, 300
for paragonimiasis, 65
Skunk, *Capillaria hepatica* in, 487
Slit-lamp examination, for angiostrongyliasis cantonensis, 382
for ocular onchocerciasis, 317
for onchocerciasis, 300
Slug, *Angiostrongylus cantonensis* in, 376
Angiostrongylus costaricensis in, 388
Snail, *Angiostrongylus cantonensis* in, 376
as intermediate host for *Alaria* sp, 108
as intermediate host for *Angiostrongylus costaricensis*, 388
as intermediate host for *Eurytrema pancreaticum*, 110
as intermediate host for *Philophthalmus* sp, 113
clonorchiasis and opisthorchiasis in, 70
fascioliasis in, 84
intestinal trematodiases in, 96
Paragonimus sp in, 55
schistosomes in, 28
Snake, as intermediate host for *Gnathostoma* sp, 454
as paratenic host for *Alaria* sp, 108
intestinal trematodiases in, 97
Mesocestoides lineatus in, 201
Solomon Islands, lymphatic filariasis in, 219
paragonimiasis in, 52
South Africa, coenurosis in, 186
dipylidiasis in, 138
Inermicapsifer madagascariensis in, 201
paragonimiasis in, 52
South America, American brugian filariasis in, 320
ancylostomiasis in, 354
ascariasis in, 397
coenurosis in, 186
gnathostomiasis in, 447
hydatidosis in, 148
Hymenolepis diminuta in, 201
Hymenolepis nana in, 201
Lagochilascaris minor in, 515
lymphatic filariasis in, 219
mansonelliasis ozzardi in, 258
mansonelliasis perstans in, 245
Southeast Asia, angiostrongyliasis cantonensis in, 374
Cyclodontostomum purvisi in, 515
dirofilariasis in, 276
gnathostomiasis in, 447
opisthorchiasis in, 71
sparganosis in, 167
Sowda (*see* Onchocerciasis)
Spain, cysticercosis in, 118
hydatidosis in, 146
intestinal trematodiases in, 96
trichinosis in, 472
Sparganosis (*see also* Diphyllobothriasis), 165-183
clinical features of, 172, 174-175
definition of, 165
diagnosis of, 177, 180
epidemiology of, 167
etiology of, 165
history of, 166
in differential diagnosis of gnathostomiasis, 458
longevity in humans, 172
pathologic features of, 175-176
synonyms for, 165-166

treatment and prevention of, 180
Sparganum proliferum, in vitro and in vivo cultivation of, 171
life cycle and transmission of, 171-172
morphology of, 171
Specimen taking and preparation, importance of, 1
Spelotrema brevicaeca, life cycle and transmission of, 103
morphology of adults of, 98, 101
morphology of eggs of, 98, 101
Speothus venaticus (*see* Bush dog)
Spinal cord, angiostrongyliasis cantonensis in, 378-379
coenurosis in, 190
halicephalobiasis in, 496
loiasis in, 271
schistosomiasis in, 42
paragonimiasis in, 42, 60, 63
sparganosis in, 175
Spiny-headed worm (*see* Acanthocephaliasis)
Spirocerca lupi, clinical features of infection by, 517
distinguishing features of, 516
distribution of, 515
history of, 511
life cycle of, 515
morphology of adults of, 516
morphology of eggs of, 516
pathologic features of infection by, 517
source of infection by, 515
synonyms for, 511
Spirometra erinacei, 165-166
Spirometra mansoni, 165-166
Spirometra mansonoides, 165-166
Spirometra ranarum, 165-166
Spirometra sp, life cycle of, 172
morphology of adults of, 169
morphology of eggs of, 169
morphology of larvae of, 169
Spiruroidea, superfamily of, 367-369, 511
Spleen, hydatidosis in, 154
hyperinfection strongyloidiasis in, 347
intestinal capillariasis in, 487
lymphatic filariasis in, 235
onchocerciasis in, 298
Spurious parasitism, in acanthocephaliasis, 521
in *Capillaria hepatica* infection, 490
Squid, ingested, as mimic of parasite, 18, 19
in transmission of anisakid larvae, 423
Sri Lanka, ancylostomiasis in, 355
angiostrongyliasis cantonensis in, 374
Bertiella sp in, 201
dipylidiasis in, 138
Pelodera sp in, 515
Philophthalmus sp trematodiasis in, 113
Stain, Delafield's hematoxylin, 271
GMS (Gomori-methenamine-silver), 155
H&E, 269
immunohistochemical, of larvae, 458
Movat, 121, 443
Russell-Movat, for worm morphology, 334
Warthin-Starry, 269
ZN (Ziehl-Neelsen), 47, 157
Steroid therapy, and hyperinfection strongyloidiasis, 345
Stichosome, 462, 472, 482, 487
Stoll's egg-counting method, 81
Stomach, anisakiasis in, 427-428
Streptocerciasis, 245, 250-256
clinical features of, 253

definition of, 250
diagnosis of, 256
epidemiology of, 251
etiology of, 245, 250
history of, 250-251
in differential diagnosis of leprosy, 253, 256
pathologic features of, 253-254
synonyms for, 250
treatment of, 256
Striated muscle, 10, 11
Strigeidae, family of, 97
Strigeidiasis (*see* Fasciolopsiasis)
Strongyloidea, superfamily of, 511
Strongyloides fuelleborni (*see also* Strongyloidiasis), 341-342, 344
  as parasite of nonhuman primates, 341
  transmission of in breast milk, 344
Strongyloides myopotami, in creeping eruption, 341
Strongyloides papillosus, 342
Strongyloides procyonis, in creeping eruption, 341
Strongyloides ransomi, 342
Strongyloides sp, in creeping eruption, 341, 367-368, 371
  parasite vs free-living forms, 1
Strongyloides stercoralis, in differential diagnosis of angiostrongyliasis cantonensis, 382
  life cycle and transmission of, 343-344
  morphology of adults of, 343
  morphology of eggs of, 343
  morphology of larvae of, 343
Strongyloides westeri, 342
Strongyloidiasis, 341-352
  clinical features of, 344-346
  definition of, 341
  diagnosis of, 347, 351
  epidemiology of, 342
  etiology of, 341
  history of, 341-342
  pathologic features of, 346-347
  synonyms for, 341
  treatment of, 351
Subcuticular glands, 10
Sub-Saharan Africa, ancylostomiasis in, 354
  mansonelliasis perstans in, 245
  onchocerciasis in, 288
  streptocerciasis in, 251
Sudan, acanthocephaliasis in, 521
  dracunculiasis in, 330
  intestinal trematodiases in, 96
  loiasis in, 262
  mansonelliasis perstans in, 245
  onchocerciasis in, 288
Suramin, in loiasis, 273
  in onchocerciasis, 303
Surgical treatment, of Achillurbainiidae trematodiasis, 112
  of American brugian filariasis, 327
  of angiostrongyliasis cantonensis, 382
  of angiostrongyliasis costaricensis, 394-395
  of anisakiasis, 430
  of ascariasis, 407
  of coenurosis, 194
  of creeping eruption, 371
  of *Dioctophyma renale* infection, 510
  of dirofilariasis, 283
  of dracunculiasis, 334
  of gnathostomiasis, 460

of hydatidosis, 159
of lymphatic filariasis, 238
of oesophagostomiasis, 506
of paragonimiasis, 65
of taeniasis, 132
Sushi, as food source of anisakiasis, 423
Swamp rabbit, American brugian filariasis in, 319
Sweden, diphyllobothriasis in, 167
Swimmer's itch (*see* Schistosomiasis)
Swine, acanthocephaliasis in, 520, 525
  as host for *Echinococcus* sp, 152
  as host for *Eurytrema pancreaticum*, 110
  as host for *Oesophagostomum bifurcum*, 502
  as host for *Taenia solium; T. saginata*-Taiwanese variant, 121-122
  as host for *Ternidens deminutus*, 502
  Capillaria hepatica in, 487
  Gnathostoma sp in, 454
  Trichinella spiralis in, 471
  Trichinella spiralis spiralis in, 471
Swine lungworm (*see Metastrongylus apri*)
Switzerland, anisakiasis in, 423
  halicephalobiasis in, 493
  hydatidosis in, 148
Sylvilagus aquaticus (*see* Swamp rabbit)
Sylvilagus floridanus alacer (*see* Eastern cottontail)
Symmers' fibrosis, 31, 42
Syndrome, Churg-Strauss, 236
  febrile fasciolitic eosinophilic, 81
  Foster-Kennedy, 60
  Frimodt-Möller's, 411
  hypereosinophilic, 236
  Löffler's, 236, 362, 402-403, 416, 419
  Meyers and Kouwenaar's, 215
  Weber-Christian, 61
  Weingarten's, 215
Syphacia muris (rat pinworm), 434, 444
Syphacia obvelata (mouse pinworm), 434
  clinical features of infection by, 517
  distinguishing features of, 516
  distribution of, 515
  history of, 511
  in the Philippines, 434
  life cycle of, 515
  morphology of adults of, 444, 516
  morphology of eggs of, 516
  pathologic features of infection by, 517
  source of infection by, 515
  synonyms for, 511
Tabanid fly, life cycle of, 263-264
  loiasis in, 261
Taenia asiatica, 117
Taenia brauni, 185
Taenia crassiceps, 197
Taenia crassicollis, 198
Taenia demarariensis, 198
Taenia echinococcus, 145
Taenia echinococcus multilocularis, 146
Taenia glomerata, 185
Taenia infantis, 198
Taenia lineatus (*see also Mesocestoides lineatus*), 197
Taenia madagascariensis (*see also Inermicapsifer madagascariensis*), 197
Taenia mucronata (*see also Bertiella mucronata*), 197
Taenia multiceps, in coenurosis, 185-196

life cycle and transmission of, 189
morphology of adults of, 187
morphology of coenurus (metacestode) of, 187-188
morphology of eggs of, 187
synonyms for, 185
vaccination of sheep for, 194
Taenia murina (*see also* Hymenolepis nana), 197
Taenia nana (*see also* Hymenolepis nana), 197
Taenia saginata, 117-136
life cycle and transmission of, 121-122
morphology of adults of, 118-119, 131
morphology of eggs of, 119-120, 131
Taenia saginata-Taiwanese variant, 117-136
life cycle and transmission of, 121-122
morphology of adults of, 118-119, 131
morphology of eggs of, 119-120, 131
Taenia serialis, in coenurosis, 185-196
life cycle and transmission of, 189
morphology of adults of, 187
morphology of coenurus (metacestode) of, 187-188
morphology of eggs of, 187
synonyms for, 185
Taeniasis and cysticercosis, 117-136
clinical features of adult tapeworm infection, 122, 125, 127
clinical features of metacestode infection (cysticercosis), 125, 127
definition of, 117
diagnosis of cysticercosis, 130-131
diagnosis of taeniasis, 130-131
epidemiology of, 118
etiology of, 117
history of, 117-118
in differential diagnosis of tuberculosis; neurocysticercosis, 131
pathologic features of adult tapeworm infection, 127-128, 130
pathologic features of cysticercosis, 128, 130
synonyms for, 117
treatment and prevention of, 131-132
Taenia solium, life cycle and transmission of, 121-122
morphology of adults of, 118-119
morphology of eggs of, 119-120
morphology of metacestode (*Cysticercus cellulosae*) of, 120-121
Taenia sp, classification of based on intermediate host, 185
life cycle and transmission of, 189
morphology of, 186-188
Taenia taeniaeformis (*see also* Hymenolepiasis), 197, 211
Taiwan, angiostrongyliasis cantonensis in, 374
intestinal capillariasis in, 481
intestinal trematodiases in, 96, 97
paragonimiasis in, 52
Raillietina celebensis in, 202
taeniasis in, 118
Taiwan taenia (*see Taenia saginata*-Taiwanese variant)
Talmud, hydatidosis in, 145
Tanzania, gnathostomiasis in, 447
onchocerciasis in, 288
T cells, in toxocariasis, 418
Temefos, in onchocerciasis, 288
Ternidens deminutus, life cycle and transmission of, 502
morphology of adults of, 501
morphology of eggs of, 501
Testicles, angiostrongyliasis costaricensis in, 390, 394

Tetrachloroethylene, in intestinal trematodiases, 104, 105
Tetramisole, in acanthocephaliasis, 528
Tetrapetalonema perstans (*see also* Mansonella perstans), 245
Tetrapetalonema streptocerca (*see also* Mansonella streptocerca), 245
Thailand, acanthocephaliasis in, 521
ancylostomiasis in, 355
angiostrongyliasis cantonensis in, 374
anisakiasis in, 423
Bertiella sp in, 201
gnathostomiasis in, 447
Hymenolepis diminuta in, 201
intestinal capillariasis in, 481
opisthorchiasis in, 70
Raillietina sp in, 202
taeniasis in, 118
trichinosis in, 472
Thelazia californiensis, clinical features of infection by, 517
distinguishing features of, 516
distribution of, 515
history of, 511
life cycle of, 515
morphology of adults of, 516
morphology of eggs of, 516
pathologic features of infection by, 517
source of infection by, 515
synonym for, 511
treatment of infection by, 514
Thelazia callipaeda, clinical features of infection by, 517
distinguishing features of, 516
distribution of, 515
history of, 511
life cycle of, 515
morphology of adults of, 516
morphology of eggs of, 516
pathologic features of infection by, 517
source of infection by, 515
synonym for, 511
treatment of infection by, 514
Thelazioidea, superfamily of, 511
Theragra chalcogramma (codfish), 369
Thermocyclops sp (*see* Cyclopidae)
Thiabendazole, in *Baylisascaris* sp infection, 514
in creeping eruption, 371
in dracunculiasis, 336
in enterobiasis, 445
in hepatic capillariasis, 490
in *Mammomonogamus laryngeus* infection, 514
in strongyloidiasis, 351
in toxocariasis, 420
in *Trichostrongylus* sp infection, 514
in trichuriasis, 467
Thiara granifera (*see also* Snail), 76
Thorny-headed worm (*see* Acanthocephaliasis)
Threadworm infection (*see* Enterobiasis; Strongyloidiasis)
Tick, 8, 9
Tiger, *Gnathostoma* sp in, 454
Timorian filariasis (*see* Lymphatic filariasis)
Timor microfilariae (*see* Lymphatic filariasis; *Brugia timori*)
Tissue processing and morphology of etiologic agents, general, 1, 2
Togo, dracunculiasis in, 330
oesophagostomiasis in, 499, 500
onchocerciasis in, 288
Tougnale (*see* Oesophagostomiasis)

Toxascaris leonina, 412
Toxocara canis, 411-421
    in differential diagnosis of angiostrongyliasis cantonensis, 382
    in differential diagnosis of angiostrongyliasis costaricensis, 392
    in differential diagnosis of gnathostomiasis, 458
    life cycle and transmission of, 414-415
    morphology of adults of, 412-413
    morphology of eggs of, 414
    morphology of larvae of, 413-414
Toxocara cati, 411-421
    life cycle and transmission of, 414-415
    morphology of adults of, 412-413
    morphology of eggs of, 414
    morphology of larvae of, 413-414
Toxocara mystax, 411
Toxocariasis, 411-421
    clinical features of, 415-418
    definition of, 411
    diagnosis of, 420
    epidemiology of, 412
    etiology of, 411
    history of, 411-412
    pathologic features of, 418-420
    synonyms for, 411
    treatment and prevention of, 420
Tracheae, 11
Transfusion, in ancylostomiasis, 364
Transplacental transmission of toxocariasis in canids, 414
Trematoda, class of, 4, 5
Trematodiases, miscellaneous, 107-115
    Achillurbainiidae trematodiasis, 111-112
    Alaria sp trematodiasis, 107-110
    Clinostomum complanatum trematodiasis, 112-113
    definition of, 107
    etiology of, 107
    Eurytrema pancreaticum trematodiasis, 110-111
    Philophthalmus sp trematodiasis, 113
Trichinella britovi, host species for, 471
Trichinella pseudospiralis, host species for, 471
    tissue reaction in, 478
Trichinella spiralis, 471-480
    in differential diagnosis of angiostrongyliasis cantonensis, 382
    life cycle and transmission of, 474
    morphology of adults of, 472
    morphology of larvae of, 472, 474
    susceptibility to freezing in meat, 480
Trichinella spiralis nativa, resistance to freezing in meat, 480
Trichinella spiralis nelsoni, 471
Trichinella spiralis spiralis, 471
Trichinellosis (*see* Trichinosis)
Trichiniasis (*see* Trichinosis)
Trichinosis, 471-480
    clinical features of, 474, 476-477
    definition of, 471
    diagnosis of, 478-479
    epidemiology of, 471-472
    etiology of, 471
    pathologic features of, 477-478
    synonyms for, 471
    treatment and prevention of, 479-480
Trichocephalus hepaticus, 487
Trichosoma tenue, 487

Trichosoma tenuissimum, 487
Trichostrongyloidea, superfamily of, 511
Trichostrongylus sp, clinical features of infection by, 517
    distinguishing features of, 516
    distribution of, 515
    history of, 511
    life cycle of, 515
    morphology of adults of, 516
    morphology of eggs of, 516
    pathologic features of infection by, 517
    source of infection by, 515
    synonyms for, 511
    treatment of infection by, 514
Trichuriasis, 461-470
    clinical features of, 465-466
    definition of, 461
    diagnosis of, 467
    epidemiology of, 461
    etiology of, 461
    pathologic features of, 466-467
    synonyms for, 461
    treatment of, 467
Trichuris trichiura, life cycle and transmission of, 465
    morphology of adults of, 461-462, 465
    morphology of eggs of, 465
    prevalence of, 397
Trichuris vulpis, 461, 467
Trichuroidea, superfamily of, 511
Triclabendazole, in fascioliasis, 90
Trinidad, mansonelliasis perstans in, 245
Trophosome, 2, 3
Tropical chlorosis (*see* Ancylostomiasis)
Tropical eosinophilia (*see* Lymphatic filariasis)
Tropical eosinophilic syndrome (*see* Lymphatic filariasis)
Tropical pulmonary eosinophilia (*see* Lymphatic filariasis)
Tropical rain forest, American brugian filariasis in, 320
Tropics and subtropics (*see also* countries and regions)
    creeping eruption in, 367
    oesophagostomiasis in, 500
    Physaloptera caucasica in, 515
    Spirocerca lupi in, 515
Tropisternus collaris, as intermediate host for acanthocephalans, 525
Tua chid (*see* Gnathostomiasis)
Tuberculosis, in differential diagnosis of dirofilariasis, 282
    in differential diagnosis of oesophagostomiasis and ternidenamiasis, 505
Tumeur de Dapaong (*see* Oesophagostomiasis)
Tuna, anisakid larvae in, 423
Tunisia, intestinal trematodiases in, 96
Tunnel anemia (*see* Ancylostomiasis)
Tunnel disease (*see* Ancylostomiasis)
Turbatrix aceti, clinical features of infection by, 517
    distinguishing features of, 516
    distribution of, 515
    history of, 511
    life cycle of, 515
    morphology of adults of, 516
    morphology of larvae of, 516
    pathologic features of infection by, 517
    source of infection by, 515
    synonyms for, 511
Turkey, hydatidosis in, 146
    intestinal trematodiases in, 96
Turtle, *Mesocestoides lineatus* in, 201

Tylenchoidea, superfamily of, 511
Type X larvae (of Spiruroidea), 367-371
Uganda, dracunculiasis in, 330
    loiasis in, 262
    mansonelliasis perstans in, 245
    onchocerciasis in, 288
    streptocerciasis in, 251
Uganda eye worm (*see* Mansonelliasis perstans)
Ukraine, *Eustrongylides* sp in, 515
Ultrasonography, 81, 159, 224, 236, 406, 428
Uncinariasis (*see* Ancylostomiasis)
United Kingdom (Great Britain), anisakiasis in, 423
    coenurosis in, 186
    dipylidiasis in, 138
    fascioliasis in, 82
    halicephalobiasis in, 493
    hydatidosis in, 146
United States, acanthocephaliasis in, 521
    Alaria sp trematodiasis in, 107
    American brugian filariasis in, 320
    ancylostomiasis in, 354
    angiostrongyliasis cantonensis in, 375
    angiostrongyliasis costaricensis in, 386
    anisakiasis in, 423
    ascariasis in, 397
    Bertiella sp in, 201
    coenurosis in, 186
    creeping eruption in, 367
    dipylidiasis in, 137-138
    dirofilariasis in, 276
    enterobiasis in, 434
    halicephalobiasis in, 493
    hydatidosis in, 146, 148
    Hymenolepis diminuta in, 201
    Hymenolepis nana in, 199
    intestinal trematodiases in, 97
    Mesocestoides lineatus in, 201
    Onchocerca gutturosa in, 515
    paragonimiasis in, 52
    Philophthalmus sp trematodiasis in, 113
    sparganosis in, 167
    trichinosis in, 472
Uruguay, hydatidosis in, 148
    trichinosis in, 472
Vaginulus plebius (*see* Slug)
Vampirolepis nana (*see* Hymenolepis nana)
Venezuela, angiostrongyliasis costaricensis in, 386
    hydatidosis in, 148
    Hymenolepis diminuta in, 201
    Inermicapsifer madagascariensis in, 201
    Microfilaria bolivarensis in, 515
    onchocerciasis in, 288
    paragonimiasis in, 52
    trichinosis in, 472
Ventriculus, 424
Vietnam, acanthocephaliasis in, 521
    angiostrongyliasis cantonensis in, 373
    Eurytrema pancreaticum in, 110
    intestinal trematodiases in, 96, 97
    opisthorchiasis in, 71
    paragonimiasis in, 52
    Raillietina celebensis in, 202
Vinegar eel (*see* Turbatrix aceti)
Visceral larva migrans, in differential diagnosis of *Alaria* sp infection, 109, 412
    in differential diagnosis of ancylostomiasis, 412
    in differential diagnosis of *Baylisascaris procyonis* infection, 412
    in differential diagnosis of gnathostomiasis, 447
    in differential diagnosis of hepatic capillariasis, 488
    in differential diagnosis of sparganosis, 412
Wallaby, as host for *Echinococcus* sp, 152
Walrus, anisakids in, 426
    Trichinella spiralis in, 471
Watasenia scintillans (squid), 369
Water buffalo, as host for *Eurytrema pancreaticum*, 110
Water plants, in intestinal trematodiases, 96
Watsonius watsoni, life cycle and transmission of, 103
    morphology of adults of, 98
    morphology of eggs of, 98
Weingarten's disease (*see* Toxocariasis)
West Africa, dracunculiasis in, 329
    mansonelliasis perstans in, 245
    Meningonema peruzzii in, 515
    oesophagostomiasis in, 499, 500
    streptocerciasis in, 251
Western Hemisphere, mansonelliasis ozzardi in, 258
Western mesocestoides worm (*see* Mesocestoides lineatus; *M. variabilis*)
West Indies, mansonelliasis ozzardi in, 258
Whale, anisakids in, 426
    as definitive host for *Bolbosoma* sp, 526
Whipworm (*see* Trichuriasis)
Wild boar, *Trichinella spiralis* in, 472
    Trichinella spiralis nelsoni in, 471
Wild cats, dirofilariasis in, 277
Windward Islands, angiostrongyliasis costaricensis in (Martinique), 386
    Hymenolepis diminuta in (Martinique), 201
Wolf, dirofilariasis in, 276
    hydatidosis in, 146
Woodbury bug (*see* Gnathostomiasis)
World Health Organization (WHO), dracunculiasis eradication program of, 329-330
Wuchereria bancrofti, life cycle and transmission of, 222-224
    morphology of adults of, 219-220
    morphology of microfilariae of, 221
Wuchereria bancrofti var *pacifica* (*see* Lymphatic filariasis)
Wuchereria filaria (*see* Lymphatic filariasis)
Wuchereria malayi (*see* Lymphatic filariasis; *Brugia malayi*)
Wuchereriasis (*see* Lymphatic filariasis)
Wucherer's filaria (*see* Lymphatic filariasis)
Yellowtail fish, anisakid larvae in, 423
Yemen, Republic of, *Bertiella* sp in, 201
    onchocerciasis in, 288
Yugoslavia, *Philophthalmus* sp trematodiasis in, 113
    trichinosis in, 472
Zambia, acanthocephaliasis in, 521
    Hymenolepis diminuta in, 201
    Inermicapsifer madagascariensis in, 201
    paragonimiasis in, 52
Zimbabwe, acanthocephaliasis in, 521
    dipylidiasis in, 138
    Hymenolepis diminuta in, 201
    Inermicapsifer madagascariensis in, 201
    mansonelliasis perstans in, 245
    ternidenamiasis in, 500
Zoonotic helminthiases (*see* American brugian filariasis; Dirofilariasis)

Figure Acknowledgments

An asterisk at the end of a figure legend indicates the contribution of a photograph, drawing, or specimen.

Anderson, J: **15**.20
Barnley, GR: **17**.20
Beaver, PC: **10**.25
Bird, AC: **18**.8,15,18
Caplan, C: **4**.30
Céspedes, R: **25**.1
Clay-Adams Medichrome Series: **7**.4,9,10; **9**.1
Connor, D: **13**.33; **15**.1,3; **17**.2,21,22,23,24,25, 26,27,28,29,30,31,32,33,41,48; **26**.1,18; **30**.1; **33**.26
Correa, CD Jr: **2**.69
Cross, JH: **5**.21; **24**.9
DePaoli, A: **35**.11
Dreyer, G: **13**.28,29
Goldschmidt, RA: **26**.39
Hadfield, J: **24**.1; **35**.8
Hopkins, A: **26**.2
Johnson, FB: **1**.34
Lane, E: **7**.59
Langbehn, H: **4**.26; **35**.1
Lehman, JS: **2**.35
Lichtenfels, JR: **10**.1,2,3; **30**.1
Low, MDW: **20**.15
McConnell, EE: **2**.23
Marcial-Rojas, RA: **30**.27,36
Marchevsky, M: **22**.24
Matia, T: **4**.31
Meleney, HE: **26**.19
Mistrey, M: **26**.1

Miyazaki, I: **30**.2,7,16,17,20,25,26,28,29
Moncrieff, RE: **7**.8
Nye, S: **24**.8,10,16
Palmieri, JR: **13**.36,37,39
Peters, M: **13**.34
Pilitt, PA: **10**.1,2,3
Poinar, G Jr: **1**.1
Qing, L: **4**.35
Raasch, F: **26**.17; **32**.10
Radke, M: **2**.5
Reddington, B: **13**.11
Rosado de Christenson, ML: **9**.40,41,42
Rothman, R: **15**.21
Savino, D: **18**.7
Schiller, EL: **13**.1; **20**.10,33
Smith, JH: **2**.34
Szechnyi, E: **20**.17
Tong-Hua, L: **9**.29,30
Yamaguchi, T: **10**.4
Young, M: **2**.57
Wolfe, MS: **30**.39
Wong, MM: **2**.6
World Health Organization: **2**.1,2; **18**.1
Zaman, V: **30**.4

Permission to reprint certain figures was granted by the following:
American Journal of Ophthalmology: **6**.7
American Journal of Tropical Medicine and Hygiene: **6**.1
British Journal of Radiology: **35**.9,10
Experimental Parasitology: **30**.2,7,16,17,20,25, 26,28,29